INTEGRATED TRACKING, CLASSIFICATION, AND SENSOR MANAGEMENT

IEEE Press
445 Hoes Lane
Piscataway, NJ 08854

INTEGRATED TRACKING, CLASSIFICATION, AND SENSOR MANAGEMENT

THEORY AND APPLICATIONS

Edited by

Mahendra Mallick
Propagation Research Associates, Inc., Marietta, GA, USA

Vikram Krishnamurthy
Department of Electrical and Computer Engineering, University of British Columbia, Vancouver, Canada

Ba-Ngu Vo
Department of Electrical and Computer Engineering, Curtin University, Western Australia, Australia

IEEE
IEEE PRESS

WILEY
A JOHN WILEY & SONS, INC., PUBLICATION

Cover Illustration: Courtesy of Ba-Ngu Vo
Cover Design: John Wiley & Sons, Inc.

Published by John Wiley & Sons, Inc., Hoboken, New Jersey.
Published simultaneously in Canada.

Library of Congress Cataloging-in-Publication Data is available.

ISBN: 978-0-470-63905-4

Printed in the United States of America

10 9 8 7 6 5 4 3 2 1

CONTENTS

PREFACE

This book has been a long time in the making, starting with a series of conversations in 2007 during the Colloquium on Information Fusion in Xi'an China, followed by many discussions at various conferences as well as phone calls from half way around the globe. These conversations were centered on the ever-growing interest in tracking and sensor management in the wider community and the accessibility of the state-of-the-art techniques to graduate students, researchers, and engineers.

The research on multitarget tracking and sensor management was driven by aerospace and space applications such as radar, sonar, guidance, navigation, air traffic control, and space exploration in the 1960s. Since then, these research areas have flourished into other diverse disciplines such as image processing, oceanography, autonomous vehicles and robotics, remote sensing, biomedical research, and space debris tracking. Recent efficient multitarget tracking techniques and advances in sensing and computing technology have opened up prospective applications in areas such as driving safety and traffic monitoring, homeland security, and surveillance of public facilities.

While interest in this discipline is rapidly growing with many exciting advances during the last decade, comprehensive and accessible account of significant developments in the field are few and far between. The focus of our book is on expository writing, clear description of theoretical developments, and real-world applications in these areas. The chapters of the book are divided into five groups under the headings: Filtering, Multitarget Multisensor Tracking, Sensor Management and Control, Estimation and Classification, and Decision Fusion and Decision Support. Each chapter is solicited from internationally renowned experts in their respective areas. By providing concise and detailed descriptions, such as pseudo codes for algorithms, we endeavor to facilitate the implementations of the state-of-the-art algorithms, thereby making a wealth of approaches and techniques accessible to a wider audience.

Chapter 1 develops three classes of filtering algorithms for the angle-only filtering problem in 3D using bearing and elevation measurements. The dynamic models used by these filtering algorithms are the nearly constant velocity model for the relative Cartesian state vector, exact discrete-time dynamic model for modified spherical coordinates (MSC), and exact continuous-time dynamic model for MSC. The extended Kalman filter (EKF), unscented Kalman filter (UKF), and particle filter (PF) are developed for each class, of which the UKF and PF based on the exact continuous-time dynamic model for MSC represent new algorithms. Finally, a comparative evaluation of their accuracy and computational complexity is presented using Monte Carlo simulations.

Chapter 2 presents a recently introduced approach called box particle filtering which emerged from the synergy between sequential Monte Carlo (SMC) methods and interval analysis. A theoretical derivation of the box particle filter is given based on mixtures of uniform probability density functions with box supports. Experiments with both simulated and real data show the advantages of the box particle filter over the conventional particle filter for certain classes of problems.

Chapter 3 presents an accessible account of developments in the random finite set approach to the multitarget tracking problem. This chapter is classified under the filtering part of the book because fundamentally, the random finite set approach poses the multitarget tracking problem as a Bayesian filtering problem (in the space of finite subsets or simple finite point patterns). In this chapter, we discuss the notion of a mathematically consistent error metric for multitarget tracking and present arguments for the finite set representation of the multitarget state. We also detail random finite set-based algorithms such as the probability hypothesis density (PHD), Cardinalized PHD (CPHD), and Multitarget Multi-Bernoulli filters.

The interacting multiple model (IMM) filter is a well-established and widely used algorithm at present for maneuvering target tracking. Currently, almost all IMM filtering algorithms used are discrete-time filtering algorithms. However, it is rather unknown that the original IMM filter was developed in a purely continuous-time setting, which subsequently led to the development of the discrete-time IMM filter. Chapter 4 presents in detail the mathematical development of exact continuous-time nonlinear filtering for jump Markov systems, including the continuous-time IMM filter as well as continuous-discrete-time IMM and particle filters.

The track-oriented multiple hypothesis tracking (MHT) for multisensor multi-target tracking is regarded as one of the most advanced tracking algorithms at present, relative to which other tracking algorithms are compared. Chapter 5 presents a hybrid-state derivation of the track-oriented MHT equations that is closely related to the original treatment by Kurien [1] with some minor modifications. The target death problem inherent in PHD filtering is also addressed and it is shown that it does not arise in the track-oriented MHT. A number of illustrative examples are considered to demonstrate the merits of MHT. In order to make the chapter self-contained, a comprehensive review of the state-of-the-art filtering and tracking algorithms are summarized in the beginning of the chapter, with extensive references.

Chapter 6 describes several strategies to improve airborne ground surveillance by enhanced tracking performance. The following topics are considered: specific sensor modeling, improved data association using signal strength measurements, exploitation of digital road maps, and detection and tracking of target groups. The proposed algorithms are shown to enhance track precision and track continuity over conventional techniques.

Chapter 7 presents a review of recent developments in the calculation of mean square error tracker performance bounds, together with examples that demonstrate how such bounds can be used as a basis for performing online sensor management. The review concentrates on the posterior Cramér–Rao lower bound (PCRLB), and describes computationally efficient formulations of the PCRLB that take account of real-world complexity. Two applications, concerned with the deployment of passive sonobuoys, and UAV trajectory planning, demonstrate that the PCRLB provides an efficient mechanism for performing sensor management in order to accurately track an evasive target.

Chapter 8 presents a review of the track-before-detect (TBD) problem, namely tracking when the measurement is an intensity map. It describes the different methods that have been applied to this problem and compares their performance on a simple scenario. A case study fusing data from an infra-red camera and microwave radar illustrate the advantages that can be gained through the improved sensitivity offered by the track-before-detect algorithm.

While centralized detection and estimation are known to outperform distributed approaches, the same is not always true when one is confronted with measurement origin uncertainty. Indeed, all known approaches to multitarget tracking are suboptimal. Thus, judicious multistage processing may outperform single-stage processing. In a sense, we are choosing between (suboptimal) distributed and (suboptimal) centralized processing. Chapter 9 identifies a number of scenarios where multistage fusion architectures lead to promising results.

Chapter 10 presents an overview of meta-level tracking algorithms for inferring target intent. Such meta-level trackers are fully compatible with existing target tracking algorithms and form the sensor–human interface. To capture the complex spatial trajectories of targets, stochastic context free grammars are used. Then Bayesian signal processing algorithms are used to estimate the target trajectory.

Chapter 11 presents an overview of stochastic control methods for radar resource management. Radar resource management is intrinsically a partially observed stochastic control problem since decisions need to be made based on the estimates provided by a tracker. Such problems are typically intractable unless the underlying structure is exploited. The chapter shows how supermodularity and lattice programming methods can be used to characterize the structure of the optimal radar scheduling policy.

Chapter 12 addresses the problem of multisensor resource management with application to multitarget tracking. Specifically, sensor selection, sensor placement, and performance evaluation are considered in detail. A particular contribution of this chapter is the derivation of the Posterior Cramér–Rao Lower Bound (PCRLB) to quantify the achievable estimation accuracy in multitarget tracking problem, which is used as the key metric for sensor management.

Chapter 13 on efficient inference in general hybrid Bayesian networks for classification introduces a probabilistic inference framework for hybrid Bayesian networks, in which both discrete and continuous variables are present and their functional relationship can be nonlinear. This type of model is very common in classification applications where discrete random variables representing entity types or situational hypotheses are to be assessed given noisy observations represented by mixed discrete and continuous variables.

Chapter 14 presents a new analytical approach for quantifying the long-run performance of a multisensor classification system modeled by a Bayesian network. The methodology has been applied to fusion performance evaluation of practical tracking and classification systems involving multiple sensor types. It illustrates the use of off-line evaluation to estimate marginal performance gains and sensor mode selection using measures and metrics derived herein.

Chapter 15 considers the problem of detecting, estimating, and searching for point and distributed sources of radiation. A Bayesian approach is adopted with the posterior density approximated using the notion of progressive correction combined with either Monte Carlo approximation or linearization.

In Chapter 16, important problems of distributed detection and decision fusion for a multisensor system are discussed. With known local sensors' performance indices, the design for optimal decision fusion rule at the fusion center and the optimal local decision rules at sensors are presented in both parallel and serial networks under either the Bayesian or Neyman–Pearson criterion. When local sensors are nonidentical and their performance indices are unknown, the counting rule is proposed and its exact as well as approximated performance are analyzed. For the challenging problem of distributed detection with correlated observations, a decision fusion framework using copula theory is described, which is shown particularly useful for non-Gaussian distributed and nonlinearly dependent sensor observations.

Chapter 17 presents the development of an automatic knowledge-based information fusion system to support the decision making process in a reliable, timely, and consistent manner even in conditions of uncertainty. This is obtained by using the framework of valuation algebra for knowledge representation and reasoning under uncertainty together with the algorithms for performing local computations in valuation algebra. These algorithms are then specialized to the theory of belief functions. Two practical examples are discussed: decision support systems for target identification and threat assessment.

ACKNOWLEDGMENTS

We are indebted to Dr. Sankar Basu of National Science Foundation who first suggested the idea of writing a book for Wiley/IEEE to Mahendra Mallick. Dr. Basu emphasized that the book should pay special attention to solving practical problems of interest with sound algorithms and examples.

The editors would like to thank their respective universities for the provision of the facilities for completing this book, namely the University of British Columbia and Curtin University. In the preparation of this book, the third editor, Professor Vo is supported in part by the Australian Research Council under the discovery grant DP0878158.

The contents of the book have greatly benefitted from interactions with numerous researchers from diverse fields. We express our sincere thanks to the late Jean-Pierre Le Cadre (IRISA/CNRS, France), Samuel S. Blackman (Raytheon Systems Company, USA), Barbara La Scala (National Australia Bank, Australia), Yvo Boers (THALES Nederland), and David Salmond (QinetiQ, UK).

Finally, we would like to acknowledge our families for their support and patience during the writing, correcting, and editing of this book.

REFERENCE

1. T. Kurien, Issues in the design of practical multitarget tracking algorithms, in: Y. Bar-Shalom (Ed.), *Multitarget-Multisensor Tracking: Advanced Applications*, Artech House, Norwood, MA, USA, 1990, Chapter 3.

CONTRIBUTORS

Fahed Abdallah, HEUDIASYC, UMR CNRS 6599, Université de Technologie de Compi'egne, France

Sanjeev Arulampalam, Submarine Combat Systems, Maritime Operations Division, Defence Science & Technology Organisation, Edinburgh, South Australia, Australia

Alessio Benavoli, Istituto "Dalle Molle" di Studi sull'Intelligenza Artificiale (IDSIA), Manno (Lugano), Switzerland

Henk Blom, National Aerospace Laboratory NLR, Amsterdam, The Netherlands

Craig Carthel, Compunetix, Inc., Monroeville, PA, USA

Kuo-Chu Chang, Systems Engineering and Operations Research, Volgenau School of Engineering, George Mason University, Fairfax, VA, USA

Qi Cheng, School of Electrical & Computer Engineering, Oklahoma State University, Stillwater, OK, USA

Daniel Clark, Department of Electrical, Electronic and Computing Engineering, Heriot-Watt University, Riccarton, Edinburgh, UK

Stefano Coraluppi, Compunetix, Inc., Monroeville, PA, USA

Samuel Davey, Intelligence Surveillance & Reconnaissance Division, Defence Science & Technology Organization, Edinburgh, South Australia, Australia

Michael Feldmann, Department Sensor Data and Information Fusion, Fraunhofer FKIE Wachtberg, Germany

Amadou Gning, Department of Computer Science, University College London, London, UK

Neil Gordon, Intelligence Surveillance & Reconnaissance Division, Defence Science & Technology Organisation, Edinburgh, South Australia, Australia

Marcel Hernandez, Hernandez Technical Solutions Ltd., Malvern, UK

Thia Kirubarajan, Electrical and Computer Engineering Department, Communications Research Laboratory, McMaster University, Hamilton, Ontario, Canada

Wolfgang Koch, Department Sensor Data and Information Fusion, Fraunhofer FKIE, Wachtberg, Germany

Vikram Krishnamurthy, Department of Electrical and Computer Engineering, University of British Columbia, Vancouver, Canada

Mahendra Mallick, Propagation Research Associates, Inc., Marietta, GA, USA

Michael Mertens, Department Sensor Data and Information Fusion, Fraunhofer FKIE Wachtberg, Germany

Lyudmila Mihaylova, School of Computing and Communications, InfoLab21, Lancaster University, Lancaster, UK

Mark Morelande, Department of Electrical and Electronic Engineering, The University of Melbourne, Parkville, Victoria, Australia

Ruixin Niu, Department of Electrical & Computer Engineering, Virginia Commonwealth University, Richmond, VA, USA

Branko Ristic, Intelligence Surveillance & Reconnaissance Division, Defence Science & Technology Organization, Fishermans Bend, Victoria, Australia

Mark Rutten, Intelligence Surveillance & Reconnaissance Division, Defence Science & Technology Organization, Edinburgh, South Australia, Australia

Eswar Sivaraman, United Airlines, Enterprise Optimization, Chicago, IL, USA

Wei Sun, SEOR and C4I Center, George Mason University, Fairfax, VA, USA

Ashok Sundaresan, GE Global Research, Niskayuna, NY, USA

Ratnasingham Tharmarasa, Department of Electrical and Computer Engineering, McMaster University, Hamilton, ON, Canada

Martin Ulmke, Department Sensor Data and Information Fusion, Fraunhofer FKIE, Wachtberg, Germany

Pramod Varshney, Department of Electrical Engineering & Computer Science, Syracuse University, Syracuse, NY, USA

Ba-Ngu Vo, Department of Electrical and Computer Engineering, Curtin University, WA, Australia

Ba-Tuong Vo, Department of Electrical and Computer Engineering, Curtin University, WA, Australia

Yanjun Yan, ARCON Corporation, Waltham, MA, USA

PART I

FILTERING

CHAPTER 1

Angle-Only Filtering in Three Dimensions

MAHENDRA MALLICK, MARK MORELANDE, LYUDMILA MIHAYLOVA,
SANJEEV ARULAMPALAM, AND YANJUN YAN

1.1 INTRODUCTION

The angle-only filtering problem in 3D using bearing and elevation measurements is an important problem in many passive tracking applications. For example, it arises in passive ranging using an infrared search and track (IRST) sensor [1, 2], passive sonar, passive radar in the presence of jamming, and satellite to satellite passive tracking [3, 4]. It is the counterpart of the bearings-only filtering problem in 2D. For the 3D case, the objective is to estimate the three-dimensional state of a target, such as position and velocity, using noisy measurements of bearing and elevation from a single maneuvering platform. A great deal of research has been carried out for the bearings-only filtering problem in 2D—see for example, [5–9] and the references therein. However, the number of publications for the angle-only filtering problem in 3D is relatively small [3, 4, 10–18].

Research in angle-only filtering in 3D began by extending the methods developed for the counterpart problem in 2D. For the 2D bearings-only filtering problem, it is well known that, for a target moving with uniform motion, target range cannot be observed without an ownship (sensor) maneuver [19]. Though the prior distribution of the initial state aids in improving observability, its contribution degrades with time. In addition, the accuracy of the state estimate is highly dependent on the nature of the maneuver and the particular target–observer geometry. Early recursive algorithms for this problem were based on the extended Kalman filter (EKF) [20–22] using Cartesian coordinates [23]. Researchers noted that the performance of these algorithms was poor due to premature collapse of the covariance matrix. This led to the formulation of the modified polar coordinates (MPC) [5, 24, 25], in which improved performance was demonstrated.

Integrated Tracking, Classification, and Sensor Management: Theory and Applications, Edited by Mahendra Mallick, Vikram Krishnamurthy, and Ba-Ngu Vo.

The state vector in MPC consists of bearing, bearing-rate, range-rate divided by range, and the inverse of range [5, 9, 24]. The important difference between the MPC and the Cartesian coordinates is that in MPC, the first three elements of the state are observable even before an ownship maneuver. By decoupling the observable and unobservable components of the state vector, this approach was demonstrated to prevent ill-conditioning of the covariance matrix which led to better filter performance [5, 24, 25]. The continuous-time dynamic model for the MPC is nonlinear and is represented by four continuous-time stochastic differential equations (SDEs). The key difficulty of using MPC is that the commonly applied nearly constant velocity model (NCVM) for nonmaneuvering targets is highly nonlinear in MPC. In fact, there has been some confusion in the literature as to how to convert the widely used NCVM from Cartesian coordinates to MPC. In the original work on bearings-only filtering in MPC [24, 25], these equations are numerically integrated to obtain the predicted state and covariance at the discrete measurement times. Subsequently, Aidala and Hammel [5] noted that exact, closed-form discrete-time stochastic difference equations in MPC can be obtained by using the nonlinear transformations between MPC and Cartesian coordinates. They proposed an EKF in these coordinates and claimed superior performance relative to its Cartesian counterpart.

Angle-only filtering in 3D is beset by the same observability issues that arise in the 2D case [19, 26]. As such, most of the research in the 3D angle-only filtering problem has focused on developing algorithms in the modified spherical coordinates (MSC) [17]—the 3D equivalent of MPC. The components of MSC are elevation, elevation-rate, bearing, bearing-rate times cosine of elevation, the inverse of range, and range-rate divided by range. As with MPC in 2D filtering, the main problem when using MSC in 3D filtering is the nonlinear dynamic model which arises when a target moves with the NCVM in Cartesian coordinates. Again, a number of ways of transforming the NCVM in Cartesian coordinates to MSC have been proposed.

As with MPC, the derivation of a dynamic model for MSC begins with a given motion model in Cartesian coordinates. The MSC dynamic model can then be obtained by transformation from MSC to relative Cartesian coordinates at time t_{k-1}, prediction using the NCVM for relative Cartesian coordinates during the time interval $[t_{k-1}, t_k]$, and then transformation from relative Cartesian coordinates to MSC at time t_k. In [17], this approach is used only to compute the predicted state estimate. The predicted covariance matrix is found by a linear, discretized approximation of the continuous-time dynamic model. The underlying Cartesian dynamic model is the Singer model [27]. A similar method is adopted in [4, 10, 11]. Li et al. [4] derived closed form analytic expressions for the discrete-time nonlinear dynamic model in MSC using an approach similar to that used by Aidala and Hammel [5] for MPC, but they do not describe calculation of the predicted covariance. In [13], the EKF is implemented using a discretized linear approximation for both the predicted state estimate and covariance matrix. A particle filter (PF) [9, 28, 29] was implemented using a multistep Euler approximation. In [14], first exact SDEs for MSC and log spherical coordinates (LSC) were derived from the NCVM in 3D for the relative Cartesian state vector. Then EKFs were implemented for MSC and LSC by numerically integrating nonlinear differential equations for the predicted state estimate and covariance matrix.

None of the aforementioned methods of constructing discrete-time dynamic models for MSC is entirely satisfactory as they include unnecessary approximations due to discretization in the calculation of either, or both of, the predicted state estimate and covariance matrix. As mentioned in the previous paragraph, transformations between the relative state vector in Cartesian coordinates and MSC can be used to construct an exact discrete-time dynamic model in MSC for any given Cartesian dynamic model. This is the approach taken here. Although the same dynamic model appears in [10, 11, 17], its usage in their EKF implementations does not properly account for the process noise. In particular, the approximate method used in [10, 11, 17] to calculate the predicted covariance matrix is only valid over time intervals for which the relative geometry between the target and ownship does not change significantly. This does not apply to the exact discrete-time dynamic model used here.

We also consider a dynamic model obtained by approximate discretization of the exact continuous-time dynamic model for MSC [14, 17]. The approximate discretization is performed using a truncated stochastic Taylor series expansion [30]. There are several reasons for considering an approximate version even though an exact discrete-time dynamic model is available. First, the exact discrete-time dynamic model is nonlinear in both the state and the process noise making implementation of the EKF and unscented Kalman filter (UKF) [9, 31, 32] more complicated. For the EKF, the gradient of the nonlinear time-evolution function with respect to the process noise must be calculated. For the UKF, both the dimension of the state and number of sigma points are increased due to using an augmented state which includes the MSC and process noise. The approximate dynamic models are nonlinear only in the predicted state estimate. Second, the EKF and UKF are only capable of performing the nonlinear dynamic model in an approximate fashion. Therefore, there are two possible sources of error; one due to approximations used in the filtering algorithm and the other due to the approximation in discretizing the continuous-time dynamic model in MSC. It is possible that lower algorithm approximations may counteract the errors incurred by using an approximate dynamic model. Third, the simpler form of the approximate dynamic model permits the use of the so-called optimal importance density (OID) [33] in the PF implementation. The opportunity to use a more efficient sampling density could potentially offset errors in the model. These algorithms belong to the class of continuous-discrete filtering [21, 34, 35], where the dynamic model is continuous and the measurement model is discrete. Our continuous-discrete filtering algorithms are more general than those in [21, 34], where the process noise is additive. In our formulation, the time-evolution function is nonlinear in MSC, process noise, and ownship motion. These algorithms find applications in ballistic target tracking [36–38], ballistic projectile tracking [39], and satellite orbit determination [20, 40], where the dynamic model is described by a SDE and the corresponding discrete-time stochastic difference equation is intractable.

In summary, three classes of filtering algorithms are considered. Each class considers the relative state vector and uses a discrete-time measurement model for bearing and elevation. The measurement model can be linear or nonlinear depending on the definition of the state vector. The first class of algorithms uses the relative Cartesian state vector with the discrete-time NCVM, and hence the measurement model is nonlinear. The second class uses the exact discrete-time dynamic model in MSC. Since the bearing and elevation are components of the MSC, the measurement model is linear. The third class uses the MSC with approximate discretization of the continuous-time dynamic model. An EKF, an UKF, and a bootstrap filter (BF) [9, 28, 29] are developed for each class. Further, an OID-based PF (OIDPF) is developed for filtering in MSC with an approximate dynamic model. Monte Carlo simulations are used to compare the algorithms with an emphasis on determining the main factors affecting performance.

The chapter is organized as follows. The statement of the problem and underlying assumptions are described in Section 1.2. Section 1.3 defines the tracker and sensor coordinate frames. Section 1.4 describes different coordinate systems for the target and ownship states. Section 1.5 presents the dynamic models of the target and ownship, whereas the measurement models for the Cartesian relative state and MSC are presented in Section 1.6. Filter initialization is described in Section 1.7.

Sections 1.8–1.10 present algorithms based on the EKF, UKF, and PF, respectively. Numerical simulations and results are described in Section 1.11 and conclusions are summarized in Section 1.12. Appendix 1A presents the derivation of the continuous-time dynamic model for MSC and Appendix 1B describes transformations between relative Cartesian coordinates and MSC. Finally, Appendix 1C presents details of the filter initialization using relative Cartesian coordinates and MSC.

1.2 STATEMENT OF PROBLEM

We consider the 3D angle-only filtering problem under the following assumptions:

1. We estimate the state of a nonmaneuvering target in 3D using bearing (β) and elevation (ϵ) angle measurements.
2. The target motion is described by an NCVM [22] in 3D.
3. The motion of the ownship or sensor is deterministic, that is, nonrandom and we know the state of the ownship precisely. The state of the target is defined in the tracker coordinate frame (T frame).

The ownship performs a number of maneuvers so that the target state becomes observable.

1.3 TRACKER AND SENSOR COORDINATE FRAMES

The origin of the T frame has geodetic longitude λ_0, geodetic latitude ϕ_0, and geodetic height h_0. The X, Y, and Z axes of the T frame are along the local east, north, and upward directions, respectively, as shown in Figure 1.1.

Traditionally, range, bearing, and elevation angles for multiple targets are defined in a single sensor frame. For the current problem, the bearing and elevation angles are defined in the T frame as shown in Figure 1.1.

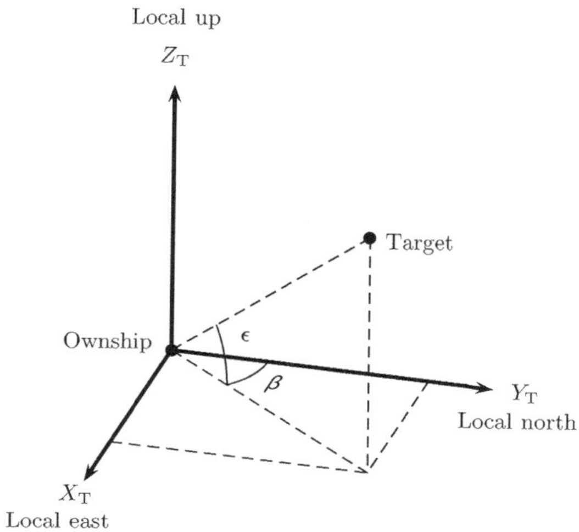

Figure 1.1 Definition of the tracker coordinate frame (T frame), bearing $\beta \in [0, 2\pi]$ and elevation angle $\epsilon \in [-\pi/2, \pi/2]$.

For a given target, a sensor coordinate frame (S frame) is defined such that the Z axis of the S frame is along the range vector. Using the bearing and elevation angles of the target, the rotational transformation matrix $\mathrm{T}_\mathrm{T}^\mathrm{S}$ from the T frame to the S frame is obtained by a (3,2) Euler sequence [41] with Euler angles $(\phi = \pi/2 - \beta, \theta = \pi/2 - \epsilon)$, respectively. The S frame is defined for obtaining decoupled SDEs for the MSC or LSC [14, 17]. The transformation matrix $\mathrm{T}_\mathrm{T}^\mathrm{S}$ is

$$\mathrm{T}_\mathrm{T}^\mathrm{S} = \begin{bmatrix} \sin\epsilon \sin\beta & \sin\epsilon\cos\beta & -\cos\epsilon \\ -\cos\beta & \sin\beta & 0 \\ \cos\epsilon\sin\beta & \cos\epsilon\cos\beta & \sin\epsilon \end{bmatrix}. \tag{1.1}$$

The rotational transformation $\mathbf{T}_\mathrm{T}^\mathrm{S}$ from the T frame to the S frame is defined differently from that in [17].

1.4 COORDINATE SYSTEMS FOR TARGET AND OWNSHIP STATES

1.4.1 Cartesian Coordinates for State Vector and Relative State Vector

The Cartesian states of the target and ownship are defined in the T frame, respectively, by

$$\mathbf{x}^\mathrm{t} := \begin{bmatrix} x^\mathrm{t} & y^\mathrm{t} & z^\mathrm{t} & \dot{x}^\mathrm{t} & \dot{y}^\mathrm{t} & \dot{z}^\mathrm{t} \end{bmatrix}', \tag{1.2}$$

and

$$\mathbf{x}^{\mathrm{o}} := \begin{bmatrix} x^{\mathrm{o}} & y^{\mathrm{o}} & z^{\mathrm{o}} & \dot{x}^{\mathrm{o}} & \dot{y}^{\mathrm{o}} & \dot{z}^{\mathrm{o}} \end{bmatrix}'. \tag{1.3}$$

The relative state vector of the target in the T frame is defined by

$$\mathbf{x} := \mathbf{x}^{\mathrm{t}} - \mathbf{x}^{\mathrm{o}}. \tag{1.4}$$

Let $\mathbf{x} = [x \quad y \quad z \quad \dot{x} \quad \dot{y} \quad \dot{z}]'$ denote the relative state vector in the T frame. Then the relative state vector can be written as

$$\begin{aligned} \mathbf{x} &:= \begin{bmatrix} x & y & z & \dot{x} & \dot{y} & \dot{z} \end{bmatrix}' \\ &= \begin{bmatrix} x^{\mathrm{t}} - x^{\mathrm{o}} & y^{\mathrm{t}} - y^{\mathrm{o}} & z^{\mathrm{t}} - z^{\mathrm{o}} & \dot{x}^{\mathrm{t}} - \dot{x}^{\mathrm{o}} & \dot{y}^{\mathrm{t}} - \dot{y}^{\mathrm{o}} & \dot{z}^{\mathrm{t}} - \dot{z}^{\mathrm{o}} \end{bmatrix}'. \end{aligned}$$

Let $\mathbf{r}^{\mathrm{T}}$ denote the range vector of the target from the sensor in the T frame. Then $\mathbf{r}^{\mathrm{T}}$ is defined by

$$\mathbf{r}^{\mathrm{T}} := \begin{bmatrix} x & y & z \end{bmatrix}' = \begin{bmatrix} x^{\mathrm{t}} - x^{\mathrm{o}} & y^{\mathrm{t}} - y^{\mathrm{o}} & z^{\mathrm{t}} - z^{\mathrm{o}} \end{bmatrix}'. \tag{1.5}$$

The slant range (or simply range) is defined as

$$r := \|\mathbf{r}^{\mathrm{T}}\| = \sqrt{x^2 + y^2 + z^2}, \quad r > 0. \tag{1.6}$$

The range vector can be expressed in terms of range, bearing, and elevation by

$$\mathbf{r}^{\mathrm{T}} = r \begin{bmatrix} \cos\epsilon \sin\beta \\ \cos\epsilon \cos\beta \\ \sin\epsilon \end{bmatrix}, \quad \beta \in [0, 2\pi], \quad \epsilon \in [-\pi/2, \pi/2]. \tag{1.7}$$

The ground range is defined by

$$\rho := \sqrt{x^2 + y^2} = r\cos\epsilon, \quad \rho > 0. \tag{1.8}$$

1.4.2 Modified Spherical Coordinates for Relative State Vector

Following Stallard's convention [17, 18], we use ω as a component of the MSC, where

$$\omega(t) := \dot{\beta}(t)\cos\epsilon(t). \tag{1.9}$$

Let $\zeta(t)$ denote the logarithm of range $r(t)$

$$\zeta(t) := \ln r(t). \tag{1.10}$$

Then

$$r(t) = \exp[\zeta(t)]. \tag{1.11}$$

Differentiating (1.10) with respect to time, we get

$$\dot{\zeta}(t) = \frac{\dot{r}(t)}{r(t)}. \tag{1.12}$$

Using the convention in [14], the relative state vector of the target in MSC is defined by

$$\begin{aligned} \boldsymbol{\xi}(t) &:= \begin{bmatrix} \xi_1(t) & \xi_2(t) & \xi_3(t) & \xi_4(t) & \xi_5(t) & \xi_6(t) \end{bmatrix}' \\ &= \begin{bmatrix} \omega(t) \ \dot{\epsilon}(t) \ \dot{\zeta}(t) \ \beta(t) \ \epsilon(t) \ \dfrac{1}{r(t)} \end{bmatrix}'. \end{aligned} \tag{1.13}$$

The components of the MSC defined in [17] and [14] are the same; however, the ordering is different.

1.5 DYNAMIC MODELS

In this section, we present the continuous-time and discrete-time models for the time evolution of the target state in each coordinate system.

1.5.1 Dynamic Model for State Vector and Relative State Vector in Cartesian Coordinates

As stated in Section 1.2 , the target is assumed to move with the NCVM in 3D. According to this motion model, the target velocity in each dimension is subject to Gaussian-distributed random perturbations. The target state in Cartesian coordinates, referenced in the T frame, then satisfies the following SDE [21, 22]

$$\frac{d\mathbf{x}^{\mathrm{t}}(t)}{dt} = \mathbf{A}\mathbf{x}^{\mathrm{t}}(t) + \mathbf{B}\mathbf{w}^{\mathrm{t}}(t), \tag{1.14}$$

where $\mathbf{w}^{\mathrm{t}}(t)$ is a zero-mean continuous-time Gaussian white noise acceleration with power spectral density matrix $\mathbf{Q}^{\mathrm{t}}(t) = \mathrm{diag}(q_x, q_y, q_z)$ and

$$\mathbf{A} = \begin{bmatrix} \mathbf{0}_3 & \mathbf{I}_3 \\ \mathbf{0}_3 & \mathbf{0}_3 \end{bmatrix}, \tag{1.15}$$

$$\mathbf{B} = \begin{bmatrix} \mathbf{0}_3 \\ \mathbf{I}_3 \end{bmatrix}, \tag{1.16}$$

with $\mathbf{0}_3$ a 3×3 matrix of zeros and $\mathbf{I}_3$ the 3×3 identity matrix.

Let $\mathbf{x}^{\mathrm{o}}(t)$ denote the nonrandom ownship state which satisfies

$$\frac{d\mathbf{x}^{\mathrm{o}}(t)}{dt} = \mathbf{A}\mathbf{x}^{\mathrm{o}}(t) + \mathbf{B}\mathbf{a}_{\mathrm{o}}^{\mathrm{T}}(t), \tag{1.17}$$

where $\mathbf{a}_{\mathrm{o}}^{\mathrm{T}}(t)$ is the ownship acceleration in the T frame. Two types of ownship motion are considered; constant velocity (CV) and a coordinated turn (CT) [22] in a plane parallel to the XY plane of the T frame. The SDE for the relative state vector of the target in the T frame is

$$\frac{d\mathbf{x}(t)}{dt} = \mathbf{A}\mathbf{x}(t) + \mathbf{B}\mathbf{w}(t), \tag{1.18}$$

where

$$\mathbf{w}(t) = \mathbf{w}^{\mathrm{t}}(t) - \mathbf{a}_{\mathrm{o}}^{\mathrm{T}}(t). \tag{1.19}$$

Since measurements are acquired at discrete times, it is desirable to have a corresponding discrete-time version of the continuous-time dynamic model. We therefore seek a stochastic difference equation in which the state at time t_k is written in terms of the state at time t_{k-1}, the ownship input, and random process noise. Let $\mathbf{x}_k := \mathbf{x}(t_k)$ denote the relative state vector of the target in Cartesian coordinates at t_k. The SDE (1.18) governing the evolution of $\mathbf{x}(t)$ corresponds to the following stochastic difference equation [21, 22]

$$\mathbf{x}_k = \mathbf{F}(\Delta_k)\mathbf{x}_{k-1} + \int_{t_{k-1}}^{t_k} \mathbf{F}(t_k - t)\mathbf{B}\mathbf{w}(t)\,dt, \tag{1.20}$$

where $\Delta_k := t_k - t_{k-1}$ is the measurement sampling interval and $\mathbf{F}(\Delta)$ is the state transition matrix for the time interval Δ

$$\mathbf{F}(\Delta) = \begin{bmatrix} 1 & \Delta \\ 0 & 1 \end{bmatrix} \otimes \mathbf{I}_3, \tag{1.21}$$

where $\otimes$ refers to the Kronecker product [42]. From (1.20) we get

$$\mathbf{x}_k = \mathbf{F}(\Delta_k)\mathbf{x}_{k-1} + \mathbf{w}_{k-1} - \mathbf{u}_{k-1}, \tag{1.22}$$

where

$$\mathbf{w}_{k-1} := \int_{t_{k-1}}^{t_k} \mathbf{F}(t_k - t)\mathbf{B}\mathbf{w}^{\mathrm{t}}(t)\,dt, \tag{1.23}$$

$$\mathbf{u}_{k-1} := \int_{t_{k-1}}^{t_k} \mathbf{F}(t_k - t)\mathbf{B}\mathbf{a}_{\mathrm{o}}^{\mathrm{T}}(t)\,dt. \tag{1.24}$$

Using the properties of $\mathbf{w}^{\mathrm{t}}(t)$, state transition matrix $\mathbf{F}$ defined in (1.21), and the definition of $\mathbf{w}_{k-1}$ in (1.23), we can show that $\mathbf{w}_{k-1}$ is a zero-mean Gaussian white integrated process noise with covariance $\mathbf{Q}(\Delta_k)$ [22]

$$\mathbf{w}_{k-1} \sim N(\cdot\,; \mathbf{0}, \mathbf{Q}(\Delta_k)), \tag{1.25}$$

where

$$\mathbf{Q}(\Delta) = \begin{bmatrix} \Delta^3/3 & \Delta^2/2 \\ \Delta^2/2 & \Delta \end{bmatrix} \otimes \mathrm{diag}(q_x, q_y, q_z). \tag{1.26}$$

The turn rate of the ownship during the interval $[t_{k-1}, t_k]$ is denoted as ω_k^{o}. Note that $\omega_k^{\mathrm{o}} = 0$ corresponds to the CV motion while the CT motion results if $\omega_k^{\mathrm{o}} \neq 0$. For the general motion of the ownship (e.g., CV, CT, or any other deterministic motion) the input vector $\mathbf{u}_{k-1}$ is given by [9]

$$\mathbf{u}_{k-1} = \mathbf{x}_k^{\mathrm{o}} - \mathbf{F}(\Delta_k)\mathbf{x}_{k-1}^{\mathrm{o}}. \tag{1.27}$$

For the current problem, we consider two types of deterministic motions for the ownship, CV and CT. Since the motion of the ownship is assumed to be deterministic, the process noise is not used for the dynamic model of the ownship and the state transition matrix is sufficient. The state transition matrix for the CV motion is given by (1.21) and that for the CT motion is

$$\mathbf{F}^{\mathrm{CT}}(\Delta, \omega) = \begin{bmatrix} 1 & 0 & 0 & \sin(\omega\Delta)/\omega & -[1-\cos(\omega\Delta)]/\omega & 0 \\ 0 & 1 & 0 & [1-\cos(\omega\Delta)]/\omega & \sin(\omega\Delta)/\omega & 0 \\ 0 & 0 & 1 & 0 & 0 & \Delta \\ 0 & 0 & 0 & \cos(\omega\Delta) & -\sin(\omega\Delta) & 0 \\ 0 & 0 & 0 & \sin(\omega\Delta) & \cos(\omega\Delta) & 0 \\ 0 & 0 & 0 & 0 & 0 & 1 \end{bmatrix}. \tag{1.28}$$

Note that when ω_k^{o} approaches zero, $\mathbf{F}^{\mathrm{CT}}(\Delta_k, \omega_k^{\mathrm{o}})$ reduces to $\mathbf{F}(\Delta_k)$ and $\mathbf{u}_{k-1}$ becomes zero.

1.5.2 Dynamic Model for Relative State Vector in Modified Spherical Coordinates

It is desired to construct a dynamic model for the relative state vector in MSC which is equivalent to the model (1.22) obtained for the relative state vector in Cartesian coordinates. Two ways of doing this are considered. The first method is to discretize a continuous-time model equivalent to (1.18). The second approach is to use the discrete-time model obtained for the relative state vector in Cartesian coordinates along with transformations between MSC and Cartesian coordinates.

We consider first discretization of a continuous-time dynamic model in MSC. We follow the approach of Stallard [17] to derive the SDEs for MSC. The key steps of the derivation, given in detail in Appendix 1A, are as follows:

1. Derive expressions for the relative velocity ($\dot{x}$, $\dot{y}$, $\dot{z}$) and relative acceleration ($\ddot{x}$, $\ddot{y}$, $\ddot{z}$) as functions of range, bearing, and elevation and their derivatives in the T frame.
2. Define an S frame such that the Z axis of the S frame is along the range vector.
3. Using the bearing and elevation angles, calculate the rotational transformation matrix $\mathrm{T}_{\mathrm{T}}^{\mathrm{S}}$ from the T frame to the S frame.
4. Transform the relative acceleration $\ddot{\mathbf{r}}^{\mathrm{T}}$ in the T frame to the S frame to yield $\ddot{\mathbf{r}}^{\mathrm{S}}$ which are functions of MSC.
5. Equate $\ddot{\mathbf{r}}^{\mathrm{S}}$ to the difference of the target acceleration (white noise acceleration) and ownship acceleration, expressed in the S frame. This produces the desired SDEs for the MSC.

This approach yields the following SDE for the target state in MSC

$$\frac{d\boldsymbol{\xi}(t)}{dt} = \mathbf{f}(\boldsymbol{\xi}(t), t) + \mathbf{G}(\boldsymbol{\xi}(t))\, \mathbf{w}^{\mathrm{t}}(t), \tag{1.29}$$

where

$$\mathbf{f}(\boldsymbol{\xi}, t) = \mathbf{s}(\boldsymbol{\xi}) - \mathbf{G}(\boldsymbol{\xi})\mathbf{a}_{\mathrm{o}}^{\mathrm{T}}(t), \tag{1.30}$$

$$\mathbf{G}(\boldsymbol{\xi}) = \xi_6 \left[\begin{array}{ccc} \cos\xi_4 & -\sin\xi_4 & 0 \\ -\sin\xi_4 \sin\xi_5 & -\cos\xi_4 \sin\xi_5 & \cos\xi_5 \\ \sin\xi_4 \cos\xi_5 & \cos\xi_4 \cos\xi_5 & \sin\xi_5 \\ \hline & \mathbf{0}_3 & \end{array}\right], \tag{1.31}$$

with

$$\mathbf{s}(\boldsymbol{\xi}) = \begin{bmatrix} \xi_1(\xi_2 \tan\xi_5 - 2\xi_3) \\ -2\xi_2\xi_3 - \xi_1^2 \tan\xi_5 \\ \xi_1^2 + \xi_2^2 - \xi_3^2 \\ \xi_1/\cos\xi_5 \\ \xi_2 \\ -\xi_3\xi_6 \end{bmatrix}. \tag{1.32}$$

An approximate discretization of (1.29) can be obtained using truncated stochastic Taylor series expansions. Stochastic Taylor series expansions are obtained by repeated application of the Itô lemma. Further details can be found in [30].

Let $\boldsymbol{\xi}_k := \boldsymbol{\xi}(t_k)$ denote the relative state vector in MSC at t_k. A first-order stochastic Taylor series approximation, also called the Euler approximation, results in the following stochastic difference equation for the time evolution of the target state in MSC

$$\boldsymbol{\xi}_k \approx \mathbf{a}_1(\boldsymbol{\xi}_{k-1}, t_{k-1}; \Delta_k) + \mathbf{v}_k, \tag{1.33}$$

where $\mathbf{v}_k \sim N(\cdot\,; \mathbf{0}, \mathbf{C}_1(\boldsymbol{\xi}_{k-1}; \Delta_k))$ and

$$\mathbf{a}_1(\boldsymbol{\xi}, t; \Delta) = \boldsymbol{\xi} + \Delta \mathbf{f}(\boldsymbol{\xi}, t), \tag{1.34}$$

$$\mathbf{C}_1(\boldsymbol{\xi}; \Delta) = \Delta \mathbf{G}(\boldsymbol{\xi}) \operatorname{diag}(q_x, q_y, q_z)\, \mathbf{G}(\boldsymbol{\xi})'. \tag{1.35}$$

A more accurate approximation can be obtained using a second-order stochastic Taylor series expansion. Define the derivative of the matrix-valued function $\mathbf{A}$ with respect to the matrix-valued argument $\mathbf{B}$ as [43]

$$D_{\mathbf{B}}\mathbf{A}(\mathbf{B}) = \mathbf{A}(\mathbf{B}) \otimes \nabla_{\mathbf{B}}, \tag{1.36}$$

where $\nabla_{\mathbf{B}} = [\partial/\partial b_{i,j}]$ with $\mathbf{B} = [b_{i,j}]$. Also, denote the lth column of $\mathbf{G}$ as $\mathbf{g}_l$. Then, the second-order stochastic Taylor series approximation of the SDE (1.29) is

$$\boldsymbol{\xi}_k \approx \mathbf{a}_2(\boldsymbol{\xi}_{k-1}, t_{k-1}; \Delta_k) + \mathbf{v}_k, \tag{1.37}$$

where $\mathbf{v}_k \sim N(\cdot\,; \mathbf{0}, \mathbf{C}_2(\boldsymbol{\xi}_{k-1}, t_{k-1}; \Delta_k))$ and

$$\mathbf{a}_2(\boldsymbol{\xi}, t; \Delta) = \boldsymbol{\xi} + \Delta \mathbf{f}(\boldsymbol{\xi}, t) + \Delta^2 \mathbf{j}(\boldsymbol{\xi}, t)/2, \tag{1.38}$$

$$\mathbf{C}_2(\boldsymbol{\xi}, t; \Delta) = \left[\mathbf{M}(\boldsymbol{\xi}, t) - \mathbf{N}(\boldsymbol{\xi}, t)\ \mathbf{G}(\boldsymbol{\xi}) + \Delta \mathbf{N}(\boldsymbol{\xi}, t)\right] \mathbf{Q}(\Delta) \begin{bmatrix} \mathbf{M}(\boldsymbol{\xi}, t)' - \mathbf{N}(\boldsymbol{\xi}, t)' \\ \mathbf{G}(\boldsymbol{\xi})' + \Delta \mathbf{N}(\boldsymbol{\xi}, t)' \end{bmatrix}, \tag{1.39}$$

with

$$\mathbf{j}(\boldsymbol{\xi}, t) = D_{\boldsymbol{\xi}'}\mathbf{f}(\boldsymbol{\xi}, t) + \sum_{l=1}^{3} (\mathbf{I}_6 \otimes \mathbf{g}_l(\boldsymbol{\xi})') D_{\boldsymbol{\xi}} D_{\boldsymbol{\xi}'} \mathbf{f}(\boldsymbol{\xi}, t), \tag{1.40}$$

$$\mathbf{M}(\boldsymbol{\xi}, t) = D_{\boldsymbol{\xi}'}\mathbf{f}(\boldsymbol{\xi}, t)\mathbf{G}(\boldsymbol{\xi}), \tag{1.41}$$

$$\mathbf{N}(\boldsymbol{\xi}, t) = D_{\boldsymbol{\xi}'}\mathbf{G}(\boldsymbol{\xi})(\mathbf{I}_3 \otimes \mathbf{f}(\boldsymbol{\xi}, t)) + \sum_{l=1}^{3} (\mathbf{I}_6 \otimes \mathbf{g}_l(\boldsymbol{\xi})') D_{\boldsymbol{\xi}} D_{\boldsymbol{\xi}'} \mathbf{G}(\boldsymbol{\xi})(\mathbf{I}_3 \otimes \mathbf{g}_l(\boldsymbol{\xi})). \tag{1.42}$$

The covariance matrix $\mathbf{Q}(\Delta)$ is given in (1.26).

Although the discrete-time dynamic equations (1.33) and (1.37) are approximate they can be made arbitrarily accurate by reducing the time over which they are applied. Therefore to increase accuracy the sampling period Δ_k can be split into subintervals with approximation (1.33) or (1.37) applied over successive subintervals. It can be expected that the second-order approximation (1.37) will require fewer subintervals than the first-order approximation (1.33) to achieve a given level of accuracy. However, this is balanced by the higher computational expense of the second-order approximation so that better accuracy with a given computational expense may possibly be achieved by performing several first-order approximations rather than relatively few second-order approximations.

We now consider a dynamic model obtained by transformation of the Cartesian discrete-time dynamic model (1.22). Let $\mathbf{f}_{\mathrm{C}}^{\mathrm{MSC}} : \mathbb{R}^6 \to \mathbb{R}^6$ denote the transformation from relative Cartesian coordinates to MSC. Similarly, let $\mathbf{f}_{\mathrm{MSC}}^{\mathrm{C}} : \mathbb{R}^6 \to \mathbb{R}^6$ denote the inverse transformation from MSC to relative Cartesian coordinates. Then,

$$\boldsymbol{\xi}_k = \mathbf{f}_{\mathrm{C}}^{\mathrm{MSC}}(\mathbf{x}_k), \tag{1.43}$$

$$\mathbf{x}_k = \mathbf{f}_{\mathrm{MSC}}^{\mathrm{C}}(\boldsymbol{\xi}_k). \tag{1.44}$$

Functional forms of $\mathbf{f}_{\mathrm{C}}^{\mathrm{MSC}}$ and $\mathbf{f}_{\mathrm{MSC}}^{\mathrm{C}}$ are presented in Appendix 1B. Details of transformations between relative Cartesian coordinates and MSC are also presented in Appendix 1B. Then, using (1.22) in (1.43) and using (1.44) for $\mathbf{x}_{k-1}$, we obtain the time evolution of the relative state vector in MSC,

$$\begin{aligned}\boldsymbol{\xi}_k &= \mathbf{f}_{\mathrm{C}}^{\mathrm{MSC}}(\mathbf{F}(\Delta_k)\mathbf{f}_{\mathrm{MSC}}^{\mathrm{C}}(\boldsymbol{\xi}_{k-1}) + \mathbf{w}_{k-1} - \mathbf{u}_{k-1}) \\ &= \mathbf{b}(\boldsymbol{\xi}_{k-1}, \mathbf{u}_{k-1}, \mathbf{w}_{k-1}),\end{aligned} \tag{1.45}$$

where $\mathbf{w}_{k-1}$ and $\mathbf{u}_{k-1}$ are specified in Section 1.5.1. The dynamic model (1.45) is exact, unlike the discretized approximations (1.33) and (1.37) of the continuous-time model (1.29). We note that a closed form analytic expression for the nonlinear function $\mathbf{b}$ is difficult to obtain. However, it is straightforward to calculate the predicted state estimate $\hat{\boldsymbol{\xi}}_{k|k-1}$ approximately given $\hat{\boldsymbol{\xi}}_{k-1|k-1}$, using nested functions in (1.45). A minor disadvantage of the exact dynamic model compared to the Taylor series approximations is that the process noise $\mathbf{w}_{k-1}$ is nonlinearly transformed. As will be seen in Sections 1.8 and 1.9 this makes the EKF and UKF slightly more complicated.

1.6 MEASUREMENT MODELS

1.6.1 Measurement Model for Relative Cartesian State

The measurement model for the bearing and elevation angles using the relative Cartesian state vector $\mathbf{x}_k$ is

$$\mathbf{z}_k = \mathbf{h}(\mathbf{x}_k) + \mathbf{n}_k, \tag{1.46}$$

where

$$\mathbf{h}(\mathbf{x}_k) = \begin{bmatrix} \beta_k \\ \epsilon_k \end{bmatrix} = \begin{bmatrix} \arctan(x_k, y_k) \\ \arctan(z_k, \rho_k) \end{bmatrix}, \quad \beta_k \in [0, 2\pi], \quad \epsilon_k \in [-\pi/2, \pi/2], \quad (1.47)$$

and

$$\mathbf{n}_k \sim N(\cdot\, ; \mathbf{0}, \mathbf{R}) \quad (1.48)$$

is the measurement noise with covariance matrix $\mathbf{R} := \mathrm{diag}(\sigma_\beta^2, \sigma_\epsilon^2)$, with σ_β and σ_ϵ the standard deviations for the bearing and elevation angle, respectively. The ground range ρ appearing in (1.47) is defined in (1.8).

1.6.2 Measurement Model for Modified Spherical Coordinates

The measurement model for the bearing and elevation angles β_k, ϵ_k using MSC is

$$\mathbf{z}_k = \mathbf{H}\boldsymbol{\xi}_k + \mathbf{n}_k, \quad (1.49)$$

$$\mathbf{H} := \begin{bmatrix} 0 & 0 & 0 & 1 & 0 & 0 \\ 0 & 0 & 0 & 0 & 1 & 0 \end{bmatrix}. \quad (1.50)$$

1.7 FILTER INITIALIZATION

Initialization is performed at time t_1 using the measurement $\mathbf{z}_1$ and prior information on the range and velocity of the target. The target state at time t_1 can be completely defined by the vector $\boldsymbol{\phi} := [\beta, \epsilon, r, s, \alpha, \gamma]'$ where β is bearing, ϵ is elevation, r is range, s is speed, α is the bearing component of heading, and γ is the elevation component of heading. The distribution of $\boldsymbol{\phi}$ given the measurement $\mathbf{z}_1$ and prior information on the range and velocity of the target is

$$\boldsymbol{\phi}|\mathbf{z}_1 \sim N(\cdot\, ; \hat{\boldsymbol{\phi}}_1, \boldsymbol{\Sigma}_1), \quad (1.51)$$

where

$$\hat{\boldsymbol{\phi}}_1 = \begin{bmatrix} \mathbf{z}_1' & \bar{r} & \bar{s} & \bar{\alpha} & \bar{\gamma} \end{bmatrix}', \quad (1.52)$$

$$\boldsymbol{\Sigma}_1 = \mathrm{diag}(\sigma_\beta^2, \sigma_\epsilon^2, \sigma_r^2, \sigma_s^2, \sigma_\alpha^2, \sigma_\gamma^2). \quad (1.53)$$

The distribution (1.51) forms the basis for initialization of the angle-only tracking algorithms in Cartesian coordinates and MSC.

1.7.1 Initialization of Relative Cartesian Coordinates

Initialization of the EKF and UKF requires a mean and covariance matrix. Given a vector $\boldsymbol{\phi}$ distributed according to (1.51) the mean and covariance matrix of the target state in relative Cartesian coordinates are

$$\hat{\mathbf{x}}_1 = \int \mathbf{c}(\boldsymbol{\phi}) N(\boldsymbol{\phi}; \hat{\boldsymbol{\phi}}_1, \boldsymbol{\Sigma}_1)\, d\boldsymbol{\phi}, \tag{1.54}$$

$$\mathbf{P}_1 = \int [\mathbf{c}(\boldsymbol{\phi}) - \hat{\mathbf{x}}_1][\mathbf{c}(\boldsymbol{\phi}) - \hat{\mathbf{x}}_1]' N(\boldsymbol{\phi}; \hat{\boldsymbol{\phi}}_1, \boldsymbol{\Sigma}_1)\, d\boldsymbol{\phi}, \tag{1.55}$$

where

$$\mathbf{c}(\boldsymbol{\phi}) = \begin{bmatrix} r\cos\epsilon\sin\beta \\ r\cos\epsilon\cos\beta \\ r\sin\epsilon \\ s\cos\gamma\sin\alpha - \dot{x}_1^{o} \\ s\cos\gamma\cos\alpha - \dot{y}_1^{o} \\ s\sin\gamma - \dot{z}_1^{o} \end{bmatrix}. \tag{1.56}$$

Note that most approaches to angle only tracking, in both 2D and 3D, use linearized approximations to the integrals (1.54) and (1.55). Here we evaluate the integrals exactly, as shown in Appendix 1C.

PFs are initiated with a collection of samples which can be found as $\mathbf{x}_1^i = \mathbf{c}(\boldsymbol{\phi}^i)$ where $\boldsymbol{\phi}^i \sim N(\cdot\,; \hat{\boldsymbol{\phi}}_1, \boldsymbol{\Sigma}_1)$ for $i = 1, \ldots, n$. The sample weights are $w_1^i = 1/n$, $i = 1, \ldots, n$.

1.7.2 Initialization of Modified Spherical Coordinates

The mean and covariance matrix of the relative target state in MSC given a vector $\boldsymbol{\phi}$ distributed according to (1.51) are

$$\hat{\boldsymbol{\xi}}_1 = \int \mathbf{d}(\boldsymbol{\phi}) N(\boldsymbol{\phi}; \hat{\boldsymbol{\phi}}_1, \boldsymbol{\Sigma}_1)\, d\boldsymbol{\phi}, \tag{1.57}$$

$$\mathbf{P}_1 = \int [\mathbf{d}(\boldsymbol{\phi}) - \hat{\boldsymbol{\xi}}_1][\mathbf{d}(\boldsymbol{\phi}) - \hat{\boldsymbol{\xi}}_1]' N(\boldsymbol{\phi}; \hat{\boldsymbol{\phi}}_1, \boldsymbol{\Sigma}_1)\, d\boldsymbol{\phi}, \tag{1.58}$$

where

$$\mathbf{d}(\boldsymbol{\phi}) = \begin{bmatrix} 0 \\ 0 \\ 0 \\ \beta \\ \epsilon \\ 1/r \end{bmatrix} + \frac{1}{r} \begin{bmatrix} \cos\beta & -\sin\beta & 0 \\ -\sin\beta\sin\epsilon & -\cos\beta\sin\epsilon & \cos\epsilon \\ \sin\beta\cos\epsilon & \cos\beta\cos\epsilon & \sin\epsilon \\ \hline & \mathbf{0}_3 & \end{bmatrix} \begin{bmatrix} s\cos\gamma\sin\alpha - \dot{x}_1^{\mathrm{o}} \\ s\cos\gamma\cos\alpha - \dot{y}_1^{\mathrm{o}} \\ s\sin\gamma - \dot{z}_1^{\mathrm{o}} \end{bmatrix}. \tag{1.59}$$

The integrals (1.57) and (1.58) can be evaluated over all variables except for the range r. The expressions are given in Appendix 1C. Integration over r can be done numerically using, for example, Monte Carlo approximation.

1.8 EXTENDED KALMAN FILTERS

The well-known EKF is based on linear approximations to nonlinear dynamic and/or measurement models [20–22]. In this section, we present the EKF recursions for the state vector in Cartesian coordinates and MSC.

After initialization using the procedure described in Section 1.7.1, a recursion of the EKF with Cartesian coordinates proceeds as shown in Algorithm 1 for $k = 2, 3, \ldots$. In this case, the dynamic model is linear and the measurement model is nonlinear, so the linearization approximation is required only in the measurement update step. This filter is denoted by CEKF. The situation is reversed in MSC, where the dynamic model

---

Algorithm 1 CEKF: A recursion of the EKF with Cartesian coordinates (CEKF).

Input: posterior mean $\hat{\mathbf{x}}_{k-1|k-1}$ and covariance matrix $\mathbf{P}_{k-1|k-1}$ at time t_{k-1} and the measurement $\mathbf{z}_k$.

Output: posterior mean $\hat{\mathbf{x}}_{k|k}$ and covariance matrix $\mathbf{P}_{k|k}$ at time t_k.

compute the predicted mean $\hat{\mathbf{x}}_{k|k-1}$ and covariance matrix $\mathbf{P}_{k|k-1}$ at time t_k

$$\begin{aligned} \hat{\mathbf{x}}_{k|k-1} &= \mathbf{F}(\Delta_k)\hat{\mathbf{x}}_{k-1|k-1} - \mathbf{u}_k, \\ \mathbf{P}_{k|k-1} &= \mathbf{F}(\Delta_k)\mathbf{P}_{k-1|k-1}\mathbf{F}(\Delta_k)' + \mathbf{Q}(\Delta_k). \end{aligned}$$

compute the Jacobian $\mathbf{H}_k = \nabla_{\mathbf{x}}\mathbf{h}(\mathbf{x})|_{\mathbf{x}=\hat{\mathbf{x}}_{k|k-1}}$.

compute the innovation covariance $\mathbf{S}_k = \mathbf{H}_k\mathbf{P}_{k|k-1}\mathbf{H}_k' + \mathbf{R}$.

compute the gain matrix $\mathbf{K}_k = \mathbf{P}_{k|k-1}\mathbf{H}_k'\mathbf{S}_k^{-1}$.

compute the posterior mean $\hat{\mathbf{x}}_{k|k}$ and covariance matrix $\mathbf{P}_{k|k}$ at time t_k

$$\begin{aligned} \hat{\mathbf{x}}_{k|k} &= \hat{\mathbf{x}}_{k|k-1} + \mathbf{K}_k(\mathbf{z}_k - \mathbf{h}(\hat{\mathbf{x}}_{k|k-1})), \\ \mathbf{P}_{k|k} &= \mathbf{P}_{k|k-1} - \mathbf{K}_k\mathbf{S}_k\mathbf{K}_k'. \end{aligned}$$

---

is nonlinear and the measurement model is linear. The approximate discrete-time dynamic model derived for MSC in Section 1.5.2 involves two integer parameters: the order of the stochastic Taylor series approximation and the number of subintervals over which the approximation is applied for each measurement sampling period. As these parameters are increased both the accuracy and computational expense of the discretization increase. A recursion of the EKF in MSC using a jth order stochastic Taylor series approximation and m subintervals per sampling period is shown in Algorithm 2. This filter is denoted by EKF-MSC(A), where "A" refers to "approximate." In order to use a common notation for the covariance matrix of the jth order stochastic Taylor series expansion $\mathbf{C}_j$, an extra time argument is added to $\mathbf{C}_1$ (see (1.35)) in Algorithm 2. Here $\mathbf{C}_1(\xi, t; \Delta)$ implies $\mathbf{C}_1(\xi; \Delta)$. This interpretation is also used in Algorithms 5 (in Section 1.9), 8 and 9 (in Section 1.10). The EKF in MSC using the exact dynamic model (1.45) is given by Algorithm 3. This filter is denoted by EKF-MSC(E), where "E" refers to "exact." Note that linearization of the dynamic equation is performed over both the previous state ξ_{k-1} and the process noise $\mathbf{w}_{k-1}$. As a result the Jacobian $\mathbf{B}$ is a 6×12 matrix rather than the 6×6 matrix obtained when the dynamic equation is nonlinear in the previous state but linear in the process noise.

Algorithm 2 EKF-MSC(A): A recursion of the EKF with MSC using the approximate dynamic model.

Input: posterior mean $\hat{\xi}_{k-1|k-1}$ and covariance matrix $\mathbf{P}_{k-1|k-1}$ at time t_{k-1} and the measurement $\mathbf{z}_k$.

Output: posterior mean $\hat{\xi}_{k|k}$ and covariance matrix $\mathbf{P}_{k|k}$ at time t_k.

set $\hat{\xi}^0_{k|k-1} = \hat{\xi}_{k-1|k-1}$ and $\mathbf{P}^0_{k|k-1} = \mathbf{P}_{k-1|k-1}$.

for $i = 1, \ldots, m$ **do**

compute the Jacobian $\mathbf{A}^i_j = D_{\xi'}\mathbf{a}_j(\xi, t_{k-1} + (i-1)\Delta_k/m; \Delta_k/m)|_{\xi=\hat{\xi}^{i-1}_{k|k-1}}$.

compute the predicted mean $\hat{\xi}^i_{k|k-1}$ and covariance matrix $\mathbf{P}^i_{k|k-1}$ at time t_k

$$\hat{\xi}^i_{k|k-1} = \mathbf{a}_j(\hat{\xi}^{i-1}_{k|k-1}, t_{k-1} + (i-1)\Delta_k/m; \Delta_k/m),$$

$$\mathbf{P}^i_{k|k-1} = \mathbf{A}^i_j \mathbf{P}^{i-1}_{k|k-1} (\mathbf{A}^i_j)' + \mathbf{C}_j(\hat{\xi}^{i-1}_{k|k-1}, t_{k-1} + (i-1)\Delta_k/m; \Delta_k/m).$$

end

compute the innovation covariance $\mathbf{S}^m_k = \mathbf{H}\mathbf{P}^m_{k|k-1}\mathbf{H}' + \mathbf{R}$.

compute the gain matrix $\mathbf{K}^m_k = \mathbf{P}^m_{k|k-1}\mathbf{H}'(\mathbf{S}^m_k)^{-1}$.

compute the posterior mean $\hat{\xi}_{k|k}$ and covariance matrix $\mathbf{P}_{k|k}$ at time t_k

$$\hat{\xi}_{k|k} = \hat{\xi}^m_{k|k-1} + \mathbf{K}^m_k(\mathbf{z}_k - \mathbf{H}\hat{\xi}^m_{k|k-1}),$$

$$\mathbf{P}_{k|k} = \mathbf{P}^m_{k|k-1} - \mathbf{K}^m_k \mathbf{S}^m_k (\mathbf{K}^m_k)'.$$

Algorithm 3 EKF-MSC(E): A recursion of the EKF with MSC using the exact dynamic model (EKF-MSC(E)).

Input: posterior mean $\hat{\boldsymbol{\xi}}_{k-1|k-1}$ and covariance matrix $\mathbf{P}_{k-1|k-1}$ at time t_{k-1} and the measurement $\mathbf{z}_k$.

Output: posterior mean $\hat{\boldsymbol{\xi}}_{k|k}$ and covariance matrix $\mathbf{P}_{k|k}$ at time t_k.

set $\hat{\boldsymbol{\xi}}^0_{k|k-1} = \hat{\boldsymbol{\xi}}_{k-1|k-1}$ and $\mathbf{P}^0_{k|k-1} = \mathbf{P}_{k-1|k-1}$.

compute the Jacobian

$\mathbf{B} = \left[D_{\boldsymbol{\xi}'}\mathbf{b}(\boldsymbol{\xi}, \mathbf{u}_{k-1}, \mathbf{w}) \; D_{\mathbf{w}'}\mathbf{b}(\boldsymbol{\xi}, \mathbf{u}_{k-1}, \mathbf{w}) \right] \big|_{\boldsymbol{\xi}=\hat{\boldsymbol{\xi}}_{k-1|k-1}, \mathbf{w}=0}$.

compute the predicted mean $\hat{\boldsymbol{\xi}}_{k-1|k}$ and covariance matrix $\mathbf{P}_{k-1|k}$ at time t_k

$$
\begin{aligned}
\hat{\boldsymbol{\xi}}_{k|k-1} &= \mathbf{b}(\hat{\boldsymbol{\xi}}^0_{k|k-1}, \mathbf{u}_{k-1}, 0), \\
\mathbf{P}_{k|k-1} &= \mathbf{B}\,\mathrm{diag}(\mathbf{P}^0_{k|k-1}, \mathbf{Q}(\Delta_k))\mathbf{B}'.
\end{aligned}
$$

compute the innovation covariance $\mathbf{S}_k = \mathbf{H}\mathbf{P}_{k|k-1}\mathbf{H}' + \mathbf{R}$.

compute the gain matrix $\mathbf{K}_k = \mathbf{P}_{k|k-1}\mathbf{H}'\mathbf{S}_k^{-1}$.

compute the posterior mean $\hat{\boldsymbol{\xi}}_{k|k}$ and covariance matrix $\mathbf{P}_{k|k}$ at time t_k

$$
\begin{aligned}
\hat{\boldsymbol{\xi}}_{k|k} &= \hat{\boldsymbol{\xi}}_{k|k-1} + \mathbf{K}_k(\mathbf{z}_k - \mathbf{H}\hat{\boldsymbol{\xi}}_{k|k-1}), \\
\mathbf{P}_{k|k} &= \mathbf{P}_{k|k-1} - \mathbf{K}_k\mathbf{S}_k\mathbf{K}_k'.
\end{aligned}
$$

1.9 UNSCENTED KALMAN FILTERS

The UKF is, like the EKF, an extension of the Kalman filter (KF) for nonlinear filtering problems. However, instead of linearizing, the UKF uses the so-called unscented transformation to approximate the moments which arise in the KF recursion [31, 32]. This approach has two advantages over linearization: it avoids the need to derive the Jacobian and it provides a more accurate approximation. In this section we briefly summarize the unscented transformation before showing how it is used as part of the UKF recursion for angle-only tracking.

Consider a random variable $\mathbf{y} \in \mathbb{R}^h$ with probability density function (PDF) p_Y and a nonlinear function $\mathbf{g} : \mathbb{R}^h \to \mathbb{R}^d$. The integral which must be evaluated to find the mean of $\mathbf{g}(\mathbf{y})$ is rarely available in closed form. The unscented transformation is an approximation of the form

$$
\int \mathbf{g}(\mathbf{y}) p_Y(\mathbf{y})\, d\mathbf{y} \approx \sum_{b=1}^{s} w_b \mathbf{g}(\mathbf{Y}_b), \tag{1.60}
$$

where $w_1, \ldots, w_s$ are weights and $\mathbf{Y}_1, \ldots, \mathbf{Y}_s$ are sigma points. There are numerous variants of the unscented transformation which differ in the number of sigma points, the locations of the sigma points, and/or the weights [32]. Here we use the form originally proposed in [31]. This form uses $s = 2h + 1$ sigma points for an h-dimensional random variable with weights and sigma points determined by a real-valued parameter, denoted as κ, and the mean and covariance matrix of the random variable.

Recursions of the UKF for angle-only tracking are given by Algorithms 4 for Cartesian coordinates and Algorithms 5 and 6 for MSC with approximate and exact dynamic models, respectively. Following our previous convention, the Algorithms 4,

---

Algorithm 4 CUKF: A recursion of the UKF with Cartesian coordinates.

Input: posterior mean $\hat{\mathbf{x}}_{k-1|k-1}$ and covariance matrix $\mathbf{P}_{k-1|k-1}$ at time t_{k-1} and the measurement $\mathbf{z}_k$.

Output: posterior mean $\hat{\mathbf{x}}_{k|k}$ and covariance matrix $\mathbf{P}_{k|k}$ at time t_k.

compute the predicted mean $\hat{\mathbf{x}}_{k|k-1}$ and covariance matrix $\mathbf{P}_{k|k-1}$ at time t_k

$$\hat{\mathbf{x}}_{k|k-1} = \mathbf{F}(\Delta_k)\hat{\mathbf{x}}_{k-1|k-1} - \mathbf{u}_{k-1},$$
$$\mathbf{P}_{k|k-1} = \mathbf{F}(\Delta_k)\mathbf{P}_{k-1|k-1}\mathbf{F}(\Delta_k)' + \mathbf{Q}(\Delta_k).$$

construct the matrix $\boldsymbol{\Sigma} = \begin{bmatrix} \mathbf{0} & \sqrt{(6+\kappa)\mathbf{P}_{k|k-1}} & -\sqrt{(6+\kappa)\mathbf{P}_{k|k-1}} \end{bmatrix}$.

for $b = 0, \ldots, 12$ **do**

compute the sigma point $\mathbf{X}_{k,b} = \hat{\mathbf{x}}_{k|k-1} + \boldsymbol{\sigma}_b$ where $\boldsymbol{\sigma}_b$ is the $(b+1)$th column of $\boldsymbol{\Sigma}$,

compute the transformed sigma point $\mathbf{Z}_{k,b} = \mathbf{h}(\mathbf{X}_{k,b})$.

end

compute the moment approximations

$$\hat{\mathbf{z}}_{k|k-1} = \sum_{b=0}^{12} w_b \mathbf{Z}_{k,b},$$
$$\mathbf{S}_k = \mathbf{P}_{\mathbf{zz},k} = \mathbf{R} + \sum_{b=0}^{12} w_b(\mathbf{Z}_{k,b} - \hat{\mathbf{z}}_{k|k-1})(\mathbf{Z}_{k,b} - \hat{\mathbf{z}}_{k|k-1})',$$
$$\mathbf{P}_{\mathbf{xz},k} = \sum_{b=0}^{12} w_b(\mathbf{X}_{k,b} - \hat{\mathbf{x}}_{k|k-1})(\mathbf{Z}_{k,b} - \hat{\mathbf{z}}_{k|k-1})'.$$

compute the gain matrix $\mathbf{K}_k = \mathbf{P}_{\mathbf{xz},k}\mathbf{S}_k^{-1}$.

compute the posterior mean $\hat{\mathbf{x}}_{k|k}$ and covariance matrix $\mathbf{P}_{k|k}$ at time t_k

$$\hat{\mathbf{x}}_{k|k} = \hat{\mathbf{x}}_{k|k-1} + \mathbf{K}_k(\mathbf{z}_k - \hat{\mathbf{z}}_{k|k-1}),$$
$$\mathbf{P}_{k|k} = \mathbf{P}_{k|k-1} - \mathbf{K}_k\mathbf{S}_k\mathbf{K}_k'.$$

5, and 6 are denoted by CUKF, UKF-MSC(A), and UKF-MSC(E), respectively. The weights used in the moment calculation are

$$w_b = \begin{cases} \kappa/(6+\kappa), & b = 0, \\ 1/[2(6+\kappa)], & b = 1, \ldots, 12. \end{cases} \tag{1.61}$$

Algorithm 5 UKF-MSC(A): A recursion of the UKF with MSC using the approximate dynamic model.

Input: posterior mean $\hat{\boldsymbol{\xi}}_{k-1|k-1}$ and covariance matrix $\mathbf{P}_{k-1|k-1}$ at time t_{k-1} and the measurement $\mathbf{z}_k$

Output: posterior mean $\hat{\boldsymbol{\xi}}_{k|k}$ and covariance matrix $\mathbf{P}_{k|k}$ at time t_k

set $\hat{\boldsymbol{\xi}}^0_{k|k-1} = \hat{\boldsymbol{\xi}}_{k-1|k-1}$ and $\mathbf{P}^0_{k|k-1} = \mathbf{P}_{k-1|k-1}$.

for $i = 1, \ldots, m$ **do**

construct the matrix $\boldsymbol{\Sigma}_i = \left[\mathbf{0} \quad \sqrt{(6+\kappa)\mathbf{P}^{i-1}_{k|k-1}} \quad -\sqrt{(6+\kappa)\mathbf{P}^{i-1}_{k|k-1}}\right]$.

for $b = 0, \ldots, 12$ **do**

compute the sigma point $\boldsymbol{\Xi}^i_{k,b} = \hat{\boldsymbol{\xi}}^{i-1}_{k|k-1} + \boldsymbol{\sigma}_b$ where $\boldsymbol{\sigma}_b$ be the $(b+1)$th column of $\boldsymbol{\Sigma}_i$.

compute the transformed sigma point and covariance matrix

$$\mathbf{E}^i_{k,b} = \mathbf{a}_j(\boldsymbol{\Xi}^i_{k,b}, t_{k-1} + (i-1)\Delta_k/m; \Delta_k/m),$$
$$\mathbf{C}^i_{k,b} = \mathbf{C}_j(\boldsymbol{\Xi}^i_{k,b}, t_{k-1} + (i-1)\Delta_k/m; \Delta_k/m).$$

end

compute the predicted mean $\hat{\boldsymbol{\xi}}^i_{k|k-1}$ and covariance matrix $\mathbf{P}^i_{k|k-1}$ at time t_k

$$\hat{\boldsymbol{\xi}}^i_{k|k-1} = \sum_{b=0}^{12} w_b \mathbf{E}^i_{k,b},$$
$$\mathbf{P}^i_{k|k-1} = \sum_{b=0}^{12} w_b [\mathbf{C}^i_{k,b} + (\mathbf{E}^i_{k,b} - \hat{\boldsymbol{\xi}}^i_{k|k-1})(\mathbf{E}^i_{k,b} - \hat{\boldsymbol{\xi}}^i_{k|k-1})'].$$

end

compute the innovation covariance $\mathbf{S}^m_k = \mathbf{H}\mathbf{P}^m_{k|k-1}\mathbf{H}' + \mathbf{R}$.

compute the gain matrix $\mathbf{K}^m_k = \mathbf{P}^m_{k|k-1}\mathbf{H}'(\mathbf{S}^m_k)^{-1}$.

compute the posterior mean $\hat{\boldsymbol{\xi}}_{k|k}$ and covariance matrix $\mathbf{P}_{k|k}$ at time t_k

$$\hat{\boldsymbol{\xi}}_{k|k} = \hat{\boldsymbol{\xi}}^m_{k|k-1} + \mathbf{K}^m_k(\mathbf{z}_k - \mathbf{H}\hat{\boldsymbol{\xi}}^m_{k|k-1}),$$
$$\mathbf{P}_{k|k} = \mathbf{P}^m_{k|k-1} - \mathbf{K}^m_k \mathbf{S}^m_k (\mathbf{K}^m_k)'.$$

Algorithm 6 UKF-MSC(E): A recursion of the UKF with MSC using the exact dynamic model.

Input: posterior mean $\hat{\boldsymbol{\xi}}_{k-1|k-1}$ and covariance matrix $\mathbf{P}_{k-1|k-1}$ at time t_{k-1} and the measurement $\mathbf{z}_k$

Output: posterior mean $\hat{\boldsymbol{\xi}}_{k|k}$ and covariance matrix $\mathbf{P}_{k|k}$ at time t_k

construct the matrices

$$\boldsymbol{\Sigma} = \left[\mathbf{0}_{6,1} \quad \sqrt{(12+\kappa)\mathbf{P}_{k-1|k-1}} \quad \mathbf{0}_{6,6} \quad -\sqrt{(12+\kappa)\mathbf{P}_{k-1|k-1}} \quad \mathbf{0}_{6,6}\right],$$

$$\boldsymbol{\Omega} = \left[\mathbf{0}_{6,1} \quad \mathbf{0}_{6,6} \quad \sqrt{(12+\kappa)\mathbf{Q}(\Delta_k)} \quad \mathbf{0}_{6,6} \quad -\sqrt{(12+\kappa)\mathbf{Q}(\Delta_k)}\right].$$

for $b = 0, \ldots, 24$ **do**

compute the sigma points $\boldsymbol{\Xi}_{k,b} = \hat{\boldsymbol{\xi}}_{k-1|k-1} + \boldsymbol{\sigma}_b$, $\mathbf{W}_{k,b} = \boldsymbol{\omega}_b$ where $\boldsymbol{\sigma}_b$ and $\boldsymbol{\omega}_b$ are the $(b+1)$th columns of $\boldsymbol{\Sigma}_i$ and $\boldsymbol{\Omega}$, respectively.

compute the transformed sigma point

$$\mathbf{E}_{k,b} = \mathbf{b}(\boldsymbol{\Xi}_{k,b}, \mathbf{u}_{k-1}, \mathbf{W}_{k,b}).$$

end

compute the predicted mean $\hat{\boldsymbol{\xi}}_{k|k-1}$ and covariance matrix $\mathbf{P}_{k|k-1}$ at time t_k

$$\hat{\boldsymbol{\xi}}_{k|k-1} = \sum_{b=0}^{24} w_b \mathbf{E}_{k,b},$$

$$\mathbf{P}_{k|k-1} = \sum_{b=0}^{24} w_b (\mathbf{E}_{k,b} - \hat{\boldsymbol{\xi}}_{k|k-1})(\mathbf{E}_{k,b} - \hat{\boldsymbol{\xi}}_{k|k-1})'.$$

compute the innovation covariance $\mathbf{S}_k = \mathbf{H}\mathbf{P}_{k|k-1}\mathbf{H}' + \mathbf{R}$.

compute the gain matrix $\mathbf{K}_k = \mathbf{P}_{k|k-1}\mathbf{H}'\mathbf{S}_k^{-1}$.

compute the posterior mean $\hat{\boldsymbol{\xi}}_{k|k}$ and covariance matrix $\mathbf{P}_{k|k}$ at time t_k

$$\hat{\boldsymbol{\xi}}_{k|k} = \hat{\boldsymbol{\xi}}_{k|k-1} + \mathbf{K}_k(\mathbf{z}_k - \mathbf{H}\hat{\boldsymbol{\xi}}_{k|k-1}),$$

$$\mathbf{P}_{k|k} = \mathbf{P}_{k|k-1} - \mathbf{K}_k\mathbf{S}_k\mathbf{K}_k'.$$

In Algorithm 5, the discrete-time dynamic model is approximated with a jth order stochastic Taylor series approximation which is applied over m subintervals per sampling period. Since the random variable being subject to a nonlinear transformation is six-dimensional, there are $2 \times 6 + 1 = 13$ sigma points. In Algorithm 6, where the exact dynamic model (1.45) is used, the nonlinear transformation is applied to a 12-dimensional random variable so $2 \times 12 + 1 = 25$ sigma points are required.

1.10 PARTICLE FILTERS

Particle filters are a class of sequential Monte Carlo methods for approximating the posterior density of the target state. The most common form of PF adopts a sequential importance sampling (SIS) [33] approach in which samples of the target state are drawn from an importance density and weighted appropriately each time a measurement is acquired. In contrast to the EKF and UKF, PFs offer the possibility of asymptotically, in the number of samples, optimal Bayesian inference, albeit at considerable computational expense. In this section we provide a brief overview of the class of auxiliary PFs [44] and describe particular implementations for angle-only filtering in relative Cartesian coordinates and MSC.

The following development uses a state vector in relative Cartesian coordinates but applies equally well to other coordinate systems. Assume the availability of a collection of n samples and weights which form an approximation to the posterior density of the target state at time t_{k-1}. We denote the weighted sample set as $\{w_{k-1}^i, \mathbf{x}_{k-1}^i\}$, where w_{k-1}^i is the weight of the ith sample $\mathbf{x}_{k-1}^i$ for $i = 1, \ldots, n$. The sum of the weights is unity. Using Bayes rule, the posterior density at time t_k is given by

$$p(\mathbf{x}_k|\mathbf{z}_{1:k}) \propto p(\mathbf{z}_k|\mathbf{x}_k) \int p(\mathbf{x}_k|\mathbf{x}_{k-1}) p(\mathbf{x}_{k-1}|\mathbf{z}_{1:k-1})\, d\mathbf{x}_{k-1}. \tag{1.62}$$

The PF approximation to the posterior density at time t_{k-1} can be written as

$$p(\mathbf{x}_{k-1}|\mathbf{z}_{1:k-1}) \approx \sum_{i=1}^{n} w_{k-1}^i \delta(\mathbf{x}_{k-1} - \mathbf{x}_{k-1}^i), \tag{1.63}$$

where δ is the Dirac delta measure. Substituting this approximation into the Bayes recursion (1.62) gives

$$p(\mathbf{x}_k|\mathbf{z}_{1:k}) \mathrel{\tilde{\propto}} p(\mathbf{z}_k|\mathbf{x}_k) \sum_{i=1}^{n} w_{k-1}^i p(\mathbf{x}_k|\mathbf{x}_{k-1}^i), \tag{1.64}$$

where $\overset{\sim}{\propto}$ means "is approximately proportional to." A marginalization over the sample index i at time t_{k-1} is present in (1.64). Reversing this marginalization gives the joint density

$$p(\mathbf{x}_k, i|\mathbf{z}_{1:k}) \overset{\sim}{\propto} w_{k-1}^i p(\mathbf{z}_k|\mathbf{x}_k) p(\mathbf{x}_k|\mathbf{x}_{k-1}^i). \tag{1.65}$$

The desired sample approximation to the posterior is obtained by sampling both the target state $\mathbf{x}_k$ and index i from $p(\mathbf{x}_k, i|\mathbf{z}_{1:k})$ and then retaining only the target state samples. Framing the sampling process in this way actually results in a simpler procedure and aids in the design of suitable sampling strategies. This idea was first proposed in [44]. Note that the process of sampling the auxiliary variable i is equivalent to a resampling step [28] in which highly weighted particles are emphasized at the expense of lowly weighted particles.

Ideally, samples would be drawn directly from (1.65). This can be done by drawing a sample index from

$$p(i|\mathbf{z}_{1:k}) \propto w_{k-1}^i \int p(\mathbf{z}_k|\mathbf{x}_k) p(\mathbf{x}_k|\mathbf{x}_{k-1}^i)\, d\mathbf{x}_k \tag{1.66}$$

and then sampling the state vector conditional on the sample index according to the density

$$p(\mathbf{x}_k|i, \mathbf{z}_{1:k}) \propto p(\mathbf{z}_k|\mathbf{x}_k) p(\mathbf{x}_k|\mathbf{x}_{k-1}^i). \tag{1.67}$$

Sampling from (1.66) and (1.67) is often not possible because the integral in (1.66) is intractable and the product in (1.67) cannot be simplified so as to obtain a sampling density for the state. An important special case when sampling from (1.66) and (1.67) is possible if the transition density is Gaussian and the likelihood is linear/Gaussian. This is not so for angle-only filtering in relative Cartesian coordinates or if the exact dynamic model (1.45) is used for filtering in MSC. However, if either of the stochastic Taylor series approximations (1.33) or (1.37) to the dynamic model are used for filtering in MSC, then sampling from (1.66) and (1.67) is possible. This is a potential advantage of using the approximate dynamic model when filtering in MSC.

More generally, when sampling directly from (1.65) is not possible, samples can instead be drawn from an importance density q and then assigned appropriate weights. A popular choice of importance density is the prior, that is,

$$q(\mathbf{x}_k, i) = w_{k-1}^i p(\mathbf{x}_k|\mathbf{x}_{k-1}^i). \tag{1.68}$$

The sample weights are found as the ratio of the posterior (1.65) and the importance density (1.68)

$$w_k^i \propto p(\mathbf{z}_k|\mathbf{x}_k^i). \tag{1.69}$$

The resulting PF is commonly called the BF [28]. While simple to implement and asymptotically convergent, the BF may not always perform well for finite sample sizes. Its chief weakness is that the importance density (1.68) does not use the current measurement $\mathbf{z}_k$. As a result, the sampling procedure can be quite inefficient with many samples having a low likelihood. In contrast, direct sampling from (1.65) via (1.66) and (1.67) ensures the drawing of samples which are in agreement with the latest available information. Sampling from (1.66) and (1.67) is referred to as using the OID [33].

Recursions of the BF in Cartesian coordinates and MSC the approximate and exact dynamic models are given in Algorithms 7, 8, and 10, respectively. The recursion of the OID PF for MSC with the approximate dynamic model is given in Algorithm 9. Following our previous convention, Algorithms 7, 8, 9, and 10 are denoted by CBF, BF-MSC(A), OIDPF-MSC(A), and BF-MSC(E), respectively. There are several points of interest. The PFs in MSC approximate the target dynamic model using a jth order stochastic Taylor series approximation applied over m subintervals per sampling period. All PFs have been regularized, as suggested in [45, 46], using a Gaussian kernel with covariance matrix

$$\mathbf{\Omega}_k = b(n)\hat{\mathbf{P}}_{k-1}, \tag{1.70}$$

Algorithm 7 CBF: A recursion of the bootstrap filter with Cartesian coordinates.

Input: a weighted sample $\{w_{k-1}^i, \mathbf{x}_{k-1}^i\}$ from the posterior at time t_{k-1} and the measurement $\mathbf{z}_k$.

Output: a weighted sample $\{w_k^i, \mathbf{x}_k^i\}$ from the posterior at time t_k.

select indices $c(1), \ldots, c(n)$ according to the weights $w_{k-1}^1, \ldots, w_{k-1}^n$.

for $i = 1, \ldots, n$ **do**

draw $\mathbf{x}_k^i \sim N(\cdot\,; \mathbf{F}(\Delta_k)\mathbf{x}_{k-1}^{c(i)}, \mathbf{Q}(\Delta_k) + \mathbf{\Omega}_k)$,

compute the un-normalized weight $\tilde{w}_k^i = N(\mathbf{z}_k; \mathbf{h}(\mathbf{x}_k^i), \mathbf{R})$.

end

compute the normalized weights

$$w_k^i = \tilde{w}_k^i \Bigg/ \sum_{j=1}^{n} \tilde{w}_k^j, \quad i = 1, \ldots, n.$$

compute the state estimate $\hat{\mathbf{x}}_k$ at time t_k

$$\hat{\mathbf{x}}_k = \sum_{i=1}^{n} w_k^i \mathbf{x}_k^i.$$

Algorithm 8 BF-MSC(A): A recursion of the bootstrap filter in MSC with the approximate dynamic model.

Input: a weighted sample $\{w_{k-1}^i, \boldsymbol{\xi}_{k-1}^i\}$ from the posterior at time t_{k-1} and the measurement $\mathbf{z}_k$.
Output: a weighted sample $\{w_k^i, \boldsymbol{\xi}_k^i\}$ from the posterior at time t_k.
select indices $c(1), \ldots, c(n)$ according to the weights $w_{k-1}^1, \ldots, w_{k-1}^n$.
for $i = 1, \ldots, n$ **do**
  set $\boldsymbol{\xi}_{k,0}^i = \boldsymbol{\xi}_{k-1}^{c(i)}$.
  for $d = 0, \ldots, m-2$ **do**
    draw
    $\boldsymbol{\xi}_{k,d+1}^i \sim N(\cdot\,; \mathbf{a}_j(\boldsymbol{\xi}_{k,d}^i, t_{k-1} + d\Delta_k/m; \Delta_k/m), \mathbf{C}_j(\boldsymbol{\xi}_{k,d}^i, t_{k-1} + d\Delta_k/m; \Delta_k/m))$.
  end
  draw $\boldsymbol{\xi}_k^i \sim N(\cdot\,; \mathbf{a}_j(\boldsymbol{\xi}_{k,m-1}^i, t_k - \Delta_k/m; \Delta_k/m), \mathbf{C}_j(\boldsymbol{\xi}_{k,m-1}^i, t_k - \Delta_k/m; \Delta_k/m) + \boldsymbol{\Omega}_k$,
  compute the un-normalized weight $\tilde{w}_k^i = N(\mathbf{z}_k; \mathbf{H}\boldsymbol{\xi}_k^i, \mathbf{R})$.
end
compute the normalized weights

$$w_k^i = \tilde{w}_k^i \Big/ \sum_{j=1}^{n} \tilde{w}_k^j, \quad i = 1, \ldots, n.$$

compute the state estimate $\hat{\mathbf{x}}_k$ at time t_k

$$\hat{\mathbf{x}}_k = \sum_{i=1}^{n} w_k^i \mathbf{f}_{\mathrm{MSC}}^{\mathrm{C}}(\boldsymbol{\xi}_k^i).$$

where $b(n)$ is a scaling factor and $\hat{\mathbf{P}}_{k-1}$ is the weighted sample covariance matrix. For samples drawn from a Gaussian distribution, the mean integrated squared error of the kernel density estimator is minimized by selecting $b(n) = (2n)^{-1/5}$ [47]. We have found that better results are obtained using a somewhat smaller scaling factor. In particular, we use $b(n) = (2n)^{-1/5}/16$. Such a reduction is suggested in [47] for samples from a multimodal distribution. Finally, the PFs in MSC provide state estimates in relative Cartesian coordinates by transforming each sample from MSC to relative Cartesian coordinates and computing the weighted mean.

Algorithm 9 OIDPF-MSC(A): A recursion of the OID PF in MSC with the approximate dynamic model.

Input: a weighted sample $\{w_{k-1}^i, \boldsymbol{\xi}_{k-1}^i\}$ from the posterior at time t_{k-1} and the measurement $\mathbf{z}_k$.
Output: a weighted sample $\{w_k^i, \boldsymbol{\xi}_k^i\}$ from the posterior at time t_k.
for $i = 1, \ldots, n$ **do**
  set $\boldsymbol{\xi}_{k,0}^i = \boldsymbol{\xi}_{k-1}^i$.
  for $d = 0, \ldots, m-2$ **do**
    draw $\boldsymbol{\xi}_{k,d+1}^i \sim N(\cdot\,; \mathbf{a}_j(\boldsymbol{\xi}_{k,d}^i, t_{k-1} + d\Delta_k/m; \Delta_k/m),$
    $\mathbf{C}_j(\boldsymbol{\xi}_{k,d}^i, t_{k-1} + d\Delta_k/m; \Delta_k/m))$.
  end
  compute the prior statistics

$$\boldsymbol{\xi}_{k|k-1}^i = \mathbf{a}_j(\boldsymbol{\xi}_{k,m-1}^i, t_k - \Delta_k/m; \Delta_k/m),$$
$$\mathbf{C}_k^i = \mathbf{C}_j(\boldsymbol{\xi}_{k,m-1}^i, t_k - \Delta_k/m; \Delta_k/m) + \boldsymbol{\Omega}_k.$$

  compute the prior measurement statistics $\mathbf{z}_k^i = \mathbf{H}\boldsymbol{\xi}_{k|k-1}^i$ and $\mathbf{S}_k^i = \mathbf{H}\mathbf{C}_k^i\mathbf{H}' + \mathbf{R}$.
  compute the un-normalized weight $\tilde{\psi}_k^i = w_{k-1}^i N(\mathbf{z}_k; \mathbf{z}_k^i, \mathbf{S}_k^i)$.
end
compute the normalized weights

$$\psi_k^i = \tilde{\psi}_k^i \Big/ \sum_{j=1}^n \tilde{\psi}_k^j, \quad i = 1, \ldots, n.$$

select indices $c(1), \ldots, c(n)$ according to the weights $\psi_k^1, \ldots, \psi_k^n$.
for $i = 1, \ldots, n$ **do**
  compute the gain matrix $\mathbf{K}^i = \mathbf{C}_k^{c(i)}\mathbf{H}'(\mathbf{S}_k^{c(i)})^{-1}$.
  compute the posterior mean $\boldsymbol{\xi}_{k|k}^i$ and covariance matrix $\boldsymbol{\Sigma}_k^i$ at time t_k

$$\boldsymbol{\xi}_{k|k}^i = \boldsymbol{\xi}_{k|k-1}^{c(i)} + \mathbf{K}^i(\mathbf{z}_k - \mathbf{z}_k^{c(i)}),$$
$$\boldsymbol{\Sigma}_k^i = \mathbf{C}_k^{c(i)} - \mathbf{K}^i\mathbf{S}_k^{c(i)}(\mathbf{K}^i)'.$$

  draw $\boldsymbol{\xi}_k^i \sim N(\cdot\,; \boldsymbol{\xi}_{k|k}^i, \boldsymbol{\Sigma}_k^i)$ and set $w_k^i = 1/n$.
end
compute the state estimate $\hat{\mathbf{x}}_k$ at time t_k

$$\hat{\mathbf{x}}_k = \sum_{i=1}^n w_k^i \mathbf{f}_{\mathrm{MSC}}^{\mathrm{C}}(\boldsymbol{\xi}_k^i).$$

Algorithm 10 BF-MSC(E): A recursion of the bootstrap filter in MSC with the exact dynamic model.

Input: a weighted sample $\{w_{k-1}^i, \boldsymbol{\xi}_{k-1}^i\}$ from the posterior at time t_{k-1} and the measurement $\mathbf{z}_k$.
Output: a weighted sample $\{w_k^i, \boldsymbol{\xi}_k^i\}$ from the posterior at time t_k.
select indices $c(1), \ldots, c(n)$ according to the weights $w_{k-1}^1, \ldots, w_{k-1}^n$.
for $i = 1, \ldots, n$ **do**
  draw $\tilde{\boldsymbol{\xi}}_{k-i}^i \sim N(\boldsymbol{\xi}_{k-1}^{c(i)}, \boldsymbol{\Omega}_k)$
  draw $\mathbf{w}_k^i \sim N(0, \mathbf{Q}(\Delta_k))$
  set $\boldsymbol{\xi}_k^i = \mathbf{b}(\tilde{\boldsymbol{\xi}}_{k-1}^i, \mathbf{u}_k, \mathbf{w}_k^i)$,
  compute the un-normalized weight $\tilde{w}_k^i = N(\mathbf{z}_k; \mathbf{H}\boldsymbol{\xi}_k^i, \mathbf{R})$.
end
compute the normalized weights

$$w_k^i = \tilde{w}_k^i \bigg/ \sum_{j=1}^{n} \tilde{w}_k^j, \quad i = 1, \ldots, n.$$

compute the state estimate $\hat{\mathbf{x}}_k$ at time t_k

$$\hat{\mathbf{x}}_k = \sum_{i=1}^{n} w_k^i \mathbf{f}_{\mathrm{MSC}}^{\mathrm{C}}(\boldsymbol{\xi}_k^i).$$

1.11 NUMERICAL SIMULATIONS AND RESULTS

In this section a performance analysis of the algorithms described in Sections 1.8–1.10 is presented. The scenario used in our simulation is similar to that used in [2].

We have made some changes to make the scenario three dimensional in nature. The target's initial ground range ρ_1, bearing β_1, target height z_1^{t}, speed s_1, course c_1 in the XY plane, and Z component $\dot{z}_1^{\mathrm{t}}$ of target velocity are shown in Table 1.1. The initial position and velocity in Cartesian coordinates can be found from Table 1.1

TABLE 1.1 Initial Parameters of Target

Variable	Value
ρ_1 (km)	138.0
β_1 (deg)	45.0
z_1^{t} (km)	9.0
s_1 (m/s)	297.0
c_1 (deg)	−135.0
$\dot{z}_1^{\mathrm{t}}$ (m/s)	0.0

TABLE 1.2 Motion Profile of Ownship

Interval (s)	Δt (s)	$\Delta\phi$ (rad)	Motion Type	ω^{o} (rad/s)
[0, 15]	15	0	CV	0
[15, 31]	16	$-\pi/4$	CT	$-\pi/64$
[31, 43]	12	0	CV	0
[43, 75]	32	$\pi/2$	CT	$\pi/64$
[75, 86]	11	0	CV	0
[86, 102]	16	$-\pi/4$	CT	$-\pi/64$
[102, 210]	108	0	CV	0

as $(138/\sqrt{2}, 138/\sqrt{2}, 9)$ km and $297/\sqrt{2}(-1, -1, 0)$ m/s, respectively. Given these initial parameters the target moves according to the NCV model given in (1.22) with the power spectral densities of the process noise, $q_x = q_y = 0.01$ m^2/s^3 and $q_z = 10^{-4}$ m^2/s^3. Note that, subject to the variations produced by the process noise, the target moves nearly in a plane parallel to the XY plane with a nearly constant height of 9 km.

The motion of the ownship is deterministic. The ownship moves in a plane parallel to the XY plane at a fixed height of 10 km with segments of CV and CT. The profile of the ownship motion is presented in Table 1.2. In Table 1.2, Δt represents the duration of the segment, $\Delta\phi$ is the total angular change during the segment, and ω^{o} is the angular velocity of the ownship during the segment. The ownship trajectory and the target trajectory from the first Monte Carlo run are shown in Figure 1.2.

The filters are initialized as described in Section 1.7 with the parameters shown in Table 1.3. All PFs are implemented with a sample size of 10,000. The UKF-MSC(A) and EKF-MSC(A) are implemented using the first-order stochastic Taylor

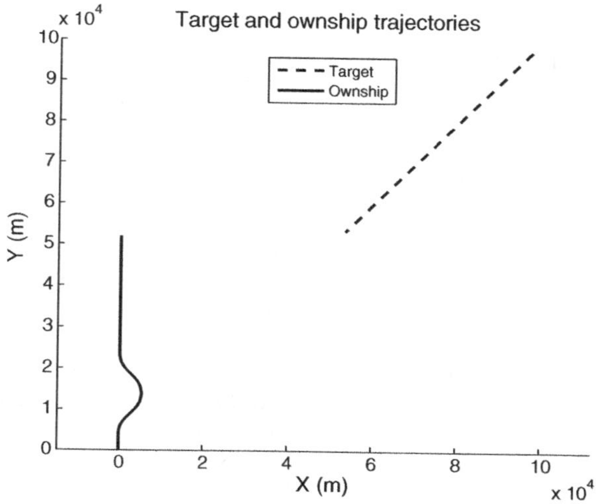

Figure 1.2 Target and ownship trajectories.

TABLE 1.3 Parameters of Prior Distribution

Parameter	Mean	Standard Deviation
Range (km)	150	13.6
Speed (m/s)	258	41.6
Bearing heading (rad/s)	$\beta_1 + \pi$	$\pi/\sqrt{12}$
Elevation heading (rad/s)	0	$\pi/60$

series approximation with $m = 8$ subintervals per sampling period. The BF-MSC(A) and OIDPF-MSC(A) are implemented using the second-order Taylor series approximation over the whole sampling period, that is, $m = 1$.

Two measures of performance are used to compare the algorithms described in Sections 1.8–1.10. The first is the root mean square (RMS) position error averaged from the end of the last observer maneuver to the end of the surveillance period. The second performance measure is the RMS position error at the end of the surveillance period. These statistics are found by averaging over 1000 realizations for bearings and elevation measurement noise standard deviations of 1, 15, and 35 mrad, or 0.057, 0.857, and 2 degrees. The results are shown in Table 1.4. The execution times of the algorithms are given in Table 1.5.

There are several points of interest to discuss. The best performance is achieved by the EKF and UKF in MSC. The degree of their superiority over the other algorithms depends on the level of the measurement noise. For low measurement noise, the EKF and UKF in MSC are only marginally better than the EKF and UKF in relative Cartesian coordinates. For high measurement noise, the performance improvements of the EKF and UKF in MSC are more pronounced. These performance improvements are offset somewhat by the significantly larger computational cost of using MSC compared to relative Cartesian coordinates, as seen in Table 1.5. Even so, the EKF

TABLE 1.4 Simulation Results for the Scenario of Figure 1.2

(a) Time-Averaged RMS Position Error

Algorithm	Measurement Noise Standard Deviation (mrad) 1	15	35
CEKF	0.650	5.786	12.00
CUKF	0.644	5.743	12.02
CBF	0.694	6.183	12.09
EKF-MSC(E)	0.629	4.773	10.65
EKF-MSC(A)	0.723	4.954	10.81
UKF-MSC(E)	0.629	4.827	10.48
UKF-MSC(A)	0.735	5.011	10.65
BF-MSC(E)	1.310	5.689	11.71
BF-MSC(A)	3.728	7.804	13.53
OIDPF-MSC(A)	3.710	7.158	11.38

(b) Final RMS Position Error

Algorithm	Measurement Noise Standard Deviation (mrad) 1	15	35
CEKF	0.472	3.363	4.780
CUKF	0.468	3.308	4.658
CBF	0.527	4.173	5.384
EKF-MSC(E)	0.462	2.413	3.333
EKF-MSC(A)	0.569	2.549	3.447
UKF-MSC(E)	0.462	2.533	3.499
UKF-MSC(A)	0.583	2.679	3.615
BF-MSC(E)	1.355	3.575	5.286
BF-MSC(A)	3.833	5.462	6.535
OIDPF-MSC(A)	3.827	4.820	5.166

TABLE 1.5 CPU Times in Seconds of Filtering Algorithms for the Measurement Standard Deviation of 15 mrad

Filtering Algorithm	CPU Time (s)	Relative CPU
CEKF	0.036	1.00
CUKF	0.11	2.97
CBF	6.28	175.22
EKF-MSC(E)	0.98	27.39
EKF-MSC(A)	1.32	36.77
UKF-MSC(E)	0.82	22.86
UKF-MSC(A)	1.14	31.73
BF-MSC(E)	15.60	434.90
BF-MSC(A)	26.86	748.82
OIDPF-MSC(A)	23.03	637.52

and UKF in MSC do not pose a severe computational burden. This is certainly true when compared to the various PF implementations. In addition to having a much greater computational cost, the PFs also perform worse than the EKF and UKF in MSC, much worse in some cases. The poor performance of the PFs could be improved by using a larger sample size but this would obviously increase their already large computational cost. An interesting aspect of the results is that the BF does not benefit from the use of MSC in the same way as the EKF and UKF. When the measurement accuracy is high (e.g., a standard deviation of 1 mrad), the RMS position error for the CBF is about half of that for the BF-MSC(E). For higher measurement standard deviations (e.g., 15, 35 mrad), the BF-MSC(E) performs slightly better than the CBF with nearly the same computational cost.

Finally, it can be noted that using the approximate dynamic model instead of the exact dynamic model provides no benefits in these examples. In addition to being computationally more expensive, using the approximate dynamic model results in worse performance than using the exact dynamic model. This is particularly so for the PFs with precise measurements. Further, the fact that the approximate dynamic model allows the OID to be used is of little consequence here as the OID does not perform much better than the BF. This can be attributed to the large initial uncertainty in target position and speed. At the expense of extra computations, the performance of the approximate dynamic model could be improved by applying the approximation over more subintervals per sampling period or by using high-order approximations. However, both of these options would increase the computational expense of an approach which is already more burdensome than using the exact dynamic model.

1.12 CONCLUSIONS

We have considered the problem of tracking a nonmaneuvering target moving in three dimensions using angle-only measurements of bearing and elevation. Three classes of algorithms based on relative Cartesian coordinates, MSC with exact dynamic model, and MSC with approximate dynamic model were developed. MSC are the three-dimensional analog of the well-known MPC. These coordinate systems have the important property of decoupling the potentially unobservable range from other observable elements of the state vector.

As in the case of 2D angle-only filtering, it appears from our results that using a coordinate system which decouples observable and nonobservable components is desirable, although the benefits of doing this were marginal in our scenario. Of the algorithms considered here, computationally efficient Gaussian approximations, such as the EKF and the UKF, were the most effective. They provided accurate state estimates at a fraction of the expense of the various PF implementations. Finally, the results obtained using MSC with the exact dynamic model were better than those obtained using MSC with the approximate dynamic model, significantly so for precise measurements. Although such an outcome is perhaps not surprising, it could not be assumed to hold since all filters only approximately implement the nonlinear dynamic model.

For the current problem, the continuous-discrete filtering algorithms using MSC, such as the EKF-MSC(A), UKF-MSC(A), BF-MSC(A), and OIDPF-MSC(A) do not provide any advantage over the corresponding algorithms which use the exact discrete-time dynamic model for MSC. The exact discrete-time dynamic model for MSC is easily derived from the discrete-time NCVM for the relative Cartesian state vector using transformations between these coordinate systems. In many real-world problems such as the ballistic target tracking [36–38], ballistic projectile tracking [39], and satellite orbit determination [20, 40], the dynamic model is derived from a force model and is described by a SDE for which the corresponding exact discrete-time stochastic difference equation cannot be obtained in closed form. As a result, the number of publications for the continuous-discrete filtering problem using the UKF [34] and PF [35] is limited. The UKF and PF algorithms developed for the continuous-discrete filtering problem in this chapter can be directly applied to these problems [48].

Our future work will consider the realistic scenario of a maneuvering target. We shall also calculate the posterior Cramér Rao lower bound (PCRLB) for the filtering problem which represents the best achievable accuracy [49]. Then the accuracy of various filtering algorithms can be compared with the PCRLB.

APPENDIX 1A DERIVATIONS FOR STOCHASTIC DIFFERENTIAL EQUATIONS IN MSC

Differentiating (1.7) with respect to the time, we get

$$\dot{r}_x^{\mathrm{T}} = \dot{r}\cos\epsilon\sin\beta - r\dot{\epsilon}\sin\epsilon\sin\beta + r\dot{\beta}\cos\epsilon\cos\beta, \tag{1A.1}$$

$$\dot{r}_y^{\mathrm{T}} = \dot{r}\cos\epsilon\cos\beta - r\dot{\epsilon}\sin\epsilon\cos\beta - r\dot{\beta}\cos\epsilon\sin\beta, \tag{1A.2}$$

$$\dot{r}_z^{\mathrm{T}} = \dot{r}\sin\epsilon + r\dot{\epsilon}\cos\epsilon. \tag{1A.3}$$

Differentiating (1A.1), (1A.2), and (1A.3) with respect to the time, we get

$$\begin{aligned}\ddot{r}_x^{\mathrm{T}} =&(\ddot{r} - r\dot{\epsilon}^2 - r\dot{\beta}^2)\cos\epsilon\sin\beta - (2\dot{r}\dot{\epsilon} + r\ddot{\epsilon})\sin\epsilon\sin\beta \\ &+ (2\dot{r}\dot{\beta} + r\ddot{\beta})\cos\epsilon\cos\beta - 2r\dot{\beta}\dot{\epsilon}\sin\epsilon\cos\beta,\end{aligned} \tag{1A.4}$$

$$\begin{aligned}\ddot{r}_y^{\mathrm{T}} =&(\ddot{r} - r\dot{\epsilon}^2 - r\dot{\beta}^2)\cos\epsilon\cos\beta - (2\dot{r}\dot{\epsilon} + r\ddot{\epsilon})\sin\epsilon\cos\beta \\ &- (2\dot{r}\dot{\beta} + r\ddot{\beta})\cos\epsilon\sin\beta + 2r\dot{\beta}\dot{\epsilon}\sin\epsilon\sin\beta,\end{aligned} \tag{1A.5}$$

$$\ddot{r}_z^{\mathrm{T}} = (\ddot{r} - r\dot{\epsilon}^2)\sin\epsilon + (2\dot{r}\dot{\epsilon} + r\ddot{\epsilon})\cos\epsilon. \tag{1A.6}$$

The acceleration of the target relative to the ownship is defined by

$$\ddot{\mathbf{r}}^{\mathrm{T}}(t) := \left[\ddot{x}(t)\ \ddot{y}(t)\ \ddot{z}(t)\right]' = \mathbf{a}^{\mathrm{T}}(t) - \mathbf{a}_{\mathrm{o}}^{\mathrm{T}}(t). \tag{1A.7}$$

We transform the relative acceleration in the T frame to the S frame by

$$\ddot{\mathbf{r}}^{\mathrm{S}}(t) = \mathrm{T}_{\mathrm{T}}^{\mathrm{S}}\ddot{\mathbf{r}}^{\mathrm{T}}(t). \tag{1A.8}$$

From (1A.8) we get

$$\ddot{r}_x^{\mathrm{S}} = \sin\epsilon\sin\beta\ddot{r}_x^{\mathrm{T}} + \sin\epsilon\cos\beta\ddot{r}_y^{\mathrm{T}} - \cos\epsilon\ddot{r}_z^{\mathrm{T}}, \tag{1A.9}$$

$$\ddot{r}_y^{\mathrm{S}} = -\cos\beta\ddot{r}_x^{\mathrm{T}} + \sin\beta\ddot{r}_y^{\mathrm{T}}, \tag{1A.10}$$

$$\ddot{r}_z^{\mathrm{S}} = \cos\epsilon\sin\beta\ddot{r}_x^{\mathrm{T}} + \cos\epsilon\cos\beta\ddot{r}_y^{\mathrm{T}} + \sin\epsilon\ddot{r}_z^{\mathrm{T}}. \tag{1A.11}$$

Using (1A.7) in (1A.8), we get

$$\ddot{\mathbf{r}}^{\mathrm{S}}(t) = \mathbf{a}^{\mathrm{S}}(t) - \mathbf{a}_{\mathrm{o}}^{\mathrm{S}}(t). \tag{1A.12}$$

Simplification of (1A.10) gives

$$\ddot{r}_y^{\mathrm{S}} = -(2\dot{r}\dot{\beta} + r\ddot{\beta})\cos\epsilon + 2r\dot{\beta}\dot{\epsilon}\sin\epsilon. \tag{1A.13}$$

Simplification of (1A.13) and substitution in the Y component of (1A.8) gives

$$\dot{\omega} = \omega(\dot{\epsilon}\tan\epsilon - 2\zeta) - (a_y^{\mathrm{S}} - \ddot{r}_y^{\mathrm{S}})/r. \tag{1A.14}$$

Writing (1A.14) using MSC, we get

$$\dot{\xi}_1 = \xi_1(\xi_2 \tan \xi_5 - 2\xi_3) - \xi_6(a_y^{\mathrm{S}} - \ddot{y}_{\mathrm{o}}^{\mathrm{S}}). \tag{1A.15}$$

Simplification of (1A.9) gives

$$\ddot{r}_x^{\mathrm{S}} = -2\dot{r}\dot{\epsilon} - r\dot{\beta}^2 \cos\epsilon \sin\epsilon - r\ddot{\epsilon}. \tag{1A.16}$$

Simplification of (1A.16) and substitution in the X component of (1A.8) gives

$$\ddot{\epsilon} = -2\zeta\dot{\epsilon} - \omega^2 \tan\epsilon - (a_x^{\mathrm{S}} - \ddot{r}_x^{\mathrm{S}})/r. \tag{1A.17}$$

Writing (1A.17) using MSC, we get

$$\dot{\xi}_2 = -2\xi_2\xi_3 - \xi_1^2 \tan \xi_5 - \xi_6(a_x^{\mathrm{S}} - \ddot{x}_{\mathrm{o}}^{\mathrm{S}}). \tag{1A.18}$$

Simplification of (1A.11) gives

$$\ddot{r}_z^{\mathrm{S}} = \ddot{r} - r\dot{\epsilon}^2 - r\dot{\beta}^2 \cos^2\epsilon. \tag{1A.19}$$

Simplification of (1A.19) and substitution in the Z component of (1A.8) gives

$$\ddot{r} = r(\dot{\epsilon}^2 + \omega^2) + (a_z^{\mathrm{S}} - \ddot{r}_z^{\mathrm{S}}). \tag{1A.20}$$

We have

$$\dot{\xi}_3 = \frac{d}{dt}(\dot{r}/r) = \ddot{r}/r - (\dot{r}/r)^2. \tag{1A.21}$$

Substitution of (1A.20) in (1A.21) and using MSC, we get

$$\dot{\xi}_3 = \xi_1^2 + \xi_2^2 - \xi_3^2 + \xi_6(a_z^{\mathrm{S}} - \ddot{z}_{\mathrm{o}}^{\mathrm{S}}). \tag{1A.22}$$

From (1.9), we get

$$\dot{\xi}_4 = \dot{\beta} = \omega/\cos\epsilon = \xi_1/\cos\xi_5. \tag{1A.23}$$

From definitions of MSC

$$\dot{\xi}_5 = \dot{\epsilon} = \xi_2, \tag{1A.24}$$

$$\dot{\xi}_6 = d/dt(1/r) = -(1/r)(\dot{r}/r) = -\xi_3\xi_6. \tag{1A.25}$$

The SDEs derived in this appendix are represented by the nonlinear vector differential equation (1.29).

APPENDIX 1B TRANSFORMATIONS BETWEEN RELATIVE CARTESIAN COORDINATES AND MSC

The transformations between relative Cartesian coordinates and MSC are given by

$$\mathbf{f}_{\mathrm{C}}^{\mathrm{MSC}}(\mathbf{x}) := \begin{bmatrix} (y\dot{x} - x\dot{y})/\sqrt{(x^2+y^2+z^2)(x^2+y^2)} \\ [\dot{z}(x^2+y^2) - z(x\dot{x}+y\dot{y})]/\sqrt{(x^2+y^2)}(x^2+y^2+z^2) \\ (x\dot{x}+y\dot{y}+z\dot{z})/(x^2+y^2+z^2) \\ \arctan(x, y) \\ \arctan(z, \sqrt{x^2+y^2}) \\ 1/\sqrt{x^2+y^2+z^2} \end{bmatrix}. \tag{1B.1}$$

$$\mathbf{f}_{\mathrm{MSC}}^{\mathrm{C}}(\boldsymbol{\xi}) := \frac{1}{\xi_6} \begin{bmatrix} \sin\xi_4 \cos\xi_5 \\ \cos\xi_4 \cos\xi_5 \\ \sin\xi_5 \\ \sin\xi_4(\xi_3\cos\xi_5 - \xi_2\sin\xi_5) + \xi_1\cos\xi_4 \\ \cos\xi_4(\xi_3\cos\xi_5 - \xi_2\sin\xi_5) - \xi_1\sin\xi_4 \\ \xi_2\cos\xi_5 + \xi_3\sin\xi_5 \end{bmatrix}. \tag{1B.2}$$

APPENDIX 1C FILTER INITIALIZATION FOR RELATIVE CARTESIAN COORDINATES AND MSC

We consider first the initial moments for relative Cartesian coordinates. The integrals (1.54) and (1.55) can be evaluated using the result

$$\int \exp(jkx) N(x; \mu, \sigma^2)\, dx = \exp(-k^2\sigma^2/2)\exp(jk\mu) \tag{1C.1}$$

and Euler's formula with integer values of k

$$\exp(jkx) = \cos(kx) + j\sin(kx), \tag{1C.2}$$

where $j = \sqrt{-1}$. We use the following notation for the moments of trigonometric functions of Gaussian random variables

$$m_{i,j}(\mu, \sigma^2) = \int \cos^i(x)\sin^j(x) N(x; \mu, \sigma^2)\, dx \tag{1C.3}$$

Then, the first-order moments are

$$\hat{\mathbf{x}}_1 = \begin{bmatrix} \bar{r}\, m_{1,0}(\hat{\epsilon}_1, \sigma_\epsilon^2)\, m_{0,1}(\hat{\beta}_1, \sigma_\beta^2) \\ \bar{r}\, m_{1,0}(\hat{\epsilon}_1, \sigma_\epsilon^2)\, m_{1,0}(\hat{\beta}_1, \sigma_\beta^2) \\ \bar{r}\, m_{0,1}(\hat{\epsilon}_1, \sigma_\epsilon^2) \\ \bar{s}\, m_{1,0}(\bar{\gamma}, \sigma_\gamma^2)\, m_{0,1}(\bar{\alpha}, \sigma_\alpha^2) - \dot{x}_1^o \\ \bar{s}\, m_{1,0}(\bar{\gamma}, \sigma_\gamma^2)\, m_{1,0}(\bar{\alpha}, \sigma_\alpha^2) - \dot{y}_1^o \\ \bar{s}\, m_{0,1}(\bar{\gamma}, \sigma_\gamma^2) - \dot{z}_1^o \end{bmatrix} \tag{1C.4}$$

where

$$m_{1,0}(\mu, \sigma^2) = \cos\mu \exp(-\sigma^2/2), \tag{1C.5}$$

$$m_{0,1}(\mu, \sigma^2) = \sin\mu \exp(-\sigma^2/2). \tag{1C.6}$$

Let $\mathbf{P}_1(i, j)$ denote the (i, j)th element of the covariance matrix. The nonzero elements of the upper triangular part of the covariance matrix $\mathbf{P}_1$ are

$$\mathbf{P}_1(1,1) = (\sigma_r^2 + \bar{r}^2)\, m_{2,0}(\hat{\epsilon}_1, \sigma_\epsilon^2)\, m_{0,2}(\hat{\beta}_1, \sigma_\beta^2) - \bar{r}^2\, m_{1,0}^2(\hat{\epsilon}_1, \sigma_\epsilon^2)\, m_{0,1}^2(\hat{\beta}_1, \sigma_\beta^2), \tag{1C.7}$$

$$\begin{aligned}\mathbf{P}_1(1,2) = {} & (\sigma_r^2 + \bar{r}^2)\, m_{2,0}(\hat{\epsilon}_1, \sigma_\epsilon^2)\, m_{1,1}(\hat{\beta}_1, \sigma_\beta^2) \\ & - \bar{r}^2\, m_{1,0}^2(\hat{\epsilon}_1, \sigma_\epsilon^2)\, m_{1,0}(\hat{\beta}_1, \sigma_\beta^2)\, m_{0,1}(\hat{\beta}_1, \sigma_\beta^2),\end{aligned} \tag{1C.8}$$

$$\mathbf{P}_1(1,3) = [(\sigma_r^2 + \bar{r}^2)\, m_{1,1}(\hat{\epsilon}_1, \sigma_\epsilon^2) - \bar{r}^2\, m_{1,0}(\hat{\epsilon}_1, \sigma_\epsilon^2)\, m_{0,1}(\hat{\epsilon}_1, \sigma_\epsilon^2)]\, m_{0,1}(\hat{\beta}_1, \sigma_\beta^2), \tag{1C.9}$$

$$\mathbf{P}_1(2,2) = (\sigma_r^2 + \bar{r}^2)\, m_{2,0}(\hat{\epsilon}_1, \sigma_\epsilon^2)\, m_{2,0}(\hat{\beta}_1, \sigma_\beta^2) - \bar{r}^2\, m_{1,0}^2(\hat{\epsilon}_1, \sigma_\epsilon^2)\, m_{1,0}^2(\hat{\beta}_1, \sigma_\beta^2), \tag{1C.10}$$

$$\mathbf{P}_1(2,3) = [(\sigma_r^2 + \bar{r}^2)\, m_{1,1}(\hat{\epsilon}_1, \sigma_\epsilon^2) - \bar{r}^2\, m_{1,0}(\hat{\epsilon}_1, \sigma_\epsilon^2)\, m_{0,1}(\hat{\epsilon}_1, \sigma_\epsilon^2)]\, m_{1,0}(\hat{\beta}_1, \sigma_\beta^2), \tag{1C.11}$$

$$\mathbf{P}_1(3,3) = (\sigma_r^2 + \bar{r}^2)\, m_{0,2}(\hat{\epsilon}_1, \sigma_\epsilon^2) - \bar{r}^2\, m_{0,1}^2(\hat{\epsilon}_1, \sigma_\epsilon^2), \tag{1C.12}$$

$$\begin{aligned}\mathbf{P}_1(4,4) = {} & (\sigma_s^2 + \bar{s}^2)\, m_{2,0}(\bar{\gamma}, \sigma_\gamma^2)\, m_{0,2}(\bar{\alpha}, \sigma_\alpha^2) \\ & - \bar{s}^2\, m_{1,0}^2(\bar{\gamma}, \sigma_\gamma^2)\, m_{0,1}^2(\bar{\alpha}, \sigma_\alpha^2),\end{aligned} \tag{1C.13}$$

$$\begin{aligned}\mathbf{P}_1(4,5) = {} & (\sigma_s^2 + \bar{s}^2)\, m_{2,0}(\bar{\gamma}, \sigma_\gamma^2)\, m_{1,1}(\bar{\alpha}, \sigma_\alpha^2) \\ & - \bar{s}^2\, m_{1,0}^2(\bar{\gamma}, \sigma_\gamma^2)\, m_{1,0}(\bar{\alpha}, \sigma_\alpha^2)\, m_{0,1}(\bar{\alpha}, \sigma_\alpha^2),\end{aligned} \tag{1C.14}$$

$$\mathbf{P}_1(4,6) = [(\sigma_s^2 + \bar{s}^2)\, m_{1,1}(\bar{\gamma}, \sigma_\gamma^2) - \bar{s}^2\, m_{1,0}(\bar{\gamma}, \sigma_\gamma^2)\, m_{0,1}(\bar{\gamma}, \sigma_\gamma^2)]\, m_{0,1}(\bar{\alpha}, \sigma_\alpha^2), \tag{1C.15}$$

$$\mathbf{P}_1(5,5) = (\sigma_s^2 + \bar{s}^2)\, m_{2,0}(\bar{\gamma}, \sigma_\gamma^2)\, m_{2,0}(\bar{\alpha}, \sigma_\alpha^2) \\ -\bar{s}^2\, m_{1,0}^2(\bar{\gamma}, \sigma_\gamma^2)\, m_{1,0}^2(\bar{\alpha}, \sigma_\alpha^2), \tag{1C.16}$$

$$\mathbf{P}_1(5,6) = [(\sigma_s^2 + \bar{s}^2)\, m_{1,1}(\bar{\gamma}, \sigma_\gamma^2) - \bar{s}^2\, m_{1,0}(\bar{\gamma}, \sigma_\gamma^2)\, m_{0,1}(\bar{\gamma}, \sigma_\gamma^2)]\, m_{1,0}(\bar{\alpha}, \sigma_\alpha^2), \tag{1C.17}$$

$$\mathbf{P}_1(6,6) = (\sigma_s^2 + \bar{s}^2)\, m_{0,2}(\bar{\gamma}, \sigma_\gamma^2) - \bar{s}^2\, m_{0,1}^2(\bar{\gamma}, \sigma_\gamma^2), \tag{1C.18}$$

where

$$m_{2,0}(\mu, \sigma^2) = [\sinh \sigma^2 + \cos^2 \mu \, \exp(-\sigma^2)] \exp(-\sigma^2), \tag{1C.19}$$

$$m_{0,2}(\mu, \sigma^2) = [\sinh \sigma^2 + \sin^2 \mu \, \exp(-\sigma^2)] \exp(-\sigma^2), \tag{1C.20}$$

$$m_{1,1}(\mu, \sigma^2) = \cos \mu \, \sin \mu \, \exp(-2\sigma^2). \tag{1C.21}$$

The elements of the lower triangular part of $\mathbf{P}_1$ are found using the symmetry condition $\mathbf{P}_1(i, j) = \mathbf{P}_1(j, i)$.

We now consider the initial moments for MSC, as given by (1.57) and (1.58). Let $\dot{x} = s \cos\gamma \sin\alpha - \dot{x}_1^{\mathrm{o}}$, $\dot{y} = s \cos\gamma \cos\alpha - \dot{y}_1^{\mathrm{o}}$ and $\dot{z} = s \sin\gamma - \dot{z}_1^{\mathrm{o}}$ denote the components of the relative velocity. Let $\hat{\xi}_1(i)$, $i = 1, \ldots, 6$, denote the ith element of the initial mean vector $\hat{\boldsymbol{\xi}}_1$. We have

$$\hat{\xi}_1(1) = \mathrm{E}(1/r)[m_{1,0}(\hat{\beta}_1, \sigma_\beta^2)\mathrm{E}(\dot{x}) - m_{0,1}(\hat{\beta}_1, \sigma_\beta^2)\mathrm{E}(\dot{y})] \tag{1C.22}$$

$$\hat{\xi}_1(2) = -\mathrm{E}(1/r)\{m_{0,1}(\hat{\epsilon}_1, \sigma_\epsilon^2)[m_{0,1}(\hat{\beta}_1, \sigma_\beta^2)\mathrm{E}(\dot{x}) \\ + m_{1,0}(\hat{\beta}_1, \sigma_\beta^2)\mathrm{E}(\dot{y})] - m_{1,0}(\hat{\epsilon}_1, \sigma_\epsilon^2)\mathrm{E}(\dot{z})\} \tag{1C.23}$$

$$\hat{\xi}_1(3) = \mathrm{E}(1/r)\{m_{1,0}(\hat{\epsilon}_1, \sigma_\epsilon^2)[m_{0,1}(\hat{\beta}_1, \sigma_\beta^2)\mathrm{E}(\dot{x}) \\ + m_{1,0}(\hat{\beta}_1, \sigma_\beta^2)\mathrm{E}(\dot{y})] + m_{0,1}(\hat{\epsilon}_1, \sigma_\epsilon^2)\mathrm{E}(\dot{z})\} \tag{1C.24}$$

$$\hat{\xi}_1(4) = \hat{\beta}_1 \tag{1C.25}$$

$$\hat{\xi}_1(5) = \hat{\epsilon}_1 \tag{1C.26}$$

$$\hat{\xi}_1(6) = \mathrm{E}(1/r) \tag{1C.27}$$

where the means of the relative velocity components are

$$\mathrm{E}(\dot{x}) = \bar{s}\, m_{1,0}(\bar{\gamma}, \sigma_\gamma^2)\, m_{0,1}(\bar{\alpha}, \sigma_\alpha^2) - \dot{x}_1^{\mathrm{o}} \tag{1C.28}$$

$$\mathrm{E}(\dot{y}) = \bar{s}\, m_{1,0}(\bar{\gamma}, \sigma_\gamma^2) m_{1,0}(\bar{\alpha}, \sigma_\alpha^2) - \dot{y}_1^{\mathrm{o}} \tag{1C.29}$$

$$\mathrm{E}(\dot{z}) = \bar{s}\, m_{0,1}(\bar{\gamma}, \sigma_\gamma^2) - \dot{z}_1^{\mathrm{o}} \tag{1C.30}$$

The upper triangular elements of the initial covariance matrix are

$$\mathbf{P}_1(1,1) = \mathrm{E}(1/r^2)[m_{2,0}(\hat{\beta}_1,\sigma_\beta^2)\mathrm{E}(\dot{x}^2) - 2m_{1,1}(\hat{\beta}_1,\sigma_\beta^2)\mathrm{E}(\dot{x}\dot{y}) + m_{0,2}(\hat{\beta}_1,\sigma_\beta^2)\mathrm{E}(\dot{y}^2)] - \hat{\xi}_1(1)^2, \tag{1C.31}$$

$$\begin{aligned}\mathbf{P}_1(1,2) = \mathrm{E}(1/r^2)\big(&\{m_{1,1}(\hat{\beta}_1,\sigma_\beta^2)[\mathrm{E}(\dot{y}^2) - \mathrm{E}(\dot{x}^2)] + [m_{0,2}(\hat{\beta}_1,\sigma_\beta^2)\\ &-m_{2,0}(\hat{\beta}_1,\sigma_\beta^2)]\mathrm{E}(\dot{x}\dot{y})\}m_{0,1}(\hat{\epsilon}_1,\sigma_\epsilon^2)\\ &+[m_{1,0}(\hat{\beta}_1,\sigma_\beta^2)\mathrm{E}(\dot{x}\dot{z}) - m_{0,1}(\hat{\beta}_1,\sigma_\beta^2)\mathrm{E}(\dot{y}\dot{z})]m_{1,0}(\hat{\epsilon}_1,\sigma_\epsilon^2)\big) - \hat{\xi}_1(1)\hat{\xi}_1(2),\end{aligned} \tag{1C.32}$$

$$\begin{aligned}\mathbf{P}_1(1,3) = \mathrm{E}(1/r^2)\big(&\{m_{1,1}(\hat{\beta}_1,\sigma_\beta^2)[\mathrm{E}(\dot{x}^2) - \mathrm{E}(\dot{y}^2)] + [m_{2,0}(\hat{\beta}_1,\sigma_\beta^2)\\ &-m_{0,2}(\hat{\beta}_1,\sigma_\beta^2)]\mathrm{E}(\dot{x}\dot{y})\}m_{1,0}(\hat{\epsilon}_1,\sigma_\epsilon^2)\\ &+[m_{1,0}(\hat{\beta}_1,\sigma_\beta^2)\mathrm{E}(\dot{x}\dot{z}) - m_{0,1}(\hat{\beta}_1,\sigma_\beta^2)\mathrm{E}(\dot{y}\dot{z})]m_{0,1}(\hat{\epsilon}_1,\sigma_\epsilon^2)\big) - \hat{\xi}_1(1)\hat{\xi}_1(3),\end{aligned} \tag{1C.33}$$

$$\mathbf{P}_1(1,4) = -\mathrm{E}(1/r)\sigma_\beta^2[\sin\hat{\beta}_1\,\mathrm{E}(\dot{x}) + \cos\hat{\beta}_1\,\mathrm{E}(\dot{y})]\exp(-\sigma_\beta^2/2), \tag{1C.34}$$

$$\mathbf{P}_1(1,6) = [\mathrm{E}(1/r^2) - \mathrm{E}(1/r)^2][m_{1,0}(\hat{\beta}_1,\sigma_\beta^2)\mathrm{E}(\dot{x}) - m_{0,1}(\hat{\beta}_1,\sigma_\beta^2)\mathrm{E}(\dot{y})], \tag{1C.35}$$

$$\begin{aligned}\mathbf{P}_1(2,2) = \mathrm{E}(1/r^2)\{&[m_{0,2}(\hat{\beta}_1,\sigma_\beta^2)\mathrm{E}(\dot{x}^2) + 2\,m_{1,1}(\hat{\beta}_1,\sigma_\beta^2)\mathrm{E}(\dot{x}\dot{y})\\ &+m_{2,0}(\hat{\beta}_1,\sigma_\beta^2)\mathrm{E}(\dot{y}^2)]m_{0,2}(\hat{\epsilon}_1,\sigma_\epsilon^2)\\ &-2[m_{0,1}(\hat{\beta}_1,\sigma_\beta^2)\mathrm{E}(\dot{x}\dot{z}) + m_{1,0}(\hat{\beta}_1,\sigma_\beta^2)\mathrm{E}(\dot{y}\dot{z})]m_{1,1}(\hat{\epsilon}_1,\sigma_\epsilon^2)\\ &+m_{2,0}(\hat{\epsilon}_1,\sigma_\epsilon^2)\mathrm{E}(\dot{z}^2)\} - \hat{\xi}_1(2)^2,\end{aligned} \tag{1C.36}$$

$$\begin{aligned}\mathbf{P}_1(2,3) = \mathrm{E}(1/r^2)\{&[\mathrm{E}(\dot{z}^2) - m_{0,2}(\hat{\beta}_1,\sigma_\beta^2)\mathrm{E}(\dot{x}^2) - 2m_{1,1}(\hat{\beta}_1,\sigma_\beta^2)\mathrm{E}(\dot{x}\dot{y})\\ &+m_{2,0}(\hat{\beta}_1,\sigma_\beta^2)\mathrm{E}(\dot{y}^2)]m_{1,1}(\hat{\epsilon}_1,\sigma_\epsilon^2)\\ &+[m_{0,1}(\hat{\beta}_1,\sigma_\beta^2)\mathrm{E}(\dot{x}\dot{z}) + m_{1,0}(\hat{\beta}_1,\sigma_\beta^2)\mathrm{E}(\dot{y}\dot{z})][m_{2,0}(\hat{\epsilon}_1,\sigma_\epsilon^2)\\ &-m_{0,2}(\hat{\epsilon}_1,\sigma_\epsilon^2)]\} - \hat{\xi}_1(2)\hat{\xi}_1(3),\end{aligned} \tag{1C.37}$$

$$\mathbf{P}_1(2,4) = \mathrm{E}(1/r)\sigma_\beta^2[\sin\hat{\beta}_1\,\mathrm{E}(\dot{y}) - \cos\hat{\beta}_1\,\mathrm{E}(\dot{x})]\exp(-\sigma_\beta^2/2)m_{0,1}(\hat{\epsilon}_1,\sigma_\epsilon^2), \tag{1C.38}$$

$$\begin{aligned}\mathbf{P}_1(2,5) = -\mathrm{E}(1/r)\sigma_\epsilon^2\{&\cos\hat{\epsilon}_1\,[m_{0,1}(\hat{\beta}_1,\sigma_\beta^2)\mathrm{E}(\dot{x}) + m_{1,0}(\hat{\beta}_1,\sigma_\beta^2)\mathrm{E}(\dot{y})]\\ &+\sin\hat{\epsilon}_1\,\mathrm{E}(\dot{z})\}\exp(-\sigma_\epsilon^2/2),\end{aligned} \tag{1C.39}$$

$$\begin{aligned}\mathbf{P}_1(2,6) = -[\mathrm{E}(1/r^2) - \mathrm{E}(1/r)^2][&m_{0,1}(\hat{\beta}_1,\sigma_\beta^2)\,m_{0,1}(\hat{\epsilon}_1,\sigma_\epsilon^2)\mathrm{E}(\dot{x})\\ &+m_{1,0}(\hat{\beta}_1,\sigma_\beta^2)\,m_{0,1}(\hat{\epsilon}_1,\sigma_\epsilon^2)\mathrm{E}(\dot{y}) - m_{1,0}(\hat{\epsilon}_1,\sigma_\epsilon^2)\mathrm{E}(\dot{z})],\end{aligned} \tag{1C.40}$$

$$\begin{aligned}\mathbf{P}_1(3,3) = \mathrm{E}(1/r^2)\{&[m_{0,2}(\hat{\beta}_1,\sigma_\beta^2)\mathrm{E}(\dot{x}^2) + 2\,m_{1,1}(\hat{\beta}_1,\sigma_\beta^2)\mathrm{E}(\dot{x}\dot{y})\\ &+m_{2,0}(\hat{\beta}_1,\sigma_\beta^2)\mathrm{E}(\dot{y}^2)]m_{2,0}(\hat{\epsilon}_1,\sigma_\epsilon^2)\end{aligned}$$

$$+2[m_{0,1}(\hat{\beta}_1, \sigma_\beta^2)\mathrm{E}(\dot{x}\dot{z}) + m_{1,0}(\hat{\beta}_1, \sigma_\beta^2)\mathrm{E}(\dot{y}\dot{z})]m_{1,1}(\hat{\epsilon}_1, \sigma_\epsilon^2)$$
$$+m_{0,2}(\hat{\epsilon}_1, \sigma_\epsilon^2)\mathrm{E}(\dot{z}^2)\} - \hat{\xi}_1(3)^2, \tag{1C.41}$$

$$\mathbf{P}_1(3,4) = \mathrm{E}(1/r)\sigma_\beta^2[\cos\hat{\beta}_1\,\mathrm{E}(\dot{x}) - \sin\hat{\beta}_1\,E(\dot{y})]\exp(-\sigma_\beta^2/2)m_{1,0}(\hat{\epsilon}_1, \sigma_\epsilon^2), \tag{1C.42}$$

$$\mathbf{P}_1(3,5) = \mathrm{E}(1/r)\sigma_\epsilon^2\{-\sin\hat{\epsilon}_1\,[m_{0,1}(\hat{\beta}_1, \sigma_\beta^2)\mathrm{E}(\dot{x}) + m_{1,0}(\hat{\beta}_1, \sigma_\beta^2)\mathrm{E}(\dot{y})]$$
$$+\cos\hat{\epsilon}_1\,\mathrm{E}(\dot{z})\}\exp(-\sigma_\epsilon^2/2), \tag{1C.43}$$

$$\mathbf{P}_1(3,6) = [\mathrm{E}(1/r^2) - \mathrm{E}(1/r)^2][m_{0,1}(\hat{\beta}_1, \sigma_\beta^2)\,m_{1,0}(\hat{\epsilon}_1, \sigma_\epsilon^2)\mathrm{E}(\dot{x})$$
$$+m_{1,0}(\hat{\beta}_1, \sigma_\beta^2)\,m_{1,0}(\hat{\epsilon}_1, \sigma_\epsilon^2)\mathrm{E}(\dot{y}) + m_{0,1}(\hat{\epsilon}_1, \sigma_\epsilon^2)\mathrm{E}(\dot{z})], \tag{1C.44}$$

$$\mathbf{P}_1(4,4) = \sigma_\beta^2, \tag{1C.45}$$

$$\mathbf{P}_1(5,5) = \sigma_\epsilon^2, \tag{1C.46}$$

$$\mathbf{P}_1(6,6) = \mathrm{E}(1/r^2) - \mathrm{E}(1/r)^2. \tag{1C.47}$$

The means of the relative velocity components $\dot{x}$, $\dot{y}$, and $\dot{z}$ are given in (1C.28–1C.30). The second-order moments are

$$\mathrm{E}(\dot{x}^2) = (\sigma_s^2 + \bar{s}^2)\,m_{2,0}(\bar{\gamma}, \sigma_\gamma^2)\,m_{0,2}(\bar{\alpha}, \sigma_\alpha^2) - 2\bar{s}\,m_{1,0}(\bar{\gamma}, \sigma_\gamma^2)\,m_{0,1}(\bar{\alpha}, \sigma_\alpha^2)\dot{x}_1^o + (\dot{x}_1^o)^2, \tag{1C.48}$$

$$\mathrm{E}(\dot{x}\dot{y}) = (\sigma_s^2 + \bar{s}^2)\,m_{2,0}(\bar{\gamma}, \sigma_\gamma^2)\,m_{1,1}(\bar{\alpha}, \sigma_\alpha^2)$$
$$-\bar{s}\,m_{1,0}(\bar{\gamma}, \sigma_\gamma^2)[m_{1,0}(\bar{\alpha}, \sigma_\alpha^2)\dot{x}_1^o + m_{0,1}(\bar{\alpha}, \sigma_\alpha^2)\dot{y}_1^o] + \dot{x}_1^o\dot{y}_1^o, \tag{1C.49}$$

$$\mathrm{E}(\dot{x}\dot{z}) = (\sigma_s^2 + \bar{s}^2)\,m_{1,1}(\bar{\gamma}, \sigma_\gamma^2)\,m_{0,1}(\bar{\alpha}, \sigma_\alpha^2)$$
$$-\bar{s}[m_{0,1}(\bar{\gamma}, \sigma_\gamma^2)\dot{x}_1^o + m_{1,0}(\bar{\gamma}, \sigma_\gamma^2)\,m_{0,1}(\bar{\alpha}, \sigma_\alpha^2)\dot{z}_1^o] + \dot{x}_1^o\dot{z}_1^o, \tag{1C.50}$$

$$\mathrm{E}(\dot{y}^2) = (\sigma_s^2 + \bar{s}^2)\,m_{2,0}(\bar{\gamma}, \sigma_\gamma^2)\,m_{2,0}(\bar{\alpha}, \sigma_\alpha^2) - 2\bar{s}\,m_{1,0}(\bar{\gamma}, \sigma_\gamma^2)\,m_{1,0}(\bar{\alpha}, \sigma_\alpha^2)\dot{y}_1^o + (\dot{y}_1^o)^2, \tag{1C.51}$$

$$\mathrm{E}(\dot{y}\dot{z}) = (\sigma_s^2 + \bar{s}^2)\,m_{1,1}(\bar{\gamma}, \sigma_\gamma^2)\,m_{1,0}(\bar{\alpha}, \sigma_\alpha^2)$$
$$-\bar{s}[m_{0,1}(\bar{\gamma}, \sigma_\gamma^2)\dot{y}_1^o + m_{1,0}(\bar{\gamma}, \sigma_\gamma^2)\,m_{1,0}(\bar{\alpha}, \sigma_\alpha^2)\dot{z}_1^o] + \dot{y}_1^o\dot{z}_1^o. \tag{1C.52}$$

The moments $\mathrm{E}(1/r^k)$, $k = 1, 2$ cannot be found in closed form but can be accurately approximated with moderate computational expense via importance sampling. In particular, let $r_i \sim N(\bar{r}, \sigma_r^2)$, $i = 1, \ldots, n$ denote samples drawn from the initial range distribution. Then, the required moments are approximated by, for $k = 1, 2$,

$$\widehat{\mathrm{E}(1/r^k)} = 1/n\sum_{i=1}^{n} r_i^k. \tag{1C.53}$$

ACKNOWLEDGMENTS

The authors thank Sam Blackman for providing the details of the tracking scenario and useful discussions and advice. The authors also thank Anand Mallick for verifying the derivations using Mathematica, developing Matlab code, generating numerical results, and preparing the LaTex document.

REFERENCES

1. S.S. Blackman and S.H. Roszkowski, Application of IMM filtering to passive ranging, *Proc. SPIE*, Vol. 3809, 1999, pp. 270–281.
2. S.S. Blackman, R.J. Dempster, B. Blyth, and C. Durand, Integration of passive ranging with multiple hypothesis tracking (MHT) for application with angle-only measurements, *Proc. SPIE*, Orlando, Florida, USA, Vol. 7698, 2010, pp. 769815-1–769815-11.
3. Q. Li, F. Guo, Y. Zhou, and W. Jiang, Observability of satellite to satellite passive tracking from angles measurements, *Proceedings of the IEEE International Conference on Control and Automation*, 2007, pp. 1926–1931.
4. Q. Li, L. Shi, H. Wang, and F. Guo, Utilization of modified spherical coordinates for satellite to satellite bearings-only tracking, *Chin. J. Space Sci.*, 29(6), 627–634, 2009.
5. V.J. Aidala and S.E. Hammel, Utilization of modified polar coordinates for bearings-only tracking, *IEEE Trans. Automat. Contr.*, AC-28(3), 283–294, 1983.
6. S. Arulampalam and B. Ristic, Comparison of the particle filter with range-parameterised and modified polar EKFs for angle-only tracking, *Proc. SPIE*, Vol. 4048, 2000, pp. 288–299.
7. S. Arulampalam, M. Clark, and R. Vinter, Performance of the shifted rayleigh filter in single-sensor bearings-only tracking, *Proceedings of the Tenth International Conference on Information Fusion*, Québec, Canada, July 2007.
8. S. Blackman and R. Popoli, *Design and Analysis of Modern Tracking Systems*, Artech House, Norwood, MA, USA, 1999.
9. B. Ristic, S. Arulampalam, and N. Gordon, *Beyond the Kalman Filter*, Artech House, Norwood, MA, USA, 2004.
10. R.R. Allen and S.S. Blackman, Implementation of an angle-only tracking filter, *Proceedings of SPIE, Signal and Data Processing of Small Targets*, Orlando, FL, USA, Vol. 1481, April 1991, pp. 292–303.
11. R.R. Allen and S.S. Blackman, Angle-only tracking with a MSC filter, *Proceedings of Digital Avionics Systems Conference*, Los Angeles, CA, USA, October 1991, pp. 561–566.
12. P. Gurfil and N.J. Kasdin, Two-step optimal estimator for three dimensional target tracking, *IEEE Trans. Aerosp. Electron. Syst.*, AES-40(3), 780–793, 2005.
13. R. Karlsson and F. Gustafsson, Range estimation using angle-only target tracking with particle filters, *Proceedings of the American Control Conference*, 2001, pp. 3743–3748.
14. M. Mallick, L. Mihaylova, S. Arulampalam, and Y. Yan, Angle-only filtering in 3D using modified spherical and log spherical coordinates, *2011 International Conference on Information Fusion*, Chicago, USA, July 5–8, 2011, pp. 1905–1912.

15. M. Mallick, M. Morelande, L. Mihaylova, S. Arulampalam, and Y. Yan, Comparison of angle-only filtering algorithms in 3D using cartesian and modified spherical coordinates, *2012 International Conference on Information Fusion*, Singapore, July 9–12, 2012, pp. 1392–1399.

16. P.N. Robinson and M.R. Yin, Modified spherical coordinates for radar, *Proceedings of the AIAA Guidance, Navigation and Control Conference*, Scottsdale, AZ, August 1–3, 1994, pp. 55–64.

17. D.V. Stallard, An angle-only tracking filter in modified spherical coordinates, *Proceedings of the AIAA Guidance, Navigation and Control Conference*, Monterey, CA, August 1987, pp. 542–550.

18. D.V. Stallard, Angle-only tracking filter in modified spherical coordinates, *J. Guid. Control, Dyn.*, 14(3), 694–696, 1991.

19. C. Jauffret and D. Pillon, Observability in passive target motion analysis, *IEEE Trans. Aerosp. Electron. Syst.*, AES-32(4), 1290–1300, 1996.

20. A. Jazwinski, *Stochastic Processes and Filtering Theory*, Academic Press, New York, 1970.

21. A. Gelb (Ed.), *Applied Optimal Estimation*, MIT Press, Cambridge, MA, USA, 1974.

22. Y. Bar-Shalom, X.R. Li, and T. Kirubarajan, *Estimation with Applications to Tracking and Navigation*, John Wiley & Sons, New York, 2001.

23. V.J. Aidala, Kalman filter behavior in bearing-only tracking applications, *IEEE Trans. Aerosp. Electron. Syst.*, AES-15(1), 29–39, 1979.

24. H.D. Hoelzer, G.W. Johnson, and A.O. Cohen, Modified polar coordinates—the key to well behaved bearings only ranging, *IR & D Report 78-M19-OOOlA*, IBM Federal Systems Division, Shipboard and Defense Systems, Manassas, VA 22110, August 31, 1978.

25. G.W. Johnson, H.D. Hoelzer, A.O. Cohen, and E.F. Harrold, Improved coordinates for target tracking from time delay information, *Proceedings of the Time Delay Estimation and Applications Conference*, Naval Postgraduate School, Monterey, CA, Vol. 2, May 1979, pp. M1–M32.

26. S.E. Hammel and V.J. Aidala, Observability requirements for three-dimensional tracking via angle measurements, *IEEE Trans. Aerosp. Electron. Syst.*, AES-24(2), 200–207, 1985.

27. R.A. Singer, Estimating optimal tracking filter performance for manned maneuvering targets, *IEEE Trans. Aerosp. Electron. Syst.*, AES-6, 473–483, 1970.

28. N.J. Gordon, D.J. Salmond, and A.F.M. Smith, Novel approach to nonlinear/non-Gaussian Bayesian state estimation, *IEE Proc.-F*, 140(2), 107–113, 1993.

29. M. Arulampalam, S. Maskell, N. Gordon and T. Clapp, A tutorial on particle filters for online nonlinear/nonGaussian Bayesian tracking, *IEEE Trans. Signal Process.*, SP-50(2), 174–188, 2002.

30. P.E. Kloeden and E. Platen, *Numerical Solution of Stochastic Differential Equations*, Springer Verlag, Berlin, 1992.

31. S. Julier, J. Uhlmann, and H.F. Durrant-Whyte, A new method for the nonlinear transformation of means and covariances in filters and estimators, *IEEE Trans. Automat. Contr.*, AC-45(3), 477–482, 2000.

32. S.J. Julier and J.K. Uhlmann, Unscented filtering and nonlinear estimation, *Proc. IEEE*, 92(3), 401–422, 2004.

33. A. Doucet, S. Godsill, and C. Andrieu, On sequential Monte Carlo sampling methods for Bayesian filtering, *Stat. Comput.*, 10(3), 197–208, 2000.

34. S. Särkkä, On unscented Kalman filtering for state estimation of continuous-time nonlinear systems, *IEEE Trans. Automat. Contr.*, AC-52(9), 1631–1641, 2007.
35. S. Särkkä, On sequential Monte Carlo sampling of discretely observed stochastic differential equations, *Proceedings of NSSPW*, September 2006.
36. M. Athans, R.P. Wishner, and A. Bertolini, Suboptimal state estimation for continuous-time nonlinear systems from discrete noisy measurements, *IEEE Trans. Automat. Contr.*, AC-13(5), 504–514, 1968.
37. R.K. Mehra , A comparison of several nonlinear filters for reentry vehicle tracking, *IEEE Trans. Automat. Contr.*, AC-16(4), 307–319, 1971.
38. A. Farina, B. Ristic, and D. Benvenuti, Tracking a ballistic target: Comparison of several nonlinear filters, *IEEE Trans. Aerosp. Electron. Syst.*, AES-38(3), 854–867, 2002.
39. V.C. Ravindra, Y. Bar-Shalom, and P. Willett, Projectile identification and impact point prediction, *IEEE Trans. Aerosp. Electron. Syst.*, AES-46(4), 2004–2021, 2010.
40. B.D. Tapley, B.E. Schutz, and G.H. Born, *Statistical Orbit Determination*, Elsevier Academic Press, Burlington, MA, 2004.
41. J.R. Wertz (Ed.), *Spacecraft Attitude Determination and Control*, D. Reidel Publishing Company, Boston, 1978.
42. R.A. Horn and C.R. Johnson, *Topics in Matrix Analysis*, Cambridge University Press, Cambridge, UK, 1991.
43. E.C. MacRae, Matrix derivatives with an application to an adaptive decision problem, *The Ann. Stat.*, 2(2), 337–345, 1974.
44. M.K. Pitt and N. Shephard, Filtering via simulation:auxiliaryparticle filters, *J. Am. Stat. Assoc.*, 94(446), 590–599, 1999.
45. M. Hürzeler and H.R. Künsch, Monte Carlo approximations for general state-space models, *J. Comput. Graph. Stat.*, 7(2), 175–193, 1998.
46. C. Musso, N. Ouddjane, and F. Le Gland, Improving regularised particle filters, in: A. Doucet, N. De Freitas, and N. Gordon (Eds.), *Sequential Monte Carlo Methods in Practice*, Springer, New York, NY, Chapter 12, pp. 247–271, 2001.
47. B.W. Silverman, *Density Estimation for Statistics and Data Analysis*, Chapman and Hall, Boca Raton, USA, 1986.
48. M. Mallick, M. Morelande, and L. Mihaylova, Continuous-discrete filtering using EKF, UKF, and PF, *2012 International Conference on Information Fusion*, Singapore, July 9–12, 2012, pp. 1087–1094.
49. T. Bréhard and J-P. Le Cadre, Closed-form posterior Cramér–Rao bounds for bearings only tracking, *IEEE Trans. Aerosp. Electron. Syst.*, AES-42(4), 1198–1223, 2006.

CHAPTER 2

Particle Filtering Combined with Interval Methods for Tracking Applications

AMADOU GNING, LYUDMILA MIHAYLOVA, FAHED ABDALLAH, AND BRANKO RISTIC

2.1 INTRODUCTION

State estimation for complex stochastic systems, in the presence of non-Gaussian noisy measurements, is of paramount importance for many applications and has been actively investigated during the last decades. One of the most popular state estimation approaches is the Bayesian inference approach [1, 2]. Within the Bayesian framework the posterior probability density function (pdf) of the state of interest is calculated, conditioned on the available measurements. Among the Bayesian approaches different algorithms have been used such as the Extended Kalman Filter (EKF) [3], unscented Kalman filter (UKF) [4–7], and sequential Monte Carlo methods (known as particle filters) [1, 2, 7, 8]. The Kalman Filter (KF) algorithm and its variants such as EKF rely on linearized state and measurement functions, typically up to a first-order Taylor series expansion. The posterior pdf of the state is approximated by the first two moments of the Gaussian distribution, the mean and covariance of the system state. The UKF [4] avoids linearization by propagating deterministically chosen points, called sigma points, to capture the mean and covariance with a reasonably small set of carefully selected points called *sigma points*. However, both the EKF and the UKF are limited to Gaussian and unimodal type of pdfs. Complex and multimodal pdfs can be represented by mixtures of Gaussians as shown in [9]. Using approximation theory [10], it can be shown that the Gaussian family function is dense in the space of continuous functions and therefore, most of the Kalman type filters have been extended to mixtures of Gaussians.

Integrated Tracking, Classification, and Sensor Management: Theory and Applications, Edited by Mahendra Mallick, Vikram Krishnamurthy, and Ba-Ngu Vo.

Particle filtering methods [1, 2] have recently emerged as a powerful tool for solving complex problems, in the presence of high nonlinearities or high-dimensional systems, where the propagation of Gaussian mixtures is not always feasible. The particle filtering approach represents the posterior state pdf through samples. However, the efficiency and accuracy of particle filters (PFs) depend significantly on the number of particles and on the proposal functions used for the importance sampling. In addition, the imprecision in the available information poses challenges to the choice of the importance sampling function, requires potentially a large number of particles and high computational complexity, which induces real-time implementation issues. In attempt to overcome these issues, both statistical and deterministic approaches have been proposed. Many works rely on statistical approaches [2, 11]. In [12], the sample set size is dynamically chosen during the estimation process, whereas in [13, 14] Rao-Blackwellization techniques are shown to reduce the computational complexity.

The interval analysis framework is another promising methodology to model measurements with unknown or complex statistical bounded errors. Initially introduced to propagate rounding errors in mathematical computations earlier in the 1950s [15], applications to state estimation have been recently intensively investigated. For instance, in [16, 17] and [18] bounded-error observers are proposed, based on the prediction–correction approach combined with various well-known interval analysis methods. These approaches are different from classical estimation methods, since the main objective is to estimate dynamically, in a guaranteed way, optimized boxes containing the hidden states.

In [19] the interval framework is used for state estimation, without focusing on obtaining estimates inside guaranteed boxes. A Box Particle Filter (Box-PF) is proposed consisting of propagated weighted boxes in a sequential way. Instead of using point samples, weighted boxes are used to approximate desired moments of the posterior pdfs.

In this chapter, the Box-PF is presented in its original ad hoc formulation given in [19] and also a theoretical Bayesian interpretation of the Box-PF [20]. The remaining part of this chapter is organized in the following way. Section 2.2 gives an overview of related works. Section 2.3 introduces the basic concepts of the interval analysis and interval methods. Section 2.4 provides an overview of the Bayesian inference methodology. Section 2.5 presents the Box-PF, combining particle filtering with interval analysis. Section 2.6 gives a theoretical derivation of the Box-PF as a sum of uniform pdfs. Section 2.8 demonstrates the advantages of the Box-PF over a dynamic localization example. The contributions and open issues for future works are summarized in Section 2.9.

2.2 RELATED WORKS

For a wealth of problems of target tracking, localization and in general for state estimation, the uncertainties can be significant and especially the measurement noise characteristics unknown or, at most, known only within a certain region, for example, the minimum and maximum values of the noise covariance matrices can be available.

With the only assumption that the model and measurement noises are bounded, a class of *bounded error approaches* (also known as *set-membership methods*) proposes to merge the information in such a way that all outputs are guaranteed to contain the true solutions. A particularity of these approaches is that, instead of providing point estimates, they provide optimized sets enclosing all the possible solutions.

In the bounded errors framework applied to state estimation problems, two popular families of sets have been widely studied. These are the ellipsoid methods, studied intensively earlier in the 1960s, and the box methods arising more recently in the 1990s. The ellipsoid sets are known to be well adapted for enclosing desired sets [21] but unfortunately, they are mostly restricted to linear cases (see in particular [22–25]). In contrast, a box family offers less adaptability to enclose desired solution sets (unless by using a finite union of disjoint boxes called *Subpaving* at a computation cost expense [16]). The main advantage of using boxes over ellipsoids is that boxes can be easily manipulated and propagated with nonlinear models. The estimation methods presented in this chapter belong to the bounded error methods using boxes.

One of the initial attempts to apply interval analysis methods to sequential nonlinear state estimation is in [26] using the Set Inversion Via Interval Analysis (SIVIA) algorithm. The SIVIA algorithm was later used in several other robotic applications, for example, [27, 28]. The aim of the SIVIA algorithm is to compute the reciprocal image of a subpaving by a possibly nonlinear function. The idea is to test, for each box of a given subpaving inside a search domain, if its image by a function of the observation model, has an intersection with the set of measurement box. A box for which this condition is not verified is excluded, otherwise if the box is entirely contained in the measurement box, it is memorized. The process is iterated with a bisection of the remaining boxes of the subpaving until a prespecified threshold volume is reached. Recursive guaranteed state estimation can be developed by using the SIVIA algorithm to perform the measurement update step [26]. The main drawbacks of the SIVIA algorithm are its exponential computational time (in relation to the problem dimension), in addition to the use of subpavings.

An alternative and real-time interval method, based on the so-called *constraints propagation* algorithm, has been recently studied for state estimation [29, 30] and successfully applied to robotics applications, for example [18, 31].

The Box-PF algorithm [19] presented in this chapter combines methodologies from both interval analysis and Bayesian inference and consists of propagated weighted boxes in a sequential way:

(i) The time update step is present and is done using interval analysis methods.
(ii) The measurement update step is also present and a likelihood term is calculated for each step. In addition, a constraint propagation algorithm is performed for each box to make it consistent with the measurements.

Finally, other related works can be found in [32–34]. These works have the particularity to unify bounded error methods and Bayesian methods by assuming the presence of both stochastic and bounded noises.

2.3 INTERVAL ANALYSIS

The interval analysis framework offers promising methodologies for modeling measurements with unknown or complex statistical bounded errors. Initially introduced to propagate rounding errors in mathematical computations earlier in the 1950s [15], applications to state estimation have been recently investigated. For instance, in [16, 17] and [18] bounded-error observers, based on a prediction–correction approach combined with various well-known interval analysis tools, have been proposed.

2.3.1 Basic Concepts

A real interval, denoted by $[x]$, is defined as a closed and connected subset

$$[x] = [\underline{x}, \overline{x}] = \{x \in \mathbb{R} \mid \underline{x} \leq x \leq, \overline{x}\}, \tag{2.1}$$

of the set $\mathbb{R}$ of real numbers where $\underline{x}$ and $\overline{x}$ denote, respectively, the minimal and maximal bounds of $[x]$. The length of the interval $[x]$ is further denoted by $|[x]|$. The set of intervals in $\mathbb{R}$ is commonly denoted as $\mathbb{IR}$. Set-theoretic operations such as intersection or union can be applied to intervals. The intersection of two intervals is always an interval, whereas the union is not. Let us denote [.] the *interval hull* operator of any set S in $\mathbb{R}$ that is, the smallest interval enclosing S. The *interval union* of two intervals $[x]$ and $[y]$ is defined as the *interval hull* of $[x] \cup [y]$, which is defined as follows [16]

$$[x] \sqcup [y] = \big[[x] \cup [y]\big] \,. \tag{2.2}$$

Furthermore, the binary operations $\{+, -, \times, /\}$ can be extended to intervals [16]. For any binary operator $\diamond$ and for any intervals $[x]$ and $[y]$, the resulting interval $[z] = [x] \diamond [y]$ is defined as

$$[z] = [x] \diamond [y] = \big[\{x \diamond y \mid x \in [x], y \in [y]\}\big] \,. \tag{2.3}$$

When the binary operation $\diamond$ is continuous, as for the usual arithmetic operations, the set $\{x \diamond y \in \mathbb{R} | x \in [x], y \in [y]\}$ is an interval. The definition (2.3) for $[x] \diamond [y]$ is equivalent to the following one

$$[x] \diamond [y] = \{x \diamond y \in \mathbb{R} | x \in [x], y \in [y]\}. \tag{2.4}$$

For the usual arithmetic operations $\{+, -, \times, /\}$, $[x] \diamond [y]$ can be implicitly computed from the bounds of $[x]$ and $[y]$ using simple formulas. The following equations hold [16, 35]

$$[x] + [y] = [\underline{x} + \underline{y}, \overline{x} + \overline{y}], \tag{2.5}$$
$$[x] - [y] = [\underline{x} - \overline{y}, \overline{x} - \underline{y}], \tag{2.6}$$
$$[x] \times [y] = [\min(\underline{x}\underline{y}, \underline{x}\overline{y}, \overline{x}\underline{y}, \overline{x}\overline{y}), \max(\underline{x}\underline{y}, \underline{x}\overline{y}, \overline{x}\underline{y}, \overline{x}\overline{y})]. \tag{2.7}$$

If the value 0 does not belong to the interval $[y]$, we also have

$$[x]/[y] = [x] \times [1/\overline{y}, 1/\underline{y}]. \tag{2.8}$$

In the case when 0 belongs to the interval $[y]$, the division operation can be defined by extending $\mathbb{R}$ to $\mathbb{R} \cup \{-\infty, \infty\}$ and by defining intervals in the form $[-\infty, \overline{x}]$, $[\underline{x}, \infty]$, and $[\infty, \infty]$.

Elementary functions such as exp, ln, cos, and sin can also be extended to intervals. If f is a function from $\mathbb{R}$ to $\mathbb{R}$, its interval counterpart $[f]$ is defined by

$$[f]([x]) = [\{f(x) | x \in [x]\}]. \tag{2.9}$$

In a similar way to the binary operation, if f is continuous and monotonic, then $[f]([x])$ is simply equal to $f([x])$. For instance,

$$[\exp]([x]) = [\exp(\underline{x}), \exp(\overline{x})]. \tag{2.10}$$

For nonmonotonic continuous functions, however, the computation of $[f]([x])$ is usually not straightforward.

A *box* represents a vector $[\mathbf{x}]$ in $\mathbb{R}^n$ which is defined as a Cartesian product of n intervals

$$[\mathbf{x}] = [x_1] \times \cdots \times [x_n]. \tag{2.11}$$

Interval arithmetics, recalled above, can be straightforwardly extended to boxes. For instance, the interval hull of a subset S of $\mathbb{R}^n$ (denoted by $[S]$) is the smallest box of $\mathbb{R}^n$ enclosing S and the interval union of two boxes $[\mathbf{x}]$ and $[\mathbf{y}]$, and is defined as

$$[\mathbf{x}] \sqcup [\mathbf{y}] = ([x_1] \sqcup [y_1]) \times \cdots \times ([x_n] \sqcup [y_n]). \tag{2.12}$$

In a similar way, binary operations between real vectors or function of real vectors can be extended to interval vectors. For instance,

$$[\mathbf{x}] + [\mathbf{y}] = ([x_1] + [y_1]) \times \cdots \times ([x_n] + [y_n]). \tag{2.13}$$

2.3.2 Inclusion Functions

Let $\mathbf{f}$ be a function from $\mathbb{R}^n$ to $\mathbb{R}^m$. It is often of interest to compute the image $\mathbf{f}([\mathbf{x}])$ of a box $[\mathbf{x}]$ by $\mathbf{f}$. In general, this image is not a box (see Figure 2.1 for an illustration),

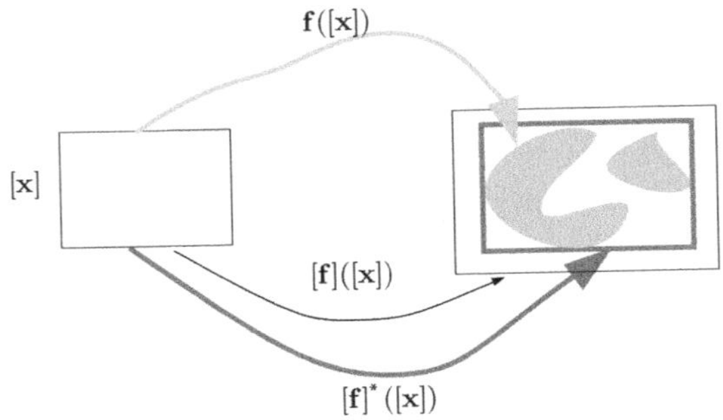

Figure 2.1 Inclusion functions obtained after applying the function **f** to a box [**x**]. The resulting image is not necessarily a box. A pessimistic inclusion function [**f**] and the minimal inclusion function [**f**]* are also presented.

and its expression may be difficult to obtain. However, it can be approximated using an *inclusion function*. An interval function [**f**] from $\mathbb{IR}^n$ to $\mathbb{IR}^m$ is said to be an *inclusion function* for **f** if

$$\mathbf{f}([\mathbf{x}]) \subseteq [\mathbf{f}]([\mathbf{x}]), \quad \forall [\mathbf{x}] \in \mathbb{IR}^n. \tag{2.14}$$

Inclusion functions may be very pessimistic, as shown in Figure 2.1. The inclusion function [**f**] is minimal if, for any **x**, [**f**]([**x**]) is the interval hull of **f**([**x**]). The minimal inclusion function for **f** is unique and can be denoted by [**f**]*.

Finding inclusion functions that can be evaluated with a convenient computational time and such that [**f**]([**x**]) is close to [**f**]*([**x**]), for most **x**, is one of the main purposes of interval analysis [16]. Different algorithms have been proposed to reduce the size of boxes enclosing **f**([**x**]) [16, 36]. Another major challenge for interval methods is to solve systems of equations involving initial conditions falling into boxes. The next section introduces the formulation of such classes of problems.

2.3.3 Constraint Satisfaction Problems

Consider n variables x_i in $\mathbb{R}$, $i \in \{1, \ldots, n\}$ linked by m relations (or constraints) of the form

$$f_j(x_1, \ldots, x_n) = 0, \quad j = 1, \ldots, m. \tag{2.15}$$

Furthermore, if **x** denotes the vector $(x_1, x_2, \ldots, x_n)^T$ (where T denotes the transpose operator) and **f** is the multivalued function whose components are f_i, that is, $\mathbf{f} = (f_1, f_2, \ldots, f_m)^T$, the m constraints (2.15) can be rewritten as

$$\mathbf{f}(\mathbf{x}) = \mathbf{0}. \tag{2.16}$$

Assuming that the vector **x** belongs to a given prior domain [**x**], the aim is to compute the smallest box [**x**′] enclosing the set of all **x** in the prior domain [**x**] satisfying

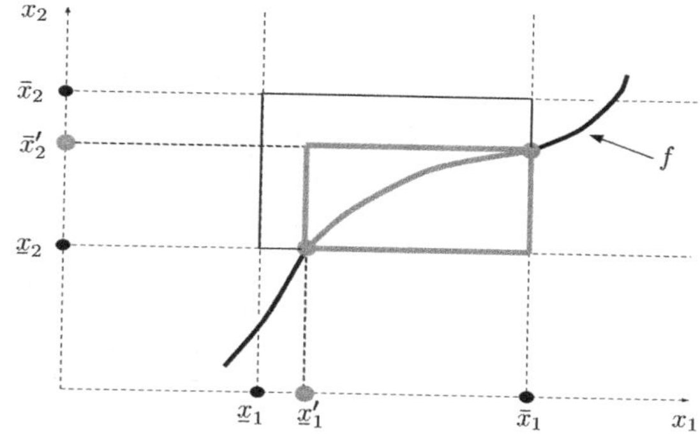

Figure 2.2 A Constraint Satisfaction Problem in $\mathbb{R}^2$ with one constraint f and an initial box $[\underline{x}_1\overline{x}_1] \times [\underline{x}_2\overline{x}_2]$. The box $[\underline{x}'_1\overline{x}_1] \times [\underline{x}'_2\overline{x}_2]$ represented in bold is the desired solution: it is included in the initial box $[\underline{x}_1\overline{x}_1] \times [\underline{x}_2\overline{x}_2]$ and it satisfies the constraint f.

the constraints $\mathbf{f}$. The so formulated problem is known as a continuous *Constraint Satisfaction Problem* (CSP), which is commonly denoted as $\mathcal{H}$, that is

$$\mathcal{H} : (\mathbf{f}(\mathbf{x}) = \mathbf{0}, \mathbf{x} \in [\mathbf{x}]). \tag{2.17}$$

Figure 2.2 illustrates a simple CSP with two variables and a single constraint. The box solution of the CSP is represented in bold.
The solution set of $\mathcal{H}$ is defined as

$$\mathbf{S} = \{\mathbf{x} \in [\mathbf{x}] \mid \mathbf{f}(\mathbf{x}) = \mathbf{0}\}. \tag{2.18}$$

Note that the set $\mathbf{S}$ is not necessarily a box. Within the interval framework, solving a CSP implies finding the smallest box $[\mathbf{x}]' \subset [\mathbf{x}]$ constituting an outer approximation of $\mathbf{S}$, such that $\mathbf{S} \subseteq [\mathbf{x}]' \subseteq [\mathbf{x}]$. In order to solve CSPs within the interval framework it is important the *consistency* condition to be satisfied [16]. There are two important types of consistency [37], as defined below and illustrated in Figure 2.3.

Definition 1 (Global consistency) *A scalar value x_i, belonging to the ith component of the box* $[\mathbf{x}]$, *is* globally consistent *[16] with a CSP $\mathcal{H}$ if it is possible to find at least one vector within* $[\mathbf{S}] = [\{\mathbf{x} \in [\mathbf{x}] \ldots \mathbf{f}\{\mathbf{x}\} = \mathbf{0}\}]$ *having x_i as its ith coordinate, that is, x_i is* globally consistent *if*

$$\exists(x_1, \ldots, x_{i-1}, x_{i+1}, \ldots, x_n) \in [x_{n_1}] \times \cdots, [x_{i-1}] \times [x_{i+1}] \times \cdots [x_n],$$
$$\mathbf{f}(x_1, \ldots, x_{i-1}, x_i, x_{i+1}, \ldots, x_n) = \mathbf{0}.$$

Definition 2 (Local consistency) *A scalar value x_i belonging to the ith component of the box* $[\mathbf{x}]$ *is* locally consistent *[16] with a CSP $\mathcal{H}$ if, for all $j = 1, \ldots, m$, it is*

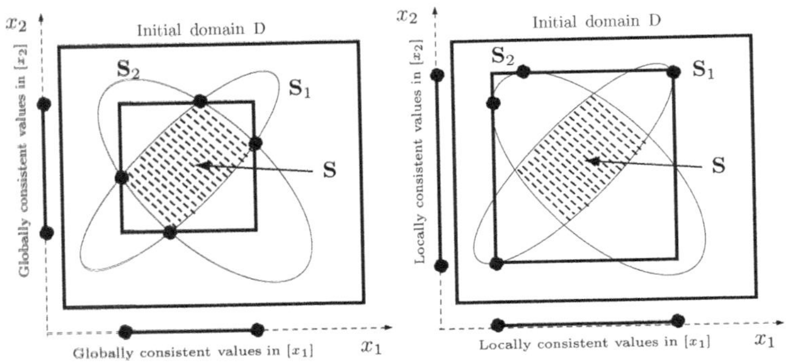

Figure 2.3 Illustration of the global (left) and local (right) consistency notions. For global consistency, any point value of $[x_1]$ corresponds at least to one vector (or solution) within $[\mathbf{S}]$. For local consistency, any punctual value of $[x_1]$ should have at least one corresponding solution in all the set $[\mathbf{S}_j]_{j=1,\ldots,m}$ taken separately.

possible to find at least one vector in $[\mathbf{S}_j] = [\{\mathbf{x} \in [\mathbf{x}] \mid f_j\{\mathbf{x}\} = \mathbf{0}\}]$ *having* x_i *as its ith coordinate, that is,* x_i *is* locally consistent *if*

$$\forall j \in \{1 \ldots m\}, \exists (x_1, \ldots, x_{i-1}, x_{i+1}, \ldots, x_n) \in [x_1] \times \cdots \times [x_{i-1}] \times [x_{i+1}] \times \cdots [x_n],$$
$$f_j(x_1, \ldots, x_{i-1}, x_i, x_{i+1}, \ldots, x_n) = 0.$$

The global consistency is equivalent to considering constraints all together while the local consistency is equivalent to taking the constraints one by one. Thus, the global consistency is stronger than the local consistency and represents the ideal solution for the CSP. Unfortunately, for most of the CSPs the existing methods can only reach the local consistency.

2.3.4 Contraction Methods

Contracting $\mathcal{H}$ means [16] replacing $[\mathbf{x}]$ by a smaller domain $[\mathbf{x}]'$ such that $\mathbf{S} \subseteq [\mathbf{x}]' \subseteq [\mathbf{x}]$. A *contractor* $\mathcal{C}$ for $\mathcal{H}$ is any operator that can be used to contract $\mathcal{H}$. Several methods for building contractors are described in [16, Chapter 4], including Gauss elimination, Gauss–Seidel algorithm, and linear programming. Each of these methods can be suitable for different types of CSP. Although the approaches presented in this chapter are not limited to any particular contractor, we will focus on a general and well-known contraction method, the *Constraints Propagation* (CP) technique [16, 37–40]. The main advantage of the CP method is its efficiency in the presence of high redundancy of data and constraints. The CP method is also known to be simple and, most importantly, to be independent of nonlinearities.

The CP method is based on the use of *primitive constraints*. A constraint is considered to be primitive if it involves a single binary operation (such as $+$, $-$, $*$ or $/$) or a single standard function (such as cos, sin, or sinh). The CP method proceeds by contracting $\mathcal{H}$ with respect to each of the primitive constraints until convergence to a minimal domain. Hereafter, a simple illustration is given though Example 1. A detailed description of the CP algorithm can be found in [16].

Example 1 Consider a three-dimensional CSP with a single constraint $z = x\exp(y)$ and an initial domain $[z] = [0, 3]$, $[x] = [1, 7]$, and $[y] = [0, 1]$. The constraint is not primitive, as it involves one arithmetic operation and one elementary function. However, it can be decomposed into two primitive constraints by introducing an auxiliary variable a

$$\begin{aligned} a &= \exp(y), \\ z &= a \cdot x. \end{aligned} \tag{2.19}$$

The domain of a can be initialized by $[a] = [0, +\infty]$. The CP algorithm alternates between two phases commonly called forward propagation and backward propagation.

In the *forward propagation* phase, the domains of a and z are contracted using (2.19). This is achieved via two steps F_1 and F_2:

- F_1: $[a] \leftarrow [a] \cap [\exp]([y]) = [0, +\infty] \cap [1, e] = [1, e]$
- F_2: $[z] \leftarrow [z] \cap [x] \cdot [a] = [0, 3] \cap [1, 7] \cdot [1, e] = [1, 3]$.

Next, a backward propagation step is performed, using an inversion of the system (2.19), which is achieved in three steps B_3, B_4, and B_5:

- B_3: $[x] \leftarrow [x] \cap ([z]/[a]) = [1, 7] \cap [1, 3]/[1, e] = [1, 3]$
- B_4: $[a] \leftarrow [a] \cap ([z]/[x]) = [1, e] \cap [1, 3]/[1, 3] = [1, e]$
- B_5: $[y] \leftarrow [y] \cap [\ln]([a]) = [0, 1] \cap [0, 1] = [0, 1]$.

After one forward–backward propagation cycle, the domains of the variables have thus been reduced to $[z] = [1, 3]$, $[x] = [1, 3]$, and $[y] = [0, 1]$. It can be checked that these domains are no longer changed after another iteration of the procedure.

In general, it can be shown that the CP contractor yields a globally consistent domain, when the order of the primitive contraction steps are carefully chosen, provided that the graph associated with the CSP contains no cycle [37]. Unfortunately, in general, cycles are often present in the CSPs and a loop is needed for the CP method (the primitive contractions are performed until no more contraction is possible). In the presence of cycles, the CP method provides locally consistent domains.

2.4 BAYESIAN FILTERING

Consider the following system:

$$\begin{cases} \mathbf{x}_{k+1} = \mathbf{f}(\mathbf{x}_k, \mathbf{v}_{k+1}), \\ \mathbf{z}_{k+1} = \mathbf{g}(\mathbf{x}_{k+1}, \mathbf{w}_{k+1}), \end{cases} \tag{2.20}$$

where $\mathbf{f} : \mathbb{R}^{n_x} \times \mathbb{R}^{n_v} \longrightarrow \mathbb{R}^{n_x}$ is in general a nonlinear transition function defining the state vector $\mathbf{x}_{k+1}$ at time $k+1$ from the previous state $\mathbf{x}_k$ and from $\mathbf{v}_k$ an independent identically distributed (iid) process noise sequence; n_x and n_v denote, respectively, the dimensions of the state, the input, and process noise vectors. The function $\mathbf{g} : \mathbb{R}^{n_x} \times \mathbb{R}^{n_w} \longrightarrow \mathbb{R}^{n_z}$ defines the relation between the state and measurement vectors, $\mathbf{w}_k$ is an iid measurement noise sequence; n_z, n_w are, respectively, dimensions of the measurement and measurement noise vectors. The states and the measurements up to time k are represented, respectively, by $\mathbf{X}_k = \{\mathbf{x}_{k_1}, k_1 = 1, \ldots, k\}$ and $\mathbf{Z}_k = \{\mathbf{z}_{k_1}, k_1 = 1, \ldots, k\}$.

Within the Bayesian framework, the posterior state pdf $p(\mathbf{X}_{k+1}|\mathbf{Z}_{k+1})$ provides a complete description of the state up to time instant $k+1$, given the measurements $\mathbf{Z}_{k+1}$ up to time $k+1$. In many applications, the marginal of the posterior pdf $p(\mathbf{x}_{k+1}|\mathbf{Z}_{k+1})$, at time instant $k+1$, also provides sufficient information and is given by

$$p(\mathbf{x}_{k+1}|\mathbf{Z}_{k+1}) = \frac{1}{\alpha_{k+1}} p(\mathbf{x}_{k+1}|\mathbf{Z}_k) p(\mathbf{z}_{k+1}|\mathbf{x}_{k+1}), \tag{2.21}$$

$$p(\mathbf{x}_{k+1}|\mathbf{Z}_k) = \int_{\mathbb{R}^{n_x}} p(\mathbf{x}_{k+1}|\mathbf{x}_k) p(\mathbf{x}_k|\mathbf{Z}_k) d\mathbf{x}_k, \tag{2.22}$$

where $\alpha_{k+1} = \int_{\mathbb{R}^{n_x}} p(\mathbf{z}_{k+1}|\mathbf{x}_{k+1}) p(\mathbf{x}_{k+1}|\mathbf{Z}_k) d\mathbf{x}_{k+1}$ is a normalization factor.

The recursion is initialized with a prior pdf $p(\mathbf{x}_0)$, for example with a uniform pdf over some region of the state space. Equation (2.22) corresponds to the time update step and equation (2.21) represents the measurement update step.

2.5 BOX PARTICLE FILTERING

In this section the original idea of the Box-PF [19] is briefly presented. The analogy and the difference with the PF are emphasized here while the Bayesian derivation of the Box-PF is shown in Section 2.6.

2.5.1 Main Steps of the Box Particle Filter

The aim of the Box-PF is to generalize particle filtering in the bounded error context. Instead of propagating weighted point samples of the posterior pdf, the key idea of the Box-PF is to propagate weighted box particles through bounded error models and using interval analysis. Figure 2.4 illustrates the main steps of the Box-PF algorithm. The Box-PF steps can be described as follows: box particle initialization, time update, measurement update, and resampling. These steps are summarized below with an emphasis to the analogy with the PF. Note that, in the next sections, *particles* are used for the PF and *box particles* for the Box-PF.

- *Box particle initialization*. For PF, this stage consists of sampling a set of N_p particles $\{\mathbf{x}_0^{(\ell)}\}_{\ell=1}^{N_p}$ in a prior bounded region of the state space. This process of

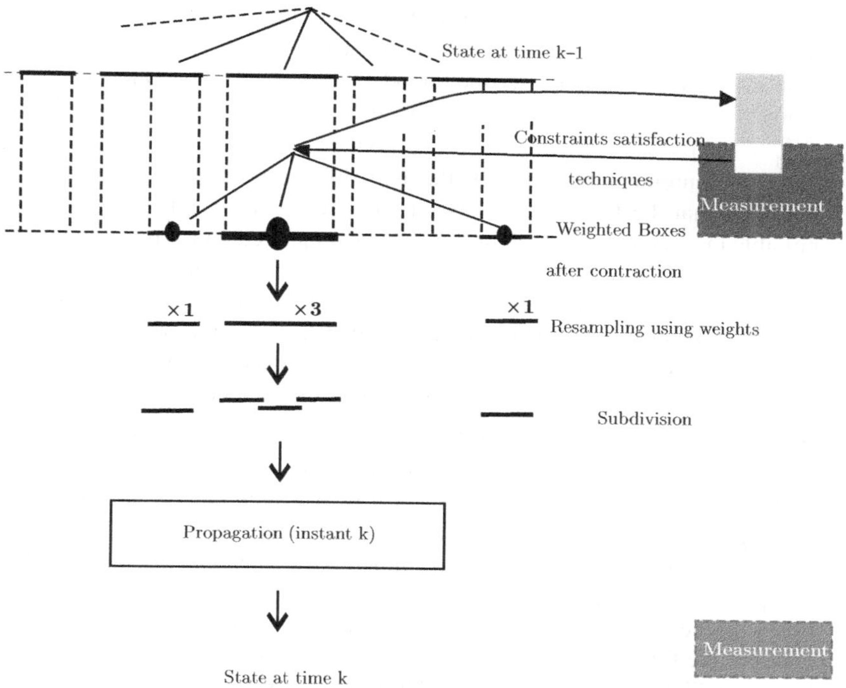

Figure 2.4 The Box Particle Filter steps.

initialization can be generalized in a natural way for boxes by splitting a prior bounded state space region into N_p equally weighted and mutually disjoint boxes $\{[\mathbf{x}_0^{(\ell)}]\}_{\ell=1}^{N_p}$. An advantage of this initialization using boxes is the possibility of exploring a large prior uncertainty region with few box particles.

- *Time update step*. In this step, for the PF, the prior pdf represented by the set of particles is propagated to the next time step. Knowing the particles $\{\mathbf{x}_k^{(l)}\}_{l=1}^{N_p}$ and the noise process $\mathbf{v}_k$ at step k, the particles are propagated at step $k+1$ according to a proposal pdf. If the proposal pdf is chosen to be the transition prior, the updated particles at time step $k+1$ can be sampled according to: $\{\mathbf{x}_{k+1}^{(\ell)} = \mathbf{f}(\mathbf{x}_k^{(\ell)}, \mathbf{v}_k^{(\ell)})\}_{\ell=1}^{N_p}$, where $\mathbf{v}_k^{(\ell)}$ is a noise realization corresponding to the particle $\mathbf{x}_k^{(\ell)}$. In the case of box particles, thanks to interval analysis tools, it is also possible to propagate boxes using the same transition function. Knowing the cloud of box particles $\{[\mathbf{x}_k^{(\ell)}]\}_{\ell=1}^{N_p}$ at time step k and assuming that the system noise is known to be enclosed in $[\mathbf{v}_{k+1}]$, the boxes at step $k+1$ can be propagated using the transition function: $\{[\mathbf{x}_{k+1}^{(\ell)}] = [\mathbf{f}]([\mathbf{x}_k^{(\ell)}], [\mathbf{v}_{k+1}])\}_{\ell=1}^{N_p}$, where $[\mathbf{f}]$ is an inclusion function for $\mathbf{f}$ (see Section 2.3). This step brings an interesting property of the Box-PF: instead of propagating each particle using one realization of the noise $\mathbf{v}_{k+1}$, the uncertainty on the noise is also propagated for each box particle.

- *Measurement update step*. Similarly to the PF, the weights of the predicted box particles in the Box-PF have to be updated using the new measurement at time $k+1$. For this purpose, likelihood factors need to be calculated using innovation quantities. For the PF, the innovations of the N_p particles are $\mathbf{r}_{k+1}^{(\ell)} = \mathbf{z}_{k+1} - \mathbf{z}_{k+1}^{(\ell)}$, for $\ell = 1, \ldots, N_p$, where $\mathbf{z}_{k+1}^{(\ell)} = \mathbf{g}(\mathbf{x}_{k+1}^{(\ell)})$ is the ℓth predicted measurement. For this step, the main difference between the Box-PF and the PF is that the Box-PF predicts for each box particle, a box enclosing the possible measurements which in turn is to be compared with a box representing the bounded measurement $[\mathbf{z}_{k+1}]$ available at time $k+1$. The innovation for the ℓth box particle indicates the proximity between the measured box and the predicted box measurement. Thus, in the bounded error framework, the innovation can be represented using the intersection between the two boxes. For all box particles, $\ell = 1, \ldots, N_p$, the predicted box measurements have the expression $[\mathbf{z}_{k+1}^{(\ell)}] = [\mathbf{g}]([\mathbf{x}_{k+1}^{(\ell)}])$, where $[\mathbf{g}]$ is an inclusion function for $\mathbf{g}$. The innovation is represented as: $[\mathbf{r}_{k+1}^{(\ell)}] = [\mathbf{z}_{k+1}^{(\ell)}] \cap [\mathbf{z}_{k+1}]$. Next for the PF, using probabilistic models $p_{\mathbf{w}}$ for the measurement noise $\mathbf{w}$, the likelihood of each particle is calculated as: $p(\mathbf{z}_{k+1}|\mathbf{x}_{k+1}^{(\ell)}) = p_{\mathbf{w}}(\mathbf{z}_{k+1} - \mathbf{z}_{k+1}^{(\ell)}) = p_{\mathbf{w}}(\mathbf{r}_{k+1}^{(\ell)})$. For the Box-PF, in the bounded error context, the likelihood is calculated using the idea that a box particle with a corresponding predicted measurement without an intersection with the measured box has a likelihood factor equal to zero. In contrast, a box particle for which the predicted measurement is included in the measured box has a likelihood close to 1. In addition, to the weights update, the Box-PF incorporates a new step. In the PF algorithm, each particle is propagated without any information about the variance of its position. In contrast, in the bounded error context, each box particle takes into account the imprecision caused by the model errors. This box correction is thus similar to the variance matrix measurement update step of Kalman filtering. Therefore, in order to preserve an appropriate size of each box, a *contraction step* (see Section 2.3.4) is performed which allows to eliminate the inconsistent part of the box particles with respect to the measured box. This leads to a measure of the *box likelihood* in the form

$$L_k^{(\ell)} = \prod_{j=1}^{n_x} L_k^{(\ell),j}, \tag{2.23}$$

where

$$L_k^{(\ell),j} = \frac{|[\tilde{\mathbf{x}}_{k+1}^{(\ell)}(j)]|}{|[\mathbf{x}_{k+1}^{(\ell)}(j)]|} \tag{2.24}$$

and where $[\tilde{\mathbf{x}}_{k+1}^{(\ell)}(j)]$ represents the new ℓth box particle after the contraction step as described above.

- *Resampling step*. As for the PF algorithm, a resampling step is also added to introduce variety in the particles. Different resampling algorithms exist [7]. In the Box-PF algorithm reported in Algorithm 1, the multinomial resampling is applied, combined with a new subdivision step.
- *Subdivision step*. For the PF algorithm, the duplicated particles after the resampling step are propagated, during the next steps, by introducing artificial noises. These noises introduce variety and help make the importance sampling more efficient. The same strategy for box particles can be used here, by adding artificial noises to the boxes. However, regarding the possibilities given by box properties, other alternative techniques can be considered in the resampling step. For instance, after the multinomial resampling step, in order to obtain smaller boxes around the regions with high likelihoods, each box can be divided by the corresponding number of realizations. The efficiency of this resampling strategy is shown in [19].
- *Estimate calculation*. For the PF, the state is usually approximated by the empirical means of the weighted particles as [14]

$$\hat{\mathbf{x}}_k = \sum_{\ell=1}^{N_p} \omega_k^{(\ell)} \mathbf{x}_k^{(\ell)}. \tag{2.25}$$

The confidence on this estimation can also be characterized using the variance matrix given by

$$\hat{\mathbf{P}}_k = \sum_{\ell=1}^{N_p} \omega_k^{(\ell)} (\hat{\mathbf{x}}_k - \mathbf{x}_k^{(\ell)})(\hat{\mathbf{x}}_k - \mathbf{x}_k^{(\ell)})^T. \tag{2.26}$$

Similarly, for the Box-PF the state can be estimated by the center of the weighted box particles according to

$$\hat{\mathbf{x}}_k = \sum_{\ell=1}^{N} \omega_k^{(\ell)} \mathbf{c}_k^{(\ell)}, \tag{2.27}$$

where $\mathbf{c}_k^{(\ell)}$ is the center of the ℓth box particle.
In addition, for each component $j = 1, \ldots, n_x$ of the state, the standard deviation $\sigma_k^2(j)$ for the estimate of the component $\hat{\mathbf{x}}_k(j)$ can be obtained according to

$$\sigma_k^2(j) = \sum_{\ell=1}^{N_p} \omega_k^{(\ell)} \left((\mathbf{c}_k^{(\ell)}(j) - \hat{\mathbf{x}}_k(j))^2 + \frac{|[\mathbf{x}_k^{(\ell)}](j)|^2}{12} \right). \tag{2.28}$$

In (2.28) the term $\frac{|[\mathbf{x}_k^{(\ell)}](j)|^2}{12}$ represents the standard deviation for the center $\mathbf{c}_k^{(\ell)}$ of each box (similarly to a uniform pdf variance).

Algorithm 1: The Box-PF Algorithm

1. *Initialisation*
   Set $k = 0$ *and generate uniformly* N_p *boxes* $\{[\mathbf{x}_k^{(\ell)}]\}_{l=1}^{N_p}$ *mutually disjoint and equally weighted with* $\frac{1}{N_p}$.
2. *Time update step*
   $[\mathbf{x}_{k+1}^{(\ell)}] = [\mathbf{f}]([\mathbf{x}_k^{(\ell)}], [\mathbf{v}_k])$, *for* $l = 1, \ldots, N_p$.
3. *Measurement update step*
   - *Predicted measurement:* $[\mathbf{z}_{k+1}^{(\ell)}] = [\mathbf{g}]([\mathbf{x}_{k+1}^{(\ell)}])$, *for* $l = 1, \ldots, N_p$.
   - *Innovation:* $[\mathbf{r}_{k+1}^{(\ell)}] = [\mathbf{z}_{k+1}^{(\ell)}] \cap [\mathbf{z}_{k+1}]$, *for* $l = 1, \ldots, N_p$.
   - *Box particle contraction: if* $[\mathbf{r}_{k+1}^{(\ell)}] \neq \emptyset$ *then contract* $[\mathbf{x}_{k+1}^{(\ell)}]$ *using* $[\mathbf{r}_{k+1}^{(\ell)}]$ *and CP algorithm to obtain* $[\tilde{\mathbf{x}}_{k+1}^{(\ell)}]$ *else* $[\tilde{\mathbf{x}}_{k+1}^{(\ell)}] = \emptyset$, *for* $l = 1, \ldots, N_p$.
   - *Likelihood:* $L^{(\ell)} = \prod_{j=1}^{n_x} L^{(\ell),j}$, *for* $l = 1, \ldots, N_p$, *where* $L_k^{(\ell),j} = \frac{|[\tilde{\mathbf{x}}_{k+1}^{(\ell)}(j)]|}{|[\mathbf{x}_{k+1}^{(\ell)}(j)]|}$
   - *Weights update:* $\omega_{k+1}^{(\ell)} = (\prod_{j=1}^{n_z} L^{(\ell),j})\omega_k^{(\ell)} = L^{(\ell)}\omega_k^{(\ell)}$, *for* $l = 1, \ldots, N_p$.
4. *Weights normalisation* $\omega_{k+1}^{(\ell)} \longleftarrow \frac{\omega_{k+1}^{(\ell)}}{\sum_{\ell=1}^{N_p} \omega_{k+1}^{\ell}}$, *for* $l = 1, \ldots, N_p$.
5. *State estimate*
   $\hat{\mathbf{x}}_k = \sum_{\ell=1}^{N_p} \omega_k^{(\ell)} \mathbf{c}_k^{(\ell)}$, $\sigma_k^2(j) = \sum_{\ell=1}^{N_p} \omega_k^{(\ell)} \left((\mathbf{c}_k^{(\ell)}(j) - \hat{\mathbf{x}}_k(j))^2 + \frac{|[\mathbf{x}_k^{(\ell)}](j)|^2}{12} \right)$.
6. *Resampling*
   $N_{eff} = \frac{1}{\sum_{\ell=1}^{N_p} (\omega_k^{\ell})^2}$. *if* $N_{eff} < N_{th}$, *then resample to create* N_p *new box particles with the same weights.*

A summary of the Box-PF algorithm is given as Algorithm 1. Here N_{eff} denotes the effective number of particles and N_{th} is a threshold below which samples are resampled.

2.6 BOX PARTICLE FILTERING DERIVED FROM THE BAYESIAN INFERENCE USING A MIXTURE OF UNIFORM PROBABILITY DENSITY FUNCTIONS

This section presents a Bayesian formulation of the Box-PF. The derivation of the Box-PF is performed by interpreting each box particle as a uniform pdf. In the linear case, with a Gaussian prior and Gaussian independent noises, the Bayesian solution corresponds to the Kalman filter. However, an exact propagation of the posterior pdf using the two Bayesian steps (2.21) and (2.22) is unfortunately not feasible in general. Approximation of the posterior pdf using a weighted combination of a chosen family of pdf is a natural solution. Among these families, the Gaussian mixture one [9] is the most popular approach since it has the advantage that each component of the mixture can be propagated using the well-established Kalman steps. However, in the presence of strong nonlinearities the propagation of the mixture is the main difficulty and is sometimes intractable.

Uniform pdfs represent another attractive family. In addition to the natural simplicity of these pdfs, by choosing box supports, the interval analysis framework offers a variety of tools that can propagate these pdfs through linear, nonlinear functions and even through differential equations. Moreover, sum of uniform functions (or piecewise constant functions) with box supports have been widely used and are, for instance, the basis of Riemann integration theory [41]. One crucial result for using a sum of uniform pdfs as approximation of a continuous function is that, as the number of components increases toward infinity and the measure of the support tends to zero, any continuous real valued function defined in a compact space can be uniformly approximated by a piecewise constant function. Hence, similarly to the Gaussian family, the piecewise functions are dense in the space of continuous functions.

2.6.1 Time Update Step

First, let us denote $U_{[\mathbf{x}]}$ the uniform pdf with the box $[\mathbf{x}]$ as support. The uniform pdf sum representation of a random variable $\mathbf{x}$ is written as

$$p(\mathbf{x}) = \sum_{i=1}^{l} w^{(i)} U_{[\mathbf{x}^{(i)}]}(\mathbf{x}), \tag{2.29}$$

where l denotes the number of mixture components, the $([\mathbf{x}^{(i)}])_{i=1}^{l}$ denote the box supports, and the $(w^{(i)})_{i=1}^{l}$ represent a set of normalized weights

$$\sum_{i=1}^{l} w^{(i)} = 1 \quad \text{and} \quad \forall i, \; w^{(i)} \geq 0. \tag{2.30}$$

Assume that, at time instant $k+1$, an approximation for the previous time pdf $p(\mathbf{x}_k|\mathbf{Z}_k)$ by a mixture of l_k uniform pdfs with box supports $[\mathbf{x}_k^{(i)}]$ is available, that is,

$$p(\mathbf{x}_k|\mathbf{Z}_k) = \sum_{i=1}^{l_k} w_k^{(i)} U_{[\mathbf{x}_k^{(i)}]}(\mathbf{x}_k). \tag{2.31}$$

Next, the time update step is aimed to be performed. Inserting (2.31) into the time update equation (2.22) (introduced in Section 2.4) gives

$$\begin{aligned} p(\mathbf{x}_{k+1}|\mathbf{Z}_k) &= \int_{\mathbb{R}^{n_x}} p(\mathbf{x}_{k+1}|\mathbf{x}_k) \sum_{i=1}^{l_k} w_k^{(i)} U_{[\mathbf{x}_k^{(i)}]}(\mathbf{x}_k) d\mathbf{x}_k \\ &= \sum_{i=1}^{l_k} w_k^{(i)} \int_{[\mathbf{x}_k^{(i)}]} p(\mathbf{x}_{k+1}|\mathbf{x}_k) U_{[\mathbf{x}_k^{(i)}]}(\mathbf{x}_k) d\mathbf{x}_k \\ &= \sum_{i=1}^{l_k} w_k^{(i)} \frac{1}{|[\mathbf{x}_k^{(i)}]|} \int_{[\mathbf{x}_k^{(i)}]} p(\mathbf{x}_{k+1}|\mathbf{x}_k) d\mathbf{x}_k. \end{aligned} \tag{2.32}$$

Consider an inclusion function $[\mathbf{f}]$ (see Section 2.3) for the transition model $\mathbf{f}$, and let us assume that the noise $\mathbf{v}_{k+1}$, at time instant $k+1$, is bounded in the box $[\mathbf{v}_{k+1}]$. Then, by definition of the inclusion functions, $\forall i = 1, \ldots, l_k$, if $\mathbf{x}_k \in [\mathbf{x}_k^{(i)}]$ then $\mathbf{x}_{k+1} \in [\mathbf{f}]([\mathbf{x}_k^{(i)}], [\mathbf{v}_{k+1}])$. Thus, for all $i = 1, \ldots, l_k$ we can write

$$p(\mathbf{x}_{k+1}|\mathbf{x}_k)U_{[\mathbf{x_k}^{(i)}]}(\mathbf{x}_k) = 0 \quad \forall \mathbf{x}_{k+1} \notin [\mathbf{f}]([\mathbf{x}_k^{(i)}], [\mathbf{v}_{k+1}]). \tag{2.33}$$

Equation (2.33) shows that for any transition function $\mathbf{f}$, using interval analysis techniques, the support for the pdf terms $\int_{[\mathbf{x}_k^{(i)}]} p(\mathbf{x}_{k+1}|\mathbf{x}_k)U_{[\mathbf{x}_k^{(i)}]}(\mathbf{x}_k)d\mathbf{x}_k$ can be approximated by $[\mathbf{f}]([\mathbf{x}_k^{(i)}], [\mathbf{v}_{k+1}])$. In addition, it can be seen that, in the Box-PF algorithm, each pdf term $\int_{[\mathbf{x}_k^{(i)}]} p(\mathbf{x}_{k+1}|\mathbf{x}_k)U_{[\mathbf{x}_k^{(i)}]}(\mathbf{x}_k)d\mathbf{x}_k$ in (2.32) is modeled using one uniform pdf component having as support the interval $[\mathbf{f}]([\mathbf{x}_k^{(i)}], [\mathbf{v}_{k+1}])$, that is,

$$\int_{[\mathbf{x}_k^{(i)}]} p(\mathbf{x}_{k+1}|\mathbf{x}_k)U_{[\mathbf{x}_k^{(i)}]}(\mathbf{x}_k)d\mathbf{x}_k = U_{[\mathbf{f}]([\mathbf{x}_k^{(i)}],[\mathbf{v}_{k+1}])}. \tag{2.34}$$

The Box-PF strategy of approximating each pdf $\int_{[\mathbf{x}_k^{(i)}]} p(\mathbf{x}_{k+1}|\mathbf{x}_k)U_{[\mathbf{x}_k^{(i)}]}(\mathbf{x}_k)d\mathbf{x}_k$ using one uniform pdf component may not be accurate enough (however, as for the PF, it is sufficient to approximate the first moments of the pdf as it is shown experimentally in Section 2.8 and in [19] with more details). Alternately, a mixture of uniform pdfs can be used to better approximate this pdf. Note that for the Gaussian mixture method, a similar suggestion is also found in [9] to properly model $p(\mathbf{x}_{k+1}|\mathbf{Z}_k)$ when the noise $\mathbf{v}_{k+1}$ variance is large (in this case each Gaussian in the mixture at time instant k is propagated and generates more than one Gaussian component for the time update). For a better approximation of the predicted pdf, for a general case, assume that the noise $\mathbf{v}_{k+1}$ pdf is also approximated with a mixture of q_k uniform pdfs

$$p(\mathbf{v}_{k+1}) = \sum_{i=1}^{q_k} \lambda_{k+1}^{(i)} U_{[\mathbf{v}_{k+1}^{(i)}]}(\mathbf{v}_{k+1}), \tag{2.35}$$

with only one component if the only information about the noise is that it is bounded. In (2.35) $\lambda_{k+1}^{(i)}$ are the q_k weighting coefficients of the uniform pdfs. The transition probability $p(\mathbf{x}_{k+1}|\mathbf{x}_k)$ can be further presented according to

$$\begin{aligned} p(\mathbf{x}_{k+1}|\mathbf{x}_k) &= \int_{[\mathbf{v}_k^{(i)}]} p(\mathbf{x}_{k+1}|\mathbf{x}_k, \mathbf{v}_{k+1}) \sum_{i=1}^{q_k} \lambda_{k+1}^{(i)} U_{[\mathbf{v}_{k+1}^i]}(\mathbf{v}_{k+1}) d\mathbf{v}_{k+1} \\ &= \sum_{i=1}^{q_k} \lambda_{k+1}^{(i)} \frac{1}{|[\mathbf{v}_k^{(i)}]|} \int_{[\mathbf{v}_{k+1}^{(i)}]} p(\mathbf{x}_{k+1}|\mathbf{x}_k, \mathbf{v}_{k+1}) d\mathbf{v}_{k+1}. \end{aligned} \tag{2.36}$$

Combining (2.32) and (2.36) leads to the expression

$$p(\mathbf{x}_{k+1}|\mathbf{Z}_k) = \sum_{i=1}^{l_k}\sum_{j=1}^{q_k} w_k^{(i)}\lambda_{k+1}^{(j)} \frac{1}{|[\mathbf{x}_k^{(i)}]|}\frac{1}{|[\mathbf{v}_{k+1}^{(j)}]|} \times \left(\int_{[\mathbf{x}_k^{(i)}]}\int_{[\mathbf{v}_{k+1}^{(j)}]} p(\mathbf{x}_{k+1}|\mathbf{x}_k, \mathbf{v}_{k+1})d\mathbf{x}_k d\mathbf{v}_{k+1}\right). \tag{2.37}$$

In order to approximate the term $\int_{[\mathbf{x}_k^{(i)}]}\int_{[\mathbf{v}_{k+1}^{(j)}]} p(\mathbf{x}_{k+1}|\mathbf{x}_k, \mathbf{v}_{k+1})d\mathbf{x}_k d\mathbf{v}_{k+1}$ in (2.37), let Φ[1] denote the real valued function of $\mathbf{y}$

$$\Phi(\mathbf{y}) = \int_{[\mathbf{x}]}\int_{[\mathbf{v}]} p(\mathbf{y}|\mathbf{x}, \mathbf{v})d\mathbf{x}d\mathbf{v} \tag{2.38}$$

defined on $\mathbb{R}^{n_x}$ (here $\mathbf{y}$ denotes one value in the state space). Using interval analysis tools, the function Φ can be approximated by a sum of constant functions with box supports. For this purpose, let us introduce two claims.

Claim 1 *Consider a set* $\mathbb{E}$ *of N mutually disjoint boxes* $([\mathbf{y}_i])_{i=1,\ldots,N}$ *on* $\mathbb{R}^{n_x}$ *such that*

$$\bigcup_{i=1}^{N}[\mathbf{y}_i] \supseteq \mathbf{f}([\mathbf{x}_k], [\mathbf{v}_{k+1}]) \quad \textit{and such that} \quad \bigcup_{i=1}^{N}[\mathbf{y}_i] \subseteq [\mathbf{f}]([\mathbf{x}_k], [\mathbf{v}_{k+1}]),$$

where $[\mathbf{f}]$ *is an inclusion function of* $\mathbf{f}$. *A real valued function* $\Phi^{\mathbb{E}}(\mathbf{y})$ *can be defined on* $\mathbb{R}^{n_x}$

$$\Phi^{\mathbb{E}}(\mathbf{y}) = \begin{cases} 0, & \forall\ \mathbf{y} \notin \bigcup_{i=1}^{N}[\mathbf{y}_i], \\ \frac{1}{|\text{box}(\mathbf{y})|}\int_{[\mathbf{x}]}\int_{[\mathbf{v}]} p(\boldsymbol{\xi} \in \text{box}(\mathbf{y})|\mathbf{x}, \mathbf{v})d\mathbf{x}d\mathbf{v}, & \text{otherwise}, \end{cases} \tag{2.39}$$

where box$(\mathbf{y})$ *denotes the only box* $[\mathbf{y}_i]$*, for an* $i \in \{1, \ldots, N\}$*, containing* $\mathbf{y}$ *and* $|\ |$ *is an operator for the volume of the box defined in Section 2.3.1. When* $\max_{i=1,\ldots,N}(|[\mathbf{y}_i]|)$ *tends to zero (and, consequently, N also tends to infinity), the function* $\Phi^{\mathbb{E}}$ *defined in* (2.39) *tends to* Φ.

Proof First, remark that

$$p(\boldsymbol{\xi} \in \text{box}(\mathbf{y})|\mathbf{x}, \mathbf{v}) = \int_{\text{box}(\mathbf{y})} p(\boldsymbol{\xi}|\mathbf{x}, \mathbf{v})d\boldsymbol{\xi}. \tag{2.40}$$

[1]Notation chosen for simplicity instead of $\Phi^{[\mathbf{x}]\times[\mathbf{v}]}$.

Since Φ is defined as, (2.38), $\Phi(\mathbf{y}) = \int_{[\mathbf{x}]}\int_{[\mathbf{v}]} p(\mathbf{y}|\mathbf{x}, \mathbf{v})d\mathbf{x}d\mathbf{v}$, by combining the expression of $\Phi^{\mathbb{IE}}$ (2.39) and equation (2.40), we can write

$$\Phi^{\mathbb{IE}}(\mathbf{y}) = \begin{cases} 0 & \forall\, \mathbf{y} \notin \bigcup_{i=1}^{N}[\mathbf{y}_i], \\ \frac{1}{|\text{box}(\mathbf{y})|}\int_{\text{box}(\mathbf{y})} \Phi(\boldsymbol{\xi})d\boldsymbol{\xi}, & \text{otherwise.} \end{cases} \tag{2.41}$$

In addition, Φ is a continuous function (see equation (2.38) which is an integration of the transition probability). Since the support $\mathbf{f}([\mathbf{x}], [\mathbf{v}])$ of Φ is compact on $\mathbb{R}^{n_x}$, Φ is also uniformly continuous and $\forall\,\epsilon, \exists\,\eta\,|\,\forall\,\mathbf{y}_1, \mathbf{y}_2 \in \mathbb{R}^{n_x}$ such that $||\mathbf{y}_1, \mathbf{y}_2|| < \eta$, $||\Phi(\mathbf{y}_1), \Phi(\mathbf{y}_2)|| < \epsilon$. Here $||.||$ is the Euclidian norm. Hence, if the $[\mathbf{y}_i]$ are chosen such that $\max_{i=1,\ldots,N}(|[\mathbf{y}_i]| < \eta)$, for any $j \in \{1, \ldots, N\}$, and $\forall\,\mathbf{y} \in [\mathbf{y}_j]$, we have

$$\begin{aligned} \|\Phi^{\mathbb{IE}}(\mathbf{y}) - \Phi(\mathbf{y})\| &= \left\| \frac{1}{|[\mathbf{y}_j]|}\int_{[\mathbf{y}_j]} (\Phi(\boldsymbol{\xi}) - \Phi(\mathbf{y}))d\boldsymbol{\xi} \right\| \\ &\leq \frac{1}{|[\mathbf{y}_j]|}\int_{[\mathbf{y}_j]} \|\Phi(\boldsymbol{\xi}) - \Phi(\mathbf{y})\|d\boldsymbol{\xi} \\ &\leq \frac{1}{|[\mathbf{y}_j]|}\int_{[\mathbf{y}_j]} \epsilon\, d\boldsymbol{\xi} = \epsilon. \end{aligned}$$

Finally, considering the hypothesis that the sets $([\mathbf{y}_i])_{i=1,\ldots,N}$ are mutually disjoint and that the support $f([\mathbf{x}_k], [\mathbf{v}_{k+1}])$ of Φ is included in the support $\bigcup_{i=1}^{N}[\mathbf{z}_i]$ of $\Phi^{\mathbb{IE}}$, we can write

$$\begin{aligned} \|\Phi^{\mathbb{IE}} - \Phi\| &= \int_{\bigcup_{i=1}^{N}[\mathbf{y}_i]} \left\|\Phi^{\mathbb{IE}}(\mathbf{y}) - \Phi(\mathbf{y})\right\| d\mathbf{y} \\ &= \sum_{i=1}^{N}\int_{[\mathbf{y}_i]} \|\Phi^{\mathbb{IE}}(\mathbf{y}) - \Phi(\mathbf{y})\|d\mathbf{y} \\ &\leq \epsilon \sum_{i=1}^{N} |[\mathbf{y}_i]| \\ &\leq \epsilon\, |[\mathbf{f}]([\mathbf{x}_k], [\mathbf{v}_{k+1}])|. \end{aligned}$$

This proves Claim 1. ■

Claim 1 is the first step allowing the approximation of the function Φ with piecewise constant functions $\Phi^{\mathbb{IE}}$. Next, Claim 2 is introduced allowing the design of an algorithm, based on interval analysis, to approximate the function $\Phi^{\mathbb{IE}}$.

Claim 2 *Consider a set* $\mathbb{IE}$ *of mutually disjoint boxes* $([\mathbf{y}_i])_{i=1,\ldots,N}$ *on* $\mathbb{R}^{n_x}$ *such that*

$$\bigcup_{i=1}^{N}[\mathbf{y}_i] \supseteq \mathbf{f}([\mathbf{x}_k], [\mathbf{v}_{k+1}])$$

and consider a set $\mathbb{IF}$ *of mutually disjoint boxes* $([\mathbf{x}_i] \times [\mathbf{v}_i])_{i=1,\ldots,M}$ *on* $\mathbb{R}^{n_x+n_v}$ *such that*

$$\bigcup_{i=1}^{M}([\mathbf{x}_i] \times [\mathbf{v}_i]) = [\mathbf{x}] \times [\mathbf{v}].$$

Then, let us define on $\mathbb{R}^{n_x+n_v}$ *the real valued function* $\Phi^{\mathbb{IE},\mathbb{IF}}$ *such that*

$$\Phi^{\mathbb{IE},\mathbb{IF}}(\mathbf{y}) = \begin{cases} 0 & \forall\, \mathbf{y} \notin \bigcup_{i=1}^{N}[\mathbf{y}_i], \\ \frac{1}{|\mathrm{box}(\mathbf{y})|} \sum\limits_{j \in I^{\mathrm{box}(\mathbf{y})}} |[\mathbf{x}_j] \times [\mathbf{v}_j]|, & \text{otherwise}, \end{cases} \tag{2.42}$$

where $box(\mathbf{y})$ *is the only box* $[\mathbf{y}_i]$*, for an* $i \in \{1, \ldots, N\}$*, containing* $\mathbf{y}$ *and* $I^{box(\mathbf{y})}$ *is a set of indexes defined as*

$$I^{box(\mathbf{y})} = \{j \in \{1, \ldots, M\} \,|\, \mathbf{f}([\mathbf{x}_j], [\mathbf{v}_j]) \subseteq box(\mathbf{y})\}.$$

When $\max_{j=1,\ldots,M}(|[\mathbf{x}_j] \times [\mathbf{v}_j]|)$ *tends to zero (and consequently M also tends to infinity), the functions* $\Phi^{\mathbb{IE},\mathbb{IF}}$ *defined in* (2.42) *tend to* $\Phi^{\mathbb{IE}}$

Proof For simplicity, let us prove the claim using two points.

- First, assume that the set $\mathbb{IE}$ is constituted by only one box $[\mathbf{y}]$ and that, consequently, $\Phi^{\mathbb{IE}}$ is defined as

$$\Phi^{\mathbb{IE}}(\mathbf{y}) = \begin{cases} 0 & \forall\, \mathbf{y} \notin [\mathbf{y}], \\ \frac{1}{|[\mathbf{y}]|} \int_{[\mathbf{x}]} \int_{[\mathbf{v}]} p(\boldsymbol{\xi} \in [\mathbf{y}] | \mathbf{x}, \mathbf{v}) d\mathbf{x} d\mathbf{v} & \text{otherwise.} \end{cases}$$

  Consider the set $\mathbf{S}$ denoting the intersection between the box $[\mathbf{x}] \times [\mathbf{v}]$ and the reciprocal image of $[\mathbf{y}]$ via the function $\mathbf{f}$, that is

$$\mathbf{S} = ([\mathbf{x}] \times [\mathbf{v}]) \cap \mathbf{f}^{-1}([\mathbf{y}]). \tag{2.43}$$

The function $\Phi^{\mathbb{IE}}$ can be rewritten, using the set $\mathbf{S}$ in the following way

$$\Phi^{\mathbb{IE}}(\mathbf{y}) = \begin{cases} 0 & \forall\, \mathbf{y} \notin [\mathbf{y}], \\ \frac{1}{\|[\mathbf{y}]\|} \int_{[\mathbf{x}]} \int_{[\mathbf{v}]} p(\boldsymbol{\xi} \in [\mathbf{y}]|\mathbf{x}, \mathbf{v}) d\mathbf{x} d\mathbf{v} & \text{otherwise,} \end{cases}$$

$$= \begin{cases} 0 & \forall\, \mathbf{y} \notin [\mathbf{y}], \\ \frac{1}{\|[\mathbf{y}]\|} \int_{\mathbf{S}} p(\boldsymbol{\xi} \in [\mathbf{y}]|\mathbf{x}, \mathbf{v}) d\mathbf{x} d\mathbf{v} + \frac{1}{\|[\mathbf{y}]\|} \int_{[\mathbf{x}]\times[\mathbf{v}] \setminus \mathbf{S}} p(\boldsymbol{\xi} \in [\mathbf{y}]|\mathbf{x}, \mathbf{v}) d\mathbf{x} d\mathbf{v} & \text{otherwise.} \end{cases} \tag{2.44}$$

By definition of $\mathbf{S}$ (2.43),

$$p(\boldsymbol{\xi} \in [\mathbf{y}]|\mathbf{x}, \mathbf{v}) = \begin{cases} 0 & \forall\, (\mathbf{x} \times \mathbf{v}) \in ([\mathbf{x}] \times [\mathbf{v}] \setminus \mathbf{S}), \\ 1 & \forall\, (\mathbf{x} \times \mathbf{v}) \in \mathbf{S}. \end{cases} \tag{2.45}$$

From equation (2.45), we can deduce that

$$\int_{\mathbf{S}} p(\boldsymbol{\xi} \in [\mathbf{y}]|\mathbf{x}, \mathbf{v}) d\mathbf{x} d\mathbf{v} = \mu(\mathbf{S}), \tag{2.46}$$

and that

$$\int_{[\mathbf{x}]\times[\mathbf{v}] \setminus \mathbf{S}} p(\boldsymbol{\xi} \in [\mathbf{y}]|\mathbf{x}, \mathbf{v}) d\mathbf{x} d\mathbf{v} = 0, \tag{2.47}$$

where μ denotes the Lebesque measure on $\mathbb{R}^{n_x+n_v}$.
Next, using equations (2.44), (2.46), and (2.47), the function $\Phi^{\mathbb{IE}}$ can be rewritten in the following way

$$\Phi^{\mathbb{IE}}(\mathbf{y}) = \begin{cases} 0 & \forall\, \mathbf{y} \notin [\mathbf{y}], \\ \frac{1}{\|[\mathbf{y}]\|} \mu(\mathbf{S}) & \text{otherwise.} \end{cases} \tag{2.48}$$

Furthermore, for $\mathbf{y} \in [\mathbf{y}]$, the set of indexes $I^{\text{box}(\mathbf{y})}$ is defined as

$$\begin{aligned} I^{\text{box}(\mathbf{y})} &= \{j \in \{1, \ldots, M\} \,|\, \mathbf{f}([\mathbf{x}_j], [\mathbf{v}_j]) \subseteq [\mathbf{y}]\} \\ &= \{j \in \{1, \ldots, M\} \,|\, ([\mathbf{x}_j], [\mathbf{v}_j]) \subseteq \mathbf{f}^{-1}[\mathbf{y}]\}. \end{aligned}$$

Hence, the following equivalence can be written

$$j \in I^{\text{box}(\mathbf{y})} \Leftrightarrow [\mathbf{x}_j] \times [\mathbf{v}_j] \subseteq \mathbf{S},$$

since $\Phi^{\mathbb{IE},\mathbb{IF}}$ is defined as

$$\Phi^{\mathbb{IE},\mathbb{IF}}(\mathbf{y}) = \begin{cases} 0 & \forall\, \mathbf{y} \notin [\mathbf{y}], \\ \frac{1}{|\mathrm{box}(\mathbf{y})|} \sum_{j \in I^{\mathrm{box}(\mathbf{y})}} |[\mathbf{x}_j] \times [\mathbf{v}_j]|, & \text{otherwise.} \end{cases}$$

To prove Claim 2, it suffices to prove that $\sum_{j \in I^{\mathrm{box}(\mathbf{y})}} |[\mathbf{x}_j] \times [\mathbf{v}_j]|$ converges towards $\mu(\mathbf{S})$ when $\max_{j=1,\ldots,M}(|[\mathbf{x}_j] \times [\mathbf{v}_j]|)$ tends to zero. This later convergence can be proven by using previous results introduced in [26]. The well-known SIVIA algorithm[16] is used for inverting sets through large classes of functions including the continuous functions. From a given prior box, and an image box, SIVIA provides two subpavings (union of disjoints intervals), inside the prior box, both converging to the true solution set. An inner subpaving included in the solution set and outer subpaving enclosing the solution set can be obtained by SIVIA. In [26] the convergence of SIVIA has been derived. When the size of the boxes inside the subpaving tends to 0, the inner subpaving tends toward the solution set (see [26] Sections 4 and 5). For our problem, the set $([\mathbf{x}_j] \times [\mathbf{v}_j])_{j \in I^{[\mathbf{y}]}}$ constitutes the inner subpaving, of an initial prior $[\mathbf{x}] \times [\mathbf{v}]$, approximating the solution set $\mathbf{f}^{-1}([\mathbf{y}]) \cap [\mathbf{x}] \times [\mathbf{v}]$. We can conclude that when $\max_{j=1,\ldots,M}(|[\mathbf{x}_j] \times [\mathbf{v}_j]|)$ tends to 0, $\sum_{j \in I^{[\mathbf{y}]}} |[\mathbf{x}_j] \times [\mathbf{v}_j]|$ tends to $\mu(\mathbf{f}^{-1}([\mathbf{y}]) \cap [\mathbf{x}] \times [\mathbf{v}]$ and consequently $\Phi^{\mathbb{IE},\mathbb{IF}}$ tends to $\Phi^{\mathbb{IE}}$.

- Next, if we assume that $\mathbb{IE}$ is a set of mutually disjoints $([\mathbf{y}_i])_{i=1,\ldots,N}$ on $\mathbb{R}^{n_x}$ on $\mathbb{R}^{n_x+n_v}$, the convergence is proven (from above) for each $[\mathbf{y}_i]$ taken separately. Consequently, since M is a finite number, this proves Claim 2. ■

By appropriately choosing first the set $\mathbb{IE}$ and secondly the set $\mathbb{IF}$, the combination of Claim 1 and Claim 2 allows the approximation of the desired function Φ at a level of accuracy that will increase with the number of boxes. As an illustration, consider a simple summation of two random variables distributed according to respectively $U_{[1,2]}$ and $U_{[4,6]}$. Figure 2.5 shows the approximation using $\Phi^{\mathbb{IE},\mathbb{IF}}$ introduced in the two claims, with respectively 20 and 2000 intervals in $\mathbb{IE}$ and $\mathbb{IF}$.

2.6.2 Measurement Update Step

Assume that, at time instant $k+1$, an approximation of the time update pdf $p(\mathbf{x}_{k+1}|\mathbf{Z}_k)$ by a mixture of $l_{k+1|k}$ uniform pdfs with interval supports $[\mathbf{x}_{k+1|k}^{(i)}]$ and weights $w_{k+1|k}$ is available and that the measurement update step is to be performed. For the Box-PF algorithm, the measurement likelihood function is taken to be one component with uniform distribution. However, for a general case, a mixture of uniform pdfs can be considered. Assume that the likelihood function has the expression (2.49) with s_{k+1}

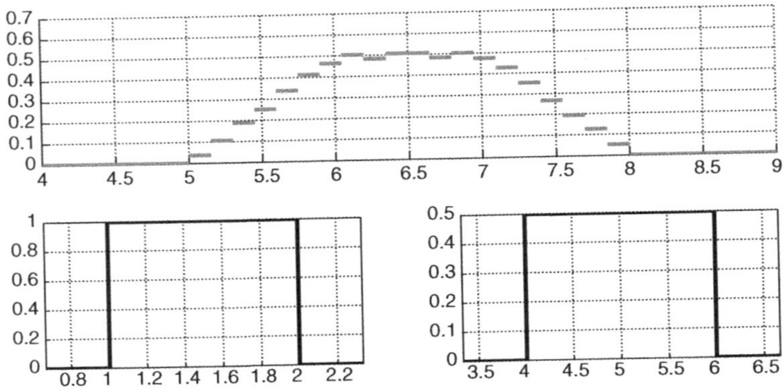

Figure 2.5 A simple illustration of the time update step performed using the combination of Claim 1 and Claim 2. The respective uniform pdfs, for the state $\mathbf{x}$, $U_{[4,6]}$, and the system noise $\mathbf{v}$, $U_{[4,2]}$, are presented at the bottom. At the top, the approximation of the resulting convolution of the two uniform pdfs is displayed.

components weighted with the normalized coefficient $\beta_{k+1}^{(j)}$

$$p(\mathbf{z}_{k+1}|\mathbf{x}_{k+1}) = \sum_{j=1}^{s_{k+1}} \beta_{k+1}^{(j)} U_{[\mathbf{z}_{k+1}^{(j)}]}(\mathbf{g}(\mathbf{x}_{k+1})), \tag{2.49}$$

where the $[\mathbf{z}_{k+1}^{(j)}]$ denotes a set of box supports satisfying

$$\cup_{j=1}^{s_{k+1}} [\mathbf{z}_{k+1}^{(j)}] = [\mathbf{z}_{k+1}].$$

The measurement update has the expression

$$\begin{aligned} p(\mathbf{x}_{k+1}|\mathbf{Z}_{k+1}) &= \frac{1}{\alpha_{k+1}} p(\mathbf{z}_{k+1}|\mathbf{x}_{k+1}) p(\mathbf{x}_{k+1}|\mathbf{Z}_k) \\ &= \frac{1}{\alpha_{k+1}} \sum_{j=1}^{s_{k+1}} \beta_{k+1}^{j} U_{[\mathbf{z}_{k+1}^{j}]}(\mathbf{g}(\mathbf{x}_{k+1})) \sum_{i=1}^{l_{k+1|k}} w_{k+1|k}^{i} U_{[\mathbf{x}_{k+1|k}^{i}]}(\mathbf{x}_{k+1}) \\ &= \frac{1}{\alpha_{k+1}} \sum_{j=1}^{s_{k+1}} \sum_{i=1}^{l_{k+1|k}} \beta_{k+1}^{j} w_{k+1|k}^{i} U_{[\mathbf{z}_{k+1}^{j}]}(\mathbf{g}(\mathbf{x}_{k+1})) U_{[\mathbf{x}_{k+1|k}^{i}]}(\mathbf{x}_{k+1}). \end{aligned} \tag{2.50}$$

The term $U_{[\mathbf{z}_{k+1}^{j}]}(\mathbf{g}(\mathbf{x}_{k+1}))U_{[\mathbf{x}_{k+1|k}^{i}]}(\mathbf{x}_{k+1})$ is also a constant function with a support being the following set

$$\left\{\mathbf{x}_{k+1} \in [\mathbf{x}_{k+1|k}^{i}] \mid \exists\, \mathbf{w}_{k+1} \in [\mathbf{w}_{k+1}] \text{ such that } \mathbf{g}(\mathbf{x}_{k+1}, \mathbf{w}_{k+1}) \in [\mathbf{z}_{k+1}^{j}]\right\}.$$

We can deduce that, using consistency algorithms (see 2.3.4 for an introduction to these techniques and [16] for more details), the predicted supports $[\mathbf{x}_{k+1|k}^{(i)}]$ for the time update pdf $p(\mathbf{x}_{k+1}|\mathbf{Z}_k)$ approximation have to be contracted with respect to the new measurement. These contraction steps give the new support for the posterior pdf $p(\mathbf{x}_{k+1}|\mathbf{Z}_{k+1})$ at time instant $k+1$. This is an interesting result since the contraction steps which have been heuristically introduced in the Box-PF are derived theoretically through the posterior pdf expression (2.50). We can see that, using the Bayesian formulation combined with a mixtures of uniform pdfs, the Box-PF procedures are theoretically justified. Furthermore, these procedures can be extended to derive a better approximation of the posterior pdfs. The weights update summarized by (2.50) is more correct and more elaborate than the Box-PF likelihood heuristic approach presented in Section (2.5). The time update step is, surprisingly, the most complicated step to derive and can be performed using the two claims introduced in Section 2.6.1.

2.7 BOX-PF ILLUSTRATION OVER A TARGET TRACKING EXAMPLE

2.7.1 Simulation Set-Up

Consider the problem of tracking a target in the 2D plane. The available measurements are range, range rate, and azimuth. The target state vector $\mathbf{x}$ consists of the target positions (x, y) and velocities $(\dot{x}, \dot{y})$ in Cartesian frame, that is $\mathbf{x} = (x, \dot{x}, y, \dot{y})^T$. The target moves according to the nearly constant velocity model [42] with a transition density function $p(\mathbf{x}_{k+1}|\mathbf{x}_k) = \mathcal{N}(\mathbf{x}_{k+1}, \mathbf{F}\mathbf{x}_k, \mathbf{Q})$, where

$$\mathbf{F} = \mathbf{I}_2 \otimes \begin{pmatrix} 1 & T \\ 0 & 1 \end{pmatrix}, \quad \mathbf{Q} = \mathbf{I}_2 \otimes \begin{pmatrix} \frac{T^3}{3} & \frac{T^2}{2} \\ \frac{T^2}{2} & T \end{pmatrix} .\vartheta, \tag{2.51}$$

$\mathbf{Q}$ is the covariance matrix of the system noise, $\mathbf{I}_2$ is the 2×2 dimensional unit matrix, $\otimes$ is the Kronecker product, $T = t_{k+1} - t_k$ the sampling interval, and ϑ the process noise intensity. Figure 2.6 shows the tracking scenario. Initially, at $k = 0$ the target is located at position $(-100, 300)$m and is moving with the velocity $(-9.5, -8.8)$m/s. The sensor is static, located at the origin of the coordinate frame. The simulation is run until scan $k = 60$. Other parameters are $\vartheta = 0.05$ and $T = 1$ s.

The measurement function $\mathbf{g}$ is defined as

$$\mathbf{g}(x) = \begin{pmatrix} \sqrt{x^2+y^2} \\ \frac{x\dot{x}+y\dot{y}}{\sqrt{x^2+y^2}} \\ \arctan(\frac{y}{x}) \end{pmatrix} + \mathbf{w}, \tag{2.52}$$

where the measurement noise $\mathbf{w}$ is a zero-mean white Gaussian process with a covariance matrix $\Sigma = \text{diag}(\sigma_r^2, \sigma_{\dot{r}}^2, \sigma_\theta^2)$, with $\sigma_r = 2.5$ m, $\sigma_{\dot{r}} = 0.01$ m/s, $\sigma_\theta = 0.25$ rad.

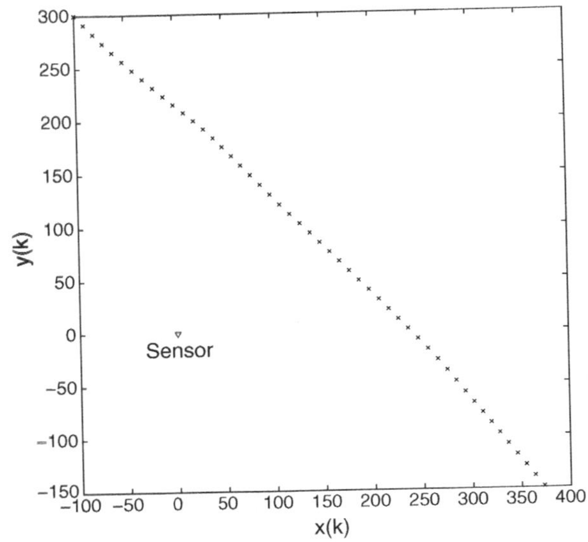

Figure 2.6 Tracking scenario.

Figure 2.7 shows a global view of the Box-PF performance, with 32 box particles and for one single run with measurements generated from (2.51). Circle marks represent the true target trajectory, while the "plus" represent the estimated trajectory. Algorithm 2 shows the steps of the constraints propagation algorithm

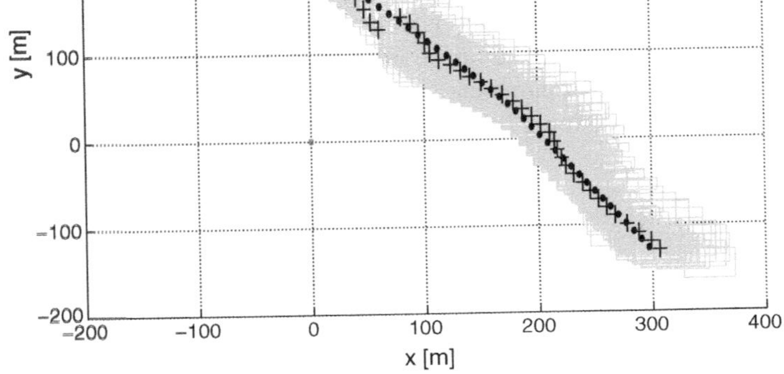

Figure 2.7 Trajectory generated using a Box-PF with 32 box particles.

Algorithm 2: Constraints propagation algorithm

1. ***Input:*** $[\mathbf{x}] = [x] \times [\dot{x}] \times [y] \times [\dot{y}]$, $\mathbf{z} = [r] \times [\dot{r}] \times [\beta]$
   constraint 1
2. $[x] = [x] \cap \sqrt{([r]^2 - [y]^2)}$
3. $[y] = [y] \cap \sqrt{([r]^2 - [x]^2)}$
4. $[r] = [r] \cap \sqrt{([x]^2 + [y]^2)}$
   constraint 2
5. $[x] = [x] \cap \dfrac{[\dot{x}][\dot{y}][y]}{[\dot{r}]^2 - [\dot{x}]} + \sqrt{\dfrac{([\dot{y}]^2 - [\dot{r}]^2)y^2}{[\dot{r}]^2 - [\dot{x}]^2} + \dfrac{[\dot{x}][\dot{y}][y]}{[\dot{r}]^2 - [\dot{x}]^2}}$
6. $[y] = [y] \cap \dfrac{[\dot{y}][\dot{x}][x]}{[\dot{r}]^2 - [\dot{y}]} + \sqrt{\dfrac{([\dot{x}]^2 - [\dot{r}]^2)x^2}{[\dot{r}]^2 - [\dot{y}]^2} + \dfrac{[\dot{y}][\dot{x}][x]}{[\dot{r}]^2 - [\dot{y}]^2}}$
7. $[\dot{x}] = [\dot{x}] \cap \dfrac{[\dot{r}].(\sqrt{[x]^2 + [y]^2}) - [y].[\dot{y}]}{[x]}$
8. $[\dot{y}] = [\dot{y}] \cap \dfrac{[\dot{r}].(\sqrt{[x]^2 + [y]^2}) - [x].[\dot{x}]}{[y]}$
9. $[\dot{r}] = [\dot{r}] \cap \dfrac{[x].[\dot{x}] + [y].[\dot{y}]}{\sqrt{[x]^2 + [y]^2}}$
   constraint 3
10. $[x] = [x] \cap \dfrac{[y]}{[\tan](\beta}$
11. $[y] = [y] \cap \tan(\beta).[x]$
12. $[\beta] = [\beta] \cap [\arctan](\dfrac{[y]}{[x]}$
13. ***Output:*** $[\mathbf{x}] = [x] \times [\dot{x}] \times [y] \times [\dot{y}]$,

(see Section 2.3.4) used to contract each box particle at the measurement update step. The original algorithm performs the contractions until convergence (there is no more contraction after a specified threshold). For this illustration we are using a loop of three iterations (we observed that more contractions do not lead to a significant improvement).

Figures 2.8 and 2.9 show the average performance of the Box-PF over 100 Monte Carlo simulation when applied to this problem.

2.8 APPLICATION FOR A VEHICLE DYNAMIC LOCALIZATION PROBLEM

In this section, a real experiment for localization of a land vehicle is considered (see [19] for further details about the experiment). The mobile frame origin M was chosen at the middle of rear axle. The elementary rotation $\delta_{\theta,k}$ and the elementary

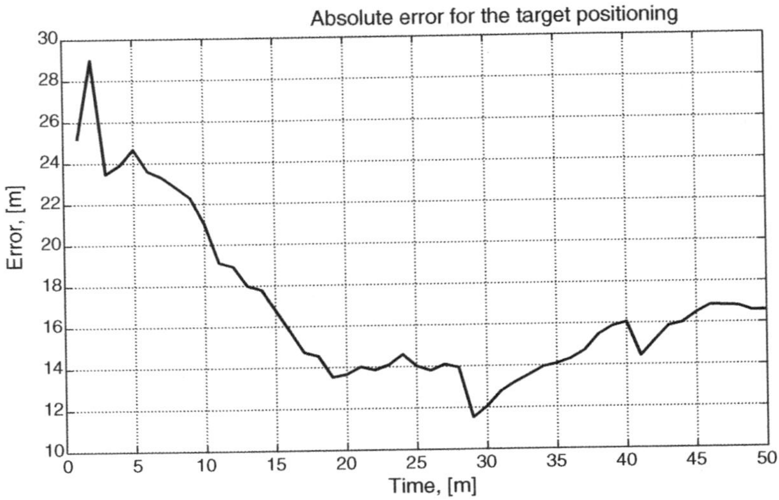

Figure 2.8 Absolute positioning error generated using a Box-PF with 32 box particles.

displacement $\delta_{S,k}$ measurements between two time steps at the point M were obtained from a fiber optic gyro and the two rear wheels odometry sensors

$$\begin{cases} \delta_{S,k} = \dfrac{\delta_{RR,k} + \delta_{RL,k}}{2}, \\ \delta_{\theta,k} = \delta_{\theta,k}^{\text{gyro}}, \end{cases} \tag{2.53}$$

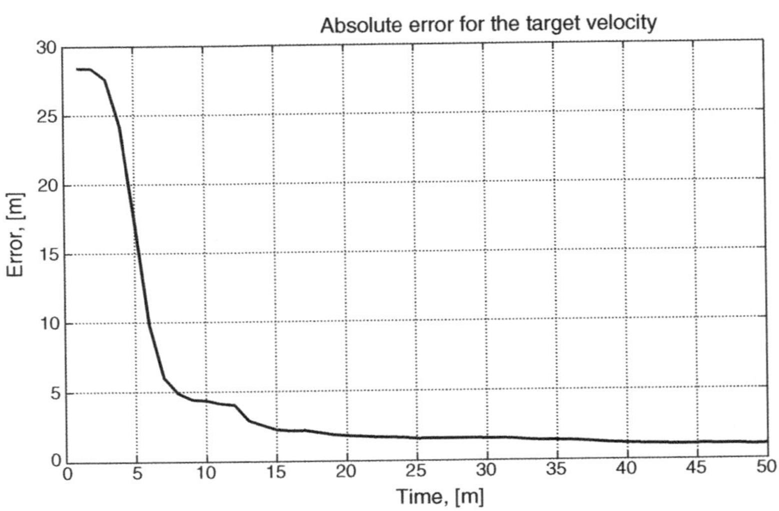

Figure 2.9 Absolute error for the velocity generated using a Box-PF with 32 box particles.

where $\delta_{RR,k}$ and $\delta_{RL,k}$ denote respectively the measured right and left rear wheels displacements between two samples, and $\delta_{\theta,k}^{\text{gyro}}$ is a measure of the elementary rotation given by the gyro. The state $\mathbf{x}_k = x_k \times y_k \times \theta_k$ constituted by the position and the heading angle of the vehicle at time instant k is propagated through the model

$$\begin{cases} x_{k+1} = x_k + \delta_{S,k} \cos\left(\theta_k + \dfrac{\delta_{\theta,k}}{2}\right), \\ y_{k+1} = y_k + \delta_{S,k} \sin\left(\theta_k + \dfrac{\delta_{\theta,k}}{2}\right), \\ \theta_{k+1} = \theta_k + \delta_{\theta,k}. \end{cases} \tag{2.54}$$

The measurement consists of 2D position provided by a Global Position System (GPS) that is $(x_{\text{GPS}}, y_{\text{GPS}})$. The longitude–latitude estimated point of the GPS was converted to a Cartesian local frame. The GPS measurement box can be quantified using the standard deviation σ_{GPS} provided in real time by the GPS receiver. Thus,

$$\begin{cases} [x_{\text{GPS}}] = & [x_{\text{GPS}} - 3\sigma_{\text{GPS}}, x_{\text{GPS}} + 3\sigma_{\text{GPS}}] \\ [y_{\text{GPS}}] = & [y_{\text{GPS}} - 3\sigma_{\text{GPS}}, y_{\text{GPS}} + 3\sigma_{\text{GPS}}] \end{cases} \tag{2.55}$$

The GPS measurement $([x_{\text{GPS}}], [y_{\text{GPS}}])$ was used to initialize the box state position $([x_1], [y_1])$ at instant t_1. The heading angle of the vehicle was initialized by $[\theta_1] = [-\infty, +\infty]$. In addition, a ground truth was available through the Thales Navigation GPS receiver used in a Post-Processed Kinematic mode, working at 1 Hz sampling rate with a local base (a Trimble 7400) and providing a few centimeters of accuracy. The data of the sensors were time stamped and logged during several tests. We report the analysis of 4.7 km path with a mean speed of 50 km/h using a 3 GHz Pentium 4 and a Matlab implementation. The two filters provide outputs at the frequency of the GPS (5 Hz).

Table 2.1 shows the mean square error for the GPS alone, the Box-PF, and PF. As a conclusion for this problem, the Box-PF and the PF give equivalent filtering performances (The PF is slightly better overall due to a faster convergence in practice). It is worth noting that for running the Box-PF, we use only 10 box particles compared with

TABLE 2.1 Comparison between the PF and Box-PF

	GPS	PF	Box-PF
Mean square error for x(m)	0.134	0.129	0.119
Mean square error for y(m)	0.374	0.217	0.242
Particle number	–	3000	10
One step running time (ms)	–	666	149

The table shows the mean square error for GPS, PF and Box-PF. We also give the mean of the running time of one step for each algorithm

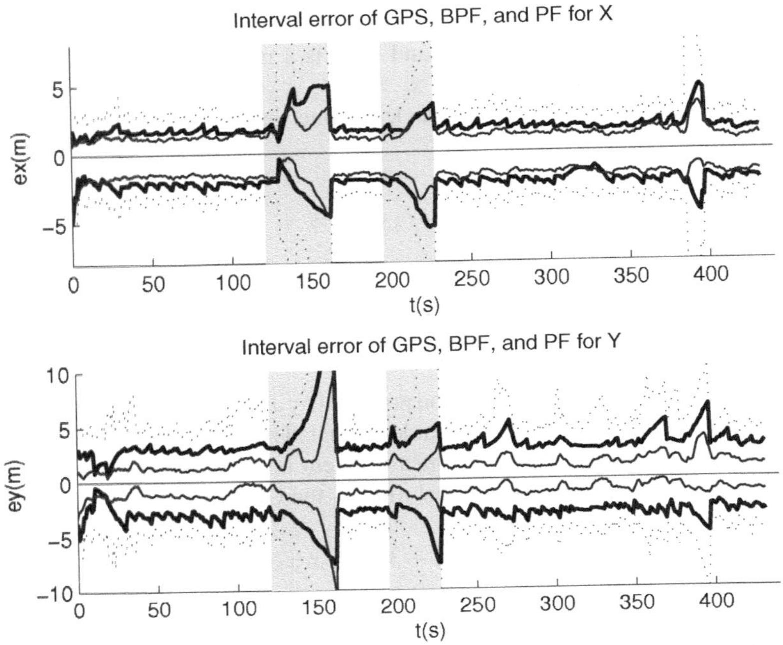

Figure 2.10 The interval error *ex* for *x* coordinate and *ey* for *y* coordinate, in meters, for the GPS (dashed), Box-PF (bold), and PF (solid).

3000 particles for the PF. This is a good motivation for further developing the Box-PF since one can reduce significantly the particles' numbers (for this application, the factor is about 300). For distributed state estimation problem, this reduction is a very attractive outcome since it can drastically reduce the amount of data to be exchanged.

Table 2.1 gives the average one step computational time for each algorithm. Since the output frequency of each filter is 5 Hz, the running time for Box-PF satisfies real-time constraints. In addition, in this example the Box-PF is approximately 4.5 times faster than the PF.

Figure 2.10 shows the interval error for x and y from the GPS (dashed black), the estimates from the Box-PF (bold), and PF (solid). The gray colored regions correspond to the period of time when the differential GPS corrections were unavailable. For the PF, the interval error is calculated by using 3σ error bounds around the point estimate. It can be seen that for this nonlinear problem, the two filters are consistent with the ground truth: the interval errors contain the value "0" which is equivalent to say that the estimated intervals contain the ground truth values. Figure 2.11 plots the heading estimated error in addition to the interval errors, in degrees, for Box-PF (bold) and PF (solid). The errors on the heading estimation angles provided by the Box-PF and the PF are of the same magnitude. One can conclude that the Box-PF is also able to estimate a variable not directly measured by sensors.

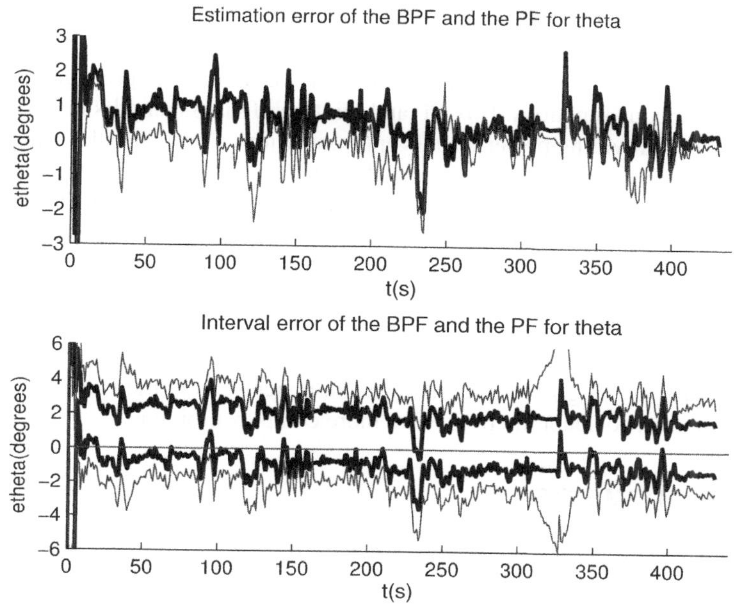

Figure 2.11 The estimated heading error e_θ and the interval errors, in degrees, for Box-PF (bold) and PF.

2.9 CONCLUSIONS

This chapter introduced a new approach for sequential nonlinear estimation based on a combination of particle filtering and interval analysis. The method was first presented within the interval framework, by introducing a new concept of box particles for the purposes of drastically reducing the number of particles.

The Box-PF algorithm was derived through the Bayesian inference based on mixtures of uniform pdfs with box supports. This study provides theoretical justifications of the Box-PF procedures and proves that mixture of uniform pdfs can be used to approximate the posteriors pdf for a state estimation problem. Furthermore, this study also provides more sophisticated procedures to sequentially propagate and correct the mixture of uniform pdfs in comparison with the original Box-PF procedures. Experiments both with simulated and real data show the feasibility and the effectiveness of the Box-PF in comparison with the PF. The Box-PF algorithm seems to be more adapted to real-time applications in comparison with the PF. Furthermore, the significant reduction of the number of particles opens interesting perspectives for distributed nonlinear and nonparametric state estimation problems.

The Box-PF can be classified between the PF method and the Gaussian mixture method in terms of number of samples or components needed. In general, a mixture of uniform pdfs needs more components than a mixture of Gaussians to approximate one given pdf but less components than a set of samples for a PF.

In addition, the Box-PF share common properties with both PFs and Gaussian mixture methods. For instance, for both PF and Box-PF, linearization of the model is not necessary since the interval analysis offers powerful tools to propagate intervals through continuous functions. Box-PF and mixture of Gaussian pdfs methods share the property of providing posterior pdfs approximations (for the PF this can be done indirectly by using kernel densities for each particle).

The study of the Box-PF, in the light of the Bayesian framework, opens numerous challenges. Current works are focused on using the new framework in the prediction step and in the correction step, introduced in Section 2.6, to implement a more sophisticated nonlinear and real-time filters. We expect a more time-consuming algorithm than the Box-PF, however, with a better accuracy in addition to a direct approximation of the posterior pdfs as improvements of the Box-PF. Future researches may introduce new steps such as merging of the mixture components, as in the Gaussian mixture methods, in order to adapt the Box-PF to more complex problems.

ACKNOWLEDGMENTS

We acknowledge the support from the UK EPSRC project EP/E027253/1, from the [European Community's] Seventh Framework Programme [FP7/2007-2013] under grant agreement No 238710 (Monte Carlo based Innovative Management and Processing for an Unrivalled Leap in Sensor Exploitation) and the EU COST action TU0702.

REFERENCES

1. A. Doucet, N. Freitas, and N. Gordon (Eds.), *Sequential Monte Carlo Methods in Practice*, Springer-Verlag, New York, 2001.
2. M. Arulampalam, S. Maskell, N. Gordon, and T. Clapp, "A tutorial on particle filters for online nonlinear/non-Gaussian Bayesian tracking," *IEEE Trans. Signal Proc.*, 50(2), 174–188, 2002.
3. A. Jaswinski, *Stochastic Processes and Filtering Theory*, Academic Press, New York, 1970.
4. S. Julier, J. Uhlmann, and H. Durrant-White, A new approach for filtering nonlinear systems, *Proceedings of the American Control Conference*, Washington, DC, 1995.
5. S. Julier and J. Uhlmann, Unscented filtering and nonlinear estimation, *Proc. IEEE*, 92(3), 401–422, 2004.
6. S. Julier, The scaled unscented transformation, in *Proceedings of the IEEE American Control Conference*, 2002, pp. 4555–4559.
7. E. Wan and R. van der Merwe, Kalman Filtering and Neural Networks. in S. Haykin (Ed.), *The Unscented Kalman Filter*, Wiley, New York, 2001, Chapter 7, pp. 221–280.
8. B. Ristic, S. Arulampalam, and N. Gordon, *Beyond the Kalman Filter: Particle Filters for Tracking Applications*, Artech House, Boston, London, 2004.
9. D.L. Alspach and H.W. Sorenson, Nonlinear Bayesian estimation using Gaussian sum approximation, *IEEE Trans. Automats Contr.*, 17(4), 439–448, 1972.

10. E.W. Cheney, *Introduction to Approximation Theory*, 2nd edition Chelsea, New York, 1982.

11. O. Cappé, S. Godsill, and E. Mouline, An overview of existing methods and recent advances in sequential Monte Carlo, *IEEE Proc.*, 95(5), 899–924, 2007.

12. D. Fox, Adapting the sample size in particle filters through KLD-sampling, *Int. J. Robot. Res.*, 22(12), 985–1003, 2003.

13. L. Mihaylova, D. Angelova, S. Honary, D.R. Bull, C.N. Canagarajah, and B. Ristic, Mobility tracking in cellular networks using particle filtering, *IEEE Trans. Wireless Commun.*, 6(10), 3589–3599, 2007.

14. F. Gustafsson, Particle filter theory and practice with positioning applications, *IEEE Aerosp. Electron. Syst. Mag.*, 25(7), 53–82, 2010.

15. R.E. Moore, *Interval Analysis*, Prentice-Hall, Englewood Cliffs, NJ, 1966.

16. L. Jaulin, M. Kieffer, O. Didrit, and E. Walter, *Applied Interval Analysis*, Springer-Verlag, Paris, 2001.

17. L. Jaulin, Nonlinear bounded-error state estimation of continuous-time systems, *Automatica*, 38(6), 1079–1082, 2002.

18. A. Gning and Ph. Bonnifait, Constraints propagation techniques on intervals for a guaranteed localization using redundant data, *Automatica*, 42(7), 1167–1175, 2006.

19. F. Abdallah, A. Gning, and P. Bonnifait, Box particle filtering for nonlinear state estimation using interval analysis, *Automatica*, 44(3), 807–815, 2008.

20. A. Gning, L. Mihaylova, and F. Abdallah, Mixture of uniform probability density functions for nonlinear state estimation using interval analysis, *Proceedings of the* 13th *International Conference on Information Fusion*, Edinburgh, UK, 2010.

21. A.B. Kurzhanski and I. Valyi, *Ellipsoidal Calculus for Estimation and Control*, Birkhauser, Boston, 1997.

22. F. Schweppe, Recursive state estimation: Unknown but bounded errors and system inputs, *Automat. Contr., IEEE Trans.*, 13(1), 22–28, 1968.

23. D. Bertsekas and I. Rhodes, Recursive state estimation for a set-membership description of uncertainty, *IEEE Trans. Automat. Contr.*, 16(2), 117–128, 1971.

24. F.L. Chernousko, *State Estimation for Dynamic Systems*, CRC Press, Boca Raton, 1994.

25. A. Kurzhanski and L. Vályi, "Ellipsoidal calculus for estimation and control," *Technical Report*, Birkhäuser, 1997.

26. Jaulin L. and Walter E., Set inversion via interval analysis for nonlinear bounded-error estimation, *Automatica*, 29(4), 1053–1064, 1993.

27. L. Jaulin, E. Walter, O. Lévêque, and D. Meizel, Set inversion for χ-algorithms, with application to guaranteed robot localization, *Math. Comput. Simul.*, 52(3-4), 197–210, 2000.

28. M. Kieffer, L. Jaulin, E. Walter, and D. Meizel, Robust autonomous robot localization using interval analysis, *Rel. Comput.*, 6(3), 337–362, 2000.

29. L. Jaulin, Interval constraint propagation with application to bounded-error estimation, *Automatica*, 36(10), 1547–1552, 2000.

30. L. Jaulin, M. Kieffer, I. Braems, and E. Walter, Guaranteed nonlinear estimation using constraint propagation on sets, *Int. J. Contr.*, 74(18), 1772–1782, 2001.

31. L. Jaulin, Robust set-membership state estimation; application to underwater robotics, *Automatica*, 45(1), 202–206, 2009.

32. M. Baum and U.D. Hanebeck, Extended object tracking based on combined set-theoretic and stochastic fusion, *Proceedings of the International Conference on Information Fusion*, 2009.

33. M. Baum and U.D. Hanebeck, Random hypersurface models for extended object tracking, *Proceedings of the IEEE International Symposium on Signal Processing and Information Technology (ISSPIT)*, 2009, pp. 178–183.

34. B. Noack, V. Klumpp, N. Petkov, and U. Hannebeck, Bounding linearization errors with sets of densities in approximate Kalman filtering, in *Proceedings of the 13th International Conference on Information Fusion*, Edinburgh, UK, 2010, ISIF.

35. R.E. Moore, *Methods and Applications of Interval Analysis*, Society for Industrial Mathematics (SIAM), Philadelphia, 1979.

36. N. Nedialko, V. Kreinovich, and S. Stark, Interval arithmetic, affine arithmetic, Taylor series methods: Why, what next?, *Numer. Algorithms*, 37, 325–336, 2004.

37. L. Jaulin, M. Kieffer, I. Braems, and E. Walter, Guaranteed nonlinear estimation using constraint propagation on sets, *Int. J. Contr.*, 74(18), 1772–1782, 2001.

38. D.L. Waltz, Generating semantic descriptions from drawings of scenes with shadows, *Technical Report*, Cambridge, MA, USA, 1972.

39. J.G. Cleary, Logical arithmetic, *Fut. Comput. Syst.*, 2, 125–149, 1987.

40. E. Davis, Constraint propagation with interval labels, *Artif. Intell.*, 32, 281–331, 1987.

41. R.E. Edwards, *What is the Riemann integral?*, Dept. of Pure Mathematics, Dept. of Mathematics, Australian National University, Canberra, 1974.

42. Y. Bar-Shalom, X. Rong Li, and T. Kirubarajan, *Estimation with Applications to Tracking and Navigation*, John Wiley and Sons, New York, 2001.

CHAPTER 3

Bayesian Multiple Target Filtering Using Random Finite Sets

BA-NGU VO, BA-TUONG VO, AND DANIEL CLARK

3.1 INTRODUCTION

The random finite set (RFS) approach, introduced by Mahler as finite set statistics (FISST), is an elegant Bayesian formulation of multitarget filtering based on RFS theory. In essence, the collection of target states at any given time is treated as a set-valued multitarget state, and the corresponding collection of sensor measurements is treated as a set-valued multitarget observation [1, 2]. The rationale behind this representation traces back to a fundamental consideration in estimation theory–estimation error. Without a meaningful notion of estimation error, the output of an estimator has very little meaning. The RFS approach has generated substantial interest in recent years due to the development of the probability hypothesis density (PHD) filter [1] and the cardinalized PHD (CPHD) filter [3]. The PHD and CPHD filters are moment approximations of the Bayes multitarget filter, which operate on the single-target state space and avoid the combinatorial problem that arises from data association. Sequential Monte Carlo (SMC) implementations [2, 4–6], and closed-form solutions [7–10] to the PHD and CPHD recursions have inspired numerous novel extensions and applications.

The aim of this chapter is to describe the RFS approach to multitarget tracking and detail RFS-based algorithms such as the PHD, CPHD filters and recent developments such as the Multitarget Multi-Bernoulli (MeMBer) filters [11, 12]. An overview of the developments in the RFS approach and the PHD/CPHD filters is given in Section 3.2. To introduce the RFS formalism for Bayesian multitarget filtering, we give a brief account of the RFS fundamentals in Section 3.3 followed by the generalization of the Bayes filter to finite-set-valued states and measurements in Section 3.4. Section 3.5 examines the fundamental notion of miss distance or estimation error for multiple targets. Sections 3.6 and 3.7 describe the PHD/CPHD filters and their

Integrated Tracking, Classification, and Sensor Management: Theory and Applications, Edited by Mahendra Mallick, Vikram Krishnamurthy, and Ba-Ngu Vo.

Gaussian mixture implementations. Section 3.9 closes this chapter by outlining the MeMBer filter as another approximation approach to the multitarget filtering problem.

3.2 OVERVIEW OF THE RANDOM FINITE SET APPROACH TO MULTITARGET FILTERING

In this section we review the Bayesian filtering paradigm for single-target tracking, outline developments in the RFS approach to multitarget tracking, and discuss the motivation for this approach.

3.2.1 Single-target Filtering

In many dynamic state estimation problems, the state is assumed to follow a Markov process on the state space $\mathcal{X} \subseteq \mathbb{R}^{n_x}$, with *transition density* $f_{k|k-1}(\cdot|\cdot)$, that is given a state $\mathbf{x}_{k-1}$ at time $k-1$, the probability density of a transition to the state $\mathbf{x}_k$ at time k is[1]

$$f_{k|k-1}(\mathbf{x}_k|\mathbf{x}_{k-1}). \tag{3.1}$$

This Markov process is partially observed in the observation space $\mathcal{Z} \subseteq \mathbb{R}^{n_z}$, as modeled by the *likelihood function* $g_k(\cdot|\cdot)$, that is given a state $\mathbf{x}_k$ at time k, the probability density of receiving the observation $\mathbf{z}_k \in \mathcal{Z}$ is

$$g_k(\mathbf{z}_k|\mathbf{x}_k). \tag{3.2}$$

It is assumed that the probability density of the measurement history $\mathbf{z}_{1:k} := (\mathbf{z}_1, \ldots, \mathbf{z}_k)$ conditioned on the system trajectory $\mathbf{x}_{0:k} := (\mathbf{x}_0, \ldots, \mathbf{z}_k)$ is given by

$$p_{1:k}(\mathbf{z}_{1:k}|\mathbf{x}_{0:k}) = g_k(\mathbf{z}_k|\mathbf{x}_k)g_{k-1}(\mathbf{z}_{k-1}|\mathbf{x}_{k-1}) \ldots g_1(\mathbf{z}_1|\mathbf{x}_1)p_0(\mathbf{x}_0),$$

for a given initial density $p_0(\cdot)$.

All information about the state at time k is encapsulated in the *posterior* (or *filtering*[2]) density

$$p_k(\cdot|\mathbf{z}_{1:k}), \tag{3.3}$$

which is defined as the probability density of the state at time k given the measurement history $\mathbf{z}_{1:k}$ up to time k. From an initial density $p_0(\cdot)$, the posterior density at time k

[1] For notational simplicity, random variables and their realizations are not distinguished.

[2] Strictly speaking, the posterior density is $p_{0:k}(\mathbf{x}_{0:k}|\mathbf{z}_{1:k})$ and the filtering density can be obtained by marginalizing the posterior density.

can be computed using the Bayes recursion

$$p_{k|k-1}(\mathbf{x}_k|\mathbf{z}_{1:k-1}) = \int f_{k|k-1}(\mathbf{x}_k|\mathbf{x})p_{k-1}(\mathbf{x}|\mathbf{z}_{1:k-1})d\mathbf{x}, \tag{3.4}$$

$$p_k(\mathbf{x}_k|\mathbf{z}_{1:k}) = \frac{g_k(\mathbf{z}_k|\mathbf{x}_k)p_{k|k-1}(\mathbf{x}_k|\mathbf{z}_{1:k-1})}{\int g_k(\mathbf{z}_k|\mathbf{x})p_{k|k-1}(\mathbf{x}|\mathbf{z}_{1:k-1})d\mathbf{x}}. \tag{3.5}$$

An optimal estimate of the state vector can be obtained with respect to a prescribed criterion. A *Bayes optimal* estimate minimizes a certain objective called the Bayes risk or loss, see for example [13]. The most common estimators are the expected a posteriori (EAP) and maximum a posteriori (MAP) given respectively by

$$\hat{\mathbf{x}}_k^{\mathrm{EAP}} = \int \mathbf{x}_k p_k(\mathbf{x}_k|\mathbf{z}_{1:k})d\mathbf{x}_k,$$

$$\hat{\mathbf{x}}_k^{\mathrm{MAP}} = \arg\sup_{x_k} p_k(\mathbf{x}_k|\mathbf{z}_{1:k}).$$

These estimators minimize certain Bayes risks, and are *consistent* in the sense that they converge almost surely to the true state.

3.2.2 Random Finite Set and Multitarget Filtering

In a multitarget environment, not only do the states of the targets vary with time, but the number of targets also changes due to targets appearing and disappearing. At the sensor, existing targets may or may not be detected. Moreover, the sensor also receives a set of spurious measurements (clutter) not originating from any target. As a result, at each time step we have a collection of indistinguishable measurements, only some of which are generated by targets (see Figure 3.1).

The objective of multitarget tracking is to jointly estimate, at each time step, the number of targets and their states from a sequence of noisy and cluttered

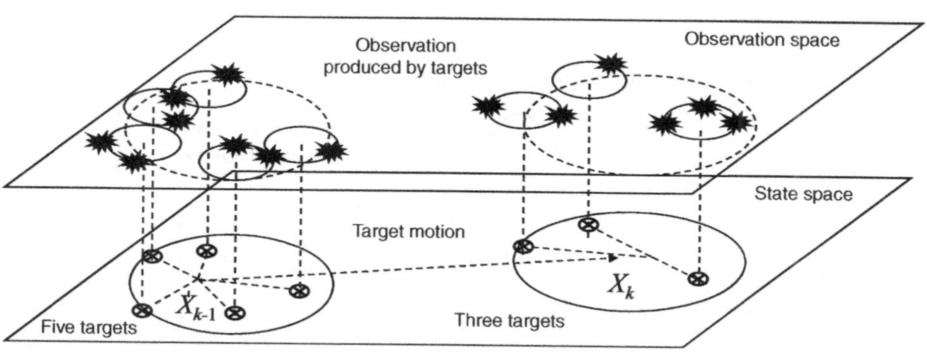

Figure 3.1 Multiple-target system model: the number of targets changes from 5 to 3, targets generate at each time a random number of measurements.

observation sets. Note that even in the special case where the sensor observes all targets and receives no false alarms, classical filtering methods are not directly applicable since there is no information on which target has generated which observation. See also Chapter 5 [14] for the classical treatment of multitarget tracking.

The earliest published work using a point process formalism for multitarget filtering appears to be [15]. A point process based filter was also proposed in [16] to estimate an unknown but fixed number of targets. In [17–19], a jump process was combined with stochastic diffusion equations on a non-Euclidean manifold to track a time varying number of targets. The same problem with continuous state evolution and marked-point process observations was considered in [20]. However, the first systematic treatment of multisensor multitarget filtering using random set theory was conceived by Mahler in 1994 [21], which later developed into FISST. Moreover, this treatment was developed as part of a unified framework for data fusion using random set theory. An early overview of this treatment appeared as a chapter in [22], while the mathematical details of the treatment are given in [23]. The 2000 monograph [24] also provides an overview of FISST and how this addresses the pitfalls of traditional Bayesian multitarget filtering techniques. A detailed treatment of FISST can be found in Mahler's latest book [25].

The FISST Bayes multitarget filter is generally intractable. In 2000 Mahler proposed to approximate the multitarget Bayes recursion by propagating the PHD of the posterior multitarget state [1, 26]. The PHD, also known in point process literature as the intensity or first moment, is a function on the single-target state space with the property that its integral over a region gives the expected number of targets in that region (see Figure 3.2). Hence the estimated number of targets is given by the integral of the PHD. The PHD value at a point $\mathbf{x}_0$ represents the expected number of targets per unit hyper-volume at $\mathbf{x}_0$, and the estimated states can be taken as the highest peaks of the PHD.

The strategy of propagating the PHD is reminiscent of the constant gain Kalman filter that propagates the mean of the posterior single-target state. The PHD filter operates on the single-target state space and avoids the combinatorial problem that arises from data association. Mahler's PHD filter is an innovative and elegant engineering approximation that captivated many researchers in multitarget tracking.

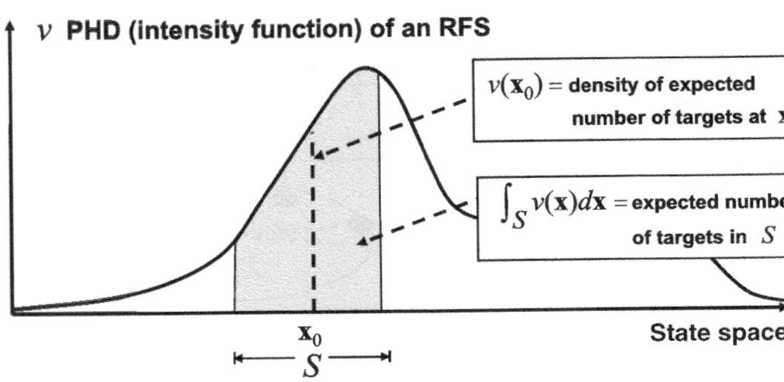

Figure 3.2 Interpreting the PHD of a random finite set.

More importantly, it provides an important step towards the practical application of FISST.

In 2003 Mahler published generalized FISST, a refined version of FISST, and the derivation of the PHD filter [1]. Additionally, the relationship between FISST set derivatives and probability density (as Radon Nikodym derivatives of probability measures) for random finite sets are established in [4], along with generic SMC implementations of the multitarget Bayes filter and PHD filter (these results were formally reported later in [2] along with the first set of convergence result for the SMC implementations). At the same time, several SMC or particle implementations of the PHD filter and multitarget Bayes filter were independently proposed in [5, 6, 27]. Convergence properties of the particle-PHD filter were later established in [2, 28, 29]. Inspired by the particle-PHD filter, important extensions to maintain track continuity have also been proposed in [30, 31]. Despite the need for additional processing such as clustering to extract multitarget state estimate, the particle-PHD filter was quickly adopted to solve a host of practical problems (see e.g., [25, 32]).

In 2005 a closed-form solution to the PHD recursion for the linear Gaussian multitarget model was discovered [7]. This result was reported in [8] together with the Gaussian Mixture PHD filter for linear and mildly nonlinear multitarget models. While more restrictive than SMC approaches, Gaussian mixture implementations are much more efficient. Moreover, they obviate the need for clustering—an expensive step in the SMC implementation. Convergence results for the GMPHD filter were established in [33]. In [34] the Gaussian mixture PHD filter is extended to linear Jump Markov multitarget model for tracking maneuvering targets, while in [35] it is extended to produce track-valued estimates.

Recently, extensions of the particle-PHD filter such as the auxiliary particle-PHD filter [36], and measurement-oriented particle labeling technique [37], partially solve the clustering problem in the extraction state estimates from the particle population. Clever use of the PHD filter with measurement-driven birth intensity was independently proposed in [37] and [38] to improve tracking performance as well as obviating exact knowledge of the birth intensity.

In 2006 Mahler published the CPHD recursion—a generalization of the PHD recursion that jointly propagates the posterior PHD and the posterior distribution of the number of targets [3, 39]. Moreover, in [9, 10] it was shown that the CPHD recursion admits a closed-form solution under linear Gaussian assumptions, and Gaussian mixture implementations were proposed for linear and mildly nonlinear multitarget models. It was demonstrated in [10] that while the GMCPHD filter is of cubic (in the number of measurement) complexity it outperforms the NP-hard JPDA filter. An intuitive interpretation of the PHD/CPHD filter in terms of bin occupancy was given in [40]. Extensions of the GMPHD and particle-PHD filters apply directly to the CPHD case.

In addition to the PHD and CPHD filters, Mahler proposed the MeMBer recursion as a tractable approximation to the Bayes multitarget recursion [25]. However, this approximation overestimates the cardinality (number of targets) and an unbiased version was proposed in [11]. Unlike the PHD and CPHD recursions, which propagate moments and cardinality distributions, the MeMBer filter propagates the parameters of a multi-Bernoulli RFS that approximates the posterior multitarget RFS. Also using multi-Bernoulli approximations, the first RFS-based multitarget Track Before Detect (TBD) filter that operates directly on the images was proposed in [12].

3.2.3 Why Random Finite Set for Multitarget Filtering?

The key feature of the RFS approach is that the multitarget state and collection of sensor measurements at any given time are treated as finite sets. *Why represent the multitarget state and multitarget measurement as finite sets*? Can't we represent them as vectors like many traditional approaches? To answer these questions, let us revisit a fundamental concept in estimation theory—estimation error. Multitarget filtering is meaningless without some notion of multitarget miss distance to measure the multitarget estimation error.

Suppose we represent the multitarget state by stacking individual states into a single vector (of dimension $n_x \times$ *number of targets*). Consider the scenarios illustrated in Figure 3.3, with the ground truth represented by the vector X and the estimate represented by the vector X'.

The estimate is correct but estimation error is $\|X - X'\| = 2$! This example reveals a fundamental inconsistency in the vector representation of the multitarget state. While this fundamental inconsistency can be remedied by taking the minimum of the distance over all permutations of individual states in the (vector) multitarget state that is

$$\min_{\text{perm}(X)} \|X - X'\| = 0, \tag{3.6}$$

there are still serious problems with the vector representation. How can the multitarget state with no target be represented by a vector? More importantly, what is the miss distance when the estimated number of targets and the true number of targets are different, like the scenarios in Figure 3.4?

Clearly, a vector representation does not admit a meaningful and mathematically consistent notion of a miss distance. A finite set, on the other hand, can

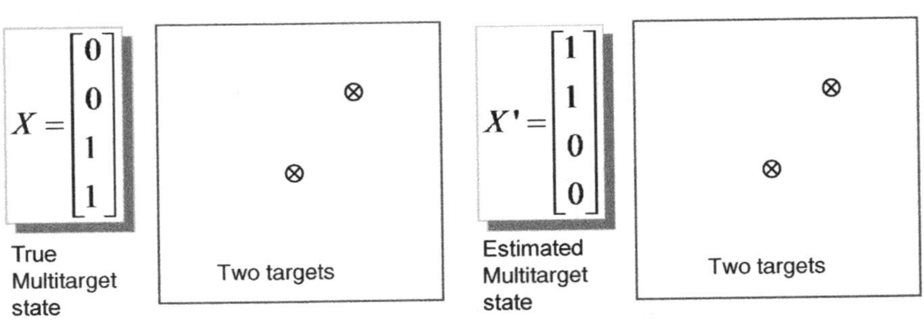

Figure 3.3 A possible vector representation of multitarget states when the estimated and true multitarget states have the same number of targets.

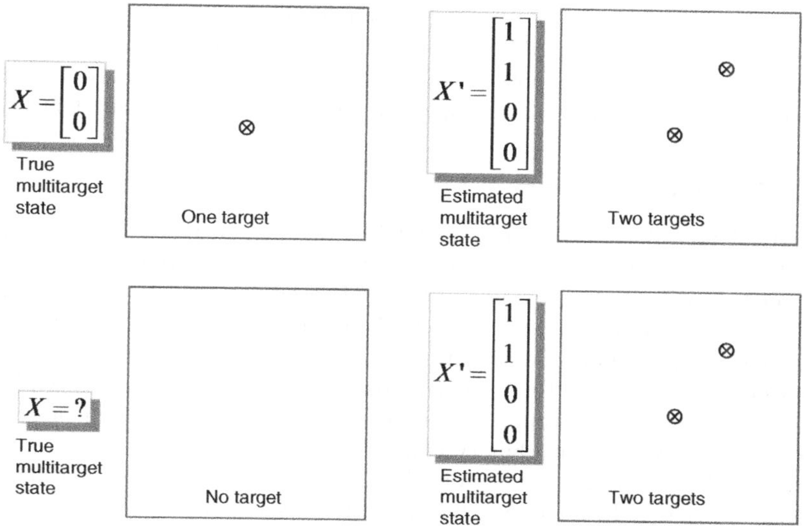

Figure 3.4 Possible scenarios where the estimated and true number of targets are different. What is the multitarget miss distance?

represent all possible occurrences of multitarget states, and distance between sets is a well understood concept (a more detailed discussion is given in Section 3.5). In fact, the miss distance defined by (3.6) is not a distance for vectors but a distance for sets.

Similarly, stacking individual measurements into a large vector is not a satisfactory representation for the multitarget measurement. The number of measurements is not fixed, and the ordering of the measurements has no significance. Hence, it is widely accepted in the multitarget tracking literature that the collection of measurements at each time is treated as a finite set [41, 42].

In the Bayesian filtering paradigm, the state and measurements are treated as realizations of random variables. Since our multitarget state and multitarget observation are finite sets, we need the concept of a random finite set to cast the multitarget filtering problem in the Bayesian framework. Moreover, we need concepts such as integration and density for set-valued variables since these concepts do not carry over from random vectors to random finite sets in a trivial manner.

3.3 RANDOM FINITE SETS

Intuitively, an RFS is a random (spatial) point pattern, for example measurements on a radar screen or positions of rain drops on a spatial area. What distinguishes an RFS from a random vector is that: the number of constituent points is random and the points themselves are random, distinct, and unordered. An RFS X on $\mathcal{X} \subseteq \mathbb{R}^d$ can be completely specified by a discrete distribution that characterizes the cardinality (number of points), and a family of joint distributions that characterizes the distribution of the points conditional on the cardinality [43]. In essence, an RFS is simply a finite-set-valued random variable. This section introduces the elements of RFS, by presenting various fundamental descriptors as well as examples of important RFSs.

3.3.1 Probability Density

At the fundamental level, like any other random variable, the randomness of an RFS is captured by its probability distribution. Similar to the random vector case, the probability density (if it exists) is a very useful descriptor of an RFS, especially in filtering and estimation. While the space $\mathcal{F}(\mathcal{X})$ of all finite subsets of $\mathcal{X}$ does not inherit the usual Euclidean notion of density from $\mathbb{R}^d$, a mathematically consistent notion of a probability density on $\mathcal{F}(\mathcal{X})$ is available from RFS or point process theory [43].

The notion of a density is intimately tied to the concept of measure and integration. Intuitively, a measure on an arbitrary space $\mathcal{Y}$ quantifies the "sizes" of subsets of $\mathcal{Y}$, for example length, area, volume, probability, and so on. The conventional choice of reference measure in RFS theory is the dimensionless measure[3] given by

$$\mu(\mathcal{T}) = \sum_{r=0}^{\infty} \frac{\lambda^r(\chi^{-1}(\mathcal{T}) \cap \mathcal{X}^r)}{r!}$$

for any subset $\mathcal{T}$ of $\mathcal{F}(\mathcal{X})$, where $\mathcal{X}^r$ is the rth Cartesian product of $\mathcal{X}$ with the convention $\mathcal{X}^0 = \{\emptyset\}$, λ^r is the rth product dimensionless Lebesgue measure on $\mathcal{X}^r$, and χ is a mapping of vectors to sets defined by $\chi(\mathbf{x}_1, \ldots, \mathbf{x}_r) = \{\mathbf{x}_i : i = 1, \ldots, r\}$. The integral of a function $f : \mathcal{F}(\mathcal{X}) \to \mathbb{R}$ over a subset $\mathcal{T}$ of $\mathcal{F}(\mathcal{X})$ with respect to μ is given as follows [2]

$$\int_{\mathcal{T}} f(X)\mu(dX) = \sum_{r=0}^{\infty} \frac{1}{r!} \int \mathbf{1}_{\mathcal{T}}(\chi(\mathbf{x}_1, \ldots, \mathbf{x}_r)) f(\{\mathbf{x}_1, \ldots, \mathbf{x}_r\}) \lambda^r(d\mathbf{x}_1 \ldots d\mathbf{x}_r),$$

where $\mathbf{1}_{\mathcal{T}}$ is the indicator function for $\mathcal{T}$. Note that we use the notation $d\mathbf{x}$ in standard integral (with respect to the hyper-volume measure that may have an associated unit) on Euclidean space, and $\lambda(d\mathbf{x})$ in integral with respect to a dimensionless measure on $\mathcal{X}$. Given an RFS X on $\mathcal{X}$, the *probability density* π with respect to the dominating measure μ satisfies

$$\mathbb{P}(X \in \mathcal{T}) = \int_{\mathcal{T}} \pi(Y)\mu(dY)$$

for any subset $\mathcal{T}$ of $\mathcal{F}(\mathcal{X})$. Unlike probability density on Euclidean space, which has the physical dimension of probability per unit hyper-volume, π is dimensionless since the reference measure is dimensionless.

[3] In fact this measure is the unnormalized probability distribution of a unit rate Poisson RFS (see subsection 3.3.5 for Poisson RFS).

3.3.2 Janossy Densities

Recall that an RFS X on $\mathcal{X} \subseteq \mathbb{R}^d$ can be completely specified by a discrete distribution $\rho(\cdot)$, and a family of symmetric[4] probability density $\{p^{(n)} : n = 1, 2, \ldots\}$. For each n, the probability $\rho(n)$ and the point distribution $p^{(n)}(\cdot)$ can be encapsulated in a function $j^{(n)}(\cdot) = \rho(n)p^{(n)}(\cdot)$ on $\mathcal{X}^n$ called the nth-order *Janossy density* (by convention, $j^{(0)} = p^{(0)}$). Moreover, if X has a probability density π (relative to the reference measure μ), then

$$\pi(\{\mathbf{x}_1, \ldots, \mathbf{x}_n\}) = j^{(n)}(\mathbf{x}_1, \ldots, \mathbf{x}_n)K^n,$$

where K is the unit in which hyper-volume on $\mathcal{X}$ is measured. Note that the Janossy density $j^{(n)}$ is not a probability density because, as a function on $\mathcal{X}^n$, it does not integrate to 1. While the nth-order Janossy density is not a probability density, $j^{(n)}(\mathbf{x}_1, \ldots, \mathbf{x}_n)d\mathbf{x}_1 \ldots d\mathbf{x}_r$ can be interpreted as the probability that the RFS X has n points, and there is exactly one point in each of the infinitesimal neighborhoods of $\mathbf{x}_1, \ldots, \mathbf{x}_n$.

3.3.3 Belief Functional and Density

Apart from the probability distribution/density, Janossy measures/densities, the *belief functional* β of an RFS X, defined by

$$\beta(S) = \mathbb{P}(X \subseteq S)$$

for all closed $S \subseteq \mathcal{X}$, is another equivalent characterization [23]. The belief functional plays important role in the FISST approach to multitarget filtering introduced by Mahler [23, 25]. For modeling of multitarget systems, the belief functional is more convenient than the probability distribution, since the former deals with closed subsets of $\mathcal{X}$ whereas the latter deals with subsets of $\mathcal{F}(\mathcal{X})$.

The belief functional is not a measure and hence the usual notion of density is not applicable. FISST provides an alternative notion of density for belief functional via the constructs of set integrals and set derivatives. The FISST density p of an RFS is obtained by taking the FISST set derivative of the belief functional β (see [23, 25] for further details). In other words, the set integral of a FISST density over a closed subset $S \subseteq \mathcal{X}$ gives the belief functional

$$\int_S p(X)\delta X = \beta(S),$$

where the integral above is a set integral defined by [1, 23, 24]

$$\int_S p(X)\delta X \equiv \sum_{i=0}^{\infty} \frac{1}{i!} \int_{S^i} p(\{\mathbf{x}_1, \ldots, \mathbf{x}_i\})d\mathbf{x}_1 \ldots d\mathbf{x}_i.$$

[4] A function is symmetric if it is invariant to the order of the arguments.

It turns out that the FISST density p is related to the probability density π by

$$\pi(X) = p(X)K^{|X|}, \tag{3.7}$$

where $|X|$ denotes the cardinality of X, and K denotes the unit in which volume on $\mathcal{X}$ is measured (see [2] for further details in this regard). Hence $p(\{\mathbf{x}_1, \ldots, \mathbf{x}_n\}) = j^{(n)}(\mathbf{x}_1, \ldots, \mathbf{x}_n)$, that is, the FISST density restricted to sets with cardinality n is the nth-order Janossy density. For the purpose of this chapter we shall not distinguish between the set derivative of an RFS's belief functional and its probability density. FISST converts the construction of multitarget densities from multitarget models into computing set derivatives of belief functionals. Procedures for analytically differentiating belief functional have been developed and described in [23, 25].

3.3.4 The Probability Hypothesis Density

The PHD also known in point process theory as an *intensity function* is a first-order statistical moment of an RFS [1]. For an RFS X on $\mathcal{X}$, its PHD is a nonnegative function v on $\mathcal{X}$ such that for each region $S \subseteq \mathcal{X}$ [43]

$$\mathbb{E}\left[|X \cap S|\right] = \int_S v(\mathbf{x})d\mathbf{x}, \tag{3.8}$$

where $\mathbb{E}$ denotes expectation. In other words, the integral of v over any region S gives the expected number of elements of X that are in S. Hence, the total mass $\hat{N} = \int v(\mathbf{x})d\mathbf{x}$ gives the expected number of elements of X. The local maxima of the PHD v are points in $\mathcal{X}$ with the highest local concentration of expected number of elements, and hence can be used to generate estimates for the elements of X. The simplest approach is to round $\hat{N}$ and choose the resulting number of highest peaks from the PHD.

3.3.5 Examples of RFS

Poisson RFS. An important class of RFSs, namely the *Poisson RFSs*, are those completely characterized by their PHDs. An RFS X on $\mathcal{X}$ is said to be *Poisson* with a given PHD v (defined on $\mathcal{X}$) if its cardinality is Poisson distributed, with mean $\bar{N} = \int v(\mathbf{x})d\mathbf{x}$, and for any finite cardinality, the elements $\mathbf{x}$ of X are independently and identically distributed (i.i.d.) according to the probability density $v(\cdot)/\bar{N}$ [43] pp. 32. A Poisson RFS with PHD v has probability density

$$\pi(\{\mathbf{x}_1, \ldots, \mathbf{x}_n\}) = e^{-\bar{N}} \prod_{i=1}^{n} v(\mathbf{x}_i),$$

with the convention $\prod_{i=1}^{0} v(\mathbf{x}) = 1$.

I.I.D. cluster RFS. An *independent and identically distributed* (i.i.d.) cluster RFS X on $\mathcal{X}$ is completely characterized by its cardinality distribution ρ and PHD v [43]. The cardinality distribution must satisfy $\bar{N} = \mathbb{E}[\rho] = \int v(\mathbf{x})d\mathbf{x}$ but can otherwise be arbitrary, and for a given cardinality the elements of X are each i.i.d with probability density $v/\bar{N}$. The probability density of an i.i.d. cluster RFS can be written as

$$\pi(\{\mathbf{x}_1, \ldots, \mathbf{x}_n\}) = n!\rho(n)\prod_{i=1}^{n} \frac{v(\mathbf{x}_i)}{\bar{N}}.$$

Note that an i.i.d. cluster process essentially captures the spatial randomness of the Poisson RFS without the restriction of Poisson cardinality.

Bernoulli RFS. A *Bernoulli* RFS on $\mathcal{X}$ has probability $1 - r$ of being empty, and probability r of being a singleton whose (only) element is distributed according to a probability density p (defined on $\mathcal{X}$). The cardinality distribution of a Bernoulli RFS is a Bernoulli distribution with parameter r. The probability density of a Bernoulli RFS is given by

$$\pi(X) = \begin{cases} 1 - r, & X = \emptyset, \\ rp(\mathbf{x}), & X = \{\mathbf{x}\}. \\ 0, & \text{otherwise} \end{cases}$$

Multi-Bernoulli RFS. A *multi-Bernoulli* RFS X on $\mathcal{X}$ is a union of a fixed number of independent Bernoulli RFSs $X^{(i)}$ with existence probability $r^{(i)} \in (0, 1)$ and probability density $p^{(i)}$ (defined on $\mathcal{X}$), $i = 1, \ldots, M$, that is

$$X = \bigcup_{i=1}^{M} X^{(i)}.$$

A multi-Bernoulli RFS is thus completely described by the multi-Bernoulli parameter set $\{(r^{(i)}, p^{(i)})\}_{i=1}^{M}$. The mean cardinality of a multi-Bernoulli RFS is $\sum_{i=1}^{M} r^{(i)}$. Moreover, the probability density π is given by $\pi(\emptyset) = \prod_{j=1}^{M}\left(1 - r^{(j)}\right)$ and

$$\pi(\{\mathbf{x}_1, \ldots, \mathbf{x}_n\}) = \pi(\emptyset) \sum_{1 \le i_1 \ne \cdots \ne i_n \le M} \prod_{j=1}^{n} \frac{r^{(i_j)} p^{(i_j)}(\mathbf{x}_j)}{1 - r^{(i_j)}}.$$

3.4 MULTIPLE TARGET FILTERING AND ESTIMATION

This section presents the multitarget filtering problem formulated in the RFS framework. We begin by stating the underlying RFS multitarget system model followed by the multitarget Bayes recursion as a solution to the multitarget filtering problem.

We also discuss the issue of multitarget state estimation and highlight the pitfalls associated with applying direct extensions of traditional estimators as well as summarize simple approaches to extracting sensible estimates.

Suppose that at time k, there are $N(k)$ targets $\mathbf{x}_{k,1}, \ldots, \mathbf{x}_{k,N(k)}$, each taking values in a state space $\mathcal{X} \subseteq \mathbb{R}^{n_x}$, and $M(k)$ measurements $\mathbf{z}_{k,1}, \ldots, \mathbf{z}_{k,M(k)}$ each taking values in an observation space $\mathcal{Z} \subseteq \mathbb{R}^{n_z}$. In the RFS approach, the collection of targets and observations, at time k, [1, 23, 25] are treated as finite sets representing the *multitarget state* and *multitarget observation*, respectively

$$X_k = \{\mathbf{x}_{k,1}, \ldots, \mathbf{x}_{k,N(k)}\} \in \mathcal{F}(\mathcal{X}),$$
$$Z_k = \{\mathbf{z}_{k,1}, \ldots, \mathbf{z}_{k,M(k)}\} \in \mathcal{F}(\mathcal{Z}),$$

where $\mathcal{F}(\mathcal{X})$ and $\mathcal{F}(\mathcal{Z})$ are the spaces finite subsets of $\mathcal{X}$ and $\mathcal{Z}$, respectively. Note that the order in which the states and measurements are listed has no significance in the RFS formulation. At each time step, some targets may die, some targets may survive and transition to a new state, and new targets may appear. Of the surviving and new born targets, each may or may not generate a measurement and those received at the sensor are also obscured by false/spurious measurements or clutter. Thus, the evolutions of the targets present and the origins of the measurements received at the sensor are not known. Such uncertainty or randomness is naturally addressed by using RFS theory to construct stochastic models for the dynamics of the multitarget state and the generation of the multitarget observations as follows.

3.4.1 Multitarget Dynamical Model

Given a multitarget state X_{k-1} at time $k-1$, each target $\mathbf{x}_{k-1} \in X_{k-1}$ either continues to exist at time k with probability $p_{S,k|k-1}(\mathbf{x}_{k-1})$ and moves to a new state $\mathbf{x}_k$ with probability density $f_{k|k-1}(\mathbf{x}_k|\mathbf{x}_{k-1})$, or dies with probability $1 - p_{S,k|k-1}(\mathbf{x}_{k-1})$ and takes on the value $\emptyset$. Thus, given a target with state $\mathbf{x}_{k-1} \in X_{k-1}$ at time $k-1$, its behavior time k is modeled by the Bernoulli RFS

$$S_{k|k-1}(\mathbf{x}_{k-1})$$

with $r = p_{S,k|k-1}(\mathbf{x}_{k-1})$ and $p(\cdot) = f_{k|k-1}(\cdot|\mathbf{x}_{k-1})$ (note that $p_{S,k|k-1}(\mathbf{x}_{k-1})$ is a probability parameterized by $\mathbf{x}_{k-1}$). This transition is commonly known in point process theory as a Markov shift [43]. The survival or death of all existing targets from time $k-1$ to k is hence modeled by the RFS

$$T_{k|k-1}(X_{k-1}) = \bigcup_{\mathbf{x}_{k-1} \in X_{k-1}} S_{k|k-1}(\mathbf{x}_{k-1}).$$

Note that conditional on X_{k-1}, the RFS of transitioned targets $T_{k|k-1}(X_{k-1})$ is a multi-Bernoulli RFS. The appearance of new targets at time k is modeled by an RFS of spontaneous births Γ_k which is usually given as either a Poisson, i.i.d. cluster, or multi-Bernoulli RFS. Consequently, the multitarget state X_k at time k is given by the union

$$X_k = T_{k|k-1}(X_{k-1}) \cup \Gamma_k. \tag{3.9}$$

It is assumed that the RFSs constituting the above union are mutually independent. The RFS multitarget transition (3.9) can be equivalently stated in the form of the multitarget transition density $\phi_{k|k-1}(\cdot|\cdot)$. The probability density that the multitarget state makes a transition from state X_{k-1} at time $k-1$ to state X_k at time k is given by

$$\phi_{k|k-1}(X_k|X_{k-1}) = \sum_{W \subseteq X_k} \pi_{T,k|k-1}(W|X_{k-1}) \cdot \pi_{\Gamma,k}(X_k - W), \tag{3.10}$$

where $\pi_{T,k|k-1}(\cdot|\cdot)$ is the probability density of the RFS $T_{k|k-1}$ of transitioned targets, $\pi_{\Gamma,k}(\cdot)$ is the probability density of the RFS Γ_k of spontaneous birth, and the difference operation denotes the set difference. Both (3.9) and (3.10) describe the time evolution of the multitarget state and incorporate the underlying models of target motion, birth and death.

3.4.2 Multitarget Observation Model

A given target $\mathbf{x}_k \in X_k$, at time k, is either detected with probability $p_{D,k}(\mathbf{x}_k)$ and generates an observation $\mathbf{z}_k$ with likelihood $g_k(\mathbf{z}_k|\mathbf{x}_k)$, or missed with probability $1 - p_{D,k}(\mathbf{x}_k)$, that is each state $\mathbf{x}_k \in X_k$ generates a Bernoulli RFS

$$D_k(\mathbf{x}_k)$$

with $r = p_{D,k}(\mathbf{x}_k)$ and $p(\cdot) = g_k(\cdot|\mathbf{x}_k)$ (note that $p_{D,k}(\mathbf{x}_k)$ is a probability parameterized by $\mathbf{x}_k$). The detection and generation of measurements for all surviving targets at time k is hence given by the RFS

$$\Theta_k(X_k) = \bigcup_{\mathbf{x}_k \in X_k} D_k(\mathbf{x}_k).$$

Note that conditional on X_k, the RFS of measurements $\Theta_k(X_k)$ generated by targets is a multi-Bernoulli RFS. In addition, the sensor also receives a set of false/spurious measurements or clutter which is modeled by an RFS K_k usually given as either a Poisson or i.i.d. cluster RFS. Consequently, at time k, the multitarget measurement Z_k generated by a multitarget state X_k is formed by the union

$$Z_k = \Theta_k(X_k) \cup K_k. \tag{3.11}$$

It is assumed that the RFSs constituting the above union are mutually independent. The RFS multitarget measurement (3.11) can be equivalently stated in the form of the multitarget likelihood $\varphi_{k|k-1}(\cdot|\cdot)$. The probability density that at time k the multitarget state X_k generates the multitarget observation Z_k is given by

$$\varphi_k(Z_k|X_k) = \sum_{W \subseteq Z_k} \pi_{\Theta,k}(W|X_k)\pi_{K,k}(Z_k - W), \tag{3.12}$$

where $\pi_{\Theta,k}(\cdot|\cdot)$ is the probability density of the RFS $\Theta_k(X_k)$ of target-generated measurements, $\pi_{K,k}(\cdot)$ is the probability density of the RFS K_k of clutter, and the difference operation denotes the set difference. Both (3.11) and (3.12) describe the generation of the multitarget observation and incorporate the underlying models of target detections, measurement generation, and clutter.

3.4.3 Multitarget Bayes Recursion

The objective of multitarget filtering is to jointly estimate the number of targets and their states from the accumulated observations. By modeling the multitarget state and observation as RFSs, the multitarget filtering problem can be posed as a Bayesian filtering problem with state space $\mathcal{F}(\mathcal{X})$ and observation space $\mathcal{F}(\mathcal{Z})$. Note that in contrast to traditional approaches, the RFS formulation avoids explicit data associations between targets and measurements. Analogous to single-target filtering, in multitarget filtering all information about the multitarget state at time k is contained in the probability density of the multitarget state conditioned on the all sets of observations received up to time k, that is the posterior density at time k. Thus, let $\pi_k(\cdot|Z_{1:k})$ denote the *multitarget posterior density* at time k. Also analogous to single-target filtering, the multitarget posterior density can be computed recursively via the multitarget Bayes recursion [1, 2, 23, 25]

$$\pi_{k|k-1}(X_k|Z_{1:k-1}) = \int \phi_{k|k-1}(X_k|X)\pi_{k-1}(X|Z_{1:k-1})\delta X, \tag{3.13}$$

$$\pi_k(X_k|Z_{1:k}) = \frac{\varphi_k(Z_k|X_k)\pi_{k|k-1}(X_k|Z_{1:k-1})}{\int \varphi_k(Z_k|X)\pi_{k|k-1}(X|Z_{1:k-1})\delta X}, \tag{3.14}$$

where $\phi_{k|k-1}(\cdot|\cdot)$ and $\varphi_k(\cdot|\cdot)$ are the multitarget transition density and multitarget likelihood, respectively. Note that the multitarget Bayes recursion was originally proposed by Mahler using FISST set integrals in [1, 23, 25], but this form is equivalent to (3.13) and (3.14) due to (3.7) [2].

Due to the combinatorial nature of multitarget densities and the multiple integrations in (3.13) and (3.14), the multitarget Bayes recursion is intractable in most practical applications and the search for tractable approximations is imperative.

3.4.4 Multitarget State Estimation

The issue of extracting point estimates for individual targets from the multitarget posterior density or an approximation to it, that is multitarget state estimation has yet to be considered. First, we summarize the main inconsistencies associated with directly extending traditional single-target EAP and MAP estimators to the multitarget case (see [25]). Then, we outline simple approaches to multitarget state estimation.

Consider a situation in which there is at most one target located in the one-dimensional interval [0, 2] with unit distances given in m. Suppose that there is 0.5 probability of the target being present/absent. Suppose also that if the target is present, it is equally likely to be found anywhere in the interval [0, 2]. Hence, the probability density of the target state is given by

$$\pi(X) = \begin{cases} 0.5 & X = \emptyset \\ 0.25 & X = \{\mathbf{x}\}, 0 \leq \mathbf{x} \leq 2. \\ 0 & \text{otherwise} \end{cases}$$

It is immediately obvious that the EAP estimate is not well defined since the addition and subtraction of sets is not meaningfully defined.

The naive MAP estimate is

$$\hat{X}^{\text{MAP}} = \arg\sup_X \pi(X) = \emptyset$$

since $\pi(\emptyset) > \pi(\{\mathbf{x}\})$ $(0.5 > 0.25)$. If the unit distance is changed from m to km, the probability density of the target state becomes

$$\pi(X) = \begin{cases} 0.5 & X = \emptyset \\ 250 & X = \{\mathbf{x}\}, 0 \leq \mathbf{x} \leq 0.002,. \\ 0 & \text{otherwise} \end{cases}$$

and hence

$$\hat{X}^{\text{MAP}} = \arg\sup_X \pi(X) = \{\mathbf{x}\}$$

for any $0 \leq \mathbf{x} \leq 0.002$ since now $\pi(\{\mathbf{x}\}) > \pi(\emptyset)$ $(250 > 0.5)$. It is immediately obvious that the MAP estimate is not well defined since a mere change in the units of measurement results in a dramatic change in the estimate.

The apparent inconsistencies with the EAP and MAP estimators as defined above are fundamentally caused by mismatches in the units of each cardinality of the multitarget density. In other words, the heart of these inconsistencies is that a comparison of sets of differing cardinalities is not meaningful, at least without a proper theoretical basis for attempting such comparisons.

We outline now some principled solutions to performing multitarget state estimation, in the form of an intuitive visualization technique and two statistical estimators with desirable properties. In the following, let X_k denote the multitarget state at time k, $\pi_k(\cdot|Z_{1:k})$ denote its posterior density, $\rho_k(\cdot|Z_{1:k})$ denote its posterior cardinality distribution, and $v_k(\cdot|Z_{1:k})$ denote its posterior intensity function.

First moment visualization. A simple approach is to exploit the physical intuition of the first moment of an RFS, its posterior intensity function. Recall that the intensity function at a point gives the density of the expected number of targets occurring at that point. The peaks of the intensity function indicate areas where there is a relatively high density of targets, in other words areas with high probability of target occurrence. Thus, state estimation proceeds by estimating the number of targets with $\hat{n} = \int v_k(\mathbf{x}|Z_{1:K})d\mathbf{x}$ or $\hat{n} = \arg\sup_n \rho_k(n|Z_{1:k})$, and extracting the $\hat{n}$ points $\hat{\mathbf{x}}_1, \ldots, \hat{\mathbf{x}}_{\hat{n}}$ such that $v_k(\hat{\mathbf{x}}_1|Z_{1:k}), \ldots, v_k(\hat{\mathbf{x}}_{\hat{n}}|Z_{1:k})$ are the $\hat{n}$ highest values of $v_k(\cdot|Z_{1:k})$. Then, the multitarget state estimate is given by $\hat{X}_k = \{\hat{\mathbf{x}}_1, \ldots, \hat{\mathbf{x}}_{\hat{n}}\}$.

Marginal multiobject estimator. The Marginal Multiobject (MaM) estimator is defined as a two step estimator. First, the number of targets is estimated using a MAP estimate on the posterior cardinality distribution (the posterior distribution of the number of targets $|X|$)

$$\hat{n} = \arg\sup_n \rho_k(n|Z_{1:k}).$$

Second, the individual target states are estimated using a MAP estimate on the posterior density given that $n = \hat{n}$ (the posterior density restricted to $|X| = \hat{n}$)

$$\hat{X}^{\text{MaM}} = \arg \sup_{X:|X|=\hat{n}} \pi_k(X|Z_{1:k}).$$

It has been shown that the MaM estimator is Bayes optimal; however convergence results are not currently known.

Joint multiobject estimator. The Joint Multiobject (JoM) estimator is defined as

$$\hat{X}_c^{\text{JoM}} = \arg\sup_X \pi_k(X|Z_{1:k})\frac{c^{|X|}}{|X|!},$$

where c is a dimensionless constant. The JoM estimator can be equivalently evaluated using a two step procedure. First, for each $n \geq 0$, determine the MAP estimate

$$\hat{X}^{(n)} = \arg \sup_{X:|X|=n} \pi_k(X|Z_{1:k}).$$

Second, set

$$\hat{X}_c^{\text{JoM}} = \hat{X}^{(\hat{n})}, \text{ where } \hat{n} = \arg\sup_n \pi_k(\hat{X}^{(n)}|Z_{1:k})\frac{c^n}{n!}.$$

It has been shown that the JoM estimator is Bayes optimal, and is a statistically consistent estimator in that it converges to the true multitarget state (in the optimal subpattern assignment (OSPA) metric discussed in the next section). Additionally, the value of c determines the desired accuracy for the state estimate and the rate of convergence of the estimator (smaller c provides better accuracy with slower convergence, and vice versa).

3.5 MULTITARGET MISS DISTANCES

The concept of a miss distance, or error, between a reference quantity and its estimated/controlled value, plays a fundamental role in any filtering/control problem. In this section we outline a number of metrics for finite sets, such as Hausdorff, Optimal Mass Transfer (OMAT), OSPA, and discuss their suitability as multitarget miss distance.

Consider a tracking application where the number of targets is not known and is to be inferred along with the positions of the targets. Suppose that the true number of targets is 2 and consider three different estimates as shown in Figure 3.5: (a) two points that are close to the true positions of the targets; (b) one point that almost coincides with the true position of one target; and (c) three points out of which two almost coincide with the true positions of the targets. The question is which of these estimates is "closest" to the truth? A satisfactory multitarget miss distance needs to capture the "difference" between two sets of vectors, namely the reference multitarget state and the estimated multitarget state, in a mathematically consistent yet physically meaningful manner. In essence, a multitarget miss distance should:

- be a metric on the space of finite sets,
- have a natural (meaningful) physical interpretation,
- capture cardinality errors and state errors meaningfully,
- be easily computed.

3.5.1 Metrics

The fundamental requirement for consistent distance measuring is that the miss distances are metrics on the space of finite sets [44]. For completeness, we

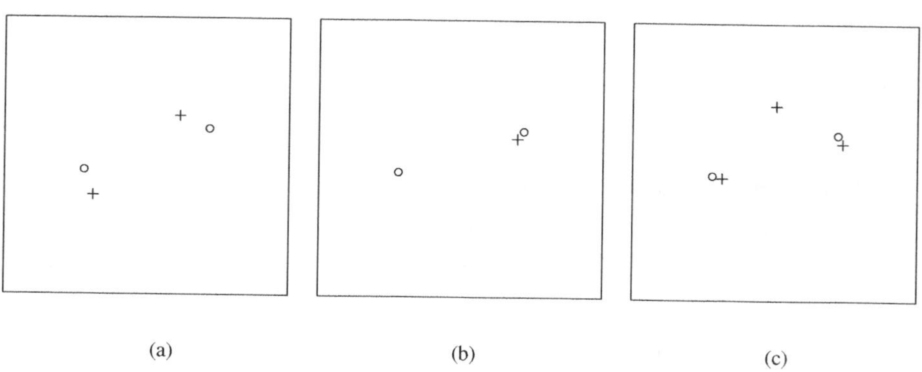

Figure 3.5 Hypothetical scenarios with ∘ representing true targets and + representing estimated targets. Which estimate is closest to the truth?

recall the definition of a metric. Let $\mathcal{S}$ be an arbitrary nonempty set. A function $d : \mathcal{S} \times \mathcal{S} \to \mathbb{R}_+ = [0, \infty)$ is called a *metric* if it satisfies the following three axioms:

1. *(identity)*: $d(\mathbf{x}, \mathbf{y}) = 0$ if and only if $\mathbf{x} = \mathbf{y}$;
2. *(symmetry)*: $d(\mathbf{x}, \mathbf{y}) = d(\mathbf{y}, \mathbf{x})$ for all $\mathbf{x}, \mathbf{y} \in \mathcal{S}$;
3. *(triangle inequality)*: $d(\mathbf{x}, \mathbf{y}) \leq d(\mathbf{x}, \mathbf{z}) + d(\mathbf{z}, \mathbf{y})$ for all $\mathbf{x}, \mathbf{y}, \mathbf{z} \in \mathcal{S}$.

In the context of multitarget miss distances, we assume a closed and bounded state-space $\mathcal{X} \subseteq \mathbb{R}^{n_x}$ equipped with a metric d (typically the Euclidean metric, $d(\mathbf{x}, \mathbf{y}) = \|\mathbf{x} - \mathbf{y}\|$). For the various metrics on $\mathcal{F}(\mathcal{X})$ appropriate indices are attached (d_H, d_p, or $\bar{d}_p^{(c)}$).

The optimal assignment paradigm of Drummond and associates [45, 46] has an intuitively appealing physical interpretation, but it can only measure the distance between multitarget states of the same cardinality in a consistent manner and is not a metric.

3.5.2 Hausdorff Metric

The *Hausdorff metric* is defined for finite nonempty subsets X and Y of $\mathcal{X}$ by

$$d_H(X, Y) = \max\left\{\max_{\mathbf{x}\in X}\min_{\mathbf{y}\in Y} d(\mathbf{x}, \mathbf{y}), \max_{\mathbf{y}\in Y}\min_{\mathbf{x}\in X} d(\mathbf{x}, \mathbf{y})\right\}.$$

Some of the Hausdorff metric's merits and difficulties in the context of multitarget filtering are discussed in [44].

The Hausdorff metric is traditionally used as a measure of dissimilarity between binary images (it typically gives a good idea of the difference in the visual perception a human would get from two images). However, it is very insensitive to differing cardinalities of finite sets, as can be seen from Scenarios C–F in Figure 3.6, which is not desirable for a performance measure of multitarget filters. Furthermore, it penalizes outliers heavily (see Scenario B of Figure 3.6), and cannot be reasonably defined if one of the sets is empty (Scenario A of Figure 3.6) though some authors set it to ∞ in this case.

3.5.3 Optimal Mass Transfer (OMAT) Metric

In [44], the *OMAT metric* was introduced to overcome some of the problems of the Hausdorff metric. The OMAT metric is defined for $1 \leq p < \infty$ and finite nonempty subsets $X = \{\mathbf{x}_1, \ldots, \mathbf{x}_m\}$ and $Y = \{\mathbf{y}_1, \ldots, \mathbf{y}_n\}$ of $\mathcal{X}$ as follows

$$d_p(X, Y) := \min_C \left(\sum_{i=1}^{m} \sum_{j=1}^{n} C_{i,j} d(\mathbf{x}_i, \mathbf{y}_j)^p \right)^{1/p},$$

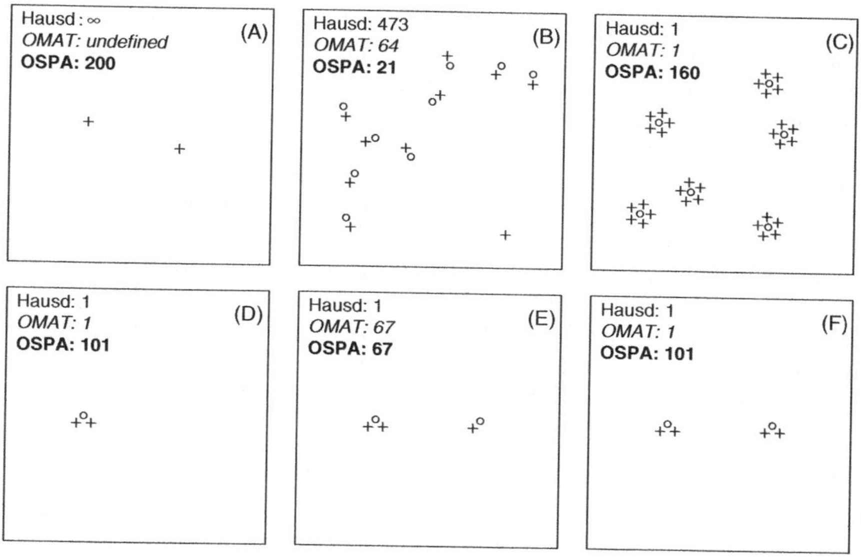

Figure 3.6 Six scenarios in a 1000×1000 m^2 window. $\circ$ actual targets, $+$ estimates. Parameters are $p = 1$ (OMAT and OSPA) and $c = 200$ (OSPA). Scenarios depict (A) two false estimates; (B) an outlier false estimate among several accurate estimates; (C) multiple estimates per target; (D–F) a comparison of balanced and unbalanced allocations of estimates to targets. Figures are not drawn to scale. The smallest distances have been considerably inflated for better viewing (from 1 m, which was assumed in the computations, to 40 m in the scale of the pictures).

where the minima are taken over all $m \times n$ transportation matrices $C = (C_{i,j})$ and where $\tilde{C}_{i,j} = 1$ if $C_{i,j} \neq 0$ and $\tilde{C}_{i,j} = 0$ otherwise. An $m \times n$ matrix C is a transportation matrix if all of its entries are nonnegative and if

$$\sum_{j=1}^{n} C_{i,j} = \frac{1}{m} \text{ for } 1 \leq i \leq m, \qquad \sum_{i=1}^{m} C_{i,j} = \frac{1}{n} \text{ for } 1 \leq j \leq n.$$

The OMAT metric partly fixes the undesirable cardinality behavior of the Hausdorff metric (see Scenario E of Figure 3.6) and is able to cope with the outlier problem by the introduction of the parameter p (Scenario B of Figure 3.6). This distance was termed the Wasserstein metric in [44] because it yields the pth order Wasserstein metric between the empirical distributions of the point patterns X and Y.

The OMAT metric is not necessarily better than the Hausdorff metric at detecting different cardinalities. Each of the examples from D to F in Figure 3.6 involves a cardinality error, and intuitively the estimation errors are roughly equal, but E is probably a bit better than F and (arguably) somewhat better than D. While the Hausdorff distance is the same in all three examples, the OMAT metric actually assigns a much larger distance in Scenario E than in the other two cases, and thus ranks the scenario that we intuitively prefer (E) as by far the worst among the three. Scenario C of Figure 3.6 gives a small distance although the cardinality of the estimated point pattern is quite far from the truth.

The OMAT metric is *geometry dependent* in the sense that a multitarget filter should be more heavily penalized for mis-estimating cardinality when the targets are far apart than when they are closely spaced. This is not a desirable property since it is virtually impossible to tell whether a multitarget filter diverges or performs well when the miss distance versus time curve increases with time. Even if this curve decreases to a small constant value, it cannot be inferred that the filter performance is good.

The distance $d_p(X, Y)$ is not defined if one of the two finite sets is empty, and in fact there is no natural extension for this case. However, when evaluating multitarget filter performance, the situation where there are no targets present but we obtain false estimates is quite common and should be captured by a reasonable metric.

3.5.4 Optimal Subpattern Assignment (OSPA) Metric

The *Optimal Subpattern Assignment (OSPA) metric* is still based on a Wasserstein construction, but eliminates most of the problems faced by the OMAT metric [47]. The OSPA metric $\bar{d}_p^{(c)}$ is defined as follows. Let $d^{(c)}(\mathbf{x}, \mathbf{y}) := \min(c, \|\mathbf{x} - \mathbf{y}\|)$ for $\mathbf{x}, \mathbf{y} \in \mathcal{X}$, and Π_k denote the set of permutations on $\{1, 2, \ldots, k\}$ for any positive integer k. Then, for $p \geq 1$, $c > 0$, and $X = \{\mathbf{x}_1, \ldots, \mathbf{x}_m\}$ and $Y = \{\mathbf{y}_1, \ldots, \mathbf{y}_n\}$ in $\mathcal{F}(\mathcal{X})$,

$$\bar{d}_p^{(c)}(X, Y) := \begin{cases} 0, & m = n = 0, \\ \left(\frac{1}{n}\left(\min_{\pi \in \Pi_n} \sum_{i=1}^{m} d^{(c)}(\mathbf{x}_i, \mathbf{y}_{\pi(i)})^p + c^p(n-m)\right)\right)^{\frac{1}{p}}, & m \leq n, \\ \bar{d}_p^{(c)}(Y, X), & m > n. \end{cases}$$

The OSPA metric is defined between any two point patterns. It is equal to c (i.e., maximal) if one of the two patterns is empty and the other is nonempty.

An alternative way of thinking about the OSPA metric is as follows: fill up the set X, which has the smaller cardinality m, with $n - m$ "dummy" points located at distance c from some points in Y and then calculate the pth order Wasserstein metric between the resulting sets. The OSPA metric can be computed efficiently by using the Hungarian method for optimal point assignment. A proof that $\bar{d}_p^{(c)}$ is indeed a metric can be found in [47].

The OSPA distance is interpreted as a pth order per-target error, comprised of a pth order per-target localization error and a pth order per-target cardinality error. The order parameter p determines the sensitivity of the metric to outliers, and the cut-off parameter c determines the relative weighting of the penalties assigned to cardinality and localization errors. A value of c which corresponds to the magnitude of a typical localization error can be considered small and has the effect of emphasizing localization errors. A value of c which corresponds to the maximal distance between targets can be considered large and has the effect of emphasizing cardinality errors.

Any value of c significantly larger than a typical localization error, but significantly smaller than the maximal distance between targets, can be considered moderate and maintains a balance between the two components.

The consistency problem of Scenarios D–F in Figure 3.6 is resolved. The OSPA metric penalizes relative differences in cardinality in an impartial way by introducing an additive component on top of the average distance in the OSPA. In this way, differences in cardinalities cannot go unnoticed in cases where an "unfortunate" positioning of the points occurs (see Scenarios C, D, and F).

For a given cut-off and a given value of p, the $\bar{d}_p^{(c)}$ distance does not substantially depend on the size of the ground truth pattern. This is because extra points are essentially penalized according to the cut-off c rather than according to their distance from other points in the pattern. While we recommend basing the choice of c only on the most fundamental a priori requirements of the test case, such as the size of the observation window, sensor accuracy, and possibly expected number of targets, more subtle adaptations may be justified in special situations, and the OSPA metric leaves this possibility open.

In the context of multitarget performance evaluation, the OSPA distance can be interpreted as a pth order "per-target" error.[5] This error is comprised of two components each separately accounting for "localization" and "cardinality" errors, which are given respectively by

$$\begin{aligned}\bar{e}_{p,\mathrm{loc}}^{(c)}(X, Y) &:= \left(\frac{1}{n}\min_{\pi\in\Pi_n}\sum_{i=1}^{m} d^{(c)}(\mathbf{x}_i, \mathbf{y}_{\pi(i)})^p\right)^{1/p},\\ \bar{e}_{p,\mathrm{card}}^{(c)}(X, Y) &:= \left(\frac{c^p(n-m)}{n}\right)^{1/p}\end{aligned} \tag{3.15}$$

if $m \le n$, and $\bar{e}_{p,\mathrm{loc}}^{(c)}(X, Y) := \bar{e}_{p,\mathrm{loc}}^{(c)}(Y, X)$, $\bar{e}_{p,\mathrm{card}}^{(c)}(X, Y) := \bar{e}_{p,\mathrm{card}}^{(c)}(Y, X)$ if $m > n$. These particular components can thus be interpreted as contributions due to localization only (within the OSPA) and cardinality only (penalized at maximal distance). Note, however, that the functions $\bar{e}_{p,\mathrm{loc}}^{(c)}$ and $\bar{e}_{p,\mathrm{card}}^{(c)}$ are themselves *not* metrics on the space $\mathcal{F}(\mathcal{X})$. Note that the decomposition of the OSPA metric into separate components is usually not necessary for performance evaluation, but may provide valuable additional information.

3.6 THE PROBABILITY HYPOTHESIS DENSITY (PHD) FILTER

The PHD filter is an approximation developed to alleviate the computational intractability in the multitarget Bayes filter. Instead of propagating the multitarget posterior density in time, the PHD filter propagates the posterior PHD, a first-order

[5] Strictly speaking, the "per-target" error is a "per-estimated-target" error when the cardinality is overestimated and a "per-true-target" error when the cardinality is underestimated.

statistical moment of the posterior multitarget state [1]. This strategy is reminiscent of the constant gain Kalman filter, which propagates the first moment (the mean) of the single-target state.

Recall the multitarget evolution and observation models from Section 3.4 and consider the following assumptions:

- Each target evolves and generates observations independently of one another.
- The birth RFS and the surviving RFSs are independent of each other.
- The clutter RFS is Poisson and independent of target-originated measurements.
- The prior and predicted multitarget RFS are Poisson.

The following two propositions show how the posterior PHD is propagated to the next time step (see [1] for the original FISST derivation or [48] for an alternative proof using classical probabilistic tools).

Proposition 1 *Suppose that, at time $k-1$, the posterior PHD v_{k-1} is given. Then, the predicted PHD $v_{k|k-1}$ is given by*

$$v_{k|k-1}(\mathbf{x}) = \int p_{S,k|k-1}(\boldsymbol{\zeta}) f_{k|k-1}(\mathbf{x}|\boldsymbol{\zeta}) v_{k-1}(\boldsymbol{\zeta}) d\boldsymbol{\zeta} + \gamma_k(\mathbf{x}), \tag{3.16}$$

where

$$\begin{aligned} p_{S,k|k-1}(\boldsymbol{\zeta}) &= \textit{probability of target existence at time } k \textit{ given previous state } \boldsymbol{\zeta}, \\ f_{k|k-1}(\cdot|\boldsymbol{\zeta}) &= \textit{single-target transition density at time } k \textit{ given previous state } \boldsymbol{\zeta}, \\ \gamma_k(\cdot) &= \textit{PHD of the birth RFS } \Gamma_k \textit{ at time } k. \end{aligned}$$

Proposition 2 *Suppose that, at time k, the predicted PHD $v_{k|k-1}$ and the finite set Z_k of measurements are given. Then, the updated PHD v_k is given by*

$$v_k(\mathbf{x}) = \left[1 - p_{D,k}(\mathbf{x}) + \sum_{\mathbf{z}\in Z_k} \frac{p_{D,k}(\mathbf{x}) g_k(z|\mathbf{x})}{\kappa_k(\mathbf{z}) + \int p_{D,k}(\boldsymbol{\zeta}) g_k(\mathbf{z}|\boldsymbol{\zeta}) v_{k|k-1}(\boldsymbol{\zeta}) d\boldsymbol{\zeta}}\right] v_{k|k-1}(\mathbf{x}). \tag{3.17}$$

where

$$\begin{aligned} p_{D,k}(\mathbf{x}) &= \textit{probability of detection given a state } \mathbf{x} \textit{ at time } k, \\ g_k(\cdot|\mathbf{x}) &= \textit{likelihood of a measurement given a state } \mathbf{x} \textit{ at time } k, \\ \kappa_k(\cdot) &= \textit{PHD of the clutter RFS } K_k \textit{ at time } k. \end{aligned}$$

It is clear from (3.16) and (3.17) that the PHD filter completely avoids the combinatorial computations arising from the unknown association of measurements with appropriate targets. Furthermore, since the posterior intensity is a function on the single-target state space $\mathcal{X}$, the PHD recursion requires much less computational power than the multitarget recursion (3.13) and (3.14), which operates on $\mathcal{F}(\mathcal{X})$. However, like the single-target Bayes recursion, the PHD recursion does not admit closed-form solutions in general, and numerical integration suffers from the "curse of dimensionality."

3.6.1 The PHD Recursion for Linear Gaussian Models

For a certain class of multitarget models, herein referred to as *linear Gaussian multitarget* models, the PHD recursion (3.16) and (3.17) admits a closed-form solution. The linear Gaussian multitarget model is summarized below:

- Each target follows a linear Gaussian dynamical model and the sensor has a linear Gaussian measurement model, that is

$$f_{k|k-1}(\mathbf{x}|\zeta) = \mathcal{N}(\mathbf{x}; \mathbf{F}_{k|k-1}\zeta, \mathbf{Q}_k), \tag{3.18}$$

$$g_k(\mathbf{z}|\mathbf{x}) = \mathcal{N}(\mathbf{z}; \mathbf{H}_k\mathbf{x}, \mathbf{R}_k), \tag{3.19}$$

  where $\mathcal{N}(\cdot; \mathbf{m}, \mathbf{P})$ denotes a Gaussian density with mean $\mathbf{m}$ and covariance $\mathbf{P}$, $\mathbf{F}_{k|k-1}$ is the state transition matrix, $\mathbf{Q}_k$ is the process noise covariance, $\mathbf{H}_k$ is the observation matrix, and $\mathbf{R}_k$ is the observation noise covariance.
- The survival and detection probabilities are state independent, that is

$$p_{S,k|k-1}(\mathbf{x}) = p_{S,k|k-1}, \tag{3.20}$$

$$p_{D,k}(\mathbf{x}) = p_{D,k}. \tag{3.21}$$

- The intensity of the birth RFS is a Gaussian mixture

$$\gamma_k(\mathbf{x}) = \sum_{i=1}^{J_{\gamma,k}} w_{\gamma,k}^{(i)} \mathcal{N}(\mathbf{x}; \mathbf{m}_{\gamma,k}^{(i)}, \mathbf{P}_{\gamma,k}^{(i)}), \tag{3.22}$$

  where $J_{\gamma,k}$, $w_{\gamma,k}^{(i)}$, $\mathbf{m}_{\gamma,k}^{(i)}$, $\mathbf{P}_{\gamma,k}^{(i)}$, $i = 1, \ldots, J_{\gamma,k}$, are given model parameters that determine the shape of the birth intensity.

The above assumptions are commonly used in many tracking algorithms [41, 42]. For clarity in the presentation, we only focus on state independent $p_{S,k}$ and $p_{D,k}$, although closed-form PHD recursions can be derived for more general cases (see [8]).

In (3.22), $\mathbf{m}_{\gamma,k}^{(i)}$, $i = 1, \ldots, J_{\gamma,k}$ are the means of the Gaussian components of the spontaneous birth intensity γ_k. These points have the highest local concentrations of expected number of spontaneous births, and represent, for example, in the context of air surveillance, airbases, or airports where targets are most likely to appear. The covariance matrix $\mathbf{P}_{\gamma,k}^{(i)}$ determines the spread of the birth intensity in the vicinity of the peak $\mathbf{m}_{\gamma,k}^{(i)}$. The weight $w_{\gamma,k}^{(i)}$ gives the expected number of new targets originating from $\mathbf{m}_{\gamma,k}^{(i)}$. Note that other forms of birth intensities can be approximated, to any desired accuracy, using Gaussian mixtures [49].

For the linear Gaussian multitarget model, the following two propositions [8] show how the Gaussian components of the posterior PHD are analytically propagated to the next time.

Proposition 3 *Suppose that the posterior PHD at time $k-1$ is a Gaussian mixture of the form*

$$v_{k-1}(\mathbf{x}) = \sum_{i=1}^{J_{k-1}} w_{k-1}^{(i)} \mathcal{N}(\mathbf{x}; \mathbf{m}_{k-1}^{(i)}, \mathbf{P}_{k-1}^{(i)}). \tag{3.23}$$

Then, under linear Gaussian multitarget assumptions, the predicted PHD for time k is also a Gaussian mixture, and is given by

$$v_{k|k-1}(\mathbf{x}) = \gamma_k(\mathbf{x}) + p_{S,k|k-1} \sum_{j=1}^{J_{k-1}} w_{k-1}^{(j)} \mathcal{N}(\mathbf{x}; \mathbf{m}_{S,k|k-1}^{(j)}, \mathbf{P}_{S,k|k-1}^{(j)}), \tag{3.24}$$

where $\gamma_k(\mathbf{x})$ is given in (3.22),

$$\mathbf{m}_{S,k|k-1}^{(j)} = \mathbf{F}_{k|k-1} \mathbf{m}_{k-1}^{(j)}, \tag{3.25}$$

$$\mathbf{P}_{S,k|k-1}^{(j)} = \mathbf{Q}_k + \mathbf{F}_{k|k-1} \mathbf{P}_{k-1}^{(j)} \mathbf{F}_{k|k-1}', \tag{3.26}$$

Proposition 4 *Suppose that the predicted PHD for time k is a Gaussian mixture of the form*

$$v_{k|k-1}(\mathbf{x}) = \sum_{i=1}^{J_{k|k-1}} w_{k|k-1}^{(i)} \mathcal{N}(\mathbf{x}; \mathbf{m}_{k|k-1}^{(i)}, \mathbf{P}_{k|k-1}^{(i)}). \tag{3.27}$$

Then, under linear Gaussian multitarget assumptions, the posterior PHD at time k is also a Gaussian mixture, and is given by

$$v_k(\mathbf{x}) = (1 - p_{D,k}) v_{k|k-1}(\mathbf{x}) + p_{D,k} \sum_{\mathbf{z} \in Z_k} \sum_{j=1}^{J_{k|k-1}} \frac{w_{k|k-1}^{(j)} q_k^{(j)}(\mathbf{z}) \mathcal{N}(\mathbf{x}; \mathbf{m}_{k|k}^{(j)}(\mathbf{z}), \mathbf{P}_{k|k}^{(j)})}{\kappa_k(\mathbf{z}) + p_{D,k} \sum_{\ell=1}^{J_{k|k-1}} w_{k|k-1}^{(\ell)} q_k^{(\ell)}(\mathbf{z})}, \tag{3.28}$$

where

$$q_k^{(j)}(\mathbf{z}) = \mathcal{N}(\mathbf{z}; \boldsymbol{\eta}_{k|k-1}^{(j)}, \mathbf{S}_{k|k-1}^{(j)}), \tag{3.29}$$

$$\boldsymbol{\eta}_{k|k-1}^{(j)} = \mathbf{H}_k \mathbf{m}_{k|k-1}^{(j)}, \tag{3.30}$$

$$\mathbf{S}_{k|k-1}^{(j)} = \mathbf{H}_k \mathbf{P}_{k|k-1}^{(j)} \mathbf{H}_k' + \mathbf{R}_k, \tag{3.31}$$

$$\mathbf{m}_{k|k}^{(j)}(\mathbf{z}) = \mathbf{m}_{k|k-1}^{(j)} + \mathbf{K}_k^{(j)}(\mathbf{z} - \eta_{k|k-1}^{(j)}), \tag{3.32}$$

$$\mathbf{P}_{k|k}^{(j)} = \mathbf{P}_{k|k-1}^{(j)} - \mathbf{P}_{k|k-1}^{(j)}\mathbf{H}_k'[\mathbf{S}_{k|k-1}^{(j)}]^{-1}\mathbf{H}_k\mathbf{P}_{k|k-1}^{(j)} \tag{3.33}$$

$$\mathbf{K}_k^{(j)} = \mathbf{P}_{k|k-1}^{(j)}\mathbf{H}_k'[\mathbf{S}_{k|k-1}^{(j)}]^{-1}. \tag{3.34}$$

It follows by induction from Propositions 3 and 4 that if the initial prior PHD v_0 is a Gaussian mixture (including the case where $v_0 = 0$), then all subsequent predicted PHDs $v_{k|k-1}$ and posterior PHDs v_k are also Gaussian mixtures. Proposition 3 provides closed-form expressions for computing the means, covariances, and weights of $v_{k|k-1}$ from those of v_{k-1}. Proposition 4 then provides closed-form expressions for computing the means, covariances, and weights of v_k from those of $v_{k|k-1}$ when a new set of measurements arrives. Propositions 3 and 4 are, respectively, the prediction and update steps of the PHD recursion for a linear Gaussian multitarget model, herein referred to as the *Gaussian mixture PHD recursion*.

The predicted PHD $v_{k|k-1}$ in Proposition 3 consists of two terms due, respectively, to the spontaneous births and existing targets. Similarly, the updated posterior PHD v_k in Proposition 4 consists of a mis-detection term, and $|Z_k|$ detection terms, one for each measurement $\mathbf{z} \in Z_k$.

Given the Gaussian mixture intensities $v_{k|k-1}$ and v_k, the corresponding expected number of targets $\hat{N}_{k|k-1}$ and $\hat{N}_k$ can be obtained by summing up the appropriate weights. Propositions 3 and 4 lead to the following closed-form recursions for $\hat{N}_{k|k-1}$ and $\hat{N}_k$:

Corollary 1 *Under the premises of Proposition 1, the mean of the predicted number of targets is*

$$\hat{N}_{k|k-1} = \hat{N}_{k-1}p_{S,k|k-1} + \sum_{j=1}^{J_{\gamma,k}} w_{\gamma,k}^{(j)}, \tag{3.35}$$

Corollary 2 *Under the premises of Proposition 2, the mean of the updated number of targets is*

$$\hat{N}_k = \hat{N}_{k|k-1}(1 - p_{D,k}) + \sum_{z\in Z_k}\sum_{j=1}^{J_{k|k-1}} \frac{p_{D,k}w_{k|k-1}^{(j)}q_k^{(j)}(\mathbf{z})}{\kappa_k(\mathbf{z}) + p_{D,k}\sum_{\ell=1}^{J_{k|k-1}} w_{k|k-1}^{(\ell)}q_k^{(\ell)}(\mathbf{z})} \tag{3.36}$$

In Corollary 1, the mean of the predicted number of targets is obtained by adding the mean number of surviving targets and the mean number of births. A similar interpretation can be drawn from Corollary 2. When there is no clutter, the mean of the updated number of targets is the number of measurements plus the mean number of targets that are not detected.

3.6.2 Implementation Issues

Managing mixture components. The Gaussian mixture PHD filter is similar to the Gaussian sum filter of [50] in the sense that they both propagate Gaussian mixtures in time. Like the Gaussian sum filter, the Gaussian mixture PHD filter also suffers from computation problems associated with the increasing number of Gaussian components as time progresses. Indeed, at time k, the Gaussian mixture PHD filter requires

$$(J_{k-1} + J_{\gamma,k})(1 + |Z_k|) = \mathcal{O}(J_{k-1}|Z_k|).$$

Gaussian components to represent v_k, where J_{k-1} is number of components of v_{k-1}. This implies the number of components in the posterior intensities increases without bound.

A simple pruning procedure can be used to reduce the number of Gaussian components propagated to the next time step. A good approximation to the Gaussian mixture posterior intensity

$$v_k(\mathbf{x}) = \sum_{i=1}^{J_k} w_k^{(i)} \mathcal{N}(\mathbf{x}; \mathbf{m}_k^{(i)}, \mathbf{P}_k^{(i)})$$

can be obtained by truncating components that have weak weights $w_k^{(i)}$. This can be done by discarding those with weights below some preset threshold, or by keeping only a certain number of components with strongest weights. Let us assume, without loss of generality, that the components with indices $i = 1, \ldots, N_P$ are those with weights, $w_k^{(i)}$, less than some specified threshold δ_1, which results in replacing intensity v_k by v_k^P, where the weights have been renormalized. Then the following error bound can be established (from Sorenson and Alspach [50])

$$\|v_k - v_k^P\|_1 \leq 2 \sum_{i=1}^{N_P} w_k^{(i)} \leq 2N_P\delta_1.$$

This shows that the L_1 error can be selected to fall within specified bounds for the pruning stage of the algorithm.

The second technique used to reduce the number of components in the mixture is to merge the Gaussians with similar means and covariances. One such approach, proposed by Salmond [51], called the Clustering algorithm is used to merge different clusters of Gaussians in a density function. Descending from the components withthe highest weights, clusters of Gaussians are selected for merging if they fall withinthe following distance determined from the covariance $\mathbf{P}_k^i$ of the ith component

$$\frac{w_k^{(i)} w_k^{(j)}}{w_k^{(i)} + w_k^{(j)}} (\mathbf{m}_k^{(i)} - \mathbf{m}_k^{(j)})' (\mathbf{P}_k^{(i)})^{-1} (\mathbf{m}_k^{(i)} - \mathbf{m}_k^{(j)}).$$

The components are merged with

$$\tilde{w}_k^{(\ell)} = \sum_{i \in L} w_k^{(i)}, \tag{3.37}$$

$$\tilde{\mathbf{x}}_k^{(\ell)} = \frac{1}{\tilde{w}_k^{(\ell)}} \sum_{i \in L} w_k^{(i)} \mathbf{x}_k^{(i)}, \tag{3.38}$$

$$\tilde{\mathbf{P}}_k^{(\ell)} = \frac{1}{\tilde{w}_k^{(\ell)}} \sum_{i \in L} w_k^{(i)} (\mathbf{P}_k^{(i)} + (\tilde{\mathbf{m}}_k^{(\ell)} - \mathbf{m}_k^{(i)})(\tilde{\mathbf{m}}_k^{(\ell)} - \mathbf{m}_k^{(i)})'). \tag{3.39}$$

Multitarget state estimation. In the Gaussian mixture representation of the posterior intensity v_k, extraction of multitarget state estimates is straightforward since the means of the constituent Gaussian components are indeed the local maxima of v_k, provided that they are reasonably well separated. Note that after pruning closely spaced Gaussian components would have been merged. Since the height of each peak depends on both the weight and covariance, selecting the $\hat{N}_k$ highest peaks of v_k may result in state estimates that correspond to Gaussians with weak weights. This is not desirable because the expected number of targets due to these peaks is small, even though the magnitudes of the peaks are large. A better alternative is to select the means of the Gaussians that have weights greater than some threshold for example 0.5.

3.6.3 Extension to Nonlinear Gaussian Models

This subsection considers extensions of the Gaussian mixture PHD filter to nonlinear target models. Specifically, the state and observation processes can be relaxed to the nonlinear model

$$\mathbf{x}_k = \boldsymbol{\varphi}_{k|k-1}(\mathbf{x}_{k-1}, \boldsymbol{\nu}_k),$$
$$\mathbf{z}_k = \mathbf{h}_k(\mathbf{x}_k, \boldsymbol{\varepsilon}_k),$$

where $\boldsymbol{\varphi}_{k|k-1}$ and $\mathbf{h}_k$ are known nonlinear functions, $\boldsymbol{\nu}_k$ and $\boldsymbol{\varepsilon}_k$ are zero-mean Gaussian process noise and measurement noise with covariances $\mathbf{Q}_k$ and $\mathbf{R}_k$, respectively.

Due to the nonlinearities of $\boldsymbol{\varphi}_{k|k-1}$ and $\mathbf{h}_k$, the posterior intensity can no longer be represented as a Gaussian mixture. Nonetheless, the Gaussian mixture PHD filter can be adapted to accommodate nonlinear Gaussian models. Analogous to the extended Kalman filter (EKF) [52], a nonlinear approximation to the Gaussian mixture PHD recursion is proposed based on applying local linearizations of $\boldsymbol{\varphi}_{k|k-1}$ and $\mathbf{h}_k$ as follows.

Linearization. In Proposition 3, the prediction step can be made to approximate nonlinear target motions by predicting the mixture components of surviving targets using first-order approximations wherever nonlinearities are encountered, that is by using the approximations (3.40) and (3.41) in place of the originals (3.25) and (3.26)

$$\mathbf{m}_{k|k-1}^{(j)} = \varphi_{k|k-1}(\mathbf{m}_{k-1}^{(j)}, \mathbf{0}), \tag{3.40}$$

$$\mathbf{P}_{k|k-1}^{(j)} = \mathbf{G}_k^{(j)}\mathbf{Q}_k[\mathbf{G}_k^{(j)}]' + \mathbf{F}_{k|k-1}^{(j)}\mathbf{P}_{k-1}^{(j)}[\mathbf{F}_{k|k-1}^{(j)}]', \tag{3.41}$$

where

$$\mathbf{F}_{k|k-1}^{(j)} = \left.\frac{\partial \varphi_{k|k-1}(\mathbf{x}, \mathbf{0})}{\partial \mathbf{x}}\right|_{\mathbf{x}=\mathbf{m}_{k-1}^{(j)}}, \qquad \mathbf{G}_k^{(j)} = \left.\frac{\partial \varphi_{k|k-1}(\mathbf{m}_{k-1}^{(j)}, \boldsymbol{\nu})}{\partial \boldsymbol{\nu}}\right|_{\boldsymbol{\nu}=\mathbf{0}}. \tag{3.42}$$

In Proposition 4, the update step can be made to approximate nonlinear measurement models by updating each of the predicted mixture components using first-order approximations wherever nonlinearities are encountered, that is by using the approximations (3.43) and (3.44) in place of the originals (3.30) and (3.31), and using the linearizations in (3.45) for the calculation of (3.33) and (3.34)

$$\eta_{k|k-1}^{(j)} = \mathbf{h}_k(\mathbf{m}_{k|k-1}^{(j)}, \mathbf{0}), \tag{3.43}$$

$$\mathbf{S}_k^{(j)} = \mathbf{U}_k^{(j)}\mathbf{R}_k[\mathbf{U}_k^{(j)}]' + \mathbf{H}_k^{(j)}\mathbf{P}_{k|k-1}^{(j)}[\mathbf{H}_k^{(j)}]', \tag{3.44}$$

where

$$\mathbf{H}_k^{(j)} = \left.\frac{\partial \mathbf{h}_k(\mathbf{x}, \mathbf{0})}{\partial \mathbf{x}}\right|_{\mathbf{x}=\mathbf{m}_{k|k-1}^{(j)}}, \quad \mathbf{U}_k^{(j)} = \left.\frac{\partial \mathbf{h}_k(\mathbf{m}_{k|k-1}^{(j)}, \boldsymbol{\varepsilon})}{\partial \boldsymbol{\varepsilon}}\right|_{\boldsymbol{\varepsilon}=\mathbf{0}}. \tag{3.45}$$

Unscented Transform. Analogous to the unscented Kalman filter (UKF) [53], a nonlinear approximation to the Gaussian mixture CPHD recursion is proposed based on the unscented transform (UT). The strategy here is to use the UT to propagate the first and second moments of each mixture component through the nonlinear transformations φ_k and $\mathbf{h}_k$ as follows.

To begin, for each mixture component of the posterior intensity, using the UT with mean $\boldsymbol{\mu}_k^{(j)}$ and covariance $\mathbf{C}_k^{(j)}$, generate a set of sigma points $\{\mathbf{y}_k^{(\ell)}\}_{\ell=0}^{L}$ and weights $\{\mathbf{u}^{(\ell)}\}_{\ell=0}^{L}$ where

$$\boldsymbol{\mu}_k^{(j)} = \left[\mathbf{m}_{k-1}^{(j)\prime}\ \mathbf{0}'\ \mathbf{0}'\right]',$$

$$\mathbf{C}_k^{(j)} = \mathrm{diag}(\mathbf{P}_{k-1}^{(j)}, \mathbf{Q}_{k-1}, \mathbf{R}_k).$$

Then, partition the sigma points into

$$\mathbf{y}_k^{(\ell)} = [(\mathbf{x}_{k-1}^{(\ell)})', (\boldsymbol{\nu}_{k-1}^{(\ell)})', (\boldsymbol{\epsilon}_k^{(\ell)})']'$$

for $\ell = 0, \ldots, L$ and proceed as follows.

For the prediction, the sigma points are propagated through the transition function according to $\mathbf{x}_{k|k-1}^{(\ell)} = \boldsymbol{\varphi}_{k|k-1}(\mathbf{x}_{k-1}^{(\ell)}, \boldsymbol{\nu}_k^{(\ell)})$ for $\ell = 0, \ldots, L$. Then, in Proposition 3, the prediction step can be made to approximate nonlinear target motions by using the approximations (3.46) and (3.47) in place of the originals (3.25) and (3.26)

$$\mathbf{m}_{k|k-1}^{(j)} = \sum_{\ell=0}^{L} \mathbf{u}^{(\ell)} \mathbf{x}_{k|k-1}^{(\ell)}, \tag{3.46}$$

$$\mathbf{P}_{k|k-1}^{(j)} = \sum_{\ell=0}^{L} \mathbf{u}^{(\ell)} (\mathbf{x}_{k|k-1}^{(\ell)} - \mathbf{m}_{k|k-1}^{(j)})(\mathbf{x}_{k|k-1}^{(\ell)} - \mathbf{m}_{k|k-1}^{(j)})'. \tag{3.47}$$

For the update, the sigma points are propagated through the measurement function according to $\mathbf{z}_{k|k-1}^{(\ell)} = \mathbf{h}_k(\mathbf{x}_{k|k-1}^{(\ell)}, \boldsymbol{\epsilon}_k^{(\ell)})$ for $\ell = 0, \ldots, L$. Then, in Proposition 4, the update step can be made to approximate nonlinear measurement models by using the approximations (3.48) and (3.49) in place of the originals (3.30) and (3.31), and using (3.50) and (3.51) in place of the originals (3.33) and (3.34)

$$\boldsymbol{\eta}_{k|k-1}^{(i)} = \sum_{\ell=0}^{L} \mathbf{u}^{(\ell)} \mathbf{z}_{k|k-1}^{(\ell)}, \tag{3.48}$$

$$\mathbf{S}_k^{(i)} = \sum_{\ell=0}^{L} \mathbf{u}^{(\ell)} (\mathbf{z}_{k|k-1}^{(\ell)} - \boldsymbol{\eta}_{k|k-1}^{(i)})(\mathbf{z}_{k|k-1}^{(\ell)} - \boldsymbol{\eta}_{k|k-1}^{(i)})', \tag{3.49}$$

$$\mathbf{P}_k^{(i)} = \mathbf{P}_{k|k-1}^{(i)} - \mathbf{G}_k^{(i)} [\mathbf{S}_k^{(i)}]^{-1} [\mathbf{G}_k^{(i)}]', \tag{3.50}$$

$$\mathbf{K}_k^{(i)} = \mathbf{G}_k^{(i)} [\mathbf{S}_k^{(i)}]^{-1}, \tag{3.51}$$

$$\mathbf{G}_k^{(i)} = \sum_{\ell=0}^{L} \mathbf{u}^{(\ell)} (\mathbf{x}_{k|k-1}^{(\ell)} - \mathbf{m}_{k|k-1}^{(i)})(\mathbf{z}_{k|k-1}^{(\ell)} - \mathbf{m}_{k|k-1}^{(i)})'. \tag{3.52}$$

Similar to its single-target counterpart, the EK-PHD filter is only applicable to differentiable nonlinear models. Moreover, calculating the Jacobian matrices may be tedious and error prone. The UK-PHD filter, on the other hand, does not suffer from these restrictions and can even be applied to models with discontinuities.

3.6.3.1 Track Continuity with the Gaussian Mixture PHD Filter. The Gaussian mixture PHD filter can be used for multiobject tracking by introducing a unique identifying label for each Gaussian component. We can then describe the Gaussian mixture PHD and the labels, at time k, as set of tuples containing the weights, means, covariances, and labels, that is

$$\tilde{v}_k = \bigcup_{i=1}^{J_k} \left\{ \left(w_k^{(i)}, \mathbf{m}_k^{(i)}, \mathbf{P}_k^{(i)}, l_k^{(i)} \right) \right\},$$

where $l_k^{(i)}$ is the track label associated with the ith component.

Prediction. Suppose that the posterior PHD at time $k-1$ is given by the mixture components and labels

$$\tilde{v}_{k-1} = \bigcup_{i=1}^{J_{k-1}} \left\{ \left(w_{k-1}^{(i)}, \mathbf{m}_{k-1}^{(i)}, \mathbf{P}_{k-1}^{(i)}, l_{k-1}^{(i)} \right) \right\}.$$

Then the predicted PHD consists of the Gaussian components surviving from the previous time step, those spawned from targets at the previous time step and those from spontaneously appearing at the current time step, which can be described by the union

$$\begin{aligned} \tilde{v}_{k|k-1} = & \left(\bigcup_{j=1}^{J_{k-1}} \left\{ \left(p_{S,k|k-1} w_{k-1}^{(j)}, \mathbf{m}_{S,k|k-1}^{(j)}, \mathbf{P}_{S,k|k-1}^{(j)}, l_{k-1}^{(j)} \right) \right\} \right) \\ & \cup \left(\bigcup_{i=1}^{J_{\gamma,k}} \left\{ \left(w_{\gamma,k}^{(i)}, \mathbf{m}_{\gamma,k}^{(i)}, \mathbf{P}_{\gamma,k}^{(i)}, l_{\gamma,k}^{(i)} \right) \right\} \right). \end{aligned}$$

Each component surviving from the previous time step maintains the same component label $l_{k-1}^{(j)}$, each new birth component is assigned a new label $l_{\gamma,k}^{(i)}$, and each spawned component has a newly introduced label $l_{\beta,k|k-1}^{(j,\ell)}$ which identifies the target which spawned it.

Update. We now suppose that the predicted PHD $v_{k|k-1}$ for time k is given by the mixture components and labels $l_{k|k-1}$

$$\bigcup_{i=1}^{J_{k|k-1}} \left\{ \left(w_{k|k-1}^{(i)}, \mathbf{m}_{k|k-1}^{(i)}, \mathbf{P}_{k|k-1}^{(i)}, l_{k|k-1}^{(i)} \right) \right\}.$$

Then the posterior components and labels can be described by the union

$$\tilde{v}_k = \bigcup_{z \in Z_k} \tilde{v}_{z,k} \cup \tilde{v}_{M,k},$$

where

$$\begin{aligned} \tilde{v}_{z,k} &= \bigcup_{j=1}^{J_{k|k-1}} \left\{ \left(w_k^{(j)}(\mathbf{z}), \mathbf{m}_{k|k}^{(j)}(\mathbf{z}), \mathbf{P}_{k|k}^{(j)}, l_{k|k}^{(j)}(\mathbf{z}) \right) \right\} \\ \tilde{v}_{M,k} &= \bigcup_{i=1}^{J_{k|k-1}} \left\{ \left((1 - p_{D,k}) w_{k|k-1}^{(i)}, \mathbf{m}_{k|k-1}^{(i)}, \mathbf{P}_{k|k-1}^{(i)}, l_{k|k-1}^{(i)} \right) \right\}. \end{aligned}$$

For each predicted component there are $1 + |Z_k|$ updated components: one from the missed detection term, $l^{(i)}_{k|k-1}$ and $|Z_k|$ for the terms updated with measurements, $l^{(j)}_{k|k}(\mathbf{z})$. In order to maintain track continuity, we need to identify which component should have the same label as in the prediction step. In most scenarios, the term with the highest PHD updated weight $w^{(j)}_k(\mathbf{z})$ is most likely to be the correct target, so we can identify this with the same label and assign new labels to the other terms. When components are merged, the label of the principal component is kept and the others discarded. More complex strategies for maintaining track continuity in more challenging scenarios are discussed in [35].

3.7 THE CARDINALIZED PHD FILTER

The primary weakness of the PHD recursion is a loss of higher order cardinality information. Since the PHD recursion is a first-order approximation, it propagates cardinality information with only a single parameter and effectively approximates the cardinality distribution by a Poisson distribution with matching mean. Since the mean and variance of a Poisson distribution are equal, when the number of targets present is high, the PHD filter estimates the cardinality with a correspondingly high variance. Additionally, the mean number of targets is effectively an EAP estimator, which can be erratic because of minor modes induced by clutter in low signal-to-noise ratio (SNR) conditions.

The CPHD recursion was proposed by Mahler in [3, 39] to address the limitations of the PHD recursion. In essence, the strategy behind the CPHD recursion is to jointly propagate the intensity function and the cardinality distribution (the probability distribution of the number of targets). An interesting interpretation of the CPHD recursion was given in [40].

The CPHD recursion rests on the following assumptions regarding the target dynamics and observations:

- Each target evolves and generates measurements independently of one another.
- The birth RFS and the surviving RFSs are independent of each other.
- The clutter RFS is an i.i.d cluster process and independent of the measurement RFSs.
- The prior and predicted multitarget RFSs are i.i.d cluster processes.

The above assumptions are similar to those in the PHD recursion, except that in this case, i.i.d cluster RFS replaces the Poisson RFS. The following propositions show how the posterior PHD and cardinality distribution can be propagated to the next time.

We also need the following notations to present the CPHD filter. Denote by C^{ℓ}_j the *binomial coefficient* $\frac{\ell!}{j!(\ell-j)!}$, P^n_j the *permutation coefficient* $\frac{n!}{(n-j)!}$, $\langle\cdot,\cdot\rangle$ the *inner product* defined between two real valued functions α and β by

$$\langle \alpha, \beta \rangle = \int \alpha(\mathbf{x})\beta(\mathbf{x})d\mathbf{x},$$

(or $\sum_{\ell=0}^{\infty} \alpha(\ell)\beta(\ell)$ when α and β are real sequences), and $e_j(\cdot)$ the *elementary symmetric function* [54] of order j defined for a finite set Z of real numbers by

$$e_j(Z) = \sum_{S \subseteq Z, |S|=j} \left(\prod_{\zeta \in S} \zeta \right),$$

with $e_0(Z) = 1$ by convention.

Proposition 5 *Suppose that, at time $k-1$, the posterior PHD v_{k-1} and posterior cardinality distribution p_{k-1} are given. Then, the predicted cardinality distribution $p_{k|k-1}$ and predicted PHD $v_{k|k-1}$ are given by [3, 10]*

$$p_{k|k-1}(n) = \sum_{j=0}^{n} p_{\Gamma,k}(n-j)\Pi_{k|k-1}[v_{k-1}, p_{k-1}](j), \tag{3.53}$$

$$v_{k|k-1}(\mathbf{x}) = \int p_{S,k|k-1}(\boldsymbol{\zeta}) f_{k|k-1}(\mathbf{x}|\boldsymbol{\zeta}) v_{k-1}(\boldsymbol{\zeta}) d\boldsymbol{\zeta} + \gamma_k(\mathbf{x}), \tag{3.54}$$

where

$$\Pi_{k|k-1}[v, p](j) = \sum_{\ell=j}^{\infty} C_j^{\ell} \frac{\langle p_{S,k|k-1}, v\rangle^j \langle 1 - p_{S,k|k-1}, v\rangle^{\ell-j}}{\langle 1, v\rangle^{\ell}} p(\ell), \tag{3.55}$$

$f_{k|k-1}(\cdot|\boldsymbol{\zeta})$ = *single-target transition density at time k given previous state $\boldsymbol{\zeta}$,*

$p_{S,k|k-1}(\boldsymbol{\zeta})$ = *probability of target existence at time k given previous state $\boldsymbol{\zeta}$,*

$\gamma_k(\cdot)$ = *PHD of spontaneous births at time k,*

$p_{\Gamma,k}(\cdot)$ = *cardinality distribution of births at time k.*

Proposition 6 *Suppose that, at time k, the predicted PHD $v_{k|k-1}$, predicted cardinality distribution $p_{k|k-1}$ and the finite set Z_k of measurements are given. Then, the updated cardinality distribution p_k and updated PHD v_k are given by [3, 10]*

$$p_k(n) = \frac{\Upsilon_k^0[v_{k|k-1}, Z_k](n) p_{k|k-1}(n)}{\langle \Upsilon_k^0[v_{k|k-1}, Z_k], p_{k|k-1}\rangle}, \tag{3.56}$$

$$\begin{aligned} v_k(\mathbf{x}) &= [1 - p_{D,k}(\mathbf{x})] \frac{\langle \Upsilon_k^1[v_{k|k-1}, Z_k], p_{k|k-1}\rangle}{\langle \Upsilon_k^0[v_{k|k-1}, Z_k], p_{k|k-1}\rangle} v_{k|k-1}(\mathbf{x}) \\ &+ p_{D,k}(\mathbf{x}) \sum_{\mathbf{z} \in Z_k} \frac{\langle \Upsilon_k^1[v_{k|k-1}, Z_k \backslash \{\mathbf{z}\}], p_{k|k-1}\rangle}{\langle \Upsilon_k^0[v_{k|k-1}, Z_k], p_{k|k-1}\rangle} \frac{g_k(\mathbf{z}|\mathbf{x})}{\kappa_k(\mathbf{z})/\langle 1, \kappa_k\rangle} v_{k|k-1}(\mathbf{x}), \end{aligned} \tag{3.57}$$

where

$$\Upsilon_k^u[v, Z](n) = \sum_{j=0}^{\min(|Z|,n)} (|Z|-j)!\,p_{K,k}(|Z|-j)P_{j+u}^n \frac{\langle 1-p_{D,k}, v\rangle^{n-(j+u)}}{\langle 1, v\rangle^n} e_j(\Xi_k(v,Z)), \tag{3.58}$$

$$\Xi_k(v, Z) = \left\{ \frac{\langle p_{D,k} g_k(\mathbf{z}|\cdot), v\rangle}{\kappa_k(\mathbf{z})/\langle 1, \kappa_k\rangle} : \mathbf{z} \in Z \right\},$$

$g_k(\cdot|\mathbf{x})$ = *single-target measurement likelihood at time k given current state* $\mathbf{x}$,

$p_{D,k}(\mathbf{x})$ = *probability of target detection at time k given current state* $\mathbf{x}$,

$\kappa_k(\cdot)$ = *PHD of clutter measurements at time k*,

$p_{K,k}(\cdot)$ = *cardinality distribution of clutter at time k*.

The CPHD cardinality prediction (3.53) is simply a convolution of the cardinality distributions of the birth and surviving targets. This is because the predicted cardinality is the sum of the cardinalities of the birth and surviving targets. The CPHD intensity prediction (3.54) is the same as the PHD prediction (3.16). Note that the cardinality and PHD in (3.53) and (3.54) are uncoupled, while the cardinality and PHD in (3.56) and (3.57) are coupled. Nonetheless, the CPHD update (3.57) is similar to the PHD update (3.17) in the sense that both have one missed detection term and $|Z_k|$ detection terms. The cardinality update (3.56) incorporates the clutter cardinality, the measurement set, the predicted intensity, and predicted cardinality distribution. Indeed (3.56) is a Bayes update, with $\Upsilon_k^0[v_{k|k-1}; Z_k](n)$ being the likelihood of the multitarget observation Z_k given that there are n targets, and $\langle \Upsilon_k^0[v_{k|k-1}; Z_k], p_{k|k-1}\rangle$ as the normalizing constant.

3.7.1 The CPHD Recursion for Linear Gaussian Models

For the linear Gaussian multitarget model, the following two propositions [10] present a closed-form solution to the CPHD recursion (3.53) and (3.54) and (3.56) and (3.57). More concisely, these propositions show how the posterior intensity (in the form of its Gaussian components) and the posterior cardinality distribution are analytically propagated in time.

Proposition 7 *Suppose at time* $k-1$ *that the posterior intensity* v_{k-1} *and posterior cardinality distribution* p_{k-1} *are given, and that* v_{k-1} *is a Gaussian mixture of the form (3.23). Then, under linear Gaussian multitarget assumptions,* $v_{k|k-1}$ *is also a Gaussian mixture, and*

$$p_{k|k-1}(n) = \sum_{j=0}^{n} p_{\Gamma,k}(n-j)\sum_{\ell=j}^{\infty} C_j^\ell p_{k-1}(\ell) p_{S,k|k-1}^j \left(1-p_{S,k}\right)^{\ell-j}, \tag{3.59}$$

$$v_{k|k-1}(\mathbf{x}) = \gamma_k(\mathbf{x}) + p_{S,k|k-1} \sum_{j=1}^{J_{k-1}} w_{k-1}^{(j)} \mathcal{N}(\mathbf{x}; \mathbf{m}_{S,k|k-1}^{(j)}, \mathbf{P}_{S,k|k-1}^{(j)}), \tag{3.60}$$

where $\gamma_k(\mathbf{x})$ *is given in (3.22),*

$$\mathbf{m}_{S,k|k-1}^{(j)} = \mathbf{F}_{k|k-1}\mathbf{m}_{k-1}^{(j)}, \tag{3.61}$$

$$\mathbf{P}_{S,k|k-1}^{(j)} = \mathbf{Q}_k + \mathbf{F}_{k|k-1}\mathbf{P}_{k-1}^{(j)}\mathbf{F}_{k|k-1}'. \tag{3.62}$$

Proposition 8 *Suppose at time k that the predicted intensity* $v_{k|k-1}$ *and predicted cardinality distribution* $p_{k|k-1}$ *are given, and that* $v_{k|k-1}$ *is a Gaussian mixture of the form (3.27). Then, under linear Gaussian multitarget assumptions,* v_k *is also a Gaussian mixture, and*

$$p_k(n) = \frac{\Psi_k^0[\mathbf{w}_{k|k-1}, Z_k](n)p_{k|k-1}(n)}{\langle \Psi_k^0[\mathbf{w}_{k|k-1}, Z_k], p_{k|k-1}\rangle}, \tag{3.63}$$

$$v_k(\mathbf{x}) = (1 - p_{D,k})\frac{\langle \Psi_k^1[\mathbf{w}_{k|k-1}, Z_k], p_{k|k-1}\rangle}{\langle \Psi_k^0[\mathbf{w}_{k|k-1}, Z_k], p_{k|k-1}\rangle}v_{k|k-1}(\mathbf{x})$$
$$+ p_{D,k}\sum_{z\in Z_k}\sum_{j=1}^{J_{k|k-1}} w_{k|k-1}^{(j)}\frac{\langle \Psi_k^1[\mathbf{w}_{k|k-1}, Z_k\backslash\{\mathbf{z}\}], p_{k|k-1}\rangle}{\langle \Psi_k^0[\mathbf{w}_{k|k-1}, Z_k], p_{k|k-1}\rangle}\frac{q_k^{(j)}(\mathbf{z})\mathcal{N}(x; \mathbf{m}_k^{(j)}(\mathbf{z}), \mathbf{P}_k^{(j)})}{\kappa_k(\mathbf{z})/\langle 1, \kappa_k\rangle}, \tag{3.64}$$

where

$$\Psi_k^u[\mathbf{w}, Z](n) = \sum_{j=0}^{\min(|Z|,n)}(|Z|-j)!p_{K,k}(|Z|-j)P_{j+u}^n\frac{\left(1-p_{D,k}\right)^{n-(j+u)}}{(\mathbf{1}'\mathbf{w})^{j+u}}e_j(\Lambda_k(\mathbf{w}, Z)), \tag{3.65}$$

$$\Lambda_k(\mathbf{w}, Z) = \left\{\frac{p_{D,k}\mathbf{w}'\mathbf{q}_k(\mathbf{z})}{\kappa_k(\mathbf{z})/\langle 1, \kappa_k\rangle} : \mathbf{z} \in Z\right\}, \tag{3.66}$$

$$\mathbf{w}_{k|k-1} = [w_{k|k-1}^{(1)}, \ldots, w_{k|k-1}^{(J_{k|k-1})}]', \tag{3.67}$$

$$\mathbf{q}_k(\mathbf{z}) = [q_k^{(1)}(\mathbf{z}), \ldots, q_k^{(J_{k|k-1})}(\mathbf{z})]', \tag{3.68}$$

$$q_k^{(j)}(\mathbf{z}) = \mathcal{N}(\mathbf{z}; \boldsymbol{\eta}_{k|k-1}^{(j)}, \mathbf{S}_{k|k-1}^{(j)}), \tag{3.69}$$

$$\boldsymbol{\eta}_{k|k-1}^{(j)} = \mathbf{H}_k\mathbf{m}_{k|k-1}^{(j)}, \tag{3.70}$$

$$\mathbf{S}_{k|k-1}^{(j)} = \mathbf{H}_k\mathbf{P}_{k|k-1}^{(j)}\mathbf{H}_k' + \mathbf{R}_k, \tag{3.71}$$

$$\mathbf{m}_k^{(j)}(\mathbf{z}) = \mathbf{m}_{k|k-1}^{(j)} + \mathbf{K}_k^{(j)}(\mathbf{z} - \boldsymbol{\eta}_{k|k-1}^{(j)}), \tag{3.72}$$

$$\mathbf{P}_k^{(j)} = [\mathbf{I} - \mathbf{K}_k^{(j)}\mathbf{H}_k]\mathbf{P}_{k|k-1}^{(j)}, \tag{3.73}$$

$$\mathbf{K}_k^{(j)} = \mathbf{P}_{k|k-1}^{(j)}\mathbf{H}_k'\left[\mathbf{S}_{k|k-1}^{(j)}\right]^{-1}. \tag{3.74}$$

It was shown in [3, 39] that the PHD recursion is a special case of the CPHD recursion. Using a similar argument, it can be shown that the Gaussian mixture PHD recursion [8] is a special case of the recursions given by Propositions 7 and 8.

Like the PHD filter, it follows by induction from Propositions 7 and 8 that if the initial intensity v_0 is a Gaussian mixture (including the case where $v_0 = 0$), then all subsequent predicted intensities $v_{k|k-1}$ and posterior intensities v_k are also Gaussian mixtures. Proposition 7 provides closed-form expressions for computing the means, covariances, and weights of $v_{k|k-1}$ from those of v_{k-1}, and also for computing the distribution $p_{k|k-1}$ from p_{k-1}. Proposition 8 then provides closed-form expressions for computing the means, covariances, and weights of v_k from those of $v_{k|k-1}$, and also for computing the distribution p_k from $p_{k|k-1}$, when a new set of measurements arrives. Propositions 7 and 8 are, respectively, the prediction and update steps of the CPHD recursion for linear Gaussian multitarget models.

3.7.2 Implementation Issues

Using the Gaussian mixture solutions above, the implementation of the Gaussian mixture CPHD filter [10] is similar to Gaussian mixture PHD filter [8]. The "pruning" and "merging" procedure described in [8], nonlinear extension via linearization and UT, as well as track continuity strategies for the PHD filter are directly applicable to the CPHD filter. A comprehensive description of implementation details for the Gaussian mixture CPHD filter is omitted since this would involve considerable repetition from the Gaussian mixture PHD filter. It is more instructive to discuss the additional steps in the implementation.

Computing Cardinality Distributions. Propagating the cardinality distribution essentially involves using (3.59) and (3.63) to recursively predict and update the weights of the distribution. However, if the cardinality distribution is infinite tailed, propagation of the entire posterior cardinality is generally not possible since this would involve propagating an infinite number of terms. In practice, if the cardinality distributions are short or moderate tailed, they can be truncated at $n = N_{\max}$ and approximated with a finite number of terms $\{p_k(n)\}_{n=0}^{N_{\max}}$. Such an approximation is reasonable when $N_{\max}$ is significantly greater than the number of targets on the scene at any time.

Computing Elementary Symmetric Functions. Evaluating the elementary symmetric functions directly from the definition is clearly intractable. Using a basic result from combinatorics theory known as the Newton–Girard formulae or equivalently Vieta's Theorem, the elementary symmetric function $e_j(\cdot)$ can be computed using the following procedure [54]. Let $\rho_1, \rho_2, \ldots, \rho_M$ be distinct roots of the polynomial

$$\alpha_M x^M + \alpha_{m-1} x^{M-1} + \cdots + \alpha_1 x + \alpha_0.$$

Then, $e_j(\cdot)$ for orders $j = 0, \ldots, M$ is given by $e_j(\{\rho_1, \rho_2, \ldots, \rho_M\}) = (-1)^j \alpha_{M-j}/\alpha_M$. The values $e_j(Z)$ can thus be evaluated by expanding out the polynomial with roots given by the elements of Z, which can be implemented using an appropriate recursion or convolution. For a finite set Z, calculation of $e_j(Z)$ requires $|Z|^2$ operations. It is shown in [55] (see Theorem 8.14) that this complexity can be reduced to $\mathcal{O}\left(|Z|\log^2|Z|\right)$ operations using a suitable decomposition and recursion.

In the CPHD recursion, each data update step requires the calculation of $|Z|+1$ elementary symmetric functions, that is one for Z and one for each set $\{Z\backslash\{\mathbf{z}\}\}$ where $\mathbf{z}\in Z$. Thus, the CPHD recursion has a complexity of $\mathcal{O}\left(|Z|^3\right)$. Furthermore, using the procedure in [55], the CPHD filter has a complexity of $\mathcal{O}\left(|Z|^2\log^2|Z|\right)$. Although this appears to be a modest saving, when $|Z|$ is large the reduction in complexity may be of some advantage. In practice, the number of measurements can be reduced by gating techniques as done in traditional tracking algorithms [41, 42].

Multitarget State Extraction. Similar to the Gaussian mixture PHD filter [8], state extraction in the Gaussian mixture CPHD filter involves first estimating the number of targets, and then extracting the corresponding number of mixture components with the highest weights from the posterior intensity as state estimates.

The number of targets can be estimated using for example an EAP estimator $\hat{N}_k=\mathbb{E}[|X_k|]$ or a MAP estimator $\hat{N}_k=\arg\max p_k(\cdot)$. Note that the EAP estimator is likely to fluctuate and be unreliable under low SNR conditions. This occurs because false alarms and target missed detections tend to induce minor modes in the posterior cardinality, and consequently the expected value is randomly shifted away from the target induced primary mode. On the other hand, the MAP estimator is likely to be more reliable since it ignores minor modes and locks directly onto the target induced primary mode. For these reasons, the MAP estimator is usually preferred over the EAP estimator [3, 39].

3.7.3 The CPHD Filter for Fixed Number of Targets

In a number of applications, there are neither target births nor deaths and the number of targets is known a priori (and fixed). While the CPHD filter is applicable in this case, it is more efficient to exploit the explicit knowledge of the number of targets.

Since there are no births nor deaths, the birth intensity is $\gamma_k(\mathbf{x})=0$ and the probability of survival is $p_{S,k|k-1}(\mathbf{x})=1$. Let $N\in\mathbb{N}$ be the fixed and known number of targets. Then, it follows from the CPHD cardinality recursion (3.53) and (3.56) that the cardinality distribution at any time must be a Dirac delta function centered on N, that is $p_{k|k-1}(\cdot)=p_k(\cdot)=\delta_N(\cdot)$. Moreover, it can be seen that the predicted and updated intensities (3.54) and (3.57) reduce to

$$v_{k|k-1}(\mathbf{x})=\int f_{k|k-1}(\mathbf{x}|\zeta)v_{k-1}(\zeta)d\zeta, \tag{3.75}$$

$$\begin{aligned} v_k(\mathbf{x}) &= [1-p_{D,k}(\mathbf{x})]\frac{\Upsilon_k^1[v_{k|k-1},Z_k](N)}{\Upsilon_k^0[v_{k|k-1},Z_k](N)}v_{k|k-1}(\mathbf{x}) \\ &\quad + p_{D,k}(\mathbf{x})\sum_{\mathbf{z}\in Z_k}\frac{\Upsilon_k^1[v_{k|k-1},Z_k\backslash\{\mathbf{z}\}](N)}{\Upsilon_k^0[v_{k|k-1},Z_k](N)}\frac{g_k(\mathbf{z}|\mathbf{x})}{\kappa_k(\mathbf{z})/\langle 1,\kappa_k\rangle}v_{k|k-1}(\mathbf{x}), \end{aligned} \tag{3.76}$$

where $\Upsilon_k^u[v,Z](\cdot)$ is given in (3.58).

A special case of the CPHD recursion for tracking a fixed number of targets in clutter (including the case of a single target) is defined by (3.75) and (3.76). The above recursion also admits a closed-form solution under linear Gaussian assumptions. The following corollaries follow directly from Propositions 7 and 8, and establish an analytic propagation of the posterior intensity given by the recursion (3.75) and (3.76).

Corollary 3 *Suppose at time $k-1$ that the posterior intensity v_{k-1} is a Gaussian mixture of the form (3.23). Then, under linear Gaussian multitarget assumptions, the predicted intensity at time k is also a Gaussian mixture and is given by*

$$v_{k|k-1}(\mathbf{x}) = \sum_{j=1}^{J_{k-1}} w_{k-1}^{(j)} \mathcal{N}(\mathbf{x}; \mathbf{m}_{S,k|k-1}^{(j)}, \mathbf{P}_{S,k|k-1}^{(j)})$$

where $\mathbf{m}_{S,k|k-1}^{(j)}$, $\mathbf{P}_{S,k|k-1}^{(j)}$ are given by (3.25) and (3.26).

Corollary 4 *Suppose at time k that the predicted intensity $v_{k|k-1}$ is a Gaussian mixture of the form (3.27). Then, under linear Gaussian multitarget assumptions, the posterior intensity at time k is also a Gaussian mixture and is given by*

$$\begin{aligned} v_k(\mathbf{x}) &= (1 - p_{D,k}) \frac{\Psi_k^1[\mathbf{w}_{k|k-1}, Z_k](N)}{\Psi_k^0[\mathbf{w}_{k|k-1}, Z_k](N)} v_{k|k-1}(\mathbf{x}) \\ &+ p_{D,k} \sum_{\mathbf{z} \in Z_k} \sum_{j=1}^{J_{k|k-1}} \frac{\Psi_k^1[\mathbf{w}_{k|k-1}, Z_k \backslash \{\mathbf{z}\}](N)}{\Psi_k^0[\mathbf{w}_{k|k-1}, Z_k](N)} \frac{w_{k|k-1}^{(j)} q_k^{(j)}(\mathbf{z})}{\kappa_k(\mathbf{z}) / \langle 1, \kappa_k \rangle} \mathcal{N}(\mathbf{x}; \mathbf{m}_k^{(j)}(\mathbf{z}), \mathbf{P}_k^{(j)}), \end{aligned}$$

where $\Psi_k^u[\mathbf{w}, Z](\cdot)$ is given by (3.65) and (3.66) and (3.68); $\mathbf{w}_{k|k-1}$ is given by (3.67); $q_k^{(j)}(\mathbf{z})$ is given by (3.69); $\mathbf{m}_k^{(j)}(\mathbf{z})$ is given by (3.72); and $\mathbf{P}_k^{(j)}$ is given by (3.73) and (3.74).

For mildly nonlinear multitarget model, the same linearization and UT technique can be applied.

3.8 NUMERICAL EXAMPLES

The performance of PHD and CPHD filters is now demonstrated with a typical tracking example. A time varying number of targets is observed in clutter on a two-dimensional surveillance region with dimensions $[-1000, 1000]\text{m} \times [-1000, 1000]\text{m}$. A maximum of 10 targets appears on the scene at any one time, and targets births and deaths occur at various times and locations. The trajectories for each target are shown in Figure 3.7.

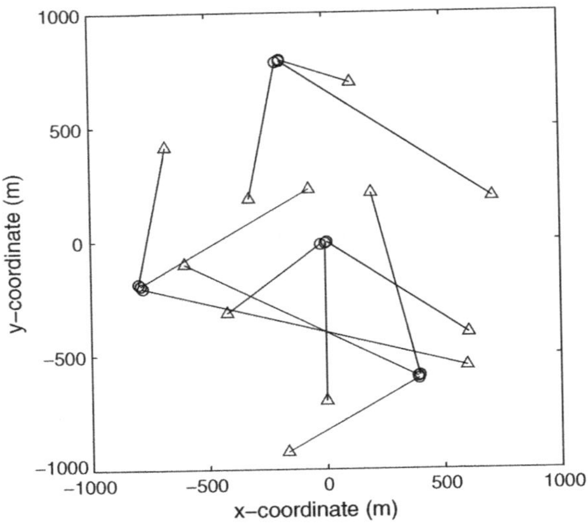

Figure 3.7 Trajectories in the xy plane. Start/Stop positions for each track are shown with ○/△.

The following dynamical and measurement models are used. The target state variable is a vector of planar position and velocity $\mathbf{x}_k = [\ p_{x,k},\ p_{y,k},\ \dot{p}_{x,k},\ \dot{p}_{y,k}\ ]'$. The single-target transition model is linear Gaussian specified by

$$\mathbf{F}_{k|k-1} = \begin{bmatrix} \mathbf{I}_2 & \Delta\mathbf{I}_2 \\ \mathbf{0}_2 & \mathbf{I}_2 \end{bmatrix}, \qquad \mathbf{Q}_k = \sigma_\nu^2 \begin{bmatrix} \frac{\Delta^4}{4}\mathbf{I}_2 & \frac{\Delta^3}{2}\mathbf{I}_2 \\ \frac{\Delta^3}{2}\mathbf{I}_2 & \Delta^2\mathbf{I}_2 \end{bmatrix},$$

where $\mathbf{I}_n$ and $\mathbf{I}_n$ denote the $n \times n$ identity and zero matrices, $\Delta = 1s$ is the sampling period, and $\sigma_\nu = 5\ \mathrm{m/s}^2$ is the standard deviation of the process noise. The probability of survival is $p_{S,k} = 0.99$. The birth process is a Poisson RFS with intensity $\gamma_k(\mathbf{x}) = \sum_{i=1}^{4} w_\gamma \mathcal{N}(\mathbf{x}; \mathbf{m}_\gamma^{(i)}, \mathbf{P}_\gamma)$ where $w_\gamma = 0.03$, $\mathbf{m}_\gamma^{(1)} = [\,0, 0, 0, 0\,]'$, $\mathbf{m}_\gamma^{(2)} = [\,400, 0, -600, 0\,]'$, $\mathbf{m}_\gamma^{(3)} = [\,-800, 0, -200, 0\,]'$, $\mathbf{m}_\gamma^{(4)} = [\,-200, 0, 800, 0\,]'$, and $\mathbf{P}_\gamma = \mathrm{diag}([\ 10,\ 10,\ 10,\ 10\ ])^2$. The probability of detection is $p_{D,k} = 0.98$. The single-target measurement model is also linear Gaussian with

$$\mathbf{H}_k = \begin{bmatrix} \mathbf{I}_2 & \mathbf{0}_2 \end{bmatrix}, \mathbf{R}_k = \sigma_\varepsilon^2 \mathbf{I}_2,$$

where $\sigma_\varepsilon = 10$ m is the standard deviation of the measurement noise. Clutter follows a Poisson RFS with intensity $\kappa_k(\mathbf{z}) = \lambda_c V u(\mathbf{z})$, where $u(\cdot)$ is a uniform probability density over the surveillance region, $V = 4 \times 10^6\ \mathrm{m}^2$ is the "volume" of the surveillance region, and $\lambda_c = 1.25 \times 10^{-5}\ \mathrm{m}^{-2}$ is the clutter intensity (giving an average of 50 clutter returns per scan).

To manage the number of mixture components, pruning and merging of components is performed at each time step using a weight threshold of $T = 10^{-5}$, a merging

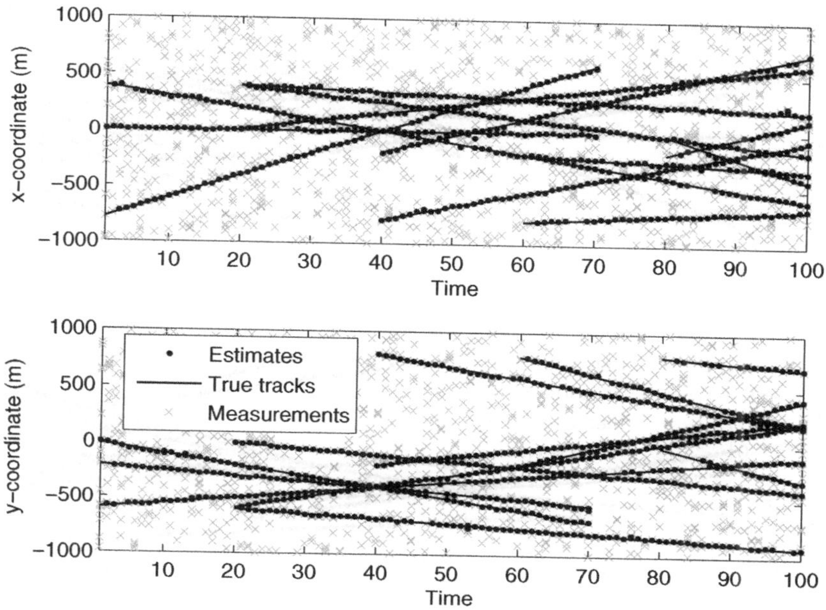

Figure 3.8 PHD filter estimates and true target tracks in x and y coordinates versus time.

threshold of $U = 4$ m, and a maximum of $J_{\max} = 100$ Gaussian components (see [8] for the exact meaning of these parameters). The number of targets is estimated using an MAP estimator on the cardinality distribution which is calculated to a maximum of $N_{\max} = 200$ terms in the CPHD filter, and state estimates are extracted as the means of corresponding components of the posterior intensity. To measure the performance of the PHD and CPHD filters, the OSPA metric of Section 3.5.4 with parameters $p = 1$ and $c = 300$ m is used on the estimated positions only (i.e., taking the positional components and excluding the velocity components).

In Figures 3.8 and 3.9, the target trajectories are shown in x and y coordinates versus time with a sample run output from the Gaussian mixture PHD and CPHD filters superimposed respectively. Note that three targets cross at time $k = 40$ whilst another two cross at time $k = 60$. It can be seen from these results that the PHD and CPHD filters are able to correctly identify target births, motions and deaths, and have no trouble handling target crossings. To give an indication of processing time, the Gaussian mixture PHD filter required 2.7 s per sample run over 100 time steps, while the Gaussian mixture CPHD filter required 10.2 s for the same data (both implemented in MATLAB on a standard notebook computer).

To verify the performance of the Gaussian mixture PHD and CPHD filters, 1000 MC runs are performed on the same target trajectories but with independently generated clutter and (target-generated) measurements for each trial. In Figure 3.10, the true number of targets at each time step is shown along with the MC average of the mean and standard deviation of the estimated cardinality distribution for both the PHD and

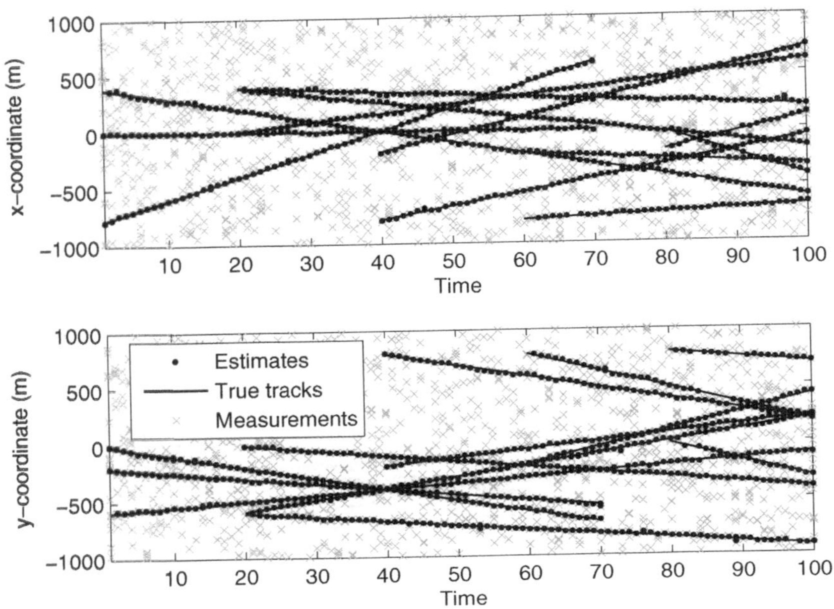

Figure 3.9 CPHD filter estimates and true target tracks in x and y coordinates versus time.

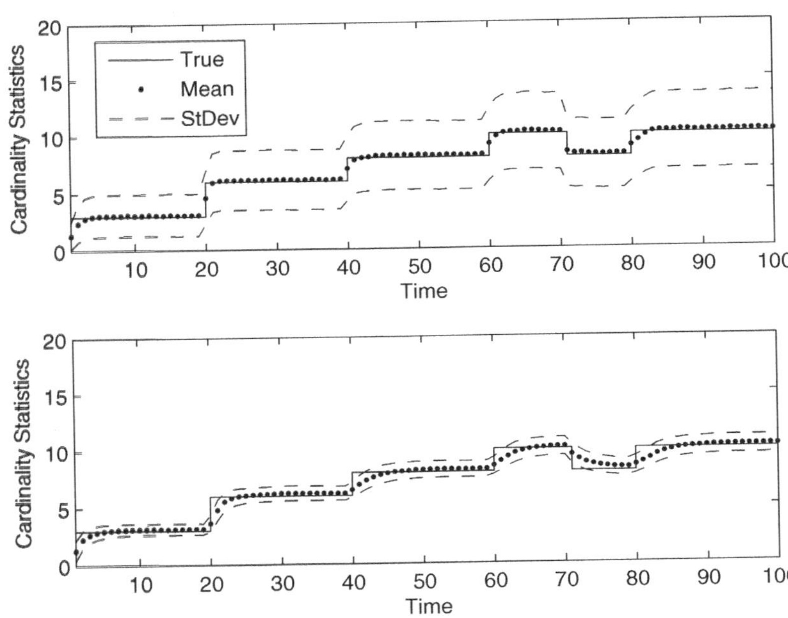

Figure 3.10 1000 MC run average of cardinality statistics versus time for (top) PHD filter and (bottom) CPHD filter.

CPHD filters. The plots demonstrate that both filters converge to the correct number of targets present, and that the variance of the cardinality distribution is much smaller in the CPHD filter than in the PHD filter (the average reduction in the variance over 100 time steps is approximately 12.5 times).

Further examination of the cardinality statistics reveals a difference in performance regarding the filters' response to changes in the number of targets. Indeed, the simulations also suggest that the average response time of the CPHD filter is slower than that of the PHD filter. An explanation for this observation is that the PHD filter's cardinality estimate has a relatively high variance, thus it has low confidence in its estimate and is easily influenced by new incoming measurement information. On the other hand, the CPHD filter's cardinality estimate has a lower variance, and as a consequence it is much more confident in its estimate and is not easily influenced by new incoming measurements.

Remark: The correct convergence of both the PHD and CPHD filters' mean number of targets is only an indication of average behavior. More importantly, it is the variance of this estimate that determines the usefulness of the filter since for any given sample path, the PHD filter's estimate of the number of targets is extremely jumpy and inaccurate, whereas the CPHD filter's estimate is far more reliable and accurate.

The MC average of the OSPA distance is shown for each time step in Figure 3.11 for both the CPHD and PHD filters. At first glance, it appears that the CPHD and PHD filters produce average errors of approximately 25 m and 40 m per-object respectively. The results of the figure also show that the OSPA distances exhibit peaks when there is a change in the number of objects. The peaks at these particular time instants can be expected since the filters are adapting to cardinality changes. In addition, the peaks in the OSPA curves are smaller in the PHD filter than in the CPHD filter. This observation can be attributed to the fact that the PHD filter has a faster response to cardinality changes than the CPHD filter and so on average the PHD filter incurs a lower penalty than the CPHD filter. Also note that the magnitude of the peaks is consistent with the magnitude of the corresponding cardinality change. Conversely, during time intervals when number of objects is steady, the PHD filter has a higher penalty than the CPHD filter. The most likely explanation here is that the CPHD filter produces more accurate state and cardinality estimates than the PHD filter in such conditions, given that both filters can be expected to have comparable localization performance.

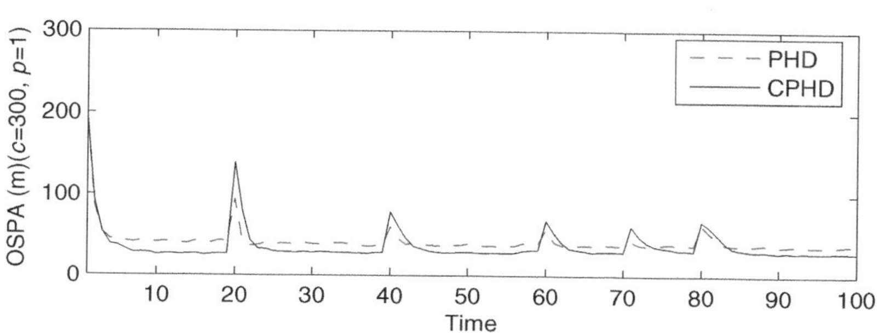

Figure 3.11 OSPA distance versus time for the GM-CPHD and GM-PHD filters.

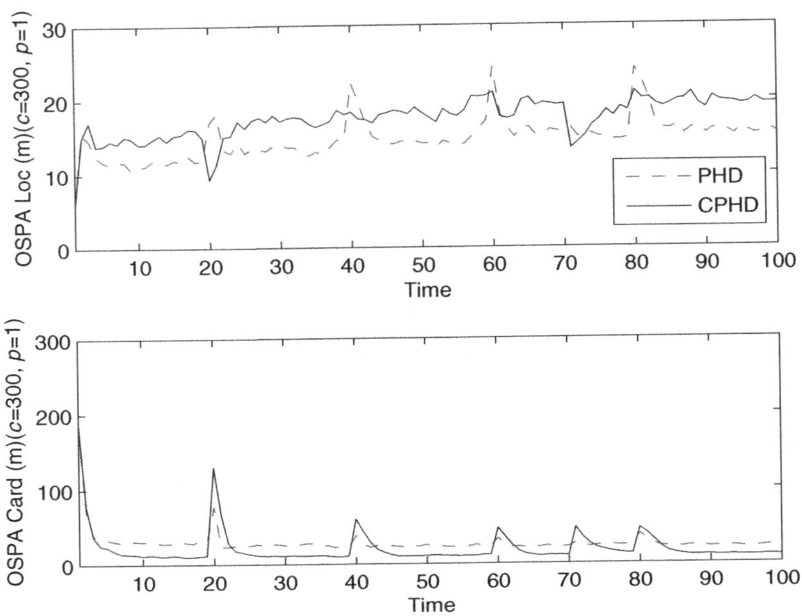

Figure 3.12 OSPA components—localization error (top) and cardinality error (bottom) versus time for the GM-CPHD and GM-PHD filters.

Indeed, inspection of the MC average of the localization and cardinality components of the OSPA distance in Figure 3.12 confirms these explanations. In terms of localization accuracy, examination of the localization component shows that the both the CPHD and PHD filters have similar performance with localization errors of roughly 15 m per-object, although the latter appears to have a slight advantage over the former due to the nature of the optimum subpattern assignment favoring consistently incorrect cardinality estimates (see the discussion at the end of Section 3.5.4 for further explanation). In terms of cardinality estimation accuracy, examination of the cardinality component shows that the CPHD filter outperforms the PHD filter when the number of objects is constant due to its more accurate propagation of cardinality information, with the filters producing cardinality errors of approximately 10 m and 25 m per-object respectively. Conversely, the cardinality component suggests that the PHD filter outperforms the CPHD filter when the number of objects changes due to its previously noted faster response. Thus, it can be seen that while both filters have comparable localization performance, the additional propagation of the complete cardinality distribution in the CPHD filter dramatically improves its cardinality performance compared to the PHD filter, and hence improves the overall performance of the CPHD filter compared to the PHD filter.

Chapter 6 presents application of the GMCPHD filter to the problem of detection and tracking of road constrained ground targets using ground moving target indicator (GMTI) radar.

3.9 MeMBer FILTER

The PHD approach is not the only approximation strategy that FISST provides. The MeMBer filter is a different approximation approach to the Bayes multitarget filter. Unlike the PHD and CPHD recursions, which propagate moments and cardinality distributions, the MeMBer filter propagates the parameters of a multi-Bernoulli RFS that approximates the posterior multitarget RFS. We conclude this chapter by providing brief summaries of the MeMBer filters proposed in [11] and [12].

3.9.1 MeMBer Recursion

The premise of the MeMBer recursion is that the multitarget RFS at each time step is approximated by a multi-Bernoulli RFS, based on the following modelling assumptions:

- Each target evolves and generates measurements independently.
- Target births follow a multi-Bernoulli RFS independent of target survivals.
- Clutter follows a Poisson RFS, not too dense, and is independent of target-generated measurements.

The MeMBer recursion propagates the multi-Bernoulli parameters of the posterior multitarget density forward in time via the following prediction step and an update step.

Prediction [25]. If at time $k-1$, the posterior multitarget density is a multi-Bernoulli of the form

$$\pi_{k-1} = \{(r_{k-1}^{(i)}, p_{k-1}^{(i)})\}_{i=1}^{M_{k-1}},$$

then the predicted multitarget density is also a multi-Bernoulli and is given by

$$\pi_{k|k-1} = \{(r_{P,k|k-1}^{(i)}, p_{P,k|k-1}^{(i)})\}_{i=1}^{M_{k-1}} \cup \{(r_{\Gamma,k}^{(i)}, p_{\Gamma,k}^{(i)})\}_{i=1}^{M_{\Gamma,k}}, \tag{3.77}$$

where

$$r_{P,k|k-1}^{(i)} = r_{k-1}^{(i)} \langle p_{S,k|l-1}, p_{k-1}^{(i)} \rangle, \tag{3.78}$$

$$p_{P,k|k-1}^{(i)}(\mathbf{x}) = \frac{\langle f_{k|k-1}(\mathbf{x}|\cdot), p_{S,k|k-1} p_{k-1}^{(i)} \rangle}{\langle p_{S,k|l-1}, p_{k-1}^{(i)} \rangle}, \tag{3.79}$$

$\{(r_{\Gamma,k}^{(i)}, p_{\Gamma,k}^{(i)})\}_{i=1}^{M_{\Gamma,k}}$ = parameters of the multi-Bernoulli RFS of births at time k.

Update [11]. If at time k the predicted multitarget density is a multi-Bernoulli of the form

$$\pi_{k|k-1} = \{(r_{k|k-1}^{(i)}, p_{k|k-1}^{(i)})\}_{i=1}^{M_{k|k-1}}$$

then the posterior multitarget density can be approximated by a multi-Bernoulli as follows

$$\pi_k \approx \{(r_{L,k}^{(i)}, p_{L,k}^{(i)})\}_{i=1}^{M_{k|k-1}} \cup \{(r_{U,k}(\mathbf{z}), p_{U,k}(\cdot;\mathbf{z}))\}_{\mathbf{z}\in Z_k}, \tag{3.80}$$

where

$$r_{L,k}^{(i)} = r_{k|k-1}^{(i)} \frac{1 - \langle p_{k|k-1}^{(i)}, p_{D,k}\rangle}{1 - r_{k|k-1}^{(i)} \langle p_{k|k-1}^{(i)}, p_{D,k}\rangle}, \tag{3.81}$$

$$p_{L,k}^{(i)}(\mathbf{x}) = p_{k|k-1}^{(i)}(x) \frac{1 - p_{D,k}(\mathbf{x})}{1 - \langle p_{k|k-1}^{(i)}, p_{D,k}\rangle}, \tag{3.82}$$

$$r_{U,k}(\mathbf{z}) = \frac{\sum_{j=1}^{M_{k|k-1}} \frac{r_{k|k-1}^{(j)}(1-r_{k|k-1}^{(j)})\langle p_{k|k-1}^{(j)}, g_k(\mathbf{z}|\cdot)p_{D,k}\rangle}{(1-r_{k|k-1}^{(j)}\langle p_{k|k-1}^{(j)}, p_{D,k}\rangle)^2}}{\kappa_k(\mathbf{z}) + \sum_{j=1}^{M_{k|k-1}} \frac{r_{k|k-1}^{(j)}\langle p_{k|k-1}^{(j)}, g_k(\mathbf{z}|\cdot)p_{D,k}\rangle}{1-r_{k|k-1}^{(j)}\langle p_{k|k-1}^{(j)}, p_{D,k}\rangle}} \tag{3.83}$$

$$p_{U,k}(\mathbf{x};\mathbf{z}) = \frac{\sum_{j=1}^{M_{k|k-1}} \frac{r_{k|k-1}^{(j)}}{1-r_{k|k-1}^{(j)}} p_{k|k-1}^{(j)}(\mathbf{x}) g_k(\mathbf{z}|\mathbf{x}) p_{D,k}(\mathbf{x})}{\sum_{j=1}^{M_{k|k-1}} \frac{r_{k|k-1}^{(j)}}{1-r_{k|k-1}^{(j)}} \langle p_{k|k-1}^{(j)}, g_k(\mathbf{z}|\cdot)p_{D,k}\rangle}. \tag{3.84}$$

In terms of complexity the MeMBer recursion is linear in the number of targets and linear in the number of measurements. This is a similar complexity to the PHD filter, but a lower complexity compared to the CPHD filter which is linear in the number of targets and cubic in the number of measurements.

3.9.2 Multitarget State Estimation

The multi-Bernoulli representation $\pi_k = \{(r_k^{(i)}, p_k^{(i)})\}_{i=1}^{M_k}$ has an intuitive interpretation that facilitates multitarget state estimation from the posterior multitarget density. The existence probability $r_k^{(i)}$ indicates how likely the ith hypothesized track is a true track, and the posterior density $p_k^{(i)}$ describes the estimated current state of the track. Hence, a multitarget state estimate can be obtained by choosing the means or modes from the posterior densities of the hypothesized tracks with existence probabilities exceeding a given threshold (e.g., 0.5). Alternatively, the following basic two-stage procedure can be used. First, we estimate the number of targets from the posterior cardinality distribution by taking its mean or mode (the mode is preferred as it is more stable than the mean). Then, we take the corresponding number of hypothesized tracks having the highest probabilities of existence and compute the individual means or modes from the individual posterior densities.

3.9.3 Extension to Track Propagation

Following the MeMBer recursions can be extended to propagate track continuity, by appropriately labeling and updating the individual Bernoulli components of the posterior multi-Bernoulli density. Thus, given a multi-Bernoulli density $\pi = \{(r^{(i)}, p^{(i)})\}_{i=1}^{M}$, to each (Bernoulli) component $(r^{(i)}, p^{(i)})$ we assign a unique (usually integer) track label $\ell^{(i)}$ to identify hypothesized tracks. The collection of triplets $\mathcal{T} = \{(\ell^{(i)}, r^{(i)}, p^{(i)})\}_{i=1}^{M}$ is referred to as a track table. While there are many possible schemes for propagating track labels, a simple scheme is presented as follows.

Prediction. If at time $k-1$ the posterior track table is $\mathcal{T}_{k-1} = \{(\ell_{k-1}^{(i)}, r_{k-1}^{(i)}, p_{k-1}^{(i)})\}_{i=1}^{M_{k-1}}$, then the predicted track table to time k is

$$\mathcal{T}_{k|k-1} = \{(\ell_{P,k|k-1}^{(i)}, r_{P,k|k-1}^{(i)}, p_{P,k|k-1}^{(i)})\}_{i=1}^{M_{k-1}} \cup \{(\ell_{\Gamma,k}^{(i)}, r_{\Gamma,k}^{(i)}, p_{\Gamma,k}^{(i)})\}_{i=1}^{M_{\Gamma,k}},$$

where $\ell_{P,k|k-1}^{(i)} = \ell_{k-1}^{(i)}$, $\ell_{\Gamma,k}^{(i)} = \mathit{new\ label}$. Thus, existing components retain their original labels, and birth components are assigned new labels.

Update. If at time k the predicted track table is $\mathcal{T}_{k|k-1} = \{(\ell_{k|k-1}^{(i)}, r_{k|k-1}^{(i)}, p_{k|k-1}^{(i)})\}_{i=1}^{M_{k|k-1}}$, then the updated track table at time k is

$$\mathcal{T}_k = \{(\ell_{L,k}^{(i)}, r_{L,k}^{(i)}, p_{L,k}^{(i)})\}_{i=1}^{M_{k|k-1}} \cup \{(\ell_{U,k}(\mathbf{z}), r_{U,k}(\mathbf{z}), p_{U,k}(\mathbf{z}))\}_{\mathbf{z}\in Z_k},$$

where $\ell_{L,k}^{(i)} = \ell_{k|k-1}^{(i)}$, $\ell_{U,k}(\mathbf{z}) = \ell_{k|k-1}^{(n)}$,

$$n = \arg\max_i \frac{r_{k|k-1}^{(i)}(1 - r_{k|k-1}^{(i)})\langle p_{k|k-1}^{(i)}, p_{D,k} g_k(\mathbf{z}|\cdot)\rangle}{(1 - r_{k|k-1}^{(i)}\langle p_{k|k-1}^{(i)}, p_{D,k}\rangle)^2}.$$

Thus, legacy components maintain their original labels, and measurement-updated components are assigned the label of the predicted track which has the largest contribution to the current measurement-updated probability of existence.

Remark: Although simple to implement, the proposed scheme is expected to perform poorly when targets are close together. Its performance can be significantly improved by adding the track association schemes considered in [35].

3.9.4 MeMBer Filter for Image Data

For image data, tracking is often performed on data that have been preprocessed into point measurements. Compressing the information on the image into a finite

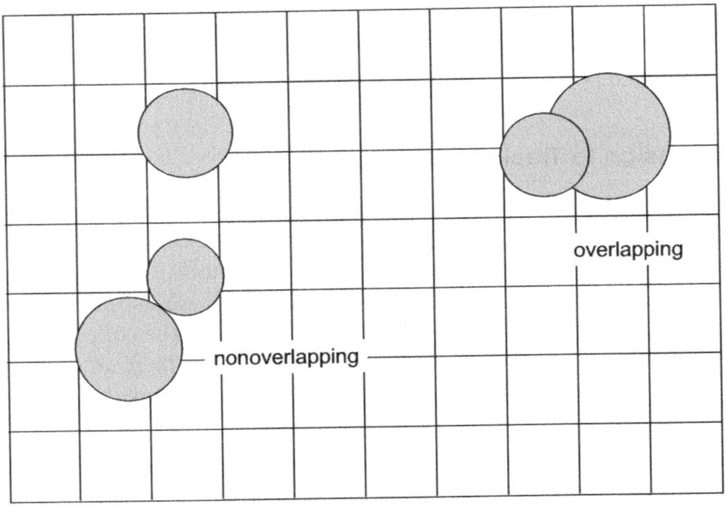

Figure 3.13 An illustration of overlapping and nonoverlapping objects.

set of points is efficient and effective for a wide range of applications. However, for applications with low SNR such as TBD, it is necessary to make use of all information contained in the images. This subsection outlines the MeMBer filter for image proposed in [12].

The type of image observation considered in this subsection is illustrated in Figure 3.13. Let $\mathbf{y} = [y_1, \ldots y_m]$ denote the image observation comprising an array of m pixel (or bin) values. The value y_i of the ith pixel can be a real number or a vector depending on the application. If the multitarget likelihood of the image observation is *separable*, that is

$$g_k(\mathbf{y}_k|X_k)=f_k(\mathbf{y}_k)\prod_{\mathbf{x}\in X_k} g_{\mathbf{y}_k}(\mathbf{x})$$

then the update step is given as follows.

Update (for image data) [12]. Given the predicted multi-Bernoulli parameters $\pi_{k|k-1} = \{(r_{k|k-1}^{(i)}, p_{k|k-1}^{(i)})\}_{i=1}^{M_{k|k-1}}$, the updated multi-Bernoulli parameters $\pi_k = \{(r_k^{(i)}, p_k^{(i)})\}_{i=1}^{M_{k|k-1}}$ are given by

$$r_k^{(i)} = \frac{r_{k|k-1}^{(i)}\left\langle p_{k|k-1}^{(i)}, g_{\mathbf{y}_k}\right\rangle}{1 - r_{k|k-1}^{(i)} + r_{k|k-1}^{(i)}\left\langle p_{k|k-1}^{(i)}, g_{\mathbf{y}_k}\right\rangle} \tag{3.85}$$

$$p_k^{(i)} = \frac{p_{k|k-1}^{(i)} g_{\mathbf{y}_k}}{\left\langle p_{k|k-1}^{(i)}, g_{\mathbf{y}_k}\right\rangle}. \tag{3.86}$$

We now describe a measurement model that has separable image observation likelihood. With reference to Figure 3.13 and omitting the time index for simplicity, a target with state $\mathbf{x}$ illuminates a set of pixels denoted by $T(\mathbf{x})$, for example $T(\mathbf{x})$ could be the set of pixels whose centers fall within certain distance from the position of the target. A pixel $i \in T(\mathbf{x})$, that is illuminated by a target with state $\mathbf{x}$, has value distributed according to $\varphi_i(\cdot, \mathbf{x})$, while a pixel $i \notin T(\mathbf{x})$, that is not illuminated by any object, has value distributed according to $\phi_i(\cdot)$. More concisely, the probability density of the value y_i of pixel i, given a state $\mathbf{x}$ is

$$p(y_i|\mathbf{x}) = \begin{cases} \varphi_i(y_i, \mathbf{x}), & i \in T(\mathbf{x}) \\ \phi_i(y_i), & i \notin T(\mathbf{x}). \end{cases} \tag{3.87}$$

For example, in TBD (see [56] Chapter 11),

$$\phi_i(y_i) = \mathcal{N}(y_i; 0, \sigma^2)$$
$$\varphi_i(y_i, \mathbf{x}) = \mathcal{N}(y_i; h_i(\mathbf{x}), \sigma^2),$$

where $\mathcal{N}(\cdot; \mu, \sigma^2)$ denotes a Gaussian density with mean μ and variance σ^2, and $h_i(x)$ is the contribution to pixel i from the state x, which depends on the point spread function, target location, and reflected energy. Note that (3.87) also holds for nonadditive models, see for example [57].

Under the following assumptions:

- conditional on the multitarget state, the values of the pixels are independently distributed;
- the regions of influences of the objects on the image do not overlap, that is $\mathbf{x} \neq \mathbf{x}' \Rightarrow T(\mathbf{x}) \cap T(\mathbf{x}') = \emptyset$,

the probability density of the observation conditional on the multiobject state X is given by

$$\begin{aligned} g(\mathbf{y}|X) &= \left(\prod_{\mathbf{x} \in X} \prod_{i \in T(\mathbf{x})} \varphi_i(y_i, \mathbf{x}) \right) \left(\prod_{i \notin \cup_{\mathbf{x} \in X} T(\mathbf{x})} \phi_i(y_i) \right), \\ &= f(\mathbf{y}) \prod_{\mathbf{x} \in X} g_{\mathbf{y}}(\mathbf{x}) \end{aligned} \tag{3.88}$$

where

$$g_{\mathbf{y}}(\mathbf{x}) = \prod\nolimits_{i \in T(\mathbf{x})} \frac{\varphi_i(y_i, \mathbf{x})}{\phi_i(y_i)},$$
$$f(\mathbf{y}) = \prod\nolimits_{i=1}^{m} \phi_i(y_i).$$

These assumptions are reasonable when targets are rigid bodies. In ground target tracking, for example, the targets would be vehicles or stationary objects that must be physically separated.

3.9.5 Implementations

Like the PHD recursions, the MeMBer filter also admits closed-form solution under linear Gaussian multitarget model. This solution can also be extended to accommodate nonlinear Gaussian dynamical and observation models with standard approximation methods. Empirical results in [11] suggest that, in Gaussian mixture implementations the CPHD filter is still superior and the MeMBer filter can only achieve similar performance to the PHD filter, while in SMC implementations the MeMBer filter outperforms both the CPHD and PHD filters. Thus, for applications where Gaussian mixture implementations are not applicable, for example highly nonlinear dynamic and observation model and/or state-dependent probability of detection, the SMC-MeMBer filter offers an attractive alternative. Moreover, for image data the measurement model is inherently nonlinear and so far, only SMC implementations are tractable [12].

ACKNOWLEDGMENTS

The first author's work is supported by the Australian Research Council under Discovery Project DP0880553. The second author's work is supported by an Australian Research Council Post Doctoral Fellowship under Discovery Project DP0989007. The third author's work is supported by the Royal Academy of Engineering/EPSRC Research Fellowship.

REFERENCES

1. R. Mahler, Multitarget Bayes filtering via first-order multitarget moments, *IEEE Trans. Aerosp. Electron. Syst.*, 39(4), 1152–1178, 2003.
2. B.-N. Vo, S. Singh, and A. Doucet, Sequential Monte Carlo methods for multitarget filtering with random finite sets, *IEEE Trans. Aerosp. Electron. Syst.*, 41(4), 1224–1245, 2005.
3. R. Mahler, PHD filters of higher order in target number, *IEEE Trans. Aerosp. & Electron. Syst.*, 43(3), 2007.
4. B.-N. Vo, S. Singh, and A. Doucet, Sequential Monte Carlo implementation of the PHD filter for multitarget tracking, *Proceedings of the International Conference on Information Fusion,* Cairns, Australia, 2003, pp. 792–799.
5. T. Zajic and R. Mahler, A particle-systems implementation of the PHD multitarget tracking filter, *Signal Processing, Sensor Fusion and Target Recognition XII, SPIE Proceedings*, Vol. 5096, 2003, pp. 291–299.
6. H. Sidenbladh, Multitarget particle filtering for the Probability Hypothesis Density, *Proceedings of the International Conferences on Information Fusion,* Cairns, Australia, 2003, pp. 800–806.

7. B.-N. Vo and W.-K. Ma, A closed-form solution to the probability hypothesis density filter, *Proceedings of the International Conference on Information Fusion,* Philadelphia, 2005.
8. B.-N. Vo and W.-K. Ma, The Gaussian mixture probability hypothesis density filter, *IEEE Trans. Signal Process.*, 54(11), 4091–4104, 2006.
9. B.-T. Vo, B.-N. Vo, and A. Cantoni, The cardinalized probability hypothesis density filter for linear Gaussian multitarget models, *Proceedings of the 40th Conference on Informations Sciences & Systems*, March 2006.
10. B.-T. Vo, B.-N. Vo, and A. Cantoni, Analytic implementations of the cardinalized probability hypothesis density filter, *IEEE Trans. Signal Process.*, 55(7), 3553–3567, 2007.
11. B.-T. Vo, B.-N. Vo, and A. Cantoni, The cardinality balanced multitarget multi-Bernoulli filter and its implementations, *IEEE Trans. Signal Process.*, 57(2), 409–423, 2009.
12. B.-N. Vo, B.-T. Vo, N.-T. Pham, and D. Suter, Joint detection and estimation of multiple objects from image observations, *IEEE Trans. Signal Proces.*, 58(10), 5129–5241, 2010.
13. C. Robert, *The Bayesian Choice: from Decision-Theoretic Motivations to Computational Implementation.*, Springer-Verlag, New York, 2001.
14. M. Mallick, S. Coraluppi, and C. Carthel, Multitarget tracking using multiple hypotheses tracking (MHT), in: M. Mallick, V. Krishnamurthy, and B-N Vo (Eds.), *Integrated Tracking, Classification, and Sensor Management: Theory and Applications*, Wiley/IEEE, 2012, chapter 5.
15. S. Mori , C. Chong, E. Tse, and R. Wishner, Tracking and identifying multiple targets without apriori identifications, *IEEE Trans. Automat. Contr.*, AC-21, 401–409, 1986.
16. R. Washburn, A random point process approach to multiobject tracking, *Proceedings of the American Control Conference*, Vol. 3, 1987, pp. 1846–1852.
17. M.I. Miller, R.S. Teichman, A. Srivastava, J.A. O'Sullivan, and D.L. Snyder, Jump-diffusion processes for automated tracking-target recognition, *Proceedings of the Conferenc on Information Science and Systems*, Baltimore, Maryland, March 1993.
18. M.I. Miller, A. Srivastava A., and U. Grenander, Conditional-mean estimation via jump-diffusion processes in multiple target tracking/recognition, *IEEE Trans. Signal Process.*, 43(11), 2678–2690, 1995.
19. A. Srivastava, M.I. Miller, and U. Grenander, Jump-diffusion processes for tracking and direction finding, *Proceedings of the 29th Allerton Conference on Communication, Control, and Computing*, University of Illinois, Urbana, 1991, pp. 563–570.
20. N. Portenko, H. Salehi, and A. Skorokhod, On optimal filtering of multitarget systems based on point process observations, *Random Operators Stochastic Equations,* 5(1), 1–34, 1997.
21. R. Mahler, Global integrated data fusion, *Proceedings of the 7th National Symposium on Sensor Fusion*, Vol. 1, (Unclassified) Sandia National Laboratories, Albuquerque, ERIM Ann Arbor MI, 1994, pp. 187–199.
22. J. Goutsias. R. Mahler, and H. Nguyen (Eds.), *Random Sets Theory and Applications*, Springer-Verlag, New York, 1997.
23. I. Goodman, R. Mahler, and H. Nguyen, *Mathematics of Data Fusion*, Kluwer Academic Publishers, Norwell, MA, USA, 1997.
24. R. Mahler, *An Introduction to Multisource-Multitarget Statistics and Applications*, Lockheed Martin Technical Monograph, March 15, 2000.

25. R. Mahler, *Statistical Multisource-Multitarget Information Fusion*, Artech House, Norwood, MA, USA, 2007.
26. R. Mahler, A theoretical foundation for the Stein–Winter probability hypothesis density (PHD) multitarget tracking approach, *Proceedings of the MSS National Symposium on Sensor and Data Fusion*, Vol. I (Unclassified), San Antonio TX, June 2000.
27. H. Sidenbladh and S.-L. Wirkander, Tracking random sets of vehicles in terrain, *Proceedings of the 2003 IEEE Workshop on Multiobject Tracking*, Madison WI, June, 2003.
28. D. Clark and J. Bell, Convergence results for the particle PHD filter, *IEEE Trans. Signal Process.*, 54(7), 2652–2661, 2006.
29. A. Johansen, S. Singh, A. Doucet, and B. Vo, Convergence of the SMC implementation of the PHD filter, *Methodol. Comput. Appl. Probab.*, 8(2), 265–291, 2006.
30. K. Panta, B. Vo, and S. Singh, Novel data association schemes for the probability hypothesis density filter, *IEEE Trans. Aerosp. Electron. Syst.*, 43(2), 556–570, 2007.
31. L. Lin, Y. Bar-Shalom, and T. Kirubarajan, Track labeling and PHD filter for multitarget tracking, *IEEE Trans., Aerosp. Electron. Syst.*, 42(3), 778–795, 2006.
32. B.-T. Vo, Random finite sets in multiobject filtering, *Ph.D Thesis*, University of Western Australia, 2008.
33. D. Clark and B. Vo, Convergence analysis of the Gaussian mixture PHD filter, *IEEE Trans. Signal Process.*, 55(4), 1204–1212, 2007.
34. A. Pasha, B.-N. Vo, H.D. Tuan, and W.K. Ma, A Gaussian mixture PHD filter for jump Markov system model, *IEEE Trans. Aerosp. Electron. Syst.*, 45(3), 919-936, 2009.
35. K. Panta, D. Clark, and B. Vo, Data association and track management for the Gaussian mixture probability hypothesis density filter, *IEEE Trans. Aerosp. Electron. Syst.*, 45(3), 1003–1016, 2009.
36. N. Whiteley, S. Singh, and S. Godsill, Auxiliary particle implementation of probability hypothesis density filter, *IEEE Trans. Aerosp. Electron. Syst.*, 46(3), 1437–1454, 2010.
37. B. Ristic, D. Clark, and B.-N.Vo, Improved SMC implementation of the PHD filter *Proceedings of the 13th Annual Conference on Information Fusion*, Edinburgh, UK, 2010.
38. J. Houssineau and D. Laneuville, PHD filter with diffuse spatial prior on the birth process with applications to GM-PHD filter, *Proceedings of the 13th Annual Conference on Information Fusion*, Edinburgh, UK, 2010.
39. R. Mahler, A theory of PHD filters of higher order in target number, *Signal Processing, Sensor Fusion, and Target Recognition XV, SPIE Defense & Security Symposium*, April 2006.
40. O. Erdinc, P. Willett, and Y. Bar-Shalom, A physical-space approach for the PHD and CPHD filters, *Signal Processing, Sensor Fusion, and Target Recognition XV, SPIE Defense & Security Symposium*, April 2006.
41. Y. Bar-Shalom and T. E. Fortmann, *Tracking and Data Association*, Academic Press, San Diego, 1988.
42. S. Blackman, *Multiple Target Tracking with Radar Applications*, Artech House, Norwood, 1986.
43. D. Daley and D. Vere-Jones, *An Introduction to the Theory of Point Processes*, Springer-Verlag, New York, NY, 1988.
44. J. Hoffman and R. Mahler, Multitarget miss distance via optimal assignment, *IEEE Trans. Syst., Man, Cybern.—Part A*, 34(3), 327–336, 2004.

45. O.E. Drummond and B.E. Fridling, Ambiguities in evaluating performance of multiple target tracking algorithms, in: O. E. Drummond (Ed.), *Signal and Data Processing of Small Targets, Proc. SPIE*, Vol. 1698, 1992, pp. 326–337.
46. R.L. Rothrock and O.E. Drummond, Performance metrics for multiple- sensor, multiple-target tracking, in: O.E. Drummond (Ed.), *Signal and Data Processing of Small Targets, Proc. SPIE*, Vol. 4048, 2000, pp. 521–531.
47. D. Schuhmacher, B.-T. Vo, and B.-N. Vo, A consistent metric for performance evaluation of multiobject filters, *IEEE Trans. Signal Process.*, 56(8), 3447–3457, 2008.
48. S. Singh, B.-N. Vo, A. Baddeley, and S. Zuev, Filters for spatial point processes *SIAM J. Contr. Optim.*, 48(4), 2275-2295, 2009.
49. J. T.-H. Lo, Finite-dimensional sensor orbits and optimal nonlinear filtering, *IEEE Trans. IT*, IT-18(5), 583–588, 1972.
50. H.W. Sorenson and D.L. Alspach, Recursive Bayesian estimation using Gaussian sum, *Automatica*, 7, 465–479, 1971.
51. D. Salmond, Tracking in uncertain environments, *Ph.D Thesis*, University of Sussex, 1989.
52. B.D. Anderson and J. B. Moore, *Optimal Filtering*, Prentice-Hall, New Jersey, 1979.
53. S.J. Julier and J.K. Uhlmann, Unscented filtering and nonlinear estimation, *Proc. IEEE*, 92(3), 401–422, 2004.
54. P. Borwein and T. Erdélyi, *Newton's Identities Section 1.1.E.2 in Polynomials and Polynomial Inequalities*, Springer-Verlag, New York, 1995.
55. J.H.A. Aho and J. Ullman, *The Design and Analysis of Computer Algorithms*, Addison-Wesley, Massachusetts, 1975.
56. B. Ristic, S. Arulampalam, and N.J. Gordon, *Beyond the Kalman Filter: Particle Filters for Tracking Applications*, Artech House, Norwood, MA, USA, 2004.
57. S. Davey, M. Rutten, and B. Cheung, A comparison of detection performance for several Track-Before-Detect algorithms, *EURASIP J. Adv. Signal Process.*, 2008(1), Article 41, 2008.

CHAPTER 4

The Continuous Time Roots of the Interacting Multiple Model Filter

HENK A.P. BLOM

4.1 INTRODUCTION

In literature, the Interacting Multiple Model (IMM) filter [1] has become one of the most popular maneuvering target tracking approaches and is often used as a benchmarking reference [2–7]. Although the IMM filter typically is known in a discrete time setting, originally it has been developed in pure continuous time [8]. The resulting continuous time IMM equations have subsequently guided the development of the discrete time IMM filter in [9] and [10]. The aim of this chapter is to present these continuous time roots of IMM, including their use for maneuvering target tracking in case of sudden continuous time mode switching events and observations that happen at nondeterministic discrete moments in time.

From a mathematical perspective, the problem considered is to recursively estimate a hybrid stochastic process $\{x_t, \theta_t\}$, where $\{x_t\}$ is an Euclidean valued stochastic process and $\{\theta_t\}$ is a discrete valued Markov process (chain). The dependence of the state process $\{x_t\}$ on the mode process $\{\theta_t\}$ poses an additional nonlinear filtering challenge to those posed by each of the two components alone. This chapter addresses nonlinear filtering for such a hybrid stochastic process evolving in continuous time. More precisely, the mode process $\{\theta_t\}$ is a continuous time finite state Markov chain, and $\{x_t\}$ is the solution of a stochastic differential equation (SDE) whose coefficients depend on $\{\theta_t\}$. By considering observations that are continuous in time, the problem is one of pure continuous filtering, and this exactly is the problem studied in [8].

The motivation of [8] for studying this continuous time filter problem was the following. For Markov jump linear systems, the prevailing view was to enumerate all potential mode history hypotheses $\{\theta_s; s \in [0, t]\}$ in a discrete time setting, and then to evaluate the conditional density of x_t as a weighted sum of mode history conditional

Integrated Tracking, Classification, and Sensor Management: Theory and Applications, Edited by Mahendra Mallick, Vikram Krishnamurthy, and Ba-Ngu Vo.

densities of x_t. This yields one density in the sum for each mode history hypothesis. For discrete time and continuous-discrete time filtering, this leads to an exponentially with time increasing number of hypotheses, for example, [11–13]. However, this prevailing view does not work the same for pure continuous time filtering [14].

In an attempt to find an escape from the prevailing view, [8] studied the pure continuous time filtering setting. In doing so, [8] could build on the great theoretical progress made in the 1970s regarding nonlinear filtering for processes involving sudden jumps [15–19]. At the same time the practical value of these theoretical results for filtering Markov jump linear systems did not meet up to expectations. For a few simple Markov jump linear cases a finite-dimensional exact nonlinear filter had been identified, that is, the Wonham filter [20] and the Björk filter [21]. For some other Markov jump linear cases several approaches had been studied [22–25], though without convincing results. Therefore, [8] explored a novel approach; that is to decompose the evolution of the joint conditional density into two coupled evolutions. One for the evolution of the conditional density of θ_t given the observations, and the other for the evolution of the mode conditional density of x_t given θ_t and the observations. The characterization of the latter evolution in [8] revealed the interaction behavior that has become known as the key characteristic of the IMM filter algorithm [1]. Because this interaction behavior is exact rather than an approximation, it may be of significant use in further studies of stochastic hybrid systems. This forms the motivation for presenting the continuous time roots of IMM in this chapter.

The filtering problems considered involve a finite state Markov chain $\{\theta_t\}$ and a solution of an SDE the coefficients of which depend of θ_t. First, Section 4.2 addresses the estimation of the Markov chain $\{\theta_t\}$ from a fully observed solution of the SDE. In literature this problem is known as Hidden Markov Model (HMM) filtering, and the basic solution is given by the Wonham filter [20]. The key contribution of Section 4.2 is to develop a transformation of this Wonham filter from an Ito type SDE to a Wiener type SDE. Next, Sections 4.3 and 4.4 address the filtering problem when both the Markov chain and the SDE solution are partially observed, that is imperfectly and/or incompletely.[1] Section 4.3 studies the structure of the joint conditional density and reveals the basic interaction behavior in the continuous time setting. Next, Section 4.4 considers the Markov jump linear case, and develops SDEs for Pre-IMM and IMM filters. Section 4.5 shows how the continuous time filter characterizations apply to filtering in case of continuous time $\{x_t, \theta_t\}$ and discrete time observations. Section 4.6 presents concluding remarks.

4.1.1 Background and Notation

The background used in this chapter consists of SDEs which may incorporate jumps [27–29], the theory of discontinuous semimartingales [19,30,31], and the fundamental filtering theorem [17–19]. These complementary developments come together in the theory of nonlinear filtering for continuous time Markov processes [15].

[1] A short version of the material in Sections 4.3 and 4.4 appears in [26].

The stochastic processes considered have an underlying complete probability space $(\Omega, \mathcal{F}, P)$ with $(\Omega, \mathcal{F})$ a measurable space and P a probability measure defined on the σ-algebra $\mathcal{F}$.[2] All stochastic processes have a time index $t \in [0, \infty)$, and have paths that are almost surely right continuous and have left-hand limits (left-hand limit is indicated by the subscript $t-$).

The following notations apply for a conditional density of a random process $\{x_t\}$ with measurable state space X. $p_{x_t}(\cdot)$ denotes the conditional density of x_t, such that $\int_U p_{x_t}(x)dx = P\{x_t \in U\}$ for any measurable $U \subset X$. $p_{x_t|\theta_s}(\cdot|\theta)$ denotes the conditional density of x_t given $\theta_t = \theta$. $p_{x_t|Y_t}(\cdot)$ denotes the conditional density of x_t given the σ-algebra Y_t, and $p_{x_t|\theta_t, Y_t}(\cdot|\theta)$ denotes the conditional density of x_t given $\theta_t = \theta$ and the σ-algebra Y_t. When there is no confusion about the σ-algebra involved, then the notation $\hat{x}_t$ stands for the conditional mean of x_t given the σ-algebra of observations up to t, that is $\hat{x}_t = E\{x_t|Y_t\}$. Similarly, $\hat{p}_{x_t}(\cdot) = p_{x_t|Y_t}(\cdot)$ and $\hat{p}_{x_t|\theta_t}(\cdot|\theta) = p_{x_t|\theta_t, Y_t}(\cdot|\theta)$.

4.2 HIDDEN MARKOV MODEL FILTER

4.2.1 Finite-State Markov Process

The process $\{\theta_t; t \geq 0\}$ is a finite state Markov chain assuming values in $\mathbb{M} = \{1, 2, \ldots, N\}$ and evolving according to the transition probabilities

$$
\begin{aligned}
P\{\theta_{t+\Delta} = \theta|\theta_t = \eta\} &= \lambda_{\eta\theta}\Delta + o(\Delta) \qquad , \text{if} \theta \neq \eta \\
&= 1 + \lambda_{\theta\theta}\Delta + o(\Delta) \quad , \text{if} \theta = \eta
\end{aligned} \tag{4.1}
$$

with $\lambda_{\theta\theta} = -\sum_{\substack{\eta \in \mathbb{M} \\ \eta \neq \theta}} \lambda_{\theta\eta}$, and $|\lambda_{\theta\theta}| < K$ for all $\theta \in \mathbb{M}$ for some finite constant K.

The process $\{\theta_t\}$ is observed through the following $\mathbb{R}^m$-valued process $\{y_t\}$ satisfying

$$
dy_t = g(\theta_t)dt + dv_t \tag{4.2}
$$

with all components of $g < \infty$ and for simplicity time invariant, $\{dv_t\}$ is an m-dimensional vector Wiener process with $E\{dv_t dv_t^T\} = Vdt$, and $\{dv_t\}$ is independent of $\{\theta_t\}$.

The filtering problem considered in this section is to characterize the evolution of the conditional probability $\hat{p}_{\theta_t}(\theta) \triangleq P\{\theta_0 = \theta|Y_t\}$. This is accomplished as follows. First, the evolution of $\{\theta_t\}$ is characterized as the solution of SDEs. Next, the exact recursive equation for $\hat{p}_{\theta_t}(\theta)$ is derived. Finally, transformations are introduced which allow a robust implementation of the optimal filter.

[2]The probability space embeds a right continuous filtration, on the positive time line, which includes all P-null sets of $\mathcal{F}$.

4.2.2 SDEs Having a Markov Chain Solution

In addition to the transition equation (4.1), for an $\mathbb{M}$-valued Markov chain $\{\theta_t\}$, we also introduce two probabilistically equivalent representations:

- $\{\theta_t\}$ as the solution of an SDE;
- an indicator representation form of $\{\theta_t\}$.

A representation of the former is

$$d\theta_t = \sum_{\substack{\theta, \eta \in \mathbb{M} \\ \theta \neq \eta}} 1_\eta(\theta_{t-})[\theta - \eta] d\rho_{\eta\theta,t}, \tag{4.3}$$

where $\{\rho_{\eta\theta,t}\}$, $\eta, \theta \in \mathbb{M}$, $\eta \neq \theta$, are $N^2 - N$ independent Poisson counting processes, which are independent of θ_0, and which have finite counting rates $\lambda_{\eta\theta}$ [32].

Equation (4.3) defines for each $\theta_0 \in \mathbb{M}$ and each realization of the Poisson counting processes a solution that can only jump at points of discontinuity of any of the Poisson counting processes. Because each has a finite counting rate, and there are a finite number of these Poisson counting processes, the number of jumps of $\{\theta_t\}$ is finite for any $t > 0$. This implies that (4.3) has a pathwise unique solution for each $\theta_0 \in \mathbb{M}$. Now it remains to show that this solution is probabilistically equivalent to (4.1). Using (4.3) yields for $\theta \neq \eta$

$$P\{\theta_{t+\Delta} = \theta | \theta_t = \eta\} = P\{(\rho_{\eta\theta,t+\Delta} - \rho_{\eta\theta,t}) \geq 1 | \theta_t = \eta\} + o(\Delta) = \lambda_{\eta\theta}\Delta + o(\Delta) \tag{4.4}$$

which implies that (4.1) holds true for the solution $\{\theta_t\}$ of (4.3).

Instead of representing $\{\theta_t\}$ as the solution of SDE (4.3), we can do so for the indicator process $\{I_t(\theta), \theta \in \mathbb{M}\}$, that is defined as follows

$$\begin{aligned} I_t(\theta) &= 1, \quad \text{if } \theta_t = \theta \\ &= 0, \quad \text{if } \theta_t \neq \theta. \end{aligned} \tag{4.5}$$

In order to assure equal distance between each two different realizations of I_t, it is common practice to consider the vector process $I_t = \mathrm{Col}\{I_t(\theta), \theta \in \mathbb{M}\}$ having N orthogonal unit vector realizations in $\{0, 1\}^N$. We develop an SDE for $\{I_t(\theta), \theta \in \mathbb{M}\}$ as follows

$$\begin{aligned} dI_t(\theta) &= I_t(\theta) - I_{t-}(\theta) \\ &= 1_\theta(\theta_t) - 1_\theta(\theta_{t-}) \\ &= \sum_{\eta \in \mathbb{M}} [1_\eta(\theta_{t-})1_\theta(\theta_t) - 1_\theta(\theta_{t-})1_\eta(\theta_t)] \end{aligned}$$

$$= \sum_{\substack{\eta \in \mathbb{M} \\ \eta \neq \theta}} [1_\eta(\theta_{t-})1_\theta(\theta_t) - 1_\theta(\theta_{t-})1_\eta(\theta_t)]$$

$$= \sum_{\substack{\eta \in \mathbb{M} \\ \eta \neq \theta}} [1_\eta(\theta_{t-})d\rho_{\eta\theta,t} - 1_\theta(\theta_{t-})d\rho_{\theta\eta,t}]$$

with $1_\theta(\theta) = 1$, and $1_\theta(\eta) = 0$, for $\eta \neq \theta$. From the last equality we get the following recursive SDE driven by the Poisson counting processes

$$dI_t(\theta) = \sum_{\substack{\eta \in \mathbb{M} \\ \eta \neq \theta}} [I_{t-}(\eta)d\rho_{\eta\theta,t} - I_{t-}(\theta)d\rho_{\theta\eta,t}]. \tag{4.6}$$

Because $p_{\theta_0}(\theta) = P\{\theta_0 = \theta\} = E\{I_0(\theta)\}$, the unconditional evolution of $\{\theta_t\}$ satisfies

$$dp_{\theta_t}(\theta) = E\{dI_t(\theta)\}$$

$$= E\left\{ \sum_{\substack{\eta \in \mathbb{M} \\ \eta \neq \theta}} [I_{t-}(\eta)d\rho_{\eta\theta,t} - I_{t-}(\theta)d\rho_{\theta\eta,t}] \right\}$$

$$= \sum_{\substack{\eta \in \mathbb{M} \\ \eta \neq \theta}} [E\{I_{t-}(\eta)d\rho_{\eta\theta,t}\} - E\{I_{t-}(\theta)d\rho_{\theta\eta,t}\}].$$

Finally, exploiting that $d\rho_{\eta\theta,t}$ is independent of $I_{t-}(\eta)$ for all $\theta, \eta \in \mathbb{M}$, yields

$$dp_{\theta_t}(\theta) = Mp_{\theta_t}(\theta)dt \tag{4.7}$$

with M being the forward generator of the finite state Markov process $\{\theta_t\}$, that is a linear operator on mappings $f : \mathbb{M} \to \mathbb{R}$, which satisfies for each such mapping

$$Mf(\theta) = \sum_{\substack{\eta \in \mathbb{M} \\ \eta \neq \theta}} [f(\eta)\lambda_{\eta\theta} - f(\theta)\lambda_{\theta\eta}]$$

$$= \sum_{\eta \in \mathbb{M}} \lambda_{\eta\theta} f(\eta). \tag{4.8}$$

4.2.3 Filtering a Hidden Markov Model (HMM)

This subsection introduces Ito differential equations for the conditional evolution of the process $\{\theta_t\}$ given the increasing sigma algebra of observations $Y_t = \sigma\{y_s; s \in [0, t]\}$, with $\{y_t\}$ being the solution of (4.2).

Theorem 1 (Wonham Equation) *The conditional evolution of the Markov process $\{\theta_t\}$ given the increasing sigma algebra of observations $Y_t = \sigma\{y_s; s \in [0, t]\}$ satisfies*

$$d\hat{p}_{\theta_t}(\theta) = M\hat{p}_{\theta_t}(\theta)dt + \hat{p}_{\theta_t}(\theta)[g(\theta) - \hat{g}_t]^T V^{-1} d\bar{v}_t \tag{4.9}$$

with M being the forward generator of Markov process $\{\theta_t\}$, and

$$\hat{g}_t = \sum_{\theta \in \mathbb{M}} g(\theta)\hat{p}_{\theta_t}(\theta) \tag{4.10}$$

$$d\bar{v}_t = dy_t - \hat{g}_t dt. \tag{4.11}$$

Proof Application of the fundamental filtering theorem [18, Theorem 8.1] to equations (4.2) and (4.6) yields the following differential for the conditional expectation of indicator $I_t(\theta)$ given Y_t

$$d\hat{I}_t(\theta) = E\left\{ \sum_{\substack{\eta \in \mathbb{M} \\ \eta \neq \theta}} [I_{t-}(\eta)d\rho_{\eta\theta,t} - I_{t-}(\theta)d\rho_{\theta\eta,t}] | Y_t \right\} + \hat{I}_t(\theta)[g(\theta) - \hat{g}_t]^T V^{-1} d\bar{v}_t.$$

Because $d\rho_{\eta\theta,t}$ is independent of Y_t and of $I_{t-}(\eta)$ for all $\theta, \eta \in \mathbb{M}$, this becomes

$$\begin{aligned} d\hat{I}_t(\theta) &= \sum_{\substack{\eta \in \mathbb{M} \\ \eta \neq \theta}} [\hat{I}_{t-}(\eta)\lambda_{\eta\theta} - \hat{I}_{t-}(\theta)\lambda_{\theta\eta}]dt + \hat{I}_t(\theta)[g(\theta) - \hat{g}_t]^T V^{-1} d\bar{v}_t \\ &= \sum_{\eta \in \mathbb{M}} \hat{I}_{t-}(\eta)\lambda_{\eta\theta} dt + \hat{I}_t(\theta)[g(\theta) - \hat{g}_t]^T V^{-1} d\bar{v}_t. \end{aligned}$$

From this equation follows that $\{\hat{I}_t(\theta)\}$ is continuous in time, hence $\hat{I}_{t-}(\theta)$ can be replaced by $\hat{I}_t(\theta)$, and we get

$$d\hat{I}_t(\theta) = M\hat{I}_t(\theta)dt + \hat{I}_t(\theta)[g(\theta) - \hat{g}_t]^T V^{-1} d\bar{v}_t.$$

Finally we write $\hat{p}_{\theta_t}(\theta)$ in terms of the conditional expectation of indicator $I_t(\theta)$

$$\hat{p}_{\theta_t}(\theta) = E\{I_t(\theta)|Y_t\} = \hat{I}_t(\theta),$$

which implies that the SDE obtained for $\hat{I}_t(\theta)$ also holds true for $\hat{p}_{\theta_t}(\theta)$. ■

Theorem 1 has originally been developed by in [20]. Part of the terms in (4.9) serves in assuring that the total weight of $\hat{p}_{\theta_t}$ sums up to unity. By dropping this condition, we get a characterization of the evolution of an unnormalized version $\hat{\pi}_{\theta_t}$ of $\hat{p}_{\theta_t}$; the best known form of this is the Zakai equation [33].

Theorem 2 (Zakai Equation) *The conditional evolution of the Markov process $\{\theta_t\}$ given the increasing sigma algebra of observations $Y_t = \sigma\{y_s; s \in [0, t]\}$ satisfies*

$$\hat{p}_{\theta_t}(\theta) = \frac{\hat{\pi}_{\theta_t}(\theta)}{\sum_{\eta \in \mathbb{M}} \hat{\pi}_{\theta_t}(\eta)} \tag{4.12}$$

with $\{\hat{\pi}_{\theta_t}(\theta)\}$ being the solution of $\hat{\pi}_{\theta_0}(\theta) = \hat{p}_{\theta_0}(\theta)$ and

$$d\hat{\pi}_{\theta_t}(\theta) = M\hat{\pi}_{\theta_t}(\theta)dt + \hat{\pi}_{\theta_t}(\theta)g(\theta)^T V^{-1}dy_t. \tag{4.13}$$

Proof Define an $\mathbb{R}$-valued process $\{\Lambda_t\}$ as the solution of

$$\Lambda_t = \Lambda_0 + \int_0^t \Lambda_s \hat{g}_s^T V^{-1} dy_s. \tag{4.14}$$

Also define $\hat{\pi}_{\theta_t}(\theta) \triangleq \Lambda_t \hat{p}_t(\theta)$, and divide both left- and right-hand sides by Λ_t, to get

$$\hat{p}_{\theta_t}(\theta) = \frac{\hat{\pi}_{\theta_t}(\theta)}{\Lambda_t}.$$

Summation of both sides over $\theta \in \mathbb{M}$ yields $\Lambda_t = \sum_{\theta \in \mathbb{M}} \hat{\pi}_{\theta_t}(\theta)$. The latter implies (4.12). Subsequent application of Ito's differentiation rule for discontinuous semi-martingales (see Appendix 4A) to $\hat{\pi}_{\theta_t}(\theta) = \Lambda_t \hat{p}_t(\theta)$ yields (4.13). ■

4.2.4 Robust Versions of the HMM Filter

Although equation (4.13) is linear $\hat{\pi}_{\theta_t}(\theta)$, its numerical evaluation easily goes wrong [34]. The key issue is that the second term in (4.13) multiplies the filtered estimate $\hat{\pi}_{\theta_t}(\theta)$ with its martingale part $\{\bar{\nu}_t\}$. In the Wonham equation, this problem even applies both to $\hat{p}_{\theta_t}(\theta)$ and $\hat{g}_t$. Such modulation of the outcome of an integral with its own driving martingale may create two problems: a mathematical problem and a physical interpretation problem. The mathematical problem is to perform the numerical integration in a way that respects the Ito nature of the SDE. For handling this mathematical problem good solutions have been developed [35]. The physical interpretation problem is that actual observations $\{y'_t\}$ may differ from the adopted $\{y_t\}$ model (4.8) at some very high frequencies that fall outside the HMM frequency range. Such a difference may cause the modulation to produce uncontrolled extra signals that subsequently influence the estimation of $\hat{\pi}_{\theta_t}(\theta)$ in some uncontrolled way, also when the numerical integration has been handled well from a mathematical perspective. Because it will be unrealistic to improve the model for arbitrarily high frequencies, this subsection shows two robust versions of the HMM filter:

- Multiplicative Functional Transformation [36];
- Log-Quotient Transformation [37].

The former approach is well known in nonlinear filtering literature, the latter not.

Theorem 3 (Multiplicative Functional Transformation) *The conditional evolution of the Markov process $\{\theta_t\}$ given the increasing sigma algebra of observations $Y_t = \sigma\{y_s; s \in [0, t]\}$ satisfies*

$$d\hat{r}_{\theta_t}(\theta) = \exp\{-g(\theta)^T V^{-1} y_t\} M[\exp\{-g(.)^T V^{-1} y_t\} \hat{r}_{\theta_t}(.)](\theta) dt$$
$$-{}^1\!/\!_2 g(\theta)^T V^{-1} g(\theta) \hat{r}_{\theta_t}(\theta) dt \tag{4.15}$$

$$\hat{p}_{\theta_t}(\theta) = \frac{\hat{r}_{\theta_t}(\theta) \exp\{g(\theta)^T V^{-1} y_t\}}{\sum_{\eta \in \mathbb{M}} \hat{r}_{\theta_t}(\eta) \exp\{g(\eta)^T V^{-1} y_t\}} \tag{4.16}$$

with initial condition $\hat{r}_{\theta_0}(\theta) = \hat{p}_{\theta_0}(\theta)$ and with M satisfying (4.10).

Proof Defining $\hat{r}_{\theta_t}(\theta) = \hat{\pi}_{\theta_t}(\theta) \exp\{-g(\theta)^T V^{-1} y_t\}$ and subsequently applying Ito's differentiation rule (Appendix 4A) yields (4.15). Due to the definition of $\hat{r}_{\theta_t}(\theta)$, we get $\hat{\pi}_{\theta_t}(\theta) = \hat{r}_{\theta_t}(\theta) \exp\{g(\theta)^T V^{-1} y_t\}$. Substituting this into (4.12) yields (4.16). ■

The differential in (4.15) no longer uses $\{dy_t\}$ as input, but $\{y_t\}$. Hence the resulting filter is robust regarding any high frequency differences between $\{y_t\}$ and $\{y'_t\}$. However, for (4.15) it is not clear what happens in case of valueless or no observations, that is when $V^{-1} = 0$. The alternative approach of [37] avoids this problem, by transforming the Zakai equation into a Wiener type equation. The transformation is taking logarithm of the un-normalized density, that is $\ln \hat{\pi}_{\theta_t}(\theta)$. Application of Ito's differentiation rule to $\ln \hat{\pi}_{\theta_t}(\theta)$, and using (4.13), yields the Log-Zakai equation.

Corollary 1 (Log-Zakai Equation) *The conditional evolution of the Markov process $\{\theta_t\}$ given the increasing sigma algebra of observations $Y_t = \sigma\{y_s; s \in [0, t]\}$ satisfies (4.12) with $\ln \hat{\pi}_{\theta_t}(\theta)$ being the solution of*

$$d \ln \hat{\pi}_{\theta_t}(\theta) = [\hat{\pi}_{\theta_t}(\theta)]^{-1} M \hat{\pi}_{\theta_t}(\theta) dt + g(\theta)^T V^{-1} [dy_t - {}^1\!/\!_2 g(\theta) dt]. \tag{4.17}$$

The Log-Zakai equation (4.17) does not involve multiplication of $\ln \hat{\pi}_{\theta_t}(\theta)$ with its martingale part. Hence, (4.17) is a Wiener type SDE. Nevertheless, straightforward numerical evaluation of (4.17) may lead to a numerical problem because the total measure $\hat{\Lambda}_t$ of $\hat{\pi}_{\theta_t}(\theta)$ increases with time. To see this, we substitute $dy_t = \hat{g}_t dt + d\bar{v}_t$ in (4.14), which yields

$$\hat{\Lambda}_t = \hat{\Lambda}_0 + \int_0^t \hat{\Lambda}_s \hat{g}_s^T V^{-1} [\hat{g}_s ds + d\bar{v}_s].$$

This means that $\{\hat{\Lambda}_t\}$ is a super-martingale, the value of which tends to grow with time. Hence, even the Log-Zakai equation lacks the kind of behavior that is needed for practical application.

According to [37], an effective way in avoiding this problem is to write the HMM filter equations for the quotient of conditional probabilities of $\theta_t = \theta$, $\theta \in \mathbb{M}$, versus the conditional probability of $\theta_t = \kappa$, for some suitably chosen reference value $\kappa \in \mathbb{M}$, that is

$$\hat{q}_{\theta_t}(\theta) = \frac{\hat{p}_{\theta_t}(\theta)}{\hat{p}_{\theta_t}(\kappa)}. \tag{4.18}$$

In such a quotient, the growing $\{\hat{\Lambda}_t\}$ effect cancels out. In order to prevent returning of estimation terms that form modulations in their own martingale part, the logarithm transformation is also maintained. The resulting filter is referred to as the Log-Quotient equation, which is a Wiener type SDE.

Theorem 4 (Log-Quotient Equation) *Assume there is a* $\kappa \in \mathbb{M}$ *for which* $g(\kappa) = 0$ *and* $p_{\theta_t}(\kappa) > 0$ *for all* t*. Then the conditional evolution of the Markov process* $\{\theta_t\}$ *given the increasing sigma algebra of observations* $Y_t = \sigma\{y_s; s \in [0, t]\}$ *satisfies*

$$\hat{p}_{\theta_t}(\theta) = \frac{\hat{q}_{\theta_t}(\theta)}{\sum_{\eta \in \mathbb{M}} \hat{q}_{\theta_t}(\eta)} \tag{4.19}$$

with $\{\hat{q}_{\theta_t}(\theta)\}$ *being the solution of* $\hat{q}_{\theta_0}(\theta) = \hat{p}_{\theta_0}(\theta)/\hat{p}_{\theta_0}(\kappa)$ *and*

$$d \ln \hat{q}_{\theta_t}(\theta) = \hat{q}_{\theta_t}(\theta)^{-1} M \hat{q}_{\theta_t}(\theta) dt - M \hat{q}_{\theta_t}(\kappa) dt + g(\theta)^T V^{-1}[dy_t - {}^1\!/\!{}_2 g(\theta) dt]. \tag{4.20}$$

Proof Substitution of (4.12) into (4.18), and subsequent evaluation yields

$$\hat{q}_{\theta_t}(\theta) = \frac{\hat{\pi}_{\theta_t}(\theta)}{\hat{\pi}_{\theta_t}(\kappa)}.$$

Taking logarithm yields $\ln \hat{q}_{\theta_t}(\theta) = \ln \hat{\pi}_{\theta_t}(\theta) - \ln \hat{\pi}_{\theta_t}(\kappa)$; hence

$$d \ln \hat{q}_{\theta_t}(\theta) = d \ln \hat{\pi}_{\theta_t}(\theta) - d \ln \hat{\pi}_{\theta_t}(\kappa).$$

Substitution of (4.17) into this equation, and subsequent evaluation yields

$$\begin{aligned} d \ln \hat{q}_{\theta_t}(\theta) = {} & \hat{q}_{\theta_t}(\theta)^{-1} M \hat{q}_{\theta_t}(\theta) dt + g(\theta)^T V^{-1}[dy_t - {}^1\!/\!{}_2 g(\theta) dt] \\ & - \hat{q}_{\theta_t}(\kappa)^{-1} M \hat{q}_{\theta_t}(\kappa) dt - g(\kappa)^T V^{-1}[dy_t - {}^1\!/\!{}_2 g(\theta) dt]. \end{aligned}$$

Taking account of $\hat{q}_{\theta_t}(\kappa) = 1$ and $g(\kappa) = 0$, this yields (4.20). ■

Remark 1 [38] has demonstrated that the Log-Quotient filter works very well for an application of quickest detection of the starting and stopping of alpha rhythm in an Electro Encephalo Gram (EEG).

4.3 SYSTEM WITH MARKOVIAN COEFFICIENTS

This section addresses the situation that the finite-state Markov chain $\{\theta_t\}$ of Section 4.2, influences the coefficients of an SDE.

4.3.1 The Filtering Problem Considered

We consider the $\mathbb{R}^n$-valued process $\{x_t\}$ as the solution of the following SDE

$$dx_t = a(x_t, \theta_t)dt + b(x_t, \theta_t)dw_t, \tag{4.21}$$

where $\{\theta_t\}$ is the process defined in subsection 4.2.1, $\{dw_t\}$ is an n-dimensional vector Wiener process with independent standard Brownian motion components independent of x_0 and $\{\theta_t\}$, and $a(x, \theta)$ and $b(x, \theta)$ are such that (4.21) has a pathwise unique solution for each initial x_0, for example [28]. Furthermore, there is an m-dimensional observation process $\{y_t\}$ satisfying

$$dy_t = h(x_t, \theta_t)dt + dv_t, \tag{4.22}$$

where all components of $h(x, \theta)$ are twice continuously differentiable in x; $\{dv_t\}$ is an m-dimensional vector Wiener process with $E\{dv_t dv_t^T\} = Vdt$, and $\{dv_t\}$ is independent of $\{x_t, \theta_t\}$ and of $\{dw_t\}$.

The filtering problem considered is to characterize the evolution of the conditional density–probability $\hat{p}_{x_t,\theta_t}(x, \theta) \triangleq p_{x_t,\theta_t|Y_t}(x, \theta)$ of the joint process $\{x_t, \theta_t\}$, given the increasing family of sigma-algebras $Y_t = \sigma\{y_s; s \in [0, t]\}$.

4.3.2 Evolution of the Joint Conditional Density

It is important to notice that the solution $\{x_t\}$ of (4.21) is not a Markov process. However, the joint process $\{x_t, \theta_t\}$ is a Markov process. The forward generator L of this joint Markov process is a Fokker–Planck–Kolmogorov operator acting upon real valued functions on $\mathbb{R}^n \times \mathbb{M}$, and satisfies

$$Lp_{x_t,\theta_t}(x, \theta) = L^\theta p_{x_t,\theta_t}(x, \theta) + Mp_{x_t,\theta_t}(x, \theta), \tag{4.23a}$$

where L^θ is the mode-conditional Fokker–Planck operator of $\{x_t\}$ given $\theta_t = \theta$, that is

$$\begin{aligned} L^\theta f(x) = &- \sum_{i=1}^{n} \frac{\partial}{\partial x_i}\{[a(x, \theta)]_i f(x, \theta)\} \\ &+ 1/2 \sum_{i,j=1}^{n} \frac{\partial^2}{\partial x_i \partial x_j}\{[b(x, \theta)b(x, \theta)^T]_{ij} f(x, \theta)\} \end{aligned} \tag{4.23b}$$

and M is the forward generator of the Markov process $\{\theta_t\}$, that is

$$Mf(x,\theta) = \sum_{\eta\in\mathbb{M}} f(x,\eta)\lambda_{\eta\theta}. \tag{4.23c}$$

The Fokker–Planck–Kolmogorov operator L determines the unconditional evolution of the joint density $p_{x_t,\theta_t}(x,\theta)$.

Proposition 1 *Let the probability measure $P\{x_t \in dx, \theta_t = \theta\}$ admit for all t a density $p_{x_t,\theta_t}(x,\theta)$, and let $L^{\theta} p_{x_t,\theta_t}(x,\theta)$ exist and be continuous for all t and all $\theta \in \mathbb{M}$. Then the evolution of density $p_{x_t,\theta_t}(x,\theta)$ satisfies the following differential*

$$dp_{x_t,\theta_t}(x,\theta) = Lp_{x_t,\theta_t}(x,\theta)dt. \tag{4.24}$$

Proof The first step is to represent the Poisson point processes in the differential (4.3) for $\{\theta_t\}$ by a Poisson random measure. Subsequently the derivation follows [27, pp. 298–299]. ■

Theorem 5 *Let the conditional probability $P\{x_t \in dx, \theta_t = \theta | Y_t\}$ admit for all t a density $\hat{p}_{x_t,\theta_t}(x,\theta)$, and let $L^{\theta}\hat{p}_{x_t,\theta_t}(x,\theta)$ exist and be continuous for all t and all $\theta \in \mathbb{M}$. Then $\hat{p}_{x_t,\theta_t}(x,\theta)$ satisfies*

$$d\hat{p}_{x_t,\theta_t}(x,\theta) = L\hat{p}_{x_t,\theta_t}(x,\theta)dt + \hat{p}_{x_t,\theta_t}(x,\theta)[h(x,\theta) - \hat{h}_t]^T V^{-1} d\bar{\nu}_t \tag{4.25}$$

with

$$\hat{h}_t = E\{h(x_t,\theta_t)|Y_t\} \tag{4.26}$$

$$d\bar{\nu}_t = dy_t - \hat{h}_t dt. \tag{4.27}$$

Proof The first step is to use the fundamental filtering theorem [19] for the characterization of the conditional expectation $\hat{f}(x_t,\theta_t) = E\{f(x_t,\theta_t)|Y_t\}$, for any mapping $f(x,\theta) : \mathbb{R}^n \times \mathbb{M} \to \mathbb{R}$ which is twice continuously differentiable in x. This yields the equation

$$d\hat{f}(\xi_t) = E\{L_B f(\xi_t)|Y_t\}dt + [E\{h(\xi_t)f(\xi_t)|Y_t\} - \hat{h}(\xi_t)\hat{f}(\xi_t)]^T V^{-1} d\bar{\nu}_t,$$

where $\xi_t \triangleq (x_t,\theta_t)$, and L_B is the backward generator of the joint Markov process $\{x_t,\theta_t\}$.

This differential is rewritten as follows

$$\sum_{\theta\in\mathbb{M}}\int_{\mathbb{R}^n} f(x,\theta)d\hat{p}_{x_t,\theta_t}(x,\theta)dx = \sum_{\theta\in\mathbb{M}}\int_{\mathbb{R}^n} \hat{p}_{x_t,\theta_t}(x,\theta)L_B f(x,\theta)dxdt$$
$$+\sum_{\theta\in\mathbb{M}}\int_{\mathbb{R}^n} \hat{p}_{x_t,\theta_t}(x,\theta)[h(x,\theta)f(x,\theta)$$
$$-\hat{h}(\xi_t)f(x,\theta)]^T dxV^{-1}d\bar{\nu}_t. \tag{4.28}$$

Next we make use of the adjoint relation that exists between the backward generator L_B and the forward generator L of a Markov process, that is

$$\sum_{\theta\in\mathbb{M}}\int_{\mathbb{R}^n} \hat{p}_{x_t,\theta_t}(x,\theta)L_B f(x,\theta)dx = \sum_{\theta\in\mathbb{M}}\int_{\mathbb{R}^n} f(x,\theta)L\hat{p}_{x_t,\theta_t}(x,\theta)dx.$$

Substituting this into (4.28) yields

$$\sum_{\theta\in\mathbb{M}}\int_{\mathbb{R}^n} f(x,\theta)d\hat{p}_{x_t,\theta_t}(x,\theta)dx = \sum_{\theta\in\mathbb{M}}\int_{\mathbb{R}^n} f(x,\theta)L\hat{p}_{x_t,\theta_t}(x,\theta)dxdt$$
$$+\sum_{\theta\in\mathbb{M}}\int_{\mathbb{R}^n} \hat{p}_{x_t,\theta_t}(x,\theta)[h(x,\theta)f(x,\theta)$$
$$-\hat{h}(\xi_t)f(x,\theta)]^T dxV^{-1}d\bar{\nu}_t.$$

Because the latter equation holds true for any twice continuously differentiable $f(x,\theta)$ this implies that (4.25) holds true. ∎

Remark 2 For $|\mathbb{M}| = 1$, equation (4.25) is well known as the optimal filter for a diffusion process [39, Theorem 6.5].

Similar to equation (4.9), some of the terms in (4.25) serve in assuring that the total weight of $\hat{p}_{x_t,\theta_t}$ sums up to unity. By dropping this condition, we get a differential characterization of the evolution of an unnormalized version $\hat{\pi}_{x_t,\theta_t}$ of $\hat{p}_{x_t,\theta_t}$; the best known form of this is the Duncan–Mortensen–Zakai equation [15].

Theorem 6 (Duncan–Mortensen–Zakai Equation) *The conditional evolution of the Markov process $\{x_t,\theta_t\}$ given the increasing sigma algebra of observations $Y_t = \sigma\{y_s; s\in[0,t]\}$ satisfies*

$$\hat{p}_{x_t,\theta_t}(x,\theta) = \frac{\hat{\pi}_{x_t,\theta_t}(x,\theta)}{\int_{\mathbb{R}^n}\sum_{\eta\in\mathbb{M}}\hat{\pi}_{x_t,\theta_t}(x',\eta)dx'} \tag{4.29c}$$

with $\{\hat{\pi}_{x_t,\theta_t}(x,\theta)\}$ *being the solution of*

$$d\hat{\pi}_{x_t,\theta_t}(x,\theta) = L\hat{\pi}_{x_t,\theta_t}(x,\theta)dt + \hat{\pi}_{x_t,\theta_t}(x,\theta)h(x,\theta)^T V^{-1} dy_t \tag{4.30}$$

under initial condition $\hat{\pi}_{x_0,\theta_0}(x,\theta) = \hat{p}_{x_0,\theta_0}(x,\theta)$.

Proof Define an $\mathbb{R}$-valued process $\{\Lambda_t\}$ as the solution of

$$\Lambda_t = \Lambda_0 + \int_0^t \Lambda_s \hat{h}_s^T dy_s.$$

Next we define $\hat{\pi}_{x_t,\theta_t}(x,\theta) \triangleq \Lambda_t \hat{p}_{x_t,\theta_t}(x,\theta)$. Dividing both left- and right-hand sides by Λ_t yields $\hat{p}_{x_t,\theta_t}(x,\theta) = \frac{\hat{\pi}_{x_t,\theta_t}(x,\theta)}{\Lambda_t}$. Integrating both sides over $\mathbb{R}^n \times \mathbb{M}$ yields $\Lambda_t = \int_{\mathbb{R}^n} \sum_{\theta\in\mathbb{M}} \hat{\pi}_{x_t,\theta_t}(x',\theta)dx'$, which implies (4.29). Subsequent application of Ito's differentiation rule to $\hat{\pi}_{x_t,\theta_t}(x,\theta) = \Lambda_t \hat{p}_{x_t,\theta_t}(x,\theta)$ yields (4.30). ■

Remark 3 For Markov jump systems, the DMZ filter equation (4.30) has originally been introduced in [23].

4.3.3 Evolution of the Conditional Density of x_t Given θ_t

This subsection decomposes the evolution of $\hat{p}_{x_t,\theta_t}(x,\theta)$ into coupled evolutions of $\hat{p}_{x_t|\theta_t}(x|\theta)$ and $\hat{p}_{\theta_t}(\theta)$. It is this decomposition in [8] which originally revealed the interaction behavior that has become so characteristic for the discrete time IMM filter algorithm.

Theorem 7 *Let the conditional probability* $P\{x_t \in dx|\theta_t = \theta, Y_t\}$ *admit for all t a density* $\hat{p}_{x_t|\theta_t}(x|\theta)$ *which is twice continuously differentiable in x, for all* $\theta \in \mathbb{M}$, *and let* $\hat{p}_{\theta_t}(\theta) > 0$, *then* $\hat{p}_{\theta_t}(\theta)$ *and* $\hat{p}_{x_t|\theta_t}(x|\theta)$ *evolve according to*

$$d\hat{p}_{\theta_t}(\theta) = M\hat{p}_{\theta_t}(\theta)dt + \hat{p}_{\theta_t}(\theta)(\hat{h}_t(\theta) - \hat{h}_t)^T V^{-1} d\bar{\nu}_t \tag{4.31}$$

$$\begin{aligned} d\hat{p}_{x_t|\theta_t}(x|\theta) &= L^\theta \hat{p}_{x_t|\theta_t}(x|\theta)dt + \hat{M}_t \hat{p}_{x_t|\theta_t}(x|\theta)dt \\ &\quad + \hat{p}_{x_t|\theta_t}(x|\theta)[h(x,\theta) - \hat{h}_t(\theta)]^T V^{-1} d\bar{\nu}_t(\theta), \end{aligned} \tag{4.32}$$

where $\hat{M}_t$ *is the interaction operator satisfying*

$$\hat{M}_t \hat{p}_{x_t|\theta_t}(x|\theta) = \sum_{\eta\in\mathbb{M}} \lambda_{\eta\theta} \frac{\hat{p}_{\theta_t}(\eta)}{\hat{p}_{\theta_t}(\theta)} [\hat{p}_{x_t|\theta_t}(x|\eta) - \hat{p}_{x_t|\theta_t}(x|\theta)] \tag{4.33}$$

and with

$$d\bar{\nu}_t = dy_t - \hat{h}_t dt \tag{4.34}$$

$$d\bar{v}_t(\theta) = dy_t - \hat{h}_t(\theta)dt \tag{4.35}$$

$$\hat{h}_t(\theta) = \int_{\mathbb{R}^n} h(x,\theta)\hat{p}_{x_t|\theta_t}(x|\theta)dx \tag{4.36}$$

$$\hat{h}_t = \sum_{\theta\in\mathbb{M}} \hat{h}_t(\theta)\hat{p}_{\theta_t}(\theta). \tag{4.37}$$

Proof By the law of total probability

$$\hat{p}_{\theta_t}(\theta) = \int_{\mathbb{R}^n} \hat{p}_{x_t,\theta_t}(x,\theta)dx.$$

Substituting (4.25) into this, and subsequent evaluation yields (4.31) with

$$\hat{h}_t(\theta) \triangleq E\{h(x_t,\theta_t)|\theta_t = \theta, Y_t\}.$$

From the latter follows (4.36) and (4.37).
Under the condition $\hat{p}_{\theta_t}(\theta) > 0$ the following holds true

$$\hat{p}_{x_t|\theta_t}(x|\theta) = \hat{p}_{\theta_t}(\theta)^{-1}\hat{p}_{x_t,\theta_t}(x,\theta).$$

Hence applying Ito's differentiation rule, using (4.25) and (4.31), yields

$$\begin{aligned} d\hat{p}_{x_t|\theta_t}(x|\theta) &= \hat{p}_{\theta_t}(\theta)^{-1}d\hat{p}_{x_t,\theta_t}(x,\theta) - \hat{p}_{x_t|\theta_t}(x|\theta)\hat{p}_{\theta_t}(\theta)^{-1}d\hat{p}_{\theta_t}(\theta)\\ &\quad+\hat{p}_{x_t|\theta_t}(x|\theta)(\hat{h}_t(\theta)-\hat{h}_t)^T V^{-1}(\hat{h}_t(\theta)-\hat{h}_t)dt. \end{aligned}$$

Substitution of (4.25) and (4.31) into this, and subsequent evaluation yields

$$\begin{aligned} d\hat{p}_{x_t|\theta_t}(x|\theta) &= \hat{p}_{\theta_t}(\theta)^{-1}L\hat{p}_{x_t,\theta_t}(x,\theta)dt - \hat{p}_{\theta_t}(\theta)^{-1}\hat{p}_{x_t|\theta_t}(x|\theta)M\hat{p}_{\theta_t}(\theta)dt\\ &\quad+\hat{p}_{x_t|\theta_t}(x|\theta)[h(x,\theta)-\hat{h}_t(\theta)]^T V^{-1}d\bar{v}_t(\theta). \end{aligned}$$

Evaluation of this, using (4.24a–c) yields (4.32) with $\hat{M}_t$ satisfying

$$\hat{M}_t\hat{p}_{x_t|\theta_t}(x|\theta) = \hat{p}_{\theta_t}(\theta)^{-1}M\hat{p}_{x_t,\theta_t}(x,\theta) - \hat{p}_{\theta_t}(\theta)^{-1}\hat{p}_{x_t|\theta_t}(x|\theta)M\hat{p}_{\theta_t}(\theta).$$

Further evaluation of the latter yields

$$\begin{aligned} \hat{M}_t\hat{p}_{x_t|\theta_t}(x|\theta) &= \hat{p}_{\theta_t}(\theta)^{-1}\sum_{\eta\in\mathbb{M}}[\hat{p}_{x_t,\theta_t}(x,\eta)\lambda_{\eta\theta} - \hat{p}_{x_t|\theta_t}(x|\theta)\hat{p}_{\theta_t}(\eta)\lambda_{\eta\theta}]\\ &= \hat{p}_{\theta_t}(\theta)^{-1}\sum_{\eta\in\mathbb{M}}\lambda_{\eta\theta}\hat{p}_{\theta_t}(\eta)[\hat{p}_{x_t|\theta_t}(x|\eta) - \hat{p}_{x_t|\theta_t}(x|\theta)]. \end{aligned}$$

■

In comparing (4.31) and (4.32) with (4.25), the appearance of the term $\hat{M}_t \hat{p}_{x_t|\theta_t}(x|\theta)$ stands out, in particular because the quotient $\frac{\hat{p}_{\theta_t}(\eta)}{\hat{p}_{\theta_t}(\theta)}$ causes $\hat{M}_t$ to be a stochastic operator, which is conditionally linear given these quotients.

4.3.4 Special Cases

In order to gain insight in the filter equations above, three special cases are considered.

Case 1. When $\{\theta_t\}$ does not switch, then $M = \hat{M}_t = 0$. Hence (4.31) and (4.32) become

$$d\hat{p}_{\theta_t}(\theta) = \hat{p}_{\theta_t}(\theta)(\hat{h}_t(\theta) - \hat{h}_t)^T V^{-1} d\bar{v}_t \tag{4.38}$$

$$\begin{aligned} d\hat{p}_{x_t|\theta_t}(x|\theta) = L^\theta \hat{p}_{x_t|\theta_t}(x|\theta)dt + \hat{p}_{x_t|\theta_t}(x|\theta)[h(x,\theta) \\ -\hat{h}_t(\theta)]^T V^{-1} d\bar{v}_t(\theta). \end{aligned} \tag{4.39}$$

Case 2. In case of $\{\theta_t\}$ observations only, that is $h(x, \theta)$ is x-invariant for $\forall\theta \in \mathbb{M}$.

Then in (4.32) the martingale term becomes zero, that is

$$d\hat{p}_{x_t|\theta_t}(x|\theta) = L^\theta \hat{p}_{x_t|\theta_t}(x|\theta)dt + \hat{M}_t \hat{p}_{x_t|\theta_t}(x|\theta)dt. \tag{4.40}$$

This means that the mode conditional evolution of $\{x_t\}$ depends on the observation process $\{y_t\}$ in an indirect way only, that is through the evolution of the conditional mode probabilities $\hat{p}_{\theta_t}(\theta)$.

Case 3. In case of no $\{\theta_t\}$ observations at all, that is $h(x, \theta) = h'(x)$.

Then $\hat{h}_t(\theta) = E\{h'(x_t)|\theta_t = \theta, Y_t\} = \int h'(x)\hat{p}_{x_t|\theta_t}(x|\theta)dx$. This implies that $\hat{h}_t(\theta)$ in (4.31) still is a function of θ (unless both $a(x, \theta)$ and $b(x, \theta)$ would be θ-invariant also). Hence, both the last term in (4.31) and the last term in (4.32) do not vanish.

4.4 MARKOV JUMP LINEAR SYSTEM

In this section, we study the situation that the finite-state Markov chain $\{\theta_t\}$ influences the coefficients of a linear system.

4.4.1 The Filtering Problem Considered

In order to study the situation that the finite-state Markov chain $\{\theta_t\}$ influences the coefficients of a linear system, we assume $h(x, \theta)$ and $a(x, \theta)$ are affine in x, and

$b(x, \theta)$ is x-invariant, that is

$$\begin{aligned} h(x,\theta) &= H(\theta)x + G(\theta) \\ a(x,\theta) &= A(\theta)x + C(\theta) \\ b(x,\theta) &= B(\theta). \end{aligned}$$

The above means we consider the $\mathbb{R}^n$-valued process $\{x_t\}$ as the solution of the following SDE

$$dx_t = A(\theta_t)x_t dt + C(\theta_t)dt + B(\theta_t)dw_t, \tag{4.41}$$

where all components of A, B, and C are $< \infty$ and for simplicity time invariant; $\{dw_t\}$ is an n-dimensional vector Wiener process with independent standard Brownian motion components, and independent of x_0 and $\{\theta_t\}$.

Furthermore, we assume m-dimensional observations satisfying the jump linear SDE

$$dy_t = H(\theta_t)x_t dt + G(\theta_t)dt + dv_t, \tag{4.42}$$

where all components of h are $< \infty$ and for simplicity time invariant, $\{dv_t\}$ is a vector Wiener process with $E\{dv_t dv_t^T\} = Vdt$, and $\{dv_t\}$ is independent of $\{x_t, \theta_t\}$ and of $\{dw_t\}$.

The remainder of this section develops for this problem differential equations for the conditional evolution of first- and second-order central moments (subsection 4.4.2), develops the continuous-time IMM filter (subsection 4.4.3), and shows the relation with the filter of Björk [21] (subsections 4.4.4 and 4.4.5).

4.4.2 Pre-IMM Filter Equations

The derivation starts with the first two mode conditional central moments of the process x_t given Y_t, under the assumption that $\hat{p}_{\theta_t}(\theta) > 0$. These mode conditional central moments are defined as follows

$$\begin{aligned} \hat{x}_t(\theta) &\triangleq E\{x_t | \theta_t = \theta, Y_t\} \\ \hat{R}_t(\theta) &\triangleq E\{[x_t - \hat{x}_t(\theta)][x_t - \hat{x}_t(\theta)]^T | \theta_t = \theta, Y_t\}. \end{aligned}$$

Theorem 8 *Let $\hat{p}_{\theta_t}(\theta) > 0$, then the mode conditional first- and second-order central moments satisfy the following set of equations*

$$d\hat{p}_{\theta_t}(\theta) = M\hat{p}_{\theta_t}(\theta)dt + \hat{p}_{\theta_t}(\theta)[H(\theta)\hat{x}_t(\theta) + G(\theta) - \hat{h}_t)]^T V^{-1} d\bar{v}_t \quad (4.43)$$

$$\begin{aligned} d\hat{x}_t(\theta) &= A(\theta)\hat{x}_t(\theta)dt + C(\theta)dt + \hat{R}_t(\theta)\mathrm{H}(\theta)^{\mathrm{T}} V^{-1} d\bar{v}_t(\theta) \\ &+ \sum_{\substack{\eta \in \mathbb{M} \\ \eta \neq \theta}} \lambda_{\eta\theta} \frac{\hat{p}_{\theta_t}(\eta)}{\hat{p}_{\theta_t}(\theta)} [\hat{x}_t(\eta) - \hat{x}_t(\theta)]dt \end{aligned} \quad (4.44)$$

$$\begin{aligned} d\hat{R}_t(\theta) &= A(\theta)\hat{R}_t(\theta)dt + \hat{R}_t(\theta)A(\theta)^T dt + B(\theta)B(\theta)^T dt \\ &- \hat{R}_t(\theta)\mathrm{H}(\theta)^{\mathrm{T}} V^{-1} H(\theta)\hat{R}_t(\theta)dt + d\varsigma_t(\theta) \\ &+ \sum_{\substack{\eta \in \mathbb{M} \\ \eta \neq \theta}} \lambda_{\eta\theta} \frac{\hat{p}_{\theta_t}(\eta)}{\hat{p}_{\theta_t}(\theta)} [\hat{R}_t(\eta) - \hat{R}_t(\theta) + [\hat{x}_t(\eta) - \hat{x}_t(\theta)][..]^T]dt \end{aligned} \quad (4.45a)$$

$$d\varsigma_t(\theta) = \sum_{i=1}^{n} E\{(x_t - \hat{x}_t(\theta))(..)^{\mathrm{T}}[(..)]_i | \theta_t = \theta, Y_t\}[H(\theta)^{\mathrm{T}} V^{-1} d\bar{v}_t(\theta)]_i \quad (4.45b)$$

with

$$d\bar{v}_t = dy_t - \hat{h}_t dt \quad (4.46)$$

$$d\bar{v}_t(\theta) = dy_t - \hat{h}_t(\theta)dt \quad (4.47)$$

$$\hat{h}_t(\theta) = H(\theta)\hat{x}_t(\theta) + G(\theta) \quad (4.48)$$

$$\hat{h}_t = \sum_{\theta \in \mathbb{M}} \hat{h}_t(\theta)\hat{p}_{\theta_t}(\theta). \quad (4.49)$$

Proof Due to (4.41) and (4.42), (4.31) yields (4.43). Moreover (4.32) becomes

$$\begin{aligned} d\hat{p}_{x_t|\theta_t}(x|\theta) &= L^{\theta}\hat{p}_{x_t|\theta_t}(x|\theta)dt + \hat{M}_t\hat{p}_{x_t|\theta_t}(x|\theta)dt \\ &+ \hat{p}_{x_t|\theta_t}(x|\theta)[x - \hat{x}_t(\theta)]H(\theta)^T V^{-1} d\bar{v}_t(\theta). \end{aligned} \quad (4.50)$$

From this we derive differentials for first and second conditional central moments of x_t given θ_t and Y_t, as follows. By the definition of $\hat{x}_t(\theta)$ we get

$$\hat{x}_t(\theta) = \int_{\mathbb{R}^n} x\hat{p}_{x_t|\theta_t}(x|\theta)dx.$$

Hence

$$d\hat{x}_t(\theta) = \int_{\mathbb{R}^n} x d\hat{p}_{x_t|\theta_t}(x|\theta)dx.$$

Substitution of (4.50) into this, and subsequent evaluation yields (4.44).
The derivation of (4.45a,b) is given in Appendix 4B. ■

The set of equations (4.43)–(4.45) forms a nonlinear filter to which we refer as the Pre-IMM filter equations. Equations (4.44) and (4.45) incorporate a specific adaptive mechanism in the form of the interaction terms

$$\sum_{\substack{\eta \in \mathbb{M} \\ \eta \neq \theta}} \lambda_{\eta\theta} \frac{\hat{p}_{\theta_t}(\eta)}{\hat{p}_{\theta_t}(\theta)} [\hat{x}_t(\eta) - \hat{x}_t(\theta)]$$

$$\sum_{\substack{\eta \in \mathbb{M} \\ \eta \neq \theta}} \lambda_{\eta\theta} \frac{\hat{p}_{\theta_t}(\eta)}{\hat{p}_{\theta_t}(\theta)} [\hat{R}_t(\eta) - \hat{R}_t(\theta) + [\hat{x}_t(\eta) - \hat{x}_t(\theta)][..]^T].$$

These interaction terms involve the quotient $\frac{\hat{p}_{\theta_t}(\eta)}{\hat{p}_{\theta_t}(\theta)}$. It has been shown in Section 4.2 that these quotients can easily be evaluated numerically using logarithm transform. The only complication is occurrence of $d\varsigma_t(\theta)$ in (4.45), by which the Pre-IMM filter is not finite dimensional.

Remark 4 If the mode process $\{\theta_t\}$ does not switch, that is $M\hat{p}_{\theta_t}(\theta) = 0$ for all $\theta \in M$, then the set of equations (4.43)–(4.45) simplifies to the continuous-time Multiple Model (MM) filter [40,41].

4.4.3 Continuous-Time IMM Filter

A sufficient condition for a set of differentials to form a finite-dimensional nonlinear filter is that all stochastic processes that appear at the right-hand sides of these differential equations, are defined by the process differentials at the left-hand sides. The set of Pre-IMM equations (4.43)–(4.45) almost satisfies this condition; the only exception is formed by equation (4.45b). Hence (4.43)–(4.45) forms a finite-dimensional filter if $d\varsigma_t(\theta) = 0$ for all $\theta \in \mathbb{M}$, in which case (4.45a) becomes

$$\begin{aligned} d\hat{R}_t(\theta) = {} & A(\theta)\hat{R}_t(\theta)dt + \hat{R}_t(\theta)A(\theta)^T dt + B(\theta)B(\theta)^T dt \\ & -\hat{R}_t(\theta)H(\theta)^{\mathrm{T}} V^{-1} H(\theta)\hat{R}_t(\theta)dt \\ & +\hat{p}_{\theta_t}(\theta)^{-1} \sum_{\substack{\eta \in \mathbb{M} \\ \eta \neq \theta}} \lambda_{\eta\theta}\hat{p}_{\theta_t}(\eta)[\hat{R}_t(\eta) - \hat{R}_t(\theta) + [\hat{x}_t(\eta) - \hat{x}_t(\theta)][..]^T]dt. \end{aligned} \tag{4.51}$$

Equations (4.43), (4.44) and (4.51) form the continuous time IMM filter. The basic structure consists of a Wonham filter (4.43) for $\hat{p}_{\theta_t}(\theta)$ plus a bank of mode-directed filters (4.44) and (4.51) for $\hat{x}_t(\theta)$ and $\hat{R}_t(\theta)$, respectively. Each mode-directed filter in this bank combines mode-directed Kalman–Bucy equations with interactions between mode conditional state estimates.

In general, $d\varsigma_t(\theta) \neq 0$, and then the Pre-IMM equations (4.43)–(4.45) do not form a closed system of equations. From information theory it is well known that, when the first and second central moments of $\hat{p}_{x_t|\theta_\tau}$ are given, then the entropy is maximized by assuming a Gaussian-shaped density, which comes down to approximating the third order central moment of $\hat{p}_{x_t|\theta_\tau}$ to zero. From (4.45b) it can be seen that this implies $d\varsigma_t(\theta) \simeq 0$. Thus by adopting a second central order neutral maximum entropy approximation for $\hat{p}_{x_t|\theta_\tau}$ we get $d\varsigma_t(\theta) \simeq 0$, by which (4.45a) becomes

$$\begin{aligned} d\hat{R}_t(\theta) \simeq\ & A(\theta)\hat{R}_t(\theta)dt + \hat{R}_t(\theta)A(\theta)^T dt + B(\theta)B(\theta)^T dt \\ & - \hat{R}_t(\theta)\mathrm{H}(\theta)^{\mathrm{T}} V^{-1} H(\theta)\hat{R}_t(\theta)dt \\ & + \hat{p}_{\theta_t}(\theta)^{-1} \sum_{\substack{\eta \in \mathbb{M} \\ \eta \neq \theta}} \lambda_{\eta\theta}\hat{p}_{\theta_t}(\eta)[\hat{R}_t(\eta) - \hat{R}_t(\theta) + [\hat{x}_t(\eta) - \hat{x}_t(\theta)][..]^T]dt. \end{aligned} \tag{4.51$'$}$$

Equations (4.43), (4.44), and (4.51′) define the continuous-time IMM filter as an approximation of the exact filter equations for a Markov jump linear system for which $d\varsigma_t(\theta) \neq 0$.

4.4.4 Linear Version of the Pre-IMM Equations

This subsection develops a version of the Pre-IMM filter equations (4.44) and (4.45) where all nonmartingale terms are linear. In order to avoid the condition $\hat{p}_{\theta_t}(\theta) > 0$, we follow the direct derivation in [42] rather than deriving it from the Pre-IMM equations.

First we define the processes $\{z_t(\theta)\}$ and $\{Z_t(\theta)\}$ as follows

$$\begin{aligned} z_t(\theta) &\triangleq I_t(\theta)x_t \\ Z_t(\theta) &\triangleq I_t(\theta)x_t x_t^T \end{aligned}$$

with $I_t(\theta)$ defined by (4.4).

Using these definitions of $I_t(\theta)$, $z_t(\theta)$, and $Z_t(\theta)$, we get

$$x_t = \sum_{\theta \in \mathbb{M}} z_t(\theta) \tag{4.52a}$$

$$x_t x_t^T = \sum_{\theta \in \mathbb{M}} Z_t(\theta) \tag{4.52b}$$

and

$$dy_t = \sum_{\theta \in \mathbb{M}} [H(\theta)z_t(\theta) + G(\theta)I_t(\theta)]dt + dv_t. \tag{4.53}$$

Next we define the filtered estimates $\{\hat{p}_{\theta_t}(\theta)\}$, $\{\hat{z}_t(\theta)\}$, and $\{\hat{Z}_t(\theta)\}$ as follows

$$\begin{aligned}\hat{p}_{\theta_t}(\theta) &\triangleq E\{I_t(\theta)|Y_t\}\\ \hat{z}_t(\theta) &\triangleq E\{z_t(\theta)|Y_t\}\\ \hat{Z}_t(\theta) &\triangleq E\{Z_t(\theta)|Y_t\}.\end{aligned}$$

Then from (4.52a,b) we get

$$\begin{aligned}\hat{x}_t = E\{x_t|Y_t\} &= \sum_{\theta\in\mathbb{M}} \hat{z}_t(\theta)\\ E\{x_t x_t^T|Y_t\} &= \sum_{\theta\in\mathbb{M}} \hat{Z}_t(\theta).\end{aligned}$$

Theorem 9 *The conditional first- and second-order moments of the process $z_t(\theta)$ satisfy the following set of equations*

$$d\hat{p}_{\theta_t}(\theta) = M\hat{p}_{\theta_t}(\theta)dt + [H(\theta)\hat{z}_t(\theta) + (G(\theta) - \hat{h}_t)\hat{p}_{\theta_t}(\theta)]^T V^{-1} d\bar{\upsilon}_t \tag{4.54}$$

$$\begin{aligned}d\hat{z}_t(\theta) = {}&[A(\theta)\hat{z}_t(\theta) + C(\theta)\hat{p}_{\theta_t}(\theta) + \sum_{\eta\in\mathbb{M}} \lambda_{\eta\theta}\hat{z}_t(\eta)]dt\\ &+[\hat{Z}_t(\theta)H(\theta)^T + \hat{z}_t(\theta)(G(\theta) - \hat{h}_t)^{\mathrm{T}}]V^{-1}d\bar{\upsilon}_t\end{aligned} \tag{4.55}$$

$$\begin{aligned}d\hat{Z}_t(\theta) = {}&[A(\theta)\hat{Z}_t(\theta) + \hat{Z}_t(\theta)A(\theta)^T + C(\theta)\hat{z}_t(\theta)^T\\ &+\hat{z}_t(\theta)C(\theta)^T + B(\theta)B(\theta)^T\hat{p}_{\theta_t}(\theta)]dt\\ &+\sum_{\eta\in\mathbb{M}} \lambda_{\eta\theta}\hat{Z}_t(\eta)dt\\ &+\left[\sum_{i=1}^{n} E\{Z_t(\theta)x_{t,i}|Y_t\}(H(\theta)^T)_i + \hat{Z}_t(\theta)(G(\theta) - \hat{h}_t)^T\right]V^{-1}d\bar{\upsilon}_t\end{aligned} \tag{4.56}$$

with

$$d\bar{\upsilon}_t = dy_t - \hat{h}_t dt \tag{4.57}$$

$$\hat{h}_t = \sum_{\theta\in\mathbb{M}} [H(\theta)\hat{z}_t(\theta) + G(\theta)\hat{p}_{\theta_t}(\theta)] \tag{4.58}$$

Proof We follow [42]

$$
\begin{aligned}
\hat{z}_t(\theta) &\triangleq E\{z_t(\theta)|Y_t\} \\
&= E\{I_t(\theta)x_t|Y_t\} \\
&= \sum_{\eta\in\mathbb{M}} \hat{p}_{\theta_t}(\eta)E\{I_t(\theta)x_t|\theta_t = \eta, Y_t\} \\
&= \hat{p}_{\theta_t}(\theta)E\{I_t(\theta)x_t|\theta_t = \theta, Y_t\} \\
&= \hat{p}_{\theta_t}(\theta)E\{x_t|\theta_t = \theta, Y_t\} \\
&= \hat{p}_{\theta_t}(\theta)\hat{x}_t(\theta).
\end{aligned}
$$

Substituting the latter into (4.43) yields (4.54).

Application of the differentiation rule for discontinuous semimartingales to the definitions of $\{z_t(\theta)\}$ and $\{Z_t(\theta)\}$ yields

$$
\begin{aligned}
dz_t(\theta) &= [A(\theta)z_t(\theta) + C(\theta)I_t(\theta) + \sum_{\eta\in\mathbb{M}} \lambda_{\eta\theta}z_t(\eta)]dt + I_t(\theta)B(\theta)dw_t + dm_t^z(\theta) \\
dZ_t(\theta) &= [A(\theta)Z_t(\theta) + Z_t(\theta)A(\theta)^T + C(\theta)z_t(\theta)^T + z_t(\theta)C(\theta)^T + B(\theta)B(\theta)^T I_t(\theta)]dt \\
&\quad + \sum_{\eta\in\mathbb{M}} \lambda_{\eta\theta}Z_t(\eta)dt + [B(\theta)x_t(\theta)^T + x_t(\theta)B(\theta)^T]dw_t + dm_t^Z(\theta),
\end{aligned}
$$

where $\{m_t^z(\theta)\}$ and $\{m_t^Z(\theta)\}$ are purely discontinuous martingales.

Application of the fundamental filtering theorem to the equations for $dz_t(\theta)$ and $dZ_t(\theta)$, and subsequent evaluation, yields (4.55) and (4.56), respectively. ■

All nonmartingale terms in (4.54)–(4.56) are linear in $\{\hat{p}_{\theta_t}(\theta), \hat{z}_t(\theta), \hat{Z}_t(\theta)\}$. [42] also has shown that for $\hat{p}_{\theta_t}(\theta) > 0$, equations (4.54)–(4.56) can be transformed to the Pre-IMM equations (4.44)–(4.45) for $\hat{x}_t(\theta)$ and $\hat{R}_t(\theta)$. The transformations to do so are

$$
\hat{x}_t(\theta) = \hat{p}_{\theta_t}(\theta)^{-1}\hat{z}_t(\theta) \tag{4.59a}
$$

$$
\hat{R}_t(\theta) = \hat{p}_{\theta_t}(\theta)^{-1}[\hat{Z}_t(\theta) - \hat{z}_t(\theta)\hat{x}_t(\theta)^T]. \tag{4.59b}
$$

It can be verified that application of Ito's differentiation rule to (4.59a,b), while using (4.54)–(4.56), indeed yields (4.44) and (4.45).

Remark 5 [43] exploits the linearity of the nonmartingale terms in (4.54)–(4.56) for the derivation of the Open-Loop-Optimal-Feedback (OLOF) control policy for a partially observed Markov jump linear system.

Remark 6 [44] exploits the linearity of the nonmartingale terms in (4.54) and (4.55) for the derivation of a stationary optimal linear filter.

4.4.5 Relation Between Bjork's Filter and Continuous-Time IMM

Björk [21] considered the Markov jump linear filtering problem given mode observations only, that is $H(\theta) = 0$. Under this condition, several terms in Pre-IMM equations (4.43)–(4.45) become zero, including $d\varsigma_t(\theta) = 0$. The latter leads to the following special version of the IMM filter equations

$$d\hat{p}_{\theta_t}(\theta) = M\hat{p}_{\theta_t}(\theta)dt + \hat{p}_{\theta_t}(\theta)\left[G(\theta) - \sum_{\eta\in\mathbb{M}} \hat{p}_{\theta_t}(\eta)G(\eta)\right]^T V^{-1}d\bar{v}_t \tag{4.60}$$

$$d\hat{x}_t(\theta) = A(\theta)\hat{x}_t(\theta)dt + C(\theta)dt + \hat{p}_{\theta_t}(\theta)^{-1}\sum_{\substack{\eta\in\mathbb{M}\\ \eta\neq\theta}} \lambda_{\eta\theta}\hat{p}_{\theta_t}(\eta)[\hat{x}_t(\eta) - \hat{x}_t(\theta)]dt \tag{4.61}$$

$$\begin{aligned} d\hat{R}_t(\theta) &= A(\theta)\hat{R}_t(\theta)dt + \hat{R}_t(\theta)A(\theta)^T dt + B(\theta)B(\theta)^T dt \\ &\quad + \hat{p}_{\theta_t}(\theta)^{-1}\sum_{\substack{\eta\in\mathbb{M}\\ \eta\neq\theta}} \lambda_{\eta\theta}\hat{p}_{\theta_t}(\eta)[\hat{R}_t(\eta) - \hat{R}_t(\theta) + [\hat{x}_t(\eta) - \hat{x}_t(\theta)][..]^T]dt. \end{aligned} \tag{4.62}$$

The set of equations (4.60)–(4.62) defines a finite-dimensional filter, though differs from the finite-dimensional filter equations in [21]. In order to establish the relation with these Björk filter equations, we transform (4.61) and (4.62) to the linear form of the previous subsection, which yields

$$\begin{aligned} d\hat{z}_t(\theta) &= A(\theta)\hat{z}_t(\theta)dt + C(\theta)\hat{p}_{\theta_t}(\theta)dt + \sum_{\eta\in\mathbb{M}} \lambda_{\eta\theta}\hat{z}_t(\eta)dt \\ &\quad + \hat{z}_t(\theta)\left[G(\theta) - \sum_{\eta\in\mathbb{M}} G(\eta)\hat{p}_{\theta_t}(\eta)\right]^T V^{-1}d\bar{v}_t \end{aligned} \tag{4.63}$$

$$\begin{aligned} d\hat{Z}_t(\theta) &= [A(\theta)\hat{Z}_t(\theta) + \hat{Z}_t(\theta)A(\theta)^T + C(\theta)\hat{z}_t(\theta)^T + \hat{z}_t(\theta)C(\theta)^T \\ &\quad + B(\theta)B(\theta)^T\hat{p}_{\theta_t}(\theta)]dt \\ &\quad + \sum_{\eta\in\mathbb{M}} \lambda_{\eta\theta}\hat{Z}_t(\eta)dt + \hat{Z}_t(\theta)[G(\theta) - \sum_{\eta\in\mathbb{M}} G(\eta)\hat{p}_{\theta_t}(\eta)]^T V^{-1}d\bar{v}_t. \end{aligned} \tag{4.64}$$

Equations (4.60) and (4.63) are the finite-dimensional filter equations developed in [21]. In contrast with (4.60) and (4.61), all nonmartingale terms in (4.63) are linear in $(\hat{p}_t(\theta), \hat{z}_t(\theta))$. However, (4.63) has a martingale term which is nonlinear in $(\hat{p}_t(\theta),\hat{z}_t(\theta))$ and involves multiplication with the innovation process $\{d\bar{v}_t\}$. This makes numerical evaluation of the Björk equations (4.60) and (4.63) more demanding than numerical evaluation of the IMM type of equations (4.60) and (4.61).

4.5 CONTINUOUS-DISCRETE FILTERING

This section addresses continuous-discrete filtering versions of the problems studied in the previous two sections. The key difference is that the observation process is now a point process, that is an increasing sequence of sampling times, each of which comes with an $\mathbb{R}^n$-valued observation.

4.5.1 The Continuous-Discrete Filtering Problem Considered

We consider the $\mathbb{R}^n$-valued process $\{x_t\}$ as the solution of SDE (4.21), that is

$$dx_t = a(x_t, \theta_t)dt + b(x_t, \theta_t)dw_t, \tag{4.65}$$

where $\{\theta_t\}$ is the process defined in subsection 4.2.1, $\{dw_t\}$ is an n-dimensional vector Wiener process with independent standard Brownian motion components independent of x_0 and $\{\theta_t\}$, and $a(x, \theta)$ and $b(x, \theta)$ are such that (4.65) has a pathwise unique solution for each initial x_0.

Furthermore, there is a sequence of m-dimensional observations $\{t_i, y_i; i = 1, 2, 3, \ldots\}$ satisfying

$$y_i = h(x_{t_i}, \theta_{t_i}) + g(\theta_{t_i})n_i, \tag{4.66}$$

where $\{t_i\}$ is a strictly increasing sequence of observation times, that is $t_i > t_{i-1}, \forall i \geq 1$; the sequence $\{t_i\}$ is independent of $\{x_t, \theta_t\}$; all components of $h(x, \theta)$ are twice continuously differentiable in x; $\{n_i\}$ is a sequence of of m-dimensional variables, each component of which is independent and standard Gaussian; $\{n_i\}$ also is independent of $\{x_t, \theta_t\}$ and of $\{t_i\}$.

The filtering problem is to characterize the evolution of the conditional density-probability $\hat{p}_{x_t,\theta_t}(x, \theta) \triangleq p_{x_t,\theta_t|Y_t}(x, \theta)$ of the joint Markov process $\{x_t, \theta_t\}$ given the initial joint density $p_{x_0,\theta_0}(x, \theta)$ and the increasing family of sigma-algebras $Y_t = \sigma\{t_i, y_i; \forall i \text{ such that } t_i \in [0, t]\}$.

4.5.2 Evolution of the Joint Conditional Density

Given the initial joint density $p_{x_0,\theta_0}(x, \theta)$ and an increasing family of sigma-algebras, Y_t, the evolution of the conditional density-probability $\hat{p}_{x_t,\theta_t}(x, \theta)$ of the joint Markov process $\{x_t, \theta_t\}$ is characterized through a sequence of alternating steps for $i = 1, 2, 3, \ldots$:

1. unconditional evolution of the joint conditional density on the interval $[t_{i-1}, t_i)$ according to (4.23);
2. Bayesian update step at moment t_i.

Elaboration of these steps leads to the following Proposition:

Proposition 2 *Let the conditional probability $P\{x_t \in dx, \theta_t = \theta | Y_t\}$ admit for all t a density $\hat{p}_{x_t,\theta_t}(x,\theta)$ which is twice continuously differentiable in x, then $\hat{p}_{x_t,\theta_t}(x,\theta)$ evolves on the half open interval $[t_{i-1}, t_i)$ according to*

$$d\hat{p}_{x_t,\theta_t}(x,\theta) = L\hat{p}_{x_t,\theta_t}(x,\theta)dt, \tag{4.67a}$$

where L is the forward generator of the joint Markov process $\{x_t, \theta_t\}$, satisfying (4.23a–c).

Moreover, at the ith observation moment t_i, the joint conditional density update satisfies

$$\hat{p}_{x_{t_i},\theta_{t_i}}(x,\theta) = \hat{p}_{x_{t_i-},\theta_{t_i-}}(x,\theta)N\{y_i; h(x,\theta), g(\theta)g(\theta)^T\}/c_i \tag{4.67b}$$

with c_i being a normalizing constant.

4.5.3 Continuous-Discrete SIR Particle Filtering

In general it is challenging to evaluate recursion (4.67a,b) in a numerical way. One of the most general approaches in discrete time filtering is to use a Sampling Importance Resampling (SIR) particle filter. At moment t, a SIR particle filter maintains a set of N_P weighted particles $\left\{\theta_t^j, x_t^j, \mu_t^j; j \in \{1, \ldots, N_P\}\right\}$. These weighted particles span the empirical density

$$\tilde{p}_{x_t,\theta_t}(x,\theta) = \sum_{j=1}^{N_p} \mu_t^j \chi\left(\theta, \theta_t^j\right) \delta\left(x - x_t^j\right) \tag{4.68}$$

as an approximation of the exact conditional density $\hat{p}_{x_t,\theta_t}(x,\theta)$, where χ and δ are indicator and Dirac functions, respectively. By using equation (4.68) it can be shown that the set of particles also spans the marginal empirical densities $\tilde{p}_{\theta_t}(\theta)$ and $\tilde{p}_{x_t|\theta_t}(x|\theta)$

$$\tilde{p}_{\theta_t}(\theta) = \int_{\mathbb{R}^n} \tilde{p}_{x_t,\theta_t}(x,\theta)dx = \sum_{j=1}^{N_P} \mu_t^j \chi(\theta, \theta_t^j) \tag{4.69}$$

$$\tilde{p}_{x_t|\theta_t}(x|\theta) = \tilde{p}_{x_t,\theta_t}(x,\theta)/\tilde{p}_{\theta_t}(\theta) = \sum_{j=1}^{N_P} \mu_t^j \chi(\theta, \theta_t^j)\delta\left(x - x_t^j\right)/\tilde{p}_{\theta_t}(\theta) \tag{4.70}$$

under the condition $\tilde{p}_{\theta_t}(\theta) > 0$.

A continuous-discrete SIR particle filter cycle for $\tilde{p}_{x_{t_i},\theta_{t_i}} \to \tilde{p}_{x_{t_{i+1}},\theta_{t_{i+1}}}$ is specified in Table 4.1. This particle filter can approximate the conditional density arbitrarily well by increasing the number N_p of particles. Key difference with the discrete time version of this SIR particle filter [45,46] is that the evolution step is now defined by

TABLE 4.1 Continuous-Discrete SIR Particle Filter

Initialization at $t = 0$: For $j = 1, .., N_p$:
Start with $i = 1$, $\mu_0^j = 1/N_p$ and simulate (x_0^j, θ_0^j) from $p_{x_0,\theta_0}(.,.)$

Particle Filter cycle:

- At moment t_i the set of particles is $\left\{\mu_{i-1}^j \in [0, 1], \theta_{t_{i-1}}^j \in \mathbb{M}, x_{t_{i-1}}^j \in \mathbb{R}^n; j = 1, \ldots, N_p\right\}$
- First, await measurement moment t_i: this yields realization y_i.
  If $t_i > T$, then $t_i := T$ and $y_i := \{\}$
- Evolution for $j = 1, \ldots, N_p$:
  Simulate on $[t_{i-1}, t_i)$ equations (4.3) and (4.65), which implies

$$\bar{\theta}_t^j = \theta_{t_{i-1}}^j + \int_{t_{i-1}}^t \sum_{\substack{\theta,\eta \in \mathbb{M} \\ \theta \neq \eta}} 1_\eta(\bar{\theta}_{s-}^j)[\theta - \eta]d\rho_{\eta\theta,s}^j \tag{4.71}$$

$$\bar{x}_t^j = \bar{x}_{t_{i-1}}^j + \int_{t_{i-1}}^t [a(\bar{x}_s^j, \bar{\theta}_s^j)ds + b(\bar{x}_s^j, \bar{\theta}_s^j)dw_s^j] \tag{4.72}$$

  This yields the end conditions: $\bar{x}_{t_i}^j$ and $\bar{\theta}_{t_i}^j$, $j = 1, ..., N_p$.
- Performing Bayes update for $j = 1, \ldots, N_p$ (unless $y_i = \{\}$)

$$\bar{\mu}_i^j = \mu_{i-1}^j p_{y_i|x_{t_i},\theta_{t_i}}(y_i|\bar{x}_{t_i}^j, \bar{\theta}_{t_i}^j)/c_i \tag{4.73}$$

  with c_i such that $\sum_{j=1}^{N_p} \bar{\mu}_i^j = 1$
- Particles and conditional density at moment t_i

$$\left\{\bar{\mu}_i^j \in [0, 1], \bar{\theta}_{t_i}^j \in \mathbb{M}, \bar{x}_{t_i}^j \in \mathbb{R}^n; j = 1, ..., N_p\right\}$$

$$\bar{p}_{x_{t_i},\theta_{t_i}}(x, \theta) = \sum_{j=1}^{N_p} \bar{\mu}_i^j \chi(\theta, \bar{\theta}_{t_i}^j)\delta(x - \bar{x}_{t_i}^j)$$

- Resampling for $j = 1, \ldots, N_p$

$$\mu_i^j = 1/N_p$$
$$\left(x_{t_i}^j, \theta_{t_i}^j\right) \sim \bar{p}_{x_{t_i},\theta_{t_i}}(x, \theta)$$

If $t_i < T$, then set $i := i + 1$ and repeat the Particle Filter cycle above.

a set of SDEs for $d\theta_t$ and dx_t rather than by a transition probability and a difference equation.

The set of evolution equations can be numerically solved as follows:

Step 1. Numerical evaluation of θ_t^j on the interval $[t_{i-1}, t_i)$ yields for each $j = 1, \ldots, N_p$ one or more mode switching moments;

Step 2. Numerical evaluation of x_t^j from t_{i-1} to the generated mode switching moment, and from this mode switching moment to the next mode switching moment, and so on until t_i is reached.

4.5.4 Markov Jump Linear Case

In order to study the situation that the finite-state Markov chain $\{\theta_t\}$ influences the coefficients of a linear system, we assume that $h(x, \theta)$, $a(x, \theta)$ are affine in x, and that $b(x, \theta)$ is x-invariant. Then we have

$$\begin{aligned} h(x, \theta) &= H(\theta)x + G(\theta) \\ a(x, \theta) &= A(\theta)x + C(\theta) \\ b(x, \theta) &= B(\theta). \end{aligned}$$

This means we consider the $\mathbb{R}^n$-valued process $\{x_t\}$ as the solution of the following SDE

$$dx_t = A(\theta_t)x_t dt + C(\theta_t)dt + B(\theta_t)dw_t, \tag{4.74}$$

where all components of A, B, and C are $< \infty$ and for simplicity time invariant; $\{dw_t\}$ is a vector Wiener process with independent standard Brownian motion components, and independent of x_0 and $\{\theta_t\}$.

Furthermore, the m-dimensional observation process satisfies

$$y_i = H(\theta_{t_i})x_{t_i} + G(\theta_{t_i}) + g(\theta_{t_i})n_i, \tag{4.75}$$

where all components of H are $< \infty$ and for simplicity time invariant, $\{n_i\}$ is an i.i.d. sequence of zero mean standard Gaussian variables with $E\{n_i n_i^T\} = I$ and $\{n_i\}$ is independent of $\{x_t, \theta_t\}$.

4.5.5 Continuous-Discrete IMM Filter

In [9], it is shown that the continuous-discrete IMM filter inherits evolution equations (4.43)–(4.45) with $d\bar{v}_t(\theta) = 0$ from continuous time IMM, and the measurement update equations from discrete time IMM [1]. The resulting continuous-discrete IMM filter is given in Table 4.2.

Both [47] and [48] propose to exploit the linearity of the transformed IMM equations (4.55) and (4.56) for the prediction equations (4.76)–(4.78) of the continuous-discrete IMM filter. This linear prediction step is specified in Table 4.3. Because the prediction step does not involve any martingale term, this does not lead to the numerical integration problem that applies in the pure continuous time setting.

TABLE 4.2 Continuous-Discrete IMM Filter

- Initialization. Start with initial values for $\hat{p}_{\theta_0}(\theta)$, $\hat{x}_0(\theta)$, and $\hat{R}_0(\theta)$.
- Set $i := 1$

IMM Filter cycle:

- First, await the measurement at t_i: this yields realization y_i.
  If $t_i > T$, then $t_i := T$ and $y_i := \{\}$
- Prediction from t_{i-1} to t_i

$$\bar{p}_{\theta_t}(\theta) = \hat{p}_{\theta_{t_{i-1}}}(\theta) + \int_{t_{i-1}}^{t} M\bar{p}_{\theta_s}(\theta)ds \tag{4.76}$$

$$\bar{x}_t(\theta) = \hat{x}_{t_{i-1}}(\theta) + \int_{t_{i-1}}^{t} \left[A(\theta)\bar{x}_s(\theta) + C(\theta) + \sum_{\substack{\eta \in \mathbb{M} \\ \eta \neq \theta}} \lambda_{\eta\theta} \frac{\bar{p}_{\theta_s}(\eta)}{\bar{p}_{\theta_s}(\theta)} [\bar{x}_s(\eta) - \bar{x}_s(\theta)] \right] ds \tag{4.77}$$

$$\bar{R}_t(\theta) = \hat{R}_{t_{i-1}}(\theta) + \int_{t_{i-1}}^{t} [A(\theta)\bar{R}_s(\theta) + \bar{R}_s(\theta)A(\theta)^T + B(\theta)B(\theta)^T$$

$$+ \sum_{\substack{\eta \in \mathbb{M} \\ \eta \neq \theta}} \lambda_{\eta\theta} \frac{\bar{p}_{\theta_s}(\eta)}{\bar{p}_{\theta_s}(\theta)} [\bar{R}_s(\eta) - \bar{R}_s(\theta) + [\bar{x}_s(\eta) - \bar{x}_s(\theta)][..]^T]]ds \tag{4.78}$$

- Bayes update of the state at moment t_i

$$\hat{x}_{t_i}(\theta) = \bar{x}_{t_i}(\theta) + K_i \nu_i(\theta) \tag{4.79}$$

$$\hat{R}_{t_i}(\theta) = \bar{R}_{t_i}(\theta) - K_i(\theta)H(\theta)\bar{R}_{t_i}(\theta) \tag{4.80}$$

$$\text{with} : \nu_i(\theta) = y_i - H(\theta)\bar{x}_{t_i}(\theta) - G(\theta) \tag{4.81}$$

$$Q_i(\theta) = H(\theta)\bar{R}_{t_i}(\theta)H(\theta)^T + g(\theta)g(\theta)^T \tag{4.82}$$

$$K_i(\theta) = \bar{R}_{t_i}(\theta)H(\theta)^T Q_i(\theta)^{-1} \tag{4.83}$$

- Bayes update of the mode probability at moment t_i

$$\hat{p}_{\theta_{t_i}}(\theta) = \bar{p}_{\theta_{t_i}}(\theta)\text{Det}\{Q_i(\theta)\}^{-1/2} \exp\{-\tfrac{1}{2}\nu_i(\theta)^T Q_i(\theta)^{-1}\nu_i(\theta)\}/c_i \tag{4.84}$$

- MMSE output

$$\hat{x}_{t_i} = \sum_{\theta \in \mathbb{M}} \left[\hat{p}_{\theta_{t_i}}(\theta)\hat{x}_{t_i}(\theta)\right] \tag{4.85}$$

$$\hat{R}_{t_i} = \sum_{\theta \in \mathbb{M}} \left[\hat{p}_{\theta_{t_i}}(\theta)\left[\hat{R}_{t_i}(\theta) + [\hat{x}_{t_i}(\theta) - \hat{x}_{t_i}][\hat{x}_{t_i}(\theta) - \hat{x}_{t_i}]^T\right]\right] \tag{4.86}$$

If $t_i < T$, then $i := i + 1$ and start the next IMM filter cycle.

TABLE 4.3 Linear Prediction from t_{i-1} to t_i in Continuous-Discrete IMM

- Transformation

$$\hat{z}_{t_{i=1}}(\theta) = \hat{p}_{\theta_{t=1}}(\theta)\hat{x}_{t_{i-1}}(\theta) \tag{4.87}$$

$$\hat{Z}_{t_{i=1}}(\theta) = \hat{p}_{\theta_{t_{i=1}}}(\theta)\hat{R}_{t_{i-1}}(\theta) + \hat{z}_{t_{i-1}}(\theta)\hat{x}_{t_{i-1}}(\theta)^T \tag{4.88}$$

- Linear prediction at the half open interval $[t_{i-1}, t_i)$, $i = 1, 2, \ldots$

$$\bar{p}_{\theta_t}(\theta) = \hat{p}_{\theta_{t_{i=1}}}(\theta) + \int_{t_{i-1}}^{t} M\bar{p}_{\theta_s}(\theta)ds \tag{4.89}$$

$$\bar{z}_t(\theta) = \hat{z}_{t_{i-1}}(\theta) + \int_{t_{i-1}}^{t} [A(\theta)\bar{z}_s(\theta) + C(\theta)\bar{p}_{\theta_s}(\theta) + \sum_{\eta\in\mathbb{M}} \lambda_{\eta\theta}\bar{z}_s(\eta)]ds \tag{4.90}$$

$$\bar{Z}_t(\theta) = \hat{Z}_{t_{i=1}}(\theta) + \int_{t_{i-1}}^{t} [A(\theta)\bar{Z}_s(\theta) + \bar{Z}_s(\theta)A(\theta)^T + C(\theta)\bar{z}_s(\theta)^T + \bar{z}_s(\theta)C(\theta)^T + B(\theta)B(\theta)^T\bar{p}_{\theta_s}(\theta)]ds \tag{4.91}$$

- Back transformation

$$\bar{x}_{t_i}(\theta) = \bar{p}_{\theta_{t_i}}(\theta)^{-1}\bar{z}_{t_i}(\theta) \tag{4.92}$$

$$\bar{R}_{t_i}(\theta) = \bar{p}_{\theta_{t_i}}(\theta)^{-1}[\bar{Z}_{t_i}(\theta) - \bar{z}_{t_i}(\theta)\hat{x}_{t_i}(\theta)^T] \tag{4.93}$$

Remark 7 For the numerical evaluation of the linear equations (4.89)–(4.91) in Table 4.3, [48] develops an approach to exploit the exponential form of linear prediction over time.

4.6 CONCLUDING REMARKS

This chapter has presented the development of the continuous time IMM equations in [8] that preceded the development of the discrete-time IMM filter.

Section 4.2 addressed the filtering of a HMM, with emphasis on the exact filter equations and the development of transformations which allow a robust implementation of the continuous-time HMM filter.

Section 4.3 characterized the evolution of the joint conditional density of the mode and the state of a Markov jump system. When the mode process is known then the filter equation reduces to the well-known nonlinear filter equations for diffusion process [39]. Section 4.3 also develops a decomposition of the joint density into coupled evolutions of the conditional mode probabilities and the mode conditional state. This decomposition leads to a stochastic interaction operator which is conditionally linear given the conditional probabilities of the modes.

Section 4.4 focused on Markov jump linear systems, and the development of filter equations for the conditional evolution of first- and second-order moments. This yields a Pre-IMM filter in continuous time, as well as a linear version of this Pre-IMM filter. Of these two versions, the nonlinear version of the Pre-IMM filter shows the interaction behavior that is well known from the discrete time IMM filter algorithm. Subsequently special cases are shown under which the Pre-IMM filter becomes the continuous-time IMM filter. Moreover, the continuous-time IMM filter is proposed as an elegant approximation of the Pre-IMM filter in all other cases.

Section 4.5 addressed continuous-discrete filtering for the situation that the observation moments are nondeterministic, and that mode switchings do not coincide with these observation moments. This yields a characterization of the evolution of the exact joint conditional density, the continuous-discrete IMM filter, and a continuous-discrete time SIR particle filter.

The continuous time IMM development in [8] has guided the development of the discrete time IMM filter, along two different routes:

- improving the timing of hypothesis merging [9];
- using Chapman–Kolmogorov for the prediction [10].

The former route has become most popular [1–3], because its interpretation fits in the prevailing hypotheses management paradigm. Nevertheless, the latter route showed that in discrete time setting the IMM interaction is exact [10]; recently this has been further elaborated in [4,49]. The value of the exact interaction is expected to increase with a strongly increasing interest in continuous-time stochastic hybrid systems [29,50–56]. This induces many valuable directions for further development of nonlinear filtering for continuous time stochastic hybrid systems. Examples of complementary continuous time filtering developments for stochastic hybrid systems are

- complementary developments in nonlinear estimation for HMMs [53];
- interrelation between filtering and control for Markov jump linear systems [6];
- filtering for Markov jump linear systems with hybrid jumps [56, 57];
- filtering of Markov modulated Poisson processes [58];
- state-dependent Markov switching [29, 55].

APPENDIX 4A DIFFERENTIATION RULE FOR DISCONTINUOUS SEMIMARTINGALES

The differentiation rule for discontinuous semimartingales [30, Theorem 12.21] reads as follows. Let $\{x_t\}$ be an n-dimensional semimartingale, and f is a twice continuously differentiable real valued function on $\mathbb{R}^n$. Then $\{f(x_t)\}$ is a semimartingale, which satisfies

$$
\begin{aligned}
f(x_t) = f(x_0) &+ \int_0^t \sum_{i=1}^n \frac{\partial f}{\partial x^i}(x_{s-}) dx_s^i + \frac{1}{2} \int_0^t \sum_{i,j=1}^n \frac{\partial^2 f}{\partial x^i \partial x^j}(x_{s-}) \left\langle x_s^{C,i}, x_s^{C,j} \right\rangle ds \\
&+ \sum_{0<s\le t} \left[f(x_s) - f(x_{s-}) - \sum_{i=1}^n \frac{\partial f}{\partial x^i}(x_s^i - x_{s-}^i) \right],
\end{aligned}
$$

where x_t^i denotes the ith component of x_t, and $\left\langle x_t^{C,i}, x_t^{C,j} \right\rangle$ is the predictable quadratic covariation between the the *i*th and *j*th components of the continuous martingale part $\{x_t^C\}$ of $\{x_t\}$.

APPENDIX 4B DERIVATION OF DIFFERENTIAL FOR $\hat{R}_T(\theta)$

This appendix derives equation (4.45a,b) following [8].

From the definition of $\hat{R}_t(\theta)$, that is the second conditional central moment of x_t given $\theta_t = \theta$ and Y_t, we get

$$
\hat{R}_t(\theta) = \int_{\mathbb{R}^n} [x - \hat{x}_t(\theta)][x - \hat{x}_t(\theta)]^T \hat{p}_{x_t|\theta_t}(x|\theta) dx.
$$

Let $r_{ij,t}(\theta)$, $i, j \in [1, n]$ denote the *ij*th component of $\hat{R}_t(\theta)$. Then according to the above

$$
r_{ij,t}(\theta) = \int_{\mathbb{R}^n} \phi_{ij,t}(x,\theta) dx, \tag{4B.1}
$$

where

$$
\phi_{ij,t}(x,\theta) = [x_i - \hat{x}_{i,t}(\theta)] x_j \hat{p}_{x_t|\theta_t}(x|\theta). \tag{4B.2}
$$

Applying Ito's differentiation rule to (4B.2), (4.44), and (4.50) yields

$$
\begin{aligned}
d\phi_{ij,t}(x,\theta) &= [x_i - \hat{x}_{i,t}(\theta)] x_j d\hat{p}_{x_t|\theta_t}(x|\theta) \\
&- x_j \hat{p}_{x_t|\theta_t}(x|\theta) d\hat{x}_{i,t}(\theta) - \hat{p}_{x_t|\theta_t}(x|\theta) \hat{p}_{\theta_t}(\theta) \mathrm{H}(\theta)_i^T V^{-1} H(\theta)[x - \hat{x}_t(\theta)] x_j dt.
\end{aligned} \tag{4B.3}
$$

Applying Ito's differentiation rule to (4B.1) and (4B.3) yields

$$
dr_{ij,t}(\theta) = \int_{\mathbb{R}^n} d\phi_{ij,t}(x,\theta) dx. \tag{4B.4}
$$

Substituting (4B.3) into (4B.4) and collecting the terms yields

$$d\hat{R}_t(\theta) = \int_{\mathbb{R}^n} [x - \hat{x}_t(\theta)]x^T d\hat{p}_{x_t|\theta_t}(x|\theta)dx - d\hat{x}_t(\theta)\hat{x}_t(\theta)^T$$
$$-\hat{R}_t(\theta)H(\theta)^T V^{-1} H(\theta) \int_{\mathbb{R}^n} [x - \hat{x}_t(\theta)]x^T \hat{p}_{x_t|\theta_t}(x|\theta)dxdt. \quad (4B.5)$$

Substituting into (4B.5) of

$$\hat{x}_t(\theta) = \int_{\mathbb{R}^n} x\hat{p}_{x_t|\theta_t}(x|\theta)dx$$
$$\hat{R}_t(\theta) = \int_{\mathbb{R}^n} [x - \hat{x}_t(\theta)][x - \hat{x}_t(\theta)]^T \hat{p}_{x_t|\theta_t}(x|\theta)dx$$

yields

$$d\hat{R}_t(\theta) = \int_{\mathbb{R}^n} [x - \hat{x}_t(\theta)]x^T d\hat{p}_{x_t|\theta_t}(x|\theta)dx - \int_{\mathbb{R}^n} x\hat{x}_t(\theta)^T d\hat{p}_{x_t|\theta_t}(x|\theta)dx$$
$$-\hat{R}_t(\theta)H(\theta)^T V^{-1} H(\theta)\hat{R}_t(\theta)dt$$
$$= \int_{\mathbb{R}^n} [x - \hat{x}_t(\theta)][x - \hat{x}_t(\theta)]^T d\hat{p}_{x_t|\theta_t}(x|\theta)dx - \int_{\mathbb{R}^n} \hat{x}(\theta)\hat{x}_t(\theta)^T d\hat{p}_{x_t|\theta_t}(x|\theta)dx$$
$$-\hat{R}_t(\theta)H(\theta)^T V^{-1} H(\theta)\hat{R}_t(\theta)dt. \quad (4B.6)$$

Substituting (4.50) and noticing that the second term equals zero, yields

$$d\hat{R}_t(\theta) = \int_{\mathbb{R}^n} [x - \hat{x}_t(\theta)][x - \hat{x}_t(\theta)]^T L^\theta \hat{p}_{x_t|\theta_t}(x|\theta)dxdt$$
$$+ \int_{\mathbb{R}^n} [x - \hat{x}_t(\theta)][x - \hat{x}_t(\theta)]^T \hat{M}_t \hat{p}_{x_t|\theta_t}(x|\theta)dxdt$$
$$+ \int_{\mathbb{R}^n} [x - \hat{x}_t(\theta)][x - \hat{x}_t(\theta)]^T \hat{p}_{x_t|\theta_t}(x|\theta)[x - \hat{x}_t(\theta)]H(\theta)^T V^{-1} d\bar{\nu}_t(\theta)]dx$$
$$-\hat{R}_t(\theta)H(\theta)^T V^{-1} H(\theta)\hat{R}_t(\theta)dt. \quad (4B.7)$$

Substituting (4.27) in the first term, followed by partial integration, and using

$$a(x, \theta) = A(\theta)x + C(\theta)$$
$$b(x, \theta) = B(\theta)$$

yields

$$\int_{\mathbb{R}^n} [x - \hat{x}_t(\theta)][x - \hat{x}_t(\theta)]^T L^\theta \hat{p}_{x_t|\theta_t}(x|\theta)dxdt$$
$$= A(\theta)\hat{R}_t(\theta) + \hat{R}_t(\theta)A(\theta)^T + B(\theta)B(\theta)^T, \quad (4B.8)$$

where use is made of: $\int_{\mathbb{R}^n} [x - \hat{x}_t(\theta)][x - \hat{x}_t(\theta)]^T \frac{\partial}{\partial x_i}\{\hat{p}_{x_t|\theta_t}(x|\theta)\}dx = 0, \ \forall i \in [1, n]$.

Substitution of (4.33) in the second term of (4B.7) and subsequent evaluation yields

$$\begin{aligned}
&\int_{\mathbb{R}^n} [x - \hat{x}_t(\theta)][x - \hat{x}_t(\theta)]^T \hat{M}_t \hat{p}_{x_t|\theta_t}(x|\theta)dx \\
&= \int_{\mathbb{R}^n} [x - \hat{x}_t(\theta)][x - \hat{x}_t(\theta)]^T \hat{p}_{\theta_t}(\theta)^{-1} \sum_{\eta \in \mathbb{M}} \lambda_{\eta\theta} \hat{p}_{\theta_t}(\eta)[\hat{p}_{x_t|\theta_t}(x|\eta) - \hat{p}_{x_t|\theta_t}(x|\theta)]dx \\
&= \hat{p}_{\theta_t}(\theta)^{-1} \sum_{\substack{\eta \in \mathbb{M} \\ \eta \neq \theta}} \lambda_{\eta\theta} \hat{p}_{\theta_t}(\eta)[\hat{R}_t(\eta) + [\hat{x}_t(\eta) - \hat{x}_t(\theta)][....]^T - \hat{R}_t(\theta)]. \qquad (4B.9)
\end{aligned}$$

Evaluation of the third term of (4B.7) yields

$$\begin{aligned}
&\int_{\mathbb{R}^n} [x - \hat{x}_t(\theta)][x - \hat{x}_t(\theta)]^T \hat{p}_{x_t|\theta_t}(x|\theta)[x - \hat{x}_t(\theta)]H(\theta)^T V^{-1} d\bar{v}_t(\theta)]dx \\
&= \sum_{i=1}^{n} \int_{\mathbb{R}^n} [x - \hat{x}_t(\theta)][x - \hat{x}_t(\theta)]^T \hat{p}_{x_t|\theta_t}(x|\theta)[x - \hat{x}_t(\theta)]_i [H(\theta)^T V^{-1} d\bar{v}_t(\theta)]_i dx \\
&= \sum_{i=1}^{n} E\{[x - \hat{x}_t(\theta)][x - \hat{x}_t(\theta)]^T [x - \hat{x}_t(\theta)]_i | \theta_t = \theta, Y_t\}[H(\theta)^T V^{-1} d\bar{v}_t(\theta)]_i.
\end{aligned}$$
$$(4B.10)$$

Substitution of (4B.8)–(4B.10) in (4B.7) yields

$$\begin{aligned}
d\hat{R}_t(\theta) = &[A(\theta)\hat{R}_t(\theta) + \hat{R}_t(\theta)A(\theta)^T + B(\theta)B(\theta)^T - \hat{R}_t(\theta)H(\theta)^T V^{-1} H(\theta)\hat{R}_t(\theta)]dt \\
&+ \hat{p}_{\theta_t}(\theta)^{-1} \sum_{\substack{\eta \in \mathbb{M} \\ \eta \neq \theta}} \lambda_{\eta\theta} \hat{p}_{\theta_t}(\eta)[\hat{R}_t(\eta) - \hat{R}_t(\theta) + [\hat{x}_t(\eta) - \hat{x}_t(\theta)][...]^T]dt \\
&+ \sum_{i=1}^{n} E\{[x - \hat{x}_t(\theta)][...]^T [...]_i | \theta_t = \theta, Y_t\}[H(\theta)^T V^{-1} d\bar{v}_t(\theta)]_i.
\end{aligned}$$

This completes the derivation of (4.45a,b).

REFERENCES

1. H. A. P. Blom and Y. Bar-Shalom, The Interacting Multiple Model algorithm for systems with Markovian switching coefficients, *IEEE Trans. Automat. Contr.*, 33, 780–783, 1988.
2. Y. Bar-Shalom and X. Rong Li, *Estimation and Tracking*, Artech House, Norwood, MA, 1993.
3. Y. Bar-Shalom, X. Rong Li, and T. Kirubarajan, Estimation with Applications to Tracking and Navigation, Wiley, New York, 2001.
4. H. A. P. Blom and Y. Bar-Shalom, Bayesian filtering of stochastic hybrid systems in discrete-time and interacting multiple model, in: B. Rozovsky and D. Crisan (Eds.), *Oxford Handbook on Nonlinear Filtering*, Oxford University Press, Oxford, pp. 526–557, 2011.
5. X. R. Li and V. P. Jilkov, Survey of maneuvering target tracking—Part V: Multiple model methods, *IEEE Trans. Aerosp. Electron. Syst.*, 41, pp. 1255–1321, 2005.
6. M. Mariton, *Jump Linear Systems in Automatic Control*, Marcel Dekker, New York, 1990.
7. E. Mazor, A. Averbuch, Y. Bar-Shalom, and J. Dayan, Interacting multiple model methods in target tracking: A survey, *IEEE Trans. Aerosp. Electron. Syst.*, 43, 103–123, 1998.
8. H. A. P. Blom, Detection-filter representations for Markov jump diffusions, *Report NLR* TR 82019U, National Aerospace Laboratory NLR, Amsterdam, March 1982.
9. H. A. P. Blom, A sophisticated tracking algorithm for air traffic control surveillance radar data, *Proceedings of the International Radar Conference*, Paris, 1984, pp. 393–398.
10. H. A. P. Blom, An efficient filter for abruptly changing systems, Proceedings of the 23rd IEEE CDC, Las Vegas, 1984, pp. 656–658.
11. C. B. Chang and M. Athans, State estimation for discrete systems with switching parameters, *IEEE Trans. Aerosp. Electron. Syst.*, 14, 418–425, 1978.
12. R. R. Tenney, R. S. Hebbert, and N. R. Sandell, A tracking filter for maneuvering sources, *IEEE Trans. Automat. Contr.*, 22, 246–261, 1977.
13. A. S. Willsky, A survey of design methods for failure detection in dynamic systems, *Automatica*, 12, 601–611, 1976.
14. S. P. Au and A. H. Haddad, Suboptimal sequential estimation-detection scheme for Poisson driven linear systems, *Inform. Sci.*, 16, 95–113, 1978.
15. M. H. A. Davis and S. I. Marcus, An introduction to nonlinear filtering, in: M. Hazewinkel and J. C. Willems (Eds.), *Stochastic Systems: The Mathematics of Filtering and Identification and Applications*, NATO Advanced Study Institute Series, Reidel, Dordrecht, pp. 53–75, 1981.
16. M. Fujisaki, G. Kallianpur, and H. Kunita, Stochastic differential equations for the nonlinear filtering problem, *Osaka J. Math.*, 9, 19–40, 1972.
17. G. Kallianpur, *Stochastic Filtering Theory*, Springer, New York, 1980.
18. R. S. Liptser and A. N. Shiryayev, Statistics of random processes, *General Theory*, Vol. I, Springer, New York, 1977.
19. J. H. Van Schuppen, Stochastic filtering theory: A discussion of concepts, methods, and results, in: M. Kohlmann and W. Vogel (Eds.), *Stochastic Control Theory and Stochastic Differential Systems*, Springer, New York, 1979.
20. W. M. Wonham, Some applications of stochastic differential equations to optimum nonlinear filtering, *SIAM J. Contr., Serie A*, 2, 347–369, 1965.

21. T. Björk, Finite dimensional optimal filters for a class of Ito processes with jumping parameters, *Stochastics,* 4, 167–183, 1980.
22. M. H. A. Davis, The application of nonlinear filtering to fault detection in linear systems, *IEEE Trans. Automat. Contr.,* 20, 257–259, 1975.
23. P. S. Krishnaprasad and S. I. Marcus, Identification and tracking: A class of nonlinear filtering problems, *Proceedings of the 1981 Joint Automatic Control Conference*, Charlotteville, VA, American Automatic Control Council, Philadelphia, 1981, pp. FP-5A. 1–4.
24. H. Kwakernaak, Filtering for systems excited by Poisson white noise, in: A. Bensoussan and J. L. Lions (Eds.), *Lecture Notes in Economics and Mathematical Systems,* Vol. 107, Springer, Berlin, pp. 468–492, 1975.
25. K. A. Loparo and Z. S. Roth, Nonlinear filtering for process diagnostics, *Proceedings of the 1981 Joint Automatic Control Conference*, Charlotteville, VA, American Automatic Control Council, Philadelphia, 1981, pp. FA-2B. 1–5.
26. H. A. P. Blom, The continuous time roots of the IMM filter, Proceedings *2012 IEEE Conference on Decision and Control*, Maui, December 2012.
27. I. I. Gihman and A. V. Skorohod, *Stochastic Differential Equations*, Springer, Berlin, 1972.
28. X. Mao and C. Yuan, *Stochastic Differential Equations with Markovian Switching,* Imperial College Press, London, 2006.
29. G. Yin and C. Zhu, *Hybrid Switching Diffusions,* Springer, New York, 2010.
30. R. J. Elliott, *Stochastic Calculus and Applications,* Springer, New York, 1982.
31. P. A. Meyer, Square and non-square integrable martingales, *Lecture Notes in Mathematics,* Vol. 190, Springer, Berlin, pp. 32–43, 1971.
32. R. W. Brockett and G. L. Blankenship, A representation theorem for linear differential equations with Markovian coefficients, *Proceedings of the Allerton Conference on Circuits and Systems Theory*, 1977, pp. 671–679.
33. M. Zakai, On the optimal filtering of diffusion processes, *Z. Wahrscheinligkeitstheorie verw. Gebiete,* 11, 230–243, 1969.
34. D. J. Clements and B. D. O. Anderson, Well-behaved Ito equations with simulations that always misbehave, *IEEE Trans. Automat. Contr.,* 18, 676–677. 1973.
35. P. E. Kloeden and E. Platen, *Numerical Solution of Stochastic Differential Equations,* Springer, Berlin, 1995.
36. J. M. C. Clark, The design of robust approximations to the stochastic differential equations of nonlinear filtering, in: J. K. Skwirzynski, (Ed.), Communication Systems and Random Process Theory, Sijthoff and Noordhoff, Alphen aan den Rijn, pp. 721–734, 1978.
37. H. A. P. Blom, Implementable differential equations for non-linear filtering, *Proceedings of the 2nd Symposium on Information Theory in the Benelux*, The Netherlands, May 1981, pp. 41–48.
38. B. Kemp and H. A. P. Blom, Optimal detection of the alpha state in a model of the human electroencephalogram, *Electroencephalogr. Clin. Neurophysiol.,* 52, 222–225. 1981.
39. A. H. Jazwinski, *Stochastic Processes and Filtering Theory,* Academic Press, New York, 1970.
40. A. H. Haddad and J. B. Cruz, Nonlinear filtering for systems with random models, *Proceedings of the 2nd Symposium of Nonlinear Estimation Theory and its Applications*, San Diego, September 1971.

41. D. G. Lainiotis, Partitioning: a unifying framework for adaptive systems; Part I, estimation, *Proc. IEEE,* 64, 1126–1143, 1976.

42. H. A. P. Blom, Markov jump-diffusion models and decision-making-free filtering, in: A. Bensoussan and J. L. Lions (Eds.), *Proceedings of the Conference on Analysis and Optimization of Systems*, Nice, France, Part 1, Springer, May 1984, pp. 568–580.

43. M. H. C. Everdij and H. A. P. Blom, Embedding adaptive JLQG into LQ martingale control with a completely observable stochastic control matrix, *IEEE Trans. Automat. Contr.,* 41, 424–430, 1996.

44. M. D. Fragoso and N. C. S. Rocha, Stationary filter for continuous-time Markovian jump linear systems, *SIAM J. Control Optim.,* 14, 801–815, 2005.

45. X. Koutsoukis, J. Kurien, and F. Zhao, Monitoring and diagnosis of hybrid systems using particle filtering methods, *Proceedings of the Mathematical Theory of Networks and Systems (MTNS)*, 2002, 1–15.

46. S. McGinnity and G. Irwin, Multiple model bootstrap filter for maneuvering target tracking, *IEEE Trans. Aerosp. Electron. Syst.,* 36, 1006–1012, 2000.

47. H. A. P. Blom, Continuous-discrete filtering for systems with Markovian switching coefficients and simultaneous jumps, *Proceedings of the 21st Asilomar Conference on Signals, Systems & Computation*, Pacific Grove, 1987, pp. 244–248.

48. S. Mori, J. W. Adaska, M. A. Pravia, and C. Y. Chong, Continuous-time interacting multiple model extrapolation, *Proceedings of the International Conference on Information Fusion*, 2008.

49. Y. Bar-Shalom, S. Challa, and H. A. P. Blom, IMM estimator versus optimal estimator for hybrid systems, *IEEE Trans. Aerosp. Electron. Syst.,* 41, 986–991, 2005.

50. J. Bect, A unifying framework of the Fokker–Planck–Kolmogorov equation for general stochastic hybrid systems, *Nonlin. Anal.: Hybrid Syst.,* 4, 357–370, 2010.

51. M. L. Bujorianu and J. Lygeros, General stochastic hybrid systems: Modelling and optimal control, *Proceedings of the 43rd IEEE Conference on Decision and Control*, 2004, pp. 1872–1877.

52. M. L. Bujorinau, and J. Lygeros, Toward a general theory of stochastic hybrid systems, in: H. A. P. Blom and J. Lygeros (Eds.), *Stochastic Hybrid Systems, Theory and Safety Critical Applications,* Springer, Berlin, pp. 3–30, 2006.

53. R. J. Elliott, L. Aggoun, and J. B. Moore, *Hidden Markov Models, Estimation and Control,* Springer, New York, 1995.

54. J. Hu, J. Lygeros, and S. Sastry, Towards a theory of stochastic hybrid systems, *Proceedings of the 3rd International Workshop on Hybrid Systems, Computation and Control (HSCC)*, Springer, 2000, pp. 160–173.

55. J. Krystul, H. A. P. Blom, and A. Bagchi, Stochastic differential equations on hybrid state spaces, in: C. G. Cassandras and J. FLygeros (Eds.), *Stochastic Hybrid Systems,* Taylor & Francis/CRC Press, Boca Raton, Chapter 2, pp. 15–45, 2007.

56. D. D. Sworder and J. E. Boyd, *Estimation Problems in Hybrid Systems,* Cambridge University Press, Cambridge, 1999.

57. H. A. P. Blom, Bayesian estimation for decision-directed stochastic control, *PhD Thesis*, Delft University of Technology, ISBN 90-9003366-1, 1990.

58. V. Krishnamurthy and R. J. Elliott, Filters for estimating Markov modulated Poisson processes and image-enhanced tracking, *Proceedings of the 34th IEEE Conference on Decision and Control*, New Orleans, 1995, pp. 63–68.

PART II

MULTITARGET MULTISENSOR TRACKING

CHAPTER 5

Multitarget Tracking Using Multiple Hypothesis Tracking

MAHENDRA MALLICK, STEFANO CORALUPPI, AND CRAIG CARTHEL

5.1 INTRODUCTION

Multitarget tracking (MTT) refers to estimating the state and the number of an unknown number of targets using scans of noisy measurements from one or more sensors in the presence of false alarm (FA) or clutter. MTT is an important and challenging problem which occurs in diverse fields including air traffic control, intelligence, surveillance, and reconnaissance (ISR), sonar tracking of submarines and marine animals, space object tracking, simultaneous localization and mapping (SLAM), terrorist activity monitoring, homeland security applications, combating maritime piracy, drug trafficking, biomedical applications, computer vision, remote sensing, oceanography, road traffic monitoring, guidance and navigation, sensor network, and chemical, biological, radiological, nuclear, and explosive (CBRNE) detection. The field of MTT is rapidly expanding and many current algorithms in MTT and associated track filtering are present in various papers. For the benefit of the reader, we present a survey of the current state-of-the-art research in tracking and filtering so that the chapter is self-contained. This serves two purposes. A comparative review of existing MTT algorithms helps the reader to know the strengths and weaknesses of the multiple hypothesis tracking (MHT) algorithm relative to other tracking algorithms. In a tracker, one can choose a filtering algorithm depending on the nature of the tracking problem. A review of various filtering algorithms helps the reader in choosing a suitable filtering algorithm for a given problem within the MHT framework. We have provided extensive references on various track filtering and MTT algorithms.

The chapter is organized as follows. Sections 5.2 and 5.3 present reviews of various tracking and track filtering algorithms, respectively. Section 5.4 gives a brief description of the various MHT algorithms and hybrid-state derivations of track-oriented MHT equations are presented in Section 5.5. The target death problem is described

Integrated Tracking, Classification, and Sensor Management: Theory and Applications, Edited by Mahendra Mallick, Vikram Krishnamurthy, and Ba-Ngu Vo.

in Section 5.6. Finally, Sections 5.7 and 5.8 present some applications of MHT and summary, respectively.

5.2 TRACKING ALGORITHMS

Real-world targets have finite size. In target tracking, two types of models for a target relative to target size are commonly used; point-target and extended target. In radar tracking problems, where the distance between the target and radar is very large compared with the size of the target, the target is treated as a point-target. A video sensor can image an extended target. In such a case, the center and orientation (or attitude) of the target can be estimated in a track filter. In this chapter, we consider point targets.

An MTT system uses one or more sensors such as radar, sonar, electro-optical, video, infra-red, multispectral, hyperspectral, and unattended ground sensor (acoustic and seismic). Table 5.1 presents the target type, sensor type, sensing mode, and sensor measurements of a number of commonly used sensors in tracking applications. The probability of detection of a sensor is less than unity. In general, a sensor can have multiple measurements in a dwell (or scan) or *field-of-view* (FOV). For a point target, we assume that at most one measurement can originate from a target in a scan. Thus there may be zero or one measurement from a target in a scan. The origin of the measurements from targets in a scan is unknown. This is known as the *measurement origin uncertainty*. In addition to measurements originating from a target, extraneous measurements due to FA or clutter can also be present in a scan. Therefore, the number of measurements in a scan is generally different from the number of targets in the dwell volume (the volume of the measurement space sensed by a sensor) due to FA and sensor probability of detection less than unity.

Two types of tracking architectures, the centralized and distributed tracking [1, 2], are commonly used for multitarget multisensor tracking. In the centralized tracking architecture, measurements from all sensors are sent to a central tracker or fusion center (FC) where *measurement-to-track association* (M2TA) and *measurement-to-track fusion* (M2TF) are performed. The centralized tracking architecture is suitable when a small number of sensors are used. In the distributed tracking architectures, one or more sensors are associated with a local tracker which performs M2TA and M2TF (M2TAF). A number of distributed fusion architectures [1, 2] exist. In one architecture, local trackers send track data to an FC which performs *track-to-track association* (T2TA) and *track-to-track fusion* (T2TF) [1, 2]. The track data depend on the nature of the T2TA and T2TF (T2TAF) algorithms used. For example, the track data for a track can be represented by the (i) state estimate and associated covariance, (ii) change in information state and associated information matrix [3–5], or (iii) decorrelated pseudo-measurement and associated measurement matrix and covariance [6–8]. The distributed fusion architecture is suitable when a large number of sensors are present. In general, the centralized tracking architecture has higher accuracy [9] than the distributed tracking architecture for the same scenario. However, the distributed tracking architecture is robust to single-point failure and

TABLE 5.1 Measurements of Some Typical Sensors

Target Type	Sensor Type	Sensing Mode	Sensor Measurements
Submarine	Sonar	Passive	Bearing
Aircraft	Infrared search and track (IRST)	Passive	Bearing, elevation
Ground moving target	Infrared	Passive	Intensity map (raw sensor measurements)
Ground moving target	Unattended ground sensor (acoustic, seismic)	Passive	Range, azimuth
Aircraft	2D radar	Active	Slant range, azimuth
Aircraft	3D radar, Air traffic control radar	Active	Slant range, azimuth, elevation
Missile	Long-range radar	Active	Slant range, direction cosines
Stationary ground target	Synthetic aperture radar (SAR)	Active	Slant range, Doppler
Ground moving target	Ground moving target indicator (GMTI) radar	Active	Slant range, azimuth, range rate
Ground moving target	GMTI radar	Active	Range-Doppler map (raw sensor measurements)
Air moving target	Airborne moving target indicator (AMTI) radar	Active	Slant range, azimuth, elevation, range-rate
Ground moving target	Video	Passive	Imagery
Ground moving target, Earth's surface	Multispectral, Hyperspectral sensors	Passive	Imagery
Aircraft, Ground target	Over the horizon radar (OTHR)	Active	Bistatic range, azimuth, bistatic range rate
Aircraft	Passive radar	Passive	Bistatic range, azimuth, bistatic range rate
Stationary and moving targets	LIDAR	Active	3D location
Submarine	Multistatic sonar	Active	Bistatic range, azimuth, bistatic range rate

does not require a high communication bandwidth. On the other hand, the algorithms for the centralized tracking are more mature than those for distributed tracking. We address centralized tracking in this chapter.

For centralized tracking, a number of multitarget multisensor tracking algorithms [2, 10] are currently used for tracking in the presence of FA and sensor probability of detection less than unity. These algorithms are as follows:

1. Global Nearest Neighbor (GNN) [2, 10],
2. Joint Probabilistic Data Association (JPDA) [2, 10–13],

3. Joint Integrated Probabilistic Data Association (JIPDA) and variants, [14],
4. Multiple Hypotheses Tracking (MHT) [2, 10, 15, 16],
5. Probability Hypothesis Density (PHD) filter [17–24] and Cardinalized PHD (CPHD) filter [20–25], and other filter,
6. Intensity Filter (iFilter) [26–29],
7. Monte Carlo Methods [30–39].

5.2.1 Tracking with Target Identity (or Track Label)

The first four algorithms, the GNN, JPDA, JIPDA, and MHT, which are well-established algorithms, have a common algorithmic structure whereas the PHD filter, CPHD filter, and iFilter have a completely different algorithmic structure. The first four tracking algorithms have two components: *data association* and *filtering*. The data-association process is much more complex than filtering. These algorithms differ in the way they perform data association.

The GNN algorithm [2, 10] associates at most one measurement to a track at a scan time so that a global cost is minimized. A track can have one or no associated measurement (missed detection event). Thus, the GNN algorithm makes a hard decision at each scan time. The GNN is the simplest algorithm and as a result it has the lowest tracking accuracy and the least computational cost among these four algorithms. The tracking accuracy of the GNN algorithm can degrade significantly in difficult tracking scenarios with closely spaced targets, high FA density, and low probability of detection.

In the standard JPDA algorithm [2, 10–12], a weighted sum of measurements (in the validated region) is associated with a track and data association is considered for a single scan of data. This process does not reflect the actual process by which a measurement is generated by a target. For a point target, at most one measurement can arise from a target in a scan. The averaging or weighted sum operation does not reflect the true underlying characteristics of the problem. The concept of *target existence* (a random variable) first proposed in [40] was incorporated in [41] for the Integrated Probabilistic Data Association (IPDA) for single-target tracking and in [14] for the JIPDA for multitarget tracking.

In the multiscan JPDA (MS-JPDA) [42–44], data association is considered using more than one scan of data. It is observed that use of two or three scans of data can produce significant improvement in tracking accuracy over the standard JPDA algorithm, which uses one scan of data. This approach is similar to the *multiframe assignment* (MFA) data association algorithm used in MHT, which is described next.

The MHT algorithm [2, 10, 15, 16, 45–47] allows a measurement to be associated with multiple tracks and multiple measurements to be associated with a track at a scan time. The algorithm makes a delayed decision by considering association of measurements in multiple scans (greater than 1) with a set of tracks. The s-dimensional (sD) assignment which represents an MFA [48–52] corresponds to assignment of measurements in $s-1$ scans with a set of tracks. Commonly used assignment algorithms are integer programming [53], Lagrangian relaxation [48–52], approximate linear programming (LP) [54, 55], and Markov Chain Monte Carlo (MCMC)-based data association [56–58]. The GNN, JPDA, and JIPDA are single-hypothesis-based algorithms. Thus, the MHT produces better tracking accuracy compared with single-hypothesis-based tracking algorithms for tracking scenarios with high FA density and closely spaced targets [10].

5.2.2 Tracking without Target Identity (or Track Label)

There are certain tracking problems such as group tracking and traffic flow monitoring, where it is not necessary to know the identity of a target. The *finite set statistics* (FISST) [17, 22–24, 59] was formulated to solve this class of problems. The multitarget Bayes recursion for the posterior density using the FISST set integrals was first proposed by Mahler in [18, 22, 59]. In general, the FISST Bayes multitarget filter is computationally intractable due to the complex set integral which is difficult to evaluate. Mahler first proposed the PHD of the posterior multitarget state [18, 19] as an approximation to the multitarget Bayes filter. The PHD is also known as the *intensity function* or *first moment*. The integral of the PHD over a volume element in the state space gives the expected number of targets in that volume element. Sequential Monte Carlo (SMC) implementations of the multitarget Bayes filter and PHD filter were presented in [60–64]. Numerically efficient Gaussian mixture based PHD filter for linear and mildly nonlinear multitarget models was presented in [64]. Track labeling features were added to the PHD filter for MTT in [65–67].

Mahler formulated the CPHD, which jointly propagates the posterior PHD and the posterior distribution of the number of targets [20, 21]. In [25] the Gaussian mixture CPHD (GMCPHD) was presented and it was shown that the computational complexity is cubic in the number of measurements. Mahler proposed the multitarget multi-Bernoulli (MeMBer) filter [22] in which the MeMBer filter propagates the parameters of a multi-Bernoulli RFS that approximates the posterior multitarget RFS. The algorithm for the MeMBer filter in [22] overestimates the number of targets. An unbiased version of the MeMBer filter was presented in [68].

The interacting multiple model (IMM) estimator [69, 70] is a well-known algorithm for maneuvering targets. SMC-PHD filter and GMPHD filter for the multiple model were presented in [71] and [72], respectively. A multiple model extension of the CPHD filter was proposed in [73] using the physical-space bin model of the CPHD filter [74]. A dynamic MM-based CPHD filter was formulated in [75] using a framework similar to that in [72].

A detailed description of the developments and algorithms in the PHD/CPHD and MeMBer filters is presented in [23] and Chapter 3 of this book [24].

Remark Although many papers study closely spaced targets, the effect of sensor resolution is not taken into account while generating simulated measurements. For example, if more than one target are within a resolution cell, then for a real sensor, only one detection will be generated. As a result, integrating over the PHD surface will not produce an accurate estimate of the actual number of true targets. To estimate the true number of targets, a *track-before-detect* (TBD) [76, 77] type algorithm is needed.

The iFilter was first proposed in [26, 27] and is described in detail in [29] and [78]. In contrast to the CPHD filter, which was derived using the finite set statistics, the iFilter is derived using Poisson point processes (PPPs) [26, 29]. Both the CPHD filter and iFilter estimate multitarget states and unknown number of targets. It was claimed in [78, 79] that the main difference between the PHD filter and iFilter is the clutter model; the PHD filter requires prior knowledge whereas the iFilter can estimate the clutter model from measurements. The CPHD filter which estimates the clutter model is presented in [80–82]. A careful and detailed comparative evaluation of the CPHD filter and iFilter is needed to determine the relative merits.

SMC-based multitarget tracking algorithms have received a great deal of attention in recent years because of their ability to handle nonlinear and non-Gaussian dynamic and measurement models [39, 77, 83, 84]. Two classes of SMC-based multitarget algorithms are commonly used,

- Particle filter (PF)-based algorithms and
- MCMC-based algorithms.

Vermaak et al. proposed three PF-based MTT algorithms in [39] which handle single-frame-based data association. These algorithms are Monte Carlo joint probabilistic data association filter (MC-JPDAF), sequential sampling particle filter (SSPF), and independent partition particle filter (IPPF). The SSPF samples individual targets sequentially by using a factorization of the importance weights whereas the IPPF assumes the associations to be independent for individual targets. These algorithms claim to handle the curse of dimensionality that arises due to the increased size of the state space associated with multiple targets.

For a multitarget tracking problem, the dimension of the state space can be high. In general, MCMC-based multitarget tracking algorithms are more effective than PF-based algorithms for a high-dimensional state space [34, 36]. As a result, the MCMC-based algorithms have been proposed by a number of researchers [31, 35–38, 85, 86]. In a comparative study [38], the MCMC-based algorithm was shown to produce improved tracking accuracy than the sampling importance resampling (SIR) and resample-move-based algorithms.

5.3 TRACK FILTERING

In this section, we describe the filtering algorithms used in a track filter. Two classes of filtering algorithms are commonly used, depending on the type of motion of the target:

- single model filter for a nonmaneuvering target, and
- multiple model filter for a maneuvering target.

Sections 5.3.1 and 5.3.2 summarize the dynamic and measurement models, respectively. Then we describe the single-model and multiple model filters.

5.3.1 Dynamic Models

If the target is nonmaneuvering, its motion can be modeled by a single kinematic or dynamic model. Such a filter is called a single model (or mode) filter [2, 69]. For a maneuvering target, the dynamic model of the target changes with time. Therefore, a single mode filter is not appropriate to describe the motion of a maneuvering target. A bank of filters, each with a different kinematic model, is used for a maneuvering target. A dynamic model can be a linear or nonlinear function of the target state. Dynamic models for a single mode filter can be discrete-time or continuous-time [69]. Next, we identify some commonly used dynamic models in target tracking. An excellent survey of dynamic models is presented in [87].

5.3.1.1 Discrete-time Dynamic Models.

- Nearly constant velocity model (NCVM)—linear [2, 10, 69, 87],
- Gauss–Markov velocity model (GMVM)—linear [2, 10, 69, 87],
- Nearly constant acceleration model (NCAM)—linear [2, 10, 69, 87],
- Gauss–Markov acceleration model (GMAM)—linear [2, 10, 69, 87],
- Singer model [10, 87, 88],
- Coordinated turn (CT) model—linear for constant angular velocity; nonlinear if angular velocity is estimated [2, 10, 69, 87],
- Dynamic models in modified polar coordinates (MPC) [77, 90] and log polar coordinates (LPC)—nonlinear [91, 92],
- Dynamic models in modified spherical coordinates (MSC) [93–99].

5.3.1.2 Continuous-time Dynamic Models.

- Dynamic models in MPC [89] and LPC—nonlinear,
- Dynamic models for MSC [95–96, 98] and LSC [95]—nonlinear,
- Dynamic model of a ballistic projectile experiencing Earth's gravity and drag force [100]—nonlinear,
- Dynamic model of a missile experiencing Earth's gravity and drag force [101–103]—nonlinear,
- Dynamic model of a satellite experiencing Earth's gravity, drag force, and solar radiation pressure (SRP) [104, 105]—nonlinear.

5.3.2 Measurement Models

Various measurement models commonly used in tracking applications and associated references are presented in Table 5.2. An excellent survey of measurement models is presented in [106].

TABLE 5.2 Measurement Models

Measurement Model	References
Bearing	Chapter 6 of [77] and references therein
Bearing, elevation	[93–99]
Intensity map (raw sensor measurements)	Chapter 11 of [77] and references therein, [76, 107]
Range, azimuth	[1, 2, 108]
Slant range, azimuth	[109–111]
Slant range, azimuth, elevation	[2, 10, 108]
Slant range, direction cosines	[112–114]
Slant range, range rate (Doppler)	[115]
Slant range, azimuth, range rate (GMTI)	[116–118]
Range-Doppler map (raw sensor measurements)	[119–122]
Slant range, azimuth, elevation, range rate	[123]
SAR images	[124, 125]
Video images	[126–128]
Bistatic range, azimuth, bistatic range rate (OTHR)	[129, 130]
LIDAR	[131, 132]

5.3.3 Single Model Filter for a Nonmaneuvering Target

A filtering algorithm has two components, the dynamic model and measurement model. Sensor measurements are usually collected at discrete times $\{t_k\}_{k=1}^{M}$, where M is the total number of scans or dwells. Therefore, the measurement model is a discrete-time model.

5.3.3.1 Discrete-time Dynamic Models. Two types of approaches are used for the discrete-time dynamic model,

- discretized continuous-time dynamic model [69, 133],
- direct discrete-time dynamic model [69].

5.3.3.2 Commonly Used Linear Dynamic Models in Tracking. We shall use the discretized continuous-time dynamic model, since this is derived from the continuous-time motion of a real-world object. Let $\mathbf{x}_k \in \mathbb{R}^n$ denote the kinematic Cartesian state vector of the target at time t_k. The dynamic models corresponding to the NCV, GMV, NCA, GMA, and CT with constant angular velocity are described by [2, 10, 69]

$$\mathbf{x}_k = \mathbf{F}_{k-1}\mathbf{x}_{k-1} + \mathbf{w}_{k-1}, \tag{5.1}$$

where $\mathbf{F}_{k-1}$ and $\mathbf{w}_{k-1}$ are the state transition matrix and zero-mean white Gaussian integrated process noise with covariance $\mathbf{Q}_{k-1}$, respectively, for the time interval $[t_{k-1}, t_k]$. Expressions for $\mathbf{F}_{k-1}$ and $\mathbf{Q}_{k-1}$ for various dynamic models are given in [2, 10, 69, 87].

5.3.3.3 Commonly Used Nonlinear Dynamic Models. Modified polar coordinates [77, 89–91] and log polar coordinates [91, 92] were proposed for the bearing-only filtering problem to prevent the filter divergence experienced by the *extended Kalman filter* (EKF) using relative Cartesian coordinates for the state vector. The relative state vector in MPC at time t_k is defined by [91]

$$\begin{aligned} \boldsymbol{\xi}_k &= [\,\xi_{k,1} \quad \xi_{k,2} \quad \xi_{k,3} \quad \xi_{k,4}\,]' \\ &:= [\,\dot{\beta}_k \quad \dot{r}_k/r_k \quad \beta_k \quad 1/r_k\,]', \end{aligned} \tag{5.2}$$

where $\boldsymbol{\beta}_k, r_k, \dot{r}_k$ are the bearing, range, and range-rate, respectively, at time t_k. The nonlinear dynamic model in MPC corresponding to the NCV model in Cartesian state is given by [91]

$$\boldsymbol{\xi}_k = \mathbf{f}(\boldsymbol{\xi}_{k-1}, \mathbf{S}^{\mathrm{MPC}}(\boldsymbol{\xi}_{k-1}, \mathbf{w}_{k-1} - \mathbf{U}_{k-1})), \tag{5.3}$$

where $\mathbf{f}$ is a nonlinear function of $\boldsymbol{\xi}_{k-1}$ and $\mathbf{S}^{\mathrm{MPC}}$ is a nonlinear function of ξ_{k-1}, $\mathbf{w}_{k-1}$ and $\mathbf{U}_{k-1}$. The term $\mathbf{U}_{k-1}$ is associated with the ownship motion [77, 91]. We note that the integrated process noise $\mathbf{w}_{k-1}$ is not an additive term in the dynamic model, rather $\boldsymbol{\xi}_k$ has a complex nonlinear dependence on the process noise $\mathbf{w}_{k-1}$. Let $\boldsymbol{\eta}_k$ denote the relative state vector in LPC at time t_k. The first three components of MPC and LPC are the same. The fourth component of LPC is $\log(r_k)$. Thus, the relative state vector in LPC at time t_k is defined by [91]

$$\begin{aligned} \boldsymbol{\eta}_k &= [\,\eta_1(t) \quad \eta_2(t) \quad \eta_3(t) \quad \eta_4(t)\,]' \\ &:= [\,\dot{\beta}_k \quad \dot{r}_k/r_k \quad \beta_k \quad \log(r_k)\,]'. \end{aligned} \tag{5.4}$$

Similarly, the nonlinear dynamic model in LPC corresponding to the NCV model in Cartesian state is given by [91]

$$\boldsymbol{\eta}_k = \mathbf{g}(\boldsymbol{\xi}_{k-1}, S^{\mathrm{LPC}}(\boldsymbol{\eta}_{k-1}, \mathbf{w}_{k-1} - \mathbf{U}_{k-1})), \tag{5.5}$$

where $\mathbf{g}$ is a nonlinear function of $\boldsymbol{\eta}_{k-1}$ and $\mathbf{S}^{\mathrm{LPC}}$ is a nonlinear function of η_{k-1}, $\mathbf{w}_{k-1}$ and $\mathbf{U}_{k-1}$.

5.3.3.4 Continuous-time Dynamic Models. Let r, β, ε denote the slant range, bearing, and elevation angle in 3D. Then, for the angle-only filtering in 3D, the relative state in MSC [95, 96, 98, 99] is defined by

$$\begin{aligned}\boldsymbol{\xi}(t) &:= [\xi_1(t) \quad \xi_2(t) \quad \xi_3(t) \quad \xi_4(t) \quad \xi_5(t) \quad \xi_6(t)]' \\ &= [\omega(t) \quad \dot{\varepsilon}(t) \quad \dot{r}(t)/r(t) \quad \beta(t) \quad \varepsilon(t) \quad 1/r(t)]',\end{aligned} \tag{5.6}$$

where $\omega(t) := \cos\varepsilon(t)\,\dot{\beta}(t)$. In an analogous manner, the LSC [95] is defined by

$$\begin{aligned}\boldsymbol{\eta}(t) &:= [\eta_1(t) \quad \eta_2(t) \quad \eta_3(t) \quad \eta_4(t) \quad \eta_5(t) \quad \eta_6(t)]' \\ &= [\omega(t) \quad \dot{\varepsilon}(t) \quad \dot{r}(t)/r(t) \quad \beta(t) \quad \varepsilon(t) \quad \log(r(t))]'.\end{aligned} \tag{5.7}$$

The nonlinear *stochastic differential equations* (SDEs) in MSC and LSC corresponding to the NCV model in Cartesian coordinates are given by [95]

$$\frac{d\boldsymbol{\xi}(t)}{dt} = \mathbf{f}^{\mathrm{MSC}}(\boldsymbol{\xi}(t), \mathbf{w}^t(t), \mathbf{a}_{\mathrm{o}}^{\mathrm{T}}(t)), \tag{5.8}$$

$$\frac{d\boldsymbol{\eta}(t)}{dt} = \mathbf{f}^{\mathrm{LSC}}(\boldsymbol{\eta}(t), \mathbf{w}^t(t), \mathbf{a}_{\mathrm{o}}^{\mathrm{T}}(t)), \tag{5.9}$$

where $\mathbf{f}^{\mathrm{MSC}}$ and $\mathbf{f}^{\mathrm{LSC}}$ are nonlinear functions and $\mathbf{w}^t(t) \in \mathbb{R}^3$ and $\mathbf{a}_{\mathrm{o}}^{\mathrm{T}}(t) \in \mathbb{R}^3$ are the zero-mean white acceleration process noise and ownship acceleration (deterministic) in the tracker coordinate frame (T frame) at time t, respectively.

Let $\mathbf{p}(t) \in \mathbb{R}^3$ and $\mathbf{v}(t) \in \mathbb{R}^3$ denote the position vector and velocity of ballistic projectile, respectively. The state is defined by

$$\mathbf{x}(t) := [\mathbf{p}'(t) \quad \mathbf{v}'(t) \quad \rho_1(t)]', \tag{5.10}$$

where $\rho_1(t)$ is a drag coefficient correction parameter, which is modeled as a first-order Gauss–Markov process [133]. Then the nonlinear SDE for a ballistic projectile under the influence of earth's gravity and atmospheric drag is described by [100]

$$d\mathbf{x}(t) = \mathbf{f}(\mathbf{x}(t))dt + d\mathbf{b}(t), \tag{5.11}$$

where $\mathbf{f}$ is a nonlinear time evolution function and $\mathbf{b}(t)$ is a Brownian motion [135] with power spectral density matrix $\mathbf{Q}(t)$. The function $\mathbf{f}$ is defined by

$$\mathbf{f}(\mathbf{x}(t)) := [\mathbf{v}'(t) \quad \mathbf{a}'_{\mathrm{gravity}} + \mathbf{a}'_{\mathrm{D}}(t) \quad -\rho_1(t)/\tau]', \tag{5.12}$$

where $\mathbf{a}_{\mathrm{gravity}}$ and $\mathbf{a}_{\mathrm{D}}(t)$ are accelerations due to gravity and atmospheric drag, respectively, and τ is the time-correlation constant for the drag coefficient correction parameter ρ_1. The SDE can also be expressed using the formal zero-mean white noise

$\mathbf{w}(t) = d\mathbf{b}(t)/dt$

$$\frac{d\mathbf{x}(t)}{dt} = \mathbf{f}(\mathbf{x}(t)) + \mathbf{w}(t). \tag{5.13}$$

Similarly, for satellite-orbit determination or space-object tracking, (5.11) is also valid, where the equation of motion is integrated in an *inertial reference frame* [104]. The total acceleration **a**(t) due to gravity, atmospheric drag, SRP, and thrust is given by [104, 105]

$$\mathbf{a}(t) = \mathbf{a}_{\text{gravity}}(t) + \mathbf{a}_{\text{D}}(t) + \mathbf{a}_{\text{SRP}}(t) + \mathbf{a}_{\text{thrust}}(t). \tag{5.14}$$

For precision orbit determination, the term acceleration due to gravity, $\mathbf{a}_{\text{gravity}}(t)$, includes the contributions due to nonspherical gravitational potential of the earth, moon, sun, and planets.

5.3.4 Filtering Algorithms

The Kalman filter (KF) [134] is an optimal estimator in the minimum mean square error (MMSE) sense, if the dynamic and measurement models are linear with additive Gaussian process noise and measurement noise, and a prior Gaussian distribution [69, 77, 135, 136]. If the dynamic model or/and measurement model is/are nonlinear, or if the noise distributions are non-Gaussian, then an optimal closed-form solution does not exist. Commonly used approximate nonlinear filtering algorithms are as follows:

- EKF and iterated EKF (IEKF) [69, 77, 133, 135–137],
- unscented Kalman filter (UKF), [77, 137–141],
- particle filter (PF) [77, 83, 84, 142],
- Gaussian sum filter (GSF) [143–148],
- Markov chain Monte Carlo (MCMC) [31, 37, 85, 86, 149],
- Beneš filter [150],
- Daum filter [151–153], and
- particle flow filter (PFF) [154–163].

5.3.4.1 Extended Kalman Filter. The first-order and second-order EKFs [69, 137] use Gaussian approximations to the filtering densities using the linear and quadratic approximations in the Taylor series expansion, respectively. In order to reduce the error in state prediction and the effect of nonlinearity in the measurement function an IEKF is often used [69, 133]. The Jacobian for the nonlinear time-evolution function and measurement functions are required for the first-order EKF whereas the Jacobian and Hessian of these nonlinear functions are needed for the second-order EKF. Calculation of the first and second derivatives can be quite tedious and error prone for certain types of problems such as the bearing-only filtering using MPC and LPC, and angle-only filtering in 3D using MSC and LSC.

5.3.4.2 *Unscented Kalman Filter.* Instead of using the Taylor series expansion of the nonlinear time-evolution function or/and measurement function, the UKF uses the unscented transform (UT) [138–141] to calculate the approximate first- and second-order moments which appear in the KF equations. The UKF has two advantages over the EKF. First, it avoids the analytic derivation of the Jacobian and Hessian matrices and their numerical evaluation. These derivative matrices can also be computed numerically. However, this increases the computational cost. Secondly, the UT provides a more accurate approximation than the approximation used in the EKF. However, the UKF is not necessarily more accurate than the EKF in all cases. For example, for the GMTI filtering problem [118], the accuracies of the EKF and UKF are comparable. Similarly, for the angle-only filtering problem in 3D using the relative Cartesian state vector or MSC with discrete-time dynamic and measurement models [97, 98], the EKF and UKF have nearly the same accuracy.

The EKF and UKF work reasonably well when the prior and posterior densities for the target state are unimodal and close to Gaussian. However, when these densities become multimodal or the process noise or/and measurement noises are non-Gaussian, then the EKF and UKF perform poorly. The GSF, PF, MCMC, and PFF are used for multimodal and non-Gaussian problems.

5.3.4.3 *Gaussian Sum Filter.* In the GSF [143–148], the posterior density of the state is approximated by a finite weighted sum of Gaussian densities, where the sum of the weights is unity. However, the online computation of the weights, means, and covariances is quite challenging [77]. For the nonlinear dynamic model excited by additive white Gaussian noise, the mean and covariance of the Gaussian components are propagated using a linearized model (or UKF approach) and the weights of the new Gaussian components are set to the previous weights [143, 144]. Then the mean, covariance, and weight are updated by Bayes rule using the measurement and measurement error covariance. In [146–148], the authors improved the accuracy of the GSF by updating the weights during the propagation step. The weights are updated by solving a convex optimization problem where the Gaussian sum approximation must satisfy the Fokker–Planck–Kolmogorov equation for continuous-time dynamical model and the Chapman–Kolmogorov equation for discrete-time dynamical model.

5.3.4.4 *Particle Filter.* PFs are SMC-based filters which approximate the posterior density of the target state by a set of random samples or particles and associated weights, $\{\mathbf{x}_k^i, w_k^i\}_{i=1}^N$. Excellent description of PFs is presented in two books [77, 84] and the review paper [83]. There are many versions of the PF algorithm and they are easy to implement, although the computational cost is high compared with that for the EKF and UKF. The sequential importance sampling (SIS) algorithm [83, 84, 164] is a Monte-Carlo-based algorithm that forms the basic building block of all PFs. PFs have received a great deal of interest for nonlinear and non-Gaussian filtering problems for more than a decade due to the seminal paper [142] where the PF is known as the bootstrap filter (BF). PFs face a common problem, the degeneracy problem, where the weight of all but one particle becomes negligible. A number of solutions such as the SIR [142], auxiliary sampling importance resampling (ASIRS) [165], and regularization [166] have been proposed to address the degeneracy problem.

The bearing-only filtering problem (see Chapter 6 of [77] and references therein) is a challenging and difficult nonlinear filtering problem due to the high degree of nonlinearity (DoN) [167–169]. Extensive studies in Chapter 6 of [77] show that the PF is the most accurate filtering algorithm for the bearing-only filtering problem. However, for the GMTI filtering problem [118], for which the DoN is low compared with that for the bearing-only filtering problem, the PF does not show any advantage over the EKF and UKF. Similarly, for the angle-only filtering problem in 3D [97, 98], we observe that for some commonly used scenarios, various PF-based algorithms do not perform better than the EKF and UKF.

5.3.4.5 Markov Chain Monte Carlo. The MCMC approach has received a great deal of attention in recent years in the context of PF-based MTT for nonlinear non-Gaussian problems. The standard PF cannot cope with new features of the posterior distribution of the target state since it provides no opportunity to generate new values for unknown quantities after their initial generation. After processing a few measurements, most of the particles except one have low weights are eliminated by the resampling process, which leads to serious sample impoverishment. An alternative solution to this problem is the MCMC-based resample-move algorithm [149]. This technique uses periodic MCMC steps to diversify particles in an importance sampling-based PF, which ensures the particles to asymptotically approximate samples from the posterior distribution. The development of sequential MCMC algorithms [31, 37, 85, 86] has made the MCMC-PF a suitable candidate for high dimensional nonlinear non-Gaussian problems. There are many variants of the MCMC-PF and it has been used for maneuvering targets [42, 170, 171], MTT in a multiple scan joint probabilistic data association (MS-JPDA) tracker [42], and group tracking [33].

5.3.4.6 Particle Flow Filter. Daum and Huang [155–163] have recently developed a new nonlinear filter based on *particle flow*, which shows a great deal of promise for difficult nonlinear non-Gaussian filtering problems. We shall refer to this class of filters as the PFF. They claim that the PFF is many orders of magnitude faster than the conventional PF and it is several orders of magnitude more accurate than the EKF [159, 160] for difficult nonlinear filtering problems. Secondly, it does not suffer from the common degeneracy problem experienced by the PF [159, 160]. An accessible account of the PFF for practicing engineers is presented in [154]. The PFF does not compute the posterior probability density function (pdf) of the target state; rather it calculates the logarithm of the un-normalized conditional density of the target state. Their original PFF algorithm [155–157], based on the incompressible flow of particles had problems with obstruction of particle flow and singularities in the equations for flow. The improved algorithm in [160] represents compressible flow (or exact flow) of particles and does not suffer from these two problems. As a result, the exact particle flow algorithm in [160] represents the exact flow of the conditional probability density. An EKF or UKF can be used with the PFF [154]. The PFF based on incompressible and exact flows are denoted by IF and EF, respectively.

A suffix "e" or "u" is used to denote the EKF or UKF in the PFF [154]. A number of examples are analyzed in [154] to demonstrate the superior performance of various PFF algorithms (e.g., EFe, EFu, EFe, EFu) over the EKF, UKF, BF, and auxiliary PF.

5.3.5 Multiple Switching Model Filter for a Maneuvering Target

A number of dynamic models or modes $\{a_i\}_{i=1}^r$, $r > 1$, are used for the state estimation of a maneuvering target. This class of problems is known as *jump Markov* or *hybrid state estimation* problem [69, 172] which includes estimation of a continuous-valued state (e.g., kinematic state) and discrete-valued mode. Two different types of scenarios can occur. Most papers consider scenarios where the measurement time interval $(t_{k-1}, t_k]$ remains constant for all k. For this type of scenarios, a discrete-time hybrid system can be described by the dynamic and measurement models, respectively, by [77]

$$\mathbf{x}_k = \mathbf{f}_{k-1}(\mathbf{x}_{k-1}, \mathbf{w}_{k-1}, m_k), \tag{5.15}$$

$$\mathbf{z}_k = \mathbf{h}_k(\mathbf{x}_k, \mathbf{v}_k, m_k), \tag{5.16}$$

where m_k is the mode in effect during the time interval $(t_{k-1}, t_k]$. The mode variable m_k is modeled by a time-homogeneous r-state first-order Markov chain with a transition probability matrix Π with elements [69]

$$\pi_{ij} := P\{m_k = a_j | m_{k-1} = a_i\}, \quad i, j = 1, 2, \ldots, r. \tag{5.17}$$

The general scenario can occur when the measurement time interval $(t_{k-1}, t_k]$ can vary with time t_k for a single sensor or due to multiple sensors. In this case, the mode variable cannot be modeled by a time-homogeneous Markov chain and a constant transition probability matrix is not possible. Details for this scenario are described in [173] where the mode of the target is described by a continuous-time Markov chain $\{m(t), t \geq 0\}$, where $m(t)$ is the mode of the target at time t. Transitions are assumed to occur at random times $\{\zeta_k\}$. The values

$$\chi_n = m(\zeta_n^+), \quad \chi_n \in \{a_1, a_2, \ldots, a_r\}, \tag{5.18}$$

of $m(t)$ at the times $\{\zeta_n^+\}$ form a discrete-state Markov sequence, known as the Markov chain embedded in the Markov process $\{m(t), t \geq 0\}$ [174]. Let $\pi_{ij}(t, t')$ denote the transition probability from time t to t' defined by

$$\pi_{ij}(t, t') := P\{m(t') = a_j | m(t) = a_i\}, \quad i, j = 1, 2, \ldots, r. \tag{5.19}$$

For a stationary or homogeneous Markov process, the transition probabilities depend only on the time difference $\tau = t' - t$ [174]:

$$\pi_{ij}(\tau) := P\{m(t + \tau) = a_j | m(t) = a_i\}, \quad \tau > 0, \quad i, j = 1, 2, \ldots, r. \tag{5.20}$$

In [174], it is shown that the transition probability matrix $\boldsymbol{\Pi}(\tau)$ satisfies the Kolmogorov equation

$$\frac{d\boldsymbol{\Pi}(\tau)}{d\tau} = \boldsymbol{\Pi}(\tau)\boldsymbol{\Lambda}, \quad \boldsymbol{\Pi}(0) = \mathbf{I}, \tag{5.21}$$

where $\boldsymbol{\Lambda}$ is the $r \times r$ *transition probability rate matrix*. The solution of the linear differential equation (5.21) is [174]

$$\boldsymbol{\Pi}(\tau) = e^{\boldsymbol{\Lambda}\tau}. \tag{5.22}$$

When $r = 2$, an analytic expression for $\boldsymbol{\Pi}(\tau)$ can be obtained [173, 174]. When $r > 2$, it is difficult to get an analytic expression for $\boldsymbol{\Pi}(\tau)$. However, we can evaluate (5.22) numerically using efficient matrix exponential evaluation algorithms [175].

Excellent descriptions of the IMM and VS-IMM algorithms are given in [69, 172] and [176, 177], respectively. Blom first formulated the suboptimal hybrid state estimator, the IMM in [178] with a subsequent journal version in [179]. Tables 5.3 and 5.4 summarize some well-known applications of the IMM and VS-IMM, respectively.

TABLE 5.3 Applications of IMM Algorithm

Problem	References
Multisensor tracking system for advanced air traffic control	[180]
Tracking algorithm for an advanced ATC system	[181]
Air traffic control tracking	[182]
Multitarget-multisensor air trafic surveillance	[183]
Precision large scale air traffic surveillance	[184]
Radar benchmark tracking problem	[185]
MHT using IMM	[185, 186]
IMM-IPDA-based Algorithms	[187]
Joint IMM/MHT tracking and identification for multisensor ground target tracking	[188]

TABLE 5.4 Applications of VS-IMM Algorithm

Problem	References
Ground target tracking with variable structure IMM estimator	[189]
Tracking Targets with Transportation Network Constraints	[190]
VS-IMM using Road Map Information for Ground Target Tracking,	[191]
MHT using VS-IMM	[192]

5.4 MHT ALGORITHMS

There are two types of MHT algorithms, hypothesis-oriented [2, 10, 15, 47] and track-oriented [2, 10, 16, 47]. The fundamental approach includes many variants.

Hypothesis-oriented MHT was first proposed by Reid [15]. The initial integer-programming formulation of the problem is due to Morefield [53]. The hybrid-state decomposition that allows for computationally efficient track-oriented MHT is due to Kurien [16]. An efficient solution to the optimization problem required for *N-scan* hypothesis pruning via Lagrangian relaxation is due to Poore and Rijavec [51]. A simple linear-programming-based relaxation approach to the same optimization problem was proposed independently by Coraluppi et al. [54] and by Storms and Spieksma [55].

The inherent suboptimality of all MHT approaches is based on the need to limit the number of track hypotheses. This can be achieved with gating techniques, with appropriate hypothesis generation logic, and with a number of hypothesis pruning and merging techniques. Arguably, the need for track confirmation schemes (see [10] for a discussion of several approaches) is required to handle spurious track formation that ensues from MHT suboptimality.

From software architecture development, maintenances, debugging, and cost effectiveness points of view, the track-oriented MHT is preferred over the hypothesis-oriented MHT. Therefore, most tracking groups at present use the track-oriented MHT. In the hypothesis-oriented MHT, the probability of a global hypothesis or a set of tracks is calculated [15], whereas the track-oriented MHT calculates the log-likelihood ratio (LLR) or track score for each track [2, 10, 16]. In the track-oriented MHT, each estimated target is represented by a directed-tree, where each branch represents a *local track hypothesis* or a *track*. Track-oriented MHT is well established as a paradigm for multisensor multitarget tracking. The fundamental approach includes many variants.

In Section 5.5, we provide a hybrid-state derivation of the track-oriented MHT equations that is closely related to the original treatment by Kurien [16], with some minor modifications. Then in Section 5.6 we address the so-called target death problem inherent in PHD filtering [22–24] and show that it does not arise in track-oriented MHT.

5.5 HYBRID-STATE DERIVATIONS OF MHT EQUATIONS

A key challenge in multisensor MTT is *measurement origin uncertainty*. We do not know how many objects are in the surveillance region and which measurement originated from which target or FA in a scan. New objects may be born in any given scan, and existing objects may die.

We assume that for each sensor scan, thresholded detection-level measurements are available. Signal processing techniques are applied to raw sensor data yielding measurements for which the detection and localization statistics are known. We are interested in a scan-based (or real-time) approach that, perhaps with some delay, yields an estimate of the number of objects and corresponding state estimates at any time.

Several approaches exist for detection-level scan-based tracking. In this section, we employ a hybrid-state formalism to describe the track-oriented MHT approach.

We assume Poisson distributed births at each scan with mean λ_b, object detection probability p_d, and object death probability p_χ at each scan. (We neglect the *time-dependent* nature of birth and death probabilities as would ensue from an underlying continuous-time formulation, and we neglect as well inter-scan birth and death events.)

Suppose N_k targets are present at time t_k. Let $\mathbf{x}_{k,i}$ denote the true state of the ith target at time t_k. The composite state $\mathbf{X}_k$ of N_k targets at time t_k is defined by [16]

$$\mathbf{X}_k := \{\mathbf{x}_{k,1}, \mathbf{x}_{k,2}, \ldots, \mathbf{x}_{k,N_k}\}. \tag{5.23}$$

Suppose M_k measurements are collected by a sensor at time t_k. In general, M_k is different from N_k due to sensor probability of detection less than unity and the presence of FA. Secondly, the origin of the measurements is unknown, that is, it is not known which measurement originates from which target. Let $\mathbf{z}_{k,i}$ and $\mathbf{Z}_k$ denote the ith measurement and composite measurement, respectively, at time t_k. Then $\mathbf{Z}_k$ is defined by

$$\mathbf{Z}_k := \{\mathbf{z}_{k,1}, \mathbf{z}_{k,2}, \ldots, \mathbf{z}_{k,M_k}\}. \tag{5.24}$$

Let $\mathbf{X}^k$ and $\mathbf{Z}^k$ denote the set of target states and measurements, respectively, up to time t_k,

$$\mathbf{X}^k := \{\mathbf{X}_1, \mathbf{X}_2, \ldots, \mathbf{X}_k\}, \tag{5.25}$$
$$\mathbf{Z}^k := \{\mathbf{Z}_1, \mathbf{Z}_2, \ldots, \mathbf{Z}_k\}. \tag{5.26}$$

Let q_k denote the discrete random variable which maps the composite target state $\mathbf{X}_k$ and FAs to the composite $\mathbf{Z}_k$ at time t_k. Let q^k be the set RVs $\{q_k\}$ up to time t_k,

$$q^k := \{q_1, q_2, \ldots, q_k\}. \tag{5.27}$$

We have a sequence of sets of measurements $\mathbf{Z}^k$, and we wish to estimate the state history $\mathbf{X}^k$ for all objects present in the surveillance region. The auxiliary discrete state history RV q^k represents a full interpretation of all measurements: which measurements are false, how the object-originated ones are to be associated, and when objects are born and die. There are two fundamental assumptions of note. The first is that there are no target births in the absence of a corresponding detection, that is we do not reason over new, undetected objects. The second is that there is *at most* one measurement per object per scan.

We are interested in the probability distribution $p\left(\mathbf{X}^k|\mathbf{Z}^k\right)$ for object state histories given data. This quantity can be obtained by conditioning over all possible auxiliary states histories q^k,

$$p\left(\mathbf{X}^k|\mathbf{Z}^k\right) = \sum_{q^k} p\left(\mathbf{X}^k, q^k|\mathbf{Z}^k\right) = \sum_{q^k} p\left(\mathbf{X}^k|\mathbf{Z}^k, q^k\right) p\left(q^k|\mathbf{Z}^k\right). \tag{5.28}$$

A pure MMSE approach would yield the following:

$$\hat{\mathbf{X}}_{\text{MMSE}}\left(\mathbf{Z}^k\right) = E\left[\mathbf{X}^k|\mathbf{Z}^k\right] = \sum_{q^k} E\left[\mathbf{X}^k|\mathbf{Z}^k, q^k\right] p\left(q^k|\mathbf{Z}^k\right). \tag{5.29}$$

The track-oriented MHT approach is a mixed MMSE/*maximum a posteriori* (MAP) one, whereby we identify the MAP estimate for the auxiliary state history q^k, and identify the corresponding MMSE estimate for the object state history $\mathbf{X}^k$ conditioned on the estimate for q^k,

$$\hat{\mathbf{X}}\left(\mathbf{Z}^k\right) = \hat{\mathbf{X}}_{\text{MMSE}}\left(\mathbf{Z}^k, \hat{q}^k\right), \tag{5.30}$$

$$\hat{q}^k = \hat{q}_{\text{MAP}}\left(\mathbf{Z}^k\right) = \underset{q^k}{\arg\max}\; p\left(q^k|\mathbf{Z}^k\right). \tag{5.31}$$

Each feasible q^k corresponds to a global hypothesis. (The set of global hypotheses is generally constrained via measurement gating and hypothesis generation logic.) We are interested in a *recursive* and *computationally efficient* expression for $p\left(q^k|\mathbf{Z}^k\right)$ that lends itself to functional optimization without the need for explicit enumeration of global hypotheses. We do so through repeated use of Bayes' rule. Note that, for notational simplicity, we use p for both probability density and probability mass functions. The normalizing constant c_k does not impact MAP estimation,

$$p\left(q^k|\mathbf{Z}^k\right) = \frac{p\left(\mathbf{Z}_k|\mathbf{Z}^{k-1}, q^k\right) p\left(q_k|\mathbf{Z}^{k-1}, q^{k-1}\right) p\left(q^{k-1}|\mathbf{Z}^{k-1}\right)}{c_k}, \tag{5.32}$$

$$c_k = p\left(\mathbf{Z}_k|\mathbf{Z}^{k-1}\right) = \sum_{q^k} p\left(\mathbf{Z}_k|\mathbf{Z}^{k-1}, q^k\right) p\left(q^k|\mathbf{Z}^{k-1}\right). \tag{5.33}$$

The recursive expression (5.32) involves two factors that we consider in turn, with the discrete state probability one first. It will be useful to introduce the aggregate variable ψ_k that accounts for the number of detections d for the τ existing tracks, the number of track deaths χ, the number of new tracks b, and the number of false returns

$r-d-b$, where r is the number of measurements in the current scan,

$$p\left(q_k|\mathbf{Z}^{k-1},q^{k-1}\right)=p\left(\psi_k|\mathbf{Z}^{k-1},q^{k-1}\right)p\left(q_k|\mathbf{Z}^{k-1},q^{k-1},\psi_k\right), \quad (5.34)$$

$$\begin{aligned}p\left(\psi_k|\mathbf{Z}^{k-1},q^{k-1}\right)=&\left\{\binom{\tau}{\chi}p_\chi^\chi\left(1-p_\chi\right)^{\tau-\chi}\right\}\\&\cdot\left\{\binom{\tau-\chi}{d}p_d^d\left(1-p_d\right)^{\tau-\chi-d}\right\}\\&\cdot\left\{\frac{\exp\left(-\lambda_b\right)\lambda_b^b}{b!}\right\}\cdot\left\{\frac{\exp\left(-\lambda_{fa}\right)\lambda_{fa}^{r-d-b}}{(r-d-b)!}\right\},\end{aligned} \quad (5.35)$$

$$p\left(q_k|\mathbf{Z}^{k-1},q^{k-1},\psi_k\right)=\frac{1}{\binom{\tau}{\chi}\binom{\tau-\chi}{d}\left(\frac{r!}{(r-d)!}\right)\binom{r-d}{b}}. \quad (5.36)$$

Substituting (5.35) and (5.36) into (5.34) and simplifying yields the following:

$$\begin{aligned}p\left(q_k|\mathbf{Z}^{k-1},q^{k-1}\right)=&\left\{\frac{\exp\left(-\lambda_b-\lambda_{fa}\right)\lambda_{fa}^r}{r!}\right\}p_\chi^\chi\left(\left(1-p_\chi\right)\left(1-p_d\right)\right)^{\tau-\chi-d}\\&\cdot\left(\frac{\left(1-p_\chi\right)p_d}{\lambda_{fa}}\right)^d\left(\frac{\lambda_b}{\lambda_{fa}}\right)^b.\end{aligned} \quad (5.37)$$

The first factor in (5.32) is given below, where $\mathbf{Z}_k=\left\{\mathbf{z}_{k,j},1\le j\le M_k\right\}$, $|J_d|+|J_b|+|J_{fa}|=M_k$, and the factors on the R.H.S. are derived from filter innovations, filter initiations, and the false measurement distribution (generally uniform over measurement space),

$$\begin{aligned}p\left(\mathbf{Z}_k|\mathbf{Z}^{k-1},q^k\right)=&\prod_{j\in J_d}f_d\left(\mathbf{z}_{k,j}|\mathbf{Z}^{k-1},q^k\right)\cdot\prod_{j\in J_b}f_b\left(\mathbf{z}_{k,j}|\mathbf{Z}^{k-1},q^k\right)\\&\cdot\prod_{j\in J_{fa}}f_{fa}\left(\mathbf{z}_{k,j}|\mathbf{Z}^{k-1},q^k\right).\end{aligned} \quad (5.38)$$

Substituting (5.37) and (5.38) into (5.32) and simplifying results in (5.39) and (5.40). This expression is the key enabler of track-oriented MHT. In particular, it provides a recursive expression for $p\left(q^k|\mathbf{Z}^k\right)$ that consists of a number of factors that relate to its constituent local track hypotheses,

$$p\left(q^k|\mathbf{Z}^k\right) = p_\chi^\chi\left((1-p_\chi)\left((1-p_d)\right)\right)^{\tau-\chi-d} \cdot \prod_{j\in J_d}\left[\frac{(1-p_\chi)\, p_d f_d\left(\mathbf{z}_{k,j}|\mathbf{Z}^{k-1}, q^k\right)}{\lambda_{fa} f_{fa}\left(\mathbf{z}_{k,j}|\mathbf{Z}^{k-1}, q^k\right)}\right] \cdot \prod_{j\in J_b}\left[\frac{\lambda_b f_b\left(\mathbf{z}_{k,j}|\mathbf{Z}^{k-1}, q^k\right)}{\lambda_{fa} f_{fa}\left(\mathbf{z}_{k,j}|\mathbf{Z}^{k-1}, q^k\right)}\right] \frac{p\left(q^{k-1}|\mathbf{Z}^{k-1}\right)}{\bar{c}_k}, \tag{5.39}$$

$$\bar{c}_k = \frac{c_k}{\left\{\frac{\exp\left(-\lambda_b - \lambda_{fa}\right)}{r!}\lambda_{fa}^r\right\} \prod_{j\in J_d\cup J_b\cup J_{fa}} f_{fa}\left(\mathbf{z}_{k,j}|\mathbf{Z}^{k-1}, q^k\right)}. \tag{5.40}$$

In practice, one cannot consider too large a set of scans before pruning or merging local hypotheses (or tracks) in some fashion. A popular mechanism to control these hypotheses is *N-scan pruning* [10, 16]. This amounts to solving (5.31), generally by a relaxation approach to an integer programming problem [51, 53–55], followed by pruning of all local hypotheses that differ from $\hat{q}^k$ in the first scan. This methodology is applied after each new scan of data is received, resulting in a fixed-delay solution to the tracking problem.

Often, *N-scan* pruning is referred to as a *maximum likelihood* (ML) approach to hypothesis management. ML estimation is closely related to MAP estimation. We have [69]

$$\hat{\mathbf{X}}_{\mathrm{MAP}}(\mathbf{y}) = \underset{\mathrm{X}}{\arg\max}\; f(\mathbf{y}|\mathbf{X})\, f(\mathbf{X}). \tag{5.41}$$

$$\hat{\mathbf{X}}_{\mathrm{ML}}(\mathbf{y}) = \underset{\mathrm{X}}{\arg\max}\; f(\mathbf{y}|\mathbf{X}). \tag{5.42}$$

Note that ML estimation is a *non-Bayesian* approach as it does not rely on a prior distribution on $\mathbf{X}$. ML estimation can be interpreted as MAP estimation with a uniform prior. In track-oriented MHT, *N-scan pruning* relies on a single parent global hypothesis, thus the ML and MAP interpretations are both valid.

Once hypotheses are resolved, in principle one has a state of object histories given by $\hat{\mathbf{X}}\left(\mathbf{Z}^k\right)$. In practice, it is common to apply track confirmation logic to all object histories [10]. This mechanism enables removal of spurious tracks induced by the suboptimality inherent in practical MHT implementations that include limited hypothesis generation, hypothesis pruning or merging, and real-time (sequential) track extraction.

Since track-confirmation logic follows the data-association logic in most MHT implementations, it is often simpler to use all measurements in the data-association process. That is, rather than reasoning as well over the possibility that certain measurements are false, it is reasonable to use all measurements as target originated. Indeed, spurious tracks are removed in the subsequent track-confirmation stage. This simplification amounts to replacing inequality constraints with equality constraints in the linear-programming formulation of the data-association process. The interested reader is referred to [54, 55] for details.

5.6 THE TARGET-DEATH PROBLEM

The PHD filter [22–24] avoids explicit track formation and rather asks the question: "What is the probability of target presence at a given volume of state space?" Parameterized by target state, this leads to a PHD surface for each time epoch of interest. A useful re-interpretation of the PHD filter, known as the bin-occupancy filter, is given in [74]. The paper describes as well the target death problem that the authors had earlier identified, and which in turn has led to the CPHD filter.

Consider the single-target case with no FA. In the absence of a target measurement, the PHD surface follows (5.37). Note that the PHD surface admits the interpretation that it identifies the probability of target presence at a given state,

$$D_{k|k}(\mathbf{x}) = (1 - p_d)\, D_{k|k-1}(\mathbf{x}), \tag{5.43}$$

where $D_{k|k-1}(\mathbf{x})$ and $D_{k|k}(\mathbf{x})$ are the predicted and updated PHD [22–24] and p_d is the probability of detection of a target. A problem with equation (5.43) is that it does not account properly for the impact of a missed detection, as described next and at greater length in [74]. This issue has been termed the *target death problem*.

While (5.43) may appear reasonable, it can be shown that it is inconsistent with the following simple Bayesian argument. Let $e_k \in \{0, 1\}$ denote the existence variable [14, 40, 77] for a target at scan k. The value 1 corresponds to the event that a target exists and the value 0 represents the absence of a target. The existence variable for a target is modeled by a two-state Markov chain [174]. Similarly, we define the detection variable $d_k \in \{0, 1\}$ [193] for a target at scan k by a two-state Markov chain [174]. We assume that the death probability at any scan is given by p_χ, as before. The updated probability of target existence after a missed detection is given by

$$\begin{aligned} P\{e_k = 1|d_k = 0\} &= P\{e_k = 1|e_{k-1} = 1, d_k = 0\}P\{e_{k-1} = 1\} \\ &= \frac{P\{e_k = 1, d_k = 0|e_{k-1} = 1\}P\{e_{k-1} = 1\}}{P\{d_k = 0|e_{k-1} = 1\}} \\ &= \frac{(1 - p_\chi)\,(1 - p_d)}{1 - (1 - p_\chi)\, p_d} P\{e_{k-1} = 1\}. \end{aligned} \tag{5.44}$$

Comparing (5.43) and (5.44), we see that the PHD filter penalizes missed detections too heavily; it is claimed in [74] that the CPHD *appears* to follow (5.44).

What happens with the track-oriented MHT approach? We compare the ratio of the probability associated with the track *coast* hypothesis (track is alive in the absence of a measurement) with the probability of track *coast* or *death*. Let q_i^k and q_j^k denote global hypotheses that include coast and death events, respectively, for the target of interest. From (5.33), we see that (5.39) follows immediately. Indeed, all factors in the global hypothesis probability cancel except for those associated with the (undetected) track,

$$\frac{p\left(q_i^k|\mathbf{Z}^k\right)}{p\left(q_i^k|\mathbf{Z}^k\right)+p\left(q_j^k|\mathbf{Z}^k\right)}=\frac{\left(1-p_\chi\right)\left(1-p_d\right)}{\left(1-p_\chi\right)\left(1-p_d\right)+p_\chi}=\frac{\left(1-p_\chi\right)\left(1-p_d\right)}{1-\left(1-p_\chi\right)p_d}. \tag{5.45}$$

Note that this validation is quite general, and in particular it is directly applicable to the multitarget case, under the simplifying assumption that no measurements satisfy the hypothesis gating criterion for the (undetected) object of interest here. We conclude that track-oriented MHT properly handles missed detections, and no target-death problem is observed.

It is of interest to explore more closely the connections between MHT and label-free tracking approaches based on the PHD; recently, Ulmke et al. in [194] have demonstrated equivalence between the approaches in the single-target case. This connection deserves further exploration, particularly in light of the differences exposed here.

5.7 EXAMPLES FOR MHT

MHT is particularly effective in dense multitarget settings and in highly cluttered environments. We first provide an illustrative example of hypothesis selection in track-oriented MHT. Next, we illustrate the benefits of MHT with actual sea trial data. Further examples may be found in [54, 195–197].

5.7.1 Example 1: *N*-scan Pruning in Track-Oriented MHT

The example assumes that two tracks, T1 and T2, have already been *resolved*. That is, prior data-association decisions have led to a single global hypothesis that includes two tracks. These tracks are illustrated with squares in Figure 5.1. Next, assume that a scan of data is received with two measurements, R1 and R2. Assume further that both R1 and R2 can feasibly be associated with T1, while only R1 can feasibly be associated with T2. This leads to a number of *local* hypotheses (or *track*) [16]. Note that this set of hypotheses includes track continuation in the absence of a measurement (often denoted a track *coast*), as well as new-track hypotheses. A second scan of data includes a single measurement R3. We assume that R3 provides feasible updates to track hypotheses that include R2, as well as spawning a new-track hypothesis. Note that we assume that tracks are terminated after two coasts, indicated by the triangles in Figure 5.1.

While the example includes a number of track hypotheses, it is important to note that each *global* hypothesis provides a complete set of data-association decisions that account for all resolved tracks and all sensor measurements. The number of global hypotheses is large, even for this simple example; the power of the track-oriented approach is that we do not require an explicit enumeration of global hypotheses.

Each track hypothesis has an associated log-likelihood ratio or track score [2, 10, 198] that reflects track initiation and termination penalties as well as nonlinear filtering scoring; in the case of linear Gaussian systems, this scoring is based on the filter *innovations*. The vector c includes the track-hypothesis scores. We are interested in the optimal global hypothesis, which amounts to identifying a vector x such that the global log-likelihood ratio is maximized: the ML solution. Having identified this solution through a two-stage relaxation approach based on linear programming or Lagrangian relaxation (as identified in Figure 5.1), many *conflicting* local hypotheses are removed. In particular, those track hypotheses that differ in the first scan past the resolved hypothesis layer are removed, while those that differ in the more recent past are maintained. The remaining local hypotheses are indicated by solid circles.

Having pruned the set of track hypothesis trees (with five surviving track hypotheses), we are ready for a new scan of data. In the example, the resolved layer always lags the current time by one scan: thus we have a multihypothesis example with hypothesis-tree depth (*N-scan*) of 1.

5.7.2 Example 2: Maneuvering Target in Heavy Clutter

We consider a maneuvering target in heavy clutter with active sonar returns. In the example, a submarine moves with a piecewise-linear motion and a speed of approximately 5 knots. The data are thresholded so that approximately 20 measurements per

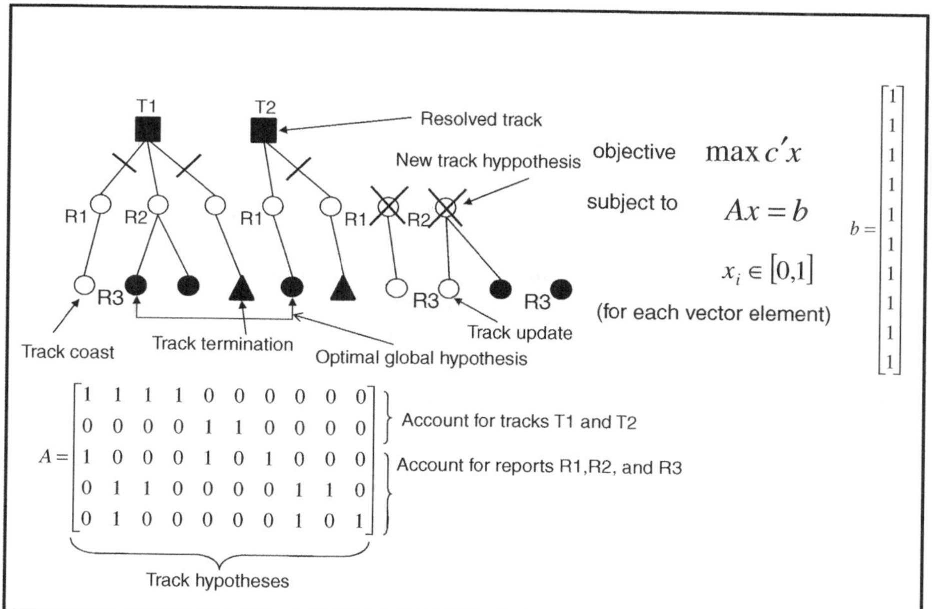

Figure 5.1 Illustration of *N*-scan pruning.

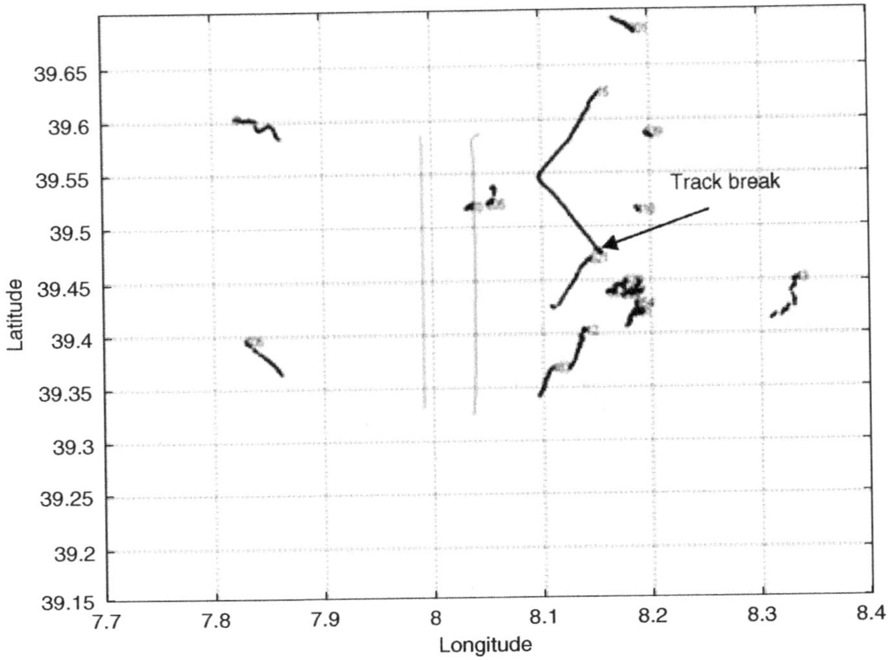

Figure 5.2 Track fragmentation on maneuver in a high-clutter environment in the absence of MHT.

scan are fed to the tracker. The scan repetition time is 60 s. The probability of target detection is approximately 0.5. The false returns are in part spurious, in part systematic in which there are wrecks, rocky outcrops, and so on, which are persistent scattering centers. Measurements include bistatic range, bearing, and bistatic range-rate.

It is good to emphasize that there is often a modeling mismatch between processing assumptions and actual target and sensor statistics. Note that while the NCV motion assumption is reasonable, it does not match actual submarine behavior. Similarly, the assumption of uniformly distributed false returns in measurement space is an optimistic simplifying assumption. Accordingly, actual sea trial data will give rise to a higher false track rate than what can be expected with simple tracker modeling.

In Figure 5.2, we see the tracking solution without the MHT. That is, the track-oriented MHT is used with no hypothesis-tree depth. We see that on the second target maneuver and close to a region of heavy clutter—as exhibited by the numerous confirmed tracks on clutter returns—the track is lost and then reacquired. This is an instance of an undesirable track fragmentation occurrence. Conversely, as shown in Figure 5.3, with MHT and a sufficiently deep hypothesis-tree depth (equal to 3 in this case), the algorithm is able to successfully disambiguate target and clutter returns, thus maintaining continuous track on the target despite a significant maneuver in heavy clutter. All other MHT parameters are kept the same. Note that Figures 5.2 and

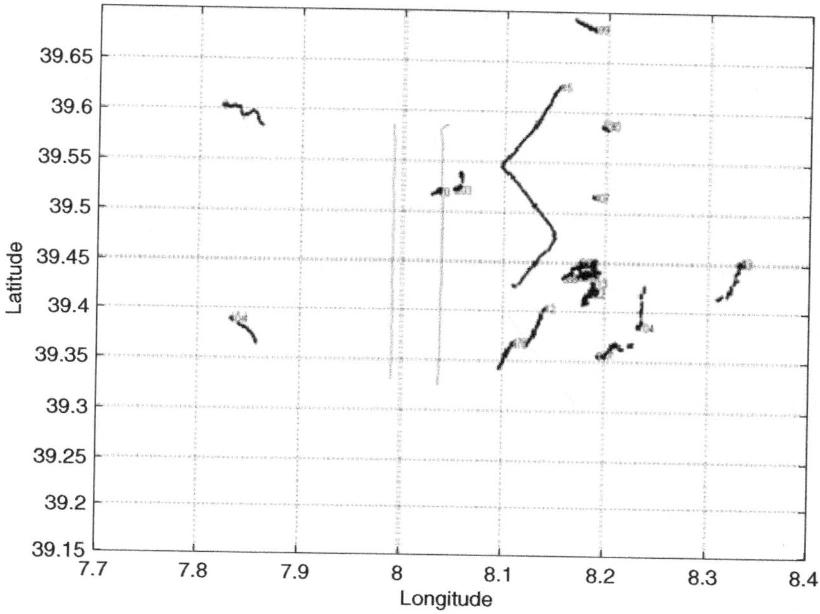

Figure 5.3 Successful tracking of a maneuvering target in a high-clutter environment with MHT.

5.3 illustrate the ground truth reconstruction for the maneuvering target as well as the nonmaneuvering source/receiver platform trajectories.

Note that, both with and without MHT, there are false tracks due to persistent, active sonar returns as noted previously. The false track rate may be reduced with subsequent use of target classification measurements or with feature-aided tracking. Also, note that the left–right disambiguation of the bistatic geometry is not perfect. Indeed, the line-array receiver induces left–right ambiguity that is more apparent in monostatic processing, and that can only be removed with ownship maneuvers. The residual ambiguity may be noted in the short-duration tracks left of the source–receiver platform trajectories, which are at the center of the display.

Of course, examples do exist in which MHT provides no appreciable gain. In fact, in some instances more complex processing may unluckily degrade tracking results. Nonetheless, in a statistical sense, there is ample theoretical and experimental evidence to confirm the benefits of MHT.

5.8 SUMMARY

The track-oriented MHT algorithm is one of the most advanced algorithms for multisensor MTT for real-world complex problems. For the benefit of the reader, we first presented a state-of-the-art overview of the current approaches used in tracking. Usually, this information is not available in any existing book. This chapter also gave a brief summary of the current nonlinear filtering algorithms. Suitable filters can be selected for the track filter given the nature and complexity of the tracking problem. A collection of dynamic and measurement models commonly used in filters are presented for a wide variety of tracking applications.

This chapter presents a detailed hybrid-state derivation of the track-oriented MHT equations following the formulation by Kurien with some minor modifications. We address the target death problem which is modeled in the PHD filter and other variants and show that it does not arise in track-oriented MHT. Finally two examples are presented which demonstrate the merits of MHT approach.

We have provided extensive references on the state-of-the-art filtering and tracking algorithms.

ACKNOWLEDGMENTS

The authors thank Sanjeev Arulampalam, Jing Liu, Yanjun Yan, and an anonymous reviewer for providing helpful comments and suggestions for improving the chapter. The authors also thank Gabriel Terejanu for useful discussions on the GSF.

REFERENCES

1. Y. Bar-Shalom and X.R. Li, *Multitarget-Multisensor Tracking: Principles and Techniques*, YBS Publishing, Storrs, CT, 1995.
2. Y. Bar-Shalom, P.K. Willett, and X. Tian, *Tracking and Data Fusion: A Handbook of Algorithms*, YBS Publishing, Storrs, CT, 2011.
3. M. Mallick and K.C. Chang, Distributed fusion using video sensors on multiple unmanned aerial vehicles, *Proc. Signal and Data Processing of Small Targets*, Vol. 6969, Orlando, FL, March 18–20, 2008.
4. S. Mori, B. Barker, C-Y Chong, and K.C. Chang, Track association and track fusion with non-deterministic target dynamics, *IEEE Trans. Aerosp. Electron. Syst.*, 38(2), 659–668, 2002.
5. A.G.O. Mutambara, *Decentralized Estimation and Control for Multisensor Systems*, CRC Press, Boca Raton, FL, 1998.
6. G. Frankel, Multisensor tracking of ballistic targets, *Proc. SPIE Signal and Data Processing of Small Targets*, Vol. 2561, 1995, pp. 337–346.
7. W. Khawsuk and L.Y. Pao, Decorrelated state estimation for distributed tracking of interacting targets in cluttered environments, *Proc. American Control Conference*, Anchorage, AK, May 8–11, 2002.
8. M. Mallick, S. Schmidt, and L.Y. Pao, Out-of-sequence track (OOST) filtering using the decorrelated pseudo measurement approach, *Proc. Signal and Data Processing of Small Targets*, Orlando, FL, April 13–15, 2004.
9. H. Chen, T. Kirubarajan, and Y. Bar-Shalom, Performance limits of track-to-track fusion versus centralized estimation: theory and application, *IEEE Trans. Aerosp. Electron. Syst.*, 39(2), 386–400, 2003.

10. S. Blackman and R. Popoli, *Design and Analysis of Modern Tracking Systems*, Artech House, Norwood, MA, 1999.

11. Y. Bar-Shalom and T.E. Fortmann, *Tracking and Data Association*, Academic Press, Waltham, MA, 1988.

12. K.C. Chang and Y. Bar-Shalom, Joint probabilistic data association for multitarget tracking with possibly unresolved measurements and maneuvers, *IEEE Trans. Automatic Control*, AC-29, 585–594, 1984.

13. T.E. Fortmann, Y. Bar-Shalom, and M. Scheffe, Sonar tracking of multiple targets using joint probabilistic data association, *IEEE Journal of Oceanic Engineering*, 8(3), 173–183, 1983.

14. D. Mušicki and R. Evans, Joint Integrated Probabilistic Data Association—JIPDA, *IEEE Trans. Aerosp. Electron. Syst.*, 40(3), 1093–1099, 2004.

15. D. Reid, An algorithm for tracking multiple targets, *IEEE Trans. Automat. Contr.*, 24(6), 843–854, 1979.

16. T. Kurien, Issues in the design of practical multitarget tracking algorithms, in: Y. Bar-Shalom (Ed.), *Multitarget-Multisensor Tracking: Advanced Applications*, Artech House, Norwood, MA, 1990.

17. R. Mahler, *An Introduction to Multisource-Multitarget Statistics and Applications*, Lockheed Martin Technical Monograph, March 15, 2000.

18. R. Mahler, A theoretical foundation for the Stein-Winter probability hypothesis density (PHD) multi-target tracking approach, *Proc. MSS Nat'l Symp. on Sensor and Data Fusion*, Vol. I (Unclassified), San Antonio, TX, June 2000.

19. R. Mahler, Multi-target Bayes filtering via first-order multi-target moments, *IEEE Trans. Aerosp. Electron. Syst.*, 39(4), 1152–1178, 2003.

20. R. Mahler, A theory of PHD filters of higher order in target number, *Signal Processing, Sensor Fusion, and Target Recognition XV, SPIE Defense & Security Symposium*, April 2006.

21. R. Mahler, PHD filters of higher order in target number, *IEEE Trans. Aerosp. Electron. Syst.*, 43(3), 2007.

22. R. Mahler, *Statistical Multisource-Multitarget Information Fusion*. Artech House, Norwood, MA, 2007.

23. B.-T. Vo, Random finite sets in multi-object filtering, *PhD Thesis*, University of Western Australia, 2008.

24. B-N Vo, B-T Vo, and D. Clark, Bayesian multiple target filtering using random finite sets, in: M. Mallick, V. Krishnamurthy, and B-N Vo (Eds.), *Integrated Tracking, Classification, and Sensor Management: Theory and Applications*, Wiley/IEEE, New York, 2012, Chapter 3.

25. B.-T. Vo, B.-N. Vo, and A. Cantoni, Analytic implementations of the cardinalized probability hypothesis density filter, *IEEE Trans. Signal Processing*, 55, 3553–3567, 2007.

26. R.L. Streit and L.D. Stone, Bayes derivation of multitarget intensity filters, *Proceedings of the 2008 International Conference on Information Fusion*, Cologne, Germany, June 30–July 3, 2008, pp. 1686–1693.

27. R.L. Streit, Multisensor multitarget intensity filters, *Proceedings of the 2008 International Conference on Information Fusion*, Cologne, Germany, June 30–July 3, 2008, pp. 1694–1701.

28. R.L. Streit, PHD intensity filtering is one step of a MAP estimation algorithm of a positron emission tomography, *Proceedings of the 2009 International Conference on Information Fusio*, Seattle, USA, July 6–9, 2009, pp. 308–315.
29. R. Streit, Multiple target tracking, in: *Poisson Point Processes: Imaging, Tracking, and Sensing*, Springer, New York, 2010, Chapter 6.
30. C. Hue, J.P. Le Cadre, and P. Perez, Tracking multiple objects with particle filtering, *IEEE Trans. Aerosp. Electron. Syst.*, 38, 791–812, 2002.
31. Z. Khan, T. Blach, and F. Dellaert, MCMC-based particle filtering for tracking a variable number of interacting targets, *IEEE Trans. Pattern Anal. Mach. Intell.*, 27, 1805–1819, 2005.
32. S.K. Pang, J. Li, and S.J. Godsill, Models and algorithms for detection and tracking of coordinated groups, *Proceedings of the IEEE Aerospace Conference*, 2008, pp. 1–17.
33. S.K. Pang, Models and algorithms for group tracking, *Ph.D. dissertation*, University of Cambridge, November 2008.
34. A. Carmi, F. Septier, and S.J. Godsill, The Gaussian mixture MCMC particle algorithm for dynamic cluster tracking, *Proceedings of the 2009 International Conference on Information Fusion*, , Seattle, USA, July 6–9, 2009.
35. F. Septier, S.K. Pang, A. Carmi, and S. Godsill, On MCMC-based particle methods for Bayesian filtering: Application to multitarget tracking, *3rd IEEE International Workshop on Computational Advances in Multi-Sensor Adaptive Processing (CAMSAP)*, 2009, pp. 360–363.
36. F. Septier, A. Carmi, and S. Godsill, Tracking of multiple contaminant clouds, *Proceedings of the 2009 International Conference on Information Fusion*, Seattle, USA, July 6–9, 2009.
37. F. Septier, A. Carmi, S.K. Pang, and S.J. Godsill, Multiple object tracking using evolutionary and hybrid MCMC-based particle algorithms, *15th IFAC Symposium on System Identification*, Saint-Malo, France, July 2009.
38. F. Septier, J. Cornebise, S. Godsill, and Y. Delignon, A comparative study of Monte-Carlo methods for multitarget tracking, *IEEE Statistical Signal Processing Workshop (SSP)*, 2011, pp. 205–208.
39. J. Vermaak, S.J. Godsill, and P. Perez, Monte Carlo filtering for multi-target tracking and data association, *IEEE Trans. Aerosp. Electron. Syst.*, 41, 309–332, 2005.
40. S.B. Colegrove and J.K. Ayliffe, An extension of Probabilistic Data Association to include track initiation and termination, *20th IREE International Convention*, Melbourne, 1985, pp. 853–856.
41. D. Mušicki, R. Evans, and S. Stankovic, Integrated Probabilistic Data Association, *IEEE Trans. Automat. Contr.*, AC-39(6), 1237–1241, 1994.
42. J. Liu, Application of improved particle filter in multiple maneuvering target tracking system, *Ph.D. dissertation*, National University of Singapore, December 2006.
43. S. Puranik and J.K. Tugnait, Tracking of multiple maneuvering targets using multiscan JPDA and IMM filtering, *IEEE Trans. Aerosp. Electron. Syst.*, 43(1), 23–35, 2007.
44. J.A. Roecker, Multiple scan joint probabilistic data association, *IEEE Trans. Aerosp. Electron. Syst.*, 31, 1204–1210, 1995.
45. S. Blackman, Multiple hypothesis tracking for multiple target tracking, *IEEE Aerosp. Electron. Syst. Magazine*, 19(1), 5–18, 2004.

46. D. Svensson, J. Wintenby, and L. Svensson, Performance evaluation of MHT and GM-CPHD in a ground target tracking scenario, *Proceedings of the 2009 International Conference on Information Fusion*, Seattle, USA, July 6–9, 2009.

47. D. Svensson, Target tracking in complex scenarios, *PhD dissertation*, Chalmers University of technology, Göteborg, Sweden, 2010.

48. S. Deb, M. Yeddanapudi, K. Pattipati, and Y. Bar-Shalom, A generalized s-d assignment algorithm for multisensor-multitarget state estimation, *IEEE Trans. Aerosp. Electron. Syst.*, 33(2), 523–538, 1997.

49. K.R. Pattipati, S. Deb, Y. Bar-Shalom, and R.B. Washburn, Passive multisensor data association using a relaxation algorithm, in: Y. Bar-Shalom (Ed.), *Multitarget-Multisensor Tracking: Advanced Applications*, Artech House, Norwood, MA, 1990, Chapter 7.

50. K.R. Pattipati, R.L. Popp, and T.T. Kirubarajan, Survey of assignment techniques for multitarget tracking, in: Y. Bar-Shalom and D. Blair (Eds.), *Multitarget-Multisensor Tracking: Applications and Advances*, Vol. III, Artech House, Norwood, MA, 2000, Chapter 2.

51. A. Poore and N. Rijavec, A Lagrangian relaxation algorithm for multidimensional assignment problems arising from multitarget tracking, *SIAM J. Optimization*, 3(3), 544–563, 1993.

52. A.B. Poore and A.J. Robertson III, A new Lagrangian relaxation based algorithm for a class of multidimensional assignment problems, *Computational Optimization and Applications*, 8(2), 129–150, 1997.

53. C. Morefield, Application of 0-1 integer programming to multitarget tracking problems, *IEEE Trans. Automat. Contr.*, 22(3), 1977.

54. S. Coraluppi, C. Carthel, M. Luettgen, and S. Lynch, All-source track and identity fusion, *Proceedings of the National Symposium on Sensor and Data Fusion*, San Antonio TX, June 2000.

55. P. Storms and F. Spieksma, An LP-based algorithm for the data association problem in multitarget tracking, *Proceedings of the 3rd International Conference on Information Fusion*, Paris, France, July 2000.

56. S. Mori and C-Y Chong, Markov chain Monte Carlo method for evaluating multi-frame data association hypotheses, *Proceedings of the 2008 International Conference on Information Fusion*, Cologne, Germany, June 30–July 3, pp. 381–388.

57. S. Oh, S. Russell, and S. Sastry, Markov chain Monte Carlo data association for general multiple-target tracking problems, *Proc. 2004 43rd IEEE Conference on Decision and Control*, Vol. 1, Paradise Island, Bahamas, December 2004, pp. 735–742.

58. S. Oh, S. Russell, and S. Sastry, Markov Chain Monte Carlo Data Association for multi-target tracking, *IEEE Trans. Automat. Contr.*, 54(3), 481–497, 2009.

59. I. Goodman, R. Mahler, and H. Nguyen, *Mathematics of Data Fusion*, Kluwer Academic Publishers, Dordrecht, The Netherlands, 1997.

60. H. Sidenbladh, and S.-L. Wirkander, Tracking random sets of vehicles in terrain, *Proc. 2003 IEEE Workshop on Multi-Object Tracking*, Madison WI, June 2003.

61. H. Sidenbladh, Multi-target particle filtering for the probability hypothesis density, *Proc. Int'l Conf. on Information Fusion*, Cairns, Australia, July 8–10, 2003, pp. 800–806.

62. B.-N. Vo, S. Singh, and A. Doucet, Sequential Monte Carlo implementation of the PHD filter for multi-target tracking, *Proc. Int'l Conf. on Information Fusion*, Cairns, Australia, July 8–10, 2003, pp. 792–799.

63. B.-N. Vo, S. Singh, and A. Doucet, Sequential Monte Carlo methods for multi-target filtering with random finite sets, *IEEE Trans. Aerosp. Electron. Syst.*, 41(4), 1224–1245, 2005.
64. T. Zajic and R. Mahler, A particle-systems implementation of the PHD multi-target tracking filter, *Proc. SPIE Signal Processing, Sensor Fusion and Target Recognition XII*, Vol. 5096, April 2003, pp. 291–299.
65. L. Lin, Y. Bar-Shalom, and T. Kirubarajan, Track labeling and PHD filter for multitarget tracking, *IEEE Trans, Aerosp. Electron. Syst.*, 42(3), 778–795, July 2006.
66. K. Panta, B. Vo, and S. Singh, Novel data association schemes for the probability hypothesis density filter, *IEEE Trans. Aerosp. Electron. Syst.*, 43(2), 556–570, 2007.
67. K. Panta, D. Clark, and B. Vo, Data association and track management for the Gaussian mixture probability hypothesis density filter, *IEEE Trans. Aerosp. Electron. Syst.*, 45(3), 1003–1016, 2009.
68. B.-T. Vo, B.-N. Vo, and A. Cantoni, The cardinality balanced multi-target multi-Bernoulli filter and its implementations, *IEEE Trans. Signal Processing*, 57(2), 409–423, 2009.
69. Y. Bar-Shalom, X. Li, and T. Kirubarajan, *Estimation with Applications to Tracking and Navigation*, Wiley & Sons, New York, 2001.
70. H.A.P. Blom, and Y. Bar-Shalom, The Interacting Multiple Model algorithm for systems with Markovian switching coefficients, *IEEE Trans. Automat. Contr.,* 33(8), 780–783, August 1988.
71. K. Punithakumar, T. Kirubarajan, and A. Sinha, Multiple-model probability hypothesis density filter for tracking maneuvering targets, *IEEE Trans. Aerosp. Electron. Syst.*, 44(1), 87–98, January 2008.
72. A. Pasha, B.-N. Vo, H. D. Tuan, and W. K. Ma, A Gaussian mixture PHD filter for jump Markov system model, *IEEE Trans. Aerosp. Electron. Syst.*, 45(3), 919–936, 2009.
73. R. Georgescu and P. Willett, Multiple model cardinalized probability hypothesis density filter, *Signal and Data Processing of Small Targets 2011*, *Proc. SPIE*, San Diego, CA, August 2011.
74. O. Erdinc, P. Willett, and Y. Bar-Shalom, The Bin-occupancy filter and its connection to the PHD filters, *IEEE Trans. Signal Processing*, 57(11), 4232–4246, November 2009.
75. X. Chen, R. Tharmarasa, M. McDonald, and T. Kirubarajan, A multiple model cardinalized probability hypothesis density filter, submitted to *IEEE Trans. Aerosp. Electron. Syst.*, 2011.
76. S. Davey, M. Rutten, and N. Gordon, Track-before-detect techniques, in: M. Mallick, V. Krishnamurthy, and B-N Vo (Eds.), *Integrated Tracking, Classification, and Sensor Management: Theory and Applications*, Wiley/IEEE, 2012, Chapter 8.
77. B. Ristic, S. Arulampalam, and N. Gordon, *Beyond the Kalman Filter*, Artech House, Norwood, MA, 2004.
78. M. Schikora, W. Koch, R. Streit, and D. Cremers, A sequential Monte Carlo method for multi-target tracking with the intensity filter, in: P. Georgieva, L. Mihaylova, L. Jain (Eds.), *Advances in Intelligent Signal Processing and Data Mining: Principles and Applications*, Springer, 2012, Chapter 2, pp. 55–88.
79. M. Schikora, W. Koch, R.L. Streit, and D. Cremers, Sequential Monte Carlo method for the iFilter, *Proc. 2011 International Conference on Information Fusion*, Chicago, USA, July 5–8, 2011, pp. 1085–1092.

80. R. Mahler and A. El-Fallah, CPHD and PHD filters for unknown backgrounds, III: Tractable multitarget filtering in dynamic clutter, in: O. Drummond (Ed.), *Signal and Data Proc. of Small Targets 2010, SPIE Proc.*, Vol. 7698, 2010.

81. R. Mahler, B.T. Vo, and B. N. Vo, CPHD filtering with unknown clutter rate and detection profile, *Proc. 14th International Conference on Information Fusion,* Chicago, Illinois, July 5–8, 2011, pp. 1834–1841.

82. R. Mahler, B.-T. Vo, and B.-N. Vo, CPHD filtering with unknown clutter rate and detection profile, *IEEE Trans. Signal Processing*, 59(8), 3497–3513, August 2011.

83. M.S. Arulampalam, S. Maskell, N. Gordon, and T. Clapp, A tutorial on particle filters for online nonlinear/non-Gaussian Bayesian tracking, *IEEE Trans. Signal Processing*, 50(2), February 2002.

84. A. Doucet, N. de Freitas, and N. Gordon, *Sequential Monte Carlo Methods in Practice*, Springer-Verlag, New York, January 2001.

85. C. Berzuini, N.G. Best, W.R. Gilks, and C. Larizza, Dynamic conditional independence models and Markov Chain Monte Carlo methods, *Journal of the American Statistical Association*, 440, 1403–1412, December 1997.

86. A. Golightly and D.J. Wilkinson, Bayesian sequential inference for nonlinear multivariate diffusions, *Statistics and Computing*, 16, 323–338, August 2006.

87. X. Rong Li and V.P. Jilkov, Survey of maneuvering target tracking Part I: Dynamic models, *IEEE Trans. Aerosp. Electron. Syst.*, 39(4), 1333–1364, October 2003.

88. R.A. Singer, Estimating optimal tracking filter performance for manned maneuvering targets, *IEEE Trans. Aerosp. Electron. Syst.*, AES-6, 473–483, July 1970.

89. H.D. Hoelzer, G.W. Johnson, and A.O. Cohen, Modified polar coordinates—The key to well behaved bearings only ranging, in: *IR & D Report* 78-M19-OOOlA, IBM Federal Systems Division, Shipboard and Defense Systems, Manassas, VA 22110, August 31, 1978.

90. V.J. Aidala and S.E. Hammel, Utilization of modified polar coordinates for bearings-only tracking, *IEEE Trans. Automat. Contr.*, 28(3), 283–294, 1983.

91. B.F. La Scala, M. Mallick, and S. Arulampalam, Differential geometry measures of nonlinearity for filtering with nonlinear dynamic and linear measurement models, *Proc. Signal and Data Processing of Small Targets*, Vol. 6699, , San Diego, CA, August 28–30, 2007.

92. T. Bréhard and J-P. Le Cadre, Closed-form Posterior Cramér-Rao bound for a manoeuvring target in the bearings only tracking context using best-fitting Gaussian distribution, *Proc. of the Ninth International Conf. on Information Fusion*, Florence, Italy, July 2006.

93. R.R. Allen and S.S. Blackman, Angle-only tracking with a MSC filter, *Proc. of Digital Avionics Systems Conference*, 1991, pp. 561–566.

94. Q. Li, F. Guo, Y. Zhou, and W. Jiang, Observability of satellite to satellite passive tracking from angles measurements, *Proc. IEEE International Conference on Control and Automation*, 2007, pp. 1926–1931.

95. M. Mallick, L. Mihaylova, S. Arulampalam, and Y. Yan, Angle-only filtering in 3D using modified spherical and log spherical coordinates, *Proc. 2011 International Conference on Information Fusion*, Chicago, USA, July 5–8, 2011, pp. 1905–1912.

96. M. Mallick, M. Morelande, and L. Mihaylova, Continuous-discrete filtering using EKF, UKF, and PF, *Proc. 2012 International Conference on Information Fusion*, Singapore, July 9–12, 2012, pp. 1087–1094.

97. M. Mallick, M. Morelande, L. Mihaylova, S. Arulampalam, and Y. Yan, Comparison of angle-only filtering algorithms in 3D using cartesian and modified spherical coordinates, *Proc. 2012 International Conference on Information Fusion*, Singapore, July 9–12, 2012, pp. 1392–1399.
98. M. Mallick, M. Morelande, L. Mihaylova, S. Arulampalam, and Y. Yan, Angle-only filtering in 3D, in: M. Mallick, V. Krishnamurthy, and B-N Vo (Eds.), *Integrated Tracking, Classification, and Sensor Management: Theory and Applications*, Wiley/IEEE, 2012, Chapter 1.
99. D.V. Stallard, An angle-only tracking filter in modified spherical coordinates, *Proc. of the AIAA Guidance, Navigation and Control Conference*, Monterey, CA, August 17–19, 1987, pp. 542–550.
100. V.C. Ravindra, Y. Bar-Shalom, and P. Willett, Projectile identification and impact point prediction, *IEEE Trans. Aerosp. Electron. Syst.*, AES-46(4), 2004–2021, October 2010.
101. M. Athans and R.P. Wishner, and A. Bertolini, Suboptimal state estimation for continuous-time nonlinear systems from discrete noisy measurements, *IEEE Trans. Automat. Contr.*, 13(5), 504–514, October 1968.
102. A. Farina, B. Ristic, and D. Benvenuti, Tracking a ballistic target: Comparison of several nonlinear filters, *IEEE Trans. Aerosp. Electron. Syst.*, AES-38(3), 854–867, July 2002.
103. R.K. Mehra, A comparison of several nonlinear filters for reentry vehicle tracking, *IEEE Trans. Automat. Contr.*, AC-16(4), 307–319, August 1971.
104. B.D. Tapley, B.E. Schutz, and G.H. Born, *Statistical Orbit Determination*, Elsevier Academic Press, New York, NY, 2004.
105. D. A. Vallado, *Fundamentals of Astrodynamics and Applications*, 3rd Edition, Microcosm Press and Springer, El Segundo, CA, USA, 2007.
106. X. Rong Li and V.P. Jilkov, A survey of maneuvering target tracking—Part III: Measurement models, *Proceedings of SPIE Conference on Signal and Data Processing of Small Targets*, San Diego, CA, USA, Vol. 4473, July–August 2001, pp. 423–446.
107. M. Taj and A. Cavallaro, Multi-camera track-before-detect, *Third ACM/IEEE International Conference on Distributed Smart Cameras (ICDSC)*, 2009, pp. 1–6.
108. M. Longbin, S. Xiaoquan, Z. Yiyu, S. Z. Kang, and Y. Bar-Shalom, Unbiased converted measurements for tracking, *IEEE Trans. Aerosp. Electron. Syst.*, 34(2), 1023–1027, July 1998.
109. E. H. Aoki, A general approach for altitude estimation and mitigation of slant range errors on target tracking using 2D radars, *Proceedings of the 2010 International Conference on Information Fusion,* Edinburgh, UK, July 26–29, 2010.
110. S. Nagaoka, Height estimation of a cruising aircraft via a radar for air traffic control, *Electron. and Commun. in Japan*, 71(11), 95–105, 1988.
111. Y. Rakvongthai, J. Ru, S. Sivananthan, and S. Oraintara, A general approach for altitude estimation and mitigation of slant range errors on target tracking using 2D radars, *Proceedings of the 2011 International Conference on Information Fusion*, Chicago, USA, July 5–8, 2011.
112. X. Tian and Y. Bar-Shalom, Robust tracking for very long range radars, Part I: Algorithm comparisons, *Proc. Signal and Data Processing of Small Targets*, Vol. 6699, San Diego, CA, August 2007.
113. X. Tian and Y. Bar-Shalom, Robust tracking for very long range radars, Part II: Measurement conversion and derivations, *Proc. Signal and Data Processing of Small Targets*, Vol. 6699, San Diego, CA, August 2007.

114. X. Tian and Y. Bar-Shalom, Coordinate conversion and tracking for very long range radars, *IEEE Trans. Aerosp. Electron. Syst.*, 45(3), 1073–1088, 2009.

115. R. Deming, J. Schindler, and L. Perlovsky, Multi-target/multi-sensor tracking using only range and Doppler measurements, *IEEE Trans. Aerosp. Electron. Syst.*, 45(2), 273–296, April 2009.

116. K. Kastella, B. Yeary, T. Zadra, R. Brouillard, and E. Frangione, Bias modeling and estimation for GMTI applications, *Proc. Third Int. Conf. on Information Fusion*, Paris, France, July 10–13, 2000.

117. M. Mallick, Maximum Likelihood geolocation using a Ground Moving Target Indicator (GMTI) Report, *Proc. 2002 IEEE Aerospace Conference*, Big Sky MT, USA, March 2002.

118. M. Mallick and S. Arulampalam, Comparison of nonlinear filtering algorithms in ground moving target indicator (GMTI) target tracking, *Proc. Signal and Data Processing of Small Targets*, San Diego, CA, August 4–7, 2003.

119. Y. Chun, W. Garber, R. Mitchell, and E. Blasch, A simple maneuver indicator from target's Range-Doppler image, *Proc. 2007 International Conference on Information Fusion*, Québec City, Canada, July 2007.

120. M.M. Hyder and K. Mahata, Range-Doppler imaging via sparse representation, *Proc. IEEE Radar Conference*, 2011, pp. 486– 491.

121. H. Schimpf and J. Billner, Robust target acquisition using consecutive range Doppler maps, MMW Advanced Target Recognition and Identification Experiment, *Meeting Proceedings RTO-MP-SET-096*, Paper 6, pp. 6-1-6-12, Neuilly-sur-Seine, France, http://www.rto.nato.int/abstracts.asp.

122. X. Tan and J. Li, A Range-Doppler imaging via forward-backward sparse Bayesian learning, *IEEE Trans. Signal Processing*, 58(4), 2421–2425, July 2010.

123. C. Cop, Ground and maritime moving target indicator radar, *Defense Today*, 58–61, September/October 2006.

124. M. Jahangir, Moving target detection for synthetic aperture radar via shadow detection, *IET International Conference on Radar Systems*, 2007, pp. 1–5.

125. X. Li and X-G Xia, Location and imaging of elevated moving target using multi-frequency velocity SAR with cross-track interferometry, *IEEE Trans. Aerosp. Electron. Syst.*, 47(2), 1203–1212, April 2011.

126. P. Arambel, J. Silver, J. Krant, M. Antone, and T. Strat, Multiple-hypothesis tracking of multiple ground targets from aerial video with dynamic sensor control, *Proc. of SPIE*, Vol. 5429, 2004, pp. 23–32.

127. P. O. Arambel and M. Antone, Multiple target tracking with a steerable airborne video sensor, *Proc. of SPIE*, Vol. 6235, 2006, pp. 623505-1–623505-11.

128. A. P. Brown, K. J. Sullivan, and D. J. Miller, Feature-aided multiple target tracking in the image plane, *Proc. of SPIE*, Vol. 6229, 2006.

129. K. Min and G. H. Wang, Research on multi-mode fusion tracking of OTHR based on Auction Algorithm, *International Conference on Computational Intelligence and Security Workshops* (CISW), December 15–19, 2007, pp. 393–396.

130. G. W. Pulford and R. J. Evans, A Multipath Data Association Tracker for over-the-horizon radar, *IEEE Trans. Aerosp. Electron. Syst.*, 34(4), 1165–1183, October 1998.

131. B. Kalyan, K. W. Lee, S. Wijesoma, and D.Moratuwage, A random finite set based detection and tracking using 3D LIDAR in dynamic environments, *IEEE International Conference on Systems Man and Cybernetics* (SMC), 2010, pp. 2288–2292.

132. T. Ogawa, H. Sakai, Y. Suzuki, K. Takagi, and K. Morikawa, Pedestrian detection and tracking using in-vehicle LIDAR for automotive application, *IEEE Intelligent Vehicles Symposium* (IV), Baden-Baden, Germany, June 5–9, 2011, pp. 734–739.

133. A. Gelb (Ed.), *Applied Optimal Estimation*, The MIT Press, Cambridge, MA, 1974.

134. R. E. Kalman. A new approach to linear filtering and prediction problems, *Transactions of the ASME—Journal of Basic Engineering*, 82, 35–45, 1960.

135. A. H. Jazwinski, *Stochastic Processes and Filtering Theory*, Academic Press, New York, 1970.

136. B. Anderson and J. Moore, *Optimal Filtering*, Prentice Hall, Englewood Cliffs, NJ, 1979.

137. S. Särkkä, Recursive Bayesian inference on stochastic differential equations, *Ph D dissertation*, Helsinki University of Technology, April 2006.

138. S. Julier, J. Uhlmann, and H. Durrant-White, A new approach for filtering nonlinear systems, *Proc. American Control Conference*, Washington, DC, 1995, pp. 1628–1632.

139. S. Julier, J. Uhlman, and H.F. Durrant-Whyte, A new method for the nonlinear transformations of means and covariances in filters and estimators, *IEEE Trans. Automat. Contr.*, 45(3), 477–482, March 2000.

140. S. Julier, The scaled unscented transformation, *Proc. of the American Control Conference*, Anchorage, AK, May 8–10, 2002.

141. S. Julier and J. Uhlmann, Unscented filtering and nonlinear estimation, *Proc. IEEE*, 92(3), 401–422, March 2004.

142. N.J. Gordon, D.J. Salmond, and A.F.M. Smith, Novel approach to nonlinear/non-Gaussian Bayesian state estimation, *IEE Proc. on Radar & Signal Processing*, 140(2), 107–113, April 1993.

143. D. L. Alspach and H. W. Sorenson. Nonlinear Bayesian estimation using Gaussian sum approximations, *IEEE Trans. Automat. Contr.*, AC-17(4), 439–448, August 1972.

144. K. Ito and K. Xiong, Gaussian filters for nonlinear filtering problems, *IEEE Trans. Automat. Contr.*, 45(5), 910–927, 2000.

145. H. W. Sorenson and D. L. Alspach. Recursive Bayesian estimation using Gaussian sums, *Automatica*, 7, 465–479, 1971.

146. G. Terejanu, P. Singla, T. Singh, and P. D. Scott, Uncertainty propagation for nonlinear dynamical systems using Gaussian mixture models, *J. Guid., Contr. Dyn.*, 3(6), 1622–1633, 2008.

147. G. Terejanu, P. Singla, T. Singh, and P. D. Scott, A novel Gaussian sum filter method for accurate solution to nonlinear filtering problem, *Proc. Eleventh International Conference on Information Fusion,* Cologne, Germany, June 30–July 3, 2008.

148. G. Terejanu, P. Singla, T. Singh, and P. D. Scott, Adaptive Gaussian sum filter for nonlinear Bayesian estimation, *IEEE Trans. Automat. Contr.*, 56(9), 2151–2156, 2011.

149. W. R. Gilks and C. Berzuini, Following a moving target-Monte Carlo, inference for dynamic Bayesian models, *J. R. Stat. Soc.* Series B, 63, 127–146, 2001.

150. V. Beněs, Exact finite dimensional filters for certain diffusions with nonlinear drift, *Stochastics*, 5, 65–92, 1981.

151. F. E. Daum, Exact finite dimensional nonlinear filters for continuous time processes with discrete time measurements, *The 23rd IEEE Conference on Decision and Control*, Vol. 23, Part 1, 1984, pp. 16–22.

152. F. E. Daum, Exact finite-dimensional nonlinear filters, *The 24th IEEE Conference on Decision and Control*, Vol. 24, Part 1, 1985, pp. 1938–1945.

153. F. Daum, Exact finite-dimensional nonlinear filters, *IEEE Trans. Automat. Contr.*, AC-31(13), 616–622, 1986.

154. S. Choi, P. Willett, F. Daum and J. Huang, Discussion and application of the homotopy filter, *Signal Processing, Sensor Fusion, and Target Recognition XX*, *Proc. SPIE*, Vol. 8050, Orlando, FL, USA, April 2011.

155. F. Daum and J. Huang, Nonlinear filters with log-homotopy, *Signal and Data Processing of Small Targets* 2007, *Proc. SPIE*, Vol. 6699, San Diego, CA, USA, August 2007.

156. F. Daum and J. Huang, Particle flow for nonlinear filters with log-homotopy, *Signal and Data Processing of Small Targets* 2008, *Proc. SPIE*, Vol. 6969, Orlando, FL, USA, March 2008.

157. F. Daum and J. Huang, Nonlinear filters with particle flow induced by log-homotopy, *Signal Processing, Sensor Fusion, and Target Recognition XVIII*, *Proc. SPIE*, Vol. 7336, Orlando, FL, USA, May 2009.

158. F. Daum and J. Huang, Nonlinear filters with particle flow, *Signal and Data Processing of Small Targets* 2009, *Proc. SPIE*, Vol. 7445, San Diego, CA, USA, August 2009.

159. F. Daum and J. Huang, Nonlinear filters with particle flow induced by log-homotopy, *Signal Processing, Sensor Fusion, and Target Recognition XVIII*, *Proc. SPIE*, Vol. 7336, Orlando, FL, USA, May 2009.

160. F. Daum, J. Huang, and A. Noushin, Exact particle flow for nonlinear filters, *Signal Processing, Sensor Fusion, and Target Recognition XIX*, *Proc. SPIE*, Vol. 7697, Orlando, FL, USA, April 2010.

161. F. Daum and J. Huang, A fresh perspective on research for nonlinear filters with particle flow induced by log-homotopy, *Signal Processing, Sensor Fusion, and Target Recognition XIX*, *Proc. SPIE*, Vol. 7697, Orlando, FL, USA, April 2010.

162. F. Daum and J. Huang, Generalized particle flow for nonlinear filters, *Signal and Data Processing of Small Targets* 2010, *Proc. SPIE*, Vol. 7698, Orlando, FL, USA, April 2010.

163. F. Daum and J. Huang, Numerical experiments for nonlinear filters with exact particle flow induced by log-homotopy, *Signal and Data Processing of Small Targets* 2010, *Proc. SPIE*, Vol. 7698, Orlando, FL, USA, April 2010.

164. A. Doucet, S. Godsill, and C. Andrieu, On sequential Monte Carlo sampling methods for Bayesian filtering, *Statistics and Computing*, 10(3), pp. 197–208, 2000.

165. M. Pitt and N. Shephard, Filtering via simulation: Auxiliary particle filters, *J. Amer. Statist. Assoc.*, 94(446), 590–599, 1999.

166. C. Musso, N. Oudjane, and F. LeGland, Improving regularised particle filters, in: A. Doucet, J.F.G. de Freitas, and N.J. Gordon (Eds.), *Sequential Monte Carlo Methods in Practice*, Springer-Verlag, New York, 2001.

167. M. Mallick, Differential geometry measures of nonlinearity with applications to ground target tracking, *Proc. 2004 International Conference on Information Fusion*, Stockholm, Sweden, June 28–July 1, 2004.

168. M. Mallick, B. F. La Scala, and S. Arulampalam, Differential geometry measures of nonlinearity for the bearing-only tracking problem, *SPIE Defense & Security Symposium*, Orlando, FL, USA, March 28–April 1, 2005.

169. M. Mallick, B. F. La Scala, and S. Arulampalam, Differential geometry measures of nonlinearity with applications to target tracking, *Fred Daum Tribute Conference*, Monterey, CA, USA, May 24, 2007.

170. J. Liu and Vadakkepat, Interacting MCMC particle filter for tracking maneuvering target, *Digital Signal Processing*, 20, 561–574, 2010.

171. J. Liu, C. Z. Han, and Y. Hu, Adaptive MCMC particle filter for tracking maneuvering target, *Proc. 30th Chinese Control Conference*, Yantai, China, July 22–24, 2011.

172. E. Mazor, A. Averbuch, Y. Bar-Shalom, and J. Dayan, Interacting multiple model methods in target tracking: A survey, *IEEE Trans. Aerosp. Electron. Syst.*, 34(1), 103–123, January 1998.

173. M. Mallick and B. F. La Scala, IMM for multi-sensor ground target tracking with variable measurement sampling intervals, *Proc. 2006 International Conference on Information Fusion*, Florence, Italy, July 10–13, 2006.

174. A. Papoulis and S. U. Pillai, *Probability, Random Variables and Stochastic Processes*, 4th Edition, McGraw-Hill, New York, 2002.

175. C. Moler and C. Van Loan, Nineteen dubious ways to compute the exponential of a matrix, *SIAM Review* 20(4), 801–836, 1978.

176. X. R. Li, Engineer's guide to variable-structure multiple-model estimation for tracking, in: Y. Bar-Shalom and W.D. Blair (Eds.), *Multitarget-Multisensor Tracking: Applications and Advances*, Vol. III, Artech House, Norwood, MA, USA, 2000, pp. 449–567.

177. X. Rong Li and V. P. Jilkov, Survey of maneuvering target tracking. Part V: Multiple-model methods, *IEEE Trans. Aerosp. Electron. Syst.*, 41(4), 1255–1321, October 2005.

178. H. A. P. Blom, An efficient filter for abruptly changing systems, *Proc. 23rd IEEE Conference on Decision and Control*, Las Vegas, NV, December 1984, pp. 656–658.

179. H. A. P. Blom and Y. Bar-Shalom, The Interacting Multiple Model algorithm for systems with Markovian switching coefficients, *IEEE Trans. Automat. Contr.*, 33(8), 780–783, August 1988.

180. H. A. P. Blom, R. A. Hogendoorn, and B. A. Van Doorn, Design of a multisensory tracking system for advanced air traffic control, in: *Multitarget-Multisensor Tracking: Application and Advances*, Vol. II, Artech House, Norwood, MA, USA, 1992, pp. 31–63, Chapter 2.

181. P. Vacher, I. Barret, and M. Gauvrit, Design of a tracking algorithm for an advanced ATC system, in: *Multitarget-Multisensor Tracking: Application and Advances*, Vol. II, Artech House, Norwood, MA, USA, 1992, pp. 1–29, Chapter 1.

182. X. R. Li and Y. Bar-Shalom, Design of an interacting multiple model algorithm for air traffic control tracking, *IEEE Trans. Contr. Syst. Technol.*, 1(3), 186–194, 1993.

183. M. Yeddanapudi, Y. Bar-Shalom, and K. Pattipati, IMM estimation for multitarget-multisensor air traffic surveillance, *Proc. of the IEEE*, 85(1), 80–96, 1997.

184. H. Wang, T. Kirubarajan, and Y. Bar-Shalom, Precision large scale air traffic surveillance using IMM/assignment estimators, *IEEE Trans. Aerosp. Electron. Syst.*, 35(1), 255–266, 1999.

185. S. S. Blackman, R. J. Dempster, M. T. Busch, and R. F. Popoli, IMM/MHT solution to radar benchmark tracking problem, *IEEE Trans. Aerosp. Electron. Syst.*, 35(2), 730–738, 1999.

186. R. J. Dempster, S. S. Blackman, and T.S. Nichols, Combining IMM filtering and MHT data association for multitarget tracking, *Proc. Twenty-Ninth Southeastern Symposium on System Theory*, 1997, pp. 123–127.

187. D. Mušicki and S. Suvorova, Tracking in clutter using IMM-IPDA based algorithms, *IEEE Trans. Aerosp. Electron. Syst.*, 44(1), 111–126, January 2008.

188. J. Lancaster and S. Blackman, Joint IMM/MHT tracking and identification for multisensor ground target tracking, *Proc. of the Ninth International Conf. on Information Fusion*, Florence, Italy, July 2006.

189. T. Kirubarajan, Y. Bar-Shalom, K. R. Pattipati, and I. Kadar, Ground target tracking with variable structure IMM estimator, *IEEE Trans. Aerosp. Electron. Syst.*, 36,(1), 26–46, January 2000.

190. B. J. Noe and N. Collins, Variable structure interacting multiple model filter (VS-IMM) for tracking targets with transportation network constraints, *Proceedings of the 2000 SPIE Conference on Signal and Data Processing of Small Targets*, Vol. 4048, Orlando, FL, April 2000, pp. 247–258.

191. B. Pannetier, K. Benameur, V. Nimier, and M. Rombaut, VS-IMM using road map information for a ground target tracking, *Proceedings of the 2005 International Conference on Information Fusion*, Philadelphia, PA, USA, July 25–29, 2005.

192. S. Blackman, K. Fong, D. E. Carroll, J. Lancaster, and R. Dempster, Integration of a road network into a radar ground moving target tracking (GMTT) system and its performance evaluation, *Proceedings of the SPIE Conference on Signal and Data Processing of Small Targets*, Vol. 7445, San Diego, CA, August 2009, pp. 74450S-1–74450S-9.

193. P. Horridge and S. Maskell, Searching for, initiating and tracking multiple targets using existence probabilities, *Proc. 2009 International Conference on Information Fusion*, July 6–9, Seattle, USA, 2009.

194. M. Ulmke, O. Erdinc, and P. Willett, Gaussian mixture cardinalized PHD filter for ground moving target tracking, *Proceedings of the 10th International Conference on Information Fusion*, Quebec City, Canada, July 2007.

195. S. Coraluppi, D. Grimmett, and O. Gérard, Distributed multi-hypothesis sonar tracking and extensions, *NURC Technical Report SR-421*, February 2005.

196. S. Coraluppi, C. Carthel, P. Willett, M. Dingboe, O. O'Neill, and T. Luginbuhl, The track repulsion effect in automatic tracking, *Proceedings of the 12th International Conference on Information Fusion*, Seattle WA, USA, July 2009.

197. S. Coraluppi and C. Carthel, Multi-stage multiple-hypothesis tracking, *ISIF J. Adv. Inform. Fusion*, 6(1), 57–67, 2011.

198. Y. Bar-Shalom, S. S. Blackman, and R. J. Fitzgerald , Dimensionless score function for multiple hypothesis tracking , *IEEE Trans. Aerosp. Electron. Syst.*, 43(1), 392–400, January 2007.

CHAPTER 6

Tracking and Data Fusion for Ground Surveillance

MICHAEL MERTENS, MICHAEL FELDMANN, MARTIN ULMKE, AND WOLFGANG KOCH

6.1 INTRODUCTION TO GROUND SURVEILLANCE

Ground surveillance comprises track extraction and maintenance of single ground moving vehicles and convoys, as well as low-flying objects such as helicopters or unmanned aerial vehicles (UAVs). Establishing and maintaining tracks of relevant ground moving objects is particularly challenged by dense target scenarios, complex target dynamics (e.g., "stop & go" behavior or strong maneuvers), masking due to the sensor's Doppler blind zone or by terrain obscuration, residual clutter, and so on. In general, these factors quickly lead to strong performance degradation or even track loss. As ground object tracking is a challenging problem, all available information sources must be exploited, that is, the sensor data themselves, as well as context knowledge about the sensor performance and the underlying scenario.

For long-range, wide-area, all-weather, and all-day surveillance operating at high data update rates, GMTI radar proves to be the sensor system of choice (GMTI: Ground Moving Target Indication). By using airborne sensor platforms in stand-off ground surveillance applications, the effect of topographical screening is alleviated, thus extending the sensors' field of view. In [1], characteristic problems of signal processing related to GMTI tracking with STAP radar are discussed (STAP: Space-Time Adaptive Processing). In this context, the following topics are of particular interest:

- *Doppler-Blindness.* Ground moving vehicles can well be masked by the clutter notch of the sensor. This physical phenomenon directly results from the low-Doppler characteristics of ground moving vehicles and causes interfering fading

Integrated Tracking, Classification, and Sensor Management: Theory and Applications, Edited by Mahendra Mallick, Vikram Krishnamurthy, and Ba-Ngu Vo.

effects that seriously affect track accuracy and track continuity. The problems are even more challenging in the presence of Doppler ambiguities (see Section 6.2).

- *Signal Strength Measurements.* Signal strength is a standard output of a modern radar system. Provided the amplitude of the target returns exceeds the false alarm background, the consideration of signal strength may lead to improved target estimates, depending on the scenario (see Section 6.2).
- *Road-Map Information.* Even military targets usually move on road networks, whose topographical coordinates are known in many cases. Digitized topographical road maps such as provided by geographical information systems (GIS) should therefore enter into the target tracking and sensor data fusion process (see Section 6.4).
- *Collectively Moving Targets.* Collectively moving convoys consisting of individual vehicles are typical of certain applications and have to be treated as aggregated entities. In some cases, the kinematic states of the individual vehicles can be treated as internal degrees of freedom. In addition, the convoy extent or the object number can become part of the object state (see Sections 6.5 and 6.6).

6.2 GMTI SENSOR MODEL

The exploitation of the sensor output relies on a correct modeling of the sensor properties. For tracking applications, there is no need to model the details of the actual signal processing and detection steps of the sensors. Rather, it is important to model the properties of the resulting measurements associated with objects of interest as well as artificial detections such as false alarms and clutter. Typical sensor properties relevant for target tracking are measurement errors, detection probabilities, false alarm distribution, and intensity. In the present context of ground target tracking, we will concentrate on the issues such as GMTI clutter notch and signal strength measurements. A more detailed discussion is provided in [1, 2].

6.2.1 Model of the GMTI Clutter Notch

For physical and technical reasons, the detection of ground moving targets by airborne radar, typically on a moving platform, is limited by strong ground clutter returns. This can be much alleviated by STAP techniques [1]. The characteristics of STAP processing, however, directly influence the GMTI tracking performance. Even after platform motion compensation by STAP filtering, low-Doppler targets can be masked by the clutter notch of the GMTI radar. Let $\mathbf{e}_k^p = (\mathbf{r}_k - \mathbf{p}_k)/|\mathbf{r}_k - \mathbf{p}_k|$ denote the unit vector pointing from the platform position $\mathbf{p}_k$ at time t_k to the target at the position $\mathbf{r}_k$ moving with the velocity $\dot{\mathbf{r}}_k$. The kinematic object state is given by $\mathbf{x}_k^T = [\mathbf{r}_k^T, \dot{\mathbf{r}}_k^T]$. Doppler blindness occurs if the radial velocities of the object as well as of the surrounding main-lobe clutter return are identical, that is, if the function

$$h_n(\mathbf{r}_k, \dot{\mathbf{r}}_k; \mathbf{p}_k) = \mathbf{e}_k^p \bullet \dot{\mathbf{r}}_k = (\mathbf{e}_k^p)^T \, \dot{\mathbf{r}}_k \tag{6.1}$$

is close to zero. In other words, $h_n(\mathbf{x}_k; \mathbf{p}_k) \approx 0$ holds if the target's velocity vector is nearly perpendicular to the sensor-to-target line-of-sight. For this reason, the equation $h_n(\mathbf{x}_k; \mathbf{p}_k) = 0$ defines the location of the GMTI clutter notch in the state space of a ground target and as such reflects a fundamental physical/technical fact without implying any further modeling assumptions.

6.2.1.1 Qualitative Discussion. Any GMTI detection model for air-to-ground radar must thus reflect the following phenomena:

1. The detection probability P_D depends on the target state and the sensor-to-target geometry.
2. P_D is small in a certain region around the clutter notch characterized by the minimum detectable velocity v_m, an important sensor parameter that must enter into the tracking process.
3. Far from the clutter notch, the detection probability depends only on the directivity pattern of the sensor and the target strength and range.
4. There exists a narrow transient region between these two domains.

GMTI models are adapted to STAP techniques in that the detection probability assumed in the tracking process is described as a function of the GMTI-specific clutter notch. While the current location of the notch is determined by the kinematic state of the target and the current sensor-to-target geometry, its width is given by a characteristic sensor parameter v_m. In this way, more detailed information on the sensor performance can be incorporated into the tracking process. This in particular permits a more appropriate treatment of missing detections. In other words, information on the potential reasons that might have caused the missing detections enters into the tracking filter. We observed that, by this measure, the number of lost tracks can significantly be reduced, while the track continuity is improved, finally leading to a more reliable ground picture. This qualitative discussion of the observed detection phenomena related to the GMTI clutter notch is similar in nature to that of resolution effects.

6.2.1.2 Quantitative Discussion. In a generic description of the detection performance of GMTI sensors, it seems plausible to write $P_D = P_D(\mathbf{x}_k)$ as a product of two factors: One factor reflects the directivity pattern and propagation effects due to the radar equation [3], that is, $P_D = P_d(r_k, \varphi_k, \vartheta_k)$ depending on range r_k, azimuth angle φ_k, and elevation angle ϑ_k. The other factor describes the clutter notch. To this end, let us consider functions of the following form

$$P_D(r_k, \varphi_k, \vartheta_k, \dot{r}_k) = P_d(r_k, \varphi_k, \vartheta_k)\left(1 - e^{-\frac{1}{2}\left(\frac{h_n}{v_m}\right)^2}\right). \tag{6.2}$$

In this expression, the sensor parameter v_m has a clear and intuitive meaning: In the region defined by $|h_n(\mathbf{x}_k)| < v_m$, we have $P_D < \frac{1}{2} P_d$. The parameter v_m is thus a quantitative measure of the minimum radial velocity with respect to the sensor platform that a ground moving target must at least have to be detectable by the sensor. The actual size of v_m depends on the particular signal processor used. Two phenomena in particular can cause problems in GMTI tracking:

1. Sensor-to-target geometries can occur where targets to be tracked are masked by the clutter notch of the sensor. This results in a series of missing detections until the geometry changes again.
2. As stopping targets are indistinguishable from ground clutter, the early detection of a stopping event itself as well as tracking of "stop & go" targets can be important to certain applications.

In Section 6.3, it is shown how the knowledge on the clutter notch of the sensor can be incorporated into a Bayesian tracking scheme. Numerical results for simulated scenarios are given in Section 6.4.

6.2.2 Signal Strength Measurements

6.2.2.1 Target Fluctuations. The realistic modeling of a target's back-scattering characteristics is difficult in general, and even more so in ground target scenarios with moving sensor platforms and different, unknown types of moving objects. Therefore, statistical models are used instead that can be handled analytically. In the following, it is assumed that the fluctuations of the input signal at the detector, resulting from fluctuations of the target radar cross section, can be described by the Swerling models [4]. The two models that are of interest in particular are Swerling-I and Swerling-III. Swerling-I holds for targets that are composed of many scatterers of comparable size, whereas Swerling-III stands for targets that are made up of mainly one dominant scattering center. Both models are based on the assumption that the fluctuation of the underlying radar cross section remains constant during a single illumination interval.

The statistical properties of chi-square target fluctuations are determined by the following general probability density, which depends on the average signal-to-noise ratio SNR_0 and the parameter m indicating the degrees of freedom ($2m$) of the associated chi-square distribution

$$\mathcal{G}^m_{\mathrm{SNR}_0}(\mathrm{SNR}) = \frac{(m/\mathrm{SNR}_0)^m}{\Gamma(m)} \mathrm{SNR}^{m-1}\, e^{-m\,\mathrm{SNR}/\mathrm{SNR}_0}. \tag{6.3}$$

Assuming independent fluctuations from scan to scan, the relevant Swerling cases are given by $m = 1$ (Swerling-I) and $m = 2$ (Swerling-III), yielding

$$\mathcal{G}^1_{\mathrm{SNR}_0}(\mathrm{SNR}) = \frac{1}{\mathrm{SNR}_0}\, e^{-\mathrm{SNR}/\mathrm{SNR}_0} \tag{6.4}$$

$$\mathcal{G}^2_{\mathrm{SNR}_0}(\mathrm{SNR}) = \frac{4\,\mathrm{SNR}}{\mathrm{SNR}_0^2}\, e^{-2\,\mathrm{SNR}/\mathrm{SNR}_0}. \tag{6.5}$$

6.2.2.2 Detection Process. The complex target signal $\mathbf{v} = (v_1, v_2)$ with orthogonal and statistically independent components is added by white Gaussian noise within the receiver unit. The detector uses the total signal $\mathbf{u} = (u_1, u_2)$ to form the signal strength $||\mathbf{u}||^2 = (u_1)^2 + (u_2)^2$ with the probability density given by the Rice distribution

$$p(||\mathbf{u}||^2|\mathbf{v}) = \frac{1}{2\sigma_n^2} e^{-\frac{||\mathbf{u}||^2 + ||\mathbf{v}||^2}{2\sigma_n^2}} I_0\left(\frac{||\mathbf{u}|| \cdot ||\mathbf{v}||}{\sigma_n^2}\right), \tag{6.6}$$

where I_0 denotes the modified Bessel function of the first kind. Normalized to the mean noise level, $\mathrm{E}\left[p(||\mathbf{u}||^2|\,0)\right] = 2\sigma_n^2$, that is, with $s = ||\mathbf{u}||^2/2\sigma_n^2$ and $\mathrm{SNR} = ||\mathbf{v}||^2/2\sigma_n^2$, this can be rewritten as

$$p(s|\mathrm{SNR}) = e^{-s-\mathrm{SNR}} I_0\left(2\sqrt{s} \cdot \sqrt{\mathrm{SNR}}\right) \tag{6.7}$$

with expectation value $\mathrm{E}[s] = 1 + \mathrm{SNR}$ and thus unit power for noise ($\mathrm{SNR} = 0$). Following the assumption from the last section, the signal-to-noise ratio SNR itself fluctuates with mean SNR_0. Thus the overall received signal strength is derived by

$$p(s|\mathrm{SNR}_0) = \int_0^\infty p(s|\mathrm{SNR})\, p(\mathrm{SNR}|\mathrm{SNR}_0)\, \mathrm{dSNR} \tag{6.8}$$

with $p(\mathrm{SNR}|\mathrm{SNR}_0)$ given by $\mathcal{G}^m_{\mathrm{SNR}_0}(\mathrm{SNR})$. Finally, we end up with the well-known distributions for the two Swerling cases

$$p(s|\mathrm{SNR}_0, m = 1) = \frac{1}{1 + \mathrm{SNR}_0} e^{-s/(1+\mathrm{SNR}_0)} \tag{6.9}$$

$$p(s|\mathrm{SNR}_0, m = 2) = \frac{1}{(1 + \mathrm{SNR}_0/2)^2} e^{-s/(1+\mathrm{SNR}_0/2)} \left(1 + \frac{s}{1 + 2/\mathrm{SNR}_0}\right). \tag{6.10}$$

As soon as signal strength is considered, the detection probability is no longer a parameter but becomes a function of the target's mean signal-to-noise ratio SNR_0 and the detection threshold λ

$$P(s > \lambda|D) = P_D(\mathrm{SNR}_0, \lambda, m) = \int_\lambda^\infty p(s|\mathrm{SNR}_0, m)\, \mathrm{d}s, \tag{6.11}$$

where D denotes detection. For the two Swerling models, the integration yields the corresponding detection probabilities, see Figure 6.1

$$P_D^{\mathbf{I}} = e^{-\lambda/(1+\mathrm{SNR}_0)} \tag{6.12}$$

$$P_D^{\mathbf{III}} = e^{-\lambda/(1+\mathrm{SNR}_0/2)} \left(1 + \frac{\lambda \cdot \mathrm{SNR}_0/2}{(1 + \mathrm{SNR}_0/2)^2}\right). \tag{6.13}$$

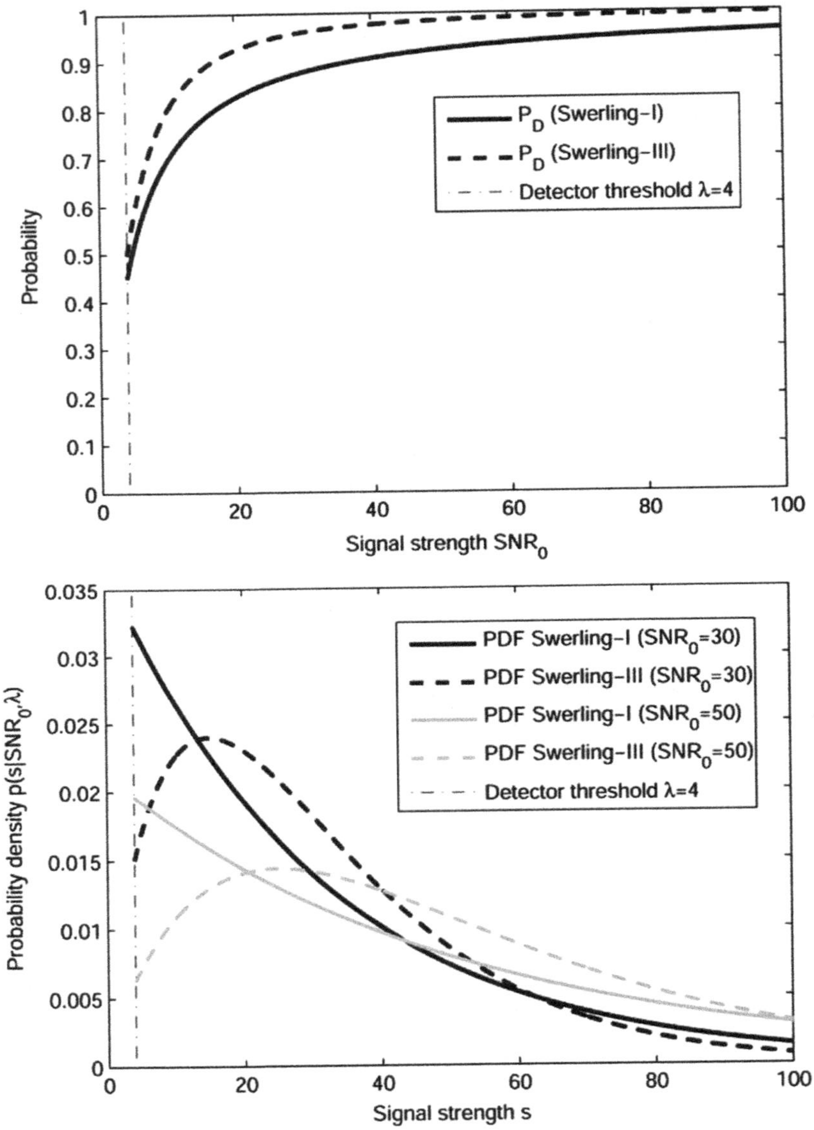

Figure 6.1 Distributions of the detection probability (top) and probability density (bottom) for the Swerling-I and Swerling-III models.

Detections are only available if the signal strength of the input signal exceeds the detection threshold. Thus the probability densities (6.9) and (6.10) need to be normalized properly, yielding for $s \geq \lambda$

$$p(s|\mathrm{SNR}_0, \lambda, \mathbf{I}) = \frac{1}{P_D^{\mathbf{I}}} \, p(s|\mathrm{SNR}_0, m = 1) \tag{6.14}$$

$$p(s|\mathrm{SNR}_0, \lambda, \mathbf{III}) = \frac{1}{P_D^{\mathbf{III}}} \, p(s|\mathrm{SNR}_0, m = 2) \, . \tag{6.15}$$

These distributions are plotted in Figure 6.1 for different values of SNR_0. Measurements, which originate from false alarms, are modeled by the Swerling-I case, that is,

$$p(s|\mathrm{CNR}_0) = \frac{1}{P_D^{\mathbf{I}}(\mathrm{CNR}_0)} \, \frac{e^{-s/(1+\mathrm{CNR}_0)}}{1 + \mathrm{CNR}_0} \tag{6.16}$$

if $s \geq \lambda$ with CNR_0 being the mean clutter strength-to-noise ratio. It is assumed that the detection threshold λ on the input signal power is high enough to suppress measurements resulting from receiver noise.

6.3 BAYESIAN APPROACH TO GROUND MOVING TARGET TRACKING

Among the data exploitation products of data fusion systems, target tracks are of particular importance. Tracking faces an omnipresent aspect in every real-world application in so far as it is dealing with fusion of data produced at different instants of time; that is, tracking is important in all applications where particular emphasis is placed on the fact that the sensor data to be exploited have the character of a time series. Tracks thus represent currently available knowledge on relevant, time-varying quantities characterizing the instantaneous *state* of individual targets or target groups of interest, such as aircraft, ships, submarines, ground vehicles, or moving persons. Quantitative measures reliably describing the quality of this knowledge are an integral part of a track. The information obtained by tracking algorithms [5–9] also includes the history of the targets. If possible, a one-to-one association between the target trajectories in the sensors' field of view and the produced tracks is to be established and has to be preserved as long as possible (track continuity). The achievable track quality does not only depend on the performance of the sensors, but also on target properties and the operational conditions within the scenario to be observed. A tracking algorithm has to perform several tasks:

1. track extraction from given sensor detections,
2. data association of sensor detections with tracks,

3. track filtering, consisting of: prediction, data processing, and, where appropriate, retrodiction,
4. track management functions, for example, track drop, split, merging.

In the following, the focus lies on track filtering algorithms as this step is mostly affected by the specific properties of ground surveillance.

6.3.1 Bayesian Tracking Filter

A Bayesian tracking algorithm is an iterative updating scheme for calculating the conditional probability density function (pdf), $p(\mathbf{x}_l|\mathbf{Z}^k)$, that represent all available knowledge on the object state $\mathbf{x}_l$ at discrete instants of time t_l. The densities are explicitly conditioned by the sensor data $\mathbf{Z}^k = \{\mathbf{Z}_1, \ldots, \mathbf{Z}_k\}$, accumulated up to some time t_k, typically the present time. Implicitly, however, they are also determined by all available context knowledge on the sensor characteristics, the dynamical object properties, the environment of the objects, topographical maps, or tactical rules governing the objects' overall behavior.

With respect to the instant of time t_l at which estimates of the object states $\mathbf{x}_l$ are required, the related density iteration process is referred to as *prediction* ($t_l > t_k$), *filtering* or *data processing* ($t_l = t_k$), or *retrodiction* ($t_l < t_k$). The propagation of the probability densities involved is given by three basic update equations, which will be derived and discussed below and illustrated by examples.

6.3.1.1 Prediction. The prediction density $p(\mathbf{x}_k|\mathbf{Z}^{k-1})$ is obtained by combining the evolution model $p(\mathbf{x}_k|\mathbf{x}_{k-1})$ with the previous filtering density $p(\mathbf{x}_{k-1}|\mathbf{Z}^{k-1})$

$$p(\mathbf{x}_{k-1}|\mathbf{Z}^{k-1}) \xrightarrow[\text{constraints}]{\text{evolution model}} p(\mathbf{x}_k|\mathbf{Z}^{k-1}) \tag{6.17}$$

$$p(\mathbf{x}_k|\mathbf{Z}^{k-1}) = \int \underbrace{p(\mathbf{x}_k|\mathbf{x}_{k-1})}_{\text{evolution model}} \underbrace{p(\mathbf{x}_{k-1}|\mathbf{Z}^{k-1})}_{\text{previous filtering}} \mathrm{d}\mathbf{x}_{k-1}. \tag{6.18}$$

6.3.1.2 Data Processing. The filtering density $p(\mathbf{x}_k|\mathbf{Z}^k)$ is obtained by combining the sensor model $p(\mathbf{Z}_k|\mathbf{x}_k)$ with the prediction density $p(\mathbf{x}_k|\mathbf{Z}^{k-1})$ according to

$$p(\mathbf{x}_k|\mathbf{Z}^{k-1}) \xrightarrow[\text{sensor model}]{\text{current sensor data}} p(\mathbf{x}_k|\mathbf{Z}^k) \tag{6.19}$$

$$p(\mathbf{x}_k|\mathbf{Z}^k) = \frac{p(\mathbf{Z}_k|\mathbf{x}_k)p(\mathbf{x}_k|\mathbf{Z}^{k-1})}{\int \underbrace{p(\mathbf{Z}_k|\mathbf{x}_k)}_{\text{sensor model}} \underbrace{p(\mathbf{x}_k|\mathbf{Z}^{k-1})}_{\text{prediction}} \mathrm{d}\mathbf{x}_k}. \tag{6.20}$$

6.3.1.3 Retrodiction. The retrodiction density $p(\mathbf{x}_l|\mathbf{Z}^k)$ is obtained by combining the object evolution model $p(\mathbf{x}_{l+1}|\mathbf{x}_l)$ with the previous prediction and filtering densities $p(\mathbf{x}_{l+1}|\mathbf{Z}^l)$, $p(\mathbf{x}_l|\mathbf{Z}^l)$ according to

$$p(\mathbf{x}_{l-1}|\mathbf{Z}^k) \xleftarrow[\text{evolution model}]{\text{filtering, prediction}} p(\mathbf{x}_l|\mathbf{Z}^k) \tag{6.21}$$

$$p(\mathbf{x}_l|\mathbf{Z}^k) = \int \frac{\overbrace{p(\mathbf{x}_{l+1}|\mathbf{x}_l)}^{\text{evolution}} \; \overbrace{p(\mathbf{x}_l|\mathbf{Z}^l)}^{\text{prev. filtering}}}{\underbrace{p(\mathbf{x}_{l+1}|\mathbf{Z}^l)}_{\text{prev. prediction}}} \; \underbrace{p(\mathbf{x}_{l+1}|\mathbf{Z}^k)}_{\text{prev. retrodiction}} \; d\mathbf{x}_{l+1} \,. \tag{6.22}$$

As a technical term, *track* is therefore either a synonym for the collection of densities $p(\mathbf{x}_l|\mathbf{Z}^k), l = 1, \ldots, k$, or of suitably chosen parameters characterizing them, such as estimates related to appropriate risk functions and the corresponding estimation error covariance matrices. According to this paradigm, an object track represents all available knowledge on the kinematic properties of a time varying object of interest, including its history and measures that describe the quality of this knowledge. One iteration of a tracking filter (without retrodiction) is depicted in Figure 6.2.

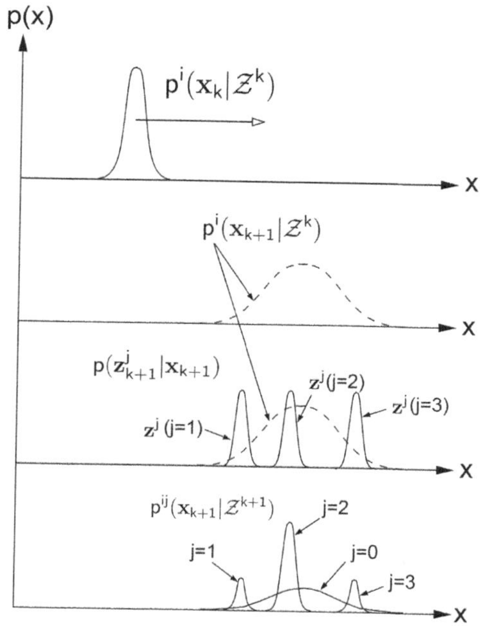

Figure 6.2 One iteration of a tracking filter. (a) density at time t_k, (b) predicted density, (c) three detections and corresponding likelihood functions (plus missed detection hypothesis), (d) updated mixture density; according to [10].

6.3.2 Essentials of GMTI Tracking

The choice of a suitable coordinate system for describing the underlying sensor-target geometry, the sensor platform trajectory, and the available a priori information on the dynamical behavior of ground moving targets are prerequisites to target tracking. In wide-area applications, a flat earth model is often not admissible. We consider three coordinate systems in which the underlying physical phenomena become transparent:

1. appropriate *ground* coordinates, typically based on a local Cartesian coordinate frame representing a tangent plane to the surface of the earth (WGS84 ellipsoid), where the descriptions of the target and platform kinematics are of a particularly simple form,
2. the moving Cartesian *antenna* coordinate system, whose x-axis is oriented along the array antenna of the GMTI radar mounted on the airborne sensor platform,
3. the *sensor* coordinate system, in which the measurements of the kinematic target parameters are described (target range, azimuth, and range rate).

In the following, we will mostly consider linear target dynamics and linear measurement models, both with normally distributed noise, leading to a Gaussian target pdf that is determined by its mean and covariance. The inclusion of the GMTI clutter notch (see Section 6.2) will lead to a Gaussian mixture structure of the pdf. The same is true for the incorporation of digital road maps as discussed in Section 6.4.

6.3.2.1 Target Dynamics. Since ground moving targets typically exhibit much less agility than, for example, military air targets, the inclusion of accelerations into the state vector may not be necessary. Therefore, the target state at time t_k is defined by

$$\mathbf{x}_k^T = [\mathbf{r}_k^T, \dot{\mathbf{r}}_k^T] = [x, y, z, v_x, v_y, v_z]. \tag{6.23}$$

As widely accepted in the tracking literature [5], the target dynamics is modeled by a linear Markov process

$$\mathbf{x}_{k+1} = \mathbf{F}_{k+1|k}\,\mathbf{x}_k + \mathbf{G}_{k+1|k}\,\mathbf{v}_{k+1} \tag{6.24}$$

which implies that the target state at time step t_{k+1} is determined by the state at the preceding time step t_k. A suitable choice of the matrices $\mathbf{F}_{k+1|k}$ and $\mathbf{G}_{k+1|k}$ is given in [11]

$$\mathbf{F}_{k+1|k} = \begin{bmatrix} \mathbf{J} & t_{k+1|k}\mathbf{J} \\ \mathbf{0} & e^{-t_{k+1|k}/\theta_t}\mathbf{J} \end{bmatrix}, \qquad \mathbf{G}_{k+1|k} = \Sigma_{k+1|k} \begin{bmatrix} \mathbf{0} \\ \mathbf{J} \end{bmatrix} \tag{6.25}$$

with time difference $t_{k+1|k} = t_{k+1} - t_k$, $\Sigma_{k+1|k} = v_t\sqrt{1 - e^{-2t_{k+1|k}/\theta_t}}$, and $\mathbf{J} = \text{diag}[1, 1, 0]$ and $\mathbf{0} = \text{diag}[0, 0, 0]$. It can be shown that the modeled target velocity is ergodic and given by

$$\mathbb{E}\left[\dot{\mathbf{r}}_k\right] = 0, \qquad \mathbb{E}\left[\dot{\mathbf{r}}_k\dot{\mathbf{r}}_l^T\right] = v_t^2 e^{-2(k-l)t_{k+1|k}/\theta_t}. \tag{6.26}$$

The parameters v_t and θ_t have to be chosen according to the expected dynamics of the targets under consideration. v_t limits the velocity standard deviation with $v_t = 20\,\text{m/s}$ being a typical value for road targets. The so-called maneuver correlation time θ_t describes the agility of the targets and typically is chosen between 30 s and 90 s. The matching of the parameter values with the true target behavior influences the quality of the track estimates.

6.3.2.2 Measurement Update. The measurement, too, is assumed to be a linear function of the target state

$$\mathbf{z}_k = \mathbf{H}_k\mathbf{x}_k + \mathbf{w}_k \tag{6.27}$$

with Gaussian measurement noise $\mathbf{w}_k$. A real GMTI sensor, however, measures in spherical coordinates, that is, azimuth φ and, possibly, elevation angle ϑ, range r, and range-rate $\dot{r}$. The corresponding Cartesian measurements are given by the nonlinear equations

$$\mathbf{z} = \begin{bmatrix} x_S + r\cos\varphi\cos\vartheta \\ y_S + r\sin\varphi\cos\vartheta \\ z_S + r\sin\vartheta \\ \dot{r} \end{bmatrix} \tag{6.28}$$

with the sensor position $\mathbf{p} = [x_S, y_S, z_S]^T$. The measurement equation (6.28) can be linearized by the following coordinate transformations:

1. shifting of the origin about the sensor position $\mathbf{p}$.
2. rotation about an angle φ around the z-axis to eliminate the y-component.
3. rotation about an angle ϑ around the y-axis to eliminate the z-component.

To be explicit, with the two-fold rotation

$$\mathbf{T}(\varphi,\vartheta) = \begin{bmatrix} \cos\vartheta & 0 & \sin\vartheta & 0 \\ 0 & 1 & 0 & 0 \\ -\sin\vartheta & 0 & \cos\vartheta & 0 \\ 0 & 0 & 0 & 1 \end{bmatrix} \cdot \begin{bmatrix} \cos\varphi & \sin\varphi & 0 & 0 \\ -\sin\varphi & \cos\varphi & 0 & 0 \\ 0 & 0 & 1 & 0 \\ 0 & 0 & 0 & 1 \end{bmatrix} \tag{6.29}$$

the transformed Cartesian measurement is given by

$$\tilde{\mathbf{z}} = \mathbf{T}(\varphi,\vartheta)\left(\mathbf{z} - [x_S, y_S, z_S, 0]^T\right) = [r, 0, 0, \dot{r}]^T. \tag{6.30}$$

In other words, in the new coordinate system, the position vector lies in the new x-direction, and thus, by definition, the radial velocity is equal to the x-component of the transformed velocity vector. The dimension of $\tilde{\mathbf{z}}$ still is 4, as the second and third components are *measured* zeros. In the new coordinate system the measurement equation (6.27) indeed is linear, and for not too large angular errors and sensor-object distances the measurement noise is approximately Gaussian with the covariance matrix

$$\tilde{\mathbf{R}} = \mathbf{T}(\varphi, \vartheta)\, \mathbf{R}\, \mathbf{T}(\varphi, \vartheta)^T \tag{6.31}$$

with the covariance $\mathbf{R}$ in spherical coordinates, given by the assumed independent measurement variances in $(r, \varphi, \vartheta, \dot{r})$. Interestingly, the *notch* function (6.1) simplifies, in the new coordinate system, to $h_n(\mathbf{r}_k, \dot{\mathbf{r}}_k; \mathbf{p}_k) = \dot{r}$ and the detection probability (6.2) becomes

$$P_D(\tilde{\mathbf{x}}_k; \tilde{\mathbf{p}}_k) = P_d(\tilde{\mathbf{x}}_k; \tilde{\mathbf{p}}_k) \left(1 - e^{-\frac{1}{2}\left(\frac{\dot{r}_k}{v_m}\right)^2}\right). \tag{6.32}$$

6.3.3 Filter Update with Clutter Notch

The data processing step of the tracking filter (6.20) requires a sensor model, defined by the likelihood function $p(\mathbf{Z}_k|\mathbf{x}_k)$. Receiving m_k detections at time step k, that is, $\mathbf{Z}_k = \{\mathbf{z}_k^1, \ldots, \mathbf{z}_k^{m_k}\}$, and taking the possibility of a missed detection into account, there are $m_k + 1$ mutually exclusive possibilities. With the standard assumption that an object leads to at most one detection[1] and a residual clutter density (false alarm rate) ρ_F, the single object likelihood function is given by

$$p(\mathbf{Z}_k, m_k|\mathbf{x}_k) = \big(1 - P_D(\mathbf{x}_k; \mathbf{p}_k)\big)\rho_F + P_D(\mathbf{x}_k; \mathbf{p}_k) \sum_{j=1}^{m_k} p\big(\mathbf{z}_k^j|\mathbf{x}_k\big) \tag{6.33}$$

with $p\big(\mathbf{z}_k^j|\mathbf{x}_k\big) = \mathcal{N}\big(\mathbf{z}_k^j; \mathbf{H}_k\mathbf{x}_k, \mathbf{R}_k\big)$ for the measurement model (6.27). When the GMTI detection model (6.2) or (6.32) is inserted into this expression, the likelihood can be split into

$$p(\mathbf{Z}_k, m_k|\mathbf{x}_k) = p_0(\mathbf{Z}_k, m_k|\mathbf{x}_k) + p_n(\mathbf{Z}_k, m_k|\mathbf{x}_k), \tag{6.34}$$

where p_0 denotes the standard likelihood without considering clutter notches, that is,

$$p_0(\mathbf{Z}_k, m_k|\mathbf{x}_k) = \big(1 - P_d(\mathbf{x}_k; \mathbf{p}_k)\big)\rho_F + P_d(\mathbf{x}_k; \mathbf{p}_k) \sum_{j=1}^{m_k} p\big(\mathbf{z}_k^j|\mathbf{x}_k\big) \tag{6.35}$$

[1]This assumption will be dropped in Section 6.5.

and the second addend p_n is the part of the overall likelihood function characteristic of the GMTI problem

$$p_n(\mathbf{Z}_k, m_k|\mathbf{x}_k) = -P_d(\mathbf{x}_k; \mathbf{p}_k)\, e^{-\frac{1}{2}\left(\frac{h_n(r_k, \phi_k, \dot{r}_k)}{v_m}\right)^2} \left(\rho_F + \sum_{j=1}^{m_k} p\left(\mathbf{z}_k^j|\mathbf{x}_k\right)\right). \qquad (6.36)$$

For a generalization to the case of Doppler-ambiguous measurements see [12].

We can immediately see that the effect of the GMTI-specific clutter notch on the likelihood function can formally be described by a measured zero of a fictitious quantity defined by the pseudo measurement function $\mathbf{h}_n$, where the minimum detectable velocity plays the role of the standard deviation of the fictitious measurements.

According to Bayes' rule (6.20), the processing of the new sensor data $\mathbf{Z}_k$ received at revisit time t_k is based on the predicted density $p(\mathbf{x}_k|\mathbf{Z}^{k-1})$ and the likelihood function $p(\mathbf{Z}_k|\mathbf{x}_k)$. Assuming a Gaussian sum representation for $p(\mathbf{x}_k|\mathbf{Z}^{k-1})$, the Gaussian sum structure of the likelihood function guarantees that also $p(\mathbf{x}_k|\mathbf{Z}^k)$ belongs to this family. According to Bayes' theorem, we obtain up to a normalizing constant

$$p(\mathbf{x}_k|\mathbf{Z}^k) \propto p(\mathbf{Z}_k|\mathbf{x}_k)\, p(\mathbf{x}_k|\mathbf{Z}^{k-1}) \propto \sum_i p_k^i\, \mathcal{N}(\mathbf{x}_k; \mathbf{x}_{k|k}^i; \mathbf{P}_{k|k}^i). \qquad (6.37)$$

The increasing number of mixture components can be treated by different mixture reduction techniques (e.g., pruning, local combining) discussed in the tracking literature [5, 6] in order to keep the number of mixture components under control. Simulations show that even a representation by only two mixture components is sufficient in many practical cases and seems to mirror the underlying physics of the detection process quite well [12, 13].

The advantage of exploiting the knowledge of the clutter notch is illustrated in the following simple example: In the considered scenario, a GMTI sensor observes a ground moving target at constant speed (15 m/s) that moves parallel to the y-axis. During the scenario, the target comes to a stop for a certain amount of time. The sensor is located in northwestern direction. Figure 6.3 presents the track results for a target tracking algorithm with and without incorporating the clutter notch information up to a time instance when the target has already stopped for several revisits. As soon as the radial speed of the target drops below the minimum detectable velocity threshold, the algorithm with clutter notch information is able to interpret the absence of detections as the result of the target being masked by the clutter notch. Therefore, the target state estimate is continued in cross-range direction as the uncertainty in range direction is small. On the other hand, the tracking algorithm without clutter notch processing simply extrapolates the track with increasing uncertainty in all directions. In Figure 6.4, the corresponding distribution of the detection probability (6.2) for the whole scenario is shown with $P_d = 0.9$ and $v_m = 4\,\text{m/s}$, and, in addition, the normalized weight of the hypothesis that the target is masked by the clutter notch. By incorporating this additional knowledge, the tracking algorithm is able to correctly

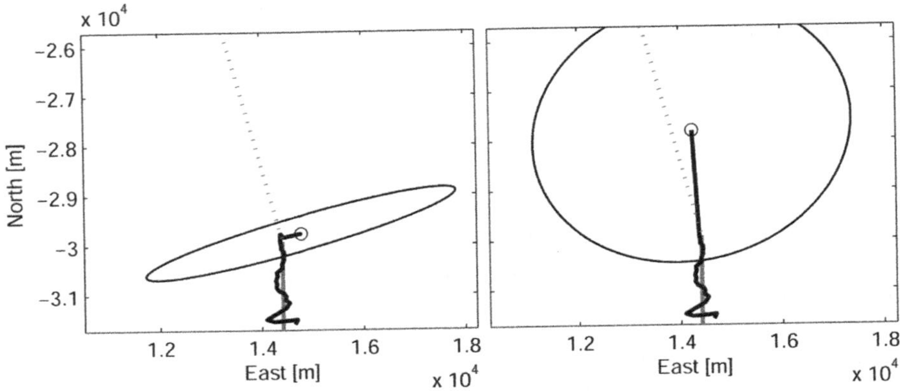

Figure 6.3 Track results for a target tracking algorithm with (left) and without (right) incorporating the clutter notch information, processing the GMTI data from a sensor which observes a target performing a move-stop-move maneuver. The presented results correspond to a time instance when the target has already stopped for several revisits. The dotted line indicates the line-of-sight between target and sensor, the latter being located in northwestern direction.

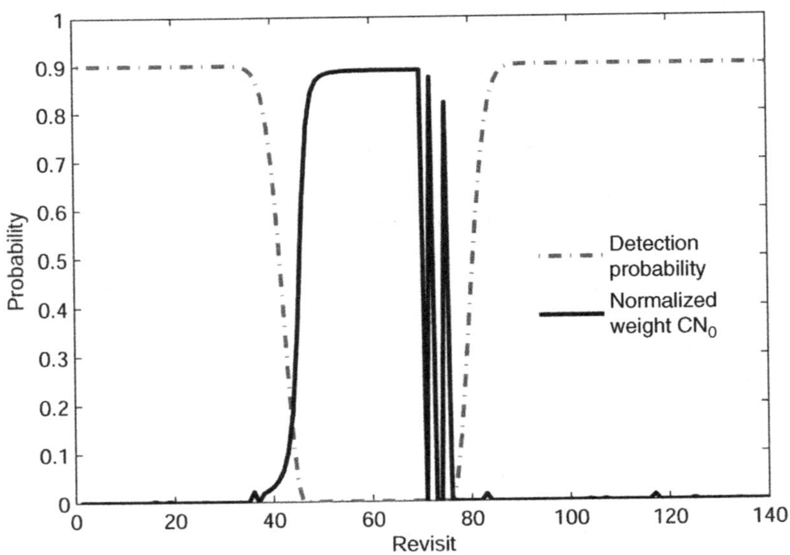

Figure 6.4 Distribution of the detection probability (6.2) with parameters $P_d = 0.9$ and $v_m = 4$ m/s and the weight of the clutter notch hypothesis, normalized to all hypothesis weights.

interpret the negative sensor output and to maintain the track due to the considerable weight of the clutter notch hypothesis.

6.3.4 Target Strength Estimation

To incorporate signal strength information, the random variable s_k is introduced, thus the target state is augmented and becomes $\hat{\mathbf{x}}_k = (\mathbf{x}_k, s_k)$ and $\mathbf{Z}_k = \{\mathbf{z}_k^i, \kappa_k^i\}_{i=1}^{m_k}$ with signal strength measurements κ_k^i. It follows

$$p(\hat{\mathbf{x}}_k|\mathbf{Z}^k) = \frac{p(\mathbf{Z}_k|\hat{\mathbf{x}}_k)\, p(\hat{\mathbf{x}}_k|\mathbf{Z}^{k-1})}{\int p(\mathbf{Z}_k|\hat{\mathbf{x}}_k)\, p(\hat{\mathbf{x}}_k|\mathbf{Z}^{k-1})\, \mathrm{d}\hat{\mathbf{x}}_k}. \tag{6.38}$$

In the following, expressions for the prior density $p(\hat{\mathbf{x}}_k|\mathbf{Z}^{k-1})$ and the likelihood function $p(\mathbf{Z}_k|\hat{\mathbf{x}}_k)$ are derived [14].

6.3.4.1 Prediction Step. First of all, the density $p(\hat{\mathbf{x}}_k|\mathbf{Z}^{k-1})$ can be split up into the kinematic and the signal strength part

$$p(\hat{\mathbf{x}}_k|\mathbf{Z}^{k-1}) = p(s_k|\mathbf{x}_k, \mathbf{Z}^{k-1})\, p(\mathbf{x}_k|\mathbf{Z}^{k-1}). \tag{6.39}$$

The kinematic prior is treated in a traditional way: the posterior density $p(\mathbf{x}_{k-1}|\mathbf{Z}^{k-1})$ is assumed to be Gaussian and is propagated to the actual time step t_k by utilizing a linear dynamics model with additive white Gaussian noise. This yields

$$p(\mathbf{x}_k|\mathbf{Z}^{k-1}) = \mathcal{N}(\mathbf{x}_k; \mathbf{x}_{k|k-1}, \mathbf{P}_{k|k-1}), \tag{6.40}$$

where the estimate $\mathbf{x}_{k|k-1}$ and covariance $\mathbf{P}_{k|k-1}$ are given by the known Kalman prediction equations. On the other hand, the prior density of the signal strength has to account for the fact that the target's mean signal-to-noise ratio SNR_0 is an a priori unknown but assumed constant value. Thus for $p(s_k|\mathbf{x}_k, \mathbf{Z}^{k-1})$ a class of densities should be chosen, which, up to a normalization constant, is invariant under the successive application of Bayes equation (6.38). Following [15], the class of inverse Gamma densities $\mathcal{I}(s; \hat{s}, \mu)$ is chosen which guarantees the aforementioned behavior

$$p(s_k|\mathbf{x}_k, \mathbf{Z}^{k-1}) = \mathcal{I}(s_k; \hat{s}_{k-1}, \mu_{k-1}) = N_{\mu_{k-1}} s_k^{-\mu_{k-1}-1} e^{-\frac{(\mu_{k-1}-1)\hat{s}_{k-1}}{s_k}} \tag{6.41}$$

with normalization constant

$$N_{\mu_{k-1}} = \frac{\left((\mu_{k-1} - 1)\hat{s}_{k-1}\right)^{\mu_{k-1}}}{\Gamma(\mu_{k-1})}. \tag{6.42}$$

The probability density $\mathcal{I}(s; \hat{s}, \mu)$ has the expectation value $\mathrm{E}[s] = \hat{s}$. If the parameter $\mu > 2$, then the variance exists with $\mathrm{Var}[s] = \frac{\hat{s}^2}{\mu-2}$. Because of $\frac{\mathrm{d}}{\mathrm{d}t}\mathrm{SNR}_0 = 0$, no *dynamics* needs to be considered for the signal strength, thus the prior density at t_k is identical to the posterior density at t_{k-1}

$$p(s_k|\mathbf{x}_x, \mathbf{Z}^{k-1}) = \mathcal{I}(s_k; \hat{s}_{k-1}, \mu_{k-1}) \tag{6.43}$$

with

$$\hat{s}_{k-1} \equiv s_{k|k-1} = s_{k-1|k-1} \qquad \text{and} \qquad \mu_{k-1} \equiv \mu_{k|k-1} = \mu_{k-1|k-1}. \tag{6.44}$$

6.3.4.2 Combined Likelihood. The combined likelihood function $p(\mathbf{Z}_k|\hat{\mathbf{x}}_k) = p(\mathbf{Z}_k|\mathbf{x}_k, s_k)$ is the probability density of the measurements and comprises all possibilities how the given sensor output $\mathbf{Z}_k$ can be interpreted, given the true target state $\hat{\mathbf{x}}_k$. Assuming independent, identically distributed false alarm measurements with the number of false alarms determined by the Poisson distribution, the likelihood function, up to a factor which does not depend on $\hat{\mathbf{x}}_k$, can be written as

$$p(\mathbf{Z}_k|\hat{\mathbf{x}}_k) \propto (1 - \mathbf{P}_D(s_k))\rho_F + \mathbf{P}_D(s_k) \sum_{i=1}^{m_k} \mathcal{N}(\mathbf{z}_k^i; \mathbf{H}_k\mathbf{x}_k, \mathbf{R}_k)\, \mathcal{LR}_S^i, \tag{6.45}$$

where ρ_F is the false alarm density and $\mathcal{N}(\mathbf{z}_k^i; \mathbf{H}_k\mathbf{x}_k, \mathbf{R}_k)$ is the normally distributed single measurement likelihood, which results from a linear measurement model with additive white Gaussian noise: $\mathbf{z}_k^i = \mathbf{H}_k\mathbf{x}_k + \mathbf{v}_k, \mathbf{v}_k \sim \mathcal{N}(\mathbf{v}_k; \mathbf{0}, \mathbf{R}_k)$ with measurement matrix $\mathbf{H}_k$ and measurement covariance $\mathbf{R}_k$. The signal strength likelihood ratio $\mathcal{LR}_S^i$ in (6.45) is given by

$$\mathcal{LR}_S^i = \frac{p(\kappa_k^i|i, D, \hat{\mathbf{x}}_k)}{p(\kappa_k^i|\neg D, \hat{\mathbf{x}}_k)}, \tag{6.46}$$

where $(D, \neg D)$ describe detection and missed detection, respectively. $p(\kappa_k^i|\neg D, \hat{\mathbf{x}}_k)$ is given by (6.16) with the clutter strength-to-noise ratio c_k and $p(\kappa_k^i|i, D, \hat{\mathbf{x}}_k)$ is distributed according to either (6.14) or (6.15). In the latter case, the detection probability is approximated by the Swerling-I case, that is, $P_D^{\mathbf{III}} \approx P_D^{\mathbf{I}}$. Finally, the combined likelihood can be calculated, yielding

$$\textit{Swerling-I:} \quad p(\mathbf{Z}_k|\hat{\mathbf{x}}_k) \propto \left(1 - e^{-\frac{\lambda}{1+s_k}}\right)\rho_F + \sum_{i=1}^{m_k} \alpha_{c_k}^i \frac{e^{-\kappa_k^i/(1+s_k)}}{1+s_k} \mathcal{N}(\mathbf{z}_k^i; \mathbf{H}_k\mathbf{x}_k, \mathbf{R}_k) \tag{6.47}$$

$$\textit{Swerling-III:} \quad p(\mathbf{Z}_k|\hat{\mathbf{x}}_k) \propto \left(1 - e^{-\frac{\lambda}{1+s_k}}\right)\rho_F + \sum_{i=1}^{m_k} \left[4\alpha_{c_k}^i \frac{e^{-2\kappa_k^i/(2+s_k)}}{(2+s_k)^2} \cdot \left(1 + \frac{\kappa_k^i s_k}{2+s_k}\right) \mathcal{N}(\mathbf{z}_k^i; \mathbf{H}_k\mathbf{x}_k, \mathbf{R}_k)\right] \tag{6.48}$$

with

$$\alpha_{c_k}^i = (1 + c_k)\, e^{\frac{\kappa_k^i - \lambda}{1+c_k}}. \tag{6.49}$$

6.3.4.3 Filter Update Step. The posterior density $p(\hat{\mathbf{x}}_k|\mathbf{Z}^k) = p(\mathbf{x}_k, s_k|\mathbf{Z}^k)$ is derived by inserting the prediction and likelihood functions into (6.38), assuming strong targets, that is, $1 + s_k \approx 2 + s_k \approx s_k$, and making use of the product formula for normal distributions.[2] The result is a weighted sum of the product of a Gaussian with an inverse Gamma density, describing the target's kinematic state and signal strength, respectively

$$p(\hat{\mathbf{x}}_k|\mathbf{Z}^k) = \sum_{i=0}^{m_k} w_k^i\, \mathcal{N}(\mathbf{x}_k; \mathbf{x}_{k|k}^i, \mathbf{P}_{k|k}^i)\, \mathcal{I}(s_k; s_{k|k}^i, \mu_{k|k}^i), \tag{6.50}$$

where the estimates $\mathbf{x}_{k|k}^i$ and covariances $\mathbf{P}_{k|k}^i$ are calculated by the known Kalman filter update equations, and weights w_k^i, signal strength estimates $s_{k|k}^i$ and $\mu_{k|k}^i$ are given by (for $i > 0$)

Swerling-I:

$$\tilde{w}_k^i = \frac{\alpha_{c_k}^i}{\rho_F} \mathcal{N}(\mathbf{z}_k^i; \mathbf{H}_k \mathbf{x}_{k|k-1}, \mathbf{S}_k^i) \frac{\mu_{k-1}}{(\mu_{k-1} - 1)\hat{s}_{k-1}} \cdot \left[\frac{(\mu_{k-1} - 1)\hat{s}_{k-1}}{(\mu_{k-1} - 1)\hat{s}_{k-1} + \kappa_k^i} \right]^{\mu_{k-1}+1} \tag{6.51}$$

$$s_{k|k}^i = \frac{\mu_{k-1} - 1}{\mu_{k-1}} \hat{s}_{k-1} + \frac{\kappa_k^i}{\mu_{k-1}} \tag{6.52}$$

$$\mu_{k|k}^i = \mu_{k-1} + 1 \tag{6.53}$$

Swerling-III:

$$\tilde{w}_k^i = \frac{4\alpha_{c_k}^i}{\rho_F} \mathcal{N}(\mathbf{z}_k^i; \mathbf{H}_k \mathbf{x}_{k|k-1}, \mathbf{S}_k^i) \frac{(1 + \kappa_k^i)\mu_{k-1}(\mu_{k-1} + 1)}{[(\mu_{k-1} - 1)\hat{s}_{k-1}]^2} \cdot \left[\frac{(\mu_{k-1} - 1)\hat{s}_{k-1}}{(\mu_{k-1} - 1)\hat{s}_{k-1} + 2\kappa_k^i} \right]^{\mu_{k-1}+2} \tag{6.54}$$

$$s_{k|k}^i = \frac{\mu_{k-1} - 1}{\mu_{k-1} + 1} \hat{s}_{k-1} + \frac{2\kappa_k^i}{\mu_{k-1} + 1} \tag{6.55}$$

$$\mu_{k|k}^i = \mu_{k-1} + 2 \tag{6.56}$$

[2] $\mathcal{N}(\mathbf{z}; \mathbf{H}\,\mathbf{x}, \mathbf{R})\mathcal{N}(\mathbf{x}; \mathbf{y}, \mathbf{P}) = \mathcal{N}(\mathbf{z}; \mathbf{H}\,\mathbf{y}, \mathbf{S})\mathcal{N}(\mathbf{x}; \mathbf{y} + \mathbf{W}(\mathbf{z} - \mathbf{H}\,\mathbf{y}), \mathbf{P} - \mathbf{W}\mathbf{S}\mathbf{W}^T)$ with $\mathbf{S} = \mathbf{H}\mathbf{P}\mathbf{H}^T + \mathbf{R}$ and $\mathbf{W} = \mathbf{P}\mathbf{H}^T\mathbf{S}^{-1}$.

and for the missed detection hypothesis ($i = 0$)

$$\tilde{w}_k^0 = 1 - \left[\frac{\hat{s}_{k-1}}{\hat{s}_{k-1} + \frac{\lambda}{\mu_{k-1}-1}} \right]^{\mu_{k-1}} \tag{6.57}$$

$$s_{k|k}^0 = \frac{1}{\tilde{w}_k^0} \left(\hat{s}_{k-1} - \left[\frac{\hat{s}_{k-1}}{\hat{s}_{k-1} + \frac{\lambda}{\mu_{k-1}-1}} \right]^{\mu_{k-1}} s_{k|k}^* \right) \tag{6.58}$$

$$\mu_{k|k}^0 = \frac{(s_{k|k}^0)^2}{\mathrm{Var}[s_{k|k}^0]} + 2 \tag{6.59}$$

with

$$w_k^i = \frac{\tilde{w}_k^i}{\sum_{j=0}^{n_k} \tilde{w}_k^i} \tag{6.60}$$

$$\alpha_{c_k}^i = (1 + c_k)\, e^{\frac{\kappa_k^i - \lambda}{1+c_k}} \tag{6.61}$$

$$s_{k|k}^* = \hat{s}_{k-1} + \frac{\lambda}{\mu_{k-1} - 1} \tag{6.62}$$

$$\begin{aligned} \mathrm{Var}[s_{k|k}^0] = {} & \frac{1}{\tilde{w}_k^0} \left[\left(\frac{(\hat{s}_{k-1})^2}{\mu_{k-1} - 2} \right) + \left(\hat{s}_{k-1} - s_{k|k}^0 \right)^2 \right] \\ & - \frac{1}{\tilde{w}_k^0} \left(\frac{\hat{s}_{k-1}}{\hat{s}_{k-1} + \frac{\lambda}{\mu_{k-1}-1}} \right)^{\mu_{k-1}} \cdot \left[\left(\frac{(s_{k|k}^*)^2}{\mu_{k-1} - 2} \right) + \left(s_{k|k}^* - s_{k|k}^0 \right)^2 \right]. \end{aligned} \tag{6.63}$$

The update equations for $s_{k|k}^i$ and $\mu_{k|k}^i$ exhibit a reasonable behavior: The expectation of the prior density $\hat{s}_{k-1}$ is modified by the measurement κ_k^i and the variance $\mathrm{Var}[s_{k|k}^i] = (s_{k|k}^i)^2/(\mu_{k|k}^i - 2)$ decreases with time as $\mu_{k|k}^i$ increases. Due to the factor $\left(1 - e^{-\lambda/(1+s_k)}\right)$ in (6.47) and (6.48), the missed detection hypothesis splits up into two components, which are processed by second-order moment matching [16] resulting in a single merged component with $\tilde{w}_k^0$, $s_{k|k}^0$, and $\mu_{k|k}^0$ as given above. In the last part of this section, the presented update scheme is implemented into the JPDAF tracking algorithm [17] and applied to a two-target scenario.

6.3.4.4 JPDAF Implementation. In the JPDAF algorithm, the complete set of possible global hypotheses is processed, that is, all possible combinations to associate measurements with tracks including missed detections and false alarms are considered. Thus it avoids the association of a certain measurement with more than one track, which is a shortcoming of single-target tracking algorithms, for example, the PDAF [16]. The individual association probabilities for each target with its relevant measurements are calculated based on the probabilities of the global hypotheses. Therefore, the calculation of the probability of the global hypotheses needs to be adapted to account for the weight factors due to the signal strength incorporation: In the original formulation of the JPDAF, the probability of the global hypothesis θ_i is given by

$$P(\theta_i|\mathbf{Z}^k) = \prod_{i=1}^{m_k} \left\{ \frac{1}{\rho_F} \mathcal{N}(\mathbf{z}_k^i; \mathbf{H}_k \mathbf{x}_{k|k-1}^{t_i}, \mathbf{S}_k^{t_i}) \right\}^{\tau_i} \cdot \prod_{t=1}^{T} \mathbf{P}_D^{\delta_t} (1 - \mathbf{P}_D)^{1-\delta_t}, \tag{6.64}$$

where $\tau_i \in \{0, 1\}$ indicates if measurement i is associated with target t_i, and $\delta_t \in \{0, 1\}$ indicates if target t is associated with a measurement. T is the number of targets. First of all, $\mathbf{P}_D^{\delta_t}$ can be shifted into the first product (dropping the index δ_t). Now the factor

$$\frac{\mathbf{P}_D}{\rho_F} \mathcal{N}\left(\mathbf{z}_k^i; \mathbf{H}_k \mathbf{x}_{k|k-1}^{t_i}, \mathbf{S}_k^{t_i}\right) \tag{6.65}$$

is replaced by (6.51) for Swerling-I and (6.54) for Swerling-III. In addition, the probability factor for the missed detection hypothesis, $1 - \mathbf{P}_D$, is replaced by (6.57). With these adjustments, $P(\theta_i|\mathbf{Z}^k)$ can be written as

$$\textit{Swerling-I:} \quad P(\theta_i|\mathbf{Z}^k) = \prod_{i=1}^{m_k} \left\{ \frac{\alpha_{c_k}^i}{\rho_F} \frac{\mu_{k-1}}{(\mu_{k-1} - 1)\hat{s}_{k-1}} \cdot \left[\frac{(\mu_{k-1} - 1)\hat{s}_{k-1}}{(\mu_{k-1} - 1)\hat{\mathbf{s}}_{k-1} + \kappa_k^i} \right]^{\mu_{k-1}+1} \right.$$
$$\left. \cdot \mathcal{N}(\mathbf{z}_k^i; \mathbf{H}_k \mathbf{x}_{k|k-1}^{t_i}, \mathbf{S}_k^{t_i}) \right\}^{\tau_i} \cdot \prod_{t=1}^{T} \left(1 - \left[\frac{\hat{s}_{k-1}}{\hat{s}_{k-1} + \frac{\lambda}{\mu_{k-1}-1}} \right]^{\mu_{k-1}} \right)^{1-\delta_t} \tag{6.66}$$

$$\textit{Swerling-III:} \quad P(\theta_i|\mathbf{Z}^k) = \prod_{i=1}^{m_k} \left\{ \frac{4\alpha_{c_k}^i}{\rho_F} \frac{(1 + \kappa_k^i)\mu_{k-1}(\mu_{k-1} + 1)}{[(\mu_{k-1} - 1)\hat{s}_{k-1}]^2} \cdot \left[\frac{(\mu_{k-1} - 1)\hat{s}_{k-1}}{(\mu_{k-1} - 1)\hat{s}_{k-1} + 2\kappa_k^i} \right]^{\mu_{k-1}+2} \right.$$
$$\left. \cdot \mathcal{N}\left(\mathbf{z}_k^i; \mathbf{H}_k \mathbf{x}_{k|k-1}^{t_i}, \mathbf{S}_k^{t_i}\right) \right\}^{\tau_i} \cdot \prod_{t=1}^{T} \left(1 - \left[\frac{\hat{s}_{k-1}}{\hat{s}_{k-1} + \frac{\lambda}{\mu_{k-1}-1}} \right]^{\mu_{k-1}} \right)^{1-\delta_t}. \tag{6.67}$$

From that, the individual association probabilities for each target are calculated in the same way as in the original JPDAF formulation.

6.3.4.5 Numerical Results. In the considered simulation scenario, see Figure 6.5, the two targets move along the same trajectory for a considerable period of time, leading to a loss of identity in case of a traditional tracking algorithm. Here, the objective is to determine the capability of target discrimination with signal strength information in the final stage of the scenario. Based on the JPDAF algorithm, the results are plotted in Figure 6.6 for Swerling-I and Swerling-III fluctuations of the two targets. Shown is the probability for correct (solid lines) and incorrect (dashed lines) association of tracks with true targets at the final revisit for different signal strength combinations, based on the two fluctuation models. The signal strength of

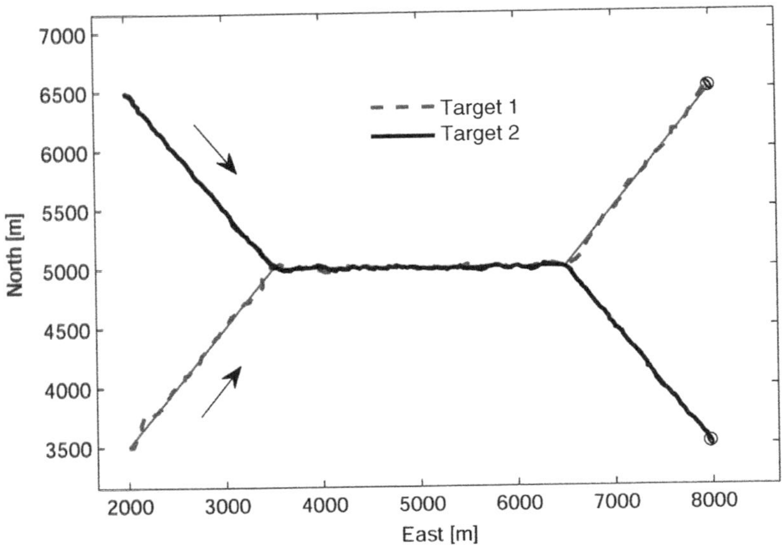

Figure 6.5 Trajectories of the two-target scenario. Shown is the snapshot at the final revisit.

one target is fix, corresponding to the minimum value of each line. The upper plots correspond to target tracking without signal strength information. In this case, the signal strength only affects the true detection probability and thus the occurrence of target detections. As expected, for all combinations of the two targets' signal strength, the probability for correct association amounts to 50 %. The lower plots illustrate the advantage of taking signal strength information into account: A clear performance gain in target discrimination is visible for both fluctuation models, with a stronger separation capability for larger differences in signal strength. Comparing these results for the two Swerling cases, apparently the discrimination performance of the algorithm is slightly higher in case of Swerling-III. This is probably due to the distinct peak in the probability density of Swerling-III fluctuations, see Figure 6.1, leading to a more obvious discrimination of the two signal strength distributions. As the results are solely based on simulation, the presented approach needs to be validated by experimental data.

6.4 EXPLOITATION OF ROAD NETWORK DATA

In many ground surveillance applications, vehicles on roads are the most important objects. This is true, for example, for traffic monitoring as well as for military reconnaissance. Digital road maps are available for many areas in a good quality, and can be exploited to enhance the tracking performance in terms of track precision and track continuity. In situation where road maps are unavailable, imprecise, or not up-to-date, tracking can, conversely, help to improve the road data base. In this section, it is shown how digital road maps can be integrated into the Bayesian tracking scheme and how

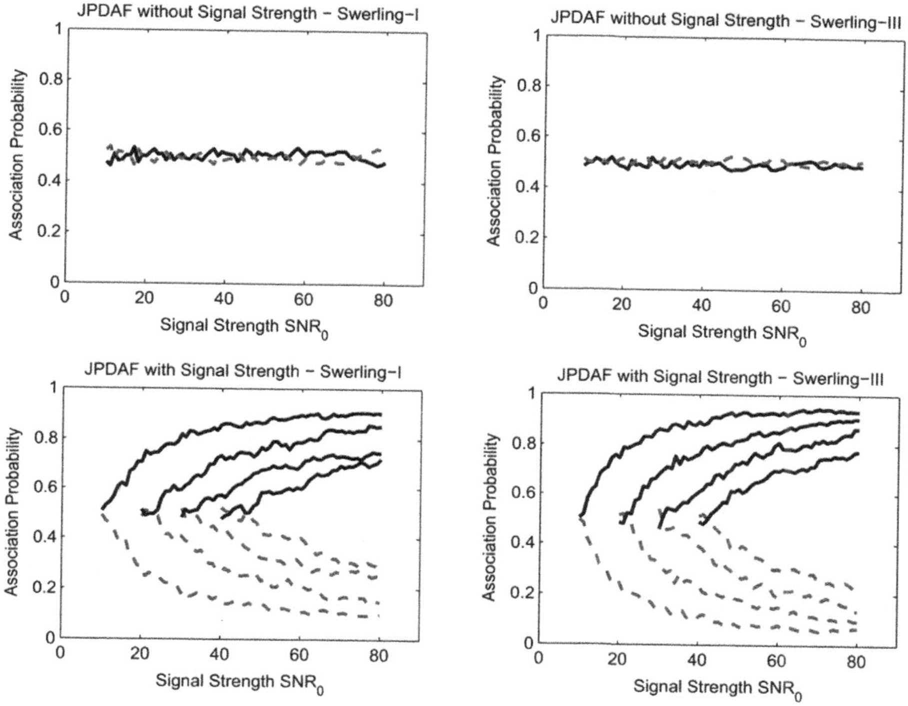

Figure 6.6 Probability for correct (solid lines) and incorrect (dashed lines) association of tracks with true targets at final revisit for different signal strength combinations, based on fluctuation model Swerling-I ((left) and Swerling-III (right). The signal strength of one target is fixed, corresponding to the minimum signal strength value of each distribution. Shown are the results for JPDAF without (top) and with (bottom) exploitation of signal strength information. The signal strength is specified in a linear scale.

they contribute to an improved tracking. We will use a Gaussian sum approach, which can seamlessly be integrated into standard Kalman filter-based tracking algorithms [10, 12, 13]. Other approaches discussed in the literature are based on the Variable Structure Interacting Multiple Model (VS-IMM) estimator [18, 19] or on particle filters [13, 20, 21].

6.4.1 Modeling of Road Networks

A given road is mathematically described by a continuous 3D curve $\mathcal{R}^*$ in Cartesian ground coordinates. Let $\mathcal{R}^*$ be parameterized by the corresponding arc length l: $\mathcal{R}^* : l \mapsto \mathcal{R}^*(l)$. In a digitized road map, $\mathcal{R}^*$ is approximated by a polygonal curve $\mathcal{R}$ defined by piecewise linear segments (see Figure 6.7). The curve $\mathcal{R}$, consisting of n_r road segments, may be characterized by $n_r + 1$ node vectors

$$\mathbf{s}_m = \mathcal{R}^*(l_m), \quad m = 1, \ldots, n_r + 1 . \tag{6.68}$$

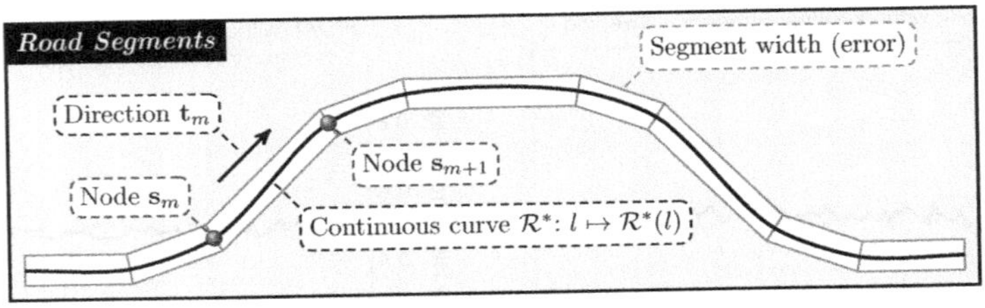

Figure 6.7 Schematic representation of a road.

From these quantities, n_r normalized tangential vectors

$$\mathbf{t}_m = \frac{\mathbf{s}_{m+1} - \mathbf{s}_m}{|\mathbf{s}_{m+1} - \mathbf{s}_m|}, \qquad m = 1, \ldots, n_r \tag{6.69}$$

can be derived. The Euclidian distance $|\mathbf{s}_{m+1} - \mathbf{s}_m|$ between two adjacent node vectors, however, is usually not identical to the distance $\lambda_m = l_{m+1} - l_m$ actually covered by a vehicle when it moves from $\mathbf{s}_m$ to $\mathbf{s}_{m+1}$ along the road. Besides the vectors $\mathbf{s}_m$, the scalar quantities $\lambda_m \geq |\mathbf{s}_{m+1} - \mathbf{s}_m|$ should therefore enter into the road model to make it more realistic. The differences $\lambda_m - |\mathbf{s}_{m+1} - \mathbf{s}_m|$ can evidently serve as a quantitative measure of the discretization errors. Using the indicator function defined by

$$\chi_m(l) = \begin{cases} 1 \text{ for } l \in [l_m, l_{m+1}) \\ 0 \qquad \text{otherwise} \end{cases} \tag{6.70}$$

with $m = 1, \ldots, n_r$, we obtain a mathematically simple description of the polygonal curve $\mathcal{R}$, by which the road $\mathcal{R}^*$ is approximated

$$\mathcal{R} : l \in [l_1, l_{n_r+1}) \mapsto \mathcal{R}(l) = \sum_{m=1}^{n_r} \left[\mathbf{s}_m + (l - l_m)\mathbf{t}_m\right] \chi_m(l)$$
$$\text{with} \quad \mathcal{R}^*(l_m) = \mathcal{R}(l_m) = \mathbf{s}_m, \quad m = 1, \ldots, n_r + 1\,. \tag{6.71}$$

The accuracy of the road can be described by a covariance matrix $\mathbf{R}_m$, that is characteristic for each node vector $\mathbf{s}_m$, which also accounts for the above-mentioned discretization error. It is assumed that the road accuracy is given for the digital road map at hand. Thus, each segment m of the polygonal road $\mathcal{R}$ is determined by the node vector $\mathbf{s}_m$ and its corresponding covariance matrix $\mathbf{R}_m$, the arc length $\lambda_m = l_{m+1} - l_m$, and the normalized tangential vector $\mathbf{t}_m$ (see Figure 6.7).

In the case of a complex road network, additional information needs to be considered, as every node vector can now have more than two neighboring nodes in the case of crossings and junctions. Therefore, a trivial association between node vectors by mere numbering—as in the case of a single road—is no longer possible. In order to deal with a complex road network consisting of a large number of nodes, it is necessary to store the knowledge of all possible associations between nodes available in, for example, a look-up table, in which for each node its location, all neighboring node

IDs, and the relative distances are saved. From this, a *local road* within the road network can be easily constructed, which is then again described by the above-mentioned quantities $\mathbf{s}_m$, $\mathbf{R}_m$, λ_m, and $\mathbf{t}_m$ for each segment.

6.4.2 Densities on Roads

When targets move on road, it seems reasonable to describe the kinematic state vector $\mathbf{x}_k^r$ of road targets at time t_k by its position on the road l_k (i.e., the arc length of the curve), and its scalar speed $\dot{l}_k$: $\mathbf{x}_k^r = [l_k, \dot{l}_k]^T$. The model for describing the dynamical behavior of road targets is therefore a 2D version of the equations presented in Section 6.3.2. We will not consider the problem of targets going on- and off-road. A possible approach for this is the use of a variable structure interaction multiple model VS-IMM scheme [18]. Alternatively, one may switch between on- and off-road tracking by a heuristic decision based on the track's proximity to a road and the alignment of the heading vector to the road direction. We will also not discuss the problem of evasive move-stop-move behavior as discussed, for example, in [22].

By making use of the related transition density $p(\mathbf{x}_k^r|\mathbf{x}_{k-1}^r)$, the predicted density in road coordinates is given by

$$p(\mathbf{x}_k^r|\mathbf{Z}^{k-1}) = \int p(\mathbf{x}_k^r|\mathbf{x}_{k-1}^r)\, p(\mathbf{x}_{k-1}^r|\mathbf{Z}^{k-1})\, \mathrm{d}\mathbf{x}_{k-1}^r. \tag{6.72}$$

Now we face the problem, that the target dynamics is given in road coordinates, while the filter update needs to be performed in Cartesian ground coordinates due to the measurements. In principle, the Bayesian formalism discussed in Section 6.3 can be applied to road targets if there exists a transformation operator $\mathcal{T}_{g\leftarrow r}$, which transforms the predicted density $p(\mathbf{x}_k^r|\mathbf{Z}^{k-1})$ from road to ground coordinates

$$\underbrace{p(\mathbf{x}_k^r|\mathbf{Z}^{k-1})}_{\text{in road coordinates}} \xrightarrow[\text{road-map errors}]{\text{road network}} \underbrace{p(\mathbf{x}_k^g|\mathbf{Z}^{k-1})}_{\text{in ground coordinates}} \ . \tag{6.73}$$

In general, such a transformation is highly nonlinear and the structure of probability densities in terms of Gaussian sums cannot be preserved. Linearity is, however, conserved if one employs linearized road segments for the mapping between road and ground coordinates, as in (6.71)

$$p(\mathbf{x}_k^g|\mathbf{Z}^{k-1}) = \int p(\mathbf{x}_k^g|\mathbf{x}_k^r)\, p(\mathbf{x}_k^r|\mathbf{Z}^{k-1})\, \mathrm{d}\mathbf{x}_k^r = \sum_{m=1}^{n_r} \int p(\mathbf{x}_k^g|\mathbf{x}_k^r, m)\, p(\mathbf{x}_k^r, m|\mathbf{Z}^{k-1})\, \mathrm{d}\mathbf{x}_k^r. \tag{6.74}$$

The density $p(\mathbf{x}_k^g|\mathbf{x}_k^r, m)$ describes the mapping from road to ground coordinates and contains the following affine transformation

$$\mathbf{t}_{g\leftarrow r}^m[\mathbf{x}_k^r] = \mathbf{T}_{g\leftarrow r}^m \mathbf{x}_k^r + \mathbf{s}_{g\leftarrow r}^m, \qquad \mathbf{T}_{g\leftarrow r}^m = \begin{bmatrix} \mathbf{t}_m & \mathbf{0} \\ \mathbf{0} & \mathbf{t}_m \end{bmatrix}, \qquad \mathbf{s}_{g\leftarrow r}^m = \begin{bmatrix} \mathbf{s}_m - l_m \mathbf{t}_m \\ \mathbf{0} \end{bmatrix}. \tag{6.75}$$

After some algebra [13], the density in ground coordinates $p(\mathbf{x}_k^g|\mathbf{Z}^{k-1})$ can be written as a sum over the road segments

$$p(\mathbf{x}_k^g|\mathbf{Z}^{k-1}) = \sum_{m=1}^{n_r} P(m|\mathbf{Z}^{k-1})\; p(\mathbf{x}_k^g|m, \mathbf{Z}^{k-1}). \tag{6.76}$$

Here, $P(m|\mathbf{Z}^{k-1})$ denotes the probability that the target moves on segment m, given the accumulated sensor data $\mathbf{Z}^{k-1}$. Its explicit form is given in [12, 13]. The density $p(\mathbf{x}_k^g|m, \mathbf{Z}^{k-1})$ can be calculated from the probability density in road coordinates and is approximately given by Gaussians. The filter update step recalculates the weight of each road segment, $P(m|\mathbf{Z}^k)$, and yields the segment-dependent densities $p(\mathbf{x}_k^g|m, \mathbf{Z}^k)$. In order to close the iterative loop with the calculation of the posterior density $p(\mathbf{x}_k^r|\mathbf{Z}^k)$, the densities in ground coordinates need to be transformed back into the road coordinate system

$$\underbrace{p(\mathbf{x}_k^g|m, \mathbf{Z}^k)}_{\text{in ground coordinates}} \xrightarrow{\text{road network}} \underbrace{p(\mathbf{x}_k^r|m, \mathbf{Z}^k)}_{\text{in road coordinates}}\;. \tag{6.77}$$

The inverse transform from Cartesian to road coordinates is simply provided by individually projecting the densities $p(\mathbf{x}_k^g|m, \mathbf{Z}^k)$ on the road, resulting in

$$p(\mathbf{x}_k^r|\mathbf{Z}^k) = \sum_{m=1}^{n_r} P(m|\mathbf{Z}^k)\; p(\mathbf{x}_k^r|m, \mathbf{Z}^k). \tag{6.78}$$

The posterior density in road coordinates is, hence, a normal mixture of segment-dependent probability densities. Before the subsequent prediction is performed (as depicted in Figure 6.8), it seems reasonable to apply a second-order approximation to the mixture densities

$$p(\mathbf{x}_k^r|\mathbf{Z}^k) \approx \mathcal{N}\big(\mathbf{x}_k^r;\, \mathbf{x}_{k|k}^r,\, \mathbf{P}_{k|k}^r\big). \tag{6.79}$$

One iteration of the iterative updating scheme is depicted in Figure 6.9. In the case of a road network, it is reasonable to introduce a local road for each target, which consists

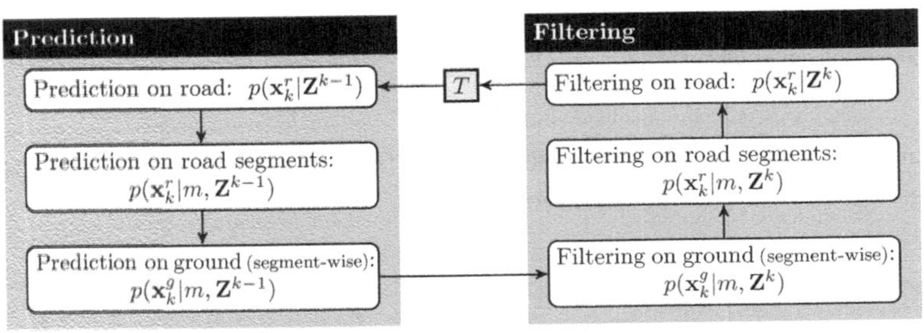

Figure 6.8 Iterative updating scheme for road-map assisted target tracking [12, 13].

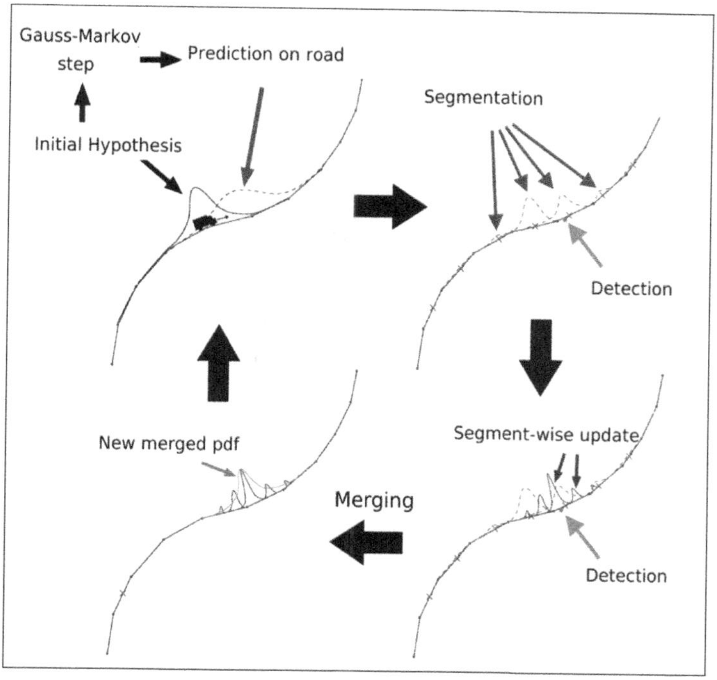

Figure 6.9 Road-map assisted target tracking schema. The pdf is predicted in road coordinates, mapped onto linear road segments, transformed to Cartesian coordinates, updated by measurements, mapped back to road segments, and finally merged into a single Gaussian pdf [10, 12, 13] (thanks to Jost Koller).

of only a limited number of segments. Depending on the target's movement, this local road needs to be adjusted continuously: Segments with vanishing probability at the tail of the road are pruned, new segments are added as the target approaches the head of the road. In addition, the orientation of the local road needs to be reversed if the target moves backwards on the road segments. In general, the target trajectory will include junctions and crossings of the road network. In such a situation, the ambiguity is resolved over time and can be described by a multiple-model approach: Every possible path at the junction or crossing leads to different road hypotheses i with $i = 1, \ldots, N_r$, each having a different continuation of the previous local road. The sum of weights $P(m|i, \mathbf{Z}^k)$ over the segments of a given road hypothesis i contributes to the road probability P_k^i at time t_k with

$$P_k^i = \frac{P_{k-1}^i \sum_{m=1}^{n_r} P(m|i, \mathbf{Z}^k)}{\sum_{j=1}^{N_r} P_{k-1}^j \sum_{m=1}^{n_r} P(m|j, \mathbf{Z}^k)} \,. \tag{6.80}$$

In Figure 6.10, a simple simulation scenario is considered in which a single target moves on a road network and passes a crossing (upper plot). The lower plot shows the calculated road probability P_k^i as a function of the revisit interval, based on the

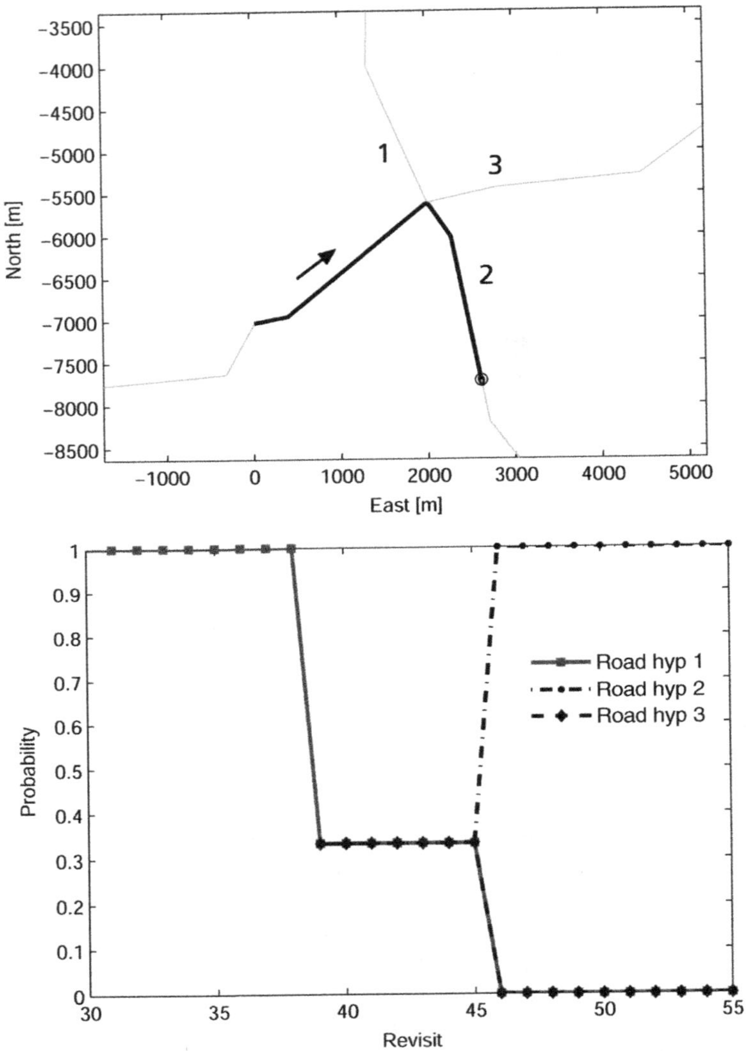

Figure 6.10 The upper plot illustrates the target trajectory (black solid line) and the road network (gray solid lines). The given numbers correspond to all possible paths after the crossing. The lower plot shows the calculated road probability P_k^i as a function of the revisit interval.

considerations above: In this scenario, the local road is described by a single road hypothesis at first. As soon as the target approaches the crossing, all possible paths after the crossing (1, 2, and 3) are considered by extending the local road of road hypothesis 1 with the associated segments in each direction. For the following seven revisits, all road probabilities are assumed equal (1/3) because the target estimate is located on a segment that is identical for the local road of all road hypotheses. The transition probabilities can be chosen differently if, for example, one road were a main highway, while the intersecting roads were secondary roads. After the crossing, the weight of the correct road hypothesis quickly dominates.

6.4.3 Application: Precision Targeting

In the following, an illustrative example for the beneficial interplay between the knowledge on the clutter notch and road-map information is briefly discussed: An easy maneuver for a ground target to prevent from being detected by a GMTI radar is to come to a stop. Such a situation is depicted in Figure 6.11. In this scenario, the target moves with a maximum speed of 15 m/s along the road, starting at the bottom. After 100 s, the target decelerates and stops. After 250 s, it accelerates again and finally moves with a constant speed of 15 m/s. The sensor is located at a fixed position in northwestern direction. Shown are the track results for three different tracking algorithms: The track of the algorithm without any extra knowledge quickly deviates from the correct target position (lower plot) as soon as the target stops by picking up false measurements. The second algorithm processes the knowledge on the clutter notch. In this case, the track is confined in cross-range direction (center plot) as there is an additional hypothesis available to interpret the sensor output, namely a target with range rate below the minimum detectable velocity of the sensor. Coincidentally, the track comes back to the correct target location as soon as the target starts moving again and new measurements are available. Finally, the additional road-map information further constrains the track in road direction and, therefore, after a few successive missed detections, a stopping target is quickly detected. The estimate remains close to the tracked stopping position until new measurements arrive, which are then picked up immediately (upper plot). In general, the interplay between clutter notch and road-map processing is optimal when the road is oriented in range direction.

6.4.4 Track-based Road-Map Extraction

In many operationally relevant scenarios, reliable digital road maps are not available or, at least, are not precise or not up-to-date. In addition, there exist fields of applications in which roads or road-like lines of communication exist only temporarily or may change over time. This can be the case in deserts or in times of a conflict. As practical evidence shows, even in typical off-road scenarios, structures similar to roads quickly evolve as vehicles usually move in the tracks of their precursor. This is especially true in a not sufficiently explored or a dangerous environment (e.g., in a mine field). In such situations, the ground vehicles moving on road networks being observed by wide-area sensors, such as GMTI radar, produce large data streams that can also be used for road-map extraction. After a suitable post-processing described below, the GMTI tracks of road targets define the corresponding road segments currently being used by the ground moving targets.

In a given sensor data fusion application, it cannot always be taken for granted that the data originating from various distributed sensors can accurately be transformed into a common coordinate system. Precisely extracted road maps with reference to the

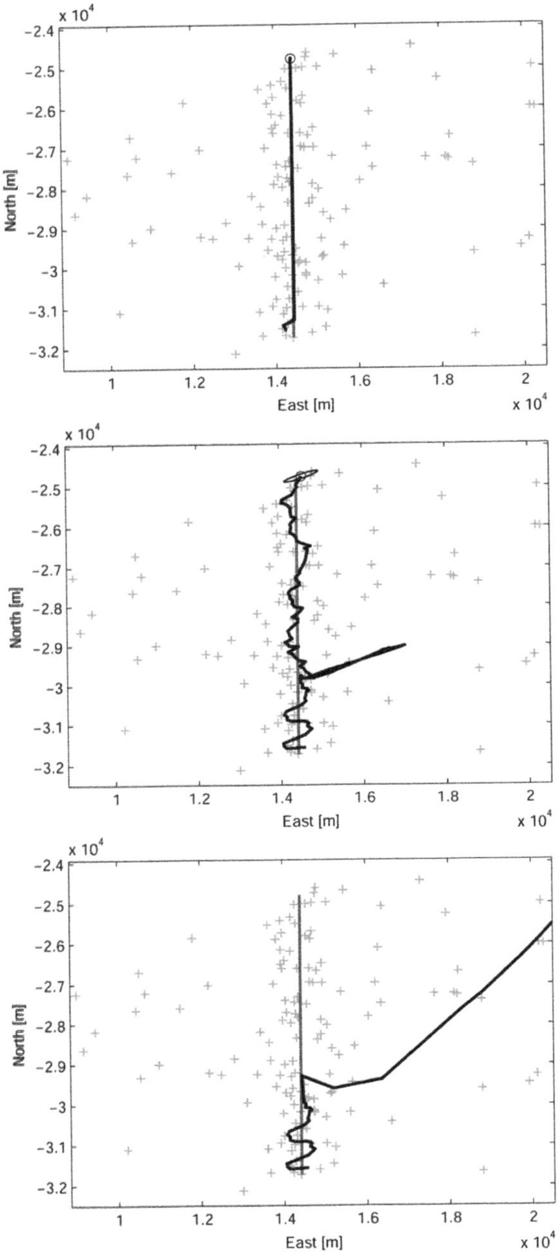

Figure 6.11 Track results (black solid lines) for a road target performing a move–stop–move maneuver: MHT with road map and clutter notch processing (upper plot), MHT with clutter notch processing (center plot), and pure MHT (lower plot). Crosses correspond to sensor measurements. The sensor is located in northwestern direction.

coordinate system of the individual sensors can be matched with each other using road-map features such as characteristic curves or crossings. By this means, a compensation of *relative* bias errors due to imprecise sensor registration or misalignment can be achieved. *Absolute* bias errors may be reduced by matching tracking-generated road maps with geo-referenced maps stored in a topographical data base.

According to the road modeling in Section 6.4.1, the accuracy by which the road is represented by the node vectors $\mathbf{s}_m$ can be described by a covariance matrix $\mathbf{R}_m$ characteristic of each node m. Typically these quantities, which are to be used in road-map assisted tracking, are not readily available as part of a geographical information database. Yet, in the case of tracking-generated road maps, reliable estimates for these quantities do exist: they are given by the position track of a target moving on road. In detail, first estimates of road segment points are defined by the collection of track position estimates and their corresponding covariance matrices.

6.4.4.1 Continuous Time Retrodiction. Due to low sensor update rates, missing detections, fading phenomena, Doppler blindness and so on, however, the accuracy and sample density of such tracking-generated road maps may be insufficient. In applications, we therefore wish to produce a suitable interpolation between adjacent *node vectors* and the related *mapping error* covariance matrices. This interpolation should take full advantage of the available knowledge of the targets' kinematic state vector and the related track accuracy as well as of background information on the vehicle's behavior, that is, the target dynamics model.

Given two adjacent nodes vectors $\mathbf{r}_{l-1|k}$, $\mathbf{r}_{l|k}$ with their related accuracies $\mathbf{R}_{l-1|k}$, $\mathbf{R}_{l|k}$, we first have to decide whether it is reasonable to create an additional node at all. Obviously additional nodes are necessary if there are curves. Conversely, for a more or less rectilinear road segment only very few nodes are required. An intuitively clear indication for the existence of a winding road can be obtained as comparing the direction of the velocity vector estimates $\dot{\mathbf{r}}_{l|k}$ with the direction of the difference vector $\mathbf{r}_{l+1|k} - \mathbf{r}_{l|k}$. See Figure 6.12 for a characteristic example. The decision also depends on the quality of the velocity estimates. Let us denote by $\phi_{l|k}$ the angle

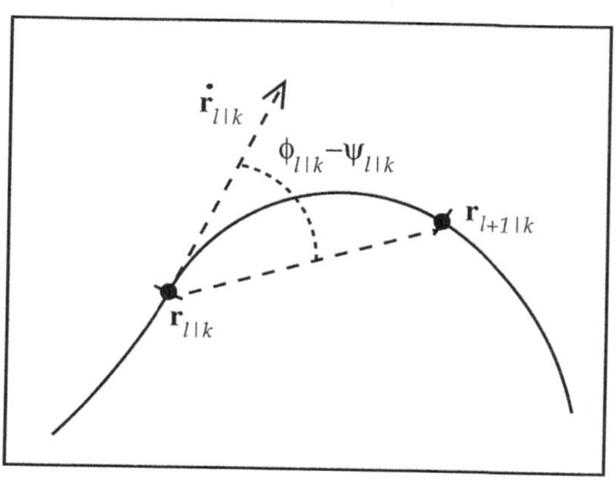

Figure 6.12 Adding a new vector node for continuous time retrodiction.

between the velocity estimate $\dot{\mathbf{r}}_{l|k}$ and one of the axes of the coordinate system. The corresponding angle for the actual velocity vector is a random variable approximately normally distributed with a variance given by $\Phi_{l|k}$. Let $\psi_{l|k}$ the corresponding angle of the difference vector $\mathbf{r}_{l+1|k} - \mathbf{r}_{l|k}$. An intuitively plausible decision criterion whether an additional node is to be introduced is thus given by

$$(\phi_{l|k} - \psi_{l|k})^2 / \Phi_{l|k} > \kappa^2. \tag{6.81}$$

If the inaccuracy of the heading estimates is large, subsequent headings are allowed to differ more than in the case of more precise estimates. It seems to be reasonable to choose the decision parameter around unity. Evidently, the estimate of the complete kinematic state vector enters into this criterion.

If an additional node is to be introduced, we consider the intermediate probability density, formally written as $p(\mathbf{x}_{l-\nu}|\mathbf{Z}^k)$ with $0 < \nu < 1$. This density expresses the available knowledge about the kinematic target state at an intermediate instant of time

$$t_{l-\nu} = t_l + \nu\,(t_{l-1} - t_l)\,, \tag{6.82}$$

where, for example, $t_{l-\nu} = (l - \nu)\Delta t$ for constant scan time Δt. From this density, an intermediate node vector and a tangential vector of the road can be derived. By considering several $\nu_1, \nu_2, \ldots$, an indication of the arc length λ of the road between the positions $\mathbf{r}_{l-1}$ and $\mathbf{r}_l$ can be obtained. Standard probability reasoning yields

$$p(\mathbf{x}_{l-\nu}|\mathbf{Z}^k) = \int p(\mathbf{x}_{l-\nu}|\mathbf{x}_l, \mathbf{Z}^k)\; p(\mathbf{x}_l|\mathbf{Z}^k)\,\mathrm{d}\mathbf{x}_l. \tag{6.83}$$

In this expression the pdf $p(\mathbf{x}_l|\mathbf{Z}^k)$ is already known and in many cases at least approximately given by

$$p(\mathbf{x}_l|\mathbf{Z}^k) = \mathcal{N}(\mathbf{x}_l;\; \mathbf{x}_{l|k},\; \mathbf{P}_{l|k}), \tag{6.84}$$

where $\mathbf{x}_{l|k}$ and $\mathbf{P}_{l|k}$ denote the expectation vector and the covariance matrix of a Gaussian pdf. For the remaining factor in the previous integrand we obtain according to Bayes' rule

$$p(\mathbf{x}_{l-\nu}|\mathbf{x}_l, \mathbf{Z}^k) = p(\mathbf{x}_{l-\nu}|\mathbf{x}_l, \mathbf{Z}^{l-1}) \tag{6.85}$$

$$= \frac{p(\mathbf{x}_l|\mathbf{x}_{l-\nu})\; p(\mathbf{x}_{l-\nu}|\mathbf{Z}^{l-1})}{\int p(\mathbf{x}_l|\mathbf{x}_{l-\nu})\; p(\mathbf{x}_{l-\nu}|\mathbf{Z}^{l-1})\,\mathrm{d}\mathbf{x}_{l-\nu}}. \tag{6.86}$$

According to (6.17), the pdf $p(\mathbf{x}_{l-\nu}|\mathbf{Z}^{l-1}) = \mathcal{N}(\mathbf{x}_{l-\nu};\, \mathbf{x}_{l-\nu|l-1},\, \mathbf{P}_{l-\nu|l-1})$ describes the knowledge available after data processing at time t_{l-1} and is also available. The transition density $p(\mathbf{x}_l|\mathbf{x}_{l-\nu})$ directly results from the target dynamics model (6.24) and is given by a Gaussian as well

$$p(\mathbf{x}_l|\mathbf{x}_{l-\nu}) = \mathcal{N}(\mathbf{x}_l;\, \mathbf{F}_{l|l-\nu}\mathbf{x}_{l-\nu},\, \mathbf{D}_{l|l-\nu}) \tag{6.87}$$

with $\mathbf{D}_{l|l-\nu} = \mathbf{G}_{l|l-\nu}\mathbf{G}_{l|l-\nu}^T$. Algebraic reasoning finally yields

$$p(\mathbf{x}_{l-\nu}|\mathbf{x}_l, \mathbf{Z}^k) = \mathcal{N}(\mathbf{x}_{l-\nu};\ \mathbf{x},\ \mathbf{P}) \tag{6.88}$$

$$\begin{aligned} \mathbf{x} &= \mathbf{x}_{l-\nu|l-1} + \mathbf{W}_{l|l-\nu}\big(\mathbf{x}_l - \mathbf{F}_{l|l-\nu}\mathbf{x}_{l-\nu|l-1}\big) \\ \mathbf{P} &= \mathbf{P}_{l-\nu|l-1} - \mathbf{W}_{l|l-\nu}\mathbf{P}_{l|l-1}\mathbf{W}_{l|l-\nu}^T \\ \mathbf{W}_{l|l-\nu} &= \mathbf{P}_{l-\nu|l-1}\mathbf{F}_{l|l-\nu}^T\mathbf{P}_{l|l-1}^{-1}\,. \end{aligned} \tag{6.89}$$

Insertion into (6.83) leads to a modified version of the well-known Rauch–Tung–Striebel formulas [5]

$$p(\mathbf{x}_{l-\nu}|\mathbf{Z}^k) = \mathcal{N}(\mathbf{x}_{l-\nu};\ \mathbf{x}_{l-\nu|k},\ \mathbf{P}_{l-\nu|k}) \tag{6.90}$$

$$\begin{aligned} \mathbf{x}_{l-\nu|k} &= \mathbf{x}_{l-\nu|l-1} + \mathbf{W}_{l|l-\nu}\big(\mathbf{x}_{l|k} - \mathbf{x}_{l|l-1}\big) \\ \mathbf{P}_{l-\nu|k} &= \mathbf{P}_{l-\nu|l-1} + \mathbf{W}_{l|l-\nu}(\mathbf{P}_{l|k} - \mathbf{P}_{l|l-1})\mathbf{W}_{l|l-\nu}^T \\ \mathbf{x}_{l|l-1} &= \mathbf{F}_{l|l-1}\mathbf{x}_{l-1|l-1} \\ \mathbf{P}_{l|l-1} &= \mathbf{F}_{l|l-1}\mathbf{P}_{l-1|l-1}\mathbf{F}_{l|l-1}^T + \mathbf{D}_{l|l-1}\,. \end{aligned} \tag{6.91}$$

The approach presented above has been applied to a simple, but realistic simulation scenario, including terrain obscuration and target stops (see Figure 6.13). The revisit interval of the simulated GMTI radar is 12 s. Its measurement accuracy (standard

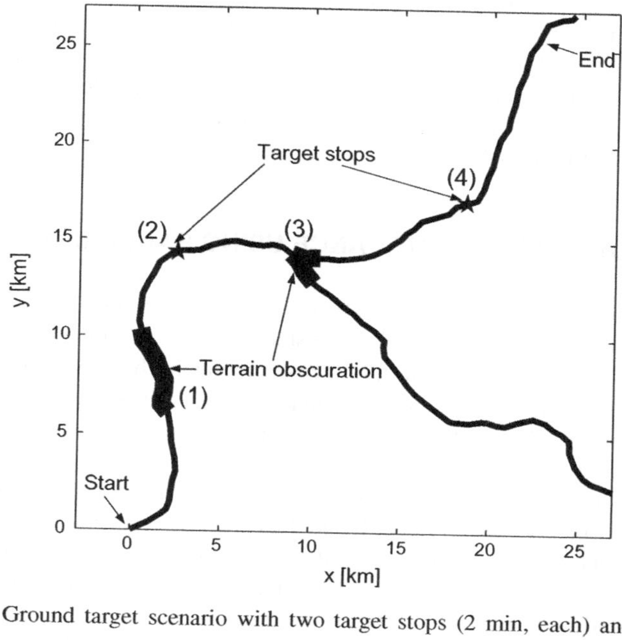

Figure 6.13 Ground target scenario with two target stops (2 min, each) and two terrain obscurations.

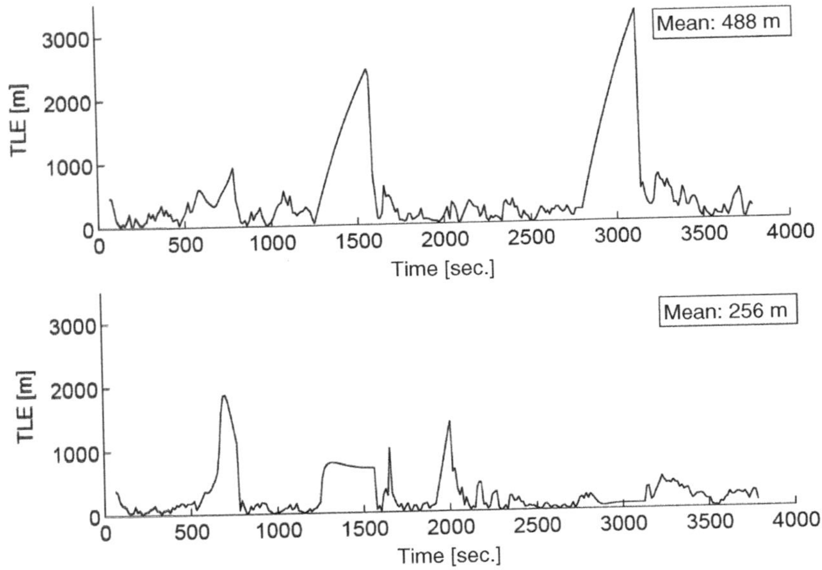

Figure 6.14 Target location error. Above: filtered track of the first vehicle; below: filtered track of the second vehicle after tracking-generated road-map extraction.

deviation) is 20 m in range (i.e., along the y-axis) and 400 m in cross-range (i.e., along the x-axis). The minimum detectable velocity is 2 m/s. The total observation time is 1 h (300 scans). The retrodicted track of a single target has been taken as a first estimate of the road to improve the tracking of the second vehicle. It has been shown in [10] that already this single iteration leads to a reduction of the target location error by 40–50 %, see Figure 6.14.

6.5 CONVOY TRACK MAINTENANCE USING RANDOM MATRICES

The explicit consideration of target extents is becoming more and more important in the development of modern tracking systems, where an object is considered as extended if it is the source of multiple measurements at the same time. In order to improve robustness and precision of estimation results, it is felt desirable to track the target extent in addition to the kinematic state of the object. Beyond that, more than these quantities cannot safely be estimated as well in the (opposite) case, where limited sensor resolution causes a fluctuating number of detections for a group of closely spaced objects (CSOs) [23–25] and thus prevents a successful tracking of (all of) the individual targets [5]. In this context, the Bayesian approach to tracking of extended objects and group targets in [26] tries to estimate both a kinematic state on the one hand and physical extent on the other so that a group of CSOs can be treated as one individual object. For the sake of simplicity, the focus here, as well as in [26, 27], is on (convoy) track maintenance, while estimation under uncertain observation-to-track association in possible presence of missed detections and false alarms is considered out of scope.

6.5.1 Object Extent within the Bayesian Framework

In order to achieve robust tracking of extended objects and group targets, the Bayesian approach in [26, 27] supplements the corresponding conventional (kinematical) track file with a random variable representing the true physical extent, which is approximated by an ellipsoid with the result that the target extent can be described by a symmetric positive definite (SPD) matrix, where an eigenvalue corresponds to the squared semi-axis length. Such a matrix-valued random variable is named a random matrix and invokes matrix-variate distributions [28]. Traditional random matrix theory is native to nuclear physics [29] and concerned with statistical properties of eigenvalues of high-dimensional matrices and so on. For a brief historical introduction beginning with [30], refer to [31], for a more detailed overview of current research topics, see [32].

Admittedly, the basic assumption to model the extent by an ellipsoid corresponds with earlier work (see, e.g., [33–35]), but the novelty of [26] is that it relies on the joint estimation of centroid kinematics and physical extent. According to the proposal in [26], the physical extent represented by an SPD random matrix $\mathbf{X}_k$ is considered besides the kinematic state of the centroid described by the sd-dimensional random vector $\mathbf{x}_k^T = [\mathbf{r}_k^T, \dot{\mathbf{r}}_k^T]$ with the spatial state component $\mathbf{r}_k$, where $d = \dim(\mathbf{r}_k)$ is also the dimension of $\mathbf{X}_k$. The decisive point for tracking of extended objects and CSOs in [26] is the special interpretation of usual sensor data as illustrated in Figure 6.15a, whereupon each measurement is considered as a measurement of the centroid scattered over object extent, that is, it is assumed that in each scan k there is

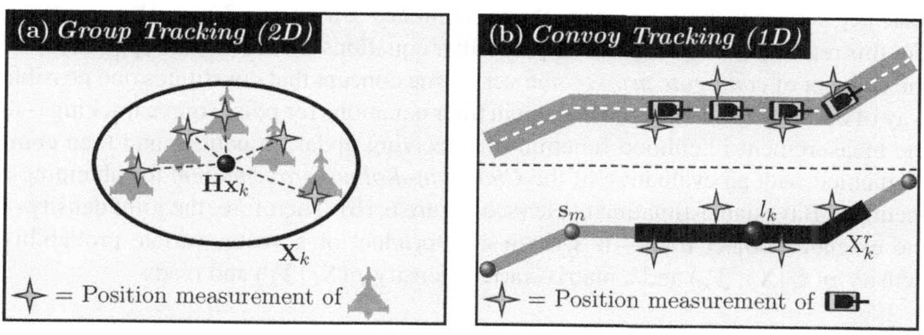

Figure 6.15 Illustration of the cluster/group tracking approach. The dashed lines in the left part of this figure symbolize the basic principle of [26], whereupon each measurement is considered as a measurement of the centroid scattered over group extent. In this context, the spatial extent in two dimensions is approximated by an ellipse. In addition, the right part of this figure shows how the scalar extent variable X_k^r in combination with the one-dimensional road coordinate system of Section 6.4.1 leads to a length-based approximation of the corresponding road-moving convoy.

a random number of n_k independent position measurements $\mathbf{y}_k^j = \mathbf{H}\mathbf{x}_k + \mathbf{w}_k^j$, where the object extent $\mathbf{X}_k$ is not directly observable. Therefore, the noise $\mathbf{w}_k^j$ is assumed to be a zero-mean normally distributed random vector with variance[3] $z\mathbf{X}_k$. On the one hand, the extent of the object can be *measured* in this way, but, on the other, this assumption implies the neglect[4] of any statistical sensor error.

With this, the likelihood to measure the set $\mathbf{Y}_k := \{\mathbf{y}_k^j\}_{j=1}^{n_k}$ (denoting the set of the n_k measurements in a particular scan k) given both kinematic state and object extent as well as the number of measurements reads $p(\mathbf{Y}_k | n_k, \mathbf{x}_k, \mathbf{X}_k) = \prod_{j=1}^{n_k} \mathcal{N}_d(\mathbf{y}_k^j; \mathbf{H}\mathbf{x}_k, z\mathbf{X}_k)$. Thereupon, the computation of the mean measurement and the measurement spread

$$\bar{\mathbf{y}}_k = \frac{1}{n_k}\sum_{j=1}^{n_k} \mathbf{y}_k^j, \qquad \overline{\mathbf{Y}}_k = \sum_{j=1}^{n_k} (\mathbf{y}_k^j - \bar{\mathbf{y}}_k)(\mathbf{y}_k^j - \bar{\mathbf{y}}_k)^T \tag{6.92}$$

is considered as a preprocessing step to rewrite the likelihood function according to

$$p(\mathbf{Y}_k | n_k, \mathbf{x}_k, \mathbf{X}_k) = \mathcal{N}_d(\bar{\mathbf{y}}_k; \mathbf{H}\mathbf{x}_k, z\mathbf{X}_k/n_k) \cdot |2\pi z\mathbf{X}_k|^{-\frac{1}{2}(n_k-1)}\, n_k^{-\frac{1}{2}d} \operatorname{etr}\left(-\tfrac{1}{2}\overline{\mathbf{Y}}_k(z\mathbf{X}_k)^{-1}\right). \tag{6.93}$$

In case of $n_k \geq d + 1$ and $\overline{\mathbf{Y}}_k > \mathbf{0}$, this likelihood function is interpreted as

$$p(\mathbf{Y}_k | n_k, \mathbf{x}_k, \mathbf{X}_k) \propto \mathcal{N}_d(\bar{\mathbf{y}}_k; \mathbf{H}\mathbf{x}_k, z\mathbf{X}_k/n_k)\, \mathcal{W}_d(\overline{\mathbf{Y}}_k; n_k - 1, z\mathbf{X}_k), \tag{6.94}$$

where $\mathcal{W}_d(\mathbf{X}; m, \mathbf{C})$ denotes the *Wishart density* [28] of an SPD random matrix $\mathbf{X}$ with expected SPD matrix $m\mathbf{C}$. In this context, a tracking algorithm is an iterative updating scheme for conditional probability densities $p(\mathbf{x}_k, \mathbf{X}_k | \mathcal{Y}_k)$ of the joint object state $(\mathbf{x}_k, \mathbf{X}_k)$ at each time t_k given the accumulated sensor data $\mathcal{Y}_k := \{\mathbf{Y}_\varkappa, n_\varkappa\}_{\varkappa=0}^{k}$. For this reason, the derivation of explicit filter equations is based on the application of the concept of *conjugate priors*—the very same concept that constitutes one possible way of deriving the well-known Kalman filter equations for point source tracking—to the measurement likelihood function for receiving update equations and then complemented with an evaluation of the *Chapman–Kolmogorov theorem* for obtaining a recursive Bayesian estimation cycle (see Figure 6.16). Therefore, the joint density—the extended object track—is written as a product of a vector-variate probability density $p(\mathbf{x}_k | \mathbf{X}_k, \mathcal{Y}_k)$ and a matrix-variate density $p(\mathbf{X}_k | \mathcal{Y}_k)$ and reads

$$p(\mathbf{x}_k, \mathbf{X}_k | \mathcal{Y}_k) = \mathcal{N}_{sd}(\mathbf{x}_k; \mathbf{x}_{k|k}, \widetilde{\mathbf{P}}_{k|k} \otimes \mathbf{X}_k)\, \mathcal{IW}_d(\mathbf{X}_k; \nu_{k|k}, \widetilde{\mathbf{X}}_{k|k}), \tag{6.95}$$

[3]The scaling factor z enables us to account for a difference between an assumed normal spread contribution of the extent and a possibly more realistic assumption such as, for example, a uniform distribution of varying scattering centers over an object of finite extent.

[4]This implicit neglect of any statistical sensor error has caused some further investigations and heuristic developments to honor the fact that both sensor error and extent contribute to the measurement spread [27].

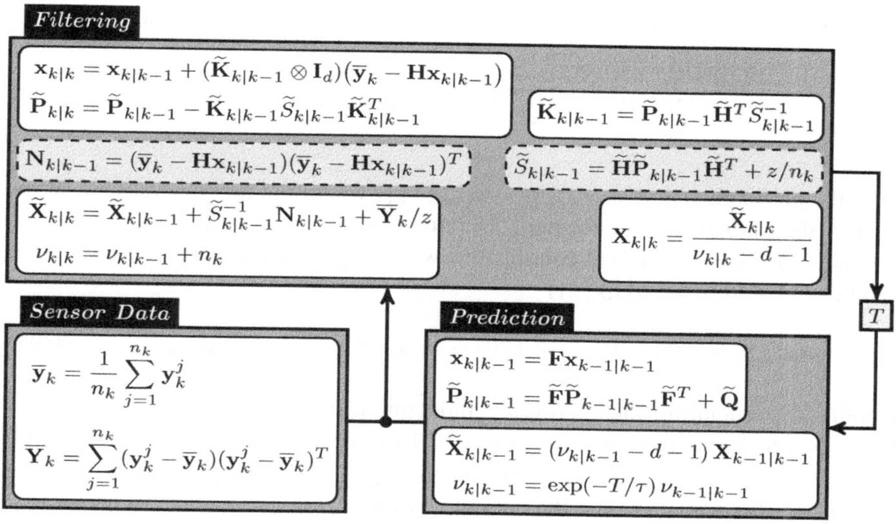

Figure 6.16 Bayesian formalism of [26].

where $\otimes$ denotes the *Kronecker product* [36] and $\mathcal{IW}_d(\mathbf{X}; m, \mathbf{C})$ the *inverse Wishart density* [28] of an SPD random matrix $\mathbf{X}$ with expected SPD matrix $\mathbf{C}/(m-d-1)$. More details on the derivation of the tracking algorithm can be found in [26].

6.5.2 Road-Map Assisted Convoy Track Maintenance

As previously mentioned, ground-moving vehicles and convoys typically move on roads, whose topographical coordinates are available up to a certain error [13, 20, 37]. Following the notation in Section 6.4.1, it seems reasonable to describe the kinematical state vector $\mathbf{x}^r_k$ of the centroid of a road-moving convoy at time t_k by its current position on the road l_k (i.e., by the arc length of the curve representing the road) and its scalar speed $\dot{l}_k$ according to $\mathbf{x}^r_k = [l_k, \dot{l}_k]^T$. As shown in Figure 6.15b, the extent of a road-moving convoy is modeled by the scalar extent variable X^r_k, which is treated as an 1×1 extent matrix, that is, the true convoy length is given by $2(\mathrm{X}^r_k)^{\frac{1}{2}}$. According to that, the conditional joint probability density of the convoy track in road coordinates reads [38, 39]

$$p(\mathbf{x}^r_k, \mathrm{X}^r_k | \mathcal{Y}_k) = \mathcal{N}_s(\mathbf{x}^r_k; \mathbf{x}^r_{k|k}, \widetilde{\mathbf{P}}^r_{k|k} \otimes \mathrm{X}^r_k)\,\mathcal{IW}_1(\mathrm{X}^r_k; \nu^r_{k|k}, \widetilde{X}^r_{k|k})\,. \tag{6.96}$$

6.5.2.1 *Time Update (Prediction).*

For sensor data processing in ground coordinates, the tracking algorithm has to provide the transformation from road into ground coordinates. In order to obtain explicit transformation formulae, the prediction in road coordinates $p(\mathbf{x}^r_k, \mathrm{X}^r_k | \mathcal{Y}_{k-1})$ (after the usual prediction step according to

Figure 6.16) is written via marginalization as a sum over the road segments

$$p(\mathbf{x}_k^r, \mathrm{X}_k^r | \mathcal{Y}_{k-1}) = \sum_{m=1}^{n_r} \theta_{k|k-1}^m \, p(\mathbf{x}_k^r, \mathrm{X}_k^r | \theta_k = m, \mathcal{Y}_{k-1}) \tag{6.97}$$

with the probability $\theta_{k|k-1}^m = P\{\theta_k = m | \mathcal{Y}_{k-1}\}$, where the random variable θ_k indicates the particular road segment. Herein, $p(\mathbf{x}_k^r, \mathrm{X}_k^r | \theta_k = m, \mathcal{Y}_{k-1})$ denotes the segment-wise density in road coordinates, which, by an application of Bayes' rule, can be related to $p(\mathbf{x}_k^r, \mathrm{X}_k^r | \mathcal{Y}_{k-1})$

$$p(\mathbf{x}_k^r, \mathrm{X}_k^r | \theta_k = m, \mathcal{Y}_{k-1}) \propto p(\theta_k = m | \mathbf{x}_k^r, \mathrm{X}_k^r) \, p(\mathbf{x}_k^r, \mathrm{X}_k^r | \mathcal{Y}_{k-1}) \,. \tag{6.98}$$

Herein, it is crucial for the appropriate modeling of the likelihood function $p(\theta_k = m | \mathbf{x}_k^r, \mathrm{X}_k^r)$ to honor not only the position of the convoy centroid $l_k = \widetilde{\mathbf{H}}_r \mathbf{x}_k^r$ with $\widetilde{\mathbf{H}}_r = [1, 0]$, but also the convoy length and the segment length $l_{m+1} - l_m$. In keeping with the methodical approach, the normal density

$$p(\theta_k = m | \mathbf{x}_k^r, \mathrm{X}_k^r) = \mathcal{N}_1(\bar{l}_m; \widetilde{\mathbf{H}}_r \mathbf{x}_k^r, \Delta l_m^2 \, \mathrm{X}_k^r) \tag{6.99}$$

with the single pseudo measurement $\bar{l}_m = (l_m + l_{m+1})/2$ and the approximation

$$\Delta l_m \approx (z_\theta)^{\frac{1}{2}} \left(1 + \frac{l_{m+1} - l_m}{2 \, (\mathrm{X}_{k|k-1}^r)^{\frac{1}{2}}} \right) \tag{6.100}$$

leads to a pseudo measurement update if, among other things, we simply replace $\bar{\mathbf{y}}_k$ with $\bar{l}_m$ and z with Δl_m^2 in Figure 6.16 (with $\overline{\mathbf{Y}}_k = \mathbf{0}$ and $n_k = 1$), that is,

$$p(\mathbf{x}_k^r, \mathrm{X}_k^r | \theta_k = m, \mathcal{Y}_{k-1}) = \mathcal{N}_s(\mathbf{x}_k^r; \mathbf{x}_{k|k-1}^{rm}, \widetilde{\mathbf{P}}_{k|k-1}^{rm} \mathrm{X}_k^r) \, \mathcal{IW}_1(\mathrm{X}_k^r; \nu_{k|k-1}^r + 1, \widetilde{X}_{k|k-1}^{rm}) \,. \tag{6.101}$$

In (6.97), $\theta_{k|k-1}^m$ denotes the probability that the convoy moves on segment m at time t_k given the sensor data $\mathcal{Y}_{k-1}$ and can be calculated as

$$\begin{aligned} \theta_{k|k-1}^m &= P\{\theta_k = m | \mathcal{Y}_{k-1}\} = \iint p(\theta_k = m | \mathbf{x}_k^r, \mathrm{X}_k^r) \, p(\mathbf{x}_k^r, \mathrm{X}_k^r | \mathcal{Y}_{k-1}) \, \mathrm{d}\mathbf{x}_k^r \, \mathrm{d}\mathrm{X}_k^r \\ &= \mathcal{T}_1(\bar{l}_m; \nu_{k|k-1}^r, \widetilde{\mathbf{H}}_r \mathbf{x}_{k|k-1}^r, \widetilde{S}_{k|k-1}^{rm} \widetilde{X}_{k|k-1}^r / \nu_{k|k-1}^r) \end{aligned} \tag{6.102}$$

under the restriction $\sum_{m=1}^{n_r} \theta_{k|k-1}^m = 1$. Herein, $\mathcal{T}_p(\mathbf{x}; \nu, \boldsymbol{\mu}, \mathbf{C})$ denotes the *Student-t density* [40] of a p-dimensional random vector $\mathbf{x}$ with mean $\boldsymbol{\mu}$ and variance $\frac{m}{m-2}\mathbf{C}$ for $m > 2$ and an SPD matrix $\mathbf{C}$. Now, the segment-wise prior density in two-dimensional ground coordinates reads

$$p(\mathbf{x}_k, \mathbf{X}_k | \theta_k = m, \mathcal{Y}_{k-1}) = \mathcal{N}_{2s}(\mathbf{x}_k; \mathbf{x}_{k|k-1}^m, \widetilde{\mathbf{P}}_{k|k-1}^m \otimes \mathbf{X}_k) \, \mathcal{IW}_2(\mathbf{X}_k; \nu_{k|k-1}, \widetilde{\mathbf{X}}_{k|k-1}^m), \tag{6.103}$$

where it seems reasonable to assume that the parameters of the normal density are implicitly defined by the following effect

$$\mathbf{x}^{m}_{k|k-1} = \mathbf{T}^{\breve{m}}_{g\leftarrow r}\mathbf{x}^{rm}_{k|k-1} + \mathbf{s}^{\breve{m}}_{g\leftarrow r} \tag{6.104}$$

$$\widetilde{\mathbf{P}}^{m}_{k|k-1} = \widetilde{\mathbf{P}}^{rm}_{k|k-1} \tag{6.105}$$

with

$$\mathbf{T}^{\breve{m}}_{g\leftarrow r} = \begin{bmatrix} \mathbf{t}_{\breve{m}} & \mathbf{0} \\ \mathbf{0} & \mathbf{t}_{\breve{m}} \end{bmatrix} \quad \text{and} \quad \mathbf{s}^{\breve{m}}_{g\leftarrow r} = \begin{bmatrix} \mathbf{s}_{\breve{m}} - l_{\breve{m}}\mathbf{t}_{\breve{m}} \\ \mathbf{0} \end{bmatrix} \tag{6.106}$$

for $m = 1, \ldots, n_r$. While $\widetilde{\mathbf{P}}^{(r)m}_{k|k-1}$ is an $s \times s$ matrix and does not depend on the dimensionality d, $\breve{m}$ denotes that one segment on which the (segment-wise) convoy centroid is located, that is, $l_{\breve{m}} \leq \widetilde{\mathbf{H}}_r\mathbf{x}^{rm}_{k|k-1} < l_{\breve{m}+1}$. This is due to the fact that $\mathbf{x}^{rm}_{k|k-1}$ is the Kalman-like update of $\mathbf{x}^{r}_{k|k-1}$ with the pseudo measurement $\bar{l}_m$ so that we always have $\widetilde{\mathbf{H}}_r\mathbf{x}^{r}_{k|k-1} \leq \widetilde{\mathbf{H}}_r\mathbf{x}^{rm}_{k|k-1} \leq \bar{l}_m$ or $\bar{l}_m \leq \widetilde{\mathbf{H}}_r\mathbf{x}^{rm}_{k|k-1} \leq \widetilde{\mathbf{H}}_r\mathbf{x}^{r}_{k|k-1}$, that is, we cannot ensure that $\mathbf{x}^{rm}_{k|k-1}$ is always located on the road segment m of the pseudo measurement $\bar{l}_m$. The transformation of the corresponding parameter matrix of the extent estimate in two-dimensional ground coordinates is performed by

$$\nu_{k|k-1} = (\nu^{r}_{k|k-1} + 1) + 1 = \nu^{r}_{k|k-1} + 2 \tag{6.107}$$

$$\widetilde{\mathbf{X}}^{m}_{k|k-1} = (\nu_{k|k-1} - 3) \cdot \mathbf{R}^{\breve{m}}_{g\leftarrow r}\, \mathrm{diag}\left(\mathrm{X}^{rm}_{k|k-1}, (\sigma^{W}_{\breve{m}}/2)^2\right) \left(\mathbf{R}^{\breve{m}}_{g\leftarrow r}\right)^T. \tag{6.108}$$

This means that the segment-wise extent estimate $\mathrm{X}^{rm}_{k|k-1} := E[\mathrm{X}^{r}_{k}|\theta_k = m, \mathcal{Y}_{k-1}]$ in road coordinates is considered and, with segment-wise road-map errors $\sigma^{W}_{\breve{m}}$ ($\sigma^{W}_{\breve{m}} > 0$), an ellipse is defined, where the two-dimensional rotation matrix $\mathbf{R}^{\breve{m}}_{g\leftarrow r}$ performs the corresponding rotation so that $E[\mathbf{X}_k|\theta_k = m, \mathcal{Y}_{k-1}]$ is aligned along the segment m. In (6.107), the number of degrees of freedom of $p(\mathrm{X}^{r}_{k}|\theta_k = m, \mathcal{Y}_{k-1})$, that is, $\nu^{r}_{k|k-1} + 1$, is incremented so that the corresponding mean square error (MSE) for the extent part [27] remains the same as $\sigma^{W}_{\breve{m}} \to 0$. This is based on the assumption that the transformation has no significant bearing on the certainty about the extent estimate.

6.5.2.2 Measurement Update (Filtering). With a measurement likelihood function as in (6.93) in mind, the posterior in road coordinates is given by

$$p(\mathbf{x}^{r}_{k}, \mathrm{X}^{r}_{k}|\mathcal{Y}_k) = \sum_{m=1}^{n_r} \theta^{m}_{k|k}\, p(\mathbf{x}^{r}_{k}, \mathrm{X}^{r}_{k}|\theta_k = m, \mathcal{Y}_k) \tag{6.109}$$

$$= \sum_{m=1}^{n_r} \iint \theta^{m}_{k|k}\, p(\mathbf{x}^{r}_{k}, \mathrm{X}^{r}_{k}|\mathbf{x}_k, \mathbf{X}_k, \theta_k = m)\, p(\mathbf{x}_k, \mathbf{X}_k|\theta_k = m, \mathcal{Y}_k)\, \mathrm{d}\mathbf{x}_k\, \mathrm{d}\mathbf{X}_k, \tag{6.110}$$

where the inverse transformation from segment-wise ground into road coordinates is formally denoted by the conditional density $p(\mathbf{x}_k^r, \mathbf{X}_k^r | \mathbf{x}_k, \mathbf{X}_k, \theta_k = m)$ and the segment-wise filtering in two-dimensional ground coordinates reads

$$p(\mathbf{x}_k, \mathbf{X}_k | \theta_k = m, \mathcal{Y}_k) = \mathcal{N}_{2s}(\mathbf{x}_k; \mathbf{x}_{k|k}^m, \widetilde{\mathbf{P}}_{k|k}^m \otimes \mathbf{X}_k) \mathcal{IW}_2(\mathbf{X}_k; \nu_{k|k}, \widetilde{\mathbf{X}}_{k|k}^m) \quad (6.111)$$

$$\propto p(\mathbf{Y}_k | n_k, \mathbf{x}_k, \mathbf{X}_k) \, p(\mathbf{x}_k, \mathbf{X}_k | \theta_k = m, \mathcal{Y}_{k-1}) \,. \quad (6.112)$$

Herein, the n_k measurements are considered with mean measurement $\overline{\mathbf{y}}_k$ and measurement spread $\overline{\mathbf{Y}}_k$ according to the filtering step in Figure 6.16, which leads to analog update equations for $\mathbf{x}_{k|k}^m$, $\widetilde{\mathbf{P}}_{k|k}^m$, $\nu_{k|k}$, and $\widetilde{\mathbf{X}}_{k|k}^m$. Afterwards, the segment-dependent measurement likelihoods $\Lambda_{k|k-1}^m := p(\mathbf{Y}_k | n_k, \theta_k = m, \mathcal{Y}_{k-1})$, that is,

$$\Lambda_{k|k-1}^m = \mathcal{T}_2\left(\overline{\mathbf{y}}_k; \nu_{k|k} - 2, \mathbf{H}\mathbf{x}_{k|k-1}^m, \widetilde{S}_{k|k-1}^m (\widetilde{\mathbf{X}}_{k|k-1}^m + \overline{\mathbf{Y}}_k / z)/(\nu_{k|k} - 2)\right)$$

$$\cdot \frac{\pi^{-(n_k-1)} \, \Gamma_2(\frac{\nu_{k|k}-1}{2})}{n_k \, \Gamma_2(\frac{\nu_{k|k-1}}{2})} \, |z\widetilde{\mathbf{X}}_{k|k-1}^m|^{\frac{1}{2}\nu_{k|k-1}} \, |z\widetilde{\mathbf{X}}_{k|k-1}^m + \overline{\mathbf{Y}}_k|^{-\frac{1}{2}(\nu_{k|k}-1)} \quad (6.113)$$

with the multivariate Gamma function $\Gamma_2(a) = \pi^{\frac{1}{2}} \Gamma(a) \Gamma(a - \frac{1}{2})$, are used to update the segment probabilities according to

$$\theta_{k|k}^m = \frac{\Lambda_{k|k-1}^m \, \theta_{k|k-1}^m}{\sum_{m'=1}^{n_r} \Lambda_{k|k-1}^{m'} \, \theta_{k|k-1}^{m'}} \,. \quad (6.114)$$

For the sake of completeness, it should be mentioned that, in case of $n_k \geq 3$ and $\overline{\mathbf{Y}}_k > \mathbf{0}$, the segment-dependent measurement likelihoods $\Lambda_{k|k-1}^m$ could also be interpreted as

$$\Lambda_{k|k-1}^m \propto \mathcal{T}_2(\overline{\mathbf{y}}_k; \nu_{k|k} - 2, \mathbf{H}\mathbf{x}_{k|k-1}^m, \widetilde{S}_{k|k-1}^m (\widetilde{\mathbf{X}}_{k|k-1}^m + \overline{\mathbf{Y}}_k / z)/(\nu_{k|k} - 2))$$

$$\cdot \, \mathcal{GB}_2^{II}(\overline{\mathbf{Y}}_k; \tfrac{n_k-1}{2}, \tfrac{\nu_{k|k-1}}{2}; z\widetilde{\mathbf{X}}_{k|k-1}^m, \mathbf{0}) \quad (6.115)$$

with the generalized matrix-variate beta type II distribution [28].

Having regard to the previous assumptions and the remarks in [13, 41], the inverse transformation from segment-wise ground to road coordinates starts with

$$\mathbf{x}_{k|k}^{rm} = \left(\mathbf{T}_{g \leftarrow r}^{\breve{m}}\right)^T \left(\mathbf{x}_{k|k}^m - \mathbf{s}_{g \leftarrow r}^{\breve{m}}\right) \quad (6.116)$$

$$\widetilde{\mathbf{P}}_{k|k}^{rm} = \widetilde{\mathbf{P}}_{k|k}^m \quad (6.117)$$

and ends with the transformation of the parameter matrix of the inverse Wishart distribution

$$\widetilde{X}_{k|k}^{rm} = (\nu_{k|k} - 3) \cdot [1, 0] \cdot \left(\mathbf{R}_{g \leftarrow r}^{\breve{m}}\right)^T \mathbf{X}_{k|k}^m \mathbf{R}_{g \leftarrow r}^{\breve{m}} \cdot [1, 0]^T . \quad (6.118)$$

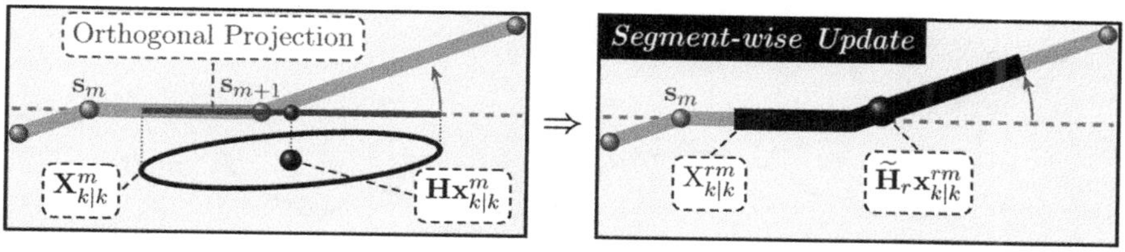

Figure 6.17 Illustration of the inverse transformation (from ground into road coordinates). It is shown how the segment-wise update of the joint object state in ground coordinates is projected onto the road in order to obtain its counterpart in segment-wise road coordinates. The orthogonal projection is evaluated for segment m or rather the straight line defined by $\mathbf{s}_m$ and $\mathbf{s}_{m+1}$ because it is assumed that the corresponding prediction of the convoy centroid $\widetilde{\mathbf{H}}_r\mathbf{x}^{rm}_{k|k-1}$ was located on segment m, that is, $\check{m} = m$. The resulting values of the projection are directly interpreted in terms of the road coordinate system with the result that protruding parts of the projected convoy track are mapped onto neighboring segments (as shown in the right part of this figure).

As illustrated in Figure 6.17, this means that the segment-wise extent estimate $\mathbf{X}^m_{k|k}$ is rotated back into the initial position. Afterwards, extracting the top left matrix element (1, 1) denotes the orthogonal projection onto the corresponding abscissa, which is equivalent to $X^{rm}_{k|k}$. Referring to (6.107) and the corresponding MSE for the extent part, the increment of the number of degrees of freedom has to be nullified, that is,

$$p(\mathbf{x}^r_k, X^r_k | \theta_k = m, \mathcal{Y}_k) = \mathcal{N}_s(\mathbf{x}^r_k; \mathbf{x}^{rm}_{k|k}, \widetilde{\mathbf{P}}^{rm}_{k|k} X^r_k)\, \mathcal{IW}_1(X^r_k; \nu_{k|k} - 1, \widetilde{X}^{rm}_{k|k}). \quad (6.119)$$

Before the subsequent prediction is performed, it is necessary to apply moment matching techniques to receive $p(\mathbf{x}^r_k, X^r_k | \mathcal{Y}_k)$ as in (6.96). Herein, the marginal densities of the joint conditional probability density are used due to the fact that the vector-variate part $p(\mathbf{x}^r_k | X^r_k, \mathcal{Y}_k)$ depends on the true extent X^r_k, that is,

$$p(X^r_k | \mathcal{Y}_k) = \sum_{m=1}^{n_r} \theta^m_{k|k}\, \mathcal{IW}_1(X^r_k; \nu_{k|k} - 1, \widetilde{X}^{rm}_{k|k}) \approx \mathcal{IW}_1(X^r_k; \nu^r_{k|k}, \widetilde{X}^r_{k|k}) \quad (6.120)$$

for the extent part of the estimation problem on the one hand and

$$\begin{aligned} p(\mathbf{x}^r_k | \mathcal{Y}_k) &= \sum_{m=1}^{n_r} \theta^m_{k|k}\, \mathcal{T}_s(\mathbf{x}^r_k; \nu_{k|k} - 1, \mathbf{x}^{rm}_{k|k}, \widetilde{\mathbf{P}}^{rm}_{k|k} \widetilde{X}^{rm}_{k|k} / (\nu_{k|k} - 1)) \\ &\approx \mathcal{T}_s\left(\mathbf{x}^r_k; \nu^r_{k|k}, \mathbf{x}^r_{k|k}, \widetilde{\mathbf{P}}^r_{k|k} \widetilde{X}^r_{k|k} / \nu^r_{k|k}\right) \end{aligned} \quad (6.121)$$

for the kinematics part on the other. By this means, the road-map assisted convoy track maintenance cycle is closed (for more details, see [38, 39]).

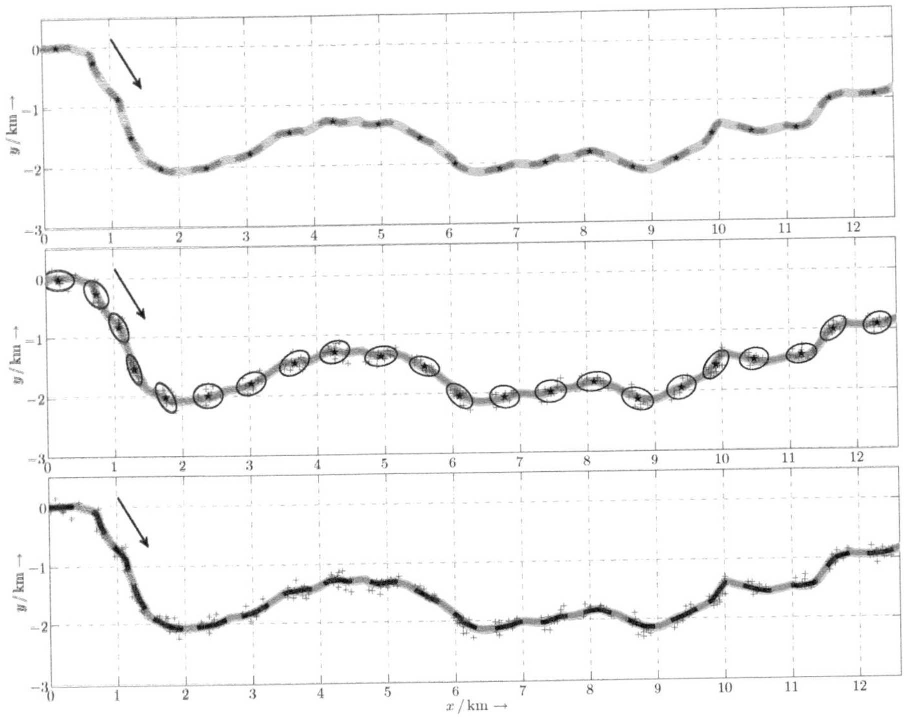

Figure 6.18 Convoy tracking scenario with 21 vehicles. Shown are, for every fifth scan k, the true target positions (top), the tracking results of the *plain* tracking algorithm (estimated centroid and ellipsoidal extent) (middle), the tracking results of the road-map assisted approach (estimated convoy track) (bottom), as well as the accumulated measurements (+).

6.5.3 Selected Numerical Examples

The single road in Figure 6.18 consisting of 100 segments was the basis of the (fictitious) tracking scenario, where the various segment lengths range from 75 m to circa 310 m. The 21 individual vehicles (point targets) of the simulated convoy moved with a constant speed of 50 km/h. The targets were arranged with 20 m distance between neighboring targets. This convoy was observed by a sensor with scan time $T = 10$ s delivering uncorrelated noisy x- and y-measurements with standard deviations $\sigma_x = 20$ m and $\sigma_y = 80$ m, respectively, where we assumed a probability of detection of 65 % for each individual vehicle (not considering the problem of limited sensor resolution here).

The middle part of Figure 6.18 shows the tracking results of the *plain* extended object tracking algorithm [26] (Section 6.5.1), while the bottom part presents the tracking results of the road-map assisted algorithm (Section 6.5.2), where only several scans are presented due to a total convoy length of 400 m. Figure 6.19 shows the estimated convoy length of the road-map assisted approach. Figure 6.20 displays the tracking error, that is, the distance between the simulated true convoy centroid

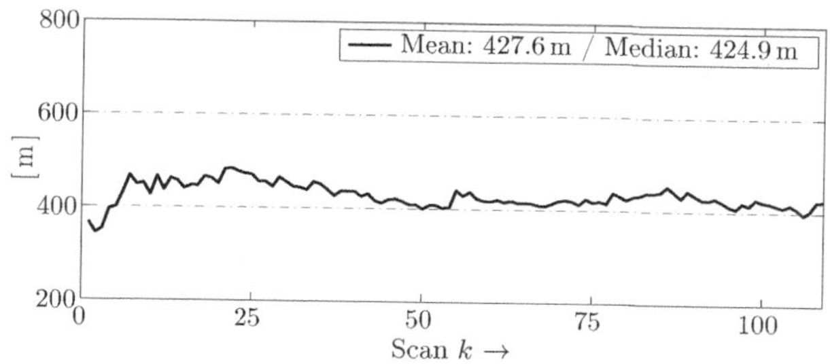

Figure 6.19 Estimated convoy length after update for the road-map assisted algorithm.

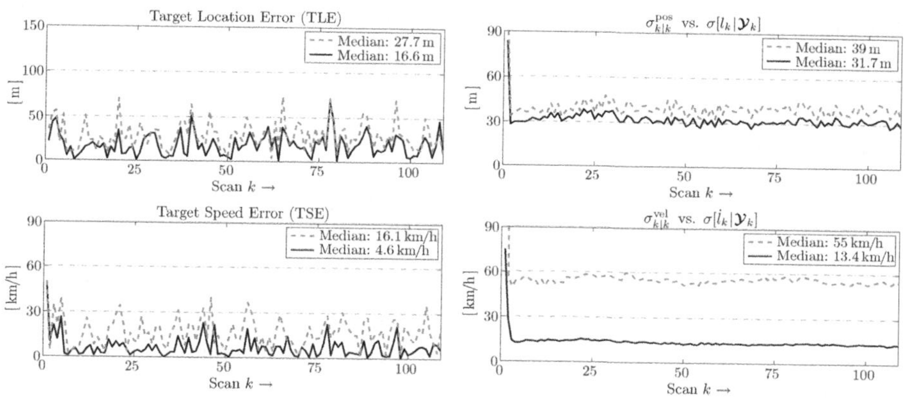

Figure 6.20 Absolute and relative errors to compare the *plain* tracking algorithm as presented in Section 6.5.1 (dashed line) with the road-map assisted approach of Section 6.5.2 (solid line).

state and the corresponding estimates, for a single run as a function of the scans. Additionally, the conditional standard deviations

$$\sigma[l_k \,|\, \mathcal{Y}_k] = \sqrt{\mathrm{Var}[l_k \,|\, \mathcal{Y}_k]} \quad \text{and} \quad \sigma[\dot{l}_k \,|\, \mathcal{Y}_k] = \sqrt{\mathrm{Var}[\dot{l}_k \,|\, \mathcal{Y}_k]} \tag{6.122}$$

for the road-map assisted algorithm and the corresponding quantities $\sigma_{k|k}^{\text{pos}} := \sqrt{\mathrm{tr}\,\mathrm{Var}[\mathbf{r}_k \,|\, \mathcal{Y}_k]}$ and $\sigma_{k|k}^{\text{vel}} := \sqrt{\mathrm{tr}\,\mathrm{Var}[\dot{\mathbf{r}}_k \,|\, \mathcal{Y}_k]}$ for the *plain* extended object tracking algorithm are presented. The comparison of both tracking approaches indicates improvement of track quality due to taking road-map information into account.

6.6 CONVOY TRACKING WITH THE CARDINALIZED PROBABILITY HYPOTHESIS DENSITY FILTER

In Section 6.5, we considered a convoy as a single entity, an *extended object*. It was assumed that the extended target consists of several scattering centers whose positions are statistically distributed over the group. If, on the other hand, the sensor resolution allows the detection of individual targets, in principle, the complete multitarget state could be tracked. Because of the exponentially growing number of data assignments, this, however, often becomes numerically intractable for larger target groups or dense clutter situations. In many practical situations, the identity of a vehicle in a convoy is in fact uninteresting. The operator typically needs to know the kinematics of the center of mass of the convoy and some information about it structure, in particular the total number of vehicles. This number cannot be reliably estimated just by the number of detections, as the latter may be strongly fluctuating due to obscuration or limited sensor resolution.

In recent years, filter algorithms have been developed that are exactly suitable to answer the above-mentioned relevant questions for convoy tracking. Based on Finite Set Theory, Mahler derived the probability hypothesis density (PHD) filter [42], which provides an estimate of the probability density of target existence and, integrated over a certain area, of the target number. An extended version, which in particular improves on the stability of the target number estimation, has been introduced as the *cardinalized* PHD (CPHD) filter [43, 44]. Alternative derivations have been given for the PHD and CPHD filters on a *physical state-space* model [45] and for the PHD filter based on Poisson point processes [46]. Details of the PHD and CPHD algorithms, their derivation and applications are given in Chapter 3. In particular, the Gaussian mixture variant (GM-CPHD) [47], which provides closed-form prediction and update equations for the filter in the case of linear Gaussian systems, is a candidate for real time multitarget tracking. Below, some important aspects, relevant to the problem of ground target tracking, are summarized. In particular, the integration of digital road maps into the algorithm (see Section 6.4) and the clutter notch model for GMTI radar sensors (see Section 6.2) are discussed.

6.6.1 Gaussian Mixture CPHD Algorithm

In the Gaussian mixture variant of the CPHD, the PHD at scan k, $v_{k|k}(\mathbf{x})$, is described as the weighted sum of $J_{k|k}$ normal distributions

$$v_{k|k}(\mathbf{x}) = \sum_{j=1}^{J_{k|k}} w_{k|k}^{(j)} \mathcal{N}\big(\mathbf{x};\, \mathbf{m}_{k|k}^{(j)},\, \mathbf{P}_{k}^{(j)}\big) \tag{6.123}$$

with $\mathbf{m}_{k|k}^{(j)}$ and $\mathbf{P}_{k|k}^{(j)}$ being the mean and covariance matrix of component j. The estimated target number is given by the integral of $v(\mathbf{x})$

$$n_{k|k} = \int v_{k|k}(\mathbf{x})\, \mathrm{d}\mathbf{x} = \sum_{j=1}^{J_{k|k}} w_{k|k}^{(j)} . \tag{6.124}$$

In the PHD, the single parameter $n_{k|k}$ is sufficient to define the cardinality distribution $p_{k|k}(n)$, that is, the probability mass function of the number of targets. The cardinality is Poisson-distributed with the parameter $n_{k|k}$, that is, $p_{k|k}(n) = e^{-n_{k|k}} \frac{n_{k|k}^n}{n!}$. However, in the CPHD, the cardinality distribution $p_{k|k}(n) = p(n|\mathbf{Z}^k)$ is arbitrary and is estimated iteratively by the Bayesian technique. Similarly to (6.124), the estimated target number can also be calculated by[5]

$$n_{k|k} = \sum_{n=1}^{\infty} n \; p_{k|k}(n) \,. \tag{6.125}$$

The prediction and update steps for each individual component's mean and covariance are performed exactly as in the Kalman filter, since it is assumed that system dynamics and the measurement process follow the linear-Gaussian assumption [6].

6.6.1.1 Prediction. Given the PHD for scan $k-1$, the probability of a target surviving P_s, the target dynamics matrix $\mathbf{F}_k$, and the process noise covariance $\mathbf{D}_k$, the predicted components are

$$w_{k|k-1}^{(j)} = P_s w_{k-1|k-1}^{(j)} \tag{6.126}$$

$$\mathbf{m}_{k|k-1}^{(j)} = \mathbf{F}_k \mathbf{m}_{k-1|k-1}^{(j)} \tag{6.127}$$

$$\mathbf{P}_{k|k-1}^{(j)} = \mathbf{F}_k \mathbf{P}_{k-1|k-1}^{(j)} \mathbf{F}_k^T + \mathbf{D}_k \,. \tag{6.128}$$

Additional components are introduced by a *birth model*, which describes potentially new targets appearing in the field of view of the sensor. A simple realization of a birth model is to introduce, at each scan, one *test component* with a certain weight w_{birth}, positioned in the middle of the field of view with zero velocity, and a large covariance $\mathbf{P}_{\text{birth}}$ covering the whole field of view of the sensor. The number of components after prediction, denoted by $J_{k|k-1}$, is equal to $J_{k-1|k-1}$ plus the number of new born components. The prediction equation for the cardinality distribution can be written in terms of a *transfer matrix* $\mathbf{M}$

$$p_{k|k-1}(n) = \sum_{n'=0}^{\infty} p_{k-1|k-1}(n') \mathbf{M}(n, n') \tag{6.129}$$

$$\mathbf{M}(n, n') = \sum_{i=0}^{\min\{n,n'\}} p_{\text{birth}}(n-i) \binom{n'}{i} (1-P_s)^{n'-i} P_s^i \,. \tag{6.130}$$

$p_{\text{birth}}(n)$ is the probability for n new targets to appear between scan $k-1$ and k, derived from the birth model. For time independent $p_{\text{birth}}(n)$, for example, for a constant scan rate, the matrix $\mathbf{M}$ is constant and can be calculated in advance.

[5]In practice, the infinite summation is replaced with a sufficiently large number, that is, a number larger than the maximum number of targets expected to be tracked at any given time.

6.6.1.2 Update. The *general* CPHD update equation for the PHD and the cardinality distribution can be written in terms of likelihood ratios [45]

$$v_{k|k}(\mathbf{x}) = \left[(1-P_d)\frac{\mathcal{L}(\mathbf{Z}_k|\neg D)}{\mathcal{L}(\mathbf{Z}_k)} + P_d\frac{\mathcal{L}(\mathbf{Z}_k|D)}{\mathcal{L}(\mathbf{Z}_k)}\right] v_{k|k-1}(\mathbf{x}) \tag{6.131}$$

$$p_{k|k}(n) = \frac{\mathcal{L}(\mathbf{Z}_k|n)}{\mathcal{L}(\mathbf{Z}_k)}\, p_{k|k-1}(n), \tag{6.132}$$

where the conditions D and $\neg D$ are short-hand notations for *target detected* and *not detected*, respectively. There are m_k measurements observed at time k. The likelihood ratios above are given by

$$\mathcal{L}(\mathbf{Z}_k|\neg D) = \frac{1}{n_{k|k-1}}\sum_{j=0}^{m_k} \alpha_k^{(j+1)}\, \beta_k^{(j)} \sigma_j(\{L_k^{(1)}, \dots, L_k^{(m_k)}\}) \tag{6.133}$$

$$\mathcal{L}(\mathbf{Z}_k|D) = \sum_{s=1}^{m_k} l(\mathbf{z}_k^{(s)}|\mathbf{x})\, \mathcal{L}(\mathbf{Z}_k|s) \tag{6.134}$$

$$\mathcal{L}(\mathbf{Z}_k|s) \equiv \frac{1}{n_{k|k-1}}\sum_{j=1}^{m_k} \beta_k^{(j)}\, \alpha_k^{(j)} \sigma_{j-1}(\{L_k^{(1)}, \dots, L_k^{(m_k)}\} \setminus \{L_k^{(s)}\}) \tag{6.135}$$

$$\mathcal{L}(\mathbf{Z}_k) = \sum_{j=0}^{m_k} \alpha_k^{(j)}\, \beta_k^{(j)} \sigma_j(\{L_k^{(1)}, \dots, L_k^{(m_k)}\}) \tag{6.136}$$

$$\mathcal{L}(\mathbf{Z}_k|n) = \sum_{j=0}^{\min\{m_k,n\}} \beta_k^{(j)} \frac{n!}{(n-j)!}(1-P_d)^{n-j}\sigma_j(\{L_k^{(1)}, \dots, L_k^{(m_k)}\}) \tag{6.137}$$

with

$$\alpha_k^{(j)} \equiv \sum_{n=j}^{\infty} \frac{n!}{(n-j)!}\, p_{k|k-1}(n)(1-P_d)^{n-j} \tag{6.138}$$

$$\beta_k^{(j)} \equiv p_c(m_k - j)\frac{(m_k - j)!}{m_k!}\,|V|^j \tag{6.139}$$

with the single detection likelihood function $l(\mathbf{z}_k^{(s)}|\mathbf{x})$, and the size of the field of view of the sensor V. We use $p_c(j)$ to denote the probability for j false alarms, and a homogeneous false alarm spatial density, V^{-1}, is assumed for simplicity. The function $\sigma_j(\mathcal{A})$ in the likelihood ratios above is defined as the sum over all possible products of elements of the set $\mathcal{A}$ with j different factors

$$\sigma_j(\{y_1, \dots y_m\}) = \sum_{1\le i_1<\dots<i_j\le m} y_{i_1}\, y_{i_2} \dots y_{i_j} \tag{6.140}$$

and $\sigma_0 \equiv 1$. The arguments of σ_j are the average likelihood functions

$$L_k^{(s)} = \frac{1}{n_{k|k-1}}\int P_d(\mathbf{x})\, v_{k|k-1}(\mathbf{x})\, l_k^{(s)}(\mathbf{z}_k^{(s)}|\mathbf{x})\, d\mathbf{x}\,. \tag{6.141}$$

The present context of Gaussian mixtures can be preserved in the case of linear measurements with Gaussian measurement errors $\mathbf{w}_k$ with (potentially detection dependent) covariance $\mathbf{R}_k^{(s)}$, that is, $\mathbf{z}_k^{(s)} = \mathbf{H}_k\mathbf{x} + \mathbf{w}_k^{(s)}$, which results in a normally distributed likelihood function

$$l_k^{(s)}(\mathbf{z}|\mathbf{x}) = \mathcal{N}\big(\mathbf{z};\, \mathbf{H}_k\mathbf{x},\, \mathbf{R}_k^{(s)}\big) \tag{6.142}$$

and an average likelihood function

$$L_k^{(s)} = \frac{1}{n_{k|k-1}} \sum_{j=1}^{J_{k|k-1}} P_d^{(j)} w_{k|k-1}^{(j)}\, \mathcal{N}\big(\mathbf{z}_k^{(s)};\, \mathbf{z}_k^{(j)},\, \mathbf{S}_k^{(j,s)}\big)\,. \tag{6.143}$$

Inserting the Gaussian mixture (6.123) into (6.131), we obtain $m_k + 1$ new components for each original component, one from the missed detection part, $(1 - P_d)$, denoted by $s = 0$, and one for each detection $\mathbf{z}_k^{(s)}$, $s = 1, \ldots, m_k$, according to (6.134). After updating, the total number of components is $J_{k|k} = (m_k + 1)\, J_{k|k-1}$.

In general, the probability of detection P_d may depend on the target state, which could be modeled by a component-dependent $P_d^{(j)}$. In that case, we also need the weighted average

$$P_{d,k-1} \equiv \frac{1}{n_{k|k-1}} \sum_{j=1}^{J_{k|k-1}} w_{k|k-1}^{(j)}\, P_d^{(j)} \tag{6.144}$$

which replaces P_d in (6.137) and (6.138); see [45].

For a missed detection, the updated component mean and covariance are equal to the predicted ones. For components j assigned to a detection s, the update follows the usual Kalman filter equations leading to a Gaussian sum of the form (6.123).

The component weights are updated by $w_{k|k}^{(j,s)} = G_k^{(j,s)} w_{k|k-1}^{(j)}$, where the updating factors are given by

$$\begin{aligned} G_k^{(j,s=0)} &= \frac{\mathcal{L}(\mathbf{Z}_k|\neg D)}{\mathcal{L}(\mathbf{Z}_k)}\,(1 - P_d^{(j)}) \\ G_k^{(j,s>0)} &= P_d^{(j)} \frac{\mathcal{L}(\mathbf{Z}_k|a_k^{(j)}{=}s)}{\mathcal{L}(\mathbf{Z}_k)}\, \mathcal{N}\big(\mathbf{z}_k^{(s)};\, \mathbf{z}_k^{(j)},\, \mathbf{S}_k^{(j,s)}\big) \end{aligned} \tag{6.145}$$

with predicted detection $\mathbf{z}_k^{(j)} = \mathbf{H}_k \mathbf{m}_{k|k-1}^{(j)}$ and innovation covariance $\mathbf{S}_k^{(j,s)}$.

For a single target, it can be shown that the GM-CPHD filter equations are identical to the standard MHT filter supplemented with a sequential likelihood ratio for track extraction [48]. The MHT likelihood ratio is equal to the component weight $w^{(j)}$, which can be regarded as a target existence probability.

6.6.2 Integration of Digital Road Maps

Digital road maps can be described by sequences of linearized segments (see Section 6.4). Road coordinates (i.e., mileage and speed) form the natural coordinate system for target state prediction, while the measurement update is done in Cartesian or sensor coordinates. In Section 6.4, the Gaussian probability density, or each component in a Gaussian mixture, is decomposed into a weighted sum of Gaussian densities on road segments. The same idea is applied here for the PHD: For a predicted PHD in road coordinates ($\tilde{\mathbf{x}}$) of the form

$$v_{k|k-1}(\tilde{\mathbf{x}}) = \sum_{j=1}^{J_{k|k-1}} w_{k|k-1}^{(j)} \, p_j(\tilde{\mathbf{x}} \,|\mathbf{Z}^{k-1}) \tag{6.146}$$

$$p_j(\tilde{\mathbf{x}} \,|\mathbf{Z}^{k-1}) = \mathcal{N}\big(\tilde{\mathbf{x}};\, \tilde{\mathbf{m}}_{k|k-1}^{(j)},\, \tilde{\mathbf{P}}_{k|k-1}^{(j)}\big) \tag{6.147}$$

each component j is decomposed into a Gaussian mixture over road segments

$$p_j(\tilde{\mathbf{x}} \,|\mathbf{Z}^{k-1}) \approx \sum_{i=1}^{N_{\text{seg}}} f_{k|k-1}^{(i,j)} \, \mathcal{N}\big(\tilde{\mathbf{x}};\, \tilde{\mathbf{m}}_{k|k-1}^{(i,j)},\, \tilde{\mathbf{P}}_{k|k-1}^{(i,j)}\big), \tag{6.148}$$

where the *normalized* weights $f_{k|k-1}^{(i,j)}$ correspond to the probability of a target to be on road segment i. The calculation of the road-segment-dependent weight, mean, and covariance is given in Section 6.4. Typically, $f_{k|k-1}^{(i,j)}$ has a sharp peak at a segment $i_{\max}$ (depending on j) and is negligible for segments far away from $i_{\max}$. Using the affine transformation given in Section 6.4, the density in Cartesian coordinates, too, is described by a Gaussian mixture over road segments

$$p_j(\mathbf{x} \,|\mathbf{Z}^{k-1}) = \sum_{i=1}^{N_{\text{seg}}} f_{k|k-1}^{(i,j)} \, \mathcal{N}\big(\mathbf{x};\, \mathbf{m}_k^{(i,j)},\, \mathbf{P}_k^{(i,j)}\big)\,. \tag{6.149}$$

$\mathbf{m}_k^{(i,j)}$ and $\mathbf{P}_k^{(i,j)}$ are calculated from their counterparts in road coordinates by Kalman-filter-like equations. In the update step, each component (i, j) is assigned to each detection s and updated according to the Kalman filter equations and (6.145). The weight update factor depends on road segment i, PHD component j, and detection s, and is denoted by $g_k^{(i,j,s)}$. The new normalized road segment weights are given by

$$f_{k|k}^{(i,j,s)} = \frac{g_k^{(i,j,s)} \, f_{k|k-1}^{(i,j)}}{G_k^{(j,s)}} \tag{6.150}$$

$$G_k^{(j,s)} = \sum_{i=1}^{N_{\text{seg}}} g_k^{(i,j,s)} \, f_{k|k-1}^{(i,j)}\,. \tag{6.151}$$

The sum $G_k^{(j,s)}$ plays the role of the update factor for the component weight (compare (6.145)): $w_{k|k}^{(j,s)} = G_k^{(j,s)} w_{k|k-1}^{(j)}$, which are, of course, not normalized. For the calculation of $g_k^{(i,j,s)}$, the weighted single detection likelihood functions are required, see (6.143)

$$L_k^{(s)} = \frac{1}{n_{k|k-1}} \sum_{j=1}^{J_{k|k-1}} P_d^{(j)} w_{k|k-1}^{(j)} \sum_{i=1}^{N_{\text{seg}}} f_{k|k-1}^{(i,j)} \mathcal{N}\big(\mathbf{z}_k^{(s)};\, \mathbf{z}_k^{(i,j)},\, \mathbf{S}_k^{(i,j,s)}\big)\,. \tag{6.152}$$

The transformation segment-wise back to road coordinates is a simple projection onto each road segment (see Section 6.4). The result, again, is a Gaussian mixture over road segments. This mixture, finally, can well be approximated by a single normal distribution in road coordinates using second-order moment matching. By this means, we obtain

$$v_{k|k}(\tilde{\mathbf{x}}) \approx \sum_{j=1}^{J_{k|k}} w_{k|k}^{(j)} \mathcal{N}\big(\tilde{\mathbf{x}};\, \tilde{\mathbf{m}}_{k|k}^{(j)},\, \tilde{\mathbf{P}}_{k|k}^{(j)}\big) \tag{6.153}$$

which now can be used to predict to the next time step.

6.6.3 Target State Dependent Detection Probability

As mentioned above, the detection probability P_d, in general, is a function of target state and sensor-target geometry. In particular for GMTI radar, due to the *clutter notch* of the sensor, P_d sharply drops to zero when the relative range rate between target and sensor falls below a certain threshold, the *minimum detectable velocity* v_m (see Section 6.2 for details). In addition, ground targets often become invisible due to terrain obscuration, for example, by hills, which also depends on the sensor-target geometry. Within the Gaussian mixture framework, such a state-dependent $P_d(\mathbf{x})$ can be approximated by a component-dependent $P_d^{(j)}$ obtained by averaging $P_d(\mathbf{x})$ over the component distribution, or simply by the value of P_d at the estimated target state

$$P_d^{(j)} \approx \int P_d(\mathbf{x}) \mathcal{N}\big(\mathbf{x};\, \mathbf{m}_{k|k-1}^{(j)},\, \mathbf{P}_{k|k-1}^{(j)}\big)\, d\mathbf{x} \tag{6.154}$$

$$\text{or} \quad P_d^{(j)} \approx P_d\left(\mathbf{m}_{k|k-1}^{(j)}\right)\,. \tag{6.155}$$

While (6.154), in general, is difficult to calculate, (6.155) is a rather crude approximation when the target state covariances increase. In the case of terrain obscuration, the situation is easier for road targets, where $P_d(\mathbf{x})$ may well be approximated by its value in the center of each road segment, that is, $P_d((\mathbf{s}_i + \mathbf{s}_{i+1})/2)$ for $i = 1, \ldots, N_{\text{seg}}$. The clutter notch of the GMTI sensor can also be treated within the same approach as in Section 6.2, leading to an additional two-fold splitting of each component of the Gaussian sum. Care has to be taken in the interpretation of components $w^{(j)}$ with negative weights, which, of course, cannot be considered as (candidate) target tracks. In particular, a Bernoulli-like interpretation [49] of $w^{(j)}$ as the probability of target existence is not meaningful for negative weights. A practical approach to treat negative weight components is to merge them with the corresponding positive weight components, as the sum of weights is positive definite.

6.6.4 Exemplary Results for Small Convoys

We use the same dynamical target and measurement model as described in Section 6.3, however restricted to a two-dimensional target motion (neglecting motion in height) and measurement of azimuth φ and range r only (see [48] for details and more results). Measurement errors are $\sigma_\varphi = 0.5°$ and $\sigma_r = 20$ m for a sensor located at $(-4000, 0)$, $P_d = 0.9$.

As an example, results for the combined tracking and target number estimation for a convoy of seven vehicles are shown in Figure 6.21, where four different time scans are overlayed. First, the group moves mostly in a direction radial to the sensor (positions a and b), implying a high P_d. The target number is correctly estimated and the structure of the group is (partially) resolved. Later, at positions c and d, the group instead moves in a cross-range direction, implying a low P_d. Here, the individual

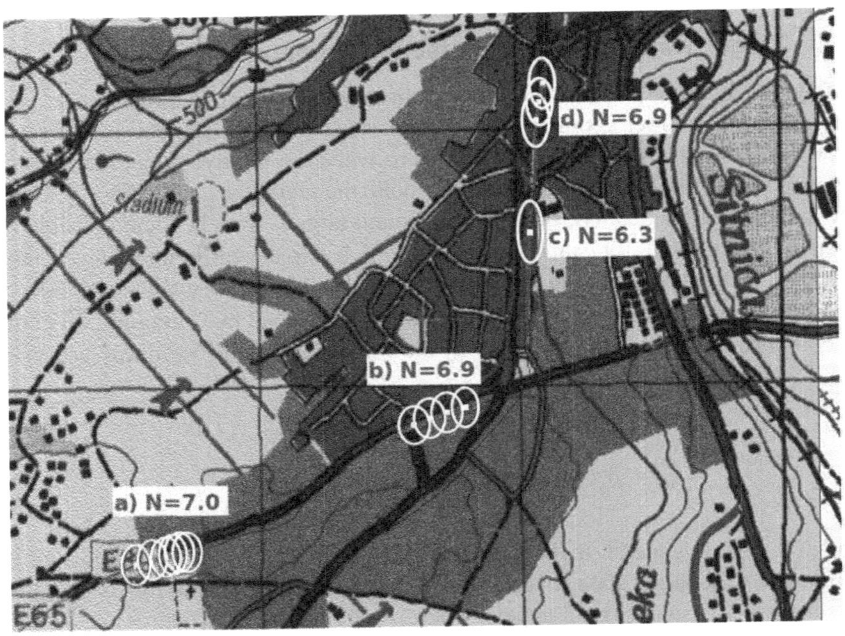

Figure 6.21 Overlay of four snapshots of tracks of a group with seven targets. At the instance c, the targets become unresolvable since they travel in cross range from the sensor, yet the clutter-notch featured GM-CPHD tracker estimates that there are 6.3 targets all co-located (i.e., represented with a single Gaussian mode—see text).

targets can no longer be resolved, but due to the clutter notch modeling, the target number estimate remains stable over a long period of time in spite of a series of missing detections. For practical purposes the position and velocity of the group's center of mass and the group size are, in fact, often more important than the exact position of each individual vehicle.

REFERENCES

1. R. Klemm, *Principles of Space-Time Adaptive Processing*, Vol. 21 of *IET Radar, Sonar, Navigation and Avionics Series*, 3rd edition, The Institution of Engineering and Technology, 2006.
2. R. Klemm (Ed.), Ground target tracking with STAP radar: The sensor, *Applications of Space-Time Adaptive Processing*, The Institution of Engineering and Technology, 2004, Chapter 14, pp. 467–500.
3. P.L. Bogler, *Radar Principles with Applications to Tracking Systems*, John Wiley & Sons, New York, 1990.
4. P. Swerling, Probability of detection for fluctuating targets, *IRE Trans. Inform. Theory*, 6(2), 269–308, 1960.
5. S.S. Blackman and R. Popoli, *Design and Analysis of Modern Tracking Systems*, Artech House, Norwood, MA, 1999.
6. Y. Bar-Shalom, X. Rong Li, and T. Kirubarajan, *Estimation with Applications to Tracking and Navigation (Theory, Algorithms and Software)*, John Wiley & Sons, New York, 2001.
7. D.L. Hall and J. Llinas, *Handbook of Multisensor Data Fusion*, Electrical Engineering & Applied Signal Processing Series. CRC Press, Boca Raton, FL, 2001.
8. W. Koch, Sensor data fusion: Methods, applications, examples, in: E. Lefebvre (Ed.), *Advances and Challenges in Multisensor Data and Information Processing*, Vol. 8 of *D: Information and Communication Security*, IOS Press, Amsterdam, Netherlands, 2007, Chapter 1, pp. 1–23.
9. J. Koller and M. Ulmke, Data fusion for ground moving target tracking, *Aerosp. Sci. Technol.*, 11(4), 261–270, 2007.
10. W. Koch, J. Koller, and M. Ulmke, Ground target tracking and road map extraction, *ISPRS J. Photogramm. Remote Sens.*, 61(3–4), 197–208, 2006.
11. G. van Keuk, Sequential track extraction, *IEEE Trans. Aerosp. Electron. Syst.*, 34(4), 1135–1148, 1998.
12. W. Koch, Ground target tracking with STAP radar: Selected tracking aspects, in: R. Klemm (Ed.), *Applications of Space-Time Adaptive Processing*, The Institution of Engineering and Technology, 2004, Chapter 15, pp. 501–540.
13. M. Ulmke and W. Koch, Road-map assisted ground moving target tracking, *IEEE Trans. Aerosp. Electron. Syst.*, 42(4), 1264–1274, 2006.
14. M. Mertens and M. Ulmke, GMTI tracking using signal strength information, *Proceedings of the International Conference on Information Fusion*, Edinburgh, UK, 2010, 8 pages.
15. W. Koch, Tracking and data fusion applications, *Proceedings of the NATO Lecture Series on Advanced Radar Signal and Data Processing. (RTO-EN-SET-086)*, Neuilly-sur-Seine, France, 2006, pp. 9/1–9/36.

16. Y. Bar-Shalom and X. Rong Li, *Estimation and Tracking: Principles, Techniques, and Software*, Artech House, Norwood, MA, 1993.
17. Y. Bar-Shalom and X. Rong Li, *Multitarget-Multisensor Tracking: Principles and Techniques*, YBS Publishing, Storrs, CT, 1995.
18. T. Kirubarajan, Y. Bar-Shalom, K.R. Pattipati, and I. Kadar, Ground target tracking with variable structure IMM estimator, *IEEE Trans. Aerosp. Electron. Syst.*, 36(1), 26–46, 2000.
19. P.J. Shea, T. Zadra, D. Klamer, E. Frangione, and R. Brouillard, Improved state estimation through use of roads in ground tracking, in: O.E. Drummond (Ed.), *Proceedings of the SPIE Conference on Signal Data Processing of Small Targets*, Vol. 4048, Orlando, FL, 2000, pp. 321–332.
20. C.S. Agate and K.J. Sullivan, Road-constrained target tracking and identification using a particle filter, in: O.E. Drummond (Ed.), *Proceedings of the SPIE Conference on Signal Data Processing of Small Targets*, Vol. 5204, San Diego, CA, 2003, pp. 532–543.
21. F. Gustafsson, Particle filter theory and practice with positioning applications, *IEEE Aerosp. Electron. Syst. Mag.*, 25(7), 53–82, 2010.
22. S. Zhang and Y. Bar-Shalom, Track segment association for GMTI tracks of evasive move-stop-move maneuvering targets, *IEEE Trans. Aerosp. Electron. Syst.*, 47(3), 1899–1914, 2011.
23. F.E. Daum and R.J. Fitzgerald, Importance of resolution in multiple-target tracking, in: O.E. Drummond (Ed.), *Proceedings of the SPIE Conference on Signal Data Processing of Small Targets*, Vol. 2235, Orlando, FL, 1994, pp. 329–338.
24. W. Koch and G. van Keuk, Multiple hypothesis track maintenance with possibly unresolved measurements, *IEEE Trans. Aerosp. Electron. Syst.*, 33(3), 883–892, 1997.
25. A.J. Perrella, Jr., Operational tracking issues in the presence of closely spaced objects (CSOs), in: O.E. Drummond (Ed.), *Proceedings of the SPIE Conference on Signal Data Processing of Small Targets*, Vol. 5204, San Diego, CA, 2003, pp. 429–435.
26. W. Koch, Bayesian approach to extended object and cluster tracking using random matrices, *IEEE Trans. Aerosp. Electron. Syst.*, 44(3), 1042–1059, 2008.
27. M. Feldmann, D. Fränken, and W. Koch, Tracking of extended objects and group targets using random matrices, *IEEE Trans. Signal Process.*, 59(4), 1409–1420, 2011.
28. A.K. Gupta and D.K. Nagar, *Matrix Variate Distributions*, Vol. 104 of *Monographs and Surveys in Pure and Applied Mathematics*, Chapman & Hall/CRC Press, Boca Raton, FL, 1999.
29. E.P. Wigner, Random matrices in physics, *SIAM Rev.*, 9(1), 1–23, 1967.
30. J. Wishart, The generalised product moment distribution in samples from a normal multivariate population, *Biometrika*, 20A(1/2), 32–52, 1928.
31. P.J. Forrester, N.C. Snaith, and J.J.M. Verbaarschot, Developments in random matrix theory, *J. Phys. A: Math. Gen.*, 36(12), R1–R10, 2003.
32. Special Issue, Random matrix theory, *J. Phys. A: Math. Gen.*, 36(12), 2859–3645, 2003.
33. S.S. Blackman, *Multiple-Target Tracking with Radar Applications*, Artech House, Norwood, MA, 1986.
34. O.E. Drummond, S.S. Blackman, and G.C. Petrisor, Tracking clusters and extended objects with multiple sensors, in: O.E. Drummond (Ed.), *Proceedings of the SPIE Conference on Signal Data Processing of Small Targets*, Orlando, FL, 1990, Vol. 1305, pp. 362–375.

35. D.J. Salmond and M. C. Parr, Track maintenance using measurements of target extent, *IEE Radar Sonar Navig.*, 150(6), 389–395, 2003.
36. D.A. Harville, *Matrix Algebra From a Statistician's Perspective*, Springer-Verlag, Springer, New York, 1997.
37. M.S. Arulampalam, N. Gordon, M. Orton, and B. Ristic, A variable structure multiple model particle filter for GMTI tracking, *Proceedings of the International Conference on Information Fusion*, Vol. 2 Annapolis, MD, 2002, pp. 927–934.
38. W. Koch and M. Feldmann, Cluster tracking under kinematical constraints using random matrices, *Robot. Autonom. Syst.*, 57(3), 296–309, 2009.
39. M. Feldmann and W. Koch, Road-map assisted convoy track maintenance using random matrices, *Proceedings of the International Conference on Information Fusion*, Cologne, Germany, 2008, pp. 250–257.
40. S. Kotz and S. Nadarajah, *Multivariate t-Distributions and Their Applications*, Cambridge University Press, Cambridge, UK, 2004.
41. M. Ulmke, Improved GMTI-tracking using road-maps and topographic information, in: O.E. Drummond (Ed.), *Proceedings of the SPIE Conference on Signal Data Processing of Small Targets*, San Diego, CA, 2003, Vol. 5204, pp. 143–154.
42. R.P.S. Mahler, Multitarget Bayes filtering via first-order multitarget moments, *IEEE Trans. Aerosp. Electron. Syst.*, 39(4), 1152–1178, 2003.
43. R.P.S. Mahler, PHD filters of higher order in target number, *IEEE Trans. Aerosp. Electron. Syst.*, 43(4), 1523–1543, 2007.
44. R.P.S. Mahler, *Statistical Multisource-Multitarget Information Fusion*, Artech House, Norwood, MA, 2007.
45. O. Erdinc, P. Willett, and Y. Bar-Shalom, The bin-occupancy filter and its connection to the PHD filters, *IEEE Trans. Signal Process.*, 57(11), 4232–4246, 2009.
46. R.L. Streit and L.D. Stone, Bayes derivation of multitarget intensity filters, *Proceedings of the International Conference on Information Fusion*, Cologne, Germany, 2008, pp. 1686–1693.
47. Ba-Tuong Vo, Ba-Ngu Vo, and A. Cantoni, Analytic implementations of the cardinalized probability hypothesis density filter, *IEEE Trans. Signal Process.*, 55(7), 3553–3567, 2007.
48. M. Ulmke, O. Erdinc, and P. Willett, GMTI tracking via the Gaussian mixture cardinalized probability hypothesis density filter, *IEEE Trans. Aerosp. Electron. Syst.*, 46(4), 1821–1833, 2010.
49. Ba-Tuong Vo, Ba-Ngu Vo, and A. Cantoni, The cardinality balanced multi-target multi-Bernoulli filter and its implementations, *IEEE Trans. Signal Process.*, 57(2), 409–423, 2009.

CHAPTER 7

Performance Bounds for Target Tracking: Computationally Efficient Formulations and Associated Applications

MARCEL HERNANDEZ

7.1 INTRODUCTION

In this chapter a review of recent developments in the calculation of mean square error (MSE) tracker performance bounds is presented, together with examples that demonstrate how such bounds can be used as the basis for performing online sensor management. In the context of target tracking, performance bounds are a powerful tool, enabling one to quantify the optimal achievable accuracy of target state estimation. This provides a mechanism for establishing the maximum degree by which suboptimal filtering algorithms could potentially be improved (e.g., [1]). Furthermore, by establishing the physical limit on system performance, these bounds allow one to identify "impossible" scenarios in which the required tracking accuracy cannot be achieved without upgrades in sensor technology.

There are a number of MSE performance bounds (see [2] and references therein), including the Bhattacharya lower bound (BLB) [3], Weiss–Weinstein lower bound (WWLB) [4], and Bobrovsky–Zakai lower bound (BZLB) [5]. However, the posterior Cramér–Rao lower bound (PCRLB) [6] is the most well-known MSE performance bound, primarily because of its low computational complexity compared to the other bounds. Indeed, there has been a great deal of interest in the PCRLB in the last decade, primarily as a result of the excellent paper [7], in which an efficient formulation of the bound was derived for the general nonlinear filtering problem. The computational simplicity of the formulation [7] has enabled the PCRLB to be used to evaluate and compare different system architectures and sensor configurations during the design phase [8], even in time-critical applications. For example, the PCRLB has been used

Integrated Tracking, Classification, and Sensor Management: Theory and Applications, Edited by Mahendra Mallick, Vikram Krishnamurthy, and Ba-Ngu Vo.

in automating the deployment of passive sonobuoys in anti-submarine warfare [9], in controlling a large network of bandwidth limited sensors in multitarget tracking [10, 11], in performing UAV trajectory planning in bearings-only tracking [12–14], and in controlling a phased array radar in order to geo-locate antiship missiles [15]. Alternative measures of performance, including information gain and entropy, have been proposed as mechanisms for sensor control in target detection, tracking, and classification applications (e.g., see [16]). However, as noted in [16], such measures typically involve evaluating difficult nonanalytical integrals of functions of the target state density, thereby making real-time sensor management difficult.

In this chapter, we detail two applications in which controlling the PCRLB provides an effective mechanism for online sensor management. The first application is concerned with the management of stationary sensors (e.g., sonobuoys tethered to the sea surface) in order to track a single, potentially manoeuvring target (e.g., a submersible). The approach used is principally that of [9], but with two minor modifications that make maximal use of the available information and limited sensor resources. We do not consider multitarget scenarios, but the reader is referred to Chapter 14 for such extensions (see also [10, 11]). The second application is UAV trajectory planning, again in tracking a single target. The general approach builds on the previous work [12–14] to provide an effective and efficient mechanism of managing the UAV platforms.

Using the PCRLB as a mechanism for performing online sensor management relies on efficient formulations of the bound existing in scenarios that are subject to real-world complexities such as measurement origin uncertainty and evasive target behavior. Fortunately, this is indeed true. In particular, efficient PCRLB formulations exist in the case of cluttered environments, in which there are both missed detections and false positives. For example, for the problem with deterministic (i.e., not randomly evolving) target dynamics, the measurement origin uncertainty was shown (in [17]) to manifest itself as a constant information reduction factor (IRF). This IRF scales the measurement contribution, such that the Cramér–Rao bound increases in magnitude (i.e., optimal performance is degraded) as the environment becomes more cluttered (i.e., as the number of false positives increases). The paper [18] (see also [19]) then derived a general expression for the PCRLB for the linear filtering problem with randomly evolving target dynamical equations. The paper [20] extended [18] to provide a PCRLB for the nonlinear filtering problem. The methodology detailed in the papers [18, 20] is referred to as the "IRF approach."

However, it has recently been shown that the IRF approach provides an optimistic bound on tracker performance [21–23]. This is primarily because the IRF approach does not take into account the exact sequence of missed detections, which can have a considerable impact on performance, particularly during the early stages of tracking. In [21], an enumeration approach was introduced for the case with missed detections (i.e., $P_d \leq 1$) but no false alarms (i.e., $P_{fa} = 0$). This enumeration approach created a bound conditional on the sequence of detections, and then created an unconditional bound as a weighted average. It was then shown in [22] that the enumeration bound was greater than the IRF bound, and therefore less optimistic. The paper [23] generalized the enumeration bound of [21] to the case in which there were also false alarms (i.e., $P_d \leq 1$ and $P_{fa} \geq 0$) by introducing the "measurement sequence conditioning" (MSC) approach. The MSC approach created a bound conditional on the number of measurements at each sampling time, and then created an unconditional bound as a weighted average (weighted by the probability of each particular sequence occurring). It was then shown (again, see [23]) that the MSC bound was also less optimistic than the corresponding IRF bound.[1] In simulations it was then shown that differences between the two bounds are greatest initially, when the impact of the actual measurement sequence is most significant.

In [24], the measurement existence sequence conditioning (MESC) approach was introduced. This is similar to the MSC approach, but rather than condition on the number of measurements, MESC conditions on at least one measurement existing at each sampling time. In simulations, the resulting MESC bound was shown to be almost identical to the MSC bound [24]. In the case with no false alarms, it has recently been shown in [25–27] that a modified Riccati (MR) equation provides an upper bound on the enumeration bound [21], but with a much lower computational overhead.[2] In simulations, the MR equation was shown to bound the enumeration PCRLB reasonably tightly, making it an effective measure of system performance. However, the MR equation cannot be guaranteed to provide a bound on tracker performance, and can only be applied in the case $P_{fa} = 0$.

Also of great interest recently has been the development of the PCRLB for the problem of tracking a manoeuvring target with Markovian model switching dynamics. This is a hybrid estimation problem where it is required to estimate sequentially both the *continuous*-valued target state and the *discrete*-valued target model motion. Direct application of the PCRLB recursive formula for nonlinear filtering [7] would lead to differentiation of terms involving the discrete-valued model variable. Since this cannot be done analytically, in one's quest for the lower bound, there are two options at one's disposal. The first is to explore alternatives to the Cramér–Rao bound, which are potentially applicable to hybrid system estimation problem (e.g., BLB, WWLB, and BZLB [28]). The difficulty with these bounds is that the derivations are extremely complex and their implementation is quite involved.

The second option is to consider *approximations* to the PCRLB for Markovian switching systems. In [29], an approach was presented in which a conditional PCRLB was calculated by conditioning on the manoeuvre sequence. The unconditional PCRLB was then determined as a weighted sum of these conditional bounds. However, this formulation typically provides an overly optimistic lower bound, because at each stage, one implicitly *knows* the sequence of manoeuvres that the filtering algorithm is required to estimate.

In [30, 31] the multimodal prior target probability density function (PDF) of the manoeuvring target was approximated by a best-fitting unimodal Gaussian (BFG) distribution. The Riccati-like recursive formula of [7] was then applied to calculate a performance measure, which is not guaranteed to provide a performance bound

[1] Which generalized the finding of [22] to the case in which $P_{fa} \geq 0$.

[2] The MR approach has a computational complexity similar to that of the IRF approach [18].

because of the approximation involved. Despite this, the performance measure is an accurate predictor of the steady-state performance of state-of-the-art tracking algorithms, such as the interacting multiple model (IMM) filter [32]. Furthermore, the mean and covariance of the unimodal BFG distribution are given by a simple recursive formula, allowing the BFG measure to be calculated extremely efficiently. This makes the approach of [30, 31] invaluable in predicting system performance in manoeuvring target scenarios.

The rest of this chapter is organized as follows. In Section 7.2 we define the basic filtering problem, and provide details of a general class of lower bounds that includes as special cases the PCRLB, BLB, WWLB, and BZLB. We also provide details of the excellent paper of [7] that introduced an efficient recursion for the PCRLB, with similar recursions also now existing for other classes of lower bound [33, 34]. In Section 7.3, we review developments in the calculation of computationally efficient expressions for the PCRLB in cluttered environments, and in Section 7.4 we detail the BFG approach that provides an approximate PCRLB for manoeuvring target tracking.

Following this, in Section 7.5 we consider the first of two sensor management applications, providing a framework for the deployment of stationary sensors in tracking manoeuvring targets. Simulations, concerned with the deployment of passive sonobuoys in submarine tracking, show the accuracy with which the PCRLB can predict system performance, and this allows the sensors to be deployed in a timely manner to effectively control the estimation error. In Section 7.6, we consider the problem of UAV trajectory planning in tracking a highly manoeuvring airborne target, and again demonstrate that the PCRLB can be effectively utilized as a measure of system performance. Finally, conclusions are given in Section 7.7.

7.2 BAYESIAN PERFORMANCE BOUNDS

7.2.1 The Estimation Problem

Let $X_k \in R^{n_x}$ denote the target state at discrete sampling time $t_k, k = 0, 1, \ldots$. The target state evolves according to a Markov process given as follows: $X_{k+1} = f_k(X_k, \omega_k)$, where $f_k(\cdot)$ is potentially a nonlinear function of the target state and w_k is an independent and identically distributed (IID) process noise sequence.

Measurements $Z_k \in R^{n_z}$ are available at the discrete sampling times t_k, $k = 1, 2, \ldots$. Each measurement is given as follows: $Z_k = h_k(X_k, \epsilon_k)$, where $h_k(\cdot)$ is also potentially a nonlinear function of the target state and ϵ_k is also an IID measurement noise sequence. The objective is to estimate the target state vector $X_{0:k} = \left\{X_0', \ldots, X_k'\right\}'$ at each sampling time k based on the sequence of measurements $Z_{1:k} = \left\{Z_1', \ldots, Z_k'\right\}'$.

7.2.2 A General Class of Lower Bounds

Let $\hat{X}_{0:k}$ denote any unbiased estimator of $X_{0:k}$, based on the sequence of measurements $Z_{1:k}$. In [28], the Schwarz inequality was invoked to determine a general class of MSE performance bounds for the nonlinear estimation problem. This class of lower bounds is given as follows

$$E\left[(X_{0:k} - \hat{X}_{0:k})(X_{0:k} - \hat{X}_{0:k})'\right] \geq V_{0:k} J_{0:k}^{-1} V_{0:k}'. \tag{7.1}$$

The inequality in (7.1) means that the difference is a nonnegative definite matrix; $J_{0:k}$ is referred to as the *Fisher information matrix* (FIM).

The $(k+1)n_x \times (k+1)n_x$ matrices $V_{0:k}$ and $J_{0:k}$ are given by

$$V_{0:k} = E\left[X_{0:k} \times \Psi(X_{0:k}, Z_{1:k})'\right], \tag{7.2}$$

$$J_{0:k} = E\left[\Psi(X_{0:k}, Z_{1:k}) \times \Psi(X_{0:k}, Z_{1:k})'\right]. \tag{7.3}$$

The expectations are with respect to $X_{0:k}$ and $Z_{1:k}$; and $\Psi(X_{0:k}, Z_{1:k})$ is a vector of dimensionality $(k+1)n_x$ of real-valued, measurable functions that satisfies

$$\int_{X_{0:k}} \Psi(X_{0:k}, Z_{1:k}) p(X_{0:k}, Z_{1:k}) dX_{0:k} = 0, \tag{7.4}$$

where $p(X_{0:k}, Z_{1:k})$ denotes the joint PDF of $X_{0:k}$ and $Z_{1:k}$.

The vector function $\Psi(X_{0:k}, Z_{1:k})$ defines the bound. For example, the general BLB is obtained by choosing $\Psi(X_{0:k}, Z_{1:k})$ as follows (e.g., [34])

$$\Psi(X_{0:k}, Z_{1:k}) = \sum_{i=1}^{M} \frac{a_i}{p(X_{0:k}, Z_{1:k})} \nabla^i_{X_{0:k}} p(X_{0:k}, Z_{1:k}), \tag{7.5}$$

where $\nabla^i_v = \left[\partial^i/\partial v_1^i, \ldots, \partial^i/\partial v_n^i\right]'$ for a vector $v = [v_1, \ldots, v_n]'$, i.e., ∇^i_v denotes the ith-order partial derivative operator with respect to v. The parameter M is an arbitrary natural number, $a_i = 1$ and a_i $(i > 1)$ are arbitrary real numbers. It can easily be shown that for this choice of Ψ (7.4) holds.[3] Of particular interest, the PCRLB (e.g., [6]) is a special case of the BLB in which $M = 1$, that is

$$\Psi(X_{0:k}, Z_{1:k}) = \frac{\nabla_{X_{0:k}} p(X_{0:k}, Z_{1:k})}{p(X_{0:k}, Z_{1:k})}. \tag{7.8}$$

[3]To provide an outline proof

$$\int_{X_{0:k}} \Psi(X_{0:k}, Z_{1:k}) p(X_{0:k}, Z_{1:k}) dX_{0:k} = \sum_{i=1}^{M} a_i \int_{X_{0:k}} \nabla^i_{X_{0:k}} p(X_{0:k}, Z_{1:k}) dX_{0:k}, \tag{7.6}$$

$$= \sum_{i=1}^{M} a_i \nabla^i_{X_{0:k}} p(Z_{1:k}) = 0. \tag{7.7}$$

In this case, it can easily be shown that $V_{0:k} = -I_{(k+1)n_x}$, where I_m denotes the m-dimensional identity matrix.[4] The PCRLB is then given as follows

$$E\left[(X_{0:k} - \hat{X}_{0:k})(X_{0:k} - \hat{X}_{0:k})'\right] \geq J_{0:k}^{-1}, \tag{7.12}$$

where $J_{0:k}$ is given by

$$J_{0:k} = E\left[\frac{\nabla_{X_{0:k}} p(X_{0:k}, Z_{1:k}) \left[\nabla_{X_{0:k}} p(X_{0:k}, Z_{1:k})\right]'}{p(X_{0:k}, Z_{1:k})^2}\right], \tag{7.13}$$

$$= E\left[-\Delta_{X_{0:k}}^{X_{0:k}} \log p(X_{0:k}, Z_{1:k})\right], \tag{7.14}$$

Δ is a second-order partial differential operator, defined as follows: $\Delta_{\Psi}^{\Theta} = \nabla_{\Psi}\nabla_{\Theta}'$.

In [28], a further class of vectors $\Psi(X_{0:k}, Z_{1:k})$ satisfying (7.4) is introduced, giving the multiple parameter WWLB. Furthermore, in [34] a general expression for the multiple parameter BZLB is provided.

In exemplar scenarios, the PCRLB has been shown to be the least tight of the performance bounds [34], although the result has not been proven in general. However, the relative optimism of the PCRLB is not surprising when one considers the fact that the PCRLB is a special case of the BLB, and is also a limiting case of the WWLB and BZLB. The BLB, WWLB, and BZLB all contain free parameters that can be optimized in order to make the bound as tight (i.e., large) as possible.

However, of critical importance the PCRLB is computationally (by far) the simplest of the performance bounds. This makes it an appealing measure of system performance in time-critical sensor management applications, in which the ability to quickly evaluate and compare sensor configurations is invaluable.

7.2.3 Efficient Fixed Dimensionality Recursions

The limitation with each of the bounds in Section 7.2.2 is that the dimensionality of the bound grows linearly as the sampling time k increases. As a result, the matrix manipulations required in calculating each bound become increasingly complex, and must be re-calculated at every time step. However, in [7], the authors considered the general nonlinear filtering problem, and were interested only in the PCRLB for the estimation of the target state at time k, that is

$$E\left[(X_k - \hat{X}_k)(X_k - \hat{X}_k)'\right] \geq J_k^{-1}, \tag{7.15}$$

[4] Again, as an outline

$$V_{0:k}' = \int_{X_{0:k}} \left\{\int_{Z_{1:k}} \nabla_{X_{0:k}} p(X_{0:k}, Z_{1:k}) dZ_{1:k}\right\} X_{0:k}' dX_{0:k} = \int_{X_{0:k}} \left[\nabla_{X_{0:k}} p(X_{0:k})\right] X_{0:k}' dX_{0:k}, \tag{7.9}$$

$$= \int_{X_{0:k}} \left\{\nabla_{X_{0:k}} \left[p(X_{0:k}) X_{0:k}'\right] - p(X_{0:k}) I_{(k+1)n_x}\right\} dX_{0:k}, \tag{7.10}$$

$$= \nabla_{X_{0:k}} \int_{X_{0:k}} p(X_{0:k}) X_{0:k}' dX_{0:k} - I_{(k+1)n_x} = -I_{(k+1)n_x}. \tag{7.11}$$

where J_k^{-1} is the bottom right $n_x \times n_x$ block of the matrix $J_{0:k}^{-1}$; with $J_{0:k}$ given by (7.14).

In [7], the Markov property of the target state transition equation was exploited in order to provide the following fixed dimensionality, Riccati-like recursion for the sequence J_k

$$J_k = D_{k-1}^{33} - (D_{k-1}^{12})'(J_{k-1} + D_{k-1}^{11})^{-1} D_{k-1}^{12} + J_Z(k), \quad k = 1, 2, \ldots \tag{7.16}$$

where

$$D_{k-1}^{11} = E\left[-\Delta_{X_{k-1}}^{X_{k-1}} \log p(X_k | X_{k-1})\right], \tag{7.17}$$

$$D_{k-1}^{12} = E\left[-\Delta_{X_{k-1}}^{X_k} \log p(X_k | X_{k-1})\right], \tag{7.18}$$

$$D_{k-1}^{33} = E\left[-\Delta_{X_k}^{X_k} \log p(X_k | X_{k-1})\right], \tag{7.19}$$

$$J_Z(k) = E\left[-\Delta_{X_k}^{X_k} \log p(Z_k | X_k)\right]. \tag{7.20}$$

The Riccati-like recursion (7.16) generalizes the CRLB of [35] to nondeterministic systems.

Calculation of the D_k matrices is (generally) straightforward. Indeed, if $f_k(\cdot)$ is a linear function of X_k (i.e., $f_k(X_k, \omega_k) = A_k X_k + \omega_k$) and $\omega_k \sim N(0, \Sigma_k)$, then it can easily be shown that (e.g., [36])

$$D_k^{11} = A_k' \Sigma_k^{-1} A_k, \tag{7.21}$$

$$D_k^{12} = -A_k' \Sigma_k^{-1}, \tag{7.22}$$

$$D_k^{33} = \Sigma_k^{-1}. \tag{7.23}$$

In this case, if we apply the matrix inversion lemma to (7.16), using relations (7.21)–(7.23), then

$$J_k = \left(\Sigma_{k-1} + A_{k-1} J_{k-1}^{-1} A_{k-1}'\right)^{-1} + J_Z(k). \tag{7.24}$$

The initial FIM is given by

$$J_0 = E\left[-\Delta_{X_0}^{X_0} \log p(X_0)\right]. \tag{7.25}$$

If the prior distribution is Gaussian with covariance C_0, then $J_0 = C_0^{-1}$.

Section 7.3 is devoted to the calculation of the matrix $J_Z(k)$ for a scenario with multiple sensors, each with independent measurement processes. Each sensor can have missed detections and spurious false alarms. In this case, the measurement likelihood, $p(Z_{k+1}|X_{k+1})$ is a mixture distribution (being a mixture of the likelihoods of target generated measurements and false alarm measurements) making $J_Z(k)$ difficult to calculate. We show that under certain assumptions a decomposition exists, and a constant IRF quantifies the effect of the measurement origin uncertainty.

In [34], the Markov property of the state transition equation was again exploited to show that for both the BLB and BZLB, the sequence of information submatrices is also given by (7.16), but with more complex D matrices in these cases. Moreover, [34] also introduced a fully recursive form for the information submatrix for the WWLB for the estimation of X_k (see also [33]). An efficient PCRLB recursion has also been determined for multitarget tracking [37].

7.3 PCRLB FORMULATIONS IN CLUTTERED ENVIRONMENTS

7.3.1 Measurement Model

Purely for notational simplicity we will focus analysis on a single sensor scenario. The extension to multisensor systems is straightforward provided that the sensors have independent measurement processes. The key expressions in the multisensor case are also included for completeness.

Let m_k denote the number of measurements (each of dimensionality n_z) available at the sensor at sampling time k, m_k can include both measurements generated by the target and false alarms. The measurement model specified in [18] in used throughout. In this model, there can be a maximum of one target generated measurement per sampling time, which occurs with constant probability P_D. Each target generated measurement, Z_k^T is a nonlinear function of the target state, corrupted by white Gaussian noise, that is

$$Z_k^T = h_k(X_k) + \epsilon_k, \tag{7.26}$$

where $\epsilon \sim N(0, R_k)$.

At each sampling time the number of false alarms has a Poisson distribution with mean λV, where V denotes the volume of the surveillance region A, and λ is the density of false alarms per unit volume. Each false alarm measurement, Z_k^F is uniformly distributed throughout the surveillance region, that is

$$Z_k^F = u_k, \tag{7.27}$$

where $p(u_k) = 1/V$ for $u_k \in A$.

In the PCRLB recursion (7.16) only the $J_Z(\cdot)$ matrix is dependent on the measurement model, with the D-matrices dependent solely on the target dynamical model. In the following sections, several different formulations of the PCRLB are presented in cluttered environments. These formulations differ in the way in which they condition on the measurement sequence.

7.3.2 Information Reduction Factor Approach

The IRF approach [18, 20] calculates the measurement contribution as follows

$$J_Z(k) = E_{m_k}\left[J_Z(k:m_k)\right], \tag{7.28}$$

$$= \sum_{m_k=1}^{\infty} Pr(m_k) J_Z(k:m_k), \tag{7.29}$$

where $Pr(m_k)$ is the probability mass function (PMF) of m_k, and $J_Z(k:m_k)$ is the measurement contribution at time k given that there are m_k measurements at that time. This is given as follows

$$J_Z(k:m_k) = E_{X_k,Z_k}\Big[\left[\nabla_{X_k} \ln p(Z_k|X_k,m_k)\right]\left[\nabla_{X_k} \ln p(Z_k|X_k,m_k)\right]'\Big]. \tag{7.30}$$

The FIM is then given by the recursion (7.16), or by (7.24) if the target dynamic are linear. The IRF PCRLB is given by J_k^{-1}.

Conditional on several unrestrictive assumptions (see Section 7.3.5), and after significant calculations [20, 23], it can be shown that

$$J_Z(k) = q_k E_{X_k}\left[H_k' R_k^{-1} H_k\right], \tag{7.31}$$

where $H_k' = \nabla_{X_k}\left[h_k(X_k)'\right]$ is the Jacobian of the measurement vector (of target generated measurements), and

$$q_k = \left[\sum_{m_k=1}^{\infty} Pr(m_k) q_k(m_k)\right]. \tag{7.32}$$

The variable $q_k(m_k) \in [0, 1]$ is the IRF conditional on there being m_k measurements at sampling time k [23] and $q_k \in [0, 1]$ is the overall IRF [20]. These IRFs are independent of the target state, and quantify the effect of missed detections and false alarms. If there is no clutter and there are no missed detections, then $q_k = 1$. In all other cases $q_k < 1$ and the IRF reduces the measurement contribution (compared to the no clutter case), and therefore increases the PCRLB.

For completeness, in Section 7.3.5, a general expression for the IRF $q_k(m_k)$ taken from [23] (see (7.46) herein) is given, along with a numerical approximation (see (7.51)). Of particular note, if the measurement, clutter, and detection models are time invariant then so too is the IRF $q_k(m_k)$, which need then only be calculated once (for each sensor). This calculation can be performed offline, enabling the PCRLB to be calculated efficiently, and implemented in real-time applications [9].

In the case of an N sensor system, with independent measurement processes at each sensor, (7.31) generalizes to

$$J_Z(k) = \sum_{i=1}^{N} q_k^i E_{X_k}\left[(H_k^i)'(R_k^i)^{-1}H_k^i\right], \tag{7.33}$$

where the superscript "i" denotes that the corresponding term is sensor specific.

An extension of the IRF approach also exists for the problem of tracking multiple targets in cluttered environments [11]. The sequence of manipulations is even more complex in that case, and the measurement origin uncertainty can be shown to manifest itself as an information reduction matrix.

7.3.3 Measurement Sequence Conditioning Approach

Let $m_{1:k}$ denote the sequence comprising of the number of measurements available at times $1, \ldots, k$, that is $m_{1:k} = \{m_1, \ldots, m_k\}$. The MSC approach [23] creates an MSE estimator performance bound as follows

$$E\left[(\hat{X}_k - X_k)(\hat{X}_k - X_k)'\right] = E_{m_{1:k}}\left[E\left[(\hat{X}_k - X_k)(\hat{X}_k - X_k)'\middle| m_{1:k}\right]\right], \tag{7.34}$$

$$\geq E_{m_{1:k}}\left[J_k^{-1}(m_{1:k})\right], \tag{7.35}$$

$$= \sum_{m_{1:k}}\left[Pr(m_{1:k}) \times J_k^{-1}(m_{1:k})\right], \tag{7.36}$$

$$\triangleq \text{MSC PCRLB}, \tag{7.37}$$

where $Pr(m_{1:k})$ is the probability of the sequence of measurements $m_{1:k}$ occurring; and $J_k(m_{1:k})$ is the FIM conditional on the measurement sequence $m_{1:k}$.

Conditioning on $m_{1:k}$, and repeating the derivation of [7], the recursive formula for the conditional FIM is given by

$$J_k(m_{1:k}) = D_{k-1}^{33} - \left(D_{k-1}^{12}\right)'\left[J_{k-1}(m_{1:k-1}) + D_{k-1}^{11}\right]^{-1} D_{k-1}^{12} + J_Z(k : m_k). \tag{7.38}$$

$J_Z(k : m_k)$ is again the measurement contribution at time k given that there are m_k measurements at that time (see (7.30)). The initial FIM is given by $J_0 = C_0^{-1}$ irrespective of the sequence $m_{1:k}$.

Clearly, $J_Z(k : 0) = 0$, that is there is no contribution if there are no measurements. For $m_k > 0$ and after some calculations (again, see [23]) it can then be shown that

$$J_Z(k : m_k) = q_k(m_k) E_{X_k}\left[H_k' R_k^{-1} H_k\right], \tag{7.39}$$

where $q_k(m_k)$ is again the IRF conditional on there being m_k measurements at sampling time k; $q_k(m_k)$ is given later in (7.46).

If the target dynamics are linear, then (7.38) can be simplified as follows (see (7.24))

$$J_k(m_{1:k}) = \left(\Sigma_{k-1} + A_{k-1}J_{k-1}(m_{1:k-1})A'_{k-1}\right)^{-1} + J_Z(k : m_k). \quad (7.40)$$

To extend the MSC formulation to an N sensor system, $m_{1:k}$ is replaced with $\mathcal{M}_{1:k} = \{M_1, \ldots, M_k\}$ in the recursions (7.38) and (7.40); where $M_k = \left\{m_k^1, \ldots, m_k^N\right\}$ and m_k^i denotes the number of measurements at sensor i at sampling time k. The measurement contribution given in (7.39) then generalizes to

$$J_Z(k : M_k) = \sum_{i=1}^{N} q_k^i(m_k^i) E_{X_k}\left[(H_k^i)'(R_k^i)^{-1}H_k^i\right], \quad (7.41)$$

again, the superscript "i" denotes the sensor-specific terms.

The MSC approach generalizes the enumeration approach of [21] to the case in which there are both false alarms and missed detections. To see this, if there are no false alarms (i.e., $\lambda = 0$), then $m_k = 0$ or 1 and it can easily be shown that $q_k(1) = 1$. Hence the approach of conditioning on the measurement sequence reduces to conditioning on the detection sequence, as in [21].

7.3.4 Measurement Existence Sequence Conditioning Approach

The MESC approach [24] is similar in principle to the MSC approach [23], but rather than condition on the number of measurements generated at each sampling time, the MESC approach conditions on the *existence* of at least one measurement. Therefore, instead of conditioning on $m_{1:k}$, the MESC approach conditions on $e_{1:k} = \{e_1, \ldots, e_k\}$ where $e_k = 1$ if $m_k \geq 1$ and $e_k = 0$ otherwise. The performance bound is then given as follows

$$\text{MESC PCRLB} \triangleq \sum_{e_{1:k}} \left[Pr(e_{1:k}) \times J_k^{-1}(e_{1:k})\right]. \quad (7.42)$$

Repeating the derivation of the recursive formula (7.16), conditional this time on $e_{1:k}$, it can easily be shown that

$$J_k(e_{1:k}) = D_{k-1}^{33} - \left(D_{k-1}^{12}\right)'\left[J_{k-1}(e_{1:k-1}) + D_{k-1}^{11}\right]^{-1} D_{k-1}^{12} + J_Z(k : e_k). \quad (7.43)$$

Again, the measurement contribution is zero if there are no measurements, that is $J_Z(k : e_k = 0) = 0$. In the case $e_k = 1$, it can then be shown that [24]

$$J_Z(k : e_k = 1) = \frac{q_k}{Pr(e_k = 1)} E_{X_k}\left[H_k'R_k^{-1}H_k\right]. \quad (7.44)$$

The IRF, q_k is again as given in (7.32). Finally, if the target dynamics are linear, (7.43) can be simplified as follows (see (7.24))

$$J_k(e_{1:k}) = \left(\Sigma_{k-1} + A_{k-1} J_{k-1}(e_{1:k-1}) A'_{k-1}\right)^{-1} + J_Z(k : e_k). \tag{7.45}$$

The extension to multisensor systems is again straightforward but is omitted for brevity.

7.3.5 Calculation of the Information Reduction Factors

7.3.5.1 *General Expression.* In calculating the IRFs, the following unrestrictive assumptions are made (e.g., see [23]) regarding the measurement model.

- There is a maximum of one target generated measurement per sampling time, which occurs with probability P_d. P_d does not have to be constant (e.g., in [9] the probability of detection was range dependent), but the IRF must be calculated for each potential value of P_d.
- False alarm measurements are independent of the target state. It is also commonly assumed (e.g., see [18, 20]) that the number of false alarms has a Poisson distribution with mean λV, where V is the volume of the surveillance region. However, it is not necessary to use a Poisson distribution to model the number of false alarms.
- The error covariance of target generated measurements is a diagonal matrix, that is $R_k = \text{diag}\left(\sigma_1^2, \ldots, \sigma_{n_z}^2\right)$. If this assumption does not hold an information reduction matrix (rather than an IRF) quantifies the effect of the measurement origin uncertainty.

It is common practice to only consider false alarms that fall in a gated region of volume V_g around the target (e.g., see (39) in [23]). Hence the average number of false alarms considered is λV_g per sampling time. The IRF, $q_k(m_k)$ is then given as follows (e.g., see (54) in [23])

$$q_k(m) = \frac{\epsilon_g(m)^2 |R_k|^{(m-2)/2}}{m V_g^{2m-2} (2\pi)^{n_z}} \left\{ \int_{Z_{mn_z}=-g}^{g} \cdots \int_{Z_{m1}=-g}^{g} \cdots \int_{Z_{1n_z}=-g}^{g} \cdots \int_{Z_{11}=-g}^{g} \right. \tag{7.46}$$

$$\left. \frac{Z_{11}^2 \exp\left\{-\sum_{i=1}^{n_z} Z_{1i}^2\right\}}{\left[\frac{(1-\epsilon_g(m))}{V_g^m} + \frac{\epsilon_g(m)}{m V_g^{m-1} \sqrt{(2\pi)^{n_z} |R_k|}} \sum_{r=1}^{m} \exp\left\{-\frac{1}{2} \sum_{j=1}^{n_z} Z_{rj}^2\right\}\right]} dZ_{11} \ldots dZ_{mn_z} \right\},$$

for $m = 1, 2, \ldots$, with $q_k(0) = 0$. The volume of the gated observation region is given by

$$V_g = (2g)^{n_z} \prod_{i=1}^{n_z} \sigma_i = (2g)^{n_z} \sqrt{|R_k|}. \tag{7.47}$$

The parameter $\epsilon_g(m_k)$ is the probability that one of the m_k measurements is target generated. This is given as follows

$$\epsilon_g(m_k) = \frac{P_d^g}{p_g(m_k)} \frac{(\lambda V_g)^{m_k - 1} \exp\left(-\lambda V_g\right)}{(m_k - 1)!}, \tag{7.48}$$

where $p_g(m_k)$ is the probability that there are m_k observations in the gated observation region

$$p_g(m_k) = (1 - P_d^g) \frac{(\lambda V_g)^{m_k} \exp\left(-\lambda V_g\right)}{m_k!} + P_d^g \frac{(\lambda V_g)^{m_k - 1} \exp\left(-\lambda V_g\right)}{(m_k - 1)!}. \tag{7.49}$$

P_d^g is the probability of obtaining a target generated measurement in the gated observation region. This is given by

$$P_d^g = P_d \left[\Phi(g) - \Phi(-g)\right]^{n_z}, \tag{7.50}$$

where $\Phi(\cdot)$ is the cumulative distribution function of a standard Normal random variable. Clearly, provided g is not small (and typically $g \geq 4$) then $P_d^g \approx P_d$.

We note that if the measurement, clutter, and detection models are time invariant then so too is the IRF, $q_k(m_k)$, which then need only be calculated once.

7.3.5.2 Numerical Approximation. We can create a Monte Carlo approximation of (7.46) as follows (see (58) in [23])

$$q_k(m_k) \approx \frac{(2g)^{n_z m_k} \epsilon_g(m_k)^2 |R_k|^{(m_k-2)/2}}{m_k V_g^{2m_k - 2} (2\pi)^{n_z}} \tag{7.51}$$
$$\times \frac{1}{N_i} \sum_{l=1}^{N_i} \frac{U_{11}[l]^2 \exp\left\{-\sum_{i=1}^{n_z} U_{1i}[l]^2\right\}}{\left[\frac{(1-\epsilon_g(m_k))}{V_g^{m_k}} + \frac{\epsilon_g(m_k)}{m_k V_g^{m_k-1} \sqrt{(2\pi)^{n_z} |R_k|}} \sum_{r=1}^{m_k} \exp\left\{-\frac{1}{2} \sum_{j=1}^{n_z} U_{rj}[l]^2\right\}\right]},$$

where $U_{ij}[l]$, $i = 1, \ldots, m_k$; $j = 1, \ldots, n_z$; $l = 1, \ldots, N_i$ are IID random variables each drawn from a uniform distribution on $[-g, g]$ (i.e., each $U_{ij}[l] \sim U[-g, g]$).

7.3.6 Relationships Between the Various Performance Bounds

The following relationship, proven in [24], holds between the three performance bounds (at all sampling times)

$$\text{IRF PCRLB} \leq \text{MESC PCRLB} \leq \text{MSC PCRLB}. \tag{7.52}$$

This extends the result of [23], in which it was shown that

$$\text{IRF PCRLB} \leq \text{MSC PCRLB}. \tag{7.53}$$

We note that the inequality (7.53) was first proven in [22], for the case with no false alarms.

It should be noted that the above inequalities hold in general, and are not reliant on the measurement model prescribed in Section (7.3.1). Indeed, the proofs of (7.52) and (7.53) are reliant only on convexity of the matrix inversion operation and Jensen's inequality (e.g., see Appendix A in [22]).

There are three special cases that are worthy of consideration.

- *Case 1*: There are no false alarms (i.e., $\lambda = 0$). In this case $m_k = e_k$ ($= 0$ or 1) for all k and the MSC and MESC approaches give identical bounds, equal to the enumeration bound of [21].
- *Case 2*: There are no missed detections (i.e., $P_d = 1$). In this case $e_k = 1$ for all k, $Pr(e_k = 1) = 1$ and the IRF and MESC approaches give identical bounds.
- *Case 3*: There are no false alarms and no missed detections. In this case, $m_k = e_k = 1$ for all k; and $q_k = q_k(1) = 1$. Clearly then all three performance bounds are identical, with each giving the no-clutter PCRLB.

The IRF PCRLB is computationally the simplest of the three performance bounds, but does not consider the impact of the actual sequence of measurements (it simply scales the measurement contribution by a constant factor at each sampling time). However, the actual sequence of measurements can have a critical impact on tracking performance, particularly after track formation, in which target state uncertainty can be high. At such times the IRF PCRLB can be optimistic compared to the MSC PCRLB [22, 23]. Differences between the two bounds are greatest when target generated measurements are accurate and the initial target state uncertainty is large.

The MESC approach has the advantage that it can be enumerated exactly, because there are precisely 2^k existence sequences $e_{1:k}$ at each sampling time k. By comparison, the MSC approach has an infinite number of potential measurement sequences. Many of these sequences have an extremely low probability of occurrence and can be instantly discounted. However, in implementing both the MSC and MESC approaches, the number of sequences that need to be considered is still prohibitively large, even for relatively small values of k. A more scalable method of estimating the MSC PCRLB (or MESC PCRLB) is to sample measurement (or existence) sequences, and determine the average value of the sequence-dependent bound (see Section VI in [23]). Although the MSC approach produces the tightest performance bound, simulations have yielded insignificant differences between the MESC and MSC bounds [24]. Furthermore, as the number of sampling times increases, all three bounds converge to the same steady-state value [23, 24].

7.4 AN APPROXIMATE PCRLB FOR MANEUERING TARGET TRACKING

7.4.1 Motion Model

Let us now consider the discrete-time linear filtering problem with multiple switching dynamic models and additive Gaussian noise. The target motion is then modeled as follows

$$X_{k+1} = F_k^{r_{k+1}} X_k + w_k(r_{k+1}), \tag{7.54}$$

where r_{k+1} specifies the target motion model (or regime) which is in effect during the time interval between sampling times k and $k+1$; $w_k(r_{k+1})$ is additive Gaussian process noise, with $w_k(r_{k+1}) \sim \mathcal{N}\big(0, Q_k^{r_{k+1}}\big)$. We assume the target motion can switch between N_m models. The evolution of the motion model sequence is modeled by a time-homogenous Markov chain with known

- transition probabilities: $p_{ij} \triangleq Pr(r_{k+1} = j | r_k = i)$ for any $i, \ j \in \{1, 2, \ldots, N_m\}$;
- initial motion model probabilities: $p_1(i) \triangleq Pr(r_1 = i)$ for $i \in \{1, 2, \ldots, N_m\}$.

7.4.2 Best-Fitting Gaussian Approach

The BFG approach [30, 31] replaces the linear jump Markov system specified by (7.54) with the BFG approximation

$$X_{k+1} \approx A_k X_k + W_k, \quad W_k \sim \mathcal{N}(0, \Sigma_k). \tag{7.55}$$

Hence, at each stage we want to replace the regime-dependent state transition equation (given by (7.54), and referred to as "model 1") with a single BFG distribution (given by (7.55), and referred to as "model 2"). A_k and Σ_k will be chosen so that at each stage the distribution of X_{k+1} has the same mean and covariance under each model, that is

$$E\left[X_{k+1}|\text{model 1}\right] = E\left[X_{k+1}|\text{model 2}\right] \triangleq \varepsilon_{k+1}, \tag{7.56}$$

$$\text{Cov}\left[X_{k+1}|\text{model 1}\right] = \text{Cov}\left[X_{k+1}|\text{model 2}\right] \triangleq C_k, \tag{7.57}$$

for all k. Finally, an approximate PCRLB is then given by the Riccati-like recursion (7.24). It has been shown [30, 31] that the BFG approach produces a measure of

TABLE 7.1 Recursive Calculation of the BFG Approximation of [30, 31]

INITIALIZATION

- Assuming the initial distribution of the state vector is Gaussian with mean $\bar{X}_0$ and covariance P_0, that is $X_0 \sim \mathcal{N}(\bar{X}_0, P_0)$, set: $\varepsilon_0 = \bar{X}_0$ and $\mathcal{C}_0 = P_0$.
- Initialize mode probabilities: $p_1(r)$, $r = 1, \ldots, N_m$.

GENERAL RECURSION (FOR $k = 0, 1, 2 \ldots$)

$$A_k = \sum_{r=1}^{N_m} F_k^r \, p_{k+1}(r), \tag{7.58}$$

$$\mathcal{C}_{k+1} = \sum_{r=1}^{N_m} p_{k+1}(r) \left[F_k^r \left(\mathcal{C}_k + \varepsilon_k \varepsilon_k' \right) (F_k^r)' + Q_k^r \right] - A_k \varepsilon_k \varepsilon_k' A_k', \tag{7.59}$$

$$\Sigma_k = \mathcal{C}_{k+1} - A_k \mathcal{C}_k A_k', \tag{7.60}$$

$$\varepsilon_{k+1} = A_k \varepsilon_k, \tag{7.61}$$

$$p_{k+2}(r) = \sum_{j=1}^{N_m} p_{jr} \, p_{k+1}(j), \qquad r = 1, \ldots, N_m. \tag{7.62}$$

performance (MOP) (clearly, no longer a bound) that is in close agreement with the RMSE performance of the IMM filter [32]. Conversely, an earlier conditional approach [29] produced an extremely optimistic bound because of its implicit assumption that the sequences of target manoeuvres were known. The BFG approach can be used in conjunction with any of the cluttered environment approaches presented in Section 7.3, in order to provide a MOP for tracking manoeuvring targets in cluttered environments.

An alternative BFG approach which matched the state *transition* mean and covariance[5] (rather than the state mean and covariance) was presented in [38]. The measures of performance calculated using the BFG approaches [30] and [38] are of similar accuracy (albeit with the BFG approach of [38] producing a lower RMSE in simulations). However, as a key discriminator, the BFG approach presented herein has a significantly lower computational expense, making it a more desirable approach, particularly when quantifying system performance in time-critical applications.

7.4.3 Recursive Computation of Best-Fitting Gaussian Approximation

A fully recursive set of equations giving ε_k, $\mathcal{C}_k$, A_k, and Σ_k for all $k \geq 0$ is given in Table 7.1. In Figures 7.1 and 7.2 an illustration of the BFG approach is provided, for a scenario (taken from [31]) in which a target can switch between constant velocity and coordinated-turn motion models (e.g., see [39] for a description of these motion models). In Figure 7.1 an initial sample of 10,000 particles is drawn from the prior

[5]The state transition mean and covariance are $E\left[X_{k+1}|X_k\right]$ and $\text{Cov}\left[X_{k+1}|X_k\right]$, respectively.

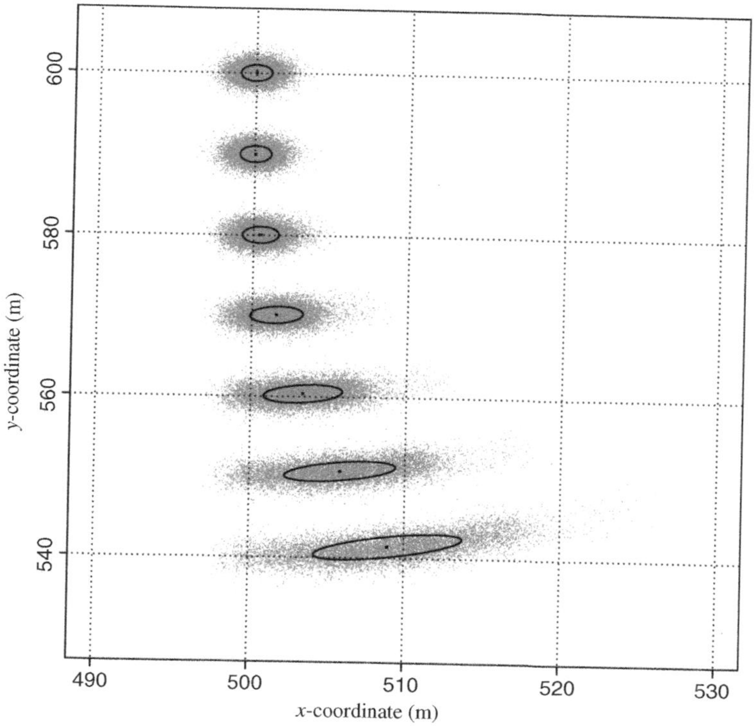

Figure 7.1 Potential target paths (gray dots) and BFG covariance ellipses (in black).

target distribution, and particle evolutions at times 1, 2, ..., 6 s are shown based on the two motion models.[6] The black ellipses show the BFG distribution covariance ($\mathcal{C}_k$) at each stage, with the center of the ellipse giving the mean (ε_k) of this distribution.

In Figure 7.2 we show the BFG PDF at sampling time $k = 6$. Also shown is an approximation of the *true* target PDF at this time. As we can see, the BFG approximation fits the true x-coordinate distribution well. It is also a reasonable fit in the case of the y-coordinate distribution, although the true distribution has heavier tails and a less pronounced maximum.

7.5 A GENERAL FRAMEWORK FOR THE DEPLOYMENT OF STATIONARY SENSORS

7.5.1 Introduction

This section is concerned with the management of stationary sensors (e.g., sonobuoys tethered to the sea surface) in order to track a single, potentially manoeuvring target (e.g., a submersible). We do not consider multitarget scenarios, but the reader is

[6]These particles are shown for illustrative purposes only (to show potential target paths) and do not form part of the calculation of ε_k or $\mathcal{C}_k$.

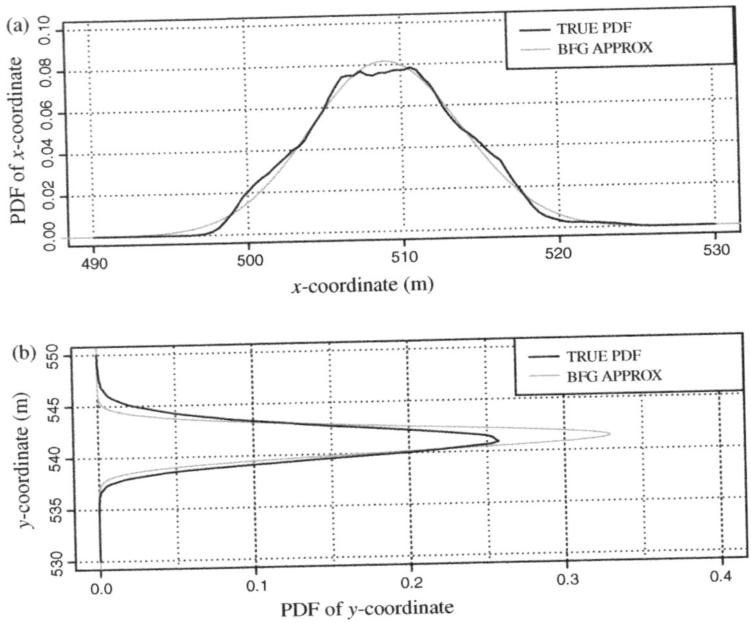

Figure 7.2 The true target state PDF and BFG approximation in the (a) x-coordinate direction, and (b) y-coordinate direction, each at sampling time $k = 6$ (corresponding to the final ellipse shown in Figure 7.1).

referred to [10, 11] for extensions to multitarget surveillance and tracking. The work of [10, 11] is presented in Chapter 12 of this book.

We allow for the possibility that there are restrictions in the number of sensors that can be utilized at any one time. For example, there may be communication bandwidth limitations, in which case only a limited number of the active sensors can transmit measurements to the fusion center at each sampling time. There are therefore three inter-related optimization problems that need to be addressed. These are as follows:

- *The Interval Between Sensor Deployments.* The issue is in determining the time at which it is likely that either: (i) the target will be too distant from the current sensor array, or (ii) the sensor/target geometry will be unfavorable. In both cases the existing sensors will no longer be able to effectively triangulate the target location.
- *The Use of Existing Sensors.* Which of the sensors already deployed should be utilized at each sampling time, or subsequent to a new deployment.
- *The Number of Sensors and Configuration of the Next Deployment.* Specifically, how many sensors should be deployed, and where should they be positioned in order that the sensor array can best estimate the target state.

The general framework presented herein is principally the approach of [9], but with two modifications. The first modification is that the time at which a new sensor deployment should be made is continually reassessed as additional information (i.e., measurements) becomes available. This is a "just-in-time" approach whereby we only commit to a deployment at the very last time instance possible. The second deviation from the approach of [9] is that the framework has the functionality to deploy new sensors in order to minimize the total predicted (and discounted) tracking performance[7] at multiple sampling times subsequent to the deployment. This is similar to the approach advocated in [40], which modified the approach of [9] to place the sensors further ahead of the target and in its likely path. Simulation results showed that the approach of [40] was still able to accurately track the target whilst significantly reducing the number of sensors required.

7.5.2 Interval Between Deployments

Measurements are available as discrete time instances t_k $(= Tk)$, $k \geq 1$; where T is the time interval between measurements. At each of these times, the algorithm predicts when (one or more) new sensors should be deployed. In calculating this deployment time, existing (i.e., already deployed) sensors will be reused until they are predicted to no longer make an effective contribution to the tracking of the target (e.g., they are too distant from the target, or the sensor/target geometry is such that the existing sensors can no longer effectively triangulate the target location).

We assume that there is an interval between the time at which a decision is made to deploy new sensors, and the time at which the sensors can be physically deployed and activated. We let t_D denote the minimum time interval (in minutes) between sensor deployments. The interval t_D should take into account both the computation time required to determine the most effective locations of the new sensors and the time required to subsequently deploy the sensors.

Let $t_k^\star$ denote the *tentative* time of the next deployment, determined at sampling time k. Let $T^\star(n)$ then denote the *confirmed* time of the nth deployment of sensors. Once a deployment time is confirmed it will not change. However, a tentative deployment time may increase or decrease depending on the impact of the subsequent measurements on the estimate of the target state.

A tentative deployment time $t_k^\star$ is calculated at the following time instances:

- *Instance* 1: Immediately following the first update of the tracking algorithm at sampling time $k = 1$. We note that initialization of the target track is assumed to occur at time $k = 0$.
- *Instance* 2: Immediately following a deployment of one or more new sensors, that is at sampling time $k = T^\star(n)$, $n = 1, 2, \ldots$.
- *Instance* 3: Provided we are not within t_D minutes of the previous tentative deployment time $t_{k-1}^\star$, that is provided $k \leq (t_{k-1}^\star - t_D)$.

[7]The MOP utilized herein is based on the PCRLB. Full details of the MOP can be found in Section 7.5.5.

At such time instances, $t_k^\star$ is then given as follows

$$t_k^\star = \inf\left\{r : b_{r|k} > L_1, r \geq (k + t_D)\right\}, \tag{7.63}$$

where $b_{r|k}$ denotes the predictive MOP at sampling time r, based on the information/measurements available up to time k. Specifically $b_{r|k}$ depends on the target state and motion model estimates at sampling time k. The method of calculating the MOP is given in Table 7.2. Therefore, $t_k^\star$ is the first occasion at which the performance measure $b_{r|k}$ exceeds the threshold L_1, with the constraint that the minimum time required to make the sensor deployment is t_D minutes. The threshold L_1 reflects the mission objective, with a small value denoting the need to accurately track the target.

TABLE 7.2 Procedure for Determining the Performance Measure $b_{k+\varsigma|k}$

STEP 1—SAMPLE POTENTIAL TARGET STATE EVOLUTIONS

- **FOR** $j = 1, \ldots, n_s$:
  - SAMPLE: $X_{k|k}^{(j)} \sim \mathcal{N}(\hat{X}_{k|k}, C_{k|k})$
    (where $\mathcal{N}(\hat{X}_{k|k}, C_{k|k})$ is the overall target state estimate at time k)
  - SAMPLE: $r_{k|k}^j \sim \hat{p}(r)$ (where $\hat{p}(r) = Pr(r_k = r|Z_{1:k})$)
  - **FOR** $r = 1, \ldots, \varsigma$
    * SAMPLE: $r_{k+r|k}^j \sim Pr(r_{k+r}|r_{k+r-1|k}^j)$
      (i.e., using the known transition probabilities between motion models)
    * SAMPLE: $X_{k+r|k}^j \sim p(X_{k+r}^j|X_{k+r-1}^j, r_{k+r|k}^j)$
      (i.e., using the state transition equation)
  - **END FOR**
- **END FOR**

STEP 2—IMPLEMENT GENERAL RECURSION

- SET $J_k = C_{k|k}^{-1}$
- **FOR** $k_1 = k + 1, \ldots, (k + \varsigma)$
  - **IF** $[k_1 < T^\star]$:
    * DETERMINE $\Phi(N, N_t, k_1, k)$
      (i.e., the subset of N out of N_t sensors perceived to be closest to the target)
    * DETERMINE $J_Z(k_1)$ using (7.66)
      (using the N sensors identified on the previous step)
  - **ELSE**:
    * DETERMINE $\Psi(N - N_n, N_t, k_1, k)$
      (i.e., the subset of $(N - N_n)$ out of N_t sensors perceived to be closest to the target)
    * DETERMINE $J_Z(k_1)$ using (7.66)
      (using the $(N - N_n)$ sensors identified on the previous step and N_n new sensors)
  - DETERMINE J_{k_1} using the recursive formula (7.65)
  - DETERMINE $b_{k_1|k}$ using (7.67)
- **END FOR**

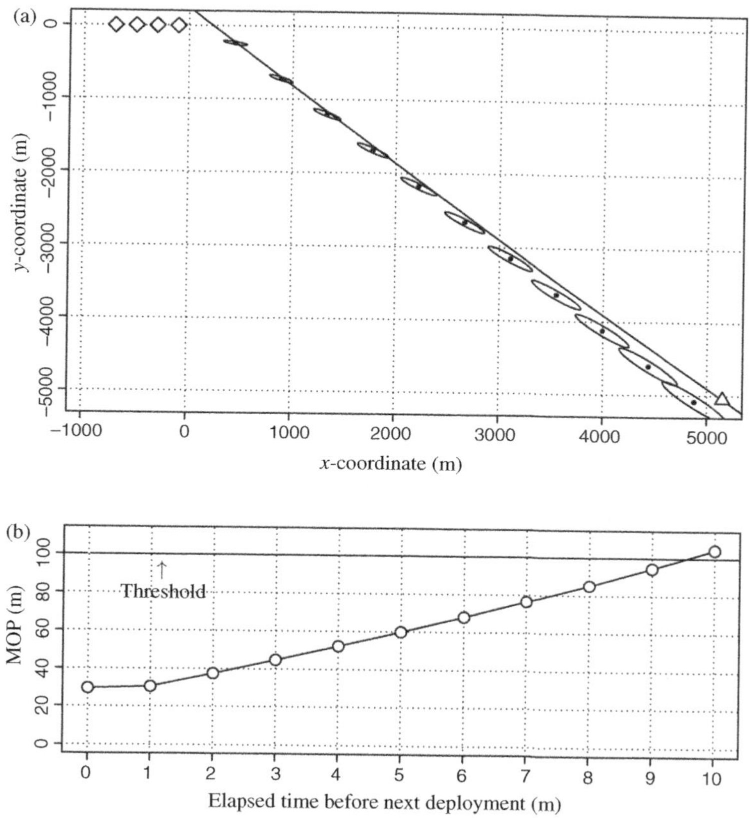

Figure 7.3 *Determining a tentative interval between deployments*. (a) The existing sensor locations are shown by diamonds, and the series of ellipses each show four times the BFG approach RMSE at intervals of 1 min. Each ellipse is centered on the estimated target location at that time. The black line shows the true target path (moving in a South-Easterly direction), and the triangle shows the target location at the proposed time of the deployment. (b) The MOP plotted against elapsed time. In this exemplar, the threshold at which we will deploy a new sensor is $L_1 = 100$ m (shown in black). Consequently, in this case it is tentatively predicted that the next sensor deployment should be made after an elapsed time of 10 min (i.e., $t_1^\star = 11$).

An example, in which $t_1^\star$ is determined can be found in Figure 7.3 (above). The scenario is as specified later in Section 7.5.7, and is also used in generating Figures 7.4–7.8. The parameter settings are given in Table 7.4 (with the maximum target acceleration, $a_T = 0.001$ m/s^2; and the probability of target nonmanoeuvre, $p_{ii} = 0.9$).

We sequentially calculate $t_k^\star$, $k = 1, 2 \ldots$, until: $t_k^\star = (k + t_D)$ at which point the next deployment is "confirmed" and we commit to making a deployment at time $T^\star(\cdot) = t_k^\star$ (e.g., see Figure 7.4). The reason that this deployment time has been confirmed is that in this case, if we were to delay in deploying and instead calculate $t_{k+1}^\star$,

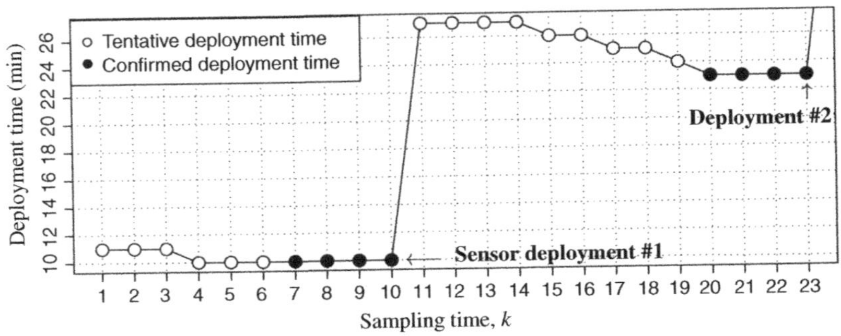

Figure 7.4 *A sequence of "tentative" and "confirmed" deployment times.* In this example, the minimum time required to make a deployment is $t_D = 3$ min.

the earliest we could then deploy would be at time $t_k^\star + 1$. However, we may find that $b_{t_k^\star|k+1} > L_1$, meaning that by delaying we would have subsequently missed an opportunity to control the performance measure adequately at time $t_k^\star$.

To summarize:

- If either $k = 1$, $k = T^\star(\cdot)$, or $k \le (t_{k-1}^\star - t_D)$, a tentative deployment time $t_k^\star$ is determined using (7.63).
- Subsequently, if $t_k^\star = (k + t_D)$, this tentative deployment time becomes a confirmed deployment time, that is $T^\star(\cdot) = t_k^\star$.

The reader should note that in [9], the time at which a deployment should be made was determined only initially (i.e., at sampling time $k = 1$) and immediately subsequent to the previous deployment (i.e., at each sampling time $k = T^\star(n), n \ge 1$). This deployment time was then fixed. Consequently, the approach of [9] did not have the flexibility to react as new information became available.

This ability to reassess and react as new information becomes available is particularly important in the manoeuvring target scenarios we will consider herein, because by delaying the decision on whether to deploy until the last time instance possible, we reduce the target state uncertainty by the maximum degree possible.

7.5.3 Use of Existing Sensors

If the total number of deployed sensors, N_t, is greater than the number of sensors that can be used at any one time, N, then at each sampling time k we must determine the subset of the sensors to be utilized. The procedure for doing this is as follows:

- *Except at a Deployment Time (i.e., for $k \ne T^\star$):* Use the N sensors closest to the perceived target location.
- *At a Deployment Time (i.e., for $k = T^\star$):* Use the N_n new sensors together with the $(N - N_n)$ previously deployed sensors closest to the perceived target location.

This procedure is used both in the acquisition of sensor measurements, and in determining the performance measure $b_{k+\varsigma|k}$ used in optimizing the deployment time, sensor locations, and number of new sensors. We let $\Phi(N, N_t, k+\varsigma, k)$ denote the subset of N sensors (out of N_t) perceived to be closest to the target location at sampling time $k+\varsigma$, based on the target state estimate at sampling time k.

We note that using the subset of sensors closest to the target can be suboptimal, because of the potential for selecting a subset of sensors with an unfavorable geometry relative to the target (i.e., a geometry that is unable to triangulate the target location effectively). However, in managing the trade-off between optimality and computational expense, selecting the closest sensors is a reasonable approximation.

At each sampling time $k+\varsigma, (\varsigma = 0, 1, \ldots)$, the "perceived" sensor/target distance is given by the average distance between the sensor and n_s potential target state evolutions sampled from the prior, that is the average distance between the sensor and $X^j_{k+\varsigma|k} \sim p(X_{k+\varsigma}|Z_{1:k})$, $j = 1, \ldots, n_s$. The procedure for calculating these samples is given in STEP 1 of Table 7.2.

The perceived sensor/target distance is therefore given by

$$\text{perceived distance} = \frac{1}{n_s}\sum_{j=1}^{n_s}\sqrt{(x^S - x^j_{k+\varsigma|k})^2 + (y^S - y^j_{k+\varsigma|k})^2 + (z^S - z^j_{k+\varsigma|k})^2},$$

where $(x^j_{k+\varsigma|k}, y^j_{k+\varsigma|k}, z^j_{k+\varsigma|k})$ denotes the geo-location of the jth sample $X^j_{k+\varsigma|k}$; and (x^S, y^S, z^S) is the sensor location.

7.5.4 Locations and Number of New Sensors

At sampling time $k = (T^\star - t_D)$ we are committed to making a sensor deployment at time $T^\star$. We then seek to minimize the following overall MOP

$$B_R = \sum_{i=0}^{R-1} \gamma^i b_{T^\star + i\Delta t|T^\star - t_D}. \tag{7.64}$$

Clearly, if $R = 1$, then $B_R = b_{T^\star|T^\star - t_D}$, that is the performance measure at the deployment time. More generally, for $R > 1$, B_R is the sum of discounted performance at R future sampling time. This is referred to as R step planning. The parameter Δt is the number of sampling times between instances at which we evaluate future performance (we take $\Delta t = 3$ herein), and $\gamma \in [0, 1]$ is a factor by which we discount future performance.

On each deployment, the minimum number of new sensors N_n is positioned so that, when used in conjunction with the existing sensors, the subsequent performance measure at the deployment time $b_{T^\star|T^\star_k - t_D}$ is no greater than a second prespecified threshold L_2. We note that only in the case of $R = 1$ step planning do we actually position the sensors in order to minimize $b_{T^\star|T^\star - t_D}$. Therefore, when using a longer planning horizon, the new sensors are typically positioned further ahead of the target in order to control performance over multiple steps. In such cases it may then be necessary to deploy a greater number of sensors than would otherwise be the case in order to control performance at time $T^\star$. Pseudo-code of the complete sensor deployment algorithm can be found in Table 7.3.

TABLE 7.3 Pseudo-code of the sensor deployment algorithm

STEP 0—INITIALIZATION

- GENERATE samples $X^j_{0|0}$, $j = 1, \ldots, n_s$, from the prior distribution $p(X_0) \sim \mathcal{N}(\hat{X}_{0|0}, C_{0|0})$
- GENERATE samples r^j_0, $j = 1, \ldots, n_s$, from the prior distribution $p(r_0)$
- GENERATE $(X^j_{1|0}, r^j_{1|0})$, $j = 1, \ldots, n_s$ using the procedure on STEP 1 of Table 7.2
- SET: $t^\star_k = -\infty$ for $k \geq 1$, $N_d = 0$, $T^\star(N_d) = -\infty$, $N_t = N_0$ and $k = 1$

STEP 1—UPDATE TARGET STATE ESTIMATE

- **IF** $k = T^\star(N_d)$—INCREMENT total sensors: $N_t = (N_t + N_n)$ **END IF**
- DETERMINE which sensors to utilize at time k: $\Phi(N, N_t, k, k-1)$
- DETERMINE $p(X_k|Z_{1:k}) \sim \mathcal{N}(\hat{X}_{k|k}, C_{k|k})$ by performing one iteration of the tracker

STEP 2—DETERMINE TENTATIVE DEPLOYMENT TIME

- **IF** $[k = 1]$ OR $[k = T^\star(N_d)]$ OR $[k \leq (t^\star_{k-1} - t_D)]$
  - SET $J_k = C^{-1}_{k|k}$ and $r = 0$
  - **DO**
    - * INCREMENT: $r = (r + 1)$
    - * GENERATE $X^j_{k+r|k}$, $j = 1, \ldots, n_s$ using the procedure on STEP 1 of Table 7.2
    - * DETERMINE which sensors to utilize at time $k + r$: $\Phi(N, N_t, k + r, k)$
    - * CALCULATE $J_Z(k + r)$ using (7.66)
    - * CALCULATE J_{k+r} using recursion (7.65)
    - * CALCULATE $b_{k+r|k}$ using (7.67)
  - **WHILE** $(b_{k+r|k} < L_1$ OR $r < t_D)$
  - SET the tentative deployment time: $t^\star_k = k + r$
- **END IF**

STEP 3—DETERMINE NUMBER/LOCATION(S) OF NEW SENSOR(S)

- **IF** $[k = (t^\star_k - t_D])$
  - SET number of new sensors: $N_n = 0$
  - INCREMENT number of deployments: $N_d = (N_d + 1)$
  - SET confirmed deployment time: $T^\star(N_d) = t^\star_k$
  - **DO**
    - * INCREMENT: $N_n = (N_n + 1)$
    - * DETERMINE N_n new sensor locations at time $t^\star_k$ using R step planning
    - * CALCULATE $b_{t^\star_k|k}$ (as in the DO loop on STEP 2)
  - **WHILE** $(b_{t^\star_k|k} > L_2$ AND $N_n < N)$
- **END IF**

STEP 4—RECURSIVE STEP

- SET $k \to (k + 1)$
- REPEAT from STEP 1

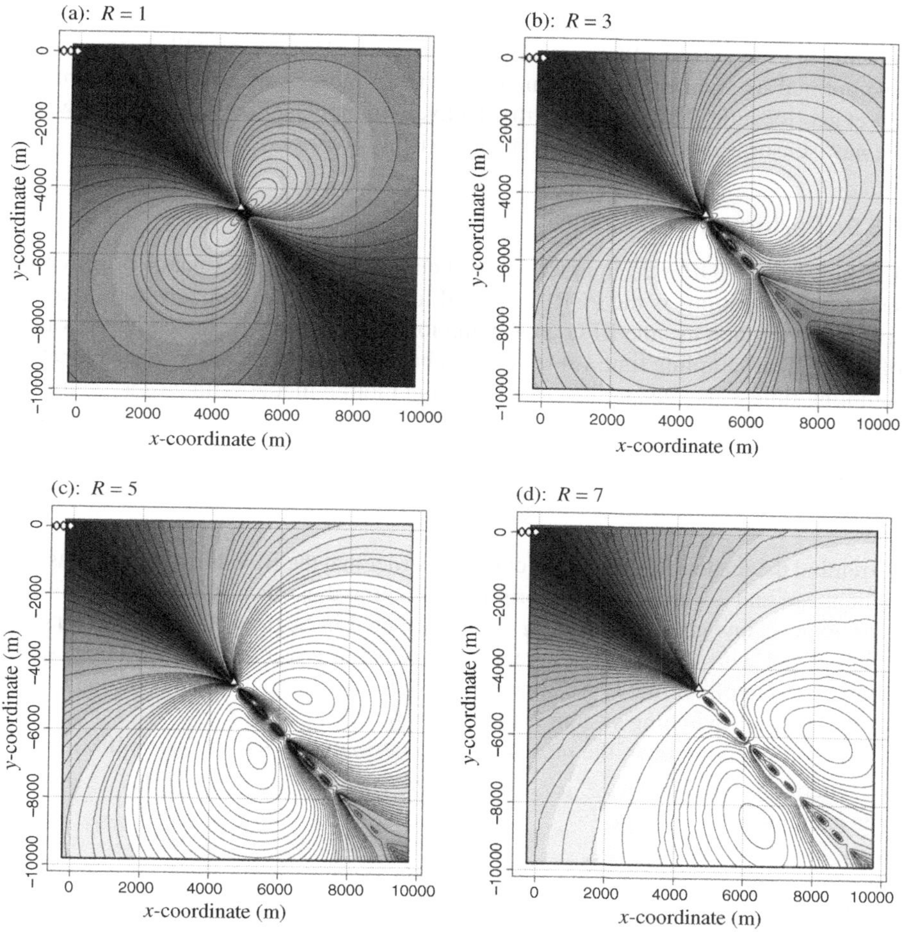

Figure 7.5 Contours plots of the overall MOP, B_R at the time of a single sensor deployment, for $R = 1, 3, 5, 7$. The lightest regions show low (i.e., desirable) values of the MOP, and the blackest regions show high values of the MOP. The positions of existing sensors are shown by white diamonds. The white triangle (close to the center of each plot) shows the true target location at the time of the sensor deployment.

Exemplar contour plots of the overall MOP, B_R for $R = 1, 3, 5, 7$ are shown in Figure 7.5. In each case, the overall MOP is based on the deployment of a single sensor. The scenario is as described later in Section 7.5.7. In particular, four sonobuoys are positioned in the top left corner of Figures 7.5a–d, and the target is moving in a South-Easterly direction at a speed of approximately 22 knots. In this case the probability of target nonmanoeuvre (in each coordinate direction), $p_{ii} = 0.9$; the maximum target acceleration in the x- and y-coordinate directions, $a_T = 0.001$ m/s^2; and all other parameter settings are given in Table 7.4. This sensor/target geometry is also shown in Figures 7.3 and 7.9.

As expected, as R increases, the optimal location for a single new sensor (shown in white in Figures 7.5a–d) moves ahead of the target along its most likely trajectory. The optimal location is also multimodal, with two modes offset from the estimated target geo-location. This is in order to effectively triangulate the target location when the new sensor is used in conjunction with existing sensors.

7.5.5 Performance Measure

The predictive MOP used in assessing and comparing the performance of different sensor configurations is the PCRLB overall location RMSE. In manoeuvring target scenarios, we then use the BFG approximation[30, 31]. Let $b_{k+\varsigma|k}$ denote the value of the MOP at sampling time $k+\varsigma$ ($\varsigma \geq 1$), based on the information available up to sampling time k (i.e., using the target state and motion model estimates at sampling time k). In order to calculate $b_{k+\varsigma|k}$ we first determine the sequence of FIMs for the BFG system. This sequence is given as follows

$$J_{k+1} = \left(A_k J_k A_k' + \Sigma_k\right)^{-1} + J_Z(k+1). \tag{7.65}$$

To remind the reader, A_k and Σ_k are chosen in order that the BFG linear system has the same mean and covariance as the original switching dynamical system. The closed-form expressions giving A_k and Σ_k were presented in Section 7.4.3. The recursion is initialized with $J_k = C_{k|k}^{-1}$, where $C_{k|k}$ is the error covariance of the target state estimate at sampling time k. An approximate MSE performance bound at time $k+\varsigma$ for the original switching system is then given by $J_{k+\varsigma}^{-1}$.

At each sampling time, the measurement contribution $J_Z(k+\varsigma)$ is approximated by generating samples $(X_{k+\varsigma|k}^j, r_{k+\varsigma|k}^j) \overset{\text{approx}}{\sim} p(X_{k+\varsigma}, r_{k+\varsigma}|Z_{1:k})$, $j = 1, \ldots, n_s$ and then averaging

$$J_Z(k+\varsigma) \approx \frac{q_k}{n_s} \sum_{j=1}^{n_s} \sum_{s=1}^{N} \left[H_{k+\varsigma}(X_{k+\varsigma|k}^j; s)' R_{k+\varsigma}^{-1} H_{k+\varsigma}(X_{k+\varsigma|k}^j; s)\right]. \tag{7.66}$$

The parameter q_k denotes the IRF (again, see Section 7.3.2). Clearly, our approach provides an approximate IRF PCRLB [18, 20] for manoeuvring target tracking. Alternatively, we could have accounted for the measurement origin uncertainty using either the MSC approach [23] or MESC approach [24], which are each known to create a less optimistic measure/bound, but at the cost of a significant increase in computational complexity.

The parameter N is the total number of sensors that be utilized at any time instance (e.g., as a result of communication bandwidth limitations). If the total number of sensors deployed (denoted by N_t) is greater than N, at each sampling time we utilize the N sensors perceived to be closest to the target (again, see Section 7.5.3).

The MOP is then given by the following approximate location RMSE performance bound

$$b_{k+\varsigma|k} = \sqrt{J_{k+\varsigma}^{-1}[x] + J_{k+\varsigma}^{-1}[y] + J_{k+\varsigma}^{-1}[z]}, \tag{7.67}$$

where $J_{k+\varsigma}^{-1}[\zeta]$ denotes the entry of $J_{k+\varsigma}^{-1}$ giving the MSE in the ζ-coordinate direction. The procedure for calculating the MOP, $b_{k+\varsigma|k}$ is summarized in Table 7.2. We note that the MOP uses the most recent tracker state estimate (i.e., at sampling time k). The accuracy with which it predicts system performance therefore depends on both the accuracy of the most recent target state estimate and the uncertainty in the future evolution of the target state.

7.5.6 Efficient Search Technique

In order to determine the sensor locations that minimize the overall MOP, B_R, we utilize an efficient search technique [12] that is a variant of the bisection method of numerical analysis (e.g., see [41]). The efficient search technique [12] enables one to quickly determine a near-optimal solution in a two-dimensional (2D) space. A *back-tracking* step is also incorporated that is effective in helping the search to break away from local optima. This is extremely important in problems for which the MOP can be multimodal—which is often the case (e.g., see Figure 7.5). It is straightforward to extend the efficient search to three dimensions and/or optimize the locations of multiple sensors [13].

The 2D search technique is now described by way of an example. Overlaid in Figures 7.6 and 7.7 is the overall MOP shown in Figure 7.5c (i.e., with $R = 5$), and the requirement is to determine the point at which this MOP is minimized. The first step in the search evaluates the MOP at nine points that divide the search region into four equally sized subquadrants, these points are shown by white diamonds in Figure 7.6a. The quadrant in which the average value of the MOP across the four corner points is lowest is then downselected. In this case, this is the bottom right quadrant. The second step (shown in Figure 7.6b) then evaluates the MOP at nine equally spaced points within this quadrant, and on this occasion determines an improved solution (in each plot a black diamond shows the best solution determined across all points evaluated).

If at any stage a nine-point division/evaluation results in an improvement in the best solution determined to that point in the search, the step number is incremented and a subdivision is then performed in the constituent quadrant which has the lowest average value of the MOP across the four corner points. In the example scenario, such divisions are shown in Figures 7.6c and 7.6d. The search continues for a prespecified number of steps (i.e., subdivisions).

A back-tracking step comes into play if a subdivision does not result in an improvement in the best solution found to that point in the search. This is the case on step 5 (shown in Figure 7.7a). The back-tracking step (step 5b) then subdivides the quadrant with second best overall performance (see Figure 7.7b). If no subsequent improvement in the best solution is found (which is the case here), further back-tracking steps will be necessary, that is subdividing the quadrants with third best (step c), and fourth best (i.e., worst) average performance (step d). These further back-tracking steps are omitted for brevity. If after completing all three back-tracking steps no improved solution has been determined, the algorithm reverts back to the nine points evaluated

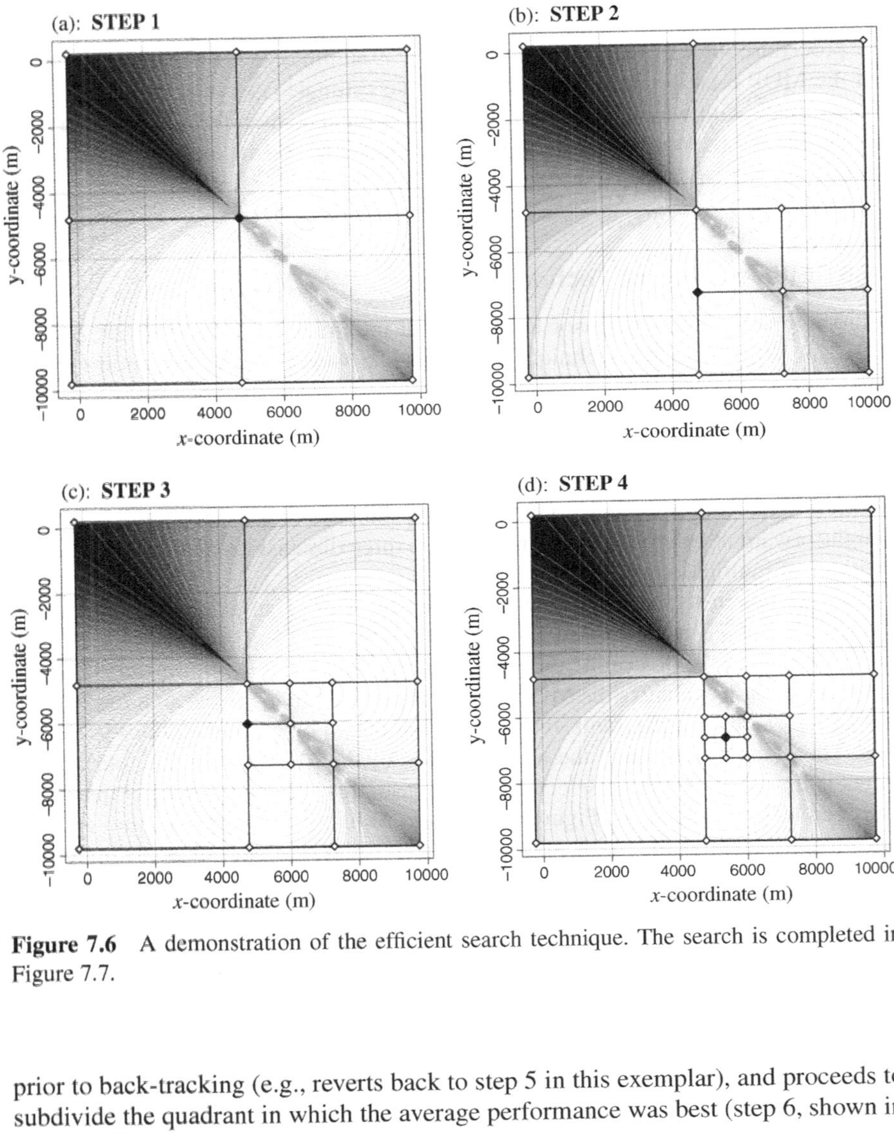

Figure 7.6 A demonstration of the efficient search technique. The search is completed in Figure 7.7.

prior to back-tracking (e.g., reverts back to step 5 in this exemplar), and proceeds to subdivide the quadrant in which the average performance was best (step 6, shown in Figure 7.7c).

For the illustrative example, the complete search (of 10 steps) and near-optimal solution are shown in Figure 7.7d, with a more detailed view shown in Figure 7.8.

7.5.7 Example—Sonobuoy Deployment in Submarine Tracking

7.5.7.1 *Measurement Model.* Passive sonobuoys are utilized. These provide angular measurements of azimuth and elevation. Hence, at sampling time k and for a

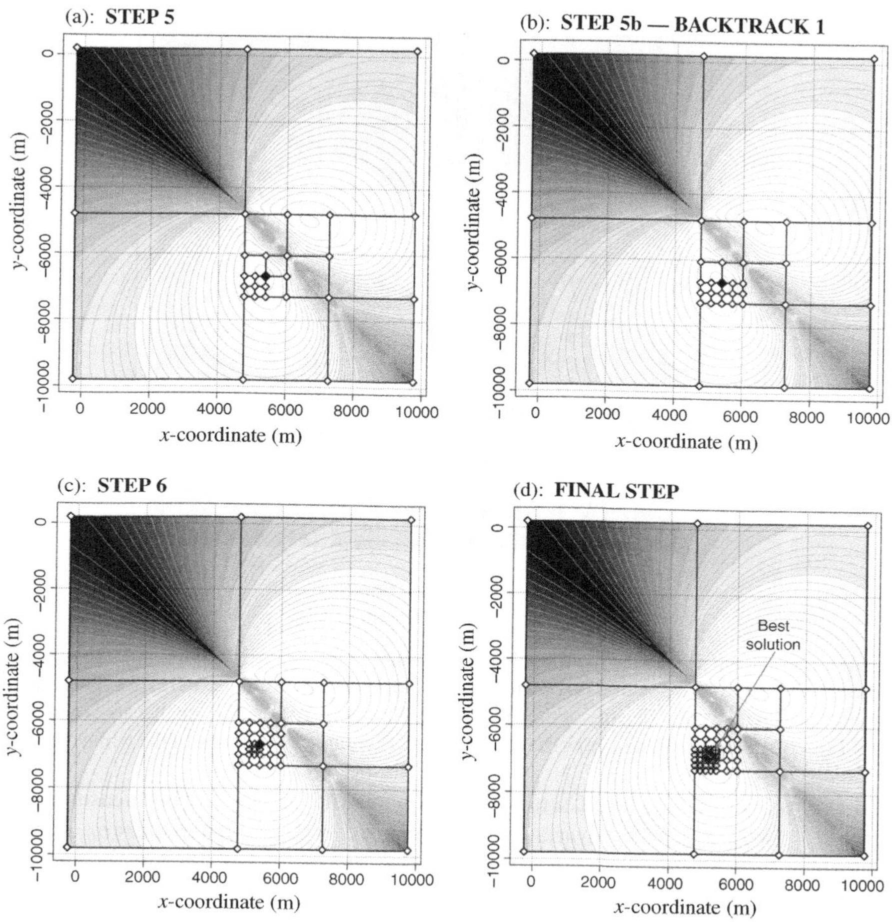

Figure 7.7 Completion of the efficient search. In subplot (d) the best solution found in the entire search is shown by a white diamond with an internal cross (located in a South–South-Easterly direction from the center of the grid).

sonobuoy positioned at (x^S, y^S, z^S), each target generated measurement is given by

$$Z_k = h_k(X_k) + \epsilon_k, \tag{7.68}$$

where

$$h_k(X_k) = \begin{pmatrix} \tan^{-1}\left(\dfrac{y_k - y^S}{x_k - x^S}\right) \\ \tan^{-1}\left(\dfrac{z_k - z^S}{\sqrt{(x_k - x^S)^2 + (y_k - y^S)^2}}\right) \end{pmatrix}, \tag{7.69}$$

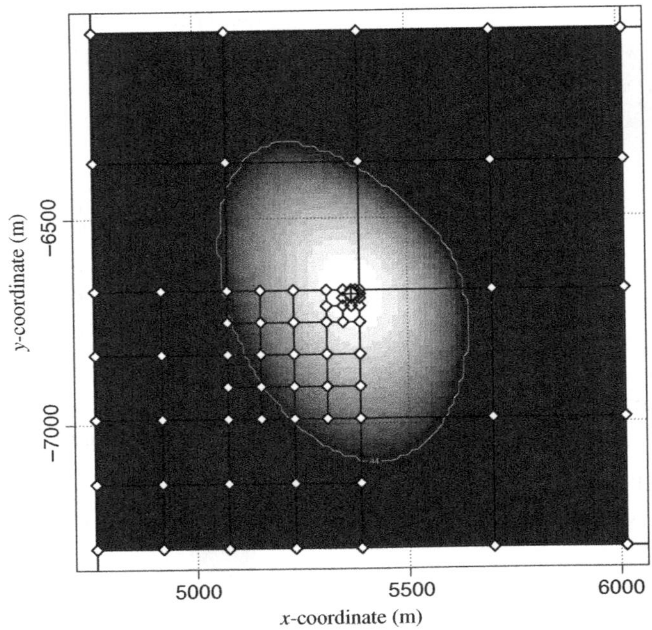

Figure 7.8 Detailed view of the complete search shown in Figure 7.7(d). Again, the best solution found in the entire search is shown by a white diamond with an internal cross. The reader should note that the color density of the overall MOP has been changed (from that shown in Figures 7.6–7.7) in order to better show the optimal value (which is at the center of the white region).

and (x_k, y_k, z_k) denotes the Cartesian coordinates of the target. The measurement error $\epsilon_k \sim \mathcal{N}(0, R_k)$, with covariance $R_k = \text{diag}(\sigma_\theta^2, \sigma_\psi^2)$. The parameters σ_θ and σ_ψ denote the azimuth and elevation measurement error standard deviations, respectively. This particular scenario is challenging, in respect that at each time instance careful utilization and deployment of the sonobuoys is required to "triangulate" the target.

Each sonobuoy has a maximum detection range of 10 km. Within this range, sonobuoys have independent detection processes, and each sonobuoy has probability P_d of detecting the target at each sampling time. False alarms have a uniform distribution in the 2D measurement space (i.e., in $[-\pi, \pi] \times [-\pi/2, \pi/2]$). The number of false alarms has a Poisson distribution with mean λV, where V is the volume of the measurement space, that is $V = 2\pi^2$.

7.5.7.2 Target Motion. The target state is denoted by: $X_k = \left[x_k\ \dot{x}_k\ y_k\ \dot{y}_k\ z_k\ \dot{z}_k\right]'$, where notation is standard. The target motion in the x- and y-coordinate directions is either nearly constant velocity (NCV) [39] or is governed by quantized accelerations (e.g., [42]). The target motion in the z-coordinate direction is always NCV. The target state evolution is therefore given by

$$X_{k+1} = F_k X_k + B_k a(r_{k+1}) + w_k. \tag{7.70}$$

The matrices F_k and B_k are given as follows

$$F_k = I_3 \otimes \begin{pmatrix} 1 & T \\ 0 & 1 \end{pmatrix}, \tag{7.71}$$

$$B_k = \begin{pmatrix} T^2/2 & 0 & 0 \\ T & 0 & 0 \\ 0 & T^2/2 & 0 \\ 0 & T & 0 \\ 0 & 0 & T^2/2 \\ 0 & 0 & T \end{pmatrix}. \tag{7.72}$$

The symbol $\otimes$ denotes the Kronecker product; w_k is additive Gaussian process noise, with $w_k \sim \mathcal{N}(0, Q_k)$; Q_k is given as follows

$$Q_k = \ell \times I_3 \otimes \begin{pmatrix} T^3/3 & T^2/2 \\ T^2/2 & T \end{pmatrix}, \tag{7.73}$$

where ℓ is the "power spectral density" of the NCV motion.

$a(r_{k+1}) = \left(a_T r^x_{k+1}\ a_T r^y_{k+1}\ 0\right)'$ is a vector of accelerations; where $r^x_{k+1},\ r^y_{k+1} = -1, 0$ or 1; and a_T is the maximum target acceleration in the x- and y-coordinate directions, respectively. Hence, in the x- and y-coordinate directions there are three potential motion models given by

$$a_T r^{\varsigma}_{k+1} \triangleq \begin{cases} \text{acceleration of } a_T \text{ m/s}^2 \text{ in } \varsigma\text{-coordinate direction}, & \text{if } r^{\varsigma}_{k+1} = 1, \\ \text{target motion is NCV}, & \text{if } r^{\varsigma}_{k+1} = 0, \\ \text{acceleration of } -a_T \text{ m/s}^2 \text{ in } \varsigma\text{-coordinate direction}, & \text{if } r^{\varsigma}_{k+1} = -1, \end{cases} \tag{7.74}$$

for $\varsigma = x, y$. There are therefore a total of nine motion model combinations. Submersible targets have only a limited capacity to accelerate, and therefore a_T has a small value.

It is assumed that the motion model in each coordinate direction is independent of the motion model in each of the other directions. The evolution of the motion model in the x- and y-coordinate directions is then modeled by a time-homogenous Markov chain with known

- transition probabilities: $p_{ij} \triangleq Pr(r^{\varsigma}_{k+1} = j | r^{\varsigma}_k = i)$, for $\varsigma = x, y$ and $i,\ j \in \{-1, 0, 1\}$;
- initial motion model probabilities: $p^{\varsigma}_1(i) \triangleq Pr(r^{\varsigma}_1 = i)$ for $\varsigma = x, y$ and $i \in \{-1, 0, 1\}$.

In later simulations, we set $p_{ij} = (1 - p_{ii})/2$ for $i \neq j$; and the target is known to be moving with NCV (in all three coordinate directions) on the first time step. Hence $p_1^{\varsigma}(0) = 1$ for $\varsigma = x, y$ (and z).

In order to implement the BFG approach [30] for the dynamical model (7.70) it is necessary to augment the target state vector to include a_T. The procedure for performing the required calculations is detailed in [31].

***7.5.7.3** Tracking Algorithm.* An IMM extended Kalman filter (IMM-EKF) [32] is used to track the manoeuvring targets. Nearest neighbor data association (e.g., [39]) is then used to deal with the measurement origin uncertainty. To provide details of the data association methodology, let $\hat{X}_{k|k-1}$ and $C_{k|k-1}$ denote the overall mean and covariance, respectively, of the (Gaussian mixture) target state distribution at sampling time k based on the measurements available up to sampling time $k - 1$. This is the IMM-EKF distribution after "mixing" [32] and immediately before measurement update at sampling time k.

For each measurement Z_k, we then calculate the following distance metric

$$\text{dist}(Z_k) \triangleq \left(Z_k - h_k(\hat{X}_{k|k-1})\right)' S_k^{-1} \left(Z_k - h_k(\hat{X}_{k|k-1})\right). \tag{7.75}$$

For brevity, the sensor index is not included in the notation. S_k is the innovation covariance, that is

$$S_k = H_k(\hat{X}_{k|k-1}) C_{k|k-1} H_k(\hat{X}_{k|k-1})' + R_k. \tag{7.76}$$

The $n_z \times n_x$ matrix H_k is the Hessian of the measurement vector $h_k(X_k)$, evaluated at $\hat{X}_{k|k-1}$, that is

$$H_k(\hat{X}_{k|k-1})' = \left[\nabla_{X_k} \left[h_k(X_k)'\right]\right]_{\hat{X}_{k|k-1}}. \tag{7.77}$$

The matrix R_k is the measurement error covariance.

We only consider measurements Z_k that are within the following gated region

$$\text{dist}(Z_k) \leq g^2. \tag{7.78}$$

We use a gate of $g = 4$ throughout. Measurements that are not within the gated region are considered to be too far from the target to have been target generated and are therefore assumed to be clutter. If two or more measurements (per sensor) are within the gated region, the measurement with the smallest distance metric value, dist($\cdot$) is associated with the target.

7.5.7.4 Parameter Settings. We consider a number of scenarios, which differ in the probability of a target nonmanoeuvre (p_{ii}) and the magnitude of the target acceleration (a_T). The maximum value of a_T considered is $a_T = 0.0125$ m/s^2, which allows the target velocity to change by a maximum of 1.5 knots (in each coordinate direction) in a one minute interval. The false alarm rate is $\lambda = 0.1$ throughout, resulting in an average of 1.97 false alarms per sensor, per time step. We consider deployment strategies with $R = 1, \ldots, 7$ step planning. At the beginning of each scenario, there are $N_0 = 4$ sonobuoys already deployed at separations of 200 m, with the initial sensor/target geometry as shown earlier in Figure 7.3. There are $N = 4$ communication channels available.

Each sonobuoy is deployed to a depth of 2 m, with the efficient search technique used to optimize the (x, y) Cartesian coordinates of each sensor. Each efficient search is within a 10 km × 10 km region centered on the estimated target location at the deployment time. The efficient search technique performs 10 subdivisions, giving a search region granularity of $10\,000/2^{10} = 9.8$ m in each coordinate direction. Sonobuoys are not permitted to be deployed within 100 m of the estimated target location at the deployment time. This constraint is incorporated into the efficient search by penalizing candidate solutions that are within the minimum distance.

The sensor management algorithm makes a new deployment on the first occasion at which the MOP (which is a predictive measure of the target location RMSE) exceeds L_1 (= 100 m herein). New sensors are then deployed in order to ensure that the subsequent MOP is no greater than L_2 $(= L_1)$ immediately subsequent to the deployment. All other parameter settings are given in Table 7.4.

7.5.7.5 Tracker Performance Metrics. Clearly, the sensor deployment algorithm prioritizes controlling the target location uncertainty. As a result, if sonobuoys were a limitless commodity, one would expect to observe similar tracking performance in each scenario considered (i.e., for each combination of a_T and p_{ii} values). Therefore, in order to compare performance across different scenarios, and also across different deployment strategies, we consider a limited supply of just six sonobuoys (with a further four sonobuoys already in position at the start of each simulation). Analysis then focuses on the following two metrics.

- The *track-life*—defined to be the period of time for which the target location RMSE is always less than 200 m.
- The overall IMM-EKF target location RMSE, based on all estimates up to the point at which each track is terminated.

The IMM-EKF target location RMSE is given as follows

$$\text{IMM-EKF RMSE} \triangleq \sqrt{\frac{1}{\mathcal{N}}\sum_i \sum_k \left\{(x_k^i - \hat{x}_k^i)^2 + (y_k^i - \hat{y}_k^i)^2 + (z_k^i - \hat{z}_k^i)^2\right\}}, \tag{7.79}$$

where $(\hat{x}_k^i, \hat{y}_k^i, \hat{z}_k^i)$ is the IMM-EKF estimate of the target location at sampling time k on run number i and (x_k, y_k, z_k) is the location of the target at that time. The parameter $\mathcal{N}$ denotes the total number of estimates considered.

7.5.7.6 Simulation Results. Simulation results are presented in Figures 7.9–7.13 and Tables 7.5–7.7. All results are averaged over 500 Monte Carlo runs, with

TABLE 7.4 Parameter Settings in the Sonobuoy Deployment Simulations

Parameter	Value
Initial (mean) target state, $\bar{X}_0$	(0 m, −7.5 m/s, 250 m, −7.5 m/s, −15 m, −0 m/s)′
Initial target state covariance, P_0	diag(1, 1, 1, 1, 1, 10^{-10})
Number of predeployed sonobuoys, N_0	4
Locations of predeployed sonobuoys	(−700 m, 0 m −2 m)′ (−500 m, 0 m, −2 m)′ (−300 m, 0 m, −2 m)′ (−100 m, 0 m, −2 m)′
Number of communication channels, N	4
Minimum time to make a deployment, t_D	180 s
Number of steps in planning horizon, R	Variable: see individual results
Number of sampling times between instances at which we evaluate performance, Δt	3
Future performance discount factor, γ	1.0 (no discounting)
Number of samples, n_s used in estimating $J_Z(\cdot)$	100
Deployment thresholds, L_1, L_2	100 m
Measurement error standard deviations: σ_θ, σ_ψ	0.5^o
Probability of detection, P_d	0.8
False alarm rate, λ	0.1
Sampling interval, T	60 s
Target acceleration, a_T (in x- and y-coordinate directions)	Variable: see individual results ($a_T = 0.001$ m/s^2 unless stated otherwise)
Transition probabilities, p_{ii} (for x-, y- coordinate motion)	Variable: see individual results ($p_{ii} = 0.9$ unless stated otherwise)
Power spectral density, ℓ (of NCV motion)	10^{-8} m^2/s^3

each run using a different target motion model realization. A sequence of six sonobuoy deployments is shown in Figure 7.9a, with the four initial sonobuoys located in the top-left corner. In Figure 7.9b, the tracker RMSE and performance measure RMSE are shown (averaged over the 500 runs), with each dip in the RMSE occurring subsequent to a sensor deployment. In Table 7.5, we show the average track life for each of the scenarios and for $R = 1, \ldots, 7$. We observe the following.

- As the magnitude of the target acceleration a_T increases, the average track life decreases, for all deployment strategies (i.e., for all values of R).
- As the magnitude of the target acceleration increases, the optimal value of R decreases.[8]

[8]The optimal value of R is the value that results in the longest average track life. This is shown in bold for each combination of a_T and p_{ii} in Table 7.5.

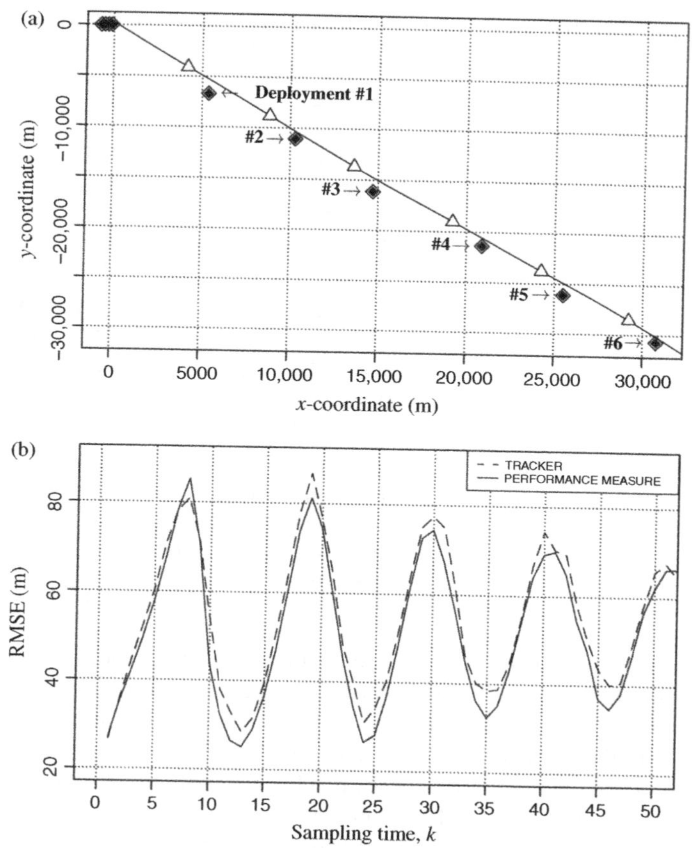

Figure 7.9 (a) *A sequence of six sonobuoy deployments*. In this case, we use a five-step planning horizon (i.e., $R = 5$), $p_{ii} = 0.9$ and $a_T = 0.001$ m/s^2. The triangles show the true target location at the time of each deployment. (b) Overall tracker and performance measure RMSEs, each plotted against sampling time (results are averaged over 500 Monte Carlo runs).

- As the probability of nonmanoeuvre, p_{ii} increases, the average track life decreases, for all values of a_T and R (see Figure 7.10).

The first two observations are as expected. Firstly, as the target manoeuvrability increases, it becomes more difficult to maintain an accurate track of the target, thereby reducing the track life. It also becomes important to deploy the sensors closer to the target (i.e., by reducing the value of R), because the increased acceleration offers the potential for the target to significantly change course. As a result, the target may not pass close to sensors that are deployed far ahead, thereby limiting their usefulness. At first, the third observation seems counter-intuitive. However, one should note that as the probability of nonmanoeuvre increases, targets tend to reach greater velocities (see Figure 7.11a), and travel greater distances as a result (see Figure 7.11b). This is

TABLE 7.5 Average Track Life (min). Results are Averaged Over 500 Runs

		Number of Planning Steps, R						
Acceleration (a_T)	Pr(no switch) (p_{ii})	1	2	3	4	5	6	7
0.001	0.1	69.8	72.5	76.9	92.4	115.6	119.4	**128.8**
	0.3	57.1	60.7	67.0	82.0	101.0	106.6	**115.8**
	0.5	52.4	56.2	58.7	75.1	92.3	98.1	**101.1**
	0.7	45.7	45.5	51.6	69.9	80.0	87.9	**91.8**
	0.9	39.7	40.3	44.0	61.0	70.0	78.8	**83.3**
0.005	0.1	32.2	54.7	64.6	74.0	77.4	**78.3**	69.6
	0.3	27.1	49.3	60.7	66.8	**67.1**	65.6	61.8
	0.5	24.8	45.9	52.8	59.2	**61.0**	57.7	50.9
	0.7	21.9	40.9	50.2	**56.8**	51.9	47.1	44.0
	0.9	24.6	41.1	50.0	**53.6**	51.6	48.4	45.4
0.0125	0.1	21.7	45.1	53.6	**54.6**	49.1	45.2	42.8
	0.3	20.8	40.4	**47.0**	44.1	39.6	37.7	34.0
	0.5	18.5	37.9	**41.8**	37.5	31.2	30.4	29.4
	0.7	18.5	32.8	**37.0**	30.7	26.6	24.1	21.8
	0.9	23.5	36.0	**37.2**	37.1	33.4	32.7	30.8

because targets that begin to accelerate remain in this model for longer. Consequently, with only a limited availability of sensors, the target tends to more quickly migrate through the sensor field and become impossible to triangulate effectively.

In Table 7.6, the average IMM-EKF location RMSE for each of the scenarios is shown. There is no obvious pattern to the results, primarily because the requirement is to maintain a tracker error of less than 100 m, and tracks are terminated as soon as this is no longer true. Hence the RMSE is similar in many of the scenarios, and notably well within the required limit. Only in the cases with the shortest track life is the tracker RMSE noticeably larger, reflecting the fact that the tracker error quickly reaches the upper limit.

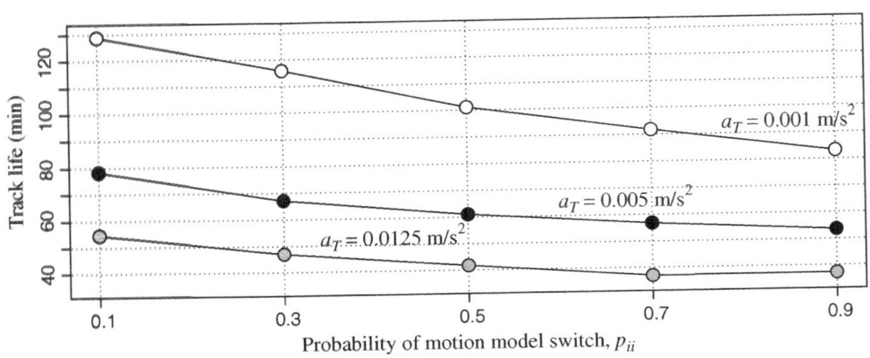

Figure 7.10 The longest average track life (i.e., the value shown in bold in each row of Table 7.5) for each value of p_{ii} and each of the three values of the maximum target acceleration, a_T.

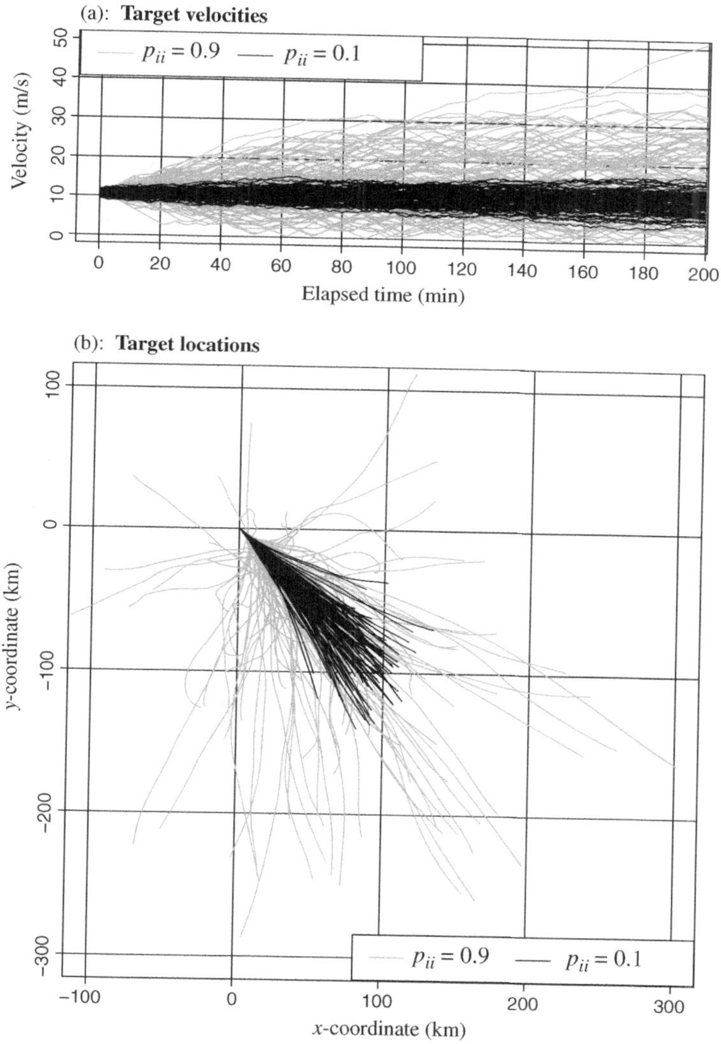

Figure 7.11 500 potential target state evolutions over a 200 min period, for $p_{ii} = 0.1$ (in black) and $p_{ii} = 0.9$ (in gray). In each case, the maximum target acceleration in the x- and y-coordinate directions is $a_T = 0.005$ m/s^2.

In Table 7.7, the percentage difference between the tracker RMSE and the BFG measure RMSE is shown in each scenario. As we can see, the BFG measure is an accurate predictor of tracker performance, with percentage differences typically no greater than 30%. Moreover, for $R \geq 5$, percentage differences are typically below 20% (see Figure 7.9b and Figure 7.13) and often much lower still. As we know, for these relatively large values of R, sensors are deployed significant distances ahead of

TABLE 7.6 Overall Tracker Location RMSE (min). Again, Results are Averaged over 500 Runs

		Number of Planning Steps, R						
Acceleration (a_T)	Pr(no switch) (p_{ii})	1	2	3	4	5	6	7
0.001	0.1	82.5	82.6	80.3	75.2	69.8	69.5	70.4
	0.3	83.4	82.6	79.7	74.4	68.9	69.3	70.9
	0.5	82.8	81.7	79.8	72.5	70.3	70.2	70.7
	0.7	82.9	82.9	76.9	72.0	69.5	69.4	71.0
	0.9	79.0	78.5	76.9	69.8	67.7	68.0	70.2
0.005	0.1	86.1	70.4	69.9	70.6	70.3	72.8	73.8
	0.3	86.5	71.0	70.1	70.3	72.7	74.8	75.6
	0.5	87.3	72.0	71.5	72.4	74.6	75.7	78.0
	0.7	89.5	73.9	72.3	73.3	75.7	78.7	79.8
	0.9	83.9	71.0	69.5	70.8	72.8	73.5	76.4
0.0125	0.1	88.1	71.1	71.3	73.2	76.5	77.8	77.6
	0.3	90.5	74.5	75.5	77.8	79.7	81.1	82.4
	0.5	93.2	76.1	76.9	80.4	84.2	84.2	85.9
	0.7	96.4	78.2	79.1	83.6	87.0	89.1	91.4
	0.9	87.2	73.1	74.1	76.2	80.2	80.9	82.1

TABLE 7.7 Percentage Difference Between the Overall Tracker Location RMSE and the Overall BFG Measure Location RMSE, that is 100 (IMM-RMSE—BFG-RMSE)/BFG-RMSE

		Number of Planning Steps, R						
Acceleration (a_T)	Pr(no switch) (p_{ii})	1	2	3	4	5	6	7
0.001	0.1	55.3	55.0	53.3	42.1	37.3	32.7	27.3
	0.3	45.6	43.5	38.3	26.8	18.6	21.0	18.6
	0.5	38.3	36.9	34.5	20.3	16.2	11.8	8.6
	0.7	34.4	33.6	24.6	14.0	9.9	0.9	3.5
	0.9	26.2	25.4	22.9	12.0	3.0	0.04	2.9
0.005	0.1	69.4	42.9	35.3	27.2	19.2	17.1	11.9
	0.3	51.4	23.8	17.6	15.6	15.9	13.6	10.9
	0.5	41.6	18.3	17.6	15.8	13.0	11.4	10.8
	0.7	43.7	21.4	18.3	16.6	15.4	16.1	12.4
	0.9	34.4	20.5	16.9	14.8	13.5	9.3	8.5
0.0125	0.1	59.4	28.5	17.8	14.9	14.0	13.7	9.3
	0.3	54.0	30.4	26.2	20.3	14.1	13.0	10.6
	0.5	53.5	31.3	25.9	23.8	17.0	12.3	10.9
	0.7	51.4	34.1	27.9	29.7	18.8	13.9	9.5
	0.9	44.0	26.0	22.3	19.7	17.4	11.6	6.9

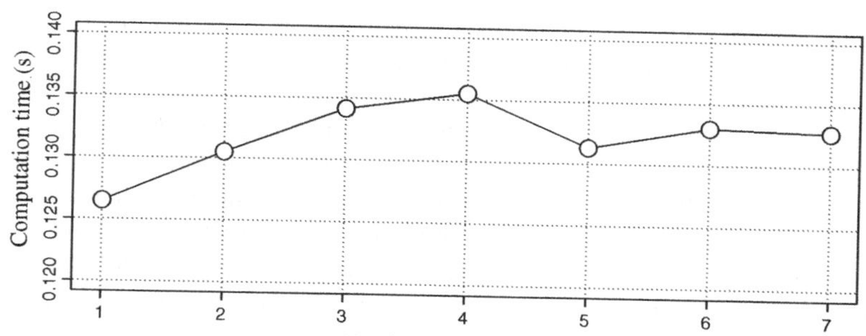

Figure 7.12 Average computation time at each sampling instance. In each case, $a_T = 0.001$ m/s^2 and results are averaged over the 500 runs at each value of p_{ii} ($= 0.1, 0.3, 0.5, 0.7, 0.9$). All simulations were programmed in *C* and run on an Intel® Core™i5-430M processor (2.26 GHz).

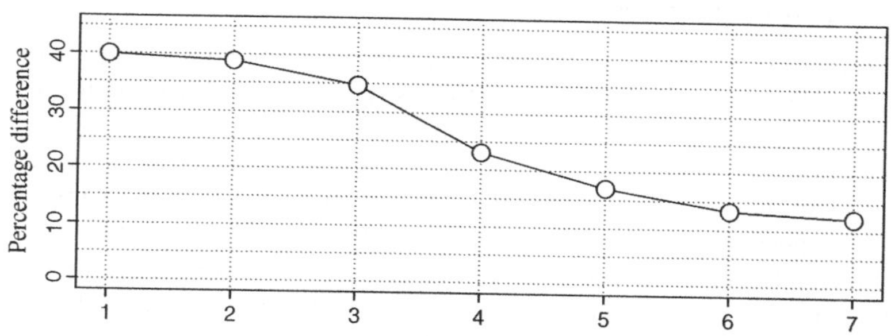

Figure 7.13 Average percentage difference between the overall tracker location RMSE and the overall BFG measure location RMSE, plotted against R. In each case, $a_T = 0.001$ m/s^2 and results are averaged over all values of p_{ii} considered (a total of 2500 runs for each value of R).

the target. In such cases, the BFG measure is less sensitive to the target state evolutions sampled in calculating the measurement contribution (7.66).[9] Consequently, for relatively large values of R, the accuracy of the BFG measure is less sensitive to the actual target state evolution, resulting in close agreement with the tracker performance.

In Figure 7.12, the average computational time at each sampling time is shown as a function of R (for $a_T = 0.001$, with similar results also obtained for the other values of a_T). We note that as R increases, the computational expense of calculating the overall MOP given in (7.64) increases. However, the track life increases as R increases from

[9] This is because with angle-only measurements, it has been shown that the measurement contribution can be strongly influenced by samples close to the sensor, to the point where the measurement contribution can show divergent behavior [43]. In such cases an alternative MOP was shown to more accurately predict sensor network performance (again, see [43]). In the current scenario, such a measure may result in improved performance for small values of R. Implementing such a modification is left for future work.

1 through 7. This results in a commensurate decrease in the proportion of sampling times at which sonobuoys need to be deployed (and optimal positions determined, which is the most computationally expensive element of the sensor deployment algorithm). As a result, the average computational time increases only gradually as R increases, and on occasions can even decrease if the corresponding increase in track life is large enough. It should also be noted that the computational time calculations herein include the expense of simulating both the measurements and target trajectories. It is clear that irrespective of the value of R, the computational burden of the sensor deployment algorithm is low enough to allow implementation in a real-time system.

7.6 UAV TRAJECTORY PLANNING

7.6.1 Scenario Overview

The primary objective is to accurately track a single manoeuvring vehicle (e.g., an aircraft) in 3D space using multiple platforms (e.g., UAVs). Each platform houses one or more sensors that provide measurements at discrete times $t_k = Tk$, $k \geq 1$. At each sampling time k, each platform performs an action/manoeuvre which controls its subsequent motion between sampling times k and $k + 1$, and most importantly, specifies the position of the sensor at the time of the next measurement.

Let a_k denote the vector prescribing the actions/manoeuvres of each platform at sampling time k. The optimization problem is to select a series of actions $\{a_k : k \geq 0\}$ in order to control a system MOP, for example $\left\{b_{k_1|k} : k > 0, k_1 > k\right\}$, at subsequent sampling times.

This optimization problem has been the subject of extensive recent analysis [12–14]. In [12, 13] the trajectory planning used only prior knowledge (i.e., did not exploit the sequence of measurements), therefore providing preflight plans that would later need to be refined online. In [14] a feedback mechanism was added that performed track update at each sampling time, prior to optimizing the subsequent manoeuvre.

7.6.2 Measure of Performance

The MOP, $b_{k_1|k}$ ($k_1 > k$) used in assessing and comparing the performance of different manoeuvre combinations is again the PCRLB overall location RMSE. Again, in manoeuvring target scenarios, the BFG approximation is exploited [30, 31]. To remind the reader, the procedure for calculating the MOP is given in Section 7.5.5. We note that in implementing the procedure in Table 7.2, the number of sensors is fixed, therefore the IF/ELSE statement on Step 2 reduces simply to calculating the measurement contribution $J_Z(k_1)$ based on the projected platform locations at sampling time k_1. We note that the MOP $b_{k_1|k}$ again exploits the target state estimate at sampling time k.

7.6.3 One-Step-Ahead Planning

In performing one-step-ahead planning, the action $a_k^\star$ is chosen in order to minimize the MOP, $b_{k+1|k}$ at the next sampling instance. The optimization problem is therefore as follows

$$a_k^\star \triangleq \arg\min_{a_k} \left[b_{k+1|k}(a_k)\right]. \tag{7.80}$$

The notation $b_{k+1|k}(a_k)$ acknowledges the fact that the MOP at sampling time $k+1$ is dependent on the action taken at sampling time k. We calculate $a_k^\star$ using the efficient search technique detailed in Section 7.5.6 (using the extension to 3D search and multiple sensors detailed in [13]).

7.6.4 Two-Step-Ahead Planning

In performing two-step-ahead planning, we determine the action $a_k^\star$ at time sampling k in order to optimize performance across the next two sampling times, that is

$$a_k^\star \triangleq \arg\min_{a_k} \left[b_{k+1|k}(a_k) + \gamma \min_{a_{k+1}} \left[b_{k+2|k}(a_k, a_{k+1})\right]\right], \tag{7.81}$$

where $b_{k+2|k}(a_k, a_{k+1})$ acknowledges the fact that the MOP at sampling time $k+2$ is dependent on the actions taken at times k and $k+1$; together with the state-of-knowledge (i.e., target state estimate) at sampling time k. The parameter $\gamma \in [0, 1]$ is a factor by which we discount the MOP at the second sampling time.

We note that in the two step-ahead approach there is a requirement to determine an action at each sensor platform at sampling time $k+1$. This action is dependent on the action at the first time step

$$\hat{a}_{k+1}(a_k) \triangleq \arg\min_{a_{k+1}} \left[b_{k+2|k}(a_k, a_{k+1})\right]. \tag{7.82}$$

The action $a_k^\star$ is then given by

$$a_k^\star \triangleq \arg\min_{a_k} \left[b_{k+1|k}(a_k) + b_{k+2|k}(a_k, \hat{a}_{k+1})\right]. \tag{7.83}$$

Again, the efficient search technique is used in order to determine $a_k^\star$ and $\hat{a}_{k+1}(a_k)$. In performing two-step-ahead planning, the second step manoeuvre $\hat{a}_{k+1}(a_k^\star)$ will never be realized. This is because upon reaching sampling time $k+1$ we will acquire measurements, update our state-of-knowledge (e.g., perform a track update), and repeat the optimization in order to calculate $a_{k+1}^\star$.

Suppose that the efficient search technique typically samples $O(n)$ actions in performing each optimization (7.80), (7.82) and (7.83). The two step-ahead approach must then typically sample $0(n^2)$ actions in performing the required optimization given in (7.81). This is a significant increase in computational expense, and therefore

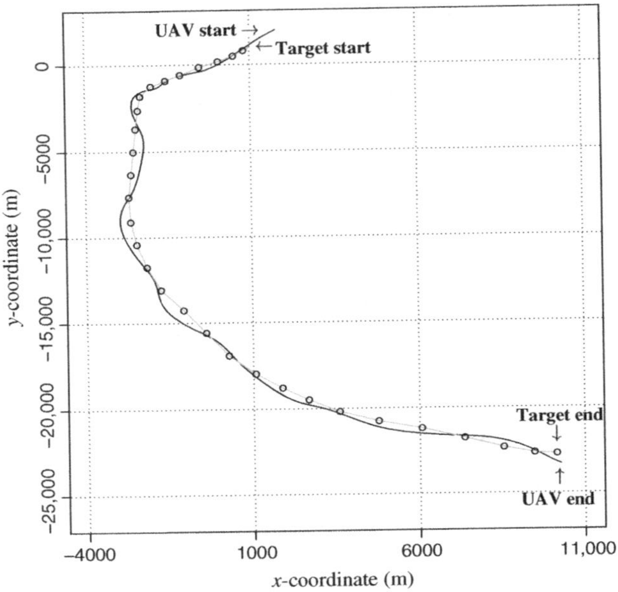

Figure 7.14 Flight path of a single UAV (in black). The true target trajectory is shown by a gray line and the tracker target state estimates are shown by circles. The probability of target nonmanoeuvre, $p_{ii} = 0.3$, with all other parameter values given in Table 7.9.

two-step-ahead planning (and indeed even longer planning horizons) should only be used if there is a clear benefit from doing so.

The flight path of a single UAV, calculated using the two-step-ahead planning approach, is shown in Figure 7.14.

7.6.5 Adaptive Horizon Planning

7.6.5.1 Motivation. Adaptive horizon planning (AHP) is motivated by the following.

- To plan effectively one must predict accurately the future impact of actions. However, this can be difficult in *highly uncertain scenarios*.[10]
- Planning based on vague, highly variable, or inaccurate predictions is ad-hoc and performance can be worse than when using simple "rules-of-thumb."

Multistep-ahead approaches take into account the future impact of actions and therefore (can) have the advantage of anticipation. However, the disadvantage of

[10] A highly uncertain scenario is one in which the underlying parameters are subject to significant random perturbations that are difficult to predict. In the context of UAV trajectory planning, uncertain and frequent target manoeuvres constitute such significant random perturbations.

multistep-ahead approaches is that, in highly uncertain scenarios, anticipation is not effective, because the future cannot be predicted with any reasonable accuracy. It may then be better to simply ignore these inaccurate predictions of the future impact of actions, and base the current action only on the immediate consequences.

The potential pitfalls of multistep-ahead planning were highlighted in [14]. In [14] it was demonstrated (via a simple toy problem and in UAV trajectory planning) that two-step-ahead planning could perform worse than one-step-ahead planning in problems with a high degree of uncertainty.

AHP [14] advocates that the length of the planning horizon should depend on the variability in predicted future performance. If this variability is high, the planning horizon should be short, and vice versa. The measures of variability that are utilized are detailed below.

7.6.5.2 Measures of the Variability in Future Performance.

At each sampling time k, we quantify the potential variability in future performance using the following two measures

$$\mathcal{V}_1(k) \triangleq \sqrt{C_{k|k}(x) + C_{k|k}(y) + C_{k|k}(z)}, \tag{7.84}$$

$$\mathcal{V}_2(k) \triangleq \max_{i=1,\ldots,N_m} \left[Pr(r_k = i | Z_{1:k}) \right]. \tag{7.85}$$

$C_{k|k}(\varsigma)$ denotes the diagonal entry of the posterior target state covariance matrix $C_{k|k}$ corresponding to the ς-component of the target state. Hence, the measure $\mathcal{V}_1(k)$ gives the uncertainty in the posterior target location estimate at each sampling time. Clearly, if $\mathcal{V}_1(k)$ takes a large value, the uncertainty in the target state estimate is high. The second measure $\mathcal{V}_2(k)$ then gives the probability associated with the most likely target motion model. If $\mathcal{V}_2(k)$ takes a relatively low value, the target motion model is also highly uncertain.

Hence $\mathcal{V}_i(\cdot)$, $i = 1, 2$ provide indicators of when future performance is highly variable and difficult to accurately predict. In such cases, caution is recommended when using multistep-ahead approaches, which cannot reliably anticipate the future consequences of actions. The AHP approach uses the variability measures as the basis for selecting the length of the planning horizon.

7.6.5.3 Implementation of Adaptive Horizon Planning.

The AHP approach is summarized in Table 7.8. The approach uses a one-step-ahead planning horizon unless both $\mathcal{V}_1(k) < \tilde{\mathcal{V}}_1^{\text{th}}$ and $\mathcal{V}_2(k) > \tilde{\mathcal{V}}_2^{\text{th}}$, where $\tilde{\mathcal{V}}_i^{\text{th}}$, $i = 1, 2$ are two prespecified thresholds. The values of the two thresholds that result in best performance are scenario specific (e.g., they depend on the parameter settings) and can only be determined by extensive simulation.

It is straightforward to extend the AHP approach in order to consider planning horizons of greater than two time steps. However, in the focal time-critical application this is impractical because the computational expense increases massively as the

TABLE 7.8 The Adaptive Horizon Planning Approach

At each sampling time, k:

- DETERMINE the two variability indicators (7.84) and (7.85).
- **IF** $\mathcal{V}_1(k) < \tilde{\mathcal{V}}_1^{\text{th}}$ and $\mathcal{V}_2(k) > \tilde{\mathcal{V}}_2^{\text{th}}$:
  - CALCULATE and USE the two-step-ahead plan (END).
- **ELSE:**
  - CALCULATE and USE the one-step-ahead plan (END).

number of time steps in the planning horizon increases.[11] A demonstration of the AHP approach is shown in Figure 7.15. In this case, we use thresholds of $\mathcal{V}_1^{\text{th}} = 50\,\text{m}$ and $\mathcal{V}_2^{\text{th}} = 0.8$, These thresholds are shown by dashed lines in Figures 7.15a and 7.15b respectively. In this example, a single UAV is utilized and the probability of target nonmanoeuvre between sampling times, $p_{ii} = 0.3$. All other parameter settings are given in Table 7.9.

In Figure 7.15a, the measure $\mathcal{V}_1(k)$ is shown (in black), together with the actual estimation error (in gray). In Figure 7.15b, the IMM model probabilities are shown (as diamonds), with the black line giving the measure $\mathcal{V}_2(k)$ at each sampling time. As we can see, when the correct motion model is highly uncertain (i.e., from sampling time 20), the variability measure $\mathcal{V}_1(k)$ is also relatively large (see Figure 7.15a). At such times, a planning horizon of one time step is used (see Figure 7.15c). At earlier sampling times, the uncertainty in the target state estimate and motion model are sufficiently low that a two step-ahead horizon is typically used.

7.6.6 Simulations

7.6.6.1 Scenarios. We utilize either one or two UAVs, and compare the performance of the three planning approaches. The probability of target nonmanoeuvre: $p_{ii} = 0.1, 0.2, 0.3$, or 0.4. This gives a total of eight scenarios. There are no missed detections and no false alarms (i.e., $P_d = 1$, $\lambda = 0$, and $q_k = 1$). We again use an IMM EKF [32] to track the target. All other parameter settings are given in Table 7.9.

7.6.6.2 Target and UAV Motion. Again, the target motion is as given in Section 7.5.7, with the target state evolution given by (7.70). Again, the target motion in the x- and y-coordinate directions is either NCV [39] or is also governed by quantized accelerations (e.g., [42]); the motion in the z-coordinate direction is always NCV. However, fast jets have a far greater capacity (than submersibles) to manoeuvre. This is reflected by the maximum target acceleration a_T taking a much larger value in this second demonstration.

[11]To reiterate, if each efficient 2D search involves $0(n)$ calculations, then with an R-step-ahead planning horizon, efficient 2D search requires $0(n^R)$ calculations. When the efficient search technique is generalized to 3D, the computational increase grows at an even greater rate with R.

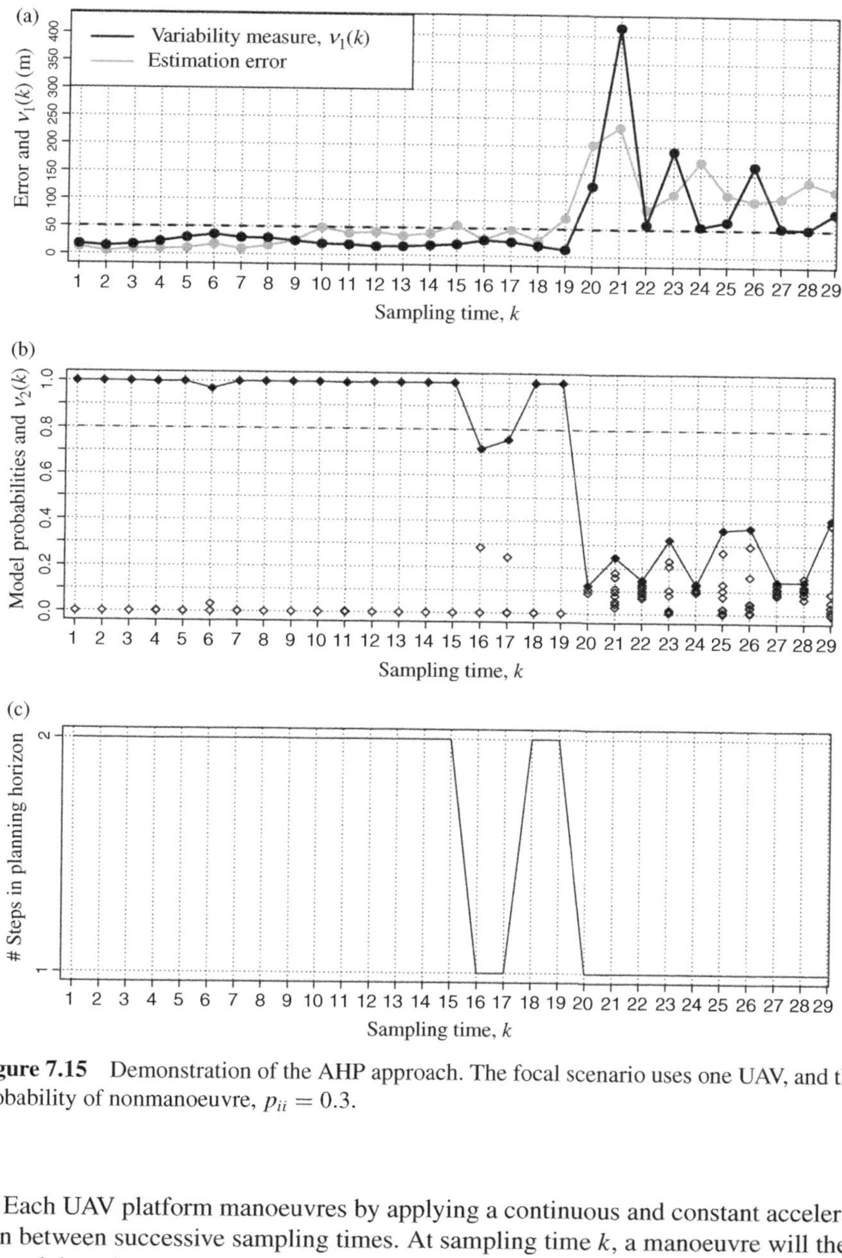

Figure 7.15 Demonstration of the AHP approach. The focal scenario uses one UAV, and the probability of nonmanoeuvre, $p_{ii} = 0.3$.

Each UAV platform manoeuvres by applying a continuous and constant acceleration between successive sampling times. At sampling time k, a manoeuvre will then control the subsequent UAV motion between times t_k and t_{k+1}, and most importantly, determine the position of the platform at the time of the next measurement (t_{k+1}). Let $X_k^{S_i} = (x_k^{S_i}\ \dot{x}_k^{S_i}\ y_k^{S_i}\ \dot{y}_k^{S_i}\ z_k^{S_i}\ \dot{z}_k^{S})'$ denote the state of a UAV platform at sampling time k, where notation is standard. Let $a_k^i = \left(a_x^{P_i}\ a_y^{P_i}\ a_z^{P_i}\right)'$ denote the accelerations

TABLE 7.9 Parameter Settings in the UAV Trajectory Planning Simulations

Parameter	Value
Initial target state, $\bar{X}_0$	(1000 m, –40 m/s 1000 m, –40 m/s 1000 m, –40 m/s)′
Initial target state covariance, P_0	diag(10^2 m^2, 1^2 m^2/s^2, 10^2 m^2, 1^2 m^2/s^2, 10^2 m^2, 1^2 m^2/s^2)
Initial state of sensor #1, $X_0^{S_1}$	(1800 m, –20 m/s 2000 m, –20 m/s 1000 m, 0 m/s)′
Initial state of sensor #2, $X_0^{S_2}$	(2000 m, –40 m/s 2000 m, –40 m/s 1000 m, 0 m/s)′
Azimuth and elevation error standard deviations: σ_θ, σ_ψ	1°
Sampling interval, T	5 s
Future performance discount factor, γ	0.5
Maximum UAV thrust (in each coordinate direction)	$2g$ m/s^2
Number of motion models (in x- and y-coordinate directions)	3
Number of motion models (in z-coordinate direction)	1
Maximum target acceleration (a_T)	10 m/s^2
Transition probabilities, p_{ii} (for x-, y-coordinate motion)	Variable: see individual results
Initial target motion model	NCV in each coordinate direction
Power spectral density, ℓ (of NCV motion)	10^{-8} m^2/s^3
AHP covariance threshold, $\mathcal{V}_1^{\text{th}}$	50 m
AHP probability threshold, $\mathcal{V}_2^{\text{th}}$	0.8

performed by this UAV in the x-, y-, and z-coordinate directions between sampling times k and $k+1$. The state of the platform then evolves according to

$$X_{k+1}^{S_i} = F_k X_k^{S_i} + B_k a_k^i. \tag{7.86}$$

The matrices F_k and B_k are again given in (7.71) and (7.72) respectively.

Each UAV has a maximum acceleration of $2g$ (in each coordinate direction) at each sampling instance, where g is the acceleration due to gravity ($g = 9.81$ m/s^2). The extension of the efficient search technique detailed in [13] is used to optimize the UAV platform accelerations. In order to reduce the computational expense of the planning approaches, the efficient search technique performs only three subdivisions. Hence, the granularity of the search region is $2g/2^3 = g/4$ m/s^2 in each coordinate direction.

7.6.6.3 Tracker Estimation Error Metrics. Simulation results are averaged over 500 runs, with each run having 30 sampling times. The 5% best and worst runs (in terms of tracker error) are removed from the analysis to reduce the potential for the results being heavily influenced by outliers. The tracker location RMSE at each sampling time is defined as follows

$$\text{IMM location RMSE at time } k$$
$$\triangleq \sqrt{\frac{1}{450}\sum_{i=1}^{450}\left\{(x_k - \hat{x}_k^i)^2 + (y_k - \hat{y}_k^i)^2 + (z_k - \hat{z}_k^i)^2\right\}}, \tag{7.87}$$

where $\hat{X}_k^i = (\hat{x}_k^i, \hat{y}_k^i, \hat{z}_k^i)$ is the IMM estimate of the target location at time k on (the included) run number i. The overall tracker location RMSE is similarly defined, but the average is over both runs and sampling times, that is we replace: $\frac{1}{450}\sum_{i=1}^{450}$ with: $\frac{1}{13{,}500}\sum_{i=1}^{450}\sum_{k=1}^{30}$ in (7.87).

7.6.6.4 Simulation Results. In Figure 7.16, we plot the IMM location RMSE for each of the planning approaches at each sampling time. In Figure 7.17, we then present the overall location RMSEs. We observe that in each of the scenarios considered, two-step-ahead planning significantly outperforms one-step-ahead planning, with decreases of 20–80% in the overall RMSE (see Figure 7.18a).

Furthermore, we observe that in the most difficult scenarios, with just one UAV and a highly manoeuvring target (i.e., with $p_{ii} \leq 0.3$), AHP significantly outperforms two-step-ahead planning, with decreases of ~20% in the overall RMSE. These improvements in performance are statistically significant[12] under both the T-test and Wilcoxon matched pairs signed rank test (e.g., see [44] for details of statistical tests). The performance of AHP and two-step-ahead planning is virtually identical in each of the other scenarios.

In Figure 7.18b, the percentage difference between the overall IMM location RMSE and the MOP is shown. We observe that in the scenarios in which AHP significantly outperforms two-step-ahead planning, the MOP is a relatively inaccurate predictor of tracking accuracy when performing two-step-ahead planning. In these cases, AHP is then able to exploit the advantages of both one-step-ahead planning (i.e., not being unduly influenced by highly uncertain estimates of future performance) and two-step-ahead planning (i.e., anticipation of the future) as appropriate. Hence, we observe that in these difficult scenarios, AHP frequently utilizes a one-step-ahead planning approach (see Figure 7.19a), rather than use a two-step-ahead approach that places undue emphasis on the potentially inaccurate second step MOP. Furthermore, AHP typically only uses a two step-ahead planning approach when the target is being

[12] Indeed, in testing the null hypothesis, H_0: the approaches have the same performance (in terms of estimation error), against H_1: the adaptive horizon approach performs better, we reject the null hypothesis in favor of H_1, with Pr(incorrectly rejected $H_0 | H_0$ true) ($\triangleq Pr$(Type I error)) typically less than 10^{-10}.

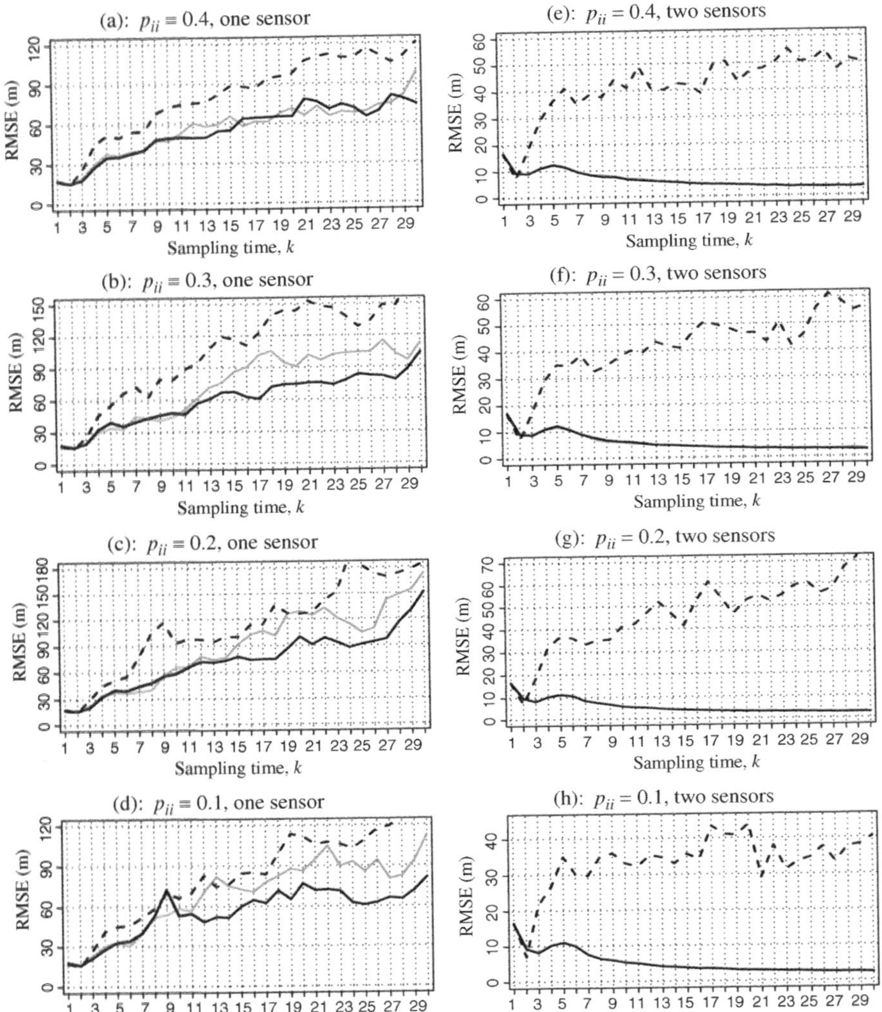

Figure 7.16 Tracker location RMSE plotted against sampling time. Key: dashed line: one-step-ahead planning, gray line: two-step-ahead planning, solid black line: adaptive horizon planning. Results are averaged over 500 runs, with the 5% best and worst runs excluded. In Figures (e)–(h) the gray line and solid black line are indistinguishable.

accurately tracked (e.g., see Figure 7.20), because an increase in the tracker covariance, or increased uncertainty in the target motion model also provide indicators of when the tracker is performing poorly.

In the scenarios with two UAVs, performing two-step-ahead planning allows the target to be accurately tracked. As a result, the MOP is an accurate predictor of

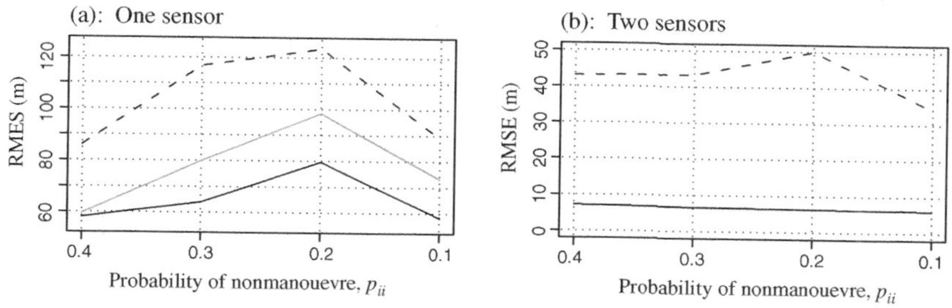

Figure 7.17 Overall tracker location RMSE for each of the scenarios. Key: dashed line: one-step-ahead planning, gray line: two-step-ahead planning, solid black line: adaptive horizon planning.

tracking performance and AHP favors a two-step-ahead planning approach at over 98% of sampling times (again, see Figure 7.19a). This explains why the tracking accuracy of the AHP approach and two-step-ahead planning approach are virtually identical in these cases.

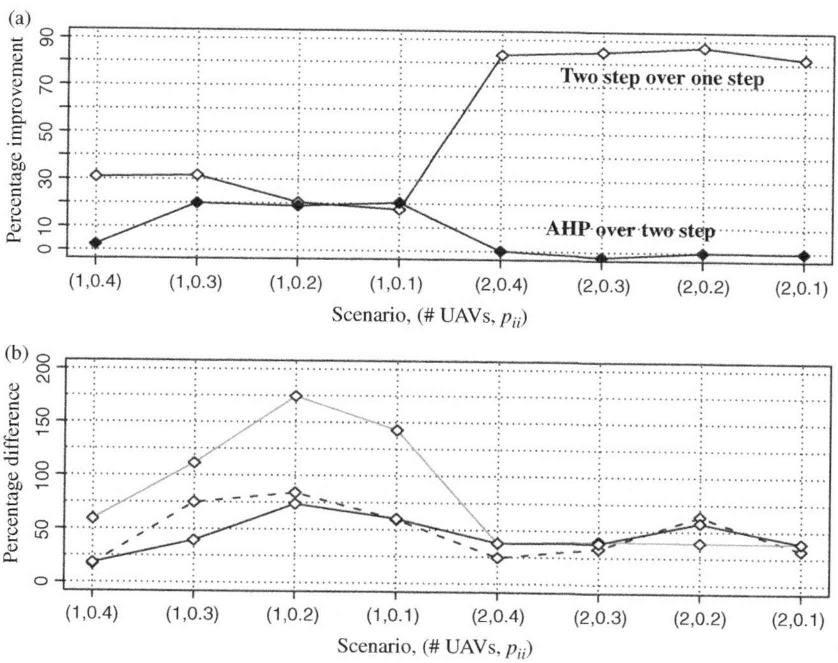

Figure 7.18 (a) The percentage improvement (in terms of the percentage reduction in RMSE) of adaptive horizon planning compared to two-step-ahead planning (solid back diamonds), and two-step-ahead planning compared to one-step-ahead planning (white diamonds). (b) Percentage difference between the IMM location RMSE and the BFG measure location RMSE, that is 100 |IMM-RMSE–BFG-RMSE|/BFG-RMSE. Key: dashed line: one-step-ahead planning, gray line: two-step-ahead planning, solid black line: AHP.

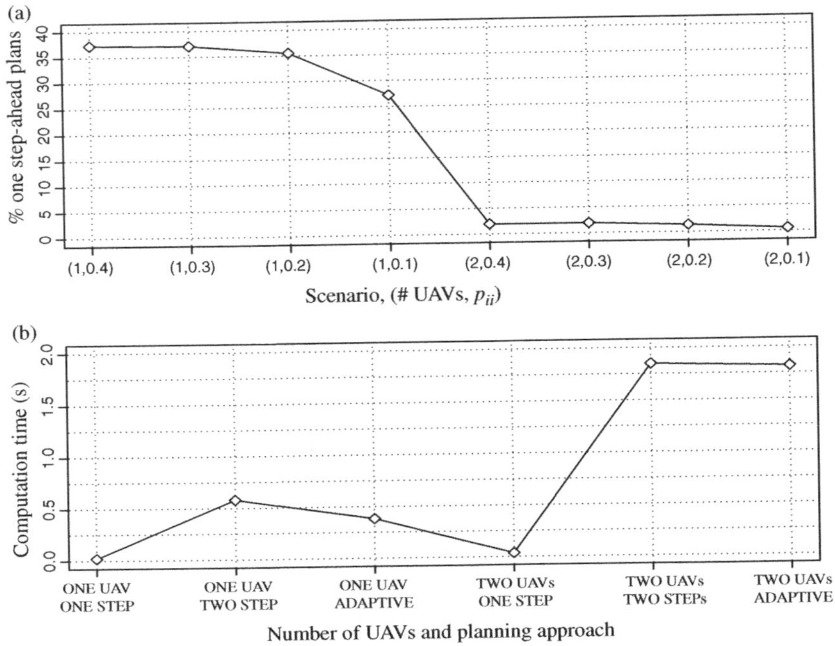

Figure 7.19 (a) The percentage of sampling times at which AHP uses a one-step-ahead planning horizon, for each of the eight scenarios. (b) The average computation time (in seconds) required to calculate a trajectory plan, for one and two UAV scenarios and each planning approach. Results are averaged over all values of p_{ii} considered (a total of 2000 runs for each number of UAVs/planning horizon combination). All simulations were programmed in *C* and run on an Intel® Core™i5-430M processor (2.26 GHz).

In Figure 7.19b, the average computation time of each planning approach is shown (with results averaged over the four p_{ii} values in each case). As we can see, the computationally simple one step-ahead approach runs much faster than both the two-step-ahead and AHP approaches. Furthermore, the AHP approach is significantly faster than two-step-ahead planning in the one UAV scenarios. This is because the AHP approach often selects the computationally simple one step-ahead plan. By comparison, in the two UAV scenarios, the AHP approach and two-step-ahead approach have virtually identical run times, simply because the AHP approach uses two-step-ahead planning on the vast majority of time steps.

In conclusion, we have demonstrated that an AHP approach can outperform more computationally expensive fixed planning horizon approaches in challenging scenarios with a high degree of uncertainty. Elaborate and computationally expensive planning strategies are widely used, but caution is recommended because they may be suboptimal if they are unduly influenced by highly uncertain measures of future performance.

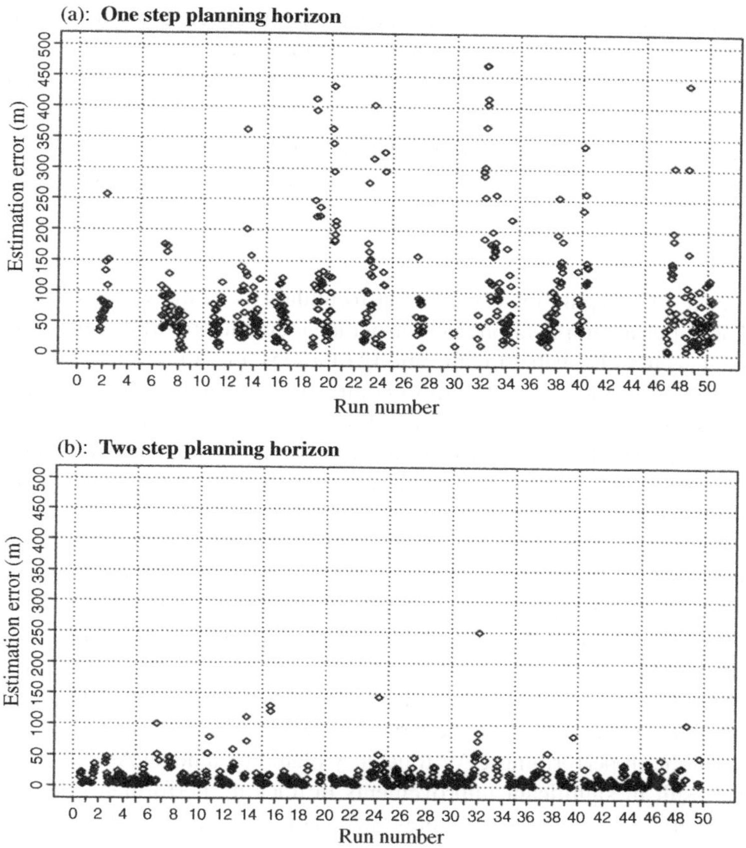

Figure 7.20 The relationship between estimation error and the number of planning steps used in the AHP approach, at each sampling time, for 50 runs of the algorithm. In each figure, on the vertical axis is the estimation error at each sampling time. In this case, a single UAV is utilized and $p_{ii} = 0.3$. It should be noted that a one-step-ahead planning horizon is used on only 37% of the time steps (see Figure 7.19a).

7.7 SUMMARY AND CONCLUSIONS

In this chapter, a review of recent developments in the calculation of MSE tracker performance bounds has been presented. These developments include the following.

- A general class of lower bounds for parameter estimation [28] that include as special cases the PCRLB, BLB, WWLB, and BBLB (see also [34]).
- An efficient recursive formula for the PCRLB [7], with similar recursions also now existing for the BLB, WWLB, and BBLB [34].
- PCRLB formulations for scenarios in which measurements can either be target generated or spurious clutter [18, 20, 23, 24]. These formulations differ in the

way in which they condition on the measurement sequence, but they all quantify the effect of the measurement origin uncertainty via an IRF [17].

- An approximate PCRLB for manoeuvring target tracking [30]. The general approach replaces the switching target dynamical system with a best-fitting linear Gaussian system. The approximate bound has been shown (in [30, 31]) to be in close agreement with the performance of state-of-the-art tracking algorithms.

The existence of computationally efficient formulations of the various performance bounds has important implications for sensor management, providing a mechanism for quickly and accurately evaluating the ability of a sensor configuration to achieve target triangulation/geo-location. The most computationally simple of the performance bounds is the PCRLB. The efficient formulations above then make the PCRLB a powerful tool in scenarios that are subject to real-world complexities such as measurement nonlinearity (e.g., only target line-of-sight), measurement origin uncertainty, and evasive target behavior.

The second half of this chapter was devoted to two applications that demonstrate how controlling the PCRLB can provide an effective mechanism for online sensor management. The first application is concerned with the deployment of stationary sensors (e.g., sonobuoys tethered to the sea surface) in order to track a single, potentially manoeuvring target (e.g., a submersible). The second application is that of UAV trajectory planning. In both cases, the PCRLB is shown to be an accurate predictor of system performance, allowing the targets to be accurately tracked.

Furthermore, in performing UAV trajectory planning, a novel AHP approach was shown to result in significant improvements in tracking performance compared to (traditional) fixed planning horizon approach. This was because the AHP approach is able to exploit the advantages of both one-step-ahead planning (i.e., not being unduly influenced by highly uncertain estimates of future performance) and two-step-ahead planning (i.e., anticipation of the future) as appropriate. Elaborate and computationally expensive planning strategies are widely used, but caution is recommended because they may be suboptimal if they are unduly influenced by highly uncertain predictions of future performance.

In both applications, the efficient formulation of the PCRLB, together with other efficiency savings (such as employing an efficient search technique), reduce, the computational overhead of the two algorithms to a minimum. As a result, real-time sensor management can be conducted, without compromising performance. We note that alternative measures of performance, including information gain and entropy, have been proposed as mechanisms for sensor control in target detection, tracking, and classification applications (e.g., see [16]). However, such measures typically involve evaluating difficult nonanalytical integrals of functions of the target state density, thereby making real-time sensor management difficult.

It is concluded that recent efficient formulations of the PCRLB offer a powerful technique for performing sensor management in target tracking, whilst allowing for real-world complexities.

ACKNOWLEDGMENTS

Dr Hernandez would like to acknowledge the support of various United Kingdom Ministry of Defence research programmes which funded the published work that motivated this chapter. These programmes include the Corporate Research Programme, the Detection and Identification of Difficult Targets Major Programme Area and the Data and Information Fusion Defence Technology Centre.

REFERENCES

1. N. Bergman, Posterior Cramér–Rao bounds for sequential estimation, in: A. Doucet, N. de Freitas, and N. Gordon (Eds.), *Sequential Monte Carlo Methods in Practice*, Springer-Verlag, New York, USA, 2001.
2. H.L. Van Trees and K.L. Bell (Eds.), *Bayesian Bounds for Parameter Estimation and Nonlinear Filtering/Tracking*, John Wiley and Sons Inc., New York, USA, 2007.
3. A. Bhattacharyya, On some analogues of the amount of information and their use in statistical estimation, SANKHYA, 8, 315–328, 1948.
4. A.J. Weiss and E. Weinstein, A lower bound on the mean square error in random parameter estimation, *IEEE Trans. Inform. Theory*, IT-31(5), 680–682, 1985.
5. B. Bobrovsky and M. Zakai, A lower bound on the estimation error for certain diffusion processes, *IEEE Trans. Inform. Theory*, IT-22(1), 45–52, 1976.
6. H.L. Van Trees, *Detection, Estimation, and Modulation Theory. Part I*, John Wiley and Sons, New York, USA, 1968.
7. P. Tichavský, C.H. Muravchik, and A. Nehorai, Posterior Cramér–Rao bounds for discrete-time nonlinear filtering, *IEEE Trans. Signal Process.*, 46(5), 1386–1396, 1998.
8. M.L. Hernandez and P.R. Horridge, Advances in the management of multisensor systems with associated applications, *Proceedings of the IEE Workshop on Target Tracking: Algorithms and Applications*, University of Sussex, UK, 2004.
9. M.L. Hernandez, T. Kirubarajan, and Y. Bar-Shalom, Multisensor resource deployment using posterior Cramér–Rao bounds, *IEEE Trans. Aerosp. Electron. Syst.*, 40(2), 399–416, 2004.
10. R. Tharmarasa, T. Kirubarajan, and M.L. Hernandez, Large-scale optimal sensor array management for multitarget tracking, *IEEE Trans. Syst., Man, Cybern.: Part C*, 37(5), 803–814, 2007.
11. R. Tharmarasa, T. Kirubarajan, M.L. Hernandez, and A. Sinha, PCRLB based multisensor array management for multitarget tracking, *IEEE Trans. Aerosp. Electron. Syst.*, 43(2), 539–555, 2007.
12. M.L. Hernandez, Optimal sensor trajectories in bearings-only tracking, *Proceedings of the 7th International Conference on Information Fusion*, Stockholm, Sweden, 2004, pp. 893–900.
13. M.I. Smith, M.L. Hernandez, and M. Cooper, Simulating and testing autonomous behaviour in multiple airborne sensor systems, *Proceedings of the SPIE Conference on Unmanned Systems Technology IX*, Orlando, Florida, USA, Vol. 6561, 2007, pp. 656100-1–656100-12.

14. M.L. Hernandez, Adaptive horizon sensor resource management: Validating the core concept, *Proceedings of the SPIE Conference on Signal and Data Processing of Small Targets*, San Diego, CA, USA, Vol. 6699, 2007, pp. 66990V-1–66990V-12.
15. B. Ristic, M.L. Hernandez, A. Farina, and H.-T. Ong, Analysis of radar allocation requirements for an IRST aided tracking of anti-ship missiles, *Proceedings of the Ninth International Conference on Information Fusion*, Florence, Italy, 2006.
16. A.O. Hero III, C.M. Kreucher, and D. Blatt, Information theoretic approaches to sensor management, in: A.O. Hero III, D. Castañón, D. Cochran, and K. Kastella (Eds.), *Foundations and Applications of Sensor Management (Signals and Communication Technology)*, Springer-Verlag, New York, USA, 2007.
17. Nui R., Willett P.K., and Y. Bar-Shalom, Matrix CRLB scaling due to measurements of uncertain origin, *IEEE Trans. Signal Process.*, 49(7), 1325–1335, 2001.
18. X. Zhang, and P.K. Willett, Cramer–Rao bounds for discrete time linear filtering with measurement origin uncertainty, *Proceedings of the Workshop on Estimation, Tracking and Fusion: A Tribute to Yaakov Bar-Shalom*, Monterey, CA, USA, 2001, pp. 546–561.
19. X. Zhang, P.K. Willett, and Y. Bar-Shalom, Dynamic Cramér–Rao bound for target tracking in clutter, *IEEE Trans. Aerosp. Electron. Syst.*, 41(4), 1154–1167, 2005.
20. M.L. Hernandez, A.D. Marrs, N.J. Gordon, S. Maskell, and C.M. Reed, Cramér–Rao bounds for nonlinear filtering with measurement origin uncertainty, *Proceedings of the 5th International Conference on Information Fusion*, Annapolis, Maryland, USA, Vol. 1, 2002, pp. 18–25.
21. A. Farina, B. Ristic, and L. Timmoneri, Cramér–Rao bounds for nonlinear filtering with $P_d < 1$ and its application to target tracking, *IEEE Trans. Signal Process.*, 50(8), 1916–1924, 2002.
22. M.L. Hernandez, B. Ristic, A. Farina, and L. Timmoneri, A comparison of two Cramér–Rao bounds for nonlinear filtering with $P_d < 1$, *IEEE Trans. Signal Process.*, 52(9), 2361–2370, 2004.
23. M.L. Hernandez, A. Farina, and B. Ristic, PCRLB for tracking in cluttered environments: Measurement sequence conditioning approach, *IEEE Trans. Aerosp. Electron. Syst.*, 42(2), 680–704, 2006.
24. H. Meng, M.L. Hernandez, Y. Liu, and X. Wang, Computationally efficient PCRLB for tracking in cluttered environments: Measurement existence conditioning approach, *IET Signal Process.*, 3(2), 133–149, 2009.
25. Y. Boers and J.N. Driessen, Modified riccati equation and its application to target tracking, *IEE Proc. Radar, Sonar Navig.*, 153(1), 7–12, 2006.
26. Y. Boers and J.N. Driessen, Results on the modified riccati equation: Target tracking applications, *IEEE Trans. Aerosp. Electron. Syst.*, 42(1), 379–384, 2006.
27. Y. Boers and J.N. Driessen, A note on bounds for target tracking with Pd<1, *IEEE Trans. Aerosp. Electron. Syst.*, 45(2), 640–646, 2009.
28. E. Weinstein and A.J. Weiss, A general class of lower bounds in parameter estimation, *IEEE Trans. Inform. Theory*, 34(2), 338–342, 1988.
29. A. Bessel, B. Ristic, A. Farina, X. Wang, and M.S. Arulampalam, Error performance bounds for tracking a manoeuvring target, *Proceedings of the 6th International Conference on Information Fusion*, Cairns, Queensland, Australia, 2003, pp. 903–910.

30. M.L. Hernandez, B. Ristic, and A. Farina, A performance bound for manoeuvring target tracking using best-fitting Gaussian distributions, *Proceedings of the 8th International Conference on Information Fusion*, Philadelphia, PA, USA, July 2005.
31. M.L. Hernandez, B. Ristic, A. Farina, T. Sathyan, and T. Kirubarajan, Performance measure for markovian switching systems using best-fitting Gaussian distributions, *IEEE Trans. Aerosp. Electron. Syst.*, 44(2), 724–747, 2008.
32. H.A.P. Blom and Y. Bar-Shalom, The interacting multiple model algorithm for systems with Markovian switching coefficients, *IEEE Trans. Automat. Contr.*, 33, 780–783, 1988.
33. I. Rapoport and Y. Oshman, Recursive Weiss–Weinstein lower bounds for discrete-time nonlinear filtering, *Proceedings of the 43rd IEEE Conference on Decision and Control*, Bahamas, 2004, pp. 2662–2667.
34. S. Reece and D. Nicholson, Tighter alternatives to the Cramér–Rao lower bound for discrete-time filtering, *Proceedings of the 8th International Conference on Information Fusion*, Philadelphia, PA, USA, 2005.
35. J. Taylor, The Cramér–Rao estimation error lower bound computation for deterministic nonlinear systems, *IEEE Trans. Automat. Contr.*, 24(2), 343–344, 1979.
36. B. Ristic, S. Zollo, and S. Arulampalam, Performance bounds for manoeuvring target tracking using asynchronous multiplatform angle-only measurements, *Proceedings of 4th International Conference on Information Fusion*, Montreal, Québec, 2001.
37. C. Hue, J.-P. Le Cadre, and P. Pérez, Posterior Cramér–Rao bounds for multitarget tracking, *IEEE Trans. Aerosp. Electron. Syst.*, 42(1), 37–49, 2006.
38. M. Morelande, B. Ristic, and M.L. Hernandez, A new best fitting gaussian performance measure for Jump Markov systems, *Proceedings of the Eleventh International Conference on Information Fusion*, Cologne, Germany, 2008, pp. 416–421.
39. Y. Bar-Shalom, X. Li, and T. Kirubarajan, *Estimation with Applications to Tracking and Navigation*, John Wiley and Sons Inc., New York, USA, 2001.
40. M.L. Hernandez and P.R. Horridge, Dynamic multisensor resource management with increased sensor utility, *Proceedings of the SPIE Conference on Signal and Data Processing of Small Targets*, Vol. 5204, San Diego, California, USA, Vol. 52904, 2003, pp. 374–389.
41. R.L. Burden and D.J. Faires, *Numerical Analysis*, 8th edition, Brooks/Cole, USA, 2000.
42. X.R. Li, Engineers guide to variable structure multiple-model estimation for tracking, in: Y. Bar-Shalom, and W. D. Blair) (Eds.), *Multitarget-Multisensor Tracking: Applications and Advances III*, Artech House, Norwood, MA, USA, 2000, pp. 499–567.
43. P.R. Horridge and M.L. Hernandez, Performance bounds for angle-only filtering with application to sensor network management, *Proceedings of the Sixth International Conference on Information Fusion*, Cairns, Queensland, Australia 2003, pp. 695–703.
44. D. Sheskin, *Handbook of Parametric and Nonparametric Statistical Procedures*, 4th edition, Chapman and Hall, London, UK, 2007.

CHAPTER 8

Track-Before-Detect Techniques

SAMUEL J. DAVEY, MARK G. RUTTEN, AND NEIL J. GORDON

8.1 INTRODUCTION

Conventional tracking algorithms assume that the sensor reports measurements that correspond to a physical point in space (*point measurements*). Some of these measurements may be due to objects of interest (*targets*) and others may be caused by noise, interference, or objects of no interest and are collectively referred to as clutter. The conventional tracker seeks to associate together the measurements from a particular target and hence infer parameters of interest for that target. This task becomes difficult when either the targets are close together and the association of measurements to targets is ambiguous, or when the spatial density of clutter measurements is very high. The first case is multitarget tracking, and is addressed in Chapters 3 and 5. The second arises as a result of poor target signal-to-noise-ratio (SNR). A high spatial density of clutter measurements usually affects the tracker in two ways: good tracks may be incorrectly associated with a clutter measurement and diverge from the correct target trajectory; or several clutter measurements may be associated together and lead to the creation of a track where no target is present, a *false track*.

Although point measurements are a standard assumption, most sensors do not intrinsically generate them. Images from optical sensors are, of course, familiar to all, but other sensors also collect images. For example, a phased array radar may simultaneously collect data in multiple azimuth and elevation beams as well as range and Doppler bins. For sensors that collect images (of one or more dimensions) a *detector* is used to extract point measurements from the sensor images. In effect, the point measurements are treated as quasi-sufficient statistics for the sensor image.

Track-Before-Detect (TBD) is the name given to a family of algorithms that may be applied to problems with low SNR targets. Most TBD algorithms operate directly on the sensor image rather than employing a point-measurement detector, as the name Track-Before-Detect naturally implies. The premise of TBD is that for low SNR

Integrated Tracking, Classification, and Sensor Management: Theory and Applications, Edited by Mahendra Mallick, Vikram Krishnamurthy, and Ba-Ngu Vo.

targets, the approximation of the sensor image by a set of points is poor and the use of a detector throws away valuable sensor information.

This chapter presents a review of many of the recent algorithms for TBD and assesses their performance. The remainder of Section 8.1 presents an outline of the historical context of TBD and summarizes the key contributions to the field. It also motivates the study of TBD with an example where conventional tracking fails. Section 8.2 defines models for targets and the measurements that are common among the alternative algorithms. Sections 8.3 through 8.7 review a number of TBD algorithms: Baum Welch; Viterbi; the Particle Filter; Maximum-Likelihood PDA; and Histogram-PMHT. These algorithms are then compared through simulated examples in Section 8.8. Section 8.9 presents an example application of TBD within a maritime surveillance context and the chapter concludes by identifying unsolved problems and future research areas for TBD in Section 8.10.

8.1.1 Historical Review of TBD Approaches

There have been broadly two types of approaches to solving TBD. Intuitively these two solution classes can be regarded as either treating TBD as a tracking problem, or treating it as a detection problem. Both perspectives have been the subject of numerous techniques and both have been used since the mid-1970s.

One possible definition of a tracker is: an automatic algorithm for determining the number of targets and their parameters from a noisy and ambiguous sequence of data by exploiting a dynamic model for target behavior. While it is more common for the sequence of data to be sets of point measurements, there is no fundamental requirement for it. Under this definition, TBD becomes a particular example of tracking where the input data are a sequence of images. In this case, the difficulty is that the function relating the observation data (images) to the target states is highly nonlinear and the noise is potentially nonGaussian. There is a progression in the TBD literature of solution methods reflecting the broader community's methods for nonlinear filtering: first Extended Kalman Filters were proposed, followed by fixed-grid methods, and more recently Sequential Monte Carlo techniques.

The second point of view is motivated by the notion of incoherent integration. It is well known that incoherently accumulating data on a stationary signal improve the SNR: if N frames are accumulated then the noise power increases at a rate of N, but the signal power increases at a rate of N^2. Thus, incoherent integration improves the SNR at the detector decision point, resulting in a higher quality decision. However, in TBD, the signal of interest is not stationary. Thus the alternative perspective on TBD is a detection problem with nuisance parameters resulting from the target motion. An approximation to incoherent integration is binary integration, where a threshold is applied to each frame and the binary result is accumulated instead. It has long been known that the penalty for this approximation is approximately 1 dB or less (e.g., [1]).

The first nonlinear filtering approach to TBD was developed by Mercier and Maybeck and published in 1978 [2, 3]. They stacked the image pixels into a long measurement vector and established a nonlinear measurement function between it and the target state. Assuming Gaussian noise, they implemented an Extended Kalman Filter (EKF) to track the target. This work was in the context of a Forward Looking Infrared (FLIR) sensor. An obvious problem with the stacked measurement vector is that the innovation covariance and Kalman gain matrices will be very large. This was mitigated by using an Information-Filter implementation which recurses the inverse of the covariance matrix instead of the covariance matrix itself. The Information-Filter implementation is linear in the number of pixels in the image. However, the EKF uses a first-order linearization of the measurement process. For TBD this was found to be a poor approximation and the EKF method was suitable only for relatively high SNR targets.

Nonlinear filtering for TBD then moved on from analytic approximation to sampling-based approximation. Barniv demonstrated that by discretizing the state space, well-known dynamic programming methods could be brought to bear. He applied the Viterbi algorithm to TBD in the mid-1980s [4–6]. Later work by Tonissen [7] sought to analytically describe the detection statistics of the Viterbi algorithm in TBD and these results were substantially improved by Johnston [8] who used extreme value theory to analyze the asymptotic statistics.

Having discretized the state space, an alternative to dynamic programming is to directly evaluate the posterior probability mass over the grid and propagate it over time. This forms the basis for the Bayesian methods described by Stone [9]. A more application orientated version of this approach was presented by Bruno who used the Baum–Welch algorithm to detect vehicles in imagery [10, 11].

A disadvantage of methods based on discretizing the state space is the high computational burden they may incur. For most problems, much of this computation is wasted effort propagating the tails of the distribution where there may be negligible probability. This leads to a desire for a dynamic grid and one way of achieving this is through particle filtering. Particle filters for TBD were introduced by Salmond [12] and Boers [13] in independent parallel research. Since this time it has been a popular method in the research community. Ristic [14] extended the target model to provide for automated initiation, and Rutten [15] introduced fluctuating target amplitude and a more efficient initiation method. The particle filter method is closely related to the Baum–Welch approach: both recursively estimate the target state posterior using samples. Whereas Baum–Welch uses a fixed grid, particle filtering uses random samples. The potential advantage of the particle filter is that it has the capability of using fewer samples by using adaptive dynamic sampling.

Target tracking usually implies data association as well as estimation. However, the above algorithms do not perform any explicit data association. The reason for this is that the data association process fundamental to Detect-Then-Track is only required when the input data are point measurements. Essentially the above methods are based on the idea that using a single frame detector degrades sensitivity. An alternative perspective is to keep the single frame detector and change the target dynamic model: rather than using a stochastic model with enough noise to capture manoeuvres, the Maximum Likelihood Probabilistic Data Association (ML-PDA) algorithm seeks to fit deterministic trajectories to point measurements [16, 17]. ML-PDA uses the PDA measurement model which is a Bayesian decomposition of the point measurement likelihood into a measurement assignment hypothesis and a conditional likelihood given known data sources. A search is performed over a grid of deterministic trajectories to maximize the PDA likelihood.

Recent research into ML-PDA has sought to reduce its computational overhead using efficient search strategies, rather than a uniformly sampled grid [18]; has extended the algorithm to handle multiple targets [19]; and has built on the extreme value theory work of Johnston to provide track validation measures [20].

In very recent work, random finite sets (RFS) were applied to TBD [21]. For the case of a single target, the RFS method is called the Bernoulli filter, since the number of targets is a Bernoulli random variable. It can be shown that the Bernoulli filter is equivalent to the particle filter TBD method in [15] and reviewed in this chapter. The multitarget RFS algorithm is the multiBernoulli filter and is described in [21]. To an extent this can be viewed as a multitarget extension of the particle filter TBD.

Histogram Probabilistic Multi-Hypothesis Tracking (H-PMHT) is a unique method for TBD [22, 23]. This method interprets the sensor image as a histogram of observations of an underlying mixture density. The states of the mixture components are linked over time via a dynamic model and Expectation-Maximization (EM) is used to determine state estimates. An advantage of H-PMHT is that it is naturally a multitarget algorithm, unlike many TBD alternatives. Also it has been shown to be a computationally efficient approach [24] largely because it does not use likelihood ratios.

In parallel to this work, other researchers developed methods based on treating TBD as a detection problem. Burczewski and Mohanty developed an incoherent integration technique for detecting moving objects against the night sky [25, 26], first published in 1978 (the same time as the EKF work of Mercier and Maybeck described earlier). The aim of this work was to detect satellites moving across a background of stars, where the target was essentially indistinguishable from the background on any one frame. They used a bank of incoherent integrators, each tuned to a different possible target dynamics hypothesis. This method is often referred to as *velocity filtering* because it employs filters tuned to a particular target velocity.

The initial velocity filtering work above assumed that the target response was a single pixel in the sensor image and explicitly enumerated all possible combinations of initial position and velocity vector. The efficiency of this method was vastly improved by Reed [27] who framed the problem in the context of filtering within a three-dimensional frame (representing two spatial dimensions and time). Reed demonstrated a frequency domain approach which avoided the need to explicitly enumerate possible target initial locations. This method was referred to as three-dimensional matched filtering. Reed extended the method to distributed targets whose response contributed to more than one pixel and considered multiple targets with velocity vectors mismatched to the velocity filters [28, 29].

In similar work, Pohlig formulated TBD as a maximum likelihood problem [30]. The aim was to avoid potential losses incurred when the velocity filters were mismatched to the target's true velocity. Tonissen extended this work to allow for correlated target amplitudes using a colored Gaussian noise process [31].

An alternative tradition in incoherent integration is the Hough Transform. As stated earlier, when binary integration is used instead of incoherent integration, the loss in sensitivity is relatively small. The Hough Transform is an image processing method for detecting lines, or other parameterized features, in an image. It thresholds the image and uses binary integration along the line to produce a transform image in the parameter domain. Bright spots in the transform domain correspond to lines in the original image. Smith and Winter applied the Hough Transform to TBD in the late 1970s [32, 33], constructing a two-dimensional image from successive samples from a one-dimensional sensor. Carlson published an extensive study of the application of the Hough Transform to radar in 1994 [34–36].

An important development in target tracking in the last decade has been the emergence of the Posterior Cramér–Rao Bound describing the best possible estimation performance as discussed in Chapter 7. While this is still relevant for TBD, a more important bound is the best possible detection performance. At the time of writing, this bound is not available. Morelande and Ristic have made some progress toward this with their work on the Barankin bound [37] but much work needs to be done.

This chapter focuses on the approach of TBD as a tracking/estimation problem rather than the detection problem perspective. All of the established methods for TBD under this umbrella are described, and a comparison of methods on some example applications is presented. Historically many TBD algorithms have been batch methods, requiring the collection of a sequence of data before the TBD algorithm is applied. However many recent methods are designed to be time recursive. The majority of the algorithms described are time recursive except when noted otherwise.

8.1.2 Limitations of Conventional Detect-then-Track

As discussed above, one way to view conventional detect-then-track is as binary integration with nuisance parameters. Similarly TBD may be considered analogous to incoherent integration with nuisance parameters. If the target state is known, then there are no nuisance parameters, and it is relatively easy to derive analytic expressions for the probability of detection and probability of false alarm for both binary and incoherent integration.

As an example, consider a batch of T frames of a known signal in complex Gaussian noise. It is well known that the output of an incoherent integrator will be chi-squared with $2T$ degrees of freedom for the noise-only case and noncentral chi-squared when the target is present. For the noncentral case, the mean depends on the target SNR. For a prescribed decision threshold, these distributions give analytic results for the detection and false alarm probabilities.

For a single frame, let $P_D(s;\beta)$ denote the probability of detecting a target with SNR s at a threshold β. The statistics of the single frame detector are the same as those of the incoherent integrator with $T = 1$, thus $P_D(s;\beta)$ is the integral of a noncentral chi-squared distribution. Similarly, let the single frame false alarm probability be $P_{FA}(\beta)$. The binary integrator is an N out of T rule based on these single frame probabilities. Thus the probability of detection and false alarm are

$$P_D^{\textbf{binary}}(s; \beta, N) = \sum_{j=N}^{T} \binom{T}{j} P_D(s; \beta)^j \left[1 - P_D(s; \beta)\right]^{T-j}, \tag{8.1}$$

$$P_{FA}^{\textbf{binary}}(\beta, N) = \sum_{j=N}^{T} \binom{T}{j} P_{FA}(\beta)^j \left[1 - P_{FA}(\beta)\right]^{T-j}. \tag{8.2}$$

Note that the binary integrator has two parameters: the single-frame threshold (which is a continuous parameter) and the detection count (which is discrete). This means that the binary integrator produces a family of Receiver Operating Characteristic (ROC) curves: one for each possible choice of N. The overall ROC is the envelope of this family.

Figure 8.1 shows the binary integration ROC curve family for $T = 20$ frames and a target SNR of 3 dB. Overlaid are two ROC curves for incoherent integration over the 20 frames: one at a target SNR of 3 dB (the solid curve) and one at a target SNR of 2 dB (the heavy solid curve). It is clear for this example that the 2 dB incoherent integration curve is very close to the envelope of the binary integration family of ROC curves: single-frame detection, as used by detect-then-track, has cost 1 dB of sensitivity.

In fact, results similar to those in Figure 8.1 can be obtained for a variety of operating conditions. This may lead to the conclusion that TBD is not needed: if the loss from single-frame detection is only around 1 dB, then it is probably tolerable. However, recall that this result assumed that the target state is known. This is a detection

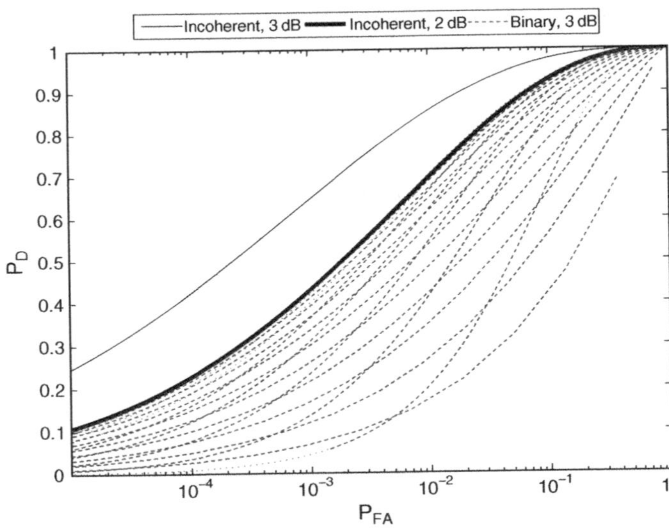

Figure 8.1 ROC, 3 dB SNR, 20 scans.

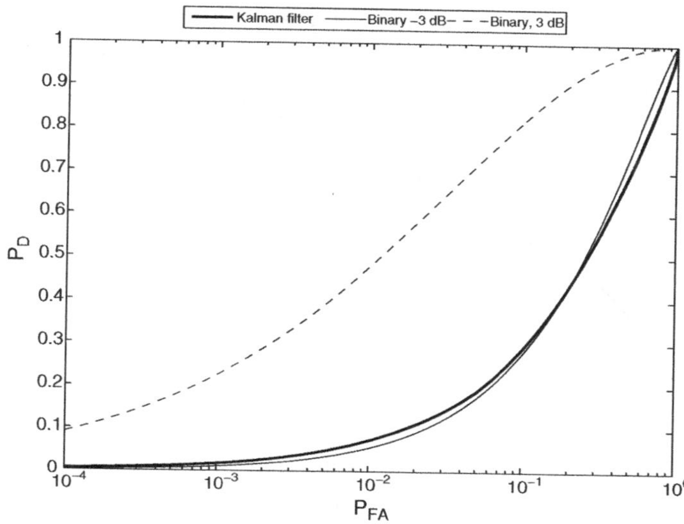

Figure 8.2 Kalman Filtering ROC, 12 frames.

result, not a joint detection and tracking result. In practice, studies have shown that the additional uncertainty due to the target dynamics changes the performance dramatically.

Figure 8.2 compares the ROC curve envelope for binary integration with a ROC curve envelope analytically derived for a Kalman-Filter-based tracker. The ROC curves were generated for a batch of 12 scans for a target with an SNR of 3 dB. The figure also shows the binary integration ROC curve for a target SNR of −3 dB, which is very similar to the Kalman-Filter curve. The uncertainty in target dynamics has translated to a 6 dB sensitivity penalty. This extreme loss comes about because the random walk target model in the Kalman Filter allows a wide family of potential false trajectories, even with very low process noise. Similar to the binary integration case, the Kalman Filter ROC is an envelope of the best achievable performance over various algorithm parameters. It assumes an N out of T rule and is optimized for N, the process noise and the measurement noise. Both noise parameters are constrained within a practical range: for zero noise, the curve is identical to the binary integration curve since this implies complete target knowledge.

In contrast, the performance loss between TBD and incoherent integration is more modest. Figure 8.3 shows the incoherent integration ROC and the binary integration ROC for a target with an SNR of 3 dB. It also shows an empirical ROC for a particle-filter-based TBD method described later in this chapter. The TBD ROC curve lies between the incoherent integration curve and the binary integration curve. For all but the very high false alarm probabilities, it is very close to the incoherent integration ROC. This demonstrates that the uncertain target dynamics are equivalent to a sensitivity loss of less than 1 dB.

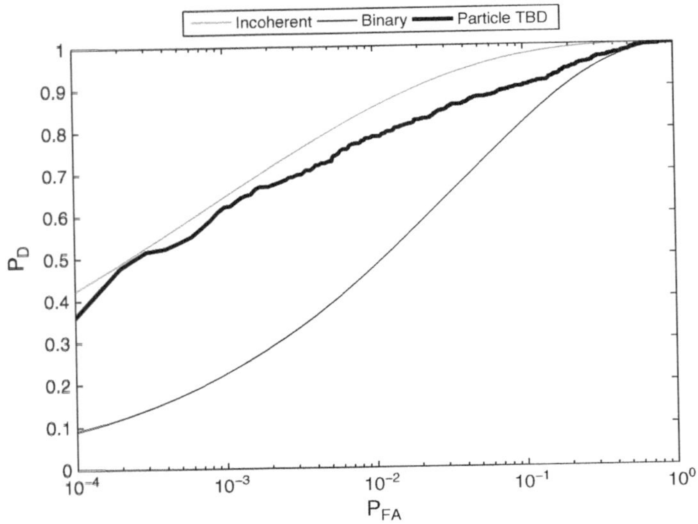

Figure 8.3 TBD ROC, 3 dB SNR, 20 frames.

Combining the results in Figures 8.2 and 8.3, we see that TBD has the potential to offer a 6 dB sensitivity improvement over conventional Kalman Filter detect-then-track methods. This is a significant improvement and acts as motivation for research into TBD methods.

8.2 MODELS

Although a variety of algorithms exist for TBD, most of them share common models and these are now introduced. Following the usual practice in target tracking, stochastic models are assumed for the movement of targets and for the sensor data collection process. In a realistic environment, targets do not follow stochastic trajectories but are subject to control inputs which are unobservable by the sensor. These unknown controls are modeled as a stochastic process.

For simplicity of notation, uniform discrete time will be assumed.

8.2.1 Target Model

In this chapter, the target shall be modeled using a state vector consisting of position and velocity in the plane and amplitude. That is,

$$\mathbf{x}_k^m \equiv \left[x_k^m, \dot{x}_k^m, y_k^m, \dot{y}_k^m, A_k^m\right]', \tag{8.3}$$

where k is a time index, and m is the target index, for the case of multiple targets. The target amplitude will be described in more detail in the next section. It is related to the target SNR and determines how easy it is to detect the target. In physical terms, it may be a function of the radar cross-section of the target, for example.

While target motion may, in general, be nonlinear, this chapter will restrict itself to a linear motion model. Chapters 2, 3, and 5 present methods for tracking with nonlinear motion models as do [38] and [14]. Thus the target state vector evolves according to

$$\boldsymbol{x}_{k+1}^{m} = \mathbf{F}_k \boldsymbol{x}_k^m + \boldsymbol{v}_k^m, \tag{8.4}$$

where $\mathbf{F}_k$ is a known matrix and $\boldsymbol{v}_k^m$ is an additive noise process. A typical assumption is that $\boldsymbol{v}_k^m$ is white and Gaussian with known covariance, $\mathbf{Q}_k$, in which case we may write

$$p\left(\boldsymbol{x}_{k+1}^{m} \middle| \boldsymbol{x}_k^m\right) \sim N\left(\boldsymbol{x}_{k+1}^{m}; \mathbf{F}_k \boldsymbol{x}_k^m, \mathbf{Q}_k\right), \tag{8.5}$$

where $N(\tau; \mu, \Sigma)$ is a multinomial Gaussian pdf with mean vector μ and covariance matrix Σ evaluated at the point τ.

For two-dimensional almost-constant-velocity motion, the matrix $\mathbf{F}_k$ is given by

$$\mathbf{F}_k = \begin{bmatrix} \mathbf{F2}_k & \mathbf{0} & \mathbf{0} \\ \mathbf{0} & \mathbf{F2}_k & \mathbf{0} \\ \mathbf{0} & \mathbf{0} & 1 \end{bmatrix}, \quad \mathbf{F2}_k = \begin{bmatrix} 1 & \Delta_k \\ 0 & 1 \end{bmatrix}, \tag{8.6}$$

where Δ_k is the elapsed time between scan k and scan $k+1$, which may not be constant in general.

The process noise in this case may be written as

$$\mathbf{Q}_k = \begin{bmatrix} \mathbf{Q2}_k & \mathbf{0} & \mathbf{0} \\ \mathbf{0} & \mathbf{Q2}_k & \mathbf{0} \\ \mathbf{0} & \mathbf{0} & q_A \end{bmatrix}, \quad \mathbf{Q2}_k = q_{XY} \begin{bmatrix} \frac{1}{3}\Delta_k^3 & \frac{1}{2}\Delta_k^2 \\ \frac{1}{2}\Delta_k^2 & \Delta_k \end{bmatrix}, \tag{8.7}$$

where q_{XY} is a scalar quantifying the variance of the accelerations, modeled as white noise, and q_A is the variance of the amplitude fluctuations, again modeled as white noise.

Although it is not typically acknowledged, the target state may be confined to a region of interest in both position and velocity: the sensor is only able to observe a limited spatial extent, and objects within it cannot travel arbitrarily fast.

The general Bayesian filter requires the convolution of the state evolution pdf with the posterior pdf from the previous observation. For the continuous state model above, this convolution is an integral that is intractable for most measurement models, even if the state evolution is linear and Gaussian.

To avoid the problem of intractable integrals a common strategy is to use a discretized state space. Whereas the continuous state space represents the set of allowable target positions and velocities as a four-dimensional volume, the discretized state space samples this volume and constrains the state to that set of samples. The advantage of this is that the integrals become summations which can be evaluated numerically even if they cannot be analytically simplified.

The simplest form of discrete state space is a uniform grid. For two-dimensional position and velocity it is intuitive to index the state space using a vector $\boldsymbol{i} \equiv [i_x, i_{\dot{x}}, i_y, i_{\dot{y}}]'$, such that

$$\boldsymbol{x}(\boldsymbol{i}) \equiv \left[\Delta_x i_x, \frac{\Delta_x}{\Delta_k} i_{\dot{x}}, \Delta_y i_y, \frac{\Delta_y}{\Delta_k} i_{\dot{y}}\right]', \tag{8.8}$$

where Δ_x and Δ_y are the grid spacing values.

In this case, the state evolution equation (8.4) is equivalent to an evolution over the index

$$\boldsymbol{i}_{k+1}^m = \mathbf{F}_{\text{grid}} \boldsymbol{i}_k^m + \tilde{\boldsymbol{v}}_k^m, \tag{8.9}$$

where

$$\mathbf{F}_{\text{grid}} = \begin{bmatrix} \mathbf{F2}_{\text{grid}} & \mathbf{0} \\ \mathbf{0} & \mathbf{F2}_{\text{grid}} \end{bmatrix}, \quad \mathbf{F2}_{\text{grid}} = \begin{bmatrix} 1 & 1 \\ 0 & 1 \end{bmatrix}, \tag{8.10}$$

and $\tilde{\boldsymbol{v}}_k^m$ is a discrete noise process with a known probability mass function (pmf). We will see that it is advantageous to limit the support of the pmf of $\tilde{\boldsymbol{v}}_k^m$ because it reduces the required computation effort. In other words, we will choose $P(\tilde{\boldsymbol{v}}_k^m)$ to be identically zero everywhere except within a small distance of $[0, 0, 0, 0]'$.

A simple example for $\tilde{\boldsymbol{v}} = \{v_j\}$ is the pmf

$$P(\tilde{\boldsymbol{v}}) = \begin{cases} \alpha & \tilde{\boldsymbol{v}} = [0, 0, 0, 0]', \\ \dfrac{1-\alpha}{80} & \tilde{\boldsymbol{v}} \neq [0, 0, 0, 0]' \text{ and } |v_j| \leq 1, \\ 0 & \text{otherwise}, \end{cases} \tag{8.11}$$

where α is a legal probability value. For $\alpha > 0.5$, this assigns the majority of the probability mass to the event that the target moves deterministically, with the remainder equally spread across the 80 elements of a $3 \times 3 \times 3 \times 3$ hypercube around it.

For the fixed grid state space, let N_x denote the number of position states in the plane, $N_{\dot{x}}$ denote the number of velocity states in the plane, and N_v denote the number of nonzero elements in the noise process pmf. The total number of elements in the state space is therefore $N_x N_{\dot{x}}$. The size of the state space and the noise process influence the computation cost of algorithms over the grid.

Note that this discrete space has not included a treatment of target amplitude. It is possible to sample amplitude in the same way that the position is sampled. However, for the purposes of this chapter, the discrete state algorithms will use the marginal measurement likelihood described in the following section. This form of the likelihood integrates over the amplitude variable and thus removes it from the model.

8.2.2 Sensor Model

The measurements considered in earlier chapters of this book consist of finite sets of points in physical space. Many sensors do not intrinsically produce such an output. An obvious example is optical sensors which collect images. Subsequent processing may be performed to extract point measurements. In this chapter, we will assume that the sensor collects an image, which may have more than two dimensions, and this image is what is supplied to the input of the TBD algorithm.

Let z_k^i denote the ith pixel in the sensor image at time k, and let $z_k = \{z_k^i\}$ denote a stacked vector of all the pixels of the image. The contribution of the target depends on the physical shape and orientation of the target, the point spread function (psf) of the sensor, and the amplitude of the signal, A_k^m. For most of this chapter, we will assume point-scatterer targets, that is the shape of the target is a single point in space. In this case, the shape of the target contribution to the sensor image is determined by the sensor psf, $\boldsymbol{h}(x, y)$, which is a property of the sensor, not the target, and so is the same for all targets.

For point-scatterer targets, the measurement function is

$$z_k = \sum_m A_k^m \boldsymbol{h}\left(x_k^m, y_k^m\right) + \boldsymbol{w}_k, \tag{8.12}$$

where we have assumed superposition of target power, and where $\boldsymbol{w}_k$ is additive sensor noise. The statistics of the sensor noise may vary depending on the physical characteristics of the sensor. The point spread function, $\boldsymbol{h}(x, y)$, may also vary between different sensors. Some common examples are discussed below.

Define the peak signal to noise ratio (peak-SNR) as

$$\gamma_k^m = \frac{\left(A_k^m\right)^2}{\sigma_w^2}, \tag{8.13}$$

where σ_w^2 is the variance of the sensor noise. This number is often expressed in a decibel scale, that is as $10\log_{10}\{\gamma_k^m\}$.

The psf is scaled such that the maximum value of $\boldsymbol{h}(x, y)$ is unity and the area under the psf is denoted $H^2 = |\boldsymbol{h}(x_k, y_k)|^2$. Note that this is assumed independent of the target state; this assumption may be violated if the target is near a boundary of the surveillance region.

8.2.2.1 Gaussian Noise. In optical sensors, it is common to subtract the average background level and then model the residual deviations as Gaussian. Thus the sensor noise is assumed to be a zero mean Gaussian random variable with covariance $\mathbf{R}$

(e.g., [4]). If no target is present, the pdf of the image z_k is therefore

$$p_G^0(z_k) = \left(|2\pi\mathbf{R}|\right)^{-1} \exp\left\{-\frac{1}{2}z_k{}'\mathbf{R}^{-1}z_k\right\}. \tag{8.14}$$

In optical sensors it is also customary to assume that the psf is Gaussian (e.g., [3]).

When a single target is present, the pdf of the image is the same Gaussian but with a nonzero mean, namely

$$p_G^1(z_k|\boldsymbol{x}_k) = \left(|2\pi\mathbf{R}|\right)^{-1} \exp\left\{-\frac{1}{2}\nu_k{}'\mathbf{R}^{-1}\nu_k\right\}, \tag{8.15}$$

with

$$\nu_k = A_k\boldsymbol{h}(x_k, y_k) - z_k. \tag{8.16}$$

Hence the likelihood ratio is

$$\begin{aligned} \mathcal{L}_G(z_k|\boldsymbol{x}_k) &\equiv \frac{p_G^1(z_k|\boldsymbol{x}_k)}{p_G^0(z_k)} \\ &= \exp\left\{-\frac{1}{2}\nu_k{}'\mathbf{R}^{-1}\nu_k + \frac{1}{2}z_k{}'\mathbf{R}^{-1}z_k\right\}. \end{aligned} \tag{8.17}$$

Assuming spatially uniform and uncorrelated noise, the noise covariance simplifies to $\mathbf{R} = \sigma_w^2\mathbf{I}$ and the likelihood ratio is

$$\mathcal{L}_G(z_k|\boldsymbol{x}_k) = \exp\left\{-\frac{A_k^2|\boldsymbol{h}(x_k, y_k)|^2}{2\sigma_w^2}\right\} \exp\left\{\frac{A_k\boldsymbol{h}(x_k, y_k)'z_k}{\sigma_w^2}\right\}. \tag{8.18}$$

Note that the assumption of uniform uncorrelated noise is not necessary, but it does simplify notation significantly and helps to illustrate the differences between the likelihoods for different noise assumptions.

When more than one target is present, the single-target contribution is replaced with the superposition as in (8.12).

Assuming that the data are scaled to unity noise variance, σ_w^2, then if a target is present the likelihood ratio can be written as

$$\mathcal{L}_G(z_k|\boldsymbol{x}_k) = \exp\left\{\frac{1}{2}\gamma_k H^2\right\} \exp\left\{-\gamma_k^{1/2}\boldsymbol{h}(x_k, y_k)'\boldsymbol{w}_k\right\}. \tag{8.19}$$

The expected value of this likelihood ratio is

$$
\begin{aligned}
E\left[\mathcal{L}_G(\cdot)\right] &= \int \exp\left\{\frac{1}{2}\gamma_k H^2\right\} \exp\left\{-\gamma_k^{1/2}\boldsymbol{h}(\cdot)'\boldsymbol{w}_k\right\} (2\pi)^{-1/2} \exp\left\{-\frac{1}{2}\boldsymbol{w}_k{}'\boldsymbol{w}_k\right\} \mathrm{d}\boldsymbol{w}_k, \\
&= \exp\left\{\frac{1}{2}\gamma_k H^2\right\} \exp\left\{\frac{1}{2}\gamma_k \boldsymbol{h}(\cdot)'\boldsymbol{h}(\cdot)\right\} \\
&\quad \times \int (2\pi)^{-1/2} \exp\left\{-\frac{1}{2}\left(\boldsymbol{w}_k{}'\boldsymbol{w}_k + 2\gamma_k^{1/2}\boldsymbol{h}(\cdot)'\boldsymbol{w}_k + \gamma_k \boldsymbol{h}(\cdot)'\boldsymbol{h}(\cdot)\right)\right\} \mathrm{d}\boldsymbol{w}_k, \\
&= \exp\left\{\gamma_k H^2\right\} \int (2\pi)^{-1/2} \exp\left\{-\frac{1}{2}\left(\boldsymbol{w}_k + \gamma_k^{1/2}\boldsymbol{h}(\cdot)\right)'\left(\boldsymbol{w}_k + \gamma_k^{1/2}\boldsymbol{h}(\cdot)\right)\right\} \mathrm{d}\boldsymbol{w}_k, \\
&= \exp\left\{\gamma_k H^2\right\}.
\end{aligned}
\tag{8.20}
$$

This exponential dependence of the likelihood ratio on target peak-SNR may be problematic: it leads to the potential requirement to deal with very high dynamic range in the TBD algorithm. For a target SNR dynamic range of 20 dB (which is far from unreasonable) the dynamic range of the likelihood would be around 10^{43}. This problem is fundamental to any detection scheme that uses the likelihood ratio and it is important to take account of this behavior in the algorithm's implementation.

8.2.2.2 Rayleigh Noise. For radar sensors, it is more common to assume complex Gaussian noise, resulting in a Rayleigh distributed noise envelope. When the phase of the target response is unknown then the standard approach [12–14] is to assume that the noise is spatially uniform and uncorrelated. This means that the envelope response in each sensor pixel is conditionally independent given the target state (or the knowledge that there is no target).

When no target is present, the pdf of the envelope at pixel i at time k, $\left|z_k^i\right|$, follows a Rayleigh distribution

$$
p_E^0\left(\left|z_k^i\right|\right) = \frac{\left|z_k^i\right|}{\sigma_w^2} \exp\left\{-\frac{\left|z_k^i\right|^2}{2\sigma_w^2}\right\}, \tag{8.21}
$$

and when a target is present, it follows a Rician distribution

$$
p_E^1\left(\left|z_k^i\right| \Big| \boldsymbol{x}_k\right) = \frac{\left|z_k^i\right|}{\sigma_w^2} \exp\left\{-\frac{\left|z_k^i\right|^2 + A_k^2\left|h^i(x_k, y_k)\right|^2}{2\sigma_w^2}\right\} \mathrm{I}_0\left(\frac{A_k\left|h^i(x_k, y_k)\, z_k^i\right|}{\sigma_w^2}\right), \tag{8.22}
$$

where $h^i(x_k, y_k)$ is the target contribution at pixel i, that is $\boldsymbol{h}(x_k, y_k) = \{h^i(x_k, y_k)\}$, and $\mathrm{I}_0(\cdot)$ is the modified Bessel function of order 0, given by

$$
\mathrm{I}_0(x) = \sum_{m=0}^{\infty} \frac{(x/2)^{2m}}{m!\,\Gamma(m+1)}. \tag{8.23}
$$

Hence the likelihood ratio for pixel i is

$$\mathcal{L}_E\left(\left|z_k^i\right|\middle|\boldsymbol{x}_k\right) = \exp\left\{-\frac{A_k^2\left|h^i(x_k, y_k)\right|^2}{2\sigma_w^2}\right\} \mathrm{I}_0\left(\frac{A_k\left|h^i(x_k, y_k)\ z_k^i\right|}{\sigma_w^2}\right). \tag{8.24}$$

Since the pixels are assumed to be conditionally independent, the likelihood of the envelope of the whole image is simply the product over the pixels

$$\begin{aligned}\mathcal{L}_E\left(|z_k|\middle|\boldsymbol{x}_k\right) &= \prod_{i=1}^{N}\mathcal{L}_E\left(\left|z_k^i\right|\middle|\boldsymbol{x}_k\right)\\ &= \exp\left\{-\frac{A_k^2\left|\boldsymbol{h}(x_k, y_k)\right|^2}{2\sigma_w^2}\right\}\prod_{i=1}^{N}\mathrm{I}_0\left(\frac{A_k\left|h^i(x_k, y_k)\, z_k^i\right|}{\sigma_w^2}\right).\end{aligned} \tag{8.25}$$

For radar sensors, the point spread function of the sensor is likely to be the spectral response of the windows used during Fourier Transforms for forming range and Doppler cells and during beam forming.

In the previous section, we found that the likelihood ratio for Gaussian noise grows exponentially with the peak-SNR. Such a relationship is more difficult to derive for the Rayleigh noise case because of the Bessel function in the likelihood. However, we may get an intuitive understanding by making the approximation

$$\mathrm{I}_0(x) \approx \frac{\exp\{x\}}{\sqrt{2\pi x}}, \tag{8.26}$$

which is accurate for large x [39]. For unity noise variance, this gives the approximate likelihood ratio

$$\mathcal{L}_E\left(|z_k|\middle|\boldsymbol{x}_k\right) \approx \exp\left\{-\frac{1}{2}\gamma_k H^2\right\}\prod_{i=1}^{N}\frac{\exp\left\{\gamma_k^{1/2}\left|h^i(x_k, y_k)\, z_k^i\right|\right\}}{\sqrt{2\pi\gamma_k^{1/2}\left|h^i(x_k, y_k)\, z_k^i\right|}}. \tag{8.27}$$

For the case where the SNR is high, the likelihood can be further approximated as

$$\mathcal{L}_E\left(|z_k|\middle|\boldsymbol{x}_k\right) \approx \exp\left\{-\frac{1}{2}\gamma_k H^2\right\}\prod_{i=1}^{N}\frac{\exp\left\{\gamma_k^{1/2}\left|h^i(x_k, y_k)\, A_k h^i(x_k, y_k)\right|\right\}}{\sqrt{2\pi\gamma_k^{1/2}\left|h^i(x_k, y_k)\, z_k^i\right|}}, \tag{8.28}$$

$$= \frac{\exp\left\{\frac{1}{2}\gamma_k H^2\right\}}{\prod_{i=1}^{N}\sqrt{2\pi\gamma_k^{1/2}\left|h^i(x_k, y_k)\, z_k^i\right|}}. \tag{8.29}$$

This expression is very similar to the Gaussian likelihood in (8.18) but has the square-root scaling term in the denominator. This scaling term slows down the growth of the likelihood ratio, but the exponential term grows more quickly. As with the Gaussian noise case, care must be taken to ensure a high dynamic range, especially for extended targets where H^2 is large.

8.2.2.3 Complex Gaussian Noise. An important feature of the Rayleigh noise model above is that the pdf has been assumed to be conditionally independent across pixels. It is intuitive that this assumption will be violated if the noise has spatial correlation. However, it is less intuitive to realize that the approach also implicitly assumes that the phase response of the target signature is spatially uncorrelated.

As stated earlier, the target signature depends on the target physical properties and the sensor psf. If the target is physically approximated by a collection of point scatterers, then the signature is the convolution of this scatterer field with the psf. For targets that are large compared with the sensor resolution, this field will contain many elements and the assumption of an uncorrelated phase response is appropriate. However, if the target is small compared with the sensor resolution, it may be appropriate to represent it as a single point scatterer, in which case the target's phase response is the phase response of the psf, which is known. For the case that the target phase response is known the likelihood must be jointly evaluated over the whole image.

For complex-Gaussian noise, the pdf of the image is

$$p_C^0(z_k) = \left(|2\pi\mathbf{R}|\right)^{-1} \exp\left\{-\frac{1}{2} z_k{}'\mathbf{R}^{-1} z_k\right\}, \tag{8.30}$$

which (unsurprisingly) looks exactly the same as the Gaussian noise case (8.14), except that the measured image, z_k, is complex valued.

When a target is present, the pdf is again very similar to the real-valued Gaussian case, except that there is an unknown phase shift, ϕ_k,

$$p_C^1(z_k|\boldsymbol{x}_k, \phi_k) = \left(|2\pi\mathbf{R}|\right)^{-1} \exp\left\{-\frac{1}{2}\nu_k{}'\mathbf{R}^{-1}\nu_k\right\}, \tag{8.31}$$

with

$$\nu_k = \exp\{j\phi_k\}A_k\boldsymbol{h}(x_k, y_k) - z_k. \tag{8.32}$$

The likelihood ratio is obtained by assuming a uniform distribution for ϕ_k over $[0, 2\pi)$ and marginalizing [40]

$$\begin{aligned}\mathcal{L}_C(z_k|\boldsymbol{x}_k) &= \int_0^{2\pi} \frac{p_C^1(z_k|\boldsymbol{x}_k, \phi_k)}{p_C^0(z_k)} p(\phi)d\phi,\\ &= \exp\left\{-\frac{A_k^2}{2}\boldsymbol{h}(x_k, y_k)^{\mathrm{H}}\mathbf{R}_w^{-1}\boldsymbol{h}(x_k, y_k)\right\} \mathrm{I}_0\left(A_k\left|\boldsymbol{h}(x_k, y_k)^{\mathrm{H}}\mathbf{R}_w^{-1}z_k\right|\right).\end{aligned} \tag{8.33}$$

Again assuming spatially uniform and uncorrelated noise, the likelihood simplifies to

$$\mathcal{L}_C(z_k|\boldsymbol{x}_k) = \exp\left\{-\frac{A_k^2\,|\boldsymbol{h}(x_k, y_k)|^2}{2\sigma_w^2}\right\} \mathrm{I}_0\left(\frac{A_k\left|\boldsymbol{h}(x_k, y_k)^{\mathrm{H}} z_k\right|}{\sigma_w^2}\right). \tag{8.34}$$

It is not surprising that the likelihood ratio for complex Gaussian noise is very similar to the Rayleigh noise case. The difference is that the product of Bessel functions in (8.25) is replaced with a single Bessel function evaluated at the output of a matched filter.

As for the Rayleigh noise case, the Bessel function can be approximated with (8.26). Assuming unity noise variance, the approximate likelihood can be written as

$$\mathcal{L}_C(z_k|\boldsymbol{x}_k) \approx \exp\left\{-\frac{1}{2}\gamma_k H^2\right\} \frac{\exp\left\{\gamma_k^{1/2}\left|\boldsymbol{h}(x_k, y_k)^{\mathrm{H}}\, z_k\right|\right\}}{\sqrt{2\pi\gamma_k^{1/2}\left|\boldsymbol{h}(x_k, y_k)^{\mathrm{H}}\, z_k\right|}}. \tag{8.35}$$

For the case where the SNR is high, the likelihood can be further approximated as

$$\mathcal{L}_C(z_k|\boldsymbol{x}_k) \approx \exp\left\{\frac{1}{2}\gamma_k H^2\right\}\left(2\pi\gamma_k H^2\right)^{-\frac{1}{2}}. \tag{8.36}$$

The complex Gaussian likelihood will be higher than the corresponding Rayleigh noise likelihood, as is expected since it also incorporates phase information. However, it will not grow as quickly as the Gaussian noise likelihood.

8.2.2.4 Point Measurements. The conventional assumption is that tracker input is a set of point measurements. Some TBD approaches also work with point measurements, such as the Hough Transform and ML-PDA. If the sensor supplies an image, as described above, then a single-frame detector is used to extract point measurements from the image. The design of this detector may require care: although rejecting false alarms is one of the tracker's primary tasks, Section 8.1.2 demonstrated that there are limits to how many false alarms the tracker can handle.

The single-frame detector consists of two parts: firstly those pixels belonging to a common object should be connected together, and secondly a threshold is applied to limit the number of false alarms at the detector output. Methods for connecting pixels are application dependent since their effectiveness depends on the expected size of the target response and the spatial correlation of the noise. A common technique is image segmentation [41].

The output of the single-frame detector is a set of measurements at each scan, ζ_k^n for $n = 1 \ldots N_k$, where N_k is the number of measurements at scan k. Each of these point measurements corresponds to a physical location in the measurement space. The dimension and nature of the measurement space is application dependent, and is usually a subspace of the target state space. For example, ζ_k^n may be a range and bearing, corresponding to a single point in the plane, or azimuth and elevation, corresponding to a radial line from the sensor.

A key difference between the point-measurement model and the image-measurement model above is that the measurement function needs to assume a source for each point measurement.

8.3 BAUM WELCH ALGORITHM

Having introduced models for the targets and the sensors, we now review some of the more recent TBD approaches. The first of these estimates the posterior pdf of the target state over a discrete grid using the Baum–Welch[1] algorithm [42, 43].

The posterior pdf of the target state can be recursively determined using the well-known Bayesian relationship

$$\begin{aligned} p(\boldsymbol{x}_k|z_{1:k}) &\propto p(z_k|\boldsymbol{x}_k) \int p(\boldsymbol{x}_k|\boldsymbol{x}_{k-1}) p(\boldsymbol{x}_{k-1}|z_{1:k-1}) d\boldsymbol{x}_{k-1}, \\ &\propto \mathcal{L}(z_k|\boldsymbol{x}_k) \int p(\boldsymbol{x}_k|\boldsymbol{x}_{k-1}) p(\boldsymbol{x}_{k-1}|z_{1:k-1}) d\boldsymbol{x}_{k-1}. \end{aligned} \tag{8.37}$$

When the processes are linear and the random elements are Gaussian, (8.37) can be implemented as a Kalman Filter, which provides a recursion for the first two moments of $p(\boldsymbol{x}_k|z_{1:k})$. However, as we have seen in the previous section, the measurement process is not linear for TBD and the measurement noise is often not Gaussian. This means that there is no analytic solution to the integral in (8.37).

A generally applicable technique for solving difficult integrals is to approximate the problem over a discrete grid and use numerical methods such as Runge–Kutta. In this case, the grid would consist of spatial sampling and also temporal sampling. Numerical integration over a fixed grid has been used for TBD [44, 45], but it tends to result in a high computation overhead.

An alternative to numerically integrating a continuous problem is to treat the problem as a discrete-state one and then use an established discrete-state estimator, such as the Baum–Welch algorithm. Baum–Welch is an algorithm that provides the posterior pmf of a state which may be *hidden*, that is not directly observed. The smoother form of the algorithm is composed of a *forwards* filter that recursively evaluates the probability $\alpha_k(\boldsymbol{x}_k) := p(\boldsymbol{x}_k, z_{1:k})$, and a *backwards* filter that recursively evaluates the forecast probability of the state and all future measurements, $\beta_k(\boldsymbol{x}_k) := p(\boldsymbol{x}_k, z_{k+1:T})$.

For the discrete-state TBD problem, $\boldsymbol{x}(\boldsymbol{i})$ is a deterministic spatial mapping from an index to physical space. Therefore work with the index only, resulting in the following

[1] Also referred to as the Hidden Markov Model filter/smoother.

forwards and backwards filters

$$\begin{aligned}\alpha_k(\boldsymbol{i}_k) &:= p(\boldsymbol{i}_k, z_{1:k}),\\ &= \sum_{\boldsymbol{i}_{k-1}} p(\boldsymbol{i}_k, \boldsymbol{i}_{k-1}, z_{1:k}),\\ &= \sum_{\boldsymbol{i}_{k-1}} p(z_k|\boldsymbol{i}_k)p(\boldsymbol{i}_k|\boldsymbol{i}_{k-1})p(\boldsymbol{i}_{k-1}, z_{1:k-1}),\\ &= p(z_k|\boldsymbol{i}_k)\sum_{\boldsymbol{i}_{k-1}} p(\boldsymbol{i}_k|\boldsymbol{i}_{k-1})\alpha_{k-1}(\boldsymbol{i}_{k-1}),\end{aligned} \tag{8.38}$$

$$\begin{aligned}\beta_k(\boldsymbol{i}_k) &:= p(\boldsymbol{i}_k, z_{k+1:T}),\\ &= \sum_{\boldsymbol{i}_{k+1}} p(\boldsymbol{i}_k, \boldsymbol{i}_{k+1}, z_{k+1:T}),\\ &= \sum_{\boldsymbol{i}_{k+1}} p(z_{k+1}|\boldsymbol{i}_{k+1})p(\boldsymbol{i}_k|\boldsymbol{i}_{k+1})p(\boldsymbol{i}_{k+1}, z_{k+2:T}),\\ &= \sum_{\boldsymbol{i}_{k+1}} p(z_{k+1}|\boldsymbol{i}_{k+1})p(\boldsymbol{i}_k|\boldsymbol{i}_{k+1})\beta_{k+1}(\boldsymbol{i}_{k+1}).\end{aligned} \tag{8.39}$$

The forwards and backwards filters are combined to produce the pdf

$$p(\boldsymbol{i}_k|z_{1:T}) = \frac{\alpha_k(\boldsymbol{i}_k)\beta_k(\boldsymbol{i}_k)}{\sum_{\boldsymbol{i}_k}\alpha_k(\boldsymbol{i}_k)\beta_k(\boldsymbol{i}_k)}. \tag{8.40}$$

8.3.1 Detection

The Baum–Welch algorithm provides a recursion for the state pdf, but does not explicitly decide whether a target is present or not. Integrating this decision with the estimator is the fundamental purpose of TBD. One option is to apply heuristic rules to the output pdf. For example, a threshold could be applied to the peaks of the pdf and a target declared when it is exceeded. Alternatively, the state space can be augmented with a nonphysical null state corresponding to the possibility that there is no target.

The transition between physical states is governed by the dynamic model and the process noise as defined in (8.9). In addition to this evolution process, assume that there is a probability of target *death*, P_{death}, which is the probability that a target will transition from any physical state to the null state. This means that the target vanishes from the sensor's field of view. We also define a probability of target *birth*, P_{birth}, which is the probability that a target spontaneously appears when there previously was none. For simplicity, assume that the pmf of the state of new targets is uniform, although this need not be the case (for example, prior knowledge about the location of airports may motivate a nonuniform distribution). Denoting the null state as $\boldsymbol{i} = \emptyset$, the augmented target transition model becomes

$$p(\boldsymbol{i}_k|\boldsymbol{i}_{k-1}) = \begin{cases} 1 - P_{\text{birth}} & \boldsymbol{i}_k = \emptyset, \boldsymbol{i}_{k-1} = \emptyset, \\ P_{\text{death}} & \boldsymbol{i}_k = \emptyset, \boldsymbol{i}_{k-1} \neq \emptyset, \\ \dfrac{P_{\text{birth}}}{N_x N_{\dot{x}}} & \boldsymbol{i}_k \neq \emptyset, \boldsymbol{i}_{k-1} = \emptyset, \\ (1 - P_{\text{death}})\, p(\tilde{\boldsymbol{v}}_k = \boldsymbol{i}_k - \mathbf{F}_{\text{grid}}\boldsymbol{i}_{k-1}) & \boldsymbol{i}_k \neq \emptyset, \boldsymbol{i}_{k-1} \neq \emptyset. \end{cases} \tag{8.41}$$

The parameters P_{birth} and P_{death} may be treated as tuning parameters and adjusted to optimize detection performance. The algorithm is initialized with $p(\boldsymbol{i}_0 = \emptyset) = 1$ and $p(\boldsymbol{i}_0) = 0\ \forall\ \boldsymbol{i}_0 \neq \emptyset$.

Once the pdf of the state has been evaluated, the Maximum A Posteriori (MAP) state estimate is by definition the state with the highest probability. In the event that this state is the null state, then the algorithm reports that there is no target.

8.3.2 Parameter Selection

The parameters of the grid filter are the measurement model, the target model, and the null-state transition probabilities. The measurement model is a description of the sensor and for most applications it is reasonable to assume that this is known, as will be done here. Typically targets do not follow stochastic trajectories, and the target model is a method to compensate for unknown control inputs. The focus of this chapter is detection performance, not estimation performance, so an almost constant velocity motion model is assumed sufficient. This leaves the birth probability, P_{birth}, and the death probability P_{death}.

Anecdotal testing indicated that the detection performance of the algorithm was insensitive to changes in P_{death}, which was therefore fixed at a small value, 0.01. Intuitively, this parameter will influence how long the filter persists with a false track or a target originated track whose target has vanished. However, this issue was not explored in this comparison.

In contrast, the false alarm rate and detection probability are strongly dependent on the birth probability, and P_{birth} may be adjusted to achieve a desired false alarm rate. When P_{birth} is close to unity, a track is quickly established whether a target is present or not. Conversely, a small value may significantly delay track establishment on a valid target.

8.3.3 Complexity Analysis

The forwards and backwards filters are evaluated for each position in the state grid and each time point, so the complexity of the Baum–Welch algorithm is order $O\left(T\, \langle N_x N_{\dot{x}} \rangle^2\right)$ where $N_x N_{\dot{x}}$ was previously defined as the number of elements in the discrete state space. This complexity can be reduced by using a compact process noise model, such as the one in (8.11). In this case, the number of nonzero terms in the summations in (8.39) and (8.39) reduces to N_v and the algorithm complexity is $O(TN_x N_{\dot{x}} N_v)$.

Efficiency can be improved by observing that the likelihood terms $p(z_k|\boldsymbol{i}_k)$ may not be independent for every state. For example, if the measurement frame is range and azimuth bins, then the likelihood is independent of target speed. In this case, the likelihood should be evaluated over the spatial grid and stored for use in calculating the forwards and backwards filters. The measurement likelihood is generally a very expensive term to calculate, especially for the Rayleigh-noise and complex-Gaussian noise models which contain a Bessel function.

In the experimental analysis, it will be seen that implementations of the grid filter using the two likelihood alternatives have substantially different computation costs. In order to remove the effects of implementation efficiency, the computation cost of the algorithm is now considered analytically.

Assume that the state space $\mathcal{X}$ is a regular grid such that $|\mathcal{X}| = A_X A_{\dot{X}}$, where A_X is the number of position states in $\mathcal{X}$ and $A_{\dot{X}}$ is the number of velocity states. Let V denote the number of elements of $\mathcal{X}$ for which the process noise has a nonzero probability mass.

The likelihood is independent of the target speed and so an efficient implementation requires A_X likelihood calculations, each of which has a cost $C_{\mathcal{L}}$.

For each state locale, (8.37) requires V products and accumulates under the sum with a further two products outside the sum. Determining the scaling constant requires a sum over the whole state space. Let $C_\times$ and C_+ denote the cost of a multiplication and a sum respectively.

Thus the total cost of a single time slice of the estimator is

$$\begin{aligned} C_{\text{GRID}} &= A_X C_{\mathcal{L}} + A_X A_{\dot{X}} C_+ + A_X A_{\dot{X}} \left\{ (V+2)C_\times + VC_+ \right\}, \\ &\approx A_X C_{\mathcal{L}} + A_X A_{\dot{X}} V (C_\times + C_+), \end{aligned} \tag{8.42}$$

since $V \gg 1$.

If the state space has a fixed physical extent in space and velocity and a symmetrical sampling spatial frequency f_s then the cost can be written as

$$C_{\text{GRID}} \approx (f_s)^2 a_1 C_{\mathcal{L}} + (f_s)^8 a_2 (C_\times + C_+). \tag{8.43}$$

So the algorithm cost contains a term that has $O(N^2)$ complexity in the grid resolution due to the likelihoods and a term that has $O(N^8)$ complexity due to the dynamic model.

It is intuitive that the cost of calculating the envelope likelihood is much higher than that of the matched likelihood. Although the second term is high in complexity, it is amenable to pipeline processing and may be relatively cheap for problems over a small spatial area.

Algorithm 1: Fixed-grid Baum–Welch

Pre-compute data likelihood ratios, $\mathcal{L}(z_k|\boldsymbol{i}_k)$
Initialize $\alpha_0(\emptyset) = 1$ and $\alpha_0(\boldsymbol{i}_0 \neq \emptyset) = 0$
for $k = 1$ to T **do**
    Forwards filter

$$\alpha_k(\boldsymbol{i}_k) = \mathcal{L}(z_k|\boldsymbol{i}_k) \sum_{\boldsymbol{i}_{k-1}} p(\boldsymbol{i}_k|\boldsymbol{i}_{k-1})\alpha_{k-1}(\boldsymbol{i}_{k-1})$$

end for
Initialize $\beta_k(\boldsymbol{i}_k) = (N_x N_{\dot{x}} + 1)^{-1}, \forall \boldsymbol{i}_k$
for $k = T - 1$ to 1 **do**
    Backwards filter

$$\beta_k(\boldsymbol{i}_k) = \sum_{\boldsymbol{i}_{k+1}} \mathcal{L}(z_{k+1}|\boldsymbol{i}_{k+1}) p(\boldsymbol{i}_k|\boldsymbol{i}_{k+1})\beta_{k+1}(\boldsymbol{i}_{k+1})$$

end for
Determine marginal posterior pmf

$$p(\boldsymbol{i}_k|z_{1:T}) = \frac{\alpha_k(\boldsymbol{i}_k)\beta_k(\boldsymbol{i}_k)}{\sum_{\boldsymbol{i}_k} \alpha_k(\boldsymbol{i}_k)\beta_k(\boldsymbol{i}_k)}$$

At each scan, choose the state with the highest probability mass

8.3.4 Summary

The Baum–Welch algorithm is summarized in Algorithm 1. The forwards–backwards filtering method is appropriate for batch processing problems. If a recursive algorithm is preferred, such as for real-time processing, then the backwards filter is omitted and the forwards filter is the recursive pdf estimator.

The main shortcoming of this method is that the computation cost required to estimate the target location to a high degree of accuracy may be excessive.

8.4 DYNAMIC PROGRAMMING: VITERBI ALGORITHM

The Baum–Welch algorithm is a method for evaluating the marginal pmf of the state at each time: it determines $p(\boldsymbol{i}_k|z_{1:T})$ and $p(\boldsymbol{i}_{k+1}|z_{1:T})$ and independently finds the MAP estimate from the marginals. A potential problem with this is that there is no guarantee that the resulting state sequence is a legitimate realization of the target model. For example, (8.11) limits the state transitions to deterministic motion with a one-cell noise process. However, the Baum–Welch estimate resulting from maximizing (8.40) is not constrained by the dynamic model and may provide jumps of more than one cell away from deterministic motion.

An alternative to estimation based on the marginal state distributions is to consider the joint pmf of the entire state sequence, $p(\boldsymbol{i}_0, \boldsymbol{i}_1 \ldots \boldsymbol{i}_k | z_{1:T})$. Although the dimension of the state sequence pmf grows exponentially with time, there are efficient algorithms that can be used to find the MAP sequence estimate that have a complexity that is linear in time. A popular algorithm for this problem is the Viterbi algorithm which has been applied to TBD by a number of authors [6, 7].

In this section, the Viterbi algorithm will be applied to the same discrete problem solved by the Baum–Welch algorithm in the previous section, namely a discretized state space augmented with a null state for target detection.

The main trick of the Viterbi algorithm is to exploit the Markov nature of the state process to avoid enumerating all of the $(N_x N_{\dot{x}})^T$ possible state sequences.

Consider a particular state position at time τ, namely $\boldsymbol{i}_\tau$. The overall sequence solution can be found by finding the best sequence through $\boldsymbol{i}_\tau$ and then comparing the best sequences for different values of $\boldsymbol{i}_\tau$. Conditioned on $\boldsymbol{i}_\tau$, the sequence pmf can be written as

$$\begin{aligned} p(\boldsymbol{i}_{0:\tau-1}, \boldsymbol{i}_{\tau+1:T} | z_{1:T}, \boldsymbol{i}_\tau) &\propto p(\boldsymbol{i}_{0:\tau-1}, \boldsymbol{i}_{\tau+1:T}, z_{1:T} | \boldsymbol{i}_\tau), \\ &= p(z_\tau | \boldsymbol{i}_\tau) p(z_{1:\tau-1} | \boldsymbol{i}_{0:\tau-1}) p(z_{\tau+1:T} | \boldsymbol{i}_{\tau+1:T}) \\ &\qquad p(\boldsymbol{i}_{0:\tau-1} | \boldsymbol{i}_\tau) p(\boldsymbol{i}_{\tau+1:T} | \boldsymbol{i}_\tau), \\ &\propto p(\boldsymbol{i}_{0:\tau-1}, z_{1:\tau-1} | \boldsymbol{i}_\tau) p(\boldsymbol{i}_{\tau+1:T}, z_{\tau+1:T} | \boldsymbol{i}_\tau). \end{aligned} \tag{8.44}$$

The conditioning breaks the optimization problem into two smaller independent optimization problems because the measurements are conditionally independent of other measurements given the current state, and the state sequence is a Markov chain. Intuitively, (8.44) says that the shortest path from A to C through B is the shortest path from A to B followed by the shortest path from B to C.

Suppose that at time k, we know the most probable path through each grid point and we know the associated path likelihood. Denote the most probable ancestor of $\boldsymbol{i}_k$ as $\theta_k(\boldsymbol{i}_k)$ and let the cost of the path through $\boldsymbol{i}_k$ be the negative path log-likelihood, $C_k(\boldsymbol{i}_k)$. The Viterbi algorithm is a recursion for $\theta_k(\boldsymbol{i}_k)$ and $C_k(\boldsymbol{i}_k)$.

The cost of a path through $\boldsymbol{i}_k$ and $\boldsymbol{i}_{k+1}$ is

$$C_k(\boldsymbol{i}_k) - \log\left\{p\left(\boldsymbol{i}_{k+1} | \boldsymbol{i}_k\right)\right\} - \log\left\{p\left(z_{k+1} | \boldsymbol{i}_{k+1}\right)\right\}, \tag{8.45}$$

so the cost of the best path through $\boldsymbol{i}_{k+1}$ is the lowest of these, namely

$$C_{k+1}(\boldsymbol{i}_{k+1}) = \min_{\boldsymbol{i}_k}\left\langle C_k(\boldsymbol{i}_k) - \log\left\{p\left(\boldsymbol{i}_{k+1} | \boldsymbol{i}_k\right)\right\}\right\rangle - \log\left\{p\left(z_{k+1} | \boldsymbol{i}_{k+1}\right)\right\}. \tag{8.46}$$

The most probable ancestor of $\boldsymbol{i}_{k+1}$ is the value of $\boldsymbol{i}_k$ that results in the lowest cost,

$$\theta_{k+1}(\boldsymbol{i}_{k+1}) = \arg\min_{\boldsymbol{i}_k}\left\langle C_k(\boldsymbol{i}_k) - \log p\left(\boldsymbol{i}_{k+1} | \boldsymbol{i}_k\right)\right\rangle. \tag{8.47}$$

At the end of the batch, the estimated state is the one with the lowest final cost, $C_k(\boldsymbol{i}_k)$, and the estimated path is found by stepping backwards through the ancestors starting with $\theta_k(\boldsymbol{i}_k)$.

The costs of all paths will grow with time since they are the accumulated log-likelihood. However, the optimization is invariant under shifts, so we can arbitrarily choose to make $C_k(\emptyset) = 0$ by shifting the costs at each time step. We can also replace the data likelihood with the likelihood ratio, which is equivalent to shifting the costs by $\log\{P(z_{k+1}|\emptyset)\}$.

8.4.1 Parameter Selection

As with the Baum–Welch algorithm, the only parameters of the Viterbi algorithm are those associated with the measurement model and the target model. Again, we assert that it is reasonable to assume that the sensor model is known. The target model will be selected based on the expected target behavior, as described in Section 8.3.2. The target birth and death parameters define a ROC curve for the detection output.

8.4.2 Complexity Analysis

The Baum–Welch algorithm accumulates probabilities over all possible state transitions and the Viterbi algorithm finds the highest probability one. Since they both operate over the same state space and its associated set of feasible transitions, it should come as no surprise that the complexity of Viterbi is also $O(TN_xN_{\dot{x}}N_v)$. Since this consists of comparisons for Viterbi as opposed to accumulates for Baum–Welch, we might expect that the Viterbi algorithm would be somewhat more expensive.

As for the Baum–Welch algorithm, observe that the likelihood calculation depends on the projection of the state into the measurement space and so it will be more efficient to determine the likelihood over the spatial grid. As before, this likelihood cost grows with the spatial area of the grid, and the overall cost of it will scale with the square of the spatial sampling frequency, f_s^2. Similarly, the state evolution component grows as f_s^8.

An important difference between the Viterbi algorithm and the Baum–Welch algorithm is the history list, $\theta_k(\boldsymbol{i}_k)$. The size of this list is the product of the size of the state space and the length of the time sequence, $A_X A_{\dot{X}} T$. Whereas Baum–Welch can be implemented as a forwards-only recursive filter, Viterbi is inherently a batch processor and the size of this history list may impact on memory requirements.

8.4.3 Summary

The Viterbi algorithm as applied to TBD is summarized in Algorithm 2. The key feature of the algorithm is that it implicitly optimizes over all possible state sequences by making locally optimal decisions.

Like Baum–Welch it is an optimal algorithm over the grid, so the potential problem with it is the cost in computation time and memory to operate over a grid of sufficient resolution and scope.

Algorithm 2: Fixed-grid Viterbi

1: Initialize $C_0(\emptyset) = 0$ and $C_0(\boldsymbol{i}_0) = \infty$ for all other states
2: **for** each scan $k = 1 \ldots T$ **do**
3: Calculate the un-normalized cost of the null state

$$c_k^0 = \min_{\boldsymbol{i}_{k-1}} \left\langle C_{k-1}(\boldsymbol{i}_{k-1}) - \log\{p(\boldsymbol{i}_k = \emptyset|\boldsymbol{i}_{k-1})\} \right\rangle$$

4: Calculate the normalized state costs

$$C_k(\boldsymbol{i}_k) = -\log \mathcal{L}(z_k|\boldsymbol{i}_k) + \min_{\boldsymbol{i}_{k-1}} \left\langle C_{k-1}(\boldsymbol{i}_{k-1}) - \log p(\boldsymbol{i}_k|\boldsymbol{i}_{k-1}) \right\rangle - c_k^0$$

5: The previous state in the most likely sequence leading to $\boldsymbol{i}_k$ is given by

$$\theta_k(\boldsymbol{i}_k) = \arg\min_{\boldsymbol{i}_{k-1}} \left\langle C_{k-1}(\boldsymbol{i}_{k-1}) - \log p(\boldsymbol{i}_k|\boldsymbol{i}_{k-1}) \right\rangle$$

6: **end for**
7: The estimated state at time T is

$$\hat{\boldsymbol{i}}_k = \arg\min_{\boldsymbol{i}_k} C_k(\boldsymbol{i}_k)$$

8: **for** each scan $k = T - 1 \ldots 1$ **do**
9: The estimated state at time k is found by backtracking

$$\hat{\boldsymbol{i}}_k = \theta_{k+1}\left(\hat{\boldsymbol{i}}_{k+1}\right)$$

10: **end for**

8.5 PARTICLE FILTER

As introduced in Section 8.3, the fundamental Bayesian filtering relationship is

$$p(\boldsymbol{x}_k|\boldsymbol{z}_{1:k}) \propto \mathcal{L}(z_k|\boldsymbol{x}_k) \int p(\boldsymbol{x}_k|\boldsymbol{x}_{k-1})p(\boldsymbol{x}_{k-1}|z_{1:k-1})d\boldsymbol{x}_{k-1}.$$

For point-measurement tracking, the densities are often assumed to be Gaussian and linear and the integral can be analytically solved, leading to the Kalman Filter. For a general nonlinear problem, such as TBD, the integral cannot be solved analytically. The previous two algorithms were based on a fixed-grid sampling approximation to the Bayesian filter. An alternative to fixed sampling is Monte Carlo sampling, that is the particle filter.

As introduced in earlier chapters, the particle filter uses a random sample approximation to the densities in the Bayesian filter, namely

$$p(\boldsymbol{x}) \approx \sum_{n=1}^{N_p} w^n \delta(\boldsymbol{x} - \boldsymbol{p}^n), \tag{8.48}$$

where N_p is the number of particles, $\delta(\cdot)$ is the Dirac delta function, $\boldsymbol{p}^n$ is the nth particle, which is a sample from the state space, and w^n is the weight of the nth particle. The weights sum to unity.

A direct implementation of particle filtering for TBD would consist of: propagating the particles forwards in time (from $k-1$ to k) using random samples from the process model; determining the weight of each particle using the frame likelihood given the state of that particle; and resampling the particles to avoid degeneracy.

The particle filter was first applied to TBD by Salmond in [12] and Boers in [13]. This was extended by Ristic and Rutten in [14] and [15]. The algorithm described in this chapter is based on the most recent version published in [15].

For the discrete state-space methods, the state space was augmented with a null state to enable automated track management decisions. This can be done for the particle filter too. However, it has been shown that this is an inefficient implementation method [46]. If the state space is augmented, then a proportion of the particles will occupy the null state, depending on the probability of a target being present, but the null state is a discrete point, so there is no value in sampling it. Only the nonnull particles provide useful information, and the effective number of samples will be diminished. Instead, it is better to directly estimate the null probability and only sample the conditional distribution $p(\boldsymbol{x}|\boldsymbol{x} \neq \emptyset)$.

Let E_k be an indicator variable such that $E_k = 1 \rightarrow \boldsymbol{x}_k \neq \emptyset$ and $E_k = 0 \rightarrow \boldsymbol{x}_k = \emptyset$. Also, let $p_k^{\emptyset}$ denote the null-state probability,

$$p_k^{\emptyset} := p\,(\boldsymbol{x}_k = \emptyset | z_{1:k}) = p(E_k = 0|z_{1:k}). \tag{8.49}$$

Recall that the transition probabilities $p(E_k|E_{k-1})$ are specified by the priors P_{birth} and P_{death}.

The nonnull marginal state distribution is approximated with particles

$$p(\boldsymbol{x}_k|\boldsymbol{x}_k \neq \emptyset, z_{1:k}) = p(\boldsymbol{x}_k|E_k = 1, z_{1:k}) \approx \sum_{n=1}^{N_p} w_k^n \delta(\boldsymbol{x}_k - \boldsymbol{p}_k^n). \tag{8.50}$$

The null-state probability recursion is simply the Bayesian filter equation

$$p\,(E_k|z_{1:k}) \propto p(z_k|E_k) \sum_{E_{k-1}=0}^{1} p\,(E_k|E_{k-1})\, p\,(E_{k-1}|z_{1:k-1}). \tag{8.51}$$

The data likelihood $p(z_k|E_k)$ can be expressed as a function of the particle weights.

Using Bayes' rule and the law of total probability, the state distribution can be written as the mixture of two components, one representing the distribution of new (birth) targets appearing at time k, and the other representing continuing targets,

$$\begin{aligned} p(\boldsymbol{x}_k|E_k = 1, z_{1:k}) \propto \mathcal{L}(z_k|\boldsymbol{x}_k)(1 - P_{\text{death}})(1 - p_{k-1}^{\emptyset}) \\ \int p(\boldsymbol{x}_k|\boldsymbol{x}_{k-1})p(\boldsymbol{x}_{k-1}|z_{1:k-1})d\boldsymbol{x}_{k-1} \\ +\mathcal{L}(z_k|\boldsymbol{x}_k)P_{\text{birth}}p_{k-1}^{\emptyset}p(\boldsymbol{x}_k|E_{k-1} = 0). \end{aligned} \tag{8.52}$$

The particle filter TBD algorithm therefore constructs two sets of particles. The first set, $\{w_b^n, \boldsymbol{p}_b^n\}$, represents the density of new (birth) targets. The second set, $\{w_c^n, \boldsymbol{p}_c^n\}$, represents the density of targets that continue from $k - 1$. The mixing proportions between these two sets are determined by the prior null-state probability, $p_{k-1}^{\emptyset}$, and by the birth and death probabilities, P_{birth} and P_{death}.

The updated null-state probability is calculated based on the particle weights, and then the two sets of particles are combined. A resampling stage reduces the total number of particles to a prescribed level and the resulting posterior particle set has uniform weights.

The algorithm declares a target detected when the null-state probability, $p_k^{\emptyset}$, is below a tunable threshold. The state estimate is then found by taking the mean of the state vectors of all particles.

8.5.1 Parameter Selection

As with the grid-based algorithm, the particle filter parameters are the measurement model, the target model, and the null-state transition probabilities. For the same reasons as given in 8.3.2, the measurement model and the target model are assumed to be fixed and known.

Studies have shown that the algorithm performance is not sensitive to the probability of target death [40]. The birth probability parameterizes an ROC curve for the algorithm output.

8.5.2 Complexity Analysis

The computation cost of the particle filter is more difficult to quantify than the cost of the grid estimators. The cost of each step is presented before giving a total cost. Let C_p^b be the cost of proposing a birth particle, C_p^c be the cost of proposing a continuing particle, C_q be the cost of evaluating the birth proposal function, and $C_R(n, m)$ be the cost of resampling n particles from a set of m.

The cost of each step is then

$$C_1 = N_b\left(C_p^b + C_{\mathcal{L}} + 3C_\times + C_q\right), \tag{8.53}$$

$$C_2 = N_c\left(C_p^c + C_{\mathcal{L}} + C_\times\right), \tag{8.54}$$

$$C_3 = (N_b + N_c + 2)C_+ + 4C_\times, \tag{8.55}$$
$$C_4 = 3C_\times + 6C_+, \tag{8.56}$$
$$C_5 = (N_b + N_c + 4)C_\times + 4C_+, \tag{8.57}$$
$$C_6 = C_R(N_c, N_b + N_c), \tag{8.58}$$

giving a total cost of

$$\begin{aligned} C_{\text{PF}} &= N_b C_p^b + N_c C_p^c + N_b C_q + C_R(N_c, N_b + N_c) + (N_b + N_c)C_{\mathcal{L}} \\ &\quad + (N_b + N_c + 12)C_+ + (4N_b + 2N_c + 11)C_\times, \\ &\approx N_b C_p^b + N_c C_p^c + N_b C_q + C_R(N_c, N_b + N_c) + (N_b + N_c)C_{\mathcal{L}}, \end{aligned} \tag{8.59}$$

where it is assumed that the cost of individual multiplications and additions is minor compared with the more complex proposal and resampling functions.

8.5.3 Summary

The particle-filter-based TBD method is summarized in Algorithm 3.

8.6 ML-PDA

The algorithms described so far have been numerical estimation methods applied to the filtering problem where the full sensor image is the measurement. However, the discussion in Section 8.1.2 highlighted that it is not necessarily the threshold process that penalizes conventional tracking but rather the combination of thresholding and a random walk dynamic model.

An alternative approach for low-SNR targets is to retain the threshold but replace the estimator. The advantage of this is that point measurements can often be treated as a linear function of the target state, or at least the nonlinearity is reduced.

The Bayesian expansion of the point-measurement likelihood enumerates a hidden assignment variable and weights the likelihood under each assignment hypothesis by its probability. In the tracking literature, this is referred to as Probabilistic Data Association (PDA) [38, 47]. Under PDA the likelihood of a scan of point measurements, $\zeta_k^1 \ldots \zeta_k^{N_k}$, is given by

$$\begin{aligned} p\left(\zeta_k^1 \ldots \zeta_k^{N_k} | \boldsymbol{x}_k\right) &= \sum_{\theta=0}^{N_k} p\left(\zeta_k^1 \ldots \zeta_k^{N_k}, \theta | \boldsymbol{x}_k\right), \\ &\propto p(\theta = 0|\boldsymbol{x}_k) + \sum_{\theta=1}^{N_k} p(\theta|\boldsymbol{x}_k)\mathcal{L}_P\left(\zeta_k^\theta | \boldsymbol{x}_k\right), \end{aligned} \tag{8.68}$$

where θ is an assignment variable indicating the measurement caused by the target and with $\theta = 0$ indicating that no measurement was due to the target. The summation

Algorithm 3: Particle Filter

1: Create a set of N_b birth particles by placing the particles in the highest intensity cells

$$\boldsymbol{p}_b^n \sim q(\boldsymbol{x}|E_{k-1}=0, z_k). \tag{8.60}$$

Note that the placement of birth particles is independent of the set of continuing particles. The un-normalized birth particle weights are calculated using the likelihood ratio and proposal density (8.60)

$$\tilde{w}_b^n = \frac{\mathcal{L}(z_k|\boldsymbol{p}_b^n)\, p(\boldsymbol{x}_k = \boldsymbol{p}_b^n|E_{k-1}=0)}{N_b\, q(\boldsymbol{x}|E_{k-1}=0, z_k)}, \tag{8.61}$$

where the particular form of the likelihood depends on the measurement model assumed for the problem.

2: Create a set of N_c continuing particles using the system dynamics (8.4) for the proposal function, with un-normalized weights

$$\tilde{w}_c^n = \frac{1}{N_c}\mathcal{L}\left(z_k|\boldsymbol{p}_c^n\right). \tag{8.62}$$

3: Determine mixing probabilities using sums of un-normalized weights

$$\tilde{M}_b = P_{\text{birth}}\, p_{k-1}^{\emptyset} \sum_{n=1}^{N_b} \tilde{w}_b^n, \tag{8.63}$$

$$\tilde{M}_c = (1 - P_{\text{death}})(1 - p_{k-1}^{\emptyset}) \sum_{n=1}^{N_c} \tilde{w}_c^n. \tag{8.64}$$

4: Update the null-state probability using the un-normalized weights

$$p_k^{\emptyset} = \frac{P_{\text{death}}(1 - p_{k-1}^{\emptyset}) + (1 - P_{\text{birth}})p_{k-1}^{\emptyset}}{\tilde{M}_b + \tilde{M}_c + P_{\text{death}}(1 - p_{k-1}^{\emptyset}) + (1 - P_{\text{birth}})p_{k-1}^{\emptyset}}. \tag{8.65}$$

5: Normalize the particle weights

$$\hat{w}_b^n = \frac{P_{\text{birth}}\, p_{k-1}^{\emptyset}}{\tilde{M}_b + \tilde{M}_c}\, \tilde{w}_b^n, \tag{8.66}$$

$$\hat{w}_c^n = \frac{(1 - P_{\text{death}})(1 - p_{k-1}^{\emptyset})}{\tilde{M}_b + \tilde{M}_c}\, \tilde{w}_c^n. \tag{8.67}$$

6: Combine the two sets of particles into one large set and resample from $N_b + N_c$ down to N_c particles with uniform weights.

above is simply a statement of the law of total probability and the factorization arises due to conditional independence.

The likelihood of the whole measurement sequence is simply the product over time since the measurements are independent conditioned on the target state

$$p(\zeta_{1:T}|\boldsymbol{x}_{0:T}) = \prod_{k=1}^{T} \left\{ p(\theta = 0|\boldsymbol{x}_k) + \sum_{\theta=1}^{N_k} p(\theta|\boldsymbol{x}_k)\mathcal{L}_P\left(\zeta_k^\theta|\boldsymbol{x}_k\right) \right\}. \tag{8.69}$$

The ML-PDA state estimate is then the deterministic target trajectory that maximizes the likelihood (8.69).

The advantage of using point measurements and the PDA likelihood is that the cost of evaluating $\mathcal{L}_P(\cdot)$ is generally much lower than the image measurement likelihood. However, the result is still a nonlinear optimization problem in the target state.

There are two key differences between the nonlinear filtering methods described earlier and the nonlinear optimization of ML-PDA: firstly, ML-PDA uses threshold detections; secondly, ML-PDA uses a deterministic target model. One consequence of the threshold is that there can be large portions of the state space where the likelihood is approximately zero and the slope of the likelihood is also approximately zero. For example, see Figure 8.4. The figure shows (8.69) evaluated over a data set with five target measurements at (600, 600) and 50 uniformly distributed clutter measurements. Empirically, it can be shown that around 70% of the state space is an approximately flat floor of zero likelihood [18]. This flat expanse in the likelihood has implications for optimization.

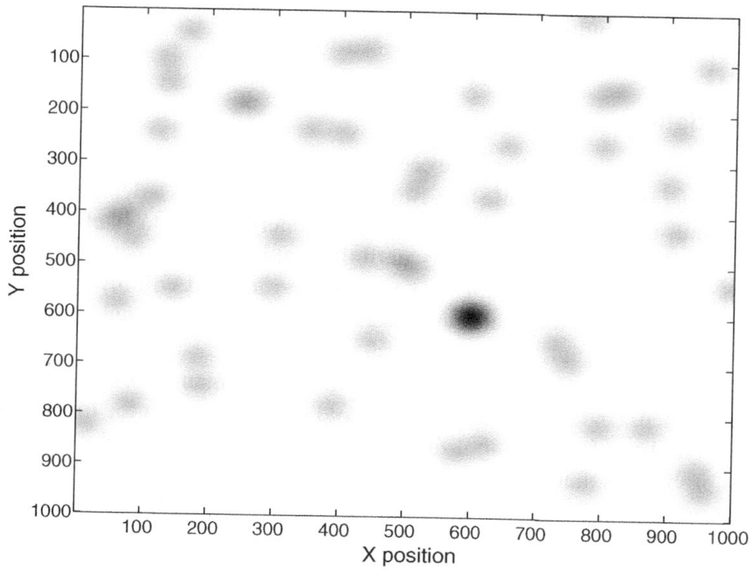

Figure 8.4 Example ML-PDA likelihood surface.

8.6.1 Optimization Methods

Extracting the state estimate for ML-PDA is a multidimensional nonlinear optimization problem. As such, various well-known optimization methods can be applied.

The original ML-PDA work of Jauffret and Bar-Shalom [16] used a coarse grid in the state space to find an approximate location for the likelihood maximum and then refined the location using the Newton–Raphson numerical method. In order to avoid local convergence to one of the many local maxima, the measurement noise was inflated, which has the effect of smoothing out the peaks. Having optimized the inflated case, the noise was gradually reduced to the proper value, with numerical optimization at each stage.

The problem with this method is that it is very costly. The multipass approach requires that the optimization problem be solved several times, and a large proportion of the effort may be spent evaluating inflated likelihoods over regions where the true likelihood is zero and flat. In part this inefficiency arises because the method tries to circumvent the structure of the likelihood. A better alternative is to exploit it.

A recent more efficient optimization strategy was proposed by Blanding [18] and is referred to as *Directed Subspace Search*. The premise of this method is that only trajectories that pass very close to at least one measurement can be local maxima of the likelihood function, and that the global maximum must be one of these local maxima. The directed subspace search constructs a grid in velocity space through each measurement and optimizes the local likelihood. Having found all of the local maxima, the global maximum is given by the highest of them.

Over a range of simulations, Blanding reported an improvement in computation speed of between 5 and 30 times. Further, the directed subspace search also produced more accurate results, with the capability of finding the global optimum at much lower target SNR than the multipass search [18].

8.6.2 Validation

The algorithms described earlier have each used a Markov chain to model the validity of tracks by incorporating a null state. Since the ML-PDA uses a deterministic target model, the null-state model is incompatible. In general, the data likelihood will be increased by adding model complexity until the number of parameters in the model reaches the number of observations. The estimation method will always result in a state estimate. However, the quality of this estimate can be quantified by the value of the likelihood at that point, $\mathcal{L}_P(\zeta_{1:T}|\hat{\boldsymbol{x}})$.

If the statistics of $\mathcal{L}_P(\zeta_{1:T}|\hat{\boldsymbol{x}})$ are known, then a Neyman–Pearson decision rule can be established. Under such a rule, a threshold is prescribed for the likelihood such that the probability of validating a false track is no more than a selected tolerance level. The problem is that these statistics are impossible to analytically determine.

Jauffret appealed to the central limit theorem and approximated the false track statistics with a Gaussian distribution [16]. However, this has been shown to be a poor fit in the tails of the distribution, which is exactly where the fit is required. Blanding extended the extreme-value work of Johnston [8] and applied it to ML-PDA [20],

Algorithm 4: ML-PDA

for $k = 1$ to T **do**
  for $j = 1$ to N_k **do**
    Construct a local grid around measurement ζ_k^j and find the optimal state, $\hat{\boldsymbol{x}}^j$.
  end for
end for
Find the best local maximum, $\hat{\boldsymbol{x}} = \max\{\hat{\boldsymbol{x}}^j\}$
Confirm the track as valid if the likelihood $\mathcal{L}_P(\zeta_{1:T}|\hat{\boldsymbol{x}})$ is greater than the prescribed threshold

achieving a high fidelity fit to empirical data. Following Blanding's approach, a Gumbel distribution can be fitted to a sample of false tracks (that is the algorithm output in the absence of a target) and this distribution can then be used to determine an appropriate likelihood threshold.

8.6.3 Summary

The ML-PDA algorithm is summarized in Algorithm 4. ML-PDA fits a deterministic target trajectory through a collection of point measurements. It is a batch algorithm and tests for the presence or absence of a single target.

The existing ML-PDA literature uses linear target models. However, the framework could easily be extended to consider nonlinear trajectories in the search space. Since a nonlinear trajectory would have more degrees of freedom, the algorithm would be expected to always prefer a nonlinear solution.

Blanding has presented a multitarget extension of ML-PDA [19] based on Joint-PDA. Although this extended version is capable of detecting and tracking multiple targets, the likelihood function has permutational complexity in the number of targets, and the states must be jointly estimated, resulting in a state space whose dimensionality is proportional to the number of targets. It is unclear how the optimization cost grows as a function of the state space dimension.

8.7 H-PMHT

The final algorithm reviewed in this chapter is the Histogram probabilistic Multi-Hypothesis Tracker (H-PMHT) [22]. H-PMHT uses a very different measurement function and applies Expectation Maximization to simplify it [48]. The derivation of H-PMHT uses quantization of the sensor image. However, the final step in the derivation takes the limit of this quantization and recovers the original data. Quantization is only a stage in the derivation of the algorithm and is not used in implementation.

Begin by quantizing the sensor image at some arbitrary scale, $\hbar^2$. The result is an integer-valued image which can be treated as a realization of a Poisson Point Process (PPP) [49]. The underlying PPP is modeled as the mixture of a PPP for each target and a PPP for false detections. Thus the PPP intensity, $f(\tau|\boldsymbol{x}_k^{1:M};\pi_k^{0:M})$, can be written as

$$f(\tau|\boldsymbol{x}_k^{1:M};\pi_k^{0:M}) = \pi_k^0 G_0(\tau) + \sum_{m=1}^{M} \pi_k^m h\left(\tau|\boldsymbol{x}_k^m\right), \tag{8.70}$$

where $\boldsymbol{x}_k^m$ is the state of the mth target and the mixing proportions form a probability vector, that is $\pi_k^m \geq 0$ and

$$\sum_{m=0}^{M} \pi_k^m = 1. \tag{8.71}$$

The mixing proportions may be interpreted as the relative power of each target. In the simplest case, the background clutter model is uniform and $G_0(\tau)$ is a constant. In spatially nonuniform clutter, mapping approaches such as [50, 51] may be used. The target component $h\left(\tau|\boldsymbol{x}_k^m\right)$ may be common across targets, for example it may represent the point spread function for a sensor observing point scatterers, or each target may have a unique signature, such as with high resolution optical sensors. For the purposes of this chapter, it is assumed to be the same for all targets and to have a known functional form with unknown parameters.

The quantized sensor image provides a count of the number of *shots* in each pixel, where each shot is an independent identically distributed random variable following a distribution defined by the intensity function, $f(\tau|\boldsymbol{x}_k^{1:M};\pi_k^{0:M})$. The sensor image does not identify which component of the mixture gave rise to each shot or the precise location of the shot within the pixel. Both of these are treated as missing data and EM is used to marginalize them out of the problem and optimize the target state parameters (i.e., the intensity map). In addition, H-PMHT allows for unobserved pixels. These are notionally sensor pixels for which no data were collected. One use for this concept is in tracking targets near the edge of the sensor frame [23]. The data from these unobserved pixels are also treated as missing data.

The probability that a shot due to target m falls in pixel i is the integral over the pixel of $h\left(\tau|\boldsymbol{x}_k^m\right)$ and was previously denoted as $h^i(\boldsymbol{x}_k^m)$ (see Section 8.2). Similarly, let $h^i(\emptyset)$ denote the probability of a clutter shot falling in pixel i, that is, the integral of $G_0(\tau)$ over pixel i.

The overall probability of a shot falling in pixel i is the mixture of the component probabilities,

$$h^i\left(\boldsymbol{x}_k^{1:M};\pi_k^{0:M}\right) = \pi_k^0 h^i(\emptyset) + \sum_{m=1}^{M} \pi_k^m h^i(\boldsymbol{x}_k^m). \tag{8.72}$$

Assuming an existing estimate of the target states, $\hat{\boldsymbol{x}}_k^{1:M}$, and the mixing proportions, $\hat{\pi}_k^{0:M}$, EM determines the probability of the missing data (E-step) and then refines the state estimate (M-step). This defines an EM algorithm at the quantization $\hbar^2$.

The derivation then takes the limit of this EM algorithm as $\hbar^2 \to 0$ and the result is the H-PMHT, which no longer uses quantization.

Define

$$||Z_k|| = \sum_i z_k^i, \tag{8.73}$$

namely the L_1 norm of the sensor image, and let

$$\bar{z}_k^i = \begin{cases} \dfrac{z_k^i}{h^i\left(\hat{\boldsymbol{x}}_k^{1:M};\hat{\pi}_k^{0:M}\right)} & i \in \mathcal{O}, \\ \dfrac{||Z_k||}{\sum_{i\in\mathcal{O}} h^i\left(\hat{\boldsymbol{x}}_k^{1:M};\hat{\pi}_k^{0:M}\right)} & l \in \bar{\mathcal{O}}, \end{cases} \tag{8.74}$$

where $\mathcal{O}$ is the set of all observed pixels and $\bar{\mathcal{O}}$ is the set of all unobserved pixels, which may be empty. $\mathcal{S}$ is the union of sets $\mathcal{O}$ and $\bar{\mathcal{O}}$.

The parameters to be estimated are the mixing proportions, π_k^m and the target states, $\boldsymbol{x}_k^m$. The updated mixing proportion estimate is given by

$$\pi_k^m = \frac{p_k^m}{\sum_{s=0}^{M} p_k^s}, \tag{8.75}$$

where

$$p_k^m = \hat{\pi}_k^m \sum_{i\in\mathcal{S}} \bar{z}_k^i h^i\left(\hat{\boldsymbol{x}}_k^m\right). \tag{8.76}$$

For the case of linear Gaussian statistics, Streit [22] demonstrated that the state maximization problem is equivalent to a point measurement filtering problem with synthetic measurements

$$\tilde{\zeta}_k^m = \frac{\hat{\pi}_k^m}{p_k^m} \sum_{i\in\mathcal{S}} \bar{z}_k^i\, h^i\left(\hat{\boldsymbol{x}}_k^m\right)\, \tilde{\zeta}_k^{m,i}, \tag{8.77}$$

where the *pixel-level centroid*, $\tilde{\zeta}_k^{m,i}$, is given by

$$\tilde{\zeta}_k^{m,i} = \frac{1}{h^i\left(\hat{\boldsymbol{x}}_k^m\right)} \int_i \tau h\left(\tau|\hat{\boldsymbol{x}}_k^{m(i)}\right) d\tau. \tag{8.78}$$

The associated synthetic measurement covariance is

$$\tilde{R}_k^m = \frac{1}{p_k^m}\Sigma, \tag{8.79}$$

and the synthetic process covariance is

$$\tilde{Q}_k^m = \frac{\sum_{i\in\mathcal{O}} h^i\left(\boldsymbol{x}_k^{1:M};\pi_k^{0:M}\right)}{||Z_k||}Q_k, \tag{8.80}$$

where Σ is the covariance matrix associated with $h(\tau|x)$, that is the spread of the target in the image.

A Kalman Filter can be used to solve this point measurement filtering problem.

8.7.1 Efficient Two-dimensional Implementation

The H-PMHT derivation uses single indexing of the sensor pixels. Single indexing does not limit the sensor dimensionality, but it makes it more difficult to decouple the sensor dimensions if they are independent. For the special case of a two-dimensional sensor with the sensor image a Cartesian product of one-dimensional cells (for example, a standard camera image), Davey developed an efficient matrix-vector formulation of H-PMHT [52].

Assume that the sensor image is collected over a rectangular grid of $N_X \times N_Y$ pixels and that the point spread function can be written as

$$h(\tau|\boldsymbol{x}) = h_X(\tau_X|\boldsymbol{x})h_Y(\tau_Y|\boldsymbol{x}). \tag{8.81}$$

Then the contribution of target m to pixel (i, j) can be written as

$$h^{i,j}\left(\boldsymbol{x}_k^m\right) = h_X^i\left(\boldsymbol{x}_k^m\right)h_Y^j\left(\boldsymbol{x}_k^m\right). \tag{8.82}$$

Define the stacked vectors

$$\boldsymbol{H}_X^m = \left[h_X^1\left(\hat{\boldsymbol{x}}_k^m\right), h_X^2\left(\hat{\boldsymbol{x}}_k^m\right), \ldots h_X^{N_X}\left(\hat{\boldsymbol{x}}_k^m\right)\right]', \tag{8.83}$$

and similarly for $\boldsymbol{H}_Y\left(\hat{\boldsymbol{x}}_k^m\right)$. The equivalent clutter constructs are $\boldsymbol{H}_X^0$ and $\boldsymbol{H}_Y^0$. These are combined across targets to form the matrices

$$\tilde{\mathbf{H}}_X = \left[\pi_k^0\boldsymbol{H}_X^0,\ \pi_k^1\boldsymbol{H}_X^1,\ \ldots\ \pi_k^M\boldsymbol{H}_X^M\right], \tag{8.84}$$

$$\mathbf{H}_Y = \left[\boldsymbol{H}_Y^0,\ \boldsymbol{H}_Y^1,\ \ldots\ \boldsymbol{H}_Y^M\right], \tag{8.85}$$

then

$$\mathbf{H} \equiv \left\{ h^{i,j} \left(\hat{\boldsymbol{x}}_k^{1:M}; \hat{\pi}_k^{0:M} \right) \right\} = \tilde{\mathbf{H}}_X \mathbf{H}_Y'. \tag{8.86}$$

The normalized measurements are $\bar{\mathbf{Z}} = z_k./\mathbf{H}$, where the Matlab-style ./ notation denotes element-wise division.

The unscaled mixing proportion estimate is given by

$$p_k^m = \hat{\pi}_k^m \{\boldsymbol{H}_X^m\}' \, \bar{\mathbf{Z}} \, \boldsymbol{H}_Y^m, \tag{8.87}$$

and the updated mixing proportions are simply scaled versions of these as in (8.75).

Thus the synthetic measurements are

$$\tilde{\zeta}_k^m = \frac{\pi^m}{p^m} \begin{bmatrix} \{\tilde{\zeta}_X^m\}' \, \bar{\mathbf{Z}} \, \boldsymbol{H}_Y^m \\ \{\boldsymbol{H}_X^m\}' \, \bar{\mathbf{Z}} \, \tilde{\zeta}_Y^m \end{bmatrix}, \tag{8.88}$$

where

$$\tilde{\zeta}_X^m = \left[\tilde{\zeta}_X^1 \left(\hat{\boldsymbol{x}}_k^m \right), \ldots \tilde{\zeta}_X^{N_X} \left(\hat{\boldsymbol{x}}_k^m \right) \right]', \tag{8.89}$$

and the pixel-level centroids are given by

$$\tilde{\zeta}_X^i(\hat{\boldsymbol{x}}_k^m) = \int_i u \, h_X \left(u | \hat{\boldsymbol{x}}_k^m \right) du. \tag{8.90}$$

with similar expressions for $\tilde{\zeta}_Y^m$ and $\tilde{\zeta}_Y^j(\hat{\boldsymbol{x}}_k^m)$.

The synthetic measurement and process covariances remain the same, and the state estimates may still be determined using a Kalman Filter.

8.7.2 Nonlinear Gaussian Measurement Function

The original presentation of the H-PMHT algorithm [22] showed that for a linear Gaussian point spread function, that is $h(\tau|\boldsymbol{x}) = \mathcal{N}(\tau; \mathbf{H}\boldsymbol{x}, \mathbf{R})$, the EM auxiliary function is equivalent to the log-likelihood of a point-measurement filtering problem. This is achieved by factorizing the auxiliary function and completing the square. The measurement component is a sum of quadratics in the target state which may be collected into a single quadratic, which is equivalent to the log of a normal distribution.

It is a simple extension of this to show that the same result holds for a Gaussian point spread function where the peak of the psf is a nonlinear function of the target state, that is $h(\tau|\boldsymbol{x}) = \mathcal{N}(\tau; f(\boldsymbol{x}), \mathbf{R})$. In this case completing the square leads to a nonlinear Gaussian noise point measurement problem. Depending on the nonlinearity, this might be solved by analytic linearization (an extended Kalman Filter) or Monte Carlo methods.

8.7.3 Track Management

The H-PMHT algorithm updates existing tracks, but does not provide a means for initiating new tracks or terminating old tracks. For the fixed-grid algorithms and particle filter introduced earlier, this function was achieved by introducing a null state. This has been done for the standard PMHT [53] but incurs an exponential complexity in the number of tracks. Since linear complexity in the number of tracks is one of the main advantages of PMHT, the null state will not be used. Instead, detection decisions will be based on the estimated target SNR, a method shown to give similar performance to the null-state model for uniform clutter [53].

A two-stage approach based on the method in [23] is used for track formation. The tracker maintains two sets of tracks: established tracks, that the tracker is confident correspond to targets, and tentative tracks, that the tracker is not confident in. The established tracks are updated first, and they vet the sensor data before it is presented to the tentative tracks. Similarly, the tentative tracks vet the data before it is passed to a new tentative track formation stage. For each track, the estimated mixing proportion may be translated into an SNR value using

$$s_k^m = 10\log_{10}\left\{\hat{\pi}_k^m/\hat{\pi}_0\right\}. \tag{8.91}$$

Track management decisions, such as track termination and tentative track promotion, may be based on the estimated track SNR. The SNR estimate is based on a single frame of data and does not include any temporal correlation, so it may be desirable to use M out of N style decision rules. An alternative is to introduce dynamics within the π_k^m priors through the Hysteresis approach [53].

The tracks vet the sensor data following the method proposed in [23]. This is done by scaling each pixel based on the tracker model

$$\acute{\mathbf{Z}} = \pi^0 N^{-1}\mathbf{Z}./\mathbf{H}. \tag{8.92}$$

The result of the vetting process is a sensor image that is suppressed in the location of the existing tracks, but unchanged in other regions. New tentative tracks are formed by finding peaks in the vetted data. When there are peaks within a threshold distance in two consecutive scans, a new tentative track is formed.

8.7.4 Summary

H-PMHT is a TBD method derived by applying EM to a Poisson Point Process mixture model. It is inherently multitarget and could easily incorporate a map for nonuniform clutter. However, initialization of the state estimates is an open problem and currently relies on conventional tracking methods.

Algorithm 5: H-PMHT

while not all tracks have converged **do**
  for each track **do**
    if this track has converged **then**
      use converged $h^i\left(\hat{\boldsymbol{x}}_k^m\right)$
    else
      calculate new $h^i\left(\hat{\boldsymbol{x}}_k^m\right)$
    end if
  end for
  calculate the overall pixel probabilities $h^i\left(\hat{\boldsymbol{x}}_k^{1:M}; \hat{\pi}_k^{0:M}\right)$
  determine expected measurements, (8.74)
  for all not-converged tracks **do**
    determine pixel-level centroids, (8.78)
    calculate synthetic measurements and covariance matrices, (8.77), (8.79) and (8.80)
    refine target state estimates by filtering the synthetic measurements and covariances
  end for
  For all tracks, determine new mixing proportion estimates, (8.75)
end while

The EM auxiliary function for H-PMHT can be expressed as a mixing proportion term and a sum of independent target components

$$Q\left(\boldsymbol{x}, \pi | \hat{\boldsymbol{x}}, \hat{\pi}\right) = \sum_{k=1}^{T} Q_{k\Pi} + \sum_{m=0}^{M} Q_X^m. \tag{8.93}$$

Ordinarily this would be iteratively optimized until $Q(\cdot)$ converges, that is all target states and the mixing proportions are estimated repeatedly until convergence. The $Q_{k\Pi}$ term couples all the targets together through normalization, so this term will not converge until all of the target states converge. However, it is likely that each target will converge at a different rate. In particular, when a track has been spuriously initiated without the support of a real target, the state estimates will usually converge very quickly, since the data association stage finds no support for the track and it essentially dead-reckons. In contrast, tracks that are initiated on a real target response may take many iterations to recover from initialization error, particularly if the target velocity is high. To reduce computation overheads, Davey proposed a partial-EM scheme where convergence is tested independently for each track [52]. This is summarized in Algorithm 5.

8.8 PERFORMANCE ANALYSIS

The previous sections reviewed five different methods for TBD. The first three methods, Baum Welch, Viterbi, and the Particle Filter, are all numerical approximations to the Bayesian filtering equations. The last two, H-PMHT and ML-PDA, use different measurement models.

This section compares these five methods using idealized simulation examples. The simulated scenarios consist of straight line trajectories in the plane observed by a cartesian image sensor. The algorithms are compared based on detection performance, estimation accuracy, and computation speed. The comparison here is an extract from [40].

8.8.1 Simulation Scenario

The algorithms described above were evaluated over an ensemble of simulation trials for an idealized scenario. The scenario considered the problem of detecting the presence or absence of a constant velocity target moving in the plane. The sensor collected frames at a constant rate, with each frame a 25×25 image. The sensor noise was Rayleigh with unit variance, and the point spread function was chosen to be the spectral response of a Hann window. This is typical of the spread expected for a radar where Fourier transforms are used for forming range cells and beam forming. Figure 8.5a shows this point spread function. Target energy is spread to the adjacent pixels around the target location, but very little spills farther. A white circle in the image is centered on the target location. Figure 8.5b shows an example of the sensor output for a target SNR of 6 dB and Figure 8.5c shows the same for a target SNR of 3 dB. At 3 dB, the target cannot be extracted from the noise based on a single frame. For illustrative purposes, the noise field is the same in Figure 8.5b and c; in practice, of course, these values vary with time and Monte Carlo iteration index.

The performance of the fixed-grid estimators (Baum–Welch, Viterbi, and ML-PDA) may be affected by the position of the target relative to the grid, that is, whether the target is close to a grid point or mid way between them. In order to average over this potential variation, the initial target position for each Monte Carlo trial was randomly sampled between 2.5 and 3 independently in X and Y. The target heading was also randomly sampled from 0° (East) to 45° (North East). Figure 8.6 shows an example of 20 Monte Carlo trials for a target moving at 1 pixel per frame. For all trials, a fixed speed of 1 pixel per frame was used. It has been found that the performance of TBD algorithms is not strongly influenced by target speed [24].

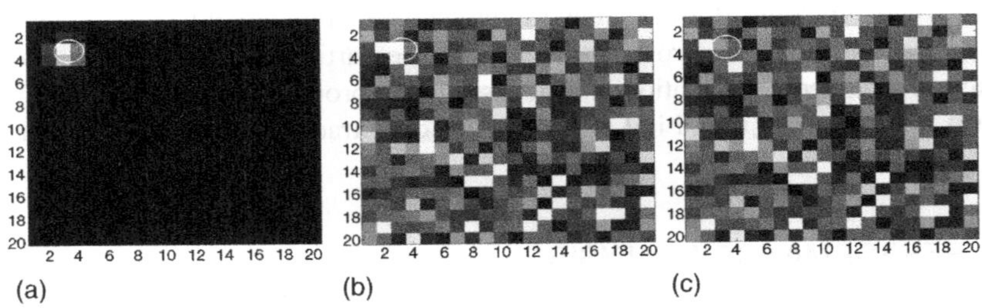

Figure 8.5 Example sensor envelope measurement images. (a) point spread function, (b) 6 dB target + noise, (c) 3dB targrt + noise.

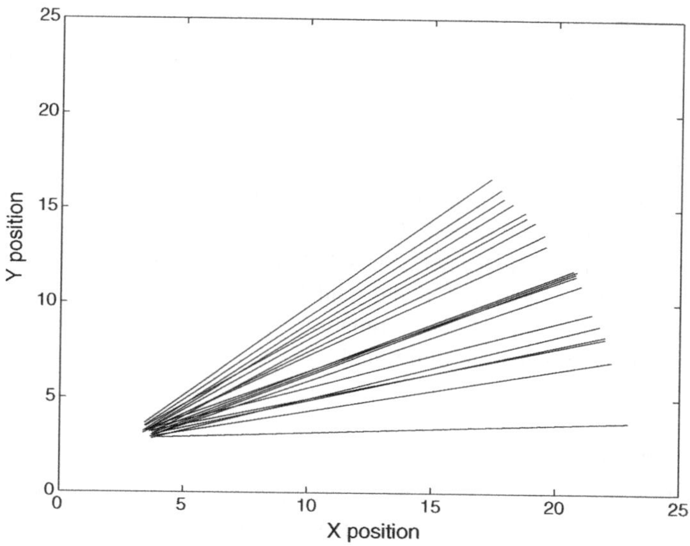

Figure 8.6 Example scenario, target speed = 1 pixel per frame.

8.8.2 Measures of Performance

A number of Measures of Performance (MOPs) will be used to compare the different TBD algorithms. To make the definitions of the MOPs crisp, the following terms will be used:

- A *detected frame* is a single frame of data for which the algorithm under test reports a track and that track is within a spatial gate of the true target location.
- The *detected-frame rate* is the proportion of frames containing a target that are detected frames, that is the relative frequency of detected frames.
- A *detected trial* is a Monte Carlo trial containing a target for which at least one frame is a detected frame.
- The *detected-trial rate* is the proportion of Monte Carlo trials containing a target that are detected trials.
- A *false report* occurs when the algorithm under test reports a track when the frame contains no target or when it reports a track that is outside the spatial gate of the true target location. If the algorithm reports multiple tracks in a single frame then the frame may contain multiple false reports.
- The *false-report rate* is the average number of false reports per frame for Monte Carlo trials that do not contain a target.
- A *false trial* is a Monte Carlo trial that does not contain a target for which at least one false report is observed.
- The *false-trial rate* is the proportion of Monte Carlo trials that are false trials.

The MOPs are then defined as

1. The *overall ROC* is the curve generated by plotting the detected-trial rate as a function of the false-trial rate.
2. The *per-frame ROC* is the curve generated by plotting the detected-frame rate as a function of the false-report rate.
3. The *estimation accuracy* is the RMS position estimation error averaged over detected frames as a function of SNR and for a prescribed false-report rate.
4. The *computation requirement* is the CPU time resource required to execute the algorithm and is measured as the total CPU time in seconds elapsed during execution of the Monte Carlo simulations. It is also expressed as a ratio with respect to the fastest algorithm tested.

8.8.3 Overall ROC

The overall ROC was evaluated for a target SNR of 6 dB and for a target SNR of 3 dB. In each case 1000 Monte Carlo trials were performed with a target and 1000 without. This allows for the estimation of very low false-trial rates which are of practical interest.

Figure 8.7a shows the overall ROC curves for each of the TBD algorithms at a target SNR of 6 dB and Figure 8.7b for a target SNR of 3 dB.

It is clear that all of the TBD algorithms do a good job of detecting the target when the SNR is 6 dB but struggle when the SNR is 3 dB. The Baum–Welch, Viterbi, and particle filter algorithms all give near perfect detection on the 6 dB case with MLPDA and H-PMHT showing some degradation. For the 3 dB case, the particle filter gives the best detection performance. In this case the scalloping loss associated with sampling the likelihood at a fixed grid location rather than at its peak causes significant performance penalty for the Baum–Welch and Viterbi algorithms.

8.8.4 Per-Frame ROC

Similar to the overall ROC, the per-frame ROC was evaluated for a target SNR of 6 dB and for a target SNR of 3 dB. Again, 1000 Monte Carlo trials were performed with a target and 1000 without. This allows for the estimation of very low false-trial rates which are of practical interest.

Figure 8.8a shows the per-frame ROC curves for each of the TBD algorithms at a target SNR of 6 dB and Figure 8.8b for a target SNR of 3 dB.

When the overall detection ROC is similar, this measure reflects the fraction of each trial for which the target was detected which is a function of the establishment delay. For this metric there is much more difference. The particle filter gives the best performance for both the 6 dB SNR target and the 3 dB one. At a false-report rate of 0.001 (i.e., 1 frame containing a false track for every 50 trials) the detected-frame rate is still around 0.75, corresponding to an average of 15 frames showing the target per trial.

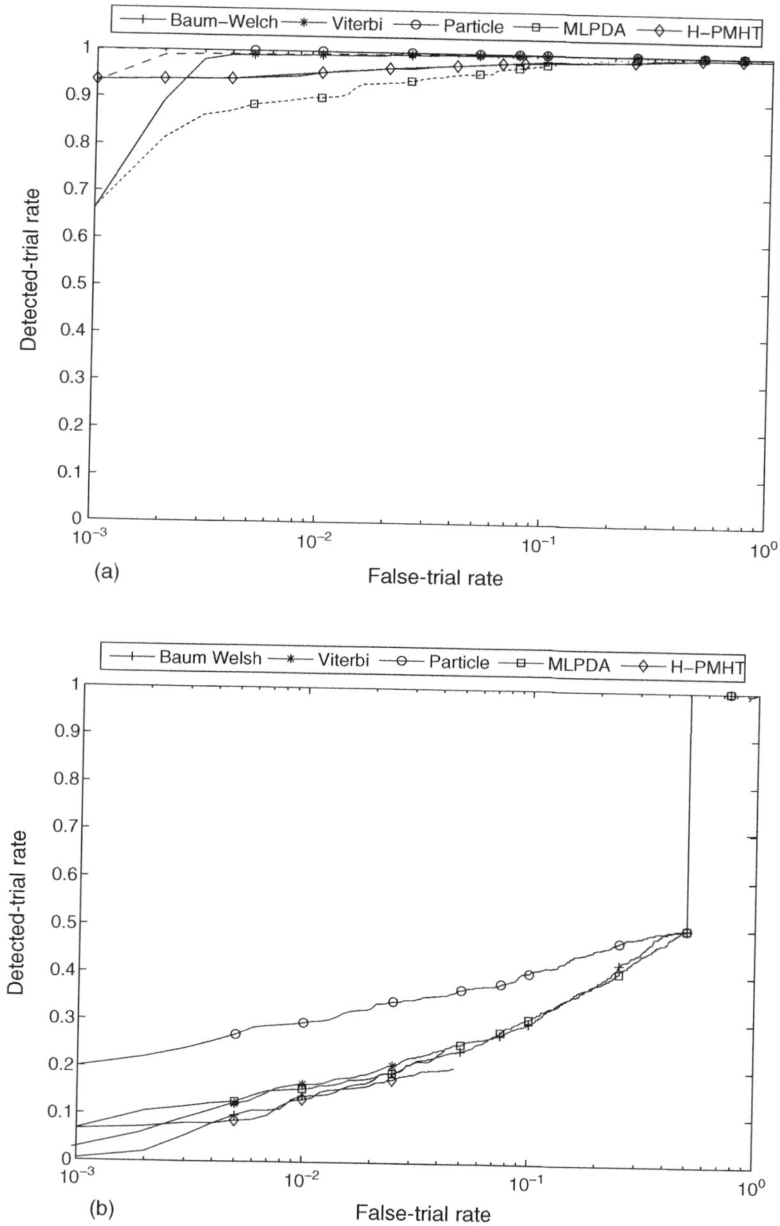

Figure 8.7 Overall ROC. (a) 6 dB target SNR, (b) 3 dB target SNR.

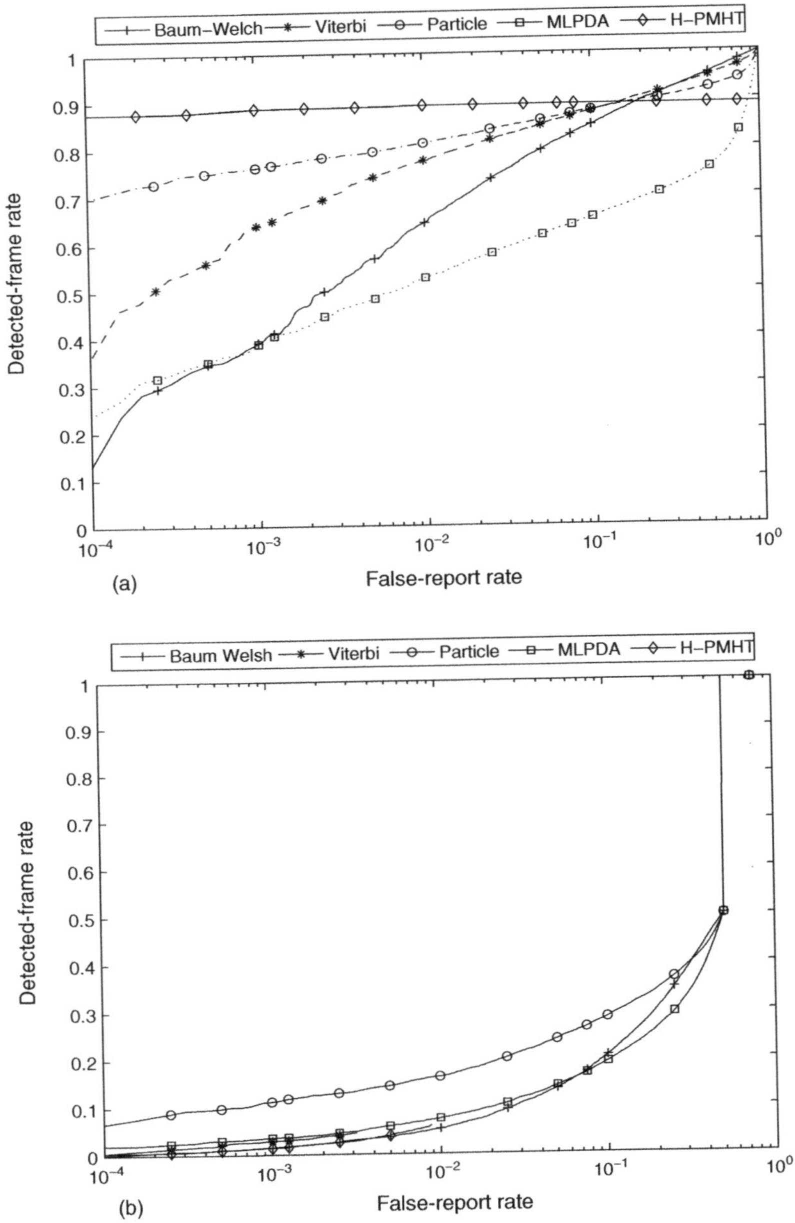

Figure 8.8 Per-frame ROC. (a) 6 dB target SNR, (b) 3 dB target SNR.

TABLE 8.1 Estimation Accuracy

Algorithm	RMS (6 dB)	RMS (3 dB)
Baum Welch	1.16	1.25
Viterbi	1.18	1.24
Particle filter	0.45	0.75
ML-PDA	1.21	1.21
H-PMHT	0.50	0.84

8.8.5 Estimation Accuracy

The estimation accuracy was also determined for a false-report rate of 0.05, that is 1 false report per 20 frames. For each algorithm, the false-report rate determines a detection threshold and target points above the threshold contribute to the calculation of RMS error.

The estimation accuracy for target SNR values of 6 dB and 3 dB are shown in Table 8.1. The grid-based algorithms (Baum–Welch, Viterbi, and ML-PDA) all give similar estimation accuracy values and are not significantly degraded by lowering the SNR. In contrast, the particle filter and H-PMHT have much better accuracy but do show degradation with lower SNR.

The accuracy of the grid algorithms would be improved by using a finer grid at the cost of computation time. Alternatively, the nonuniform grid method in Section 8.6 could be used.

8.8.6 Computation Requirements

Table 8.2 shows the computation requirements of the five different TBD algorithms. These times are expressed as CPU seconds and as a ratio relative to the fastest algorithm. Each time represents the number of CPU seconds elapsed while the algorithm executes 100 Monte Carlo trials with one parameter setting. The overall time required to evaluate the metrics was much higher, since each curve consists of many of these runs. The testing was performed for MATLAB implementations of the algorithms on a 2.8 GHz quad-core desktop machine. An operational implementation would likely give much faster execution time.

TABLE 8.2 Computation Requirement

Algorithm	cpu Time (s)	cpu Time (Ratio)
Baum Welch	1193	24
Viterbi	1185	24
Particle filter	1174	23
ML-PDA	152	3
H-PMHT	50	1

As should be expected, the numerical approximation methods take much longer to execute than the parametric approximation of H-PMHT. ML-PDA is also quite fast because it uses a threshold stage to reduce the data presented to the tracker. For all of the algorithms other than H-PMHT, the main expense is the calculation of the likelihood ratio which contains a product of Bessel functions. This function is very expensive to compute, so in a time-constrained application, it may be advantageous to use a low fidelity approximation for it.

8.9 APPLICATIONS: RADAR AND IRST FUSION

This section presents an example application of TBD where substantial improvement is obtained over conventional tracking. This example considers the fusion of microwave radar and Infra-Red Search and Track (IRST) in a ship air defence environment similar to [54]. It is advantageous to fuse these sensors because they provide complementary information. The radar provides high accuracy range and Doppler measurements but relatively poor accuracy in azimuth and elevation. In contrast, the IRST is a passive optical sensor that cannot observe range but has high accuracy in elevation and azimuth. The sensors are assumed to be co-located and observe a single inbound target. The parameters of the sensors are summarized in Table 8.3.

The nominal maximum range given in Table 8.3 is the range at which the target SNR is 12 dB. This is roughly the minimum SNR at which you might expect a conventional detect-then-track system to reliably track the target. For a sensor with nominal maximum range r_m and a target at range r, the SNR at the sensor is

$$\mathrm{SNR} = 12 + 40\log_{10}\left(\frac{r_m}{r}\right). \tag{8.94}$$

TABLE 8.3 Sensor Specifications

	Sensor 1	Sensor 2
Nominal Type	Agile Beam Radar	IR Camera
Range resolution (m)	50	–
Doppler resolution (m/s)	8	–
Azimuth resolution (mrad)	20	1
Elevation resolution (mrad)	20	1
Number of range cells	700	0
Number of doppler cells	100	0
Number of azimuth cells	1	180
Number of elevation cells	1	180
Minimum range (km)	15	–
Nominal max. range (km)	100	25
Noise type	Rayleigh	Gaussian
Update period (s)	3	1

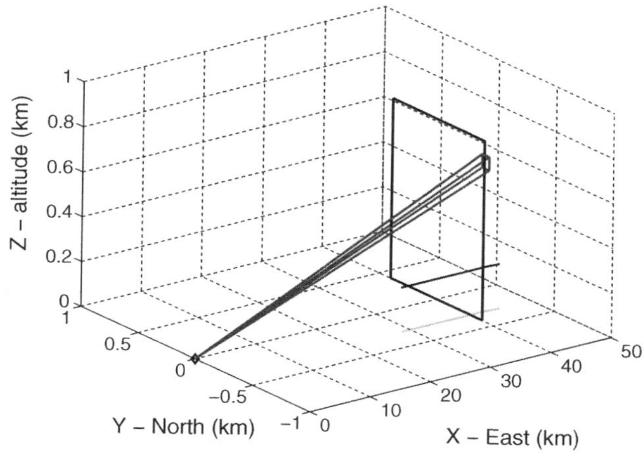

Figure 8.9 Simulated scenario.

This simple dependency assumes a drop off in power proportional to r^4 and ignores target radar cross-section considerations.

The target starts at a location 45 km East of the platform and 300 m South of the platform at an altitude of 200 m. The target moves at 278 m/s (1000 km/h) due West. The scenario consists of 90 s of observations during which time the target closes from a range of approximately 45 km to one of approximately 20 km. The target trajectory is shown in Figure 8.9. The platform is shown as a diamond at the origin and the target trajectory is plotted as a solid line. The projection of the trajectory into the X–Y plane is shown as a dashed line. The physical extent of an arbitrary single pixel for each sensor is plotted as a heavy line for sensor 1 and a gray line for sensor 2. The figure highlights that in this example the sensors are approximately orthogonal: sensor 1 provides informative measurements for the X dimension, but very little information about the Y and Z dimensions. In contrast, sensor 2 provides informative measurements for the Y and Z dimensions and no information about the X dimension. The target location in Y and Z was chosen to position the target within the beam of sensor 1 without being centered on the beam. This location was chosen to provide the opportunity for significant fusion gain.

The first sensor has a much longer detection range than the second. This means that the target is easily discernable by the first sensor for the whole scenario, whereas for the second sensor it is only likely to be detected by a conventional tracker in the second half of the scenario. Figure 8.10 shows the target SNR as a function of time for both sensors. The SNR at the second sensor does not reach the nominal conventional detection level until after approximately 70 s.

A particle filter TBD algorithm was implemented for this application and is now compared with the performance of an MHT using threshold measurements extracted from the sensor images.

The scenario was run through 100 Monte Carlo trials. Since the first sensor provides high SNR, we expect to easily detect the target, but with poor SNR in the

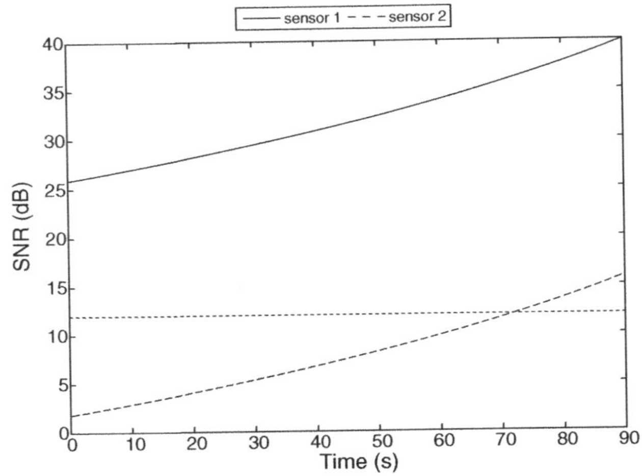

Figure 8.10 Target SNR.

imaging sensor, the estimation error in the Y–Z plane is likely to be high. To illustrate this, the estimation error was measured separately for X, Y and Z and also as a total three-dimensional distance. Due to outliers, the RMS error was found to be inappropriate, and instead the estimation performance is quantified through the root-median-squared error. The one-dimensional errors are plotted in Figure 8.11a–c and the three-dimensional error is plotted in Figure 8.11d.

It is clear from the figures that the particle TBD algorithm is able to exploit the sensor 2 frames even though the target SNR is low. The estimation error is low in all three dimensions after just a few seconds. In contrast, the conventional tracker has high estimation error in the Y and Z dimensions until the second half of the scenario.

For the X-dimension, the MHT gives very low estimation error for the whole scenario. This is because the X information is exclusively from sensor 1, for which the SNR is high and the target is always detected. The TBD estimation error is slightly higher for the X-dimension because it is fusing both sensors. When the SNR is poor for the second sensor, the algorithm injects birth particles which are uniformly distributed in range and bias the state estimate towards the mid-range region. As the SNR improves, the continuing particles dominate and this effect is reduced.

For the Y-dimension, both trackers start out with an error of 300 m because each bases its initial track on sensor 1 information which initializes the Y-dimension at 0. This initial bias is quickly overcome in the TBD output, but lasts for much longer in the MHT output. After around 40 s, the MHT error drops rapidly as the target SNR reaches a point where the target becomes detectable. Even though the target cannot be reliably detected at this point, occasional observations are enough to correct the bias and the track is sustained by the sensor 1 data.

Similarly, the Z-dimension estimate of both trackers begins with an initial bias. For sensor 1, the Z-estimate is based on the elevation of the main beam and the range. As the target moves towards the platform, the range reduces and so does the sensor

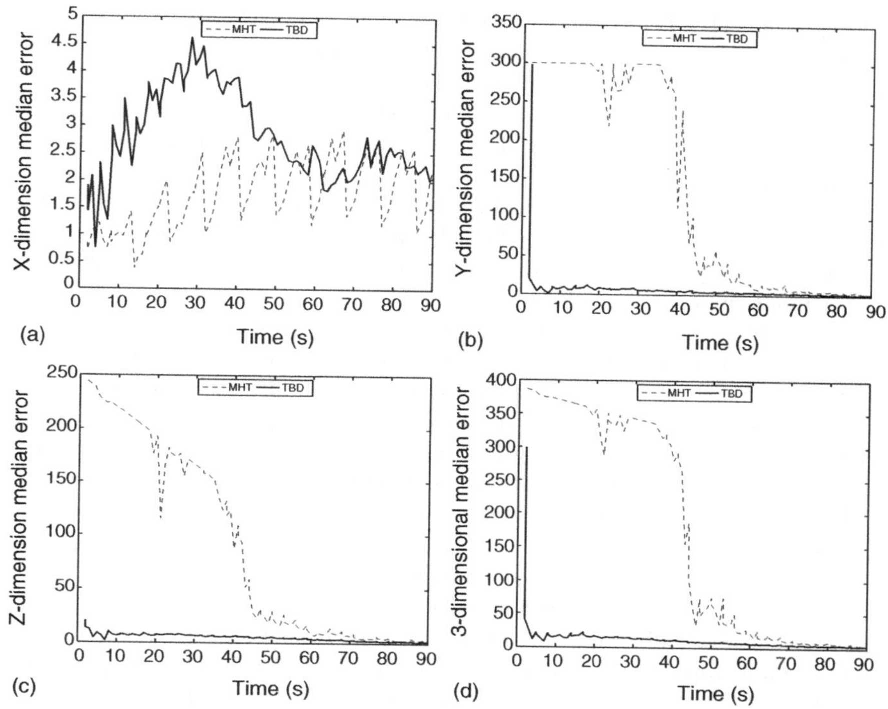

Figure 8.11 Estimation error. (a) X position, (b) Y position, (c) Z position, (d) 3D position.

1 Z-estimate. This is seen as an almost linear drop in error for the first half of the scenario for the MHT.

The three-dimensional estimation error plot is mostly dominated by the Y-dimension error which has the largest magnitude.

No attempt was made to optimize either algorithm for computation efficiency, so it is not appropriate to draw any conclusions from their execution times. Nevertheless, we observe that the MHT approach took around 25% more time to execute than the TBD approach. Most of the computation cost incurred by the MHT approach was due to the point measurement detector.

8.10 FUTURE DIRECTIONS

This chapter has reviewed some of the important algorithms in the field of TBD. Through examples, the approach of tracking without a single-frame detector has been shown to give far greater sensitivity to targets with low SNR. Anecdotal evidence also demonstrates that these algorithms are not cost-prohibitive in terms of computation requirements. However, some further improvement of the current algorithms in the literature is still required to deliver operational systems.

Many of the TBD algorithms assume that there is only one target, or none. This assumption may not be too detrimental for targets that are well separated, but for interacting targets it is a serious limitation. The multitarget extensions of the algorithms presented in this chapter all involve extending the state space to include the states of all of the targets in one vector. This incurs an exponential growth in complexity and will only be practical for small problems.

There is also very little work published on TBD for multiple sensors. This chapter presented an example of multisensor TBD, but it is one of the first applications in the area. Although in principle the extension to multiple sensors may be pedestrian, in practice things are rarely straightforward.

For operationally relevant implementations, the scaling of the algorithm will become increasingly important. Modern radars collect data over hundreds of thousands of bins, and optical sensors have millions of pixels. TBD algorithms will need to be highly efficient to scale up to high resolution sensors. Further, the high data rates required may not be feasible for distributed sensing environments, so it will be necessary to consider TBD under constrained communication. An interesting question is: what is the most valuable information to send down a communication channel for target detection?

Related to the issue of constrained communication is the notion of optimal detection performance. For position estimation accuracy, the Cramer–Rao Lower Bound provides a valuable tool for quantifying algorithm performance. For TBD, a measure for detection performance is required. This measure may also lead to a crisp analytic explanation of the performance loss between conventional tracking and TBD.

Finally, the problem of background estimation will become increasingly important. When TBD sensitivity is exploited, it will be imperative to properly model the statistical characteristics of the scene against which targets are to be detected. As was seen for the case of conventional tracking, high quality detection is often more about rejecting spurious signals than finding the real targets.

REFERENCES

1. I.S. Reed and I. Selin, A sequential test for the presence of a signal in one of k possible positions, *IEEE Trans. Inform. Theor.*, 9(4), 286–288, 1963.
2. D.E. Mercier, An extended Kalman filter for use in a shared aperture medium range tracker, *M.S. thesis*, Air Force Institute of Technology, Wright-Patterson AFB, 1978.
3. P.S. Maybeck and D.E. Mercier, A target tracker using spatially distributed infrared measurements, *IEEE trans. Automat. Contr.*, 25(2), 222–225, 1980.
4. Y. Barniv, Dynamic programming solution for detecting dim moving targets, *IEEE trans. Aerosp. Electron. Syst.*, 21(1), 144–156, 1985.
5. Y. Barniv, Dynamic programming solution for detecting dim moving targets. Part II: Analysis, *IEEE trans. Aerosp. Electron. Syst.*, 23(6), 144–156, 1987.
6. Y. Barniv, Dynamic programming algorithm for detecting dim moving targets, in: *Multitarget-Multisensor Tracking: Advanced Applications*, Artech House, Norwood, MA, Y Bar-Shalom (Ed.), Chapter 4, 1990.

7. S.M. Tonissen and R.J. Evans, Performance of dynamic programming techniques for track-before-detect, *IEEE trans. Aerosp. Electron. Syst.*, 32(4), 1440–1451, 1996.
8. L.A. Johnston and V. Krishnamurthy, Performance analysis of a dynamic programming track before detect algorithm, *IEEE trans. Aerosp. Electron. Syst.*, 38(1), 228–242, 2002.
9. L.D. Stone, C.A. Barlow, and T.L. Corwin, *Bayesian Multiple Target Tracking*, Artech House, Norwood, MA, 1999.
10. M.G.S. Bruno and J.M.F. Moura, Multiframe detector/tracker: Optimal performance, *IEEE trans. Aerosp. Electron. Syst.*, 37(3), 925–945, 2001.
11. M.G.S. Bruno, Bayesian methods for multiaspect target tracking in image sequences, *IEEE trans. Aerosp. Electron. Syst.*, 52(7), 1848–1861, 2004.
12. D.J. Salmond and H. Birch, A particle filter for track-before-detect, *Proceedings of the American Control Conference*, Arlington, VA, USA, June 2001, pp. 3755–3760.
13. Y. Boers and J.N. Driessen, Particle filter based detection for tracking, *Proceedings of the American Control Conference*, Arlington, VA, USA, June 2001, pp. 4393–4397.
14. B. Ristic, S. Arulampalam, and N.J. Gordon, *Beyond the Kalman Filter: Particle Filters for Tracking Applications*, Artech House, Norwood, MA, 2004.
15. M.G. Rutten, N.J. Gordon, and S. Maskell, Recursive track-before-detect with target amplitude fluctuations, *IEE Proc. Radar, Sonar Navig.*, 152(5), 345–352, 2005.
16. C. Jauffret and Y. Bar-Shalom, Track formation with bearing and frequency measurements in clutter, *IEEE trans. Aerosp. Electron. Syst.*, 26(6), 999–1010, 1990.
17. T. Kirubarajan and Y. Bar-Shalom, Low observable target motion analysis using amplitude information, *IEEE trans. Aerosp. Electron. Syst.*, 32(4), 1367–1384, 1996.
18. W.R. Blanding, P.K. Willett, Y. Bar-Shalom, and R. Lynch, Directed subspace search ML-PDA with application to active sonar tracking, *IEEE trans. Aerosp. Electron. Syst.*, 44(1), 201–216, 2008.
19. W.R. Blanding, P.K. Willett, and Y. Bar-Shalom, ML-PDA: Advances and a new multitarget approach, *EURASIP J. Appl. Signal Process.*, Article No. 38 , Volume 2008, January, 2008.
20. W.R. Blanding, P.K. Willett, and Y. Bar-Shalom, Offline and real-time methods for ML-PDA track validation, *IEEE trans. Signal Process.*, 55(5), 1994–2006, 2007.
21. B.-N. Vo, B.-T. Vo, N.-T. Pham, and D. Suter, Joint detection and estimation of multiple objects from image observations, *IEEE Trans. Signal Process.*, 58(10), 5129–5141, 2010.
22. R.L. Streit, Tracking on intensity-modulated data streams, *Technical Report* 11221, NUWC, Newport, Rhode Island, USA, May 2000.
23. R.L. Streit, M.L. Graham, and M.J. Walsh, Multitarget tracking of distributed targets using histogram-PMHT, *Dig. Signal Process.*, 12(2), 394–404, 2002.
24. S.J. Davey, M.G. Rutten, and B. Cheung, A comparison of detection performance for several track-before-detect algorithms, *EURASIP J. Appl. Signal Process.*, 2008.
25. R.M. Burczewski and N.C. Mohanty, Detection of moving optical objects, *International Telemetering Conference*, Los Angeles, California, 1978, pp. 325–330.
26. N.C. Mohanty, Computer tracking of moving point targets in space, *IEEE Trans. Pattern Anal. Mach. Intell.*, 3, 606–611, 1981.
27. I.S. Reed, R.M. Gagliardi, and H.B. Shao, Application of three-dimensional filtering to moving target detection, *IEEE Trans. Aerosp. Electron. Syst.*, 19(6), 898–905, 1983.
28. I.S. Reed, R.M. Gagliardi, and L.B. Stotts, Optical moving target detection with 3-D matched filtering, *IEEE Trans. Aerosp. Electron. Syst.*, 24(4), 327–336, 1988.

29. I.S. Reed, R.M. Gagliardi, and L.B. Stotts, A recursive moving-target-indication algorithm for optical image sequences, *IEEE Trans. Aerosp. Electron. Syst.*, 26(3), 434–440, 1990.

30. S.C. Pohlig, An algorithm for detection of moving optical targets, *IEEE Trans. Aerosp. Electron. Syst.*, 25(1), 56–63, 1989.

31. S.M. Tonissen and Y. Bar-Shalom, Maximum likelihood track-before-detect with fluctuating target amplitude, *IEEE Trans. Aerosp. Electron. Syst.*, 34(3), 796–809, 1998.

32. M.C. Smith and E.M. Winter, On the detection of target trajectories in a multi target environment, *IEEE Conference on Decision and Control*, San Diego, California, 1978, pp. 1189–1194.

33. M.C. Smith, Feature space transform for multitarget detection, *IEEE Conference on Decision and Control*, Alburquerque, New Mexico, 1980, pp. 835–836.

34. B.D. Carlson, E.D. Evans, and S.L. Wilson, Search radar detection and track with the Hough transform. Part I: System concept, *IEEE trans. Aerosp. Electron. Syst.*, 30(1), 102–108, 1994.

35. B.D. Carlson, E.D. Evans, and S.L. Wilson, Search radar detection and track with the Hough transform. Part II: Detection statistics, *IEEE trans. Aerosp. Electron. Syst.*, 30(1), 109–115, 1994.

36. B.D. Carlson, E.D. Evans, and S.L. Wilson, Search radar detection and track with the Hough transform. Part III: Detection performance with binary integration, *IEEE trans. Aerosp. Electron. Syst.*, 30(1), 116–125, 1994.

37. M. Morelande and B. Ristic, Signal-to-noise ratio threshold effect in track before detect, *Radar, Sonar Navig., IET*, 3(6), 601–608, 2009.

38. S.S. Blackman and R. Popoli, *Design and Analysis of Modern Tracking Systems*, Artec House, Norwood, Massachusetts, USA, 1999.

39. G.B. Arfken and H.J. Weber, *Mathematical Methods for Physicists*, Sixth Edition, Elsevier Academic Press, New York, 2005.

40. S.J. Davey, M.G. Rutten, and B. Cheung, Using phase to improve track-before-detect, 48(1), 832–849, Jan. 2012.

41. R.M. Haralick and L.G. Shapiro, *Computer and Robot Vision*, Vol. 1, Addison-Wesley, Reading, MA, 1992.

42. L.E. Baum, T. Petrie, G. Soules, and N. Weiss, A maximization technique occurring in the statistical analysis of probabilistic functions of Markov chains, *Ann. Math. Stat.*, 41(1), 394–404, 1970.

43. B. Jang and L. Rabiner, An introduction to hidden Markov models, *IEEE ASSP Mag.*, 3(1), 4–16, 1986.

44. M. McDonald and B. Balaji, Continuous-discrete filtering for dim manoeuvring maritime targets, *Proceedings of the 10th International Conference on Information Fusion*, July 2007.

45. M. McDonald and B. Balaji, Track-before-detect using Swerling 0, 1, and 3 target models for small manoeuvring maritime targets, *EURASIP J. Adv. Signal Process.*, 17 January, 2008.

46. M.G. Rutten, B. Ristic, and N.J. Gordon, A comparison of particle filters for recursive track-before-detect, *Fusion 2005: Proceedings of the 8th International Conference on Information Fusion*, Philadelphia, PA, USA, July 2005, pp. A6–2.

47. Y. Bar-Shalom and X.R. Li, *Multitarget-Multisensor Tracking: Principles and Techniques*, YBS Publishing, Storrs, Conneticut, USA, 1995.

48. A.P. Dempster, N.M. Laird, and D.B. Rubin, Maximum likelihood from incomplete data via the EM algorithm, *J. R. Stat. Soc.*, 140, 1–38, 1977.

49. R.L. Streit, *Poisson Point Processes: Imaging, Tracking, and Sensing*, Springer, New York, NY, 2010.

50. S.B. Colegrove and S.J. Davey, PDAF with multiple clutter regions and target models, *IEEE trans. Aerosp. Electron. Syst.*, 39(1), 110–124, 2003.

51. T.E. Luginbuhl, Y. Sun, and P. Willett, A track management system for the PMHT algorithm, *Fusion 2001: Proceedings of the 4th International Conference on Information Fusion*, Montreal, Canada, August 2001.

52. S.J. Davey, Detecting a small boat using histogram PMHT, *Journal of Advances in Information Fusion*, 6(2), 167–186, Dec. 2011.

53. S.J. Davey and D.A. Gray, Integrated track maintenance for the PMHT via the Hysteresis model, *IEEE Trans. Aerosp. Electron. Syst.*, 43(1), 93–111, 2007.

54. B. Ristic, M. Hernandez, A. Farina, and H.-.T Ong, Analysis of radar allocation requirements for an IRST aided tracking of anti-ship missiles, *Fusion 2006: Proceedings of the 9th International Conference on Information Fusion*, Florence, Italy, July 2006.

CHAPTER 9

Advances in Data Fusion Architectures

STEFANO CORALUPPI AND CRAIG CARTHEL

9.1 INTRODUCTION

The multisensor multitarget tracking problem is a generalization of the classical nonlinear filtering problem. In the latter, generally one seeks a recursive filtering solution for a time-ordered sequence of measurements that are known to correspond to the same object. The general tracking problem allows for missed detections, false returns, and unknown measurement associations: in short, the measurement *provenance* is unknown. The filtering problem requires a statistical description for object state evolution as well as a measurement model. The general tracking problem requires as well the statistics of the target birth–death process and of sensor detection performance.

The generalization noted above has led to a rich literature on track management and data association methodologies. A useful, up-to-date reference text for multitarget tracking has been published recently [1]. Our own research has focused on *multiple-hypothesis tracking* (MHT) techniques, and is grounded on the seminal hybrid-state formulation of the track-oriented multiple-hypothesis tracking problem [2]; see Chapter 5 of this text and references therein [3].

It is well known that distributed estimation is suboptimal in general [1]. That is, centralized processing of sensor measurements from all sensors provides optimal state estimation, while the combination of single-sensor state estimates does not accurately reflect the correlation induced by common target process noise. Since the individual sensors view the same object, their estimates are not conditionally independent.

Similarly, distributed detection is suboptimal in that in general the fusion center lacks sufficient statistics from upstream processing [4]. Accordingly, distributed tracking suffers from suboptimal track confirmation statistics [5].

Much of the track fusion work reported in the literature addresses the (batch) two-dimensional assignment problem, where one has two set of tracks from two sensors,

Integrated Tracking, Classification, and Sensor Management: Theory and Applications, Edited by Mahendra Mallick, Vikram Krishnamurthy, and Ba-Ngu Vo.

each with an associated tracking processor [6]. Though suboptimal, this processing scheme is necessitated due to the existence of inflexible legacy systems, as well as the physical, temporal, and communications bandwidth constraints in wide-area surveillance.

What if all multisensor data are available at the fusion center? Is the single-stage tracking architecture, or *track-while-fuse*, best? Some of the arguments in favor of multistage processing relate to its simplicity. For example, multistage processing allows for straightforward approaches to content with fading-target effects in undersea surveillance where target detection depends highly on scenario geometry [7]; to address track extraction in highly cluttered or dim-target environments [8]; and to content with ghosting issues in passive radar [9]. In all of these instances, single-stage tracking solutions exist but are significantly more complex.

In this chapter, we discuss a number of settings where multistage processing is beneficial. These are the following: single-sensor dense-target scenarios; multiscale sensor scenarios; large networks of low-performance sensors; and multiscale object scenarios. The discussion will be of interest to tracking and fusion practitioners whether or not an MHT paradigm is adopted. Interestingly, we find that the usual assumption is estimation and detection theory (i.e., that centralized single-stage processing is best) does not hold in target tracking. The reason is that optimal tracking is intractable; thus we are faced with either *suboptimal* centralized single-stage processing or with judiciously chosen alternatives. Thus, the unifying philosophy in the examples is that tracking at reduced complexity often can be achieved by appropriate concatenation of relatively simple processing stages.

9.2 DENSE-TARGET SCENARIOS

The *track coalescence effect* degrades the performance of probabilistic data association trackers in dense-target scenarios [10]; attempts at suboptimal Bayesian processing to combat this effect have been reported [11]. Recently, it has been observed that some trackers exhibit an opposite effect [12], including the track-oriented MHT [13]; we denote it as the *track-repulsion effect*. It is of interest to understand the track-repulsion effect at a simple analytical level, particularly the suboptimality of scan-based tracking with respect to optimal track estimation as achieved with a batch, *maximum likelihood* (ML) approach.

We consider a simple scenario. Two targets in one-dimensional Cartesian space are observed with linear measurements in Gaussian noise; for target $i \in \{1, 2\}$, we denote by $X_k^{(i)}$ the state at time k, and measurements are given by the following, where $v_k^{(i)} \sim N\left(0, \sigma^2\right)$ is a zero mean Gaussian random variable that is uncorrelated with other measurement errors. We assume that both targets are known to be stationary and at unknown locations equidistant from the coordinate system origin.

$$Y_k^{(i)} = X_k^{(i)} + v_k^{(i)}, \qquad k = 1, \ldots, N. \tag{9.1}$$

$$X_k^{(1)} = -X_k^{(2)} = \bar{X}, \qquad k = 1, \ldots, N. \tag{9.2}$$

For simplicity, we neglect false contacts and missed detections. Thus, the measurement-origin uncertainty is limited to confusion as to which target gives rise to which measurement. Assume that a large number N of data scans are available. It can be shown that, for $N \to \infty$, the ML solution is given by the following location estimates $\hat{X}^{(1)}$ and $\hat{X}^{(2)}$

$$\hat{X}^{(1)} = -\hat{X}^{(2)} = \bar{X}. \tag{9.3}$$

That is, the ML solution does not suffer track-repulsion bias. On the other hand, for $N \to \infty$, the scan-based solution as obtained, for example, with a multihypothesis tracker with a constant position kinematic motion model is the following, where $v, w \sim N\left(0, \sigma^2\right)$ are uncorrelated random variables

$$\hat{X}^{(1)} = -\hat{X}^{(2)} = E\left[\max\left\{\bar{X} + v, -\bar{X} + w\right\}\right]. \tag{9.4}$$

The track-repulsion effect is given by the displacement $\hat{X}^{(1)} - \bar{X}$, or $-\left(\hat{X}^{(2)} - \bar{X}\right)$. Figure 9.1 illustrates its magnitude for a range of values for target spacing ($2\bar{X}$), and assuming $\sigma = 10$ [m]. We see that the scan-based solution suffers a nontrivial bias for sufficiently close targets. For example, at a target spacing equal to the measurement error standard deviation of 10 m, the track displacement is approximately 2 m, that is,

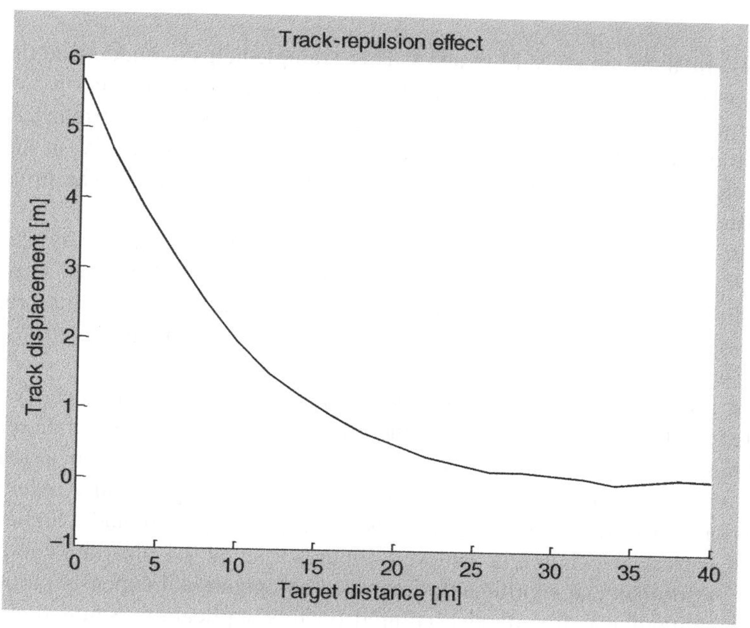

Figure 9.1 Track displacement in scan-based hard data association.

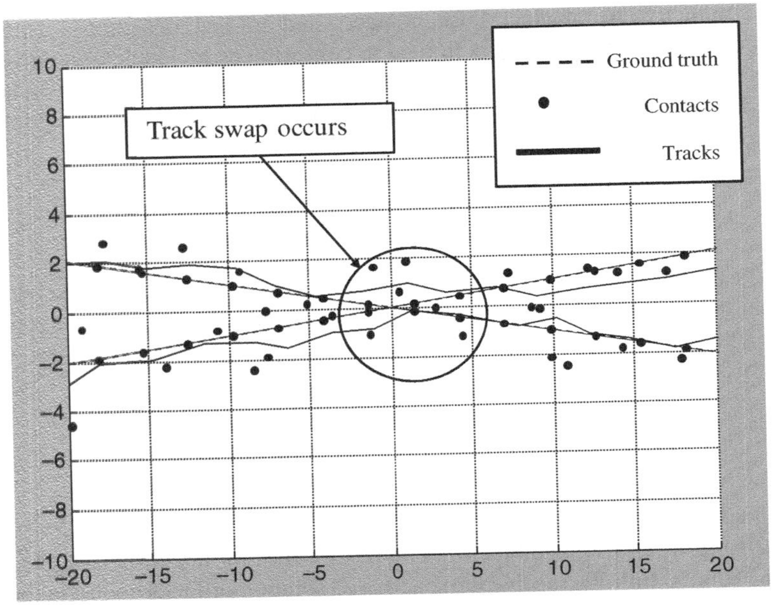

Figure 9.2 The track-repulsion effect in slowly crossing target scenarios.

the tracks are approximately 14 m apart. This example helps to understand the same phenomenon leading to the track-swapping results in the previous section.

The track-repulsion effect has a damaging effect on tracking performance. Figure 9.2 illustrates that, in slowly crossing target settings (or, alternatively, in high-rate sensor settings), the maneuvering-target solution is often selected. The problem is exacerbated as the data rate increases.

Within an MHT setting, the track-repulsion effect can be mitigated by reasoning over a long-enough time window in making data-association decisions. This, however, comes at a significant computational expense particularly when clutter rates are high. A multistage tracking approach to contend with this challenge is *track-break-track* [13].

A first tracking stage with small hypothesis tree depth serves to identify target-originated contacts. In this first stage, the probability of correct association is low. Accordingly, we then remove all track labels, or *break* all data association decisions. A second MHT processing stage is applied to the (significantly reduced) set of detections; this time, a larger hypothesis tree depth is applied, and a higher probability of correct association is achieved. Figures 9.3 and 9.4 illustrate comparable tracking performance at significantly improved computational expense, particularly at the large hypothesis tree depths required for this application. Simulation details are in [13].

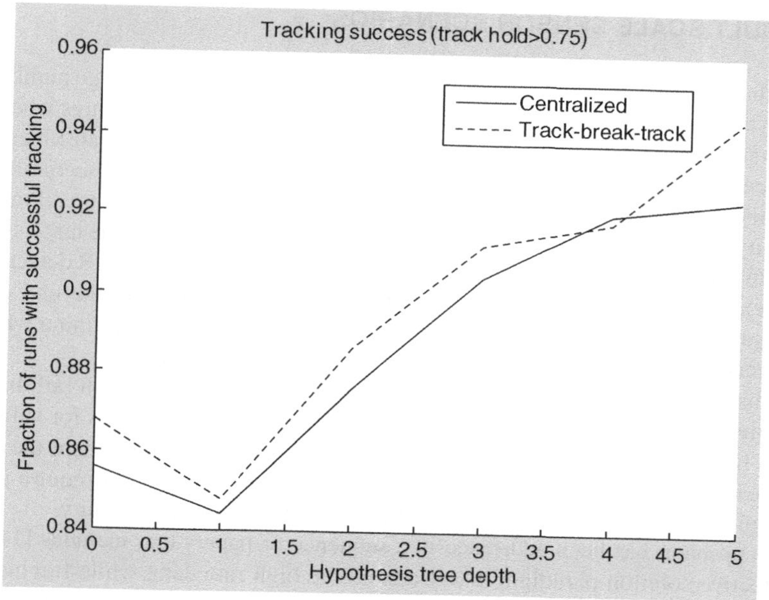

Figure 9.3 Comparable tracking performance of *track-while-fuse* and *track-break-track* architectures. In both cases, performance increases for larger hypothesis tree depths.

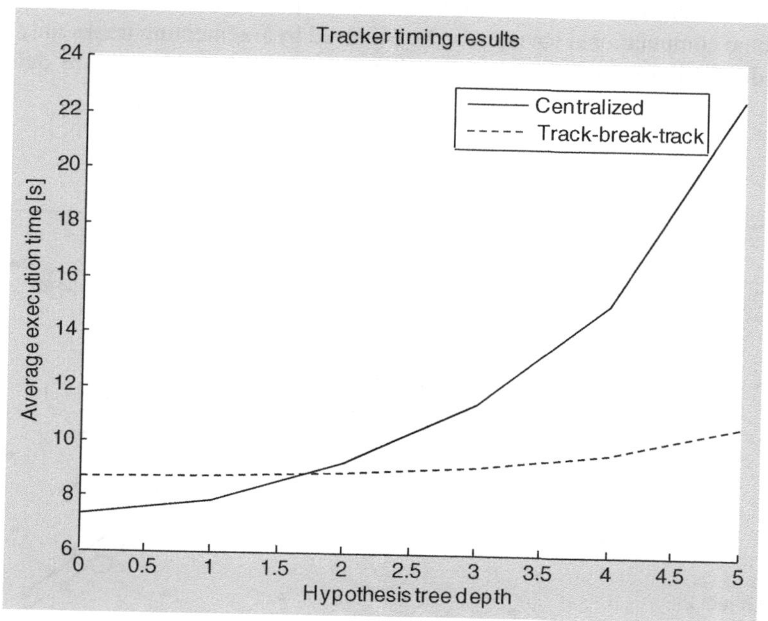

Figure 9.4 Computational savings in *track-break-track* processing for large hypothesis tree depths.

9.3 MULTISCALE SENSOR SCENARIOS

A key challenge in MHT approaches to tracking is the need to handle large numbers of track hypotheses. The problem is exacerbated in multiscale sensor settings where the feature-rich sensor measurements arrive infrequently relative to higher rate kinematic-only measurements. Figure 9.5 illustrates the problem with a notional scenario.

In this example, we have two targets that approach one another and travel in close proximity, during which numerous scans of data are received before the targets move apart. Infrequent, feature-rich data are available (e.g., satellite-based SAR detections): in the example, we see the object detections at the beginning and the end of the target trajectories. These detections provide vital information to determine whether the targets crossed or not.

In this setting, a single-stage (and track-oriented) MHT solution will fail. Indeed, generally it is computationally infeasible to maintain track hypotheses for sufficient time to benefit from the second set of SAR detections in deciding where the white and black objects have gone. However, it is also true that we do not require a high probability of correct association while the objects are in close proximity.

This problem has been addressed in a sequence of papers that includes [14–17]. An effective solution paradigm is to track on the high-rate data, while fragmenting the tracks when group composition changes. Subsequently, track stitching can be achieved with the aid of feature-rich information from a complementary sensor. This stitching is illustrated by the dotted lines in Figure 9.5.

Some further enhancements to this multistage paradigm are suggested in [18]. First, some computational savings can be achieved by fragmenting tracks only when groups divide: in Figure 9.5, we see that there is no loss of track quality by deferring

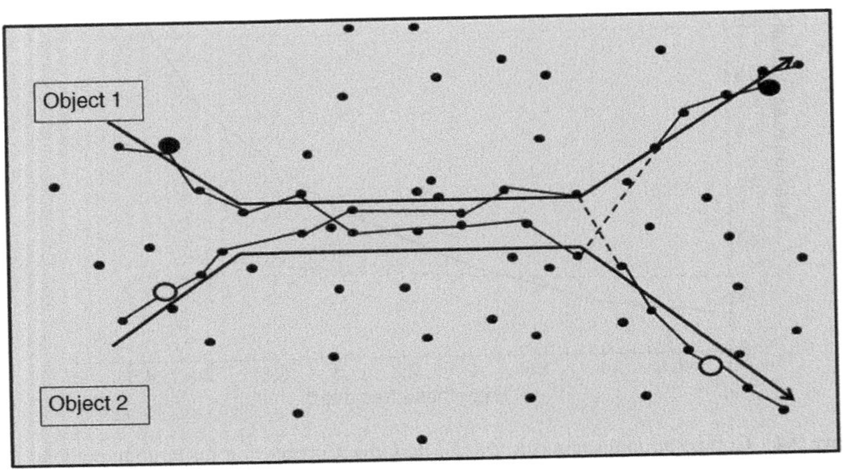

Figure 9.5 Multistage tracking with multiscale sensor data.

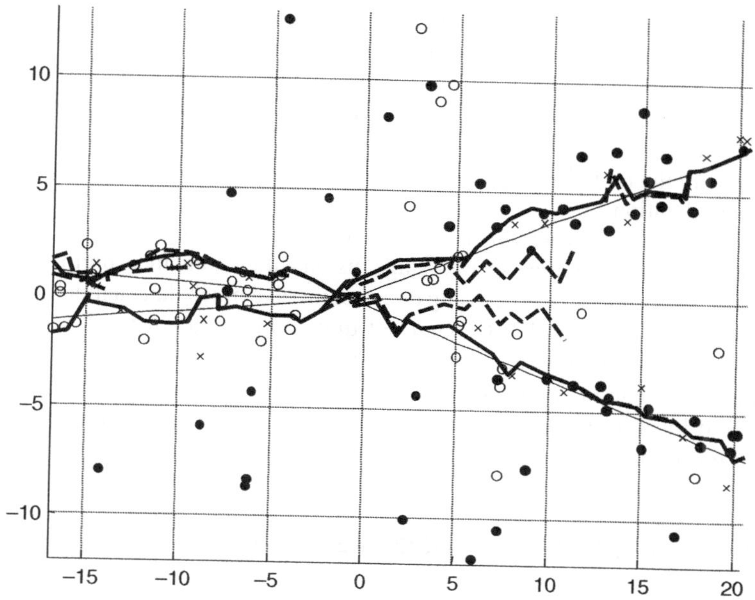

Figure 9.6 Crossing-target trajectories (thin solid lines). Single-sensor tracks for high-rate sensor data exhibit track swap. *Track-before-fuse* (with downstream logic to recover from processing errors) results in track fragmentation (dotted lines). *Track-break-fuse* successfully tracks through the ambiguity despite the upstream error (solid lines).

association decisions when the targets approach. Second, it is not necessary to maintain group statistics (centroid, etc.) for the paradigm to be used effectively. Finally, when implementing a multistage fusion process, one must define how the inevitable upstream processing errors will be dealt with in downstream processing.

Figure 9.6 illustrates a simple track-termination methodology to re-initialize processing when anomalous tracks are detected. Assume that we have a high-rate feature-poor sensor (e.g., a radar sensor), and a second, feature-rich low-rate or intermittent-coverage sensor (e.g., a signal-intelligence sensor). Assume further that two targets cross in a region where the low-rate sensor does not provide measurements. High-rate contacts are circles (target-originated returns) and dots (false returns). The feature-rich sensor provides measurements with precise identity information and no false returns; the measurements are illustrated by crosses. Target kinematic ambiguity occurs in a region that lacks feature measurements.

Single-sensor tracking often results in track swap. Indeed, in the realization shown in Figure 9.6 the single-sensor tracks do not cross. One approach to improve decision-making is to track on data from both sensors (*track-while-fuse*), with a large hypothesis tree depth to enable exploitation of feature measurements before and after the coverage gap. Unfortunately, this centralized tracking solution is unrealizable due to computational constraints.

The results of two alternate multistage schemes are shown in Figure 9.6. The first scheme, *track-before-fuse*, takes the high-rate single-sensor tracks and fuses these with the feature measurements. This processing requires fusion logic to identify and recover from upstream processing errors. Indeed, we see that the upstream track swap leads to track fragmentation once the anomaly is detected. The *track-before-fuse* results are the dotted lines in Figure 9.6: note the fragmentation and subsequent track re-initialization.

A second approach, *track-break-fuse*, removes track labels after the first single-sensor tracking stage, enabling second-stage fusion with feature measurements. In this scheme, since second-stage processing does not involve much data, a much larger hypothesis tree depth is possible (though not quite as large as in *track-before-fuse*). Correspondingly, correct association decisions are achieved with much lower computational effort than in (single-stage) *track-while-fuse*. The *track-break-fuse* results are the solid lines in Figure 9.6. This approach is computationally more costly but exhibits improved track continuity (solid line). Further details are in [18].

9.4 TRACKING IN LARGE SENSOR NETWORKS

Surveillance with large networks of cheap sensors requires the ability to maintain relevant track hypotheses over large numbers of sensor scans, and the need for effective information extraction from poor-quality sensor data. Our proposed approach is *fuse-before-track*. This two-stage approach seeks to collapse a large number poor sensor scans into a single, equivalent scan of higher quality. Techniques to do so that are insensitive to scan ordering are particularly appealing. Following data reduction, a conventional scan-based MHT tracker is employed. Details may be found in [19, 20].

To begin, we develop an effective ML approach as a first stage in *fuse-before-track* processing. There are two key aspects to the approach developed here. First, we avoid computationally costly numerical optimization schemes by evaluating the likelihood function *only* at contact locations. Having identified the top-scoring contact, one could imagine the following cumbersome methodology: (1) remove contact data due to a single target (roughly, remove a number of contacts equal to number of scans times the target detection probability); (2) collapse the extracted contacts into an equivalent fused contacts; (3) iterate the ML equations on remaining contacts. The second key aspect of our approach is to replace steps (1–3) with a simpler scheme whereby the top M contacts from the first set of ML evaluations are extracted, and each contact is kept within its original data scan. That is, we do not collapse the scans into a single, equivalent scan. In addition to the simplicity of our approach, the procedure (1–3) is potentially problematic in the case of closely spaced targets.

We consider the case of linear measurements of two-dimensional target positional perturbed by additive Gaussian noise, as given by (9.5).

$$Z = X + w, \;\; w \sim N(0, \Sigma). \tag{9.5}$$

The measurement covariance matrix Σ is assumed to be constant over all target-induced contacts. In each scan of data, all targets in the surveillance region are detected with probability p, and false contacts are uniformly distributed in this region (of area u [m^2]), with the number of false contacts Poisson distributed with mean λ. For a given sensing time epoch, assume that N is the number of synchronous sensors, and let n_i, $1 \le i \le N$ be the number of contacts from each sensor. The contacts from the ith sensor are denoted by Z_{ij}, $1 \le j \le n_i$. The likelihood function for target location is given by (9.6).

$$\Lambda(X) = \prod_{i=1}^{N} \left\{ \frac{1-p}{u^{n_i}} \mu_\lambda(n_i) + \frac{p \cdot \mu_\lambda(n_i - 1)}{u^{n_i - 1} n_i} \sum_{j=1}^{n_i} p\left(Z_{ij} \mid X\right) \right\}, \tag{9.6}$$

$$\mu_\lambda(n) = \frac{\lambda^n \exp(-\lambda)}{n!}, \tag{9.7}$$

$$p(Z|X) = \frac{1}{2\pi |\Sigma|^{0.5}} \exp\left(-\frac{1}{2} (Z - X)' \Sigma^{-1} (Z - X) \right). \tag{9.8}$$

We evaluate $\Lambda(\cdot)$ for $X \in \left\{Z_{ij}, 1 \le i \le N, 1 \le j \le n_i\right\}$. We identify the top M likelihood function evaluations; the contacts corresponding to these contacts are kept, and all others are discarded, leading to a *thinned* version of the N sets of contacts. In particular, we now have $M \ll \sum_i n_i$ contacts for a given time epoch. It is important that we select $M > p \cdot \lambda_T \cdot N$, where λ_T is the expected number of targets.

The ML approach defined here is consistent with processing paradigms that invoke hard data association, since at no stage here is there is a weighted merging of contact data. The significantly smaller contact files that result from the ML processing scheme described here constitute the input to an MHT processing stage.

It is important for MHT processing to extend over a reasonable time extent. This is quite problematic in large sensor fields, where there are many data scans in short-duration time intervals. Accordingly, it is imperative to reduce significantly the number of track hypotheses prior to MHT process with large tree depth (i.e., large *n-scan*). A straightforward methodology to enable this is described here. As a reminder, we assume that the sensors are synchronized so as to have the same sequence of scan times. (Our methodology could be extended to handle the more general case of nonsynchronous sensors, though we do not consider this here.)

In a first MHT processing stage (with small or zero *n-scan*), we associate contacts into tracks, though with track termination for nonzero time increments. That is, we perform automatic tracking separately for each (synchronous) collection of sets of contacts, with no track continuity between these collections. The resulting tracks exist over multiple sensor scans, but have zero time duration. As a byproduct of this process, tentative tracks that fail to achieve the track confirmation threshold are discarded.

The second MHT processing stage contends with a vastly simpler data-association task that associates short-duration tracks over time. In particular, this second stage can easily handle large hypothesis tree depths (*n-scan* $\gg 0$) with modest computational

Figure 9.7 A scenario with crossing-target trajectories (thin solid lines). ML–MHT (dotted lines) and ML–MHT2 (solid lines) illustrate the success of the latter to avoid track swap in scenarios with many poor-quality sensors.

expense. The large *n-scan* allows for nonzero time-depth reasoning in large sensor networks. This capability is particularly useful in dense target scenarios with nontrivial target disambiguation.

The MHT and ML–MHT (only one MHT stage) are severely challenged in multitarget scenarios in large sensor networks, due to their inability to utilize a large *n-scan*. More complex multistage processing is possible, for example, the ML–MHT2. The interested reader is referred to [20]; an illustration is given in Figure 9.7. Further work on this topic has been documented recently [21].

9.5 MULTISCALE OBJECTS

Most approaches to multisensor multitarget tracking assume at most one detection per target per scan of data. A notable exception is the *probabilistic multiple-hypothesis tracker* (PMHT) [22]. Another exception has been developed for poor-quality segmentation of closely spaced objects, motivated by image sequences of cell tissues [23]. A third approach, motivated by extend-object tracking, is to recast the object as representative of a group of smaller objects; as such, multiple measurements may be assigned to the group [24, 25].

One of the issues in treating an extended object as a group of objects is that the usual independence assumptions across objects must be relaxed. This sometimes occurs in other settings as well: see [26] for an application to particle tracking where target interactions cannot be neglected.

Like most tracking paradigms, MHTrequires at most one detection per target per scan. In [27], we propose a multistage MHT approach to contend with the challenge of active–passive data fusion where multiple passive returns may correspond to the same object with multiple onboard emitter modes. We consider multisensor fusion of radar, electronic support measures (ESM), and *global positioning system* (GPS) data. GPS data are only available for own (friendly) platforms.

A discussion of ESM sensors and references to past tracking work in this area are in [6]. A Bayesian classification approach in the absence of measurement provenance ambiguity is discussed in [28]. A track association approach based on clustering techniques is discussed in [29]. The importance of effective multisensor electronic surveillance remains an important technology requirement in security and defense; a recent discussion may be found in [30].

A principal challenge in ESM-radar fusion is the need to address multilevel object tracking, whereby each platform includes a number of independently operated emitter modes. This is in addition to the usual tracking challenges that include measurement provenance uncertainty, feature and localization errors, false returns, missed detections, and limited observability with passive sensors that risks producing ghost-tracking solutions. The multistage approach adopted in [27] is illustrated in Figure 9.8.

The key architectural insight is to perform multisensor ESM tracking at the emitter-mode level. We invoke the simplifying assumption that emitters are independent objects. Clearly, they are not: those on the same platform experience birth and death

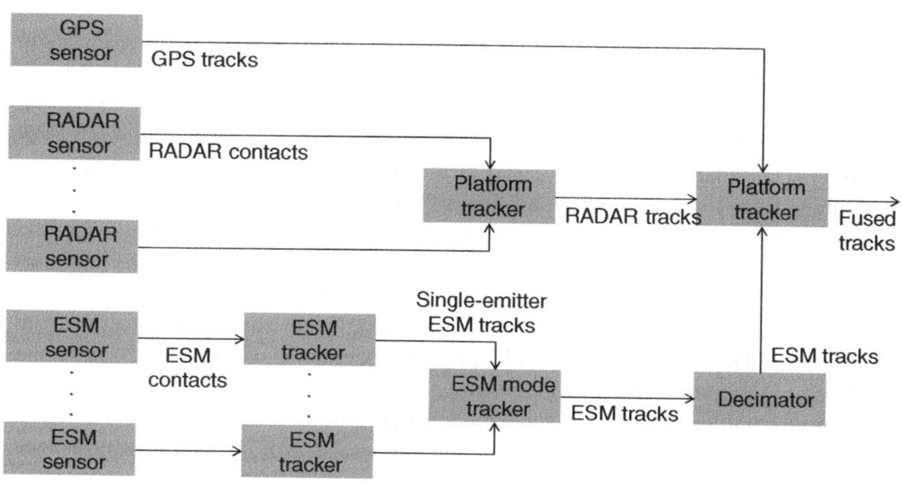

Figure 9.8 A multistage architecture enabling active–passive fusion for multiscale objects.

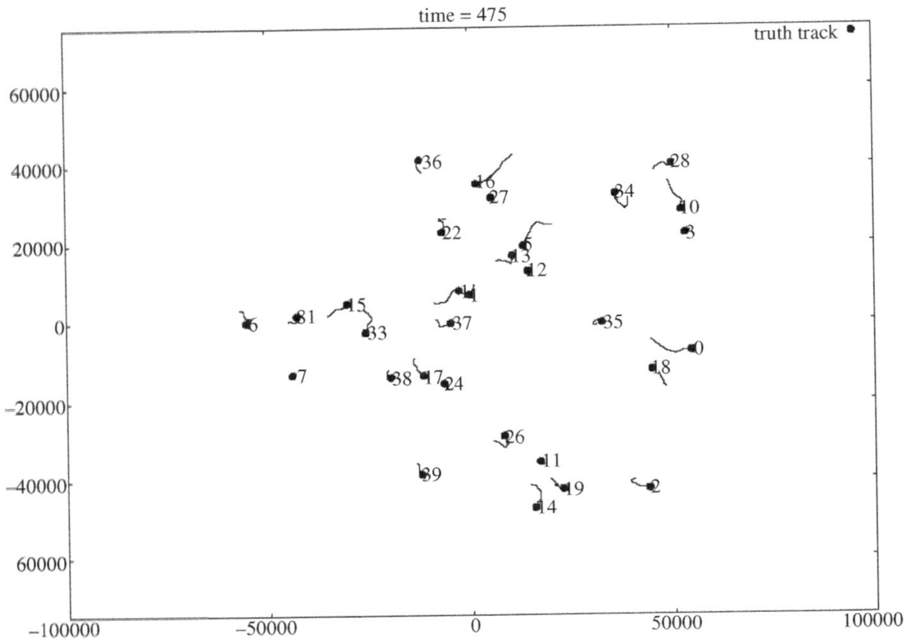

Figure 9.9 Ground truth.

at the same time, and their motion is along the same platform trajectory. Nonetheless, this approach allows us to circumvent the complexities in addressing coupled-target existence and dynamics.

Multisensor ESM tracks undergo a decimation step whereby each measurement is placed in a unique sensor scan. This is essential in that platform-level tracking will allow only one measurement per target per scan. Optimal platform filtering that allows for updates based on measurement sets is possible, but computationally prohibitive [31]. Radar data are processed in a centralized platform-level tracker. The final integration step is given by track fusion over GPS, radar, and ESM tracks.

Figures 9.9–9.16 illustrate a standard air surveillance scenario simulation and tracking solution under our multistage processing paradigm. Figure 9.16 illustrates a snapshot in time of the final output of the air surveillance data fusion. Each track may have platform type information as well, though this cannot be inferred from these figures. The illustrations provide qualitative evidence of the effectiveness and flexibility in multistage fusion.

An important aspect in multistage fusion is the need for robustness to contend with upstream data association errors as with all *track-before-fuse* architectures (see Section 9.3). The novelty addressed in this section is the multiscale object model required for active–passive data fusion.

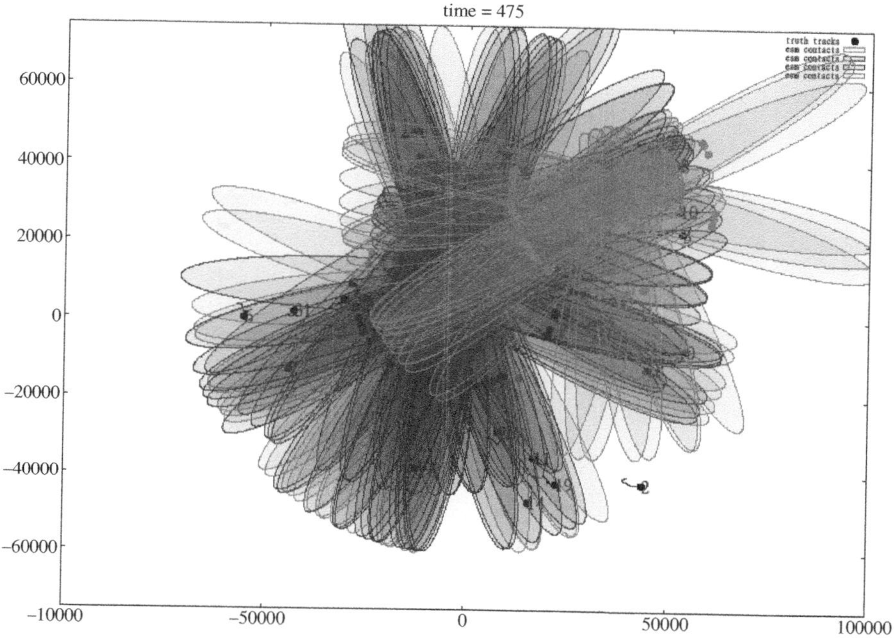

Figure 9.10 ESM measurements.

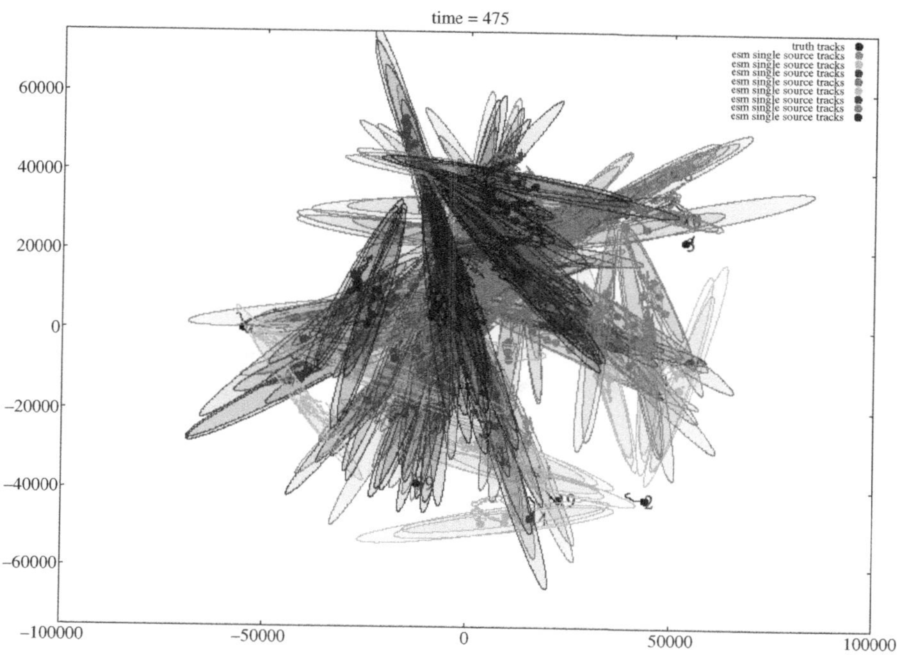

Figure 9.11 Single-sensor ESM emitter-level tracks.

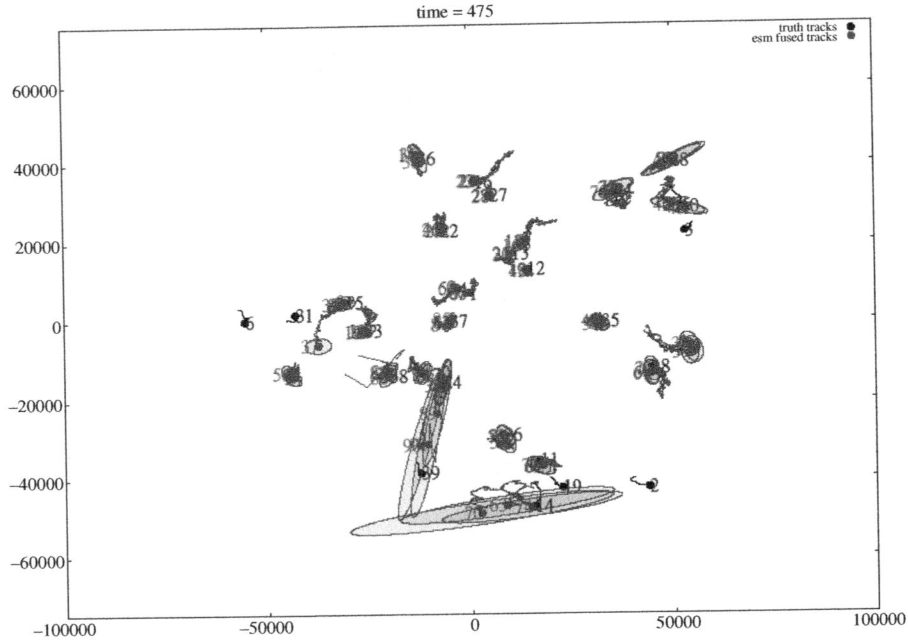

Figure 9.12 Multisensor ESM emitter-level tracks.

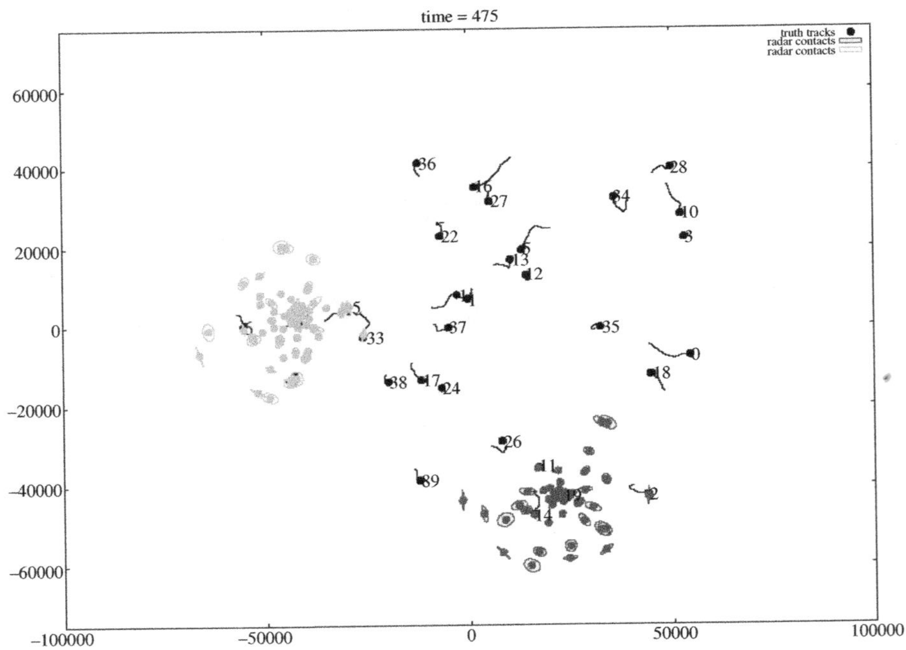

Figure 9.13 Radar measurements.

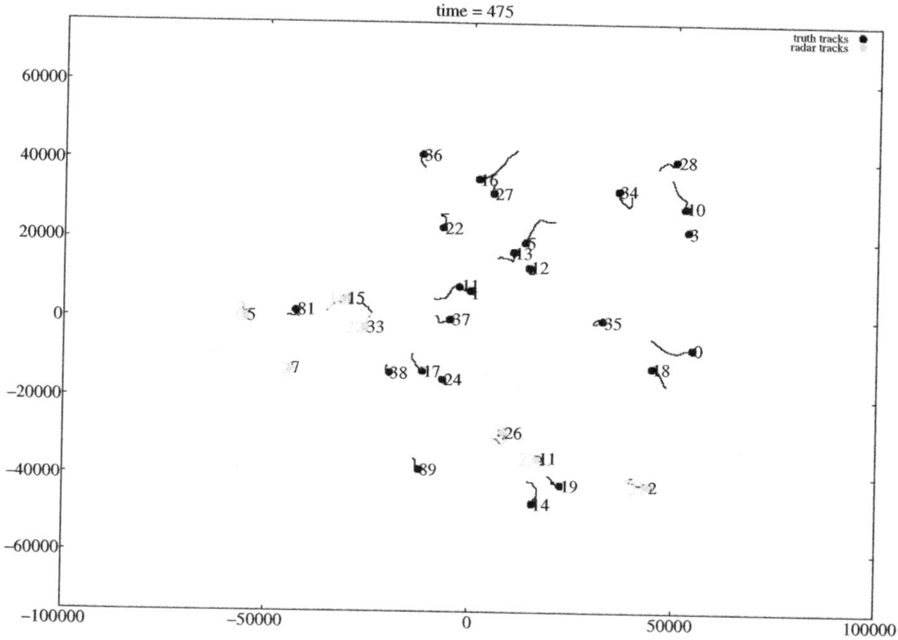

Figure 9.14 Multisensor radar platform-level tracks.

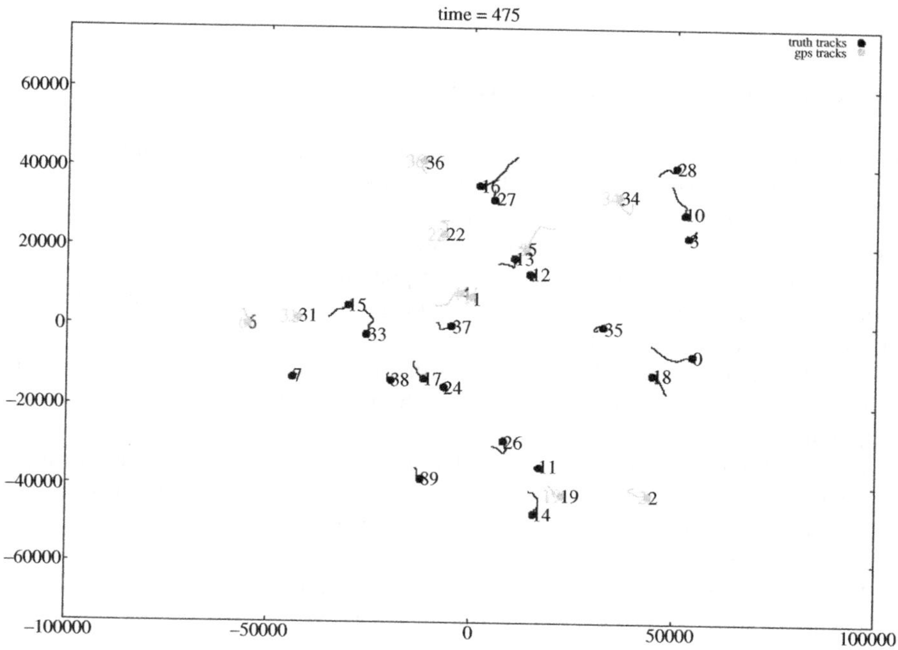

Figure 9.15 GPS tracks (on cooperative targets).

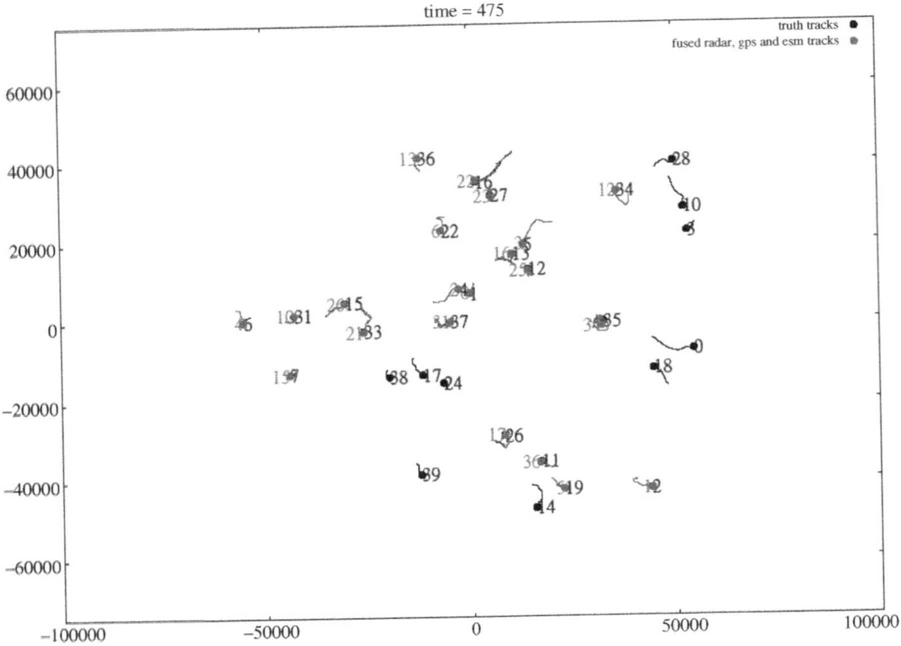

Figure 9.16 Fused multisensor (ESM, radar, GPS) platform-level tracks.

9.6 MEASUREMENT AGGREGATION

As noted in Section 9.5, the decimation step in the multistage architecture illustrated in Figure 9.8 replaces each set of emitter-level track updates with an equivalent sequence of emitter-level track updates, thus enabling downstream platform-level tracking. Indeed, this procedure allows us to apply a downstream scan-based MHT processing module where many emitter tracks may correspond to the same object. Equally important, the decimation step includes measurement-aggregation functionality. This is an essential aspect to successful multisensor fusion and is discussed next.

Complementary sensors that observe distinct portions of the target state pose a challenge to centralized fusion processing. The multistage approach allows, first, high-confidence same-sensor association decisions. Next, measurement aggregation leads to conditionally independent equivalent measurements that are more informative of the full target state than the original sensor measurements. Finally, downstream cross-sensor association of equivalent measurements is performed with improved association performance as compared with centralized measurement processing. Much of the past work on equivalent measurements for tracking is due to Drummond and colleagues, who refer to equivalent measurements as *tracklets* [6, 32]. Here, we motivate the benefit of measurement aggregation by examining the issue in an idealized two-target setting with perfect detection data.

Assume two-dimensional linear target motion according to an *Ornstein–Uhlenbeck* (OU) motion model in each dimension [33]. The discrete-time motion model is given below. Note that the initial target distribution is defined according to the target steady-state distribution; this ensures a stationary stochastic process. We assume there are known to be two targets at all times, with no target birth or death events.

$$X_{k+1} = A_k X_k + w_k, \tag{9.9}$$

$$X_0 \sim N\left(0, \bar{Q}\right), \ \bar{Q} = \begin{bmatrix} \frac{q}{2\gamma} & 0 \\ 0 & \frac{q}{2\gamma} \end{bmatrix}, \quad q > 0, \ \gamma > 0 \tag{9.10}$$

$$A_k = \begin{bmatrix} \exp\left(-\gamma \Delta t_k\right) & 0 \\ 0 & \exp\left(-\gamma \Delta t_k\right) \end{bmatrix}, \ \Delta t_k = t_{k+1} - t_k, \tag{9.11}$$

$$w_k \sim N\left(0, Q_k\right), Q_k = \begin{bmatrix} \frac{1-\exp\left(-2\gamma\Delta t_k\right)}{2\gamma} q & 0 \\ 0 & \frac{1-\exp\left(-2\gamma\Delta t_k\right)}{2\gamma} q \end{bmatrix}. \tag{9.12}$$

We assume two sensors with common revisit time Δt. The sensors are randomly staggered in time with respect to one another. In particular, for each of the two sensors, the first measurement time is defined according to the uniform distribution; that is, for each sensor $i \in \{1, 2\}$, we have $t_{i,1} \sim U(0, \Delta t)$, where we denote by $t_{i,k}$ the kth measurement from the ith sensor. Additionally, we denote by t_k the kth measurement from *either* sensor.

We assume perfect detection performance, that is, no missed detections and no false alarms. The sensor measurement model for sensor $i \in \{1, 2\}$ is as follows

$$Z_k = X_k + v_k, v_k \sim N\left(0, R\right), R = R_i. \tag{9.13}$$

The measurement time sequence differs for the two sensors, as noted above. The (time-invariant) measurement error covariance matrices are given by the following

$$R = R_1 = \begin{bmatrix} \varepsilon \cdot \sigma^2 & 0 \\ 0 & \sigma^2 \end{bmatrix}, \qquad R = R_2 = \begin{bmatrix} \sigma^2 & 0 \\ 0 & \varepsilon \cdot \sigma^2 \end{bmatrix}, \quad \varepsilon < 1. \tag{9.14}$$

For small values of ε, the sensors are increasingly complementary: one sensor provides precise measurements of target state in one dimension, and the other sensor in the second dimension.

Associated measurements are processed sequentially according to the well-known *Kalman filter* (KF). The equations are provided below, using the usual notation for predicted and updated states and covariances. Note that filter initialization reflects

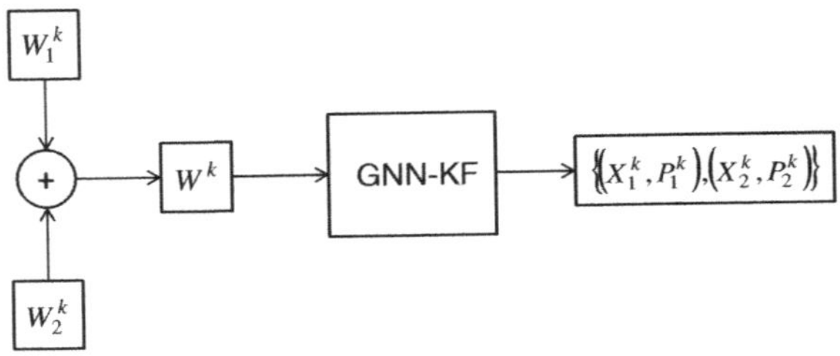

Figure 9.17 Centralized architecture for multitarget filtering.

prior state information under the OU motion model.

$$X(1|0) = 0, \tag{9.15}$$

$$P(1|0) = \bar{Q}, \tag{9.16}$$

$$X(k+1|k) = A_k X(k|k), \tag{9.17}$$

$$P(k+1|k) = A_k P(k|k) A_k' + Q_k, \tag{9.18}$$

$$L_{k+1} = P(k+1|k)(P(k+1|k) + R)^{-1}, \tag{9.19}$$

$$X(k+1|k+1) = X(k+1|k) + L_{k+1}(Z_{k+1} - X(k+1|k)), \tag{9.20}$$

$$P(k+1|k+1) = (I - L_{k+1})P(k+1|k). \tag{9.21}$$

We consider two alternative online schemes for processing available measurements. The first approach, illustrated in Figure 9.17, is the centralized approach, where the two sequences of sets of measurements are combined into a single sequence and processed with *global nearest neighbor* (GNN) data association and the KF for each track. The second approach, illustrated in Figure 9.18, is the distributed approachwith measurement aggregation. Under this approach, each of the two sequences of sets of

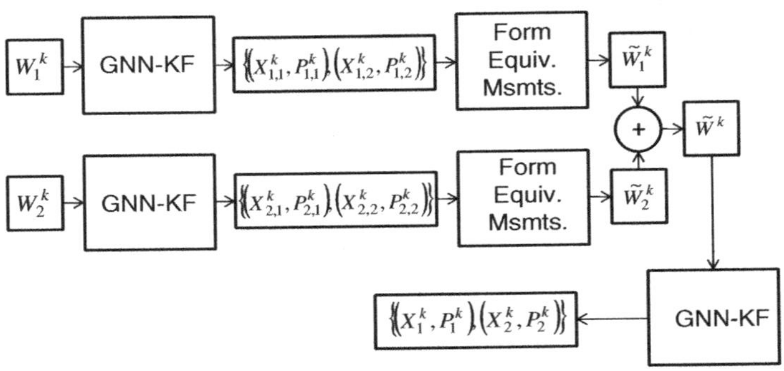

Figure 9.18 Distributed architecture for multitarget filtering.

measurements is processed separately with the GNN–KF approach. Next, measurements are processed in batches of size m, and replaced by an equivalent measurement. The two sequences of sets of equivalent measurements are sent to the downstream processing component, which consists of another GNN–KF module.

We denote by $W_{i,k}$ the set of (two) measurements from sensor i at time $t_{i,k}$. The sequence of sets of measurements from sensor i up to time t_k is denoted by $W_i^k = (W_{i,1}, W_{i,2}, ..., W_{i,\widetilde{k}})$. Note the difference in the indices k and $\widetilde{k}$ that is due to the difference between $t_{i,k}$ and t_k. We denote by W^k the time-ordered union of the subsequences of measurement sets W_1^k and W_2^k up to time t_k. Note that the measurement set W_k corresponds to time t_k. We denote by $(X_{i,j}^k, P_{i,j}^k)$ the sequence of state estimates and covariances for target j obtained by processing W_i^k.

The sequence of sets of equivalent measurements (with associated equivalent covariances) from sensor i is denoted by $\widetilde{W}_i^k = (\widetilde{W}_{i,1}, \widetilde{W}_{i,2}, \ldots, \widetilde{W}_{i,k})$. We denote by $\widetilde{W}^k$ the time-ordered union of the sub-sequences of measurement sets $\widetilde{W}_1^k$ and $\widetilde{W}_2^k$. Note that these sequences are shorter than the original sequences W^k, W_1^k, and W_2^k, in the sense that each set of measurements replaces a sequence of m sets of measurements. Indeed, the sequences of sets of measurements W_1^k, W_2^k, W^k, $\widetilde{W}_1^k$, $\widetilde{W}_2^k$, and $\widetilde{W}^k$ all have different lengths. Nonetheless, all sequences are defined over the same time interval, from zero to t_k. Processing W^k or $\widetilde{W}^k$, the sequence of actual or equivalent measurement sets from both sensors, results in (X_j^k, P_j^k) for target $j \in \{1, 2\}$.

The distributed approach would appear to be more robust to measurement origin uncertainty. Indeed, association decisions are first performed on same-sensor data. Subsequently, cross-sensor fusion is performed with better-localized equivalent measurements. On the down side, the distributed architecture delays the arrival of measurements to the fusion center. Thus, excessive measurement aggregation may degrade multitarget filtering performance.

In the distributed approach, an equivalent measurement is determined as follows, based on the state estimate and covariance that precede and follow m prediction-update steps, that is, based on $X(k|k)$, $P(k|k)$, $P(k + m|k + m)$, and $X(k + m|k + m)$. First, note that $X(k + m|k)$ and $P(k + m|k)$ are known by repeated application of equations (9.17, 9.18). We have the following relationships

$$X(k+m|k+m) = X(k+m|k) + \widetilde{L}(\widetilde{Z} - X(k+m|k)), \tag{9.22}$$

$$P(k+m|k+m) = (I - \widetilde{L})P(k+m|k), \tag{9.23}$$

$$\widetilde{L} = P(k+m|k)(P(k+m|k) + \widetilde{R})^{-1}. \tag{9.24}$$

Solving these equations leads to the following equivalent measurement and covariance

$$\widetilde{Z} = X(k+m|k) + \widetilde{L}^{-1}(X(k+m|k+m) - X(k+m|k)), \tag{9.25}$$

$$\widetilde{R}^{-1} = (\widetilde{L}^{-1} - I)P(k+m|k)P(k+m|k). \tag{9.26}$$

TABLE 9.1 Simulation Parameters for Multitarget Filtering

Parameter Description	Setting
Target process noise ($m^2 s^{-1}$)	$q = 10^{-4}$
Target drift term (s^{-1})	$\gamma = 5 \times 10^{-5}$
Sensor measurement error variance (m^2)	$\sigma^2 = 4$
Sensor variance reduction factor	$\varepsilon = 2.5 \times 10^{-5}$
Sensor revisit time (s)	$\Delta t = 1$
Scenario duration (s)	50
Number of Monte Carlo runs	100
Measurement aggregation	1–8

Table 9.1 identifies the simulation parameters that we use in our multitarget filtering study.

We consider centralized and distributed fusion architectures, the latter with a range of measurement-aggregation settings in forming equivalent measurements. We consider a minimum of no measurement aggregation (a level of one, which corresponds to centralized fusion) up to a maximum aggregation level of eight. For the purposes of this multitarget filtering study, the equivalent measurements are treated as unlabeled

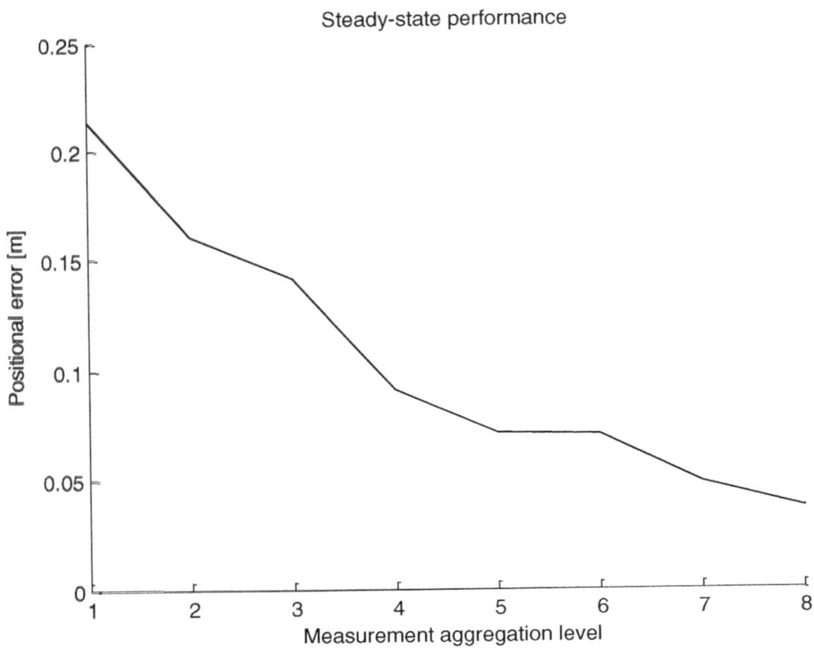

Figure 9.19 Multitarget filtering performance as a function of measurement aggregation level.

measurements, in the sense that contact-level (or detection level) data association is performed in the downstream cross-sensor GNN–KF. Later, in our discussion of air surveillance, we will retain upstream track information.

Figure 9.19 illustrates the average terminal *Optimal Sub-Pattern Assignment* (OSPA) metric (i.e., the average at time 50 s) for all choices of the measurement-aggregation level. The OSPA metric is the (smaller) RMSE based on the better (optimal) track-truth association at each time and each realization, where the RMSE averages over both targets. It is important to note that there is a tradeoff. Indeed, higher aggregation levels incur delay before precise target localization, though with a reduced instance of association errors and ghosting solutions.

As we have seen, distributed multitarget tracking outperforms centralized solutions for a number of relevant applications. One such setting, where cross-sensor data association is difficult, is explored in this section in a simplified context. Same-sensor data association and equivalent-measurement formation prior to fusion enables higher accuracy cross-sensor data association. In a multiple-hypothesis tracking context, one could in principle attack the problem directly with sufficient hypothesis depth in a centralized configuration, but the computational implications are prohibitive.

9.7 CONCLUSIONS

This chapter has provided several illustrations that motivate the use of multistage processing architectures to address difficult challenges in automatic tracking. Indeed, there are performance limitations in centralized, single-stage fusion, and some of these can be addressed through multistage processing. Interestingly, we find that multistage processing is not only useful when processing complementary, multiscale data. As shown in Section 9.5 in the context of mixed active/passive surveillance, multistage fusion enables efficient handling of multiscale targets, as discussed here data.

The traditional application domains for automatic-tracking surveillance technology have centered on physical security and defense. We believe there is significant opportunity for application of these technologies to diverse applications including cyber security and biomedical imaging; for the latter topic, an introductory survey may be found in [34].

REFERENCES

1. Y. Bar-Shalom, P. Willett, and X. Tian, *Tracking and Data Fusion: A Handbook of Algorithms*, YBS Publishing, Storrs, CT, USA, 2011.
2. T. Kurien, Issues in the design of practical multitarget tracking algorithms, *Multitarget-Multisensor Tracking: Advanced Applications*, Artech House, Norwood, MA, USA, 1990.
3. M. Mallick, S. Coraluppi, and C. Carthel, Multitarget tracking using multiple hypothesis tracking, in *Integrated Tracking, Classification, and Sensor Management: Theory and Applications*, Wiley, New York, Chapter 5, 2012.
4. J. Tsitsiklis, Decentralized detection, Vol. 2, *Advances in Signal Processing*, JAI Press, Greenwich, CT, USA, 1993.

5. W. Blanding, P. Willett, Y. Bar-Shalom, and S. Coraluppi, Multisensor track management for targets with fluctuating SNR, *IEEE Trans. Aerosp. Electron. Syst.*, 45(4), 1275–1292, 2009.
6. S. Blackman and R. Popoli, *Design and Analysis of Modern Tracking Systems*, Artech House, Norwood, MA, USA, 1999.
7. S. Coraluppi and C. Carthel, Distributed tracking in multistatic sonar, *IEEE Trans. Aerosp. Electron. Syst.*, 41(3), 1138–1147, 2005.
8. C. Carthel, S. Coraluppi, P. Willett, M. Maratea, and A. Maguer, Maximum likelihood approach to HF radar performance characterization, *Proceedings of the 12th International Conference on Information Fusion*, Seattle WA, USA, July 2009.
9. M. Daun and W. Koch, Multistatic target tracking for non-cooperative illuminating by DAB/DVB-T, *Proceedings of the IEEE OCEANS Conference*, Aberdeen, Scotland, June 2007.
10. R. Fitzgerald, Track biases and coalescence with probabilistic data association, *IEEE Trans. Aerosp. Electron. Syst.*, 21(6), 822–825, 1985.
11. H. Blom and A. Bloem, Probabilistic data association avoiding track coalescence, *IEEE Trans. Automat. Contr.*, 45(2), 247–259, 2000.
12. P. Willett, T. Luginbuhl, and E. Giannopoulos, MHT tracking for crossing sonar targets, *Proceedings of the SPIE Conference on Signal and Data Processing of Small Targets*, August 2007, Vol. 6699, pp. 66991C-1 - 66991C-12.
13. S. Coraluppi, C. Carthel, P. Willett, M. Dingboe, O. O'Neill, and T. Luginbuhl, The track repulsion effect in automatic tracking, *Proceedings of the 12th International Conference on Information Fusion*, Seattle WA, USA, July 2009, pp. 2225–2230.
14. R. Ivey, A. Waxman, D. Fay, and D. Martin, Learn-while-tracking, feature discovery and fusion of high-resolution radar range profiles, *Proceedings of the 6th International Conference on Information Fusion*, Cairns, Australia, 2003, pp. 741–748.
15. J. Lancaster and S. Blackman, Joint IMM/MHT tracking and identification with confusers and track switching, *Proc. SPIE*, 6236, 2006, pp. 6236 1A - 6236 1A-12.
16. C.-Y. Chong et al., Efficient multiple hypothesis tracking by track segment graph, *Proceedings of the 12th International Conference on Information Fusion*, Seattle WA, USA, July 2009, pp. 2177–2184.
17. R. Ivey, J. Horn, and R. Merkert, Long-duration fused feature learning aided tracking, *Proceedings of the SPIE Conference on Multisensor, Multisource Information Fusion: Architectures, Algorithms, and Applications*, Orlando FL, April 2010, Vol. 7710, pp. 77100M-1–77100M-12.
18. S. Coraluppi and C. Carthel, Multistage multiple-hypothesis tracking, *ISIF J. Adv. Inform. Fusion*, 6(1), 2011.
19. S. Coraluppi, M. Guerriero, C. Carthel, and P. Willett, Fuse-before-track in large sensor networks, *ISIF J. Adv. Inform. Fusion*, 5(1), 18–31, 2010.
20. S. Coraluppi and C. Carthel, An ML-MHT approach to tracking dim targets in large sensor networks, *Proceedings of the 13th International Conference on Information Fusion*, Edinburgh, Scotland, July 2010.
21. R. Georgescu and P. Willett, Random finite set Markov chain Monte Carlo predetection fusion, *Proceedings of the 14th International Conference on Information Fusion*, Chicago IL, USA, July 2011.

22. R. Streit and T. Luginbuhl, A probabilistic multihypothesis tracking algorithm without enumeration and pruning, *Proceedings of the Sixth Joint Service Data Fusion Symposium*, Laurel, Maryland, June 1993.

23. T. Kirubarajan, Y. Bar-Shalom, and K. Pattipati, Multiassignment for tracking a large number of overlapping objects, *IEEE Trans. Aerosp. Electron. Syst.*, 37(1), 2001.

24. W. Koch, Bayesian approach to extended object and cluster tracking using random matrices, *IEEE Trans. Aerosp. Electron. Syst.*, 44(3), 2008.

25. M. Feldman, D. Franken, and W. Koch, Tracking of extended Objects and Group Targets using random matrics, *IEEE Trans. Signal Process.*, 59(4), 1409–1420, 2011.

26. N. Oxtoby, J. Ralph, D. Samsonov, and C. Durniak, Tracking interacting dust: comparison of tracking and state estimation techniques for dusty plasmas, *Proceedings of SPIE Conference on Signal and Data Processing of Small Targets*, Orlando FL, USA, April 2010, Vol. 7698, pp. 76980C-1 to 76980C-11.

27. S. Coraluppi and C. Carthel, A hierarchical MHT approach to ESM-radar fusion, *Proceedings 15th International Conference on Information Fusion*, Singapore, July 2012, pp. 677–683.

28. S. Challa and G. Pulford, Joint target tracking and classification using radar and ESM sensors, *IEEE Trans. Aerosp. Electron. Syst.*, 37(3), 1039–1055, 2001.

29. Y. Zhou and J. Mickeal, A sequential ESM track association algorithm based on the use of information theoretic criteria, *Proceedings of the 10th International Conference on Information Fusion*, Quebec City, Canada, July 2007.

30. S. Kemkemian, M. Nouvel-Fiani, and E. Chamouard, Radar and electronic warfare cooperation: how to improve the system efficiency, *IEEE Aerosp. Electron. Syst. Magazine*, 26(8), 32–38, August 2011.

31. S. Coraluppi and C. Carthel, Bayesian filtering in electronic surveillance, *Proceedings of the SPIE Conference on Signal Processing, Sensor Fusion, and Target Recognition XXI*, Baltimore MD, USA, April 2012, Vol. 8392, pp. 839202-1 to 839202-10.

32. O. Drummond, W. Blair, G. Brown, T. Ogle, Y. Bar-Shalom, R. Cooperman, and W. Barker, Performance assessment and comparison of various tracklet methods for maneuvering targets, in *Proceedings of the SPIE*, San Diego, USA, August 2003, Vol. 5096, pp. 514–538.

33. L. Stone, C. Barlow, and T. Corwin, *Bayesian Multiple Target Tracking*, Artech House, Norwood, MA, USA, 1999.

34. E. Meijering, I. Smal, and G. Danuser, Tracking in molecular bioimaging, *IEEE Signal Process. Mag.*, 23(3), 46–53, May 2006.

CHAPTER 10

Intent Inference and Detection of Anomalous Trajectories: A Metalevel Tracking Approach

VIKRAM KRISHNAMURTHY

10.1 INTRODUCTION

Classical target tracking assumes a state-space model with target maneuvers (acceleration) modeled as a finite state Markov chain. Such models are useful on short timescales (order of several seconds) and several well-known target tracking algorithms have been developed in the literature. This chapter is motivated by *metalevel target tracking* applications on longer time scales (order of several minutes). In metalevel tracking, one is interested in devising automated procedures that assist the human operator to interpret the tracks obtained from a conventional tracking algorithm. On such longer time scales, most real-world targets are "destination-aware"—they have a well-defined destination, and they rarely move according to a "drunkard's" random walk (Markov chain). From a probabilistic modeling point of view, being destination-aware means that the initial and final target states (in terms of position) need to be chosen from a joint distribution before specifying the transition law of the target dynamics. Naturally finite state Markov chains cannot capture this long-range dependency. More sophisticated models are required to reflect the target's intent to proceed from its source location to its destination. This chapter develops metalevel tracking algorithms that assist human operators by extracting spatial patterns from target tracks to identify suspicious/anomalous spatial trajectories. The targets' spatial trajectories are modeled by two types of random processes:

- Stochastic context-free grammars (SCFGs).
- Reciprocal random processes.

Integrated Tracking, Classification, and Sensor Management: Theory and Applications, Edited by Mahendra Mallick, Vikram Krishnamurthy, and Ba-Ngu Vo.

We also present Bayesian filtering and smoothing algorithms for these processes to estimate the trajectories and hence intent of the targets.

10.1.1 Examples of Metalevel Tracking

To motivate this chapter, we start with three examples.

Example 1. Pattern of life: Consider the following forensic surveillance application called *pattern of life* analysis that involves identification of a target's daily interaction with its environment. Such analysis can be used to predict a target's behavior based on habit or schedule. Consider the street map in Figure 10.1a. Normal behavior of local traffic between points (x_1, y_1) and (x_T, y_T) is observed to flow in a straight road between the two points. However, if the local population has insider information about a new threat (such as possible installation of an improvised explosive device), then the local pattern of traffic flow changes to avoid the straight road as shown in Figure 10.1b. Such abnormal behavior deviates from routine habit and indicates an event of interest. The goal is to detect such changes in either single or aggregate target behavior. Doing so requires a parsimonious representation of target trajectories. Such a representation is provided by SCFG models that are also scalable toward dealing with trajectories of different shapes and sizes. If the terminal time T is fixed, then Markovian bridges and, more generally, reciprocal processes (RP) provide the framework for modeling such trajectories. Clearly, a Markov process is un-satisfactory since it cannot be clamped at (x_T, y_T). Developing SCFG models, reciprocal Markov models and associated signal processing and classification algorithms is the main aim of this chapter.

Example 2. Anomalous trajectory detection of destination-aware target: As a second example, consider real-time detection of an anomalous trajectory of a destination-aware target. Suppose that prior information about the intended

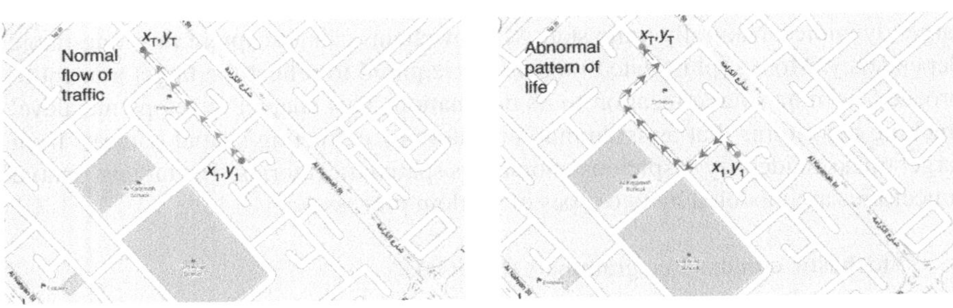

(a) (b)

Figure 10.1 (a) A map showing the normal pattern of trajectories between (x_1, y_1) and (x_T, y_T). (b) The same map showing an anomalous pattern of trajectories between (x_1, y_1) and (x_T, y_T). The deviant path signifies a strong intent by traffic to avoid the area where the embassy is located.

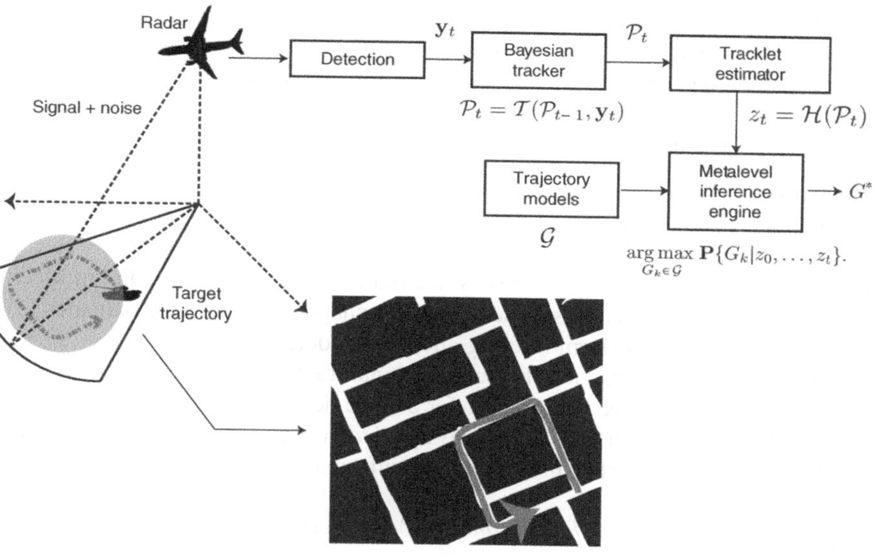

Figure 10.2 The architecture of the proposed metalevel inference for target intent.

destination (x_T, y_T) of a target starting from (x_1, y_1) is available. Knowledge about the intended destination can be obtained, for example, from flight plans or shipping registries for airplanes and ships, respectively. A trajectory model could provide predictive information about the path of the target's trajectory to enhance the accuracy of a base-level tracker. A pictorial representation of this example is shown in Figure 10.2. Further details are provided in Section 10.2. Other examples include determining if a target is circling a restricted area (perimeter surveillance), or alternatively if a vessel is loitering near the coast (for a possible smuggling attempt).

Example 3. Move-stop-move targets: As a third example, consider a surveillance problem where it is of interest to detect targets that move, stop and then move. For example, with m denoting a target moving at a specific epoch, and s denoting a stopped target, the sequence $m, \dots, m, s \dots, s, m, \dots, m$ could indicate a checkpoint where the target stops. A sequence of checkpoints can be modeled similarly. An anomalous target might have a different pattern of move and stops indicating stopping off at regions that are not checkpoints, possibly to break into a secure zone. So detecting anomalous behavior is equivalent to detecting a change in the underlying model. Other examples include determining if a target is circling a restricted area (perimeter surveillance), or alternatively if a vessel is loitering near the coast (for a possible smuggling attempt).

In the examples above, the human operator is primarily interested in determining specific patterns in target trajectories from estimated tracks. These patterns can then be used to infer the possible intent of the target [1]. Examples of such specific patterns include loops, arcs, circles, rectangles, closed trajectories, move-stop-move, and combination of these. Such trajectories exhibit complex spatial dependencies.

10.1.2 SCFGs and Reciprocal Markov Chains

The main goal is to devise SCFG and reciprocal Markov chain models and associated polynomial time Bayesian state estimation algorithms to extract spatial patterns from tracklets estimated by the conventional target tracker. In other words, we develop models and filtering algorithms to assist the human operator in determining specific target patterns. The algorithms presented in this chapter use the track estimates from an existing tracker to perform filtering and change detection. In this sense, they are at a higher layer of abstraction than conventional tracking and are fully compatible with existing trackers, see Figure 10.2 for a more detailed schematic. The methods developed in this chapter can be viewed as *middleware forming the human-sensor interface*. We are interested in interpreting information from a tracker to assist a human operator.

Why Stochastic Context-Free Grammars?: A target's trajectory can be viewed as composed of a premeditated set of subtrajectories to achieve a certain goal. However, premeditation implies noncausal and long-range time correlation between parts of the trajectory. Such complex interdependencies cannot be captured by Gauss–Markov state-space models that assume that the current part of the trajectory is only dependent on the immediate past. As will be shown in this chapter, SCFGs provide a parsimonious modeling framework for modeling complex spatial trajectories compared to Markov chains.

As depicted in Figure 10.3, in formal language theory, grammars can be classified into four different types depending on the forms of their production rules [2]. Stochastic regular grammars or finite state automata are equivalent to HMMs. SCFGs are a significant generalization of regular grammars. Of the four grammars in Figure 10.3, only stochastic regular and SCFGs have polynomial complexity estimation algorithms and are therefore of practical use in radar tracking applications. It is well known in formal language theory that SCFGs are more general than HMMs

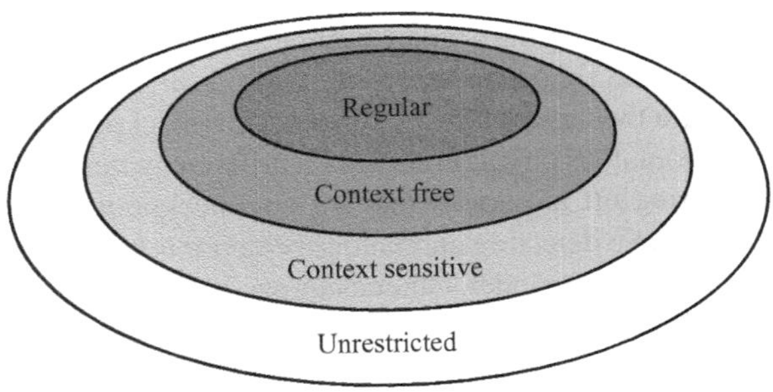

Figure 10.3 The Chomsky hierarchy of formal languages. This chapter deals with signal processing at the stochastic context-free grammar level that includes Hidden Markov Models (regular grammars) as a special case.

(stochastic finite automata) and can capture long-range dependencies and recursively embedded structures in patterns.

Modeling anomalous trajectories with SCFGs has several potential advantages:

(i) *Intuitive models* SCFGs have a compact formal representation in terms of production rules that allow human intuition to be easily codified into high-level rules, see [3] where the complex dynamics of a multifunction radar were modeled using SCFGs. It allows for modeling anomalous trajectory patterns. This, in turn, permits the design of high-level Bayesian signal processing algorithms to detect and classify anomalous trajectories. The ability for the designer to encode domain knowledge into a knowledge base is important because the lack of sufficient field data is a limiting factor in training classification systems.

(ii) *Ability to model complex trajectories* The recursive embedding structure of the possible anomalous trajectory patterns is more naturally modeled in SCFG. As mentioned earlier, the Markovian type model has dependency that has fixed length, and the growing state space is difficult to handle since the maximum range dependency must be considered. SCFGs can model arc, rectangles, closed trajectories and other anomalous trajectories like move-stop-move. A move-stop-move trajectory is a tactic used by targets to evade radar detection. Further details are provided in Section 10.3.2.

(iii) *Predictive capacity* SCFGs are more efficient in modeling hidden branching processes when compared to stochastic regular grammars or hidden Markov models with the same number of parameters. The predictive power of a SCFG measured in terms of entropy is greater than that of the stochastic regular grammar [4]. An SCFG is equivalent to a multitype Galton–Watson branching process with finite number of rewrite rules, and its entropy calculation is discussed in [5].

Similar advantages hold for reciprocal processes when modeling destination-aware targets. Reciprocal processes can be viewed as one-dimensional Markov random fields (MRFs) and take into account the joint dependence of target origin and destination. Bayesian signal processing algorithms for reciprocal Markov chains observed in noise have polynomial complexity [6].

10.1.3 Literature Survey

Trajectory modeling for intent inference is mainly approached in two ways: (a) anomaly detection [7–9] or (b) model-based inference [10, 11]. In the former, a specific trajectory is not identified. Rather, all possible trajectories of a target are categorized as either normal or anomalous. For example, in [8], a support-vector machine approach is taken to classify aberrant trajectories from normal trajectories. A key feature of model-based inference is to obtain a semantic interpretation of a complex pattern through the use of simpler subpatterns. For example, in [11], a dynamic Bayesian network is used to identify scenarios where a shopper either enters a retail store, leaves a store, or passes by the store. We consider tracklets as the simpler subpatterns comprising a trajectory (which is semantically equivalent to intent). The tracklets are explained in detail in Section 10.2.2.

Our work is related in spirit to the approach taken in [12, 13]. In [12], CFGs are used to identify two-person interactions such as hugs, hand-shakes, kicks, and punches. A stochastic context-free grammar approach is taken in [13] to recognize cheating actions in card-games at casinos. Our work departs from them significantly as we consider trajectory modeling in a tracking situation and not an action recognition system. The work presented in this chapter builds upon the work in [14] on target tracking using stochastic context-free grammars in radar tracking applications.

The use of SCFGs as a modeling tool for detecting anomalous trajectories requires the ability to compute model likelihoods. The inside–outside (IO) algorithm [4] is the conventional method used to compute the probability of an observed trajectory belonging to a given grammar model. However, the IO algorithm requires the stochastic grammar to be in a restrictive form. The Earley–Stolcke parser [15], on the other hand, is able to deal with arbitrarily structured grammars and is the algorithm used in this chapter for the Bayesian estimation of the model probabilities.

The problem of state-space realization and optimal smoothing for *Gaussian* reciprocal processes is addressed in detail in [16]. Only recently has the associated problems for a *discrete state* reciprocal process, which we shall call a reciprocal chain (RP) been addressed. The optimal smoother for a partially observed, or hidden RP (HRP), derived using a Bayesian approach, was presented in [6]. In [17], reciprocal processes are used for destination-aware tracking.

10.1.4 Main Results

This chapter uses a radar tracking framework to develop the main ideas. However, the notion of tracklets and the associated modeling of trajectories using SCFGs is scalable toward other sensing modalities like multicamera surveillance systems, geopositioning-based trackers and cellular basestation triangulation. The main results are as follows:

1. The detection of anomalous trajectories is formulated as a classification problem in Section 10.2. The SCFG models use the output from base-level tracking algorithms to either provide feedback to enhance the tracker or to perform higher level inference recognizing anomalous trajectories.
2. SCFGs are used as a novel modeling framework for anomalous trajectories like arcs, rectangles, closed paths, and so on. They are also used to model move-stop-move behavior that is a common evasive tactic used by targets. Toward this end, a quantized representation of velocity directions are used as low-level features that even allow placing probabilistic constraints on the final destination of the target.
3. A brief description of reciprocal processes is given together with a procedure for modeling the three-point transitions from a digital road map.

4. The experimental simulations carried out demonstrate the discriminative power of SCFG models in detecting anomalous trajectories. Various scenarios are described that show the flexibility of SCFG models in detecting different anomalous trajectories.

10.2 ANOMALOUS TRAJECTORY CLASSIFICATION FRAMEWORK

In this section, a system-level description of the anomalous trajectory classification problem is presented. We first describe the radar tracking framework and provide a mathematical description of trajectory classification in Section 10.2.1. In Section 10.2.2, complete details on a typical radar tracking system is presented. We also describe how to compute the velocity tracklets.

10.2.1 Trajectory Classification in Radar Tracking

In this section, an overview of the trajectory classification problem as it pertains to a radar tracking scenario is described. A diagrammatic representation of the proposed system is shown in Figure 10.2. A radar is assumed to make measurements $\mathbf{y}_t$ related to the position and velocity of targets in a particular region of interest (ROI). These measurements $\mathbf{y}_t$ are utilized by a base-level tracker $\mathcal{T}$ to estimate the actual position and velocity of the targets. Such a setup is the conventional "tracker" module used in many tracking applications. We introduce an additional module called the tracklet estimator $\mathcal{H}$ that produces quantized position estimates and velocity directions $\hat{z}_t$ using the output Π_t of a base-level tracker. These, in turn, are used by a metalevel inference engine to determine target intent.

The base-level tracker is a nonlinear Bayesian filter (such as a particle filter) that can be represented as an operator $\mathcal{T}$ that uses radar measurements $\mathbf{y}_t$ to update a posterior distribution Π_t over the position and velocity of the target by

$$\Pi_t = \mathcal{T}(\Pi_{t-1}, \mathbf{y}_t). \tag{10.1}$$

The distribution in (10.1) is then used by the tracklet estimator to obtain quantized estimates of the target position and its velocity direction. The tracklet estimator can be represented as an operator $\mathcal{H}_i$ such that

$$\hat{z}_t = \mathcal{H}_i(\Pi_t), \tag{10.2}$$

where $i \in$ {position, velocity}.

The aim is to provide models for the process $\hat{z}_t$ that can be used to classify anomalous trajectories. The target trajectory is associated with an intent depending on the context of the environment. For example, a circling behavior might be indicative of a reconnaissance operation in the vicinity of a sensitive asset like a check-post. A boat that is loitering near the shore-line could also be indicative of a smuggling operation. In [14], a pincer movement is described that is a formation maneuver often practiced by armored vehicles. In most cases, the target trajectory can be discerned from some feature (shape, destination, etc.) of its trajectory. The trajectory of a target is defined as the time evolution of its path as measured by the state variables of position, velocity, and acceleration. A geometric shape like an arc can be used to model a U-turn (or doubling back behavior). Furthermore, rectangles can be used to model circling behavior around regular man-made structures. In addition, the intended destination or even the starting point of a target can be used to infer intent if the target is known to be a friend or a foe.

The defining features of the considered trajectories are either its shape or the destination of the target. These are further characterized through velocity or position tracklets. Each type of target intent is thus assumed to be generated by a particular model $G_k \in \mathcal{G}, k = \{1, \ldots, K\}$, where there are K different types of target intent under consideration. As a target moves in a region of interest, it generates tracklets $\hat{z}_t$. The anomalous trajectory classification task is then defined as finding the model G_k that has the highest probability of explaining the observed tracklet sequence $\hat{z}_0, \ldots, \hat{z}_t$,

$$G^* = \underset{G_k \in \mathcal{G}}{\operatorname{argmax}}\, \mathbf{P}\{G_k | \hat{z}_0, \ldots, \hat{z}_t\}. \tag{10.3}$$

10.2.2 Radar Tracking System Overview

As shown in Figure 10.2, a particular instance of a target trajectory is considered that goes from its starting point (x_1, y_1) to a destination (x_T, y_T). There may be other intermediate way-points identified as well. Such a trajectory is destination constrained because the starting, intermediate, and ending destinations are known. Moreover, the target trajectory is of a rectangular shape although it is not a closed rectangle. This imposes a shape constraint on the trajectory (assumed due to existing road constraints).

A typical radar tracking application involves certain assumptions on the dynamics of the target. The target dynamics are summarized using its kinematic state in the vector $\mathbf{s}_t$ and a description of how $\mathbf{s}_t$ evolves. The general form of the target dynamics is

$$\mathbf{s}_{t+1} = f_t(\mathbf{s}_t, \mathbf{w}_t), \tag{10.4}$$

where $\mathbf{s}_t = [x_t, y_t, \dot{x}_t, \dot{y}_t]^T$ is the state vector, $f_t(\cdot)$ is a possibly time-varying function and $\mathbf{w}_t$ is a noise process with a known distribution $\mathbf{P}\{\mathbf{w}_t\}$. The state variables x_t, y_t refer to the position of the target while $\dot{x}_t, \dot{y}_t$ refer to the velocity of the target in Cartesian coordinates.

A radar cannot measure the kinematic state of a target directly. It can, however, make measurements $\mathbf{y}_t$ related to the state $\mathbf{s}_t$ that are typically of the form

$$\mathbf{y}_t = h(\mathbf{s}_t) + \mathbf{v}_t, \tag{10.5}$$

where $\mathbf{y}_t = [r_t, \dot{r}_t, \theta_t]$ is the measurement vector, $h(\cdot)$ is a nonlinear measurement function and $\mathbf{v}_t$ is a noise process with a known distribution $\mathbf{P}\{\mathbf{v}_t\}$. The measurement variable r measures the radial distance of the target to the radar (called range), $\dot{r}$ is

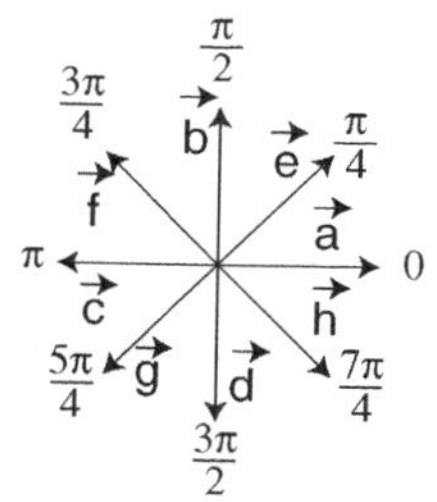

Figure 10.4 The eight angular velocity directions that are referred to as velocity tracklets. Each angle is labeled by a lowercase alphabet for convenience. The overhead $\vec{}$ refers to the fact that each lowercase alphabet denotes a unit directional vector.

the rate of change of range, and θ is the azimuth angle (angle between the position vector of the target and the x-axis).

A base-level tracker (like an extended Kalman filter) is then used to track the state of the target $\mathbf{s}_t$ using the radar measurements $\mathbf{y}_t$. The output of such a tracker consists of a posterior probability distribution $\Pi_t = \mathbf{P}\{\mathbf{s}_t|\mathbf{y}_t\}$. The state estimate can then be evaluated from the posterior distribution using a conditional expectation. The state estimates $\mathbf{E}\{\mathbf{s}_t|\mathbf{y}_t\}$ form the input to the tracklet estimator.

The tracklet estimator $\mathcal{H}$ is a quantization module that outputs either position or velocity tracklets. Tracklets are used as subunits, that comprise target trajectories. The SCFG models utilize velocity tracklets as subunits of the trajectory shape. The velocity tracklet estimator uses the posterior distribution estimates Π_t to compute conditional mean estimates of the target velocity $\dot{x}$, $\dot{y}$ in the horizontal and vertical directions. These velocity estimates are then used to find the direction of motion of the target. The possible directions of motion of the target are quantized into eight radial angular directions from the set $\mathcal{Q} = \{\vec{a} = 0, \vec{e} = \frac{\pi}{4}, \vec{b} = \frac{\pi}{2}, \vec{f} = \frac{3\pi}{4}, \vec{c} = \pi, \vec{g} = \frac{5\pi}{4}, \vec{d} = \frac{3\pi}{2}, \vec{h} = \frac{7\pi}{4}\}$. The radial directions are shown in Figure 10.4 and each is labeled for notational convenience with a lowercase alphabet and an additional $\vec{}$ to denote that is a unit directional vector. The velocity tracklet estimator thus outputs estimates of the "true" velocity vectors

$$z_t = \underset{q \in \mathcal{Q}}{\operatorname{argmin}} \left| \arctan\left(\frac{\dot{y}_t}{\dot{x}_t}\right) - q \right|. \tag{10.6}$$

Here z_t denotes the "true" velocity directions (generated by an appropriate SCFG model in $\mathcal{G}$) and the tracklet $\hat{z}_t$ denotes the (noisy) estimate of z_t.

10.3 TRAJECTORY MODELING AND INFERENCE USING STOCHASTIC CONTEXT-FREE GRAMMARS

In this section, we present SCFG models and associated signal processing algorithms for trajectory modeling and classification. SCFGs will be the main tool that we will use to model complex spatial trajectories of targets. The output of an SCFG is a string of terminal symbols. These terminal symbols are precisely the tracklets that we aim to model. Finally, the Earley–Stolcke parser is presented to perform statistical signal processing of the tracklets.

10.3.1 Review of Stochastic Context-Free Grammars

Since SCFGs are not widely used in radar signal processing, we begin with a short formal description of SCFGs A textbook treatment can be found in [18]. A context-free grammar G_{CFG} is a 4-tuple $(\mathcal{N}, \mathcal{V}, S, \mathcal{A})$, where $\mathcal{N}$ is a finite set of nonterminals $(N_i, i = 1, \ldots, |\mathcal{N}|)$, $\mathcal{V}$ is a finite set of terminals $(v_i, i = 1, \ldots, |\mathcal{V}|)$ such that $(\mathcal{N} \cap \mathcal{V} = \emptyset)$, $S \in \mathcal{N}$ is the chosen start symbol (initial nonterminal) and $\mathcal{A}$ is a finite set of production rules a_m of the form $(A \to \alpha)$, $A \in \mathcal{N}$, and $\alpha \in (\mathcal{N} \cup \mathcal{V})^+$. The set $(\mathcal{N} \cup \mathcal{V})^+$ denotes all finite length strings of symbols in $(\mathcal{N} \cup \mathcal{V})$, excluding strings of length 0 (the case where strings of length 0 is included is indicated by $(\mathcal{N} \cup \mathcal{V})^*$). The $\to$ symbol denotes a rewrite operation that replaces the nonterminal A with the string α. A stochastic context-free grammar is defined as a pair (G_{CFG}, p), where $p : \mathcal{A} \to [0, 1]$ is a probability function over the production rules $(A \to \alpha) \in \mathcal{A}$ such that $\forall A \in \mathcal{N}, \sum_{i=1}^{n_A} p(A \to \alpha_i) = 1$. The number of alternative production rules associated with A is denoted n_A.

A grammar is a generative model that produces an output sequence of terminal symbols. A symbol refers to an element of the set $(\mathcal{N} \cup \mathcal{V})$ that can be either a single nonterminal or a single terminal. A sequence of such symbols is called a string $\alpha \in (\mathcal{N} \cup \mathcal{V})^+$. When the string is complete and composed entirely of terminals (such that no further symbols can be concatenated or produced in the string), it is called a sentence. The grammar generation process begins with the start symbol $S \in \mathcal{N}$. A rule $(S \to \alpha_1)$ is then chosen from the subset of rules in $\mathcal{A}$ whose left-hand side is the start symbol S. This symbol S is then replaced by the corresponding string α_1 on the right-hand side of the chosen rule. If α_1 contains only terminal symbols, then the grammar generation process terminates. In this case, the sentence generated by the grammar G_{CFG} is the string α_1. However, if α_1 contains nonterminals, then each nonterminal is replaced by a new string according to the choice of a production rule in $\mathcal{A}$ until all nonterminals are consumed by the generation process. This can be represented as a parse tree as shown in Figure 10.5. A parse tree is a graphical representation of the string generation process in which nonterminal symbols branch out through the choice of production rules until only terminal symbols are obtained. The nonterminal symbols form the root nodes of the parse tree while the terminal symbols form the leaves of the parse tree. The resulting final string of terminals is considered to be the sentence generated by the grammar. The set of all such sentences that can be generated by a particular grammar G_{CFG} is termed a language L_{CFG}.

10.3.2 SCFG Models for Anomalous Trajectories

In this section, we model various trajectories of interest using stochastic context-free grammar models. We make the assumption that certain kinds of trajectory shapes

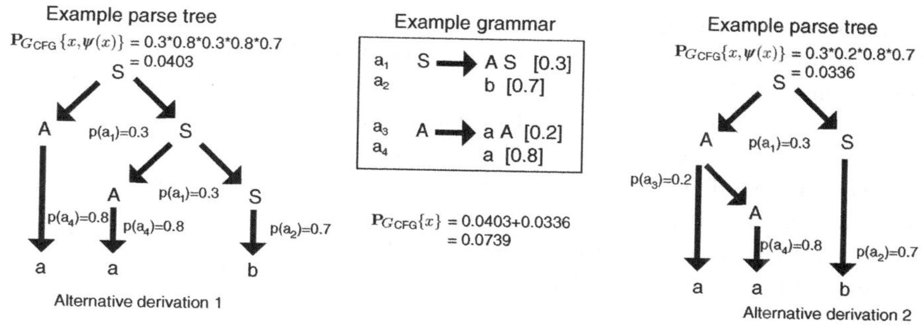

Figure 10.5 A example grammar with context-free production rules and representative alternative parse trees of a generation process for the string *aab*. The quantities in [ ] brackets denote the probability of choosing that production rule. The individual probabilities of each parse tree is shown together with the total probability of the string.

are correlated with target intent. This can be seen, for example, in the case of a target making a U-turn. The associated *U* shape (henceforth called an *arc*) of the trajectory can be approximated as a trapezoidal shape as shown in Figure 10.6a. An SCFG can be used to model all trajectories of such a characteristic shape (with

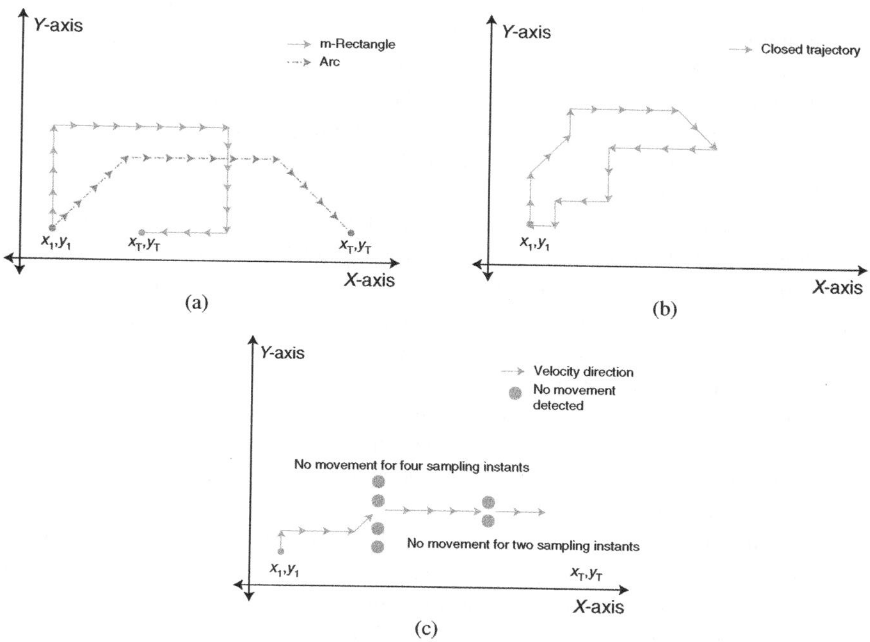

Figure 10.6 (a) A rectangular trajectory and an arc trajectory are shown with different destinations (b) A closed trajectory is shown whose beginning and ending points coincide. (c) A move-stop-move trajectory is shown that is a tactic used by targets to evade radar detection.

implicit scale-invariance) and can thus be used to recognize U-shaped trajectories. While U-turns are common trajectories to observe in civilian traffic, they generally imply anomalous behavior in military targets. Another example of target intent involving arc-shaped trajectories is a tank formation maneuver called a *pincer* operation [14]. The second shape of interest is that of rectangles. Most man-made structures like roads and buildings, constrain the path of targets to have a rectangular shape. We show that an SCFG model can be constructed for such rectangular trajectories to which we can attach intent depending on the surveillance environment. Moreover, we can also construct models for closed trajectories and the move-stop-move trajectory.

For all the SCFG models considered in this chapter, each type of trajectory has an associated grammar model G. All the grammars have a common set of terminals $\mathcal{V} = \mathcal{Q}$ and a common start symbol S. They may have different rule spaces $\mathcal{A}$ and/or nonterminal spaces $\mathcal{N}$. While modeling trajectory shapes using grammar models, we will focus on the structure of the production rules. The nonterminal space is implicitly included when writing the production rules.

Line trajectory A target traveling in a straight path creates linear trajectories with local Markov dependency, and it is characterized by rules of the form $S \rightarrow \vec{a}S \mid \vec{a}$ with $\vec{a}$ representing the target's direction of motion. An example string of a target traveling in a straight horizontal line for four sampling instants is "*aaaa*." The production rules of a line grammar generates a language that is equivalent to that of a hidden Markov model formulation [14] (or equivalently a regular grammar). A regular grammar is constrained to have only one nonterminal on either side of a production rule. The linear shape can be represented as the language $\mathcal{L}_{\text{line}} = \{x \in \vec{a}^n\}$. This notation implies that all strings generated by a line grammar G_{line} will have the form $\vec{a}^n$. The notation $\vec{a}^n$ implies that the terminal symbol $\vec{a}$ appears n times consecutively in a sequence. The two other geometric shapes of interest are arcs and rectangles. These shapes possess long-range and self-embedding dependencies that require production rules that regular grammars (and hence Markov models) cannot represent.

Arc Trajectory An arc-shaped trajectory can be expressed as a language $\mathcal{L}_{\text{arc}} = \{x \in \vec{a}^n \vec{b}^+ \vec{c}^n\}$, where there is an equal number of matching upward $\vec{a}$ and downward $\vec{c}$ tracklets and an arbitrary number of forward tracklets $\vec{b}$. The $+$ symbol denotes an arbitrary number of $\vec{b}$ symbols. The symbol x represents any arbitrary string belonging to the language. Such a language can be generated by the grammar shown in Figure 10.7a. The grammar can be constructed based on techniques reviewed in [19].

Rectangular Trajectory The *m-rectangle* language (with associated grammar shown in Figure 10.7b) is $\mathcal{L}_{\text{m-rectangle}} = \{\vec{a}^m \vec{b}^+ \vec{c}^m \vec{d}^+\}$ and it can model any trajectory comprising of four sides at right angles (not necessarily a closed curve) with at least two opposite sides being of equal length. Why do we consider m-rectangles instead of rectangles? This is because the language comprising of only rectangles is not context free. The language comprising of only rectangles can be generated by a more specific class of grammars called context-sensitive

Arc grammar

$\mathcal{N} = \{S, X, A, B, C\}$

$\mathcal{V} = \{\vec{a}, \vec{b}, \vec{c}\}$

$$\begin{aligned}
a_1 &: S \rightarrow AXC\ [p_1] \\
a_2 &: X \rightarrow AXC\ [p_2] \\
a_3 &: X \rightarrow ABC\ [p_3] \\
a_4 &: X \rightarrow B\ [p_4] \\
a_5 &: A \rightarrow \vec{a}\ [p_5] \\
a_6 &: B \rightarrow \vec{b}B\ [p_6] \\
a_7 &: B \rightarrow \vec{b}\ [p_7] \\
a_8 &: C \rightarrow \vec{c}\ [p_8]
\end{aligned}$$

(a)

m-Rectangle grammar

$\mathcal{N} = \{S, X, A, B, C, D\}$

$\mathcal{V} = \{\vec{a}, \vec{b}, \vec{c}, \vec{d}\}$

$$\begin{aligned}
a_1 &: S \rightarrow AXCD\ [p_1] \\
a_2 &: X \rightarrow AXC\ [p_2] \\
a_3 &: X \rightarrow ABC\ [p_3] \\
a_4 &: X \rightarrow B\ [p_4] \\
a_5 &: A \rightarrow \vec{a}\ [p_5] \\
a_6 &: B \rightarrow \vec{b}B\ [p_6] \\
a_7 &: B \rightarrow \vec{b}\ [p_7] \\
a_8 &: C \rightarrow \vec{c}\ [p_8] \\
a_9 &: D \rightarrow \vec{d}D\ [p_9] \\
a_{10} &: D \rightarrow \vec{d}\ [p_{10}]
\end{aligned}$$

(b)

Move-stop-move grammar

$\mathcal{N} = \{S, M, X\}$

$\mathcal{V} = \{m, x\}$

$$\begin{aligned}
a_1 &: S \rightarrow MXM\ [p_1] \\
a_2 &: X \rightarrow xX\ [p_2] \\
a_3 &: X \rightarrow x\ [p_3] \\
a_4 &: M \rightarrow mM\ [p_4] \\
a_5 &: M \rightarrow m\ [p_5]
\end{aligned}$$

(c)

Figure 10.7 An arc grammar in (a) and an m-rectangle grammar in (b). A move-stop-move grammar is shown in (c).

grammars. As a result, algorithms of polynomial complexity for recognizing such trajectories cannot be constructed. However, we can construct heuristic rules by fixing the number of $\vec{b}$'s and $\vec{d}$'s in the grammar to generate rectangles and squares. More importantly, we can quantize the tracklets more coarsely to generate closed trajectories.

Closed trajectory Consider the closed trajectory in Figure 10.6b. We can resolve each directional vector onto the unit directions represented by $\vec{a}$ and $\vec{b}$. A closed figure then comprises of an equal number of $\vec{a}$ (up) and $\vec{c}$ (down) movements together with an equal number of $\vec{b}$ (left) and $\vec{d}$ (right) movements. Such a trajectory also comprises an arc-like language where an equal number of opposing movements is represented by the language $\mathcal{L}_{\text{equal}} = \{\vec{k}^n \vec{l}^n\}$, where k and l refer to opposite movements from the set $\mathcal{Q}$.

Move-stop-move trajectory A move-stop-move trajectory results from another coarse representation of the tracklets that allows us to model a common evasion tactic used by targets. If the target stops moving (or its velocity drops below a threshold), then the tracker is unable to track it. As a result, targets seeking to evade a radar often intersperse periods of movement with periods of no movements. The sporadic stopping between two periods with movement can be modeling as a self-embedding grammar of the form shown in Figure 10.7c. A move-stop-move trajectory in which the target stops for four sampling instants would take the form "*mmmmxxxxmmm*," where each m refers to a movement in any one of the directions in $\mathcal{Q}$ and x refers to a stop.

As seen in Figure 10.7, the rules needed to generate shapes such as arcs and m-rectangles have a syntax that is more complex than a regular grammar (because they have more than one nonterminal on the right-hand side). These grammars are self-embedding context-free grammars that cannot be represented by a Markov chain. A context-free grammar is self-embedding if there exists a nonterminal A such that $A \stackrel{*}{\Rightarrow} \eta A \beta$ with $\eta, \beta \in (\mathcal{N} \cup \mathcal{Q})^+$. The relation $\stackrel{*}{\Rightarrow}$ denotes a transitive and reflexive closure of the production relation $\Rightarrow$. In simpler terms, such a notation is used to signify that even though there might not exist a production rule of the exact form $A \Rightarrow \eta A \beta$, existing production rules can be used on each other to come up with such a production rule. For the arc production rules presented, the self-embedding property can be seen by the second production rule $X \rightarrow A\ X\ C$ where X can repeatedly call itself to lengthen the output string.

10.3.3 Bayesian Signal Processing of SCFG Models

This section deals with the solution of the following sequential classification problem. Given an observation sequence of the target's estimated velocity directions $\hat{z} = \hat{z}_1, \ldots, \hat{z}_t$, can we classify the target's trajectory? The set of permissible grammar models is given by $\mathcal{G}$, which contains all the anomalous trajectories described in Section 10.3.2. Mathematically, we seek the grammar posterior probability

$$G^\star = \underset{G_k \in \mathcal{G}}{\operatorname{argmax}}\, \mathbf{P}\{G_k | \hat{z}_1, \ldots, \hat{z}_t\}, \tag{10.7}$$

where $G^\star$ is the grammar model (or corresponding anomalous trajectory) with the maximum probability given the observation sequence. The computation of the likelihoods using partial sentences $\hat{z}_1, \ldots, \hat{z}_t$ rather than a complete trajectory $\hat{z}_1, \ldots, \hat{z}_T$ is a nontrivial exercise and requires the computation of *prefix* probabilities. The prefix probability $\mathbf{P}\{\hat{z}_1, \ldots, \hat{z}_t; G_k\}$ of the string $\hat{z}_1, \ldots, \hat{z}_t$ is the probability that grammar G_k derives the string $\hat{z}_1, \ldots, \hat{z}_t, y$ that has $\hat{z}_1, \ldots, \hat{z}_t$ as its prefix and $y \in (\mathcal{N} \cup \mathcal{V})^*$ is an arbitrary suffix. The prefix probability is defined as

$$\mathbf{P}\{\hat{z}_1, \ldots, \hat{z}_t; G_k\} = \sum_{y \in (\mathcal{N} \cup \mathcal{V})^*} \mathbf{P}\{\hat{z}_1, \ldots, \hat{z}_t, y; G_k\}. \tag{10.8}$$

The computation of the prefix probabilities is carried out using the Earley–Stolcke parser that is described next.

Bayesian methods of grammatical inference The inference of SCFG models is predominantly carried out using the inside–outside algorithm [4]. The inside–outside algorithm is very similar to the forward–backward algorithm used in the inference of hidden Markov models [20]. In the case of SCFGs, the inside–outside algorithm restricts the grammar to have a particular form called the Chomsky normal form (CNF). However, any grammar can be transformed into the Chomsky normal form [21]. We begin by defining the inside probability for a stochastic grammar model. The inside probability is denoted by $\gamma(i, j, n) = \mathbf{P}\{n \overset{*}{\Rightarrow} \hat{z}_i, \ldots, \hat{z}_j\}$. This is described as the probability that nonterminal $n \in \mathcal{N}$ derives the observations (terminals) from $\hat{z}_i, \ldots, \hat{z}_j$. The inside–outside algorithm computes this probability recursively by first computing the inside probability of single observations $\gamma(i, i, n)\ \forall n, i$ and then incrementally computing the inside probability for all sequences of length 2 and so on. The recovery of the most likely parse tree $\psi(x)$ is computed using the probabilistic CYK parser [22] that is a dynamic programming algorithm similar to the Viterbi algorithm used for maximum likelihood sequence estimation in HMMs.

The Earley–Stolcke parser The Earley–Stolcke parser scans an input string $\hat{z} = \hat{z}_1, \ldots, \hat{z}_t, \ldots, \hat{z}_T$ from left to right and is able to compute the probability of the string $\mathbf{P}\{\hat{z}|G_k\}$ given the parameters of the SCFG. As each symbol $\hat{z}_t$ is scanned, a set of states u_t is created that represents the condition of the inference process at that point in the scan. Each state in u_t represents (1) a production rule $a_m \in \mathcal{A}$ such that we are currently scanning a portion of the input string that is derived from its right-hand side, (2) a point (marker) in that production rule that shows how much of that rule's right side we have recognized so far, and (3) a pointer back to the position in the input string at which we began to look for that instance of the production rule. Each state is an incomplete portion of the parse tree that generated the input string x. These states are referred to as the control structure used by the Earley–Stolcke parser to store the incomplete parse trees and are represented as ${}_i^jX \rightarrow \lambda \cdot Y\mu\ [\alpha, \gamma]$.

The upper case letters X and Y are nonterminals, λ and μ are substrings of nonterminals and terminals, "." is the marker that specifies the end position j for the partially parsed input, i is the starting index of the substring that is generated by the nonterminal X. For the purposes of dealing with the start symbol, the Earley–Stolcke uses a dummy state ${}_0^0S' \rightarrow \cdot S\ [1, 1]$ that is the initial state of the Earley–Stolcke parser.

Earley-Stolcke operations In general, we operate on a state set u_t as follows: the states in the set are processed in order, by performing one of three operations on each depending on the form of the state. These operations may add more states to u_t and may also put states in a new state set u_{t+1}. Whenever an operation attempts to add a new state, it is linked to an existing state. The *predictor* operation is applicable to a state when there is a nonterminal to the right of the dot. It causes the addition of one new state to u_t for each alternative production rule of that nonterminal. The dot is placed at the beginning of the production rule in each new state. The pointer is set to t, since the state was created in u_t. Thus, the predictor adds to u_t all productions that might generate substrings beginning at x_{t+1}. More formally, for a state ${}^{t}_{s}X \rightarrow \lambda \cdot Y\mu$ in the state set u_t, the predictor adds a new state ${}^{t}_{t}Y \rightarrow \cdot\nu$ for each of the alternative production rules $(Y \rightarrow \nu) \in \mathcal{A}$. A link is thus created between these states. The state ${}^{t}_{t}Y \rightarrow \cdot\nu$ is called a *predicted* state.

The *scanning* operation, on the other hand, is applicable just in the case when there is a terminal to the right of the dot. The scanner compares that symbol with $\hat{z}_{t+1}$, and if they match, it adds the state to $u(t+1)$, with the dot moved over one symbol in the state to indicate that terminal symbol has been scanned. If ${}^{t}_{s}X \rightarrow \lambda \cdot a\mu$ exists and $\hat{z}_{t+1} = a$, the scanning operation adds a new state ${}^{t+1}_{s}X \rightarrow \lambda a \cdot \mu$ to state set $u(t+1)$ that is called a *scanned* state. A link is also created between these states.

The third operation, the *completer*, is applicable to a state if its dot is at the end $({}^{t}_{s}X \rightarrow \lambda Y\mu.)$ of its production. Such a state is called a "complete" state. For every complete state, the completer goes back to the state set s indicated by the pointer in the complete state, and adds all states from $u(s)$ to $u(t)$ that have X (the nonterminal corresponding to that production) to the right of the dot. It moves the dot over X in these states. Intuitively, $u(t)$ is the state set we were in when we went looking for that X. We have now found it, so we go back to all the states in $u(s)$ that caused us to look for a X, and we move the dot over the X in these states to show that it has been successfully scanned. A completion operation adds a new state ${}^{t}_{s}X \rightarrow \lambda Y \cdot \mu$ (called a completed state) using ${}^{r}_{s}X \rightarrow \lambda \cdot Y\mu$ and ${}^{t}_{r}Y \rightarrow \nu\cdot$. A link pointing from ${}^{t}_{r}Y \rightarrow \nu\cdot$ to ${}^{t}_{s}X \rightarrow \lambda Y \cdot \mu$ is also created. In such a manner, the Earley–Stolcke parser continues until all the observation symbols have been scanned. If the final state set u_T contains the state ${}^{0}_{0}S' \rightarrow S\cdot$, then the algorithm terminates successfully. It represents a successful parse of the sentence $\hat{z}_1, \ldots, \hat{z}_T$.

Earley–Stolcke probabilities We mentioned earlier that each state is associated with a forward probability $\alpha({}^{j}_{i}X \rightarrow \lambda \cdot \mu)$ that is the sum of the probabilities of all paths of length i which end in the state ${}^{j}_{i}X \rightarrow \lambda \cdot \mu$ and generate observations $\hat{z}_1, \ldots, \hat{z}_i$. The inner probability $\gamma({}^{j}_{i}X \rightarrow \lambda \cdot \mu)$ of a state is defined as the sum of the probability of all paths of length $k - i$ that pass through the state ${}^{j}_{i}X \rightarrow \lambda \cdot \mu$ and hence derive the observations $\hat{z}_i, \ldots, \hat{z}_{k-1}$. A path is simply a sequence of Earley states linked through the operations of prediction, scanning, and completion as mentioned above. The length of a path is defined as the number of scanned states in it. Since a state can be scanned only when the terminal to the right of the dot matches the input symbol, there is a one-to-one correspondence between scanned states and length of the input. In Figure 10.8, we show the concept on an Earley path and the manner in which they can split and merge via the operations of prediction, scanning, and completion. The updates of the forward probability α and inside probability γ for each of the state operations is summarized in Figure 10.9.

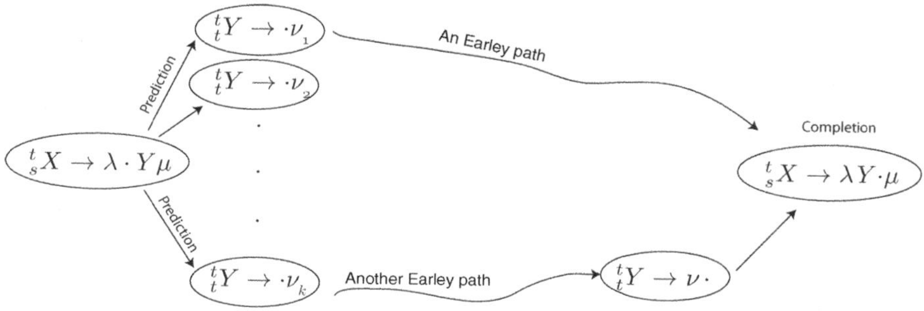

Figure 10.8 This figure shows the links created due to the prediction, scanning, and completion operations between states. The prediction creates visible splits in the path while scanning and completion merge different paths. Some paths created by the prediction operation do not lead to a viable string and hence remain incomplete.

Prediction

$$ {}_s^tX \to \lambda \cdot Y\mu \; [\alpha, \gamma] \Rightarrow {}_t^tY \to \cdot\nu \; [\alpha', \gamma'] $$

$\alpha' += \alpha\mathbf{P}\{Y \to \nu\}$ (a sum over all duplicate states),

$$ \gamma' = \mathbf{P}\{Y \to \nu\}. $$

(a)

Scanning

$$ {}_s^tX \to \lambda \cdot a\mu \; [\alpha, \gamma] \Rightarrow {}_s^{t+1}X \to \lambda a \cdot \mu \; [\alpha', \gamma'] $$

$$ \alpha' = \alpha\mathbf{P}\{a\}\,, $$

$$ \gamma' = \alpha\mathbf{P}\{a\}. $$

(b)

Completion

$$ \left.\begin{array}{l} {}_r^tY \to \nu \cdot \; [\alpha'', \gamma''] \\ {}_s^rX \to \lambda \cdot Y\mu \; [\alpha, \gamma] \end{array}\right\} \Rightarrow {}_s^tX \to \lambda Y \cdot \mu \; [\alpha', \gamma'] $$

$\alpha' += \alpha\gamma''$ (a sum over all duplicate states),

$\gamma' += \gamma\gamma''$ (a sum over all duplicate states).

(c)

Figure 10.9 The updates for the forward probability α and the inside probability γ for each of the Earley–Stolcke operations.

10.4 TRAJECTORY MODELING AND INFERENCE USING RECIPROCAL PROCESSES (RP)

Here we consider reciprocal processes for modeling target intent. They constitute and alternative model to SCFGs for metalevel tracking. Our treatment here is brief, we refer the reader to [6] for details.

How to model a digital road map or metalevel tracking? Consider the pattern-of-life problem outlined in the introduction. In the context of traffic flow, a road map such as that in Figure 10.1 can be represented as a directed graph $(\mathcal{V}, \mathcal{E})$ where the set of vertices $\mathcal{V} = \{v_1, v_2, \ldots, v_M\}$ denote road intersections and set of edges $\mathcal{E}$ denote streets between vertices. At road intersections, it is natural to model traffic by three-point transition probabilities $Q(i, j, l) = P(X_k = j | X_{k-1} = i, X_{k+1} = l)$ for edges e_{ij} and $e_{jl} \in \mathcal{E}$. Based on vehicular traffic data accumulated over time, the maximum likelihood estimate of the three-point transition probabilities can be estimated as $\hat{Q}(i, j, l) = N(e_{ij} \cap e_{jl})/N(i, l)$ where $N(e_{ij} \cap e_{jl})$ denotes the number of vehicles that travel on edge e_{ij} and edge e_{jl}, $N(i, l)$ denotes the number of vehicles that travel from vertex i to l.

In the pattern-of-life problem we are interested in vehicles that move between two specific vertices, say vertex 1 and M. (In Figure 10.1, these vertices are denoted by Cartesian coordinates (x_1, y_1) and (x_T, y_T), respectively). There may be other intermediate way-points identified as well. Such a trajectory is destination constrained because the starting, intermediate, and ending destinations are known. Random paths on the graph $(\mathcal{V}, \mathcal{E})$ between two vertices are modeled conveniently as a Markovian bridge (which is a special case of a reciprocal process). A change in the pattern-of-life corresponds to a change in the three-point transition probabilities. Detectors can be constructed for change detection of reciprocal processes observed in noise.

Reciprocal Markov Model Consider a N-state random process z_k indexed on the set $\{0, 1, \ldots, T\}$ for some integer T. The process z_k is said to be *reciprocal* [23] if

$$P(z_k | z_s, s \neq k) = P(z_k | z_{k-1}, X_{k+1}) , \tag{10.9}$$

for each $k = 1, \ldots, T-1$. Thus, z_k is conditionally independent of $z_0, \ldots, z_{k-2}, z_{k+2}, \ldots, z_T$ given its neighbors z_{k-1} and z_{k+1}. The RP model is specified by the set of three-point transition functions (10.9) together with a given joint distribution on the end points $P(z_0, z_T)$. In this chapter, we consider homogeneous three-point transitions, denoted by

$$Q(i, j, \ell) = P(z_k = j | z_{k-1} = i, z_{k+1} = \ell)$$

A reciprocal process is a one-dimensional version of a MRF, although it is not necessarily a Markov process. However, any Markov process is reciprocal ([23], Lemma 1.2).

It has been shown ([23], Lemma 1.4) that fixing the end point of a RP generates a Markov bridge (MB). A MB is a Markov process with the end point z_T fixed to a specified value. So for the finite (N) state case, a RP can be regarded as N Markov bridges, one corresponding to each of the possible final states taken by z_T. We will derive the models for each of these Markov bridges from the RP model utilizing the following important property of a reciprocal process ([23], property (a3), p. 80). Let $0 \leq s < t < u < v \leq T$, then

$$P(z_u | z_s, z_v)\, P(z_t | z_s, z_u) = P(z_t | z_s, z_v) P(z_u | z_t, z_v) . \tag{10.10}$$

Let $s = k - 1$, $u = k + 1$, and $v = T$, then, formally, for $k = 1, \ldots, T - 1$,

$$P(z_k | z_{k-1}, z_T) = \frac{P(z_k | z_{k-1}, z_{k+1})}{P(z_{k+1} | z_k, z_T)} P(z_{k+1} | z_{k-1}, z_T) . \qquad (10.11)$$

Then (10.11) yields a backward recursion that fully specifies the set of N MB transitions for $k = T - 2, T - 3, \ldots, 0$

$$B_t^m(i, j) = P(z_{k+1} = j | z_k = i, z_T = m) = \frac{Q(i, j, \ell)}{B_{k+1}^m(j, \ell)} \times \left(\sum_{m=1}^{N} \frac{Q(i, m, \ell)}{B_{k+1}^m(m, \ell)} \right)^{-1} , \qquad (10.12)$$

the last term on the right-hand side being the normalization constant. Initialization is with $B_{T-1}^m(i, j) = 1$ if $j = m$ and zero otherwise. Observe that the quantity on the RHS of (10.12) is independent of the index ℓ. As mentioned in [23] (p. 80), this important property is the RP analog to the Chapman–Kolmogorov equation for Markov processes. Now since the terms $B_{k+1}^m(j, .)$ form a probability distribution (i.e., they are nonnegative and sum to unity) for each i, m, and t, there is at least one index ℓ for which $B_{k+1}^m(j, \ell)$ is nonzero. So this index may be selected on the RHS of (10.12), and thus (10.12) is well defined. From a numerical perspective, it may be appropriate to choose a value of ℓ (for each j, m, k) that maximizes the value of $B_{k+1}^m(j, \ell)$. Determining the set of MBs from the specified three-point transitions requires complexity of $O(N^3 T)$.

The Markov bridge with final state k is given an initial probability distribution π_i^m given by the conditional distribution

$$\pi_i^m = P(z_0 = i | z_T = m) = \frac{\Pi_{i,m}}{\sum_{j=1}^{N} \Pi_{j,m}} , \qquad (10.13)$$

where $\Pi_{i,m} = P(z_0 = i, z_T = m)$ is the specified RP end points distribution. Thus, we have demonstrated that any RP may be uniquely specified by a finite set of Markov bridges with probability transition matrices given by (10.12) and initial distributions (10.13).

Remark: As described in the pattern-of-life and move-stop-move examples in the introduction, z_k denotes the underlying true tracklets that can be modeled as a Brownian bridge or more generally a reciprocal process. The three-point transitions for a road network modeled by the directed graph $(\mathcal{V}, \mathcal{E})$ are estimated as $\hat{Q}_{ijl}$ as outlined above.

10.5 EXAMPLE 1: METALEVEL TRACKING FOR GMTI RADAR

So far we have given two classes of models, SCFGs and reciprocal processes to model metalevel trajectories of targets. Here we present a brief outline of a detailed example considered in [14] on metalevel GMTI tracking and intent inference. Details of the underlying space–time adaptive processing (STAP) and IMM tracker are given in [14].

Let the set of terminals $\mathcal{V} = \{\vec{a}, \vec{b}, \vec{c}, \vec{d}, \vec{e}, \vec{f}, \vec{g}, \vec{h}\}$ denote eight possible acceleration directions of travel of the moving target depicted in Figure 10.4. At each time k, $z_k \in \mathcal{V}$ denotes *mode* of the target. The target dynamics are modeled as

$$\mathbf{s}_k = F\mathbf{s}_{k-1} + Gw_{k-1}(z_k). \tag{10.14}$$

$\mathbf{s}_k = (x_k, y_k, \dot{x}_k, \dot{y}_k)'$ denotes the ground moving target's position and velocity in Cartesian coordinates, and assuming constant velocity model, the transition matrix model, and the noise gain are, respectively,

$$F = \begin{pmatrix} 1 & 0 & T & 0 \\ 0 & 1 & 0 & T \\ 0 & 0 & 1 & 0 \\ 0 & 0 & 0 & 1 \end{pmatrix}, \quad G = \begin{pmatrix} T^2/2 & 0 \\ 0 & T^2/2 \\ T & 0 \\ 0 & T \end{pmatrix}.$$

The process noise w_k is a white Gaussian process with the covariance matrix

$$\mathcal{Q} = \rho_{z_k} \cdot \begin{pmatrix} \sigma_a^2 & 0 \\ 0 & \sigma_o^2 \end{pmatrix} \cdot \rho'_{z_k}, \quad \text{with } \rho_{z_k} = \begin{pmatrix} \sin z_k & \cos z_k \\ -\cos z_k & \sin z_k \end{pmatrix},$$

where σ_a^2 is the uncertainty along the direction indicated by z_k and σ_o^2 is orthogonal to it. Thus, the modes z_k modulate the process noise v and cause it to switch between different variance values.

The observation model describing the output of the GMTI STAP measurements is

$$\mathbf{y}_k = h(\mathbf{s}_k) + v_k$$

$$h(\mathbf{s}_k) = \begin{bmatrix} r_k \\ \dot{r}_k \\ \theta_k \end{bmatrix} = \begin{bmatrix} \sqrt{\bar{x}_k^2 + \bar{y}_k^2 + \bar{p}_k^2} \\ \frac{\bar{x}_k \dot{\bar{x}}_k + \bar{y}_k \dot{\bar{y}}_k}{\sqrt{\bar{x}_k^2 + \bar{y}_k^2 + \bar{p}_k^2}} \\ \tan^{-1}(\bar{x}_k, \bar{y}_k) \end{bmatrix}. \tag{10.15}$$

r_k is the range, $\dot{r}_k$ is the range rate, θ_k is the azimuth angle, and $v_k \sim \mathcal{N}(0, \mathcal{R})$. The covariance matrix $\mathcal{R}$ is a diagonal matrix with the diagonal elements equal to the variances of the range, range rate, and azimuth angle measurements, which are denoted as $\sigma_{r_k}^2$, $\sigma_{\dot{r}_k}^2$, and $\sigma_{\theta_k}^2$, respectively. To compensate for the radar's platform motion, we define the coordinates $\bar{x}_k = x_k - x_k^P$, where x_k^P is the x coordinate of the sensor platform at time k; similarly for $\bar{y}_k$. Finally, $\bar{p}_k$ denotes the height of the sensor typically the altitude of the aircraft on which the GMTI radar is mounted.

With the above model, if the modes $z_k \in \mathcal{V}$ in (10.14) are generated by a SCFG instead of a regular grammar, the target's trajectory exhibits sophisticated geometric patterns. For example, the line, arc, rectangular, closed, and move-stop-move trajectories. Similarly, other trajectory patterns such as extended trapeziums, and so on, can be considered, see [18] where complex patterns such as Chinese characters are considered.

Given the above setup, a tracker (e.g., IMM algorithm or sequential Markov chain Monte Carlo method) can be used to construct tracklet estimates $\hat{z}_k$ defined in (10.2). The goal then is to use these tracklet estimates in a metalevel tracking module to determine the target trajectory.

In [14], we applied the Earley–Stolcke parser to generate estimates of the trajectory. Figure 10.10 illustrates the likelihood probabilities of different geometric patterns as an m-rectangle is parsed. As it can be seen from the top panel of the figure, the correct geometric pattern maintains its high probability as the probabilities of other patterns drop because the input sequence does not support them. Some patterns such as vertical line and clockwise m-rectangle had high probabilities initially because the initial segment of the input terminal string matches their syntactic structure. However, as more terminals are parsed, their probabilities drop. This observation means that it is possible to prune a parse tree as its probability drops below a certain threshold. If the input terminal sequence does not support the syntactic rules of a syntactic pattern, the parse tree corresponding to the pattern could be pruned completely, and which could greatly reduce the computational complexity and the storage requirement.

10.6 EXAMPLE 2: DATA FUSION IN A MULTICAMERA NETWORK

Here we present a brief outline of a detailed example considered in [24] on how data fusion can be performed on SCFGs to make a decision regarding the intent of a target. For example, a camera observing only one of the linear segments of a persons trajectory is likely to classify the trajectory as being a line. However, when the different parts of the trajectory observed by different cameras are fused a more accurate inference of the trajectory can be made.

Assume that each camera transmits the posterior probability of each SCFG model given the observed string at that sensor node to a central fusion node. The central fusion node combines these posterior probabilities to obtain a fused distribution that is used to determine the intent of the target being tracked. This is done by assigning the intent to the model with the highest posterior probability. The combination formula (also called a consensus rule) $C_F : [0, 1]^J \rightarrow [0, 1]$ is a mapping from the posterior probabilities computed by all the J camera sensors in the network to a fused posterior density obtained at the central node. Consider the case of J camera sensors each computing the posterior probability of a given SCFG model among M possible models provided with the observation string $\hat{z}_1, \ldots, \hat{z}_k$. The posterior probability of the mth model computed

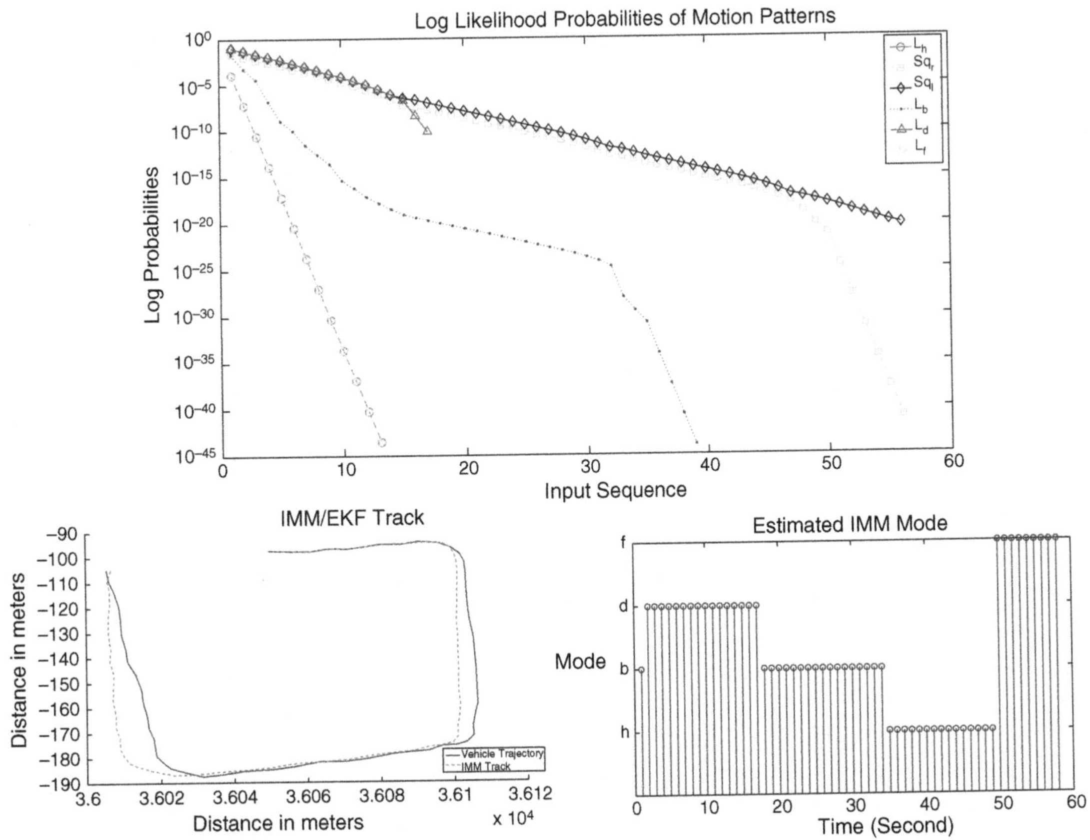

Figure 10.10 This figure from [14] illustrates the numerical result of parsing a *m*-rectangle pattern. The log-likelihood probabilities of different geometric patterns are shown in the top figure. The trajectory and its corresponding track are shown at the bottom left figure, and the estimated IMM modes are shown at the bottom right figure.

by the jth sensor is given by $p_j(\theta_m|\hat{z}_1, \ldots, \hat{z}_k)$, where the θ_m $m = 1, 2, \ldots, 7$ denote seven types of trajectories comprising of lines, arcs, and rectangles described in the figures below.

Two obvious consensus rules are the linear opinion pool and the logarithmic opinion pool [25]. The linear opinion pool is a consensus rule of the form

$$C_F(p_1, p_2, \ldots, p_J) = \sum_{i=1}^{J} \alpha_i p_i, \tag{10.16}$$

where $\sum_i \alpha_i = 1$. The logarithmic opinion pool is of the form

$$C_F(p_1, p_2, \ldots, p_J) = \frac{\prod_{i=1}^{J} p_i^{\alpha_i}}{\int \prod_{i=1}^{J} p_i^{\alpha_i}}. \tag{10.17}$$

Consider the following hallway surveillance problem depicted in Figure 10.11a. Because of the narrow hallways, the field of view of cameras used indoors have to be larger. The usefulness of the scale-invariance property of SCFGs becomes clear when we consider that a target can traverse shapes of different sizes if different corridors are used. A larger square trajectory is shown in Figure 10.11b. Alternatively, different people could traverse the hallways at different speeds thus leading to the self-embedding of the directional tracklets. This occurs because the tracklets are simply unit directional vectors without any notion of magnitude. In both cases, the underlying SCFG modeling is able to recognize the trajectory without ad hoc modifications to account for scale.

Noiseless target mode estimates The noiseless case of an m-rectangle trajectory is considered in Figure 10.12 from [24]. The intended trajectory is given by the tracklet sequence $\{aabbbccddd\}$. This scenario is different from the earlier external building scenario because of the nature of the camera placement in indoor scenarios. Some of the cameras do not merely measure linear sections of the trajectory. For example, the string observed by sensor 1 is $\{aabxxxxxxd\}$ that includes turn information as well as the tracklet at the end of the trajectory. Intuitively, we would expect a better intent inference that is apparent in the posterior probabilities in Figure 10.12.

Noisy target mode estimates Finally, a noisy version of the rectangular trajectory is simulated and the posterior probabilities are shown in Figure 10.13. The intended trajectory is the same, namely, $\{aabbbccddd\}$. However, as before, random perturbations occur due to the noisy estimates provided by the tracklet estimator. For example, the tracklets observed at sensor 3 is $\{xxxxxggcxx\}$, where the sixth and seventh symbols have been perturbed from c's to g's (which corresponds to an error of $\frac{\pi}{4}$ radians). The resulting fused posterior probabilities are shown in Figure 10.13. The fused probabilities in Figure 10.13a and b are able to incorporate information from all the sensors and correctly classify the trajectory as an m-rectangle even in the presence of noise and large observation gaps.

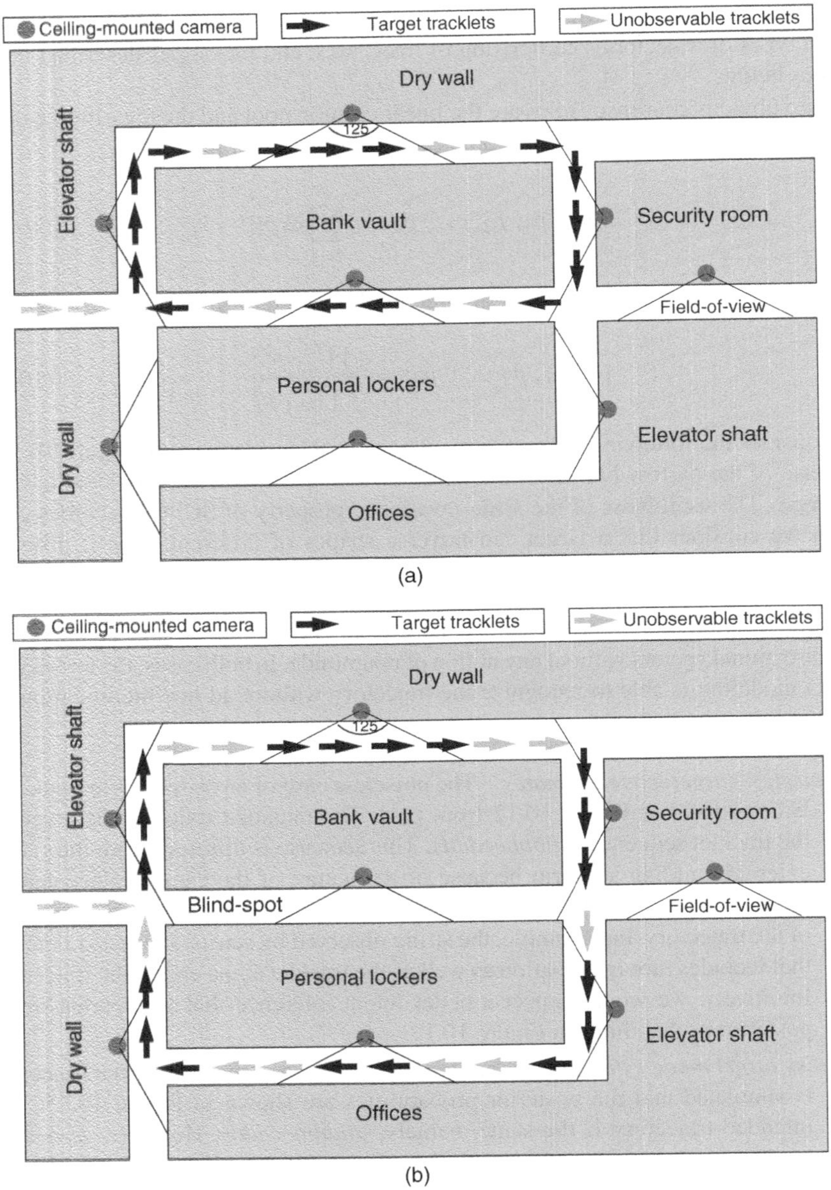

Figure 10.11 An internal hallway surveillance example from [24]. A rectangular trajectory is shown in (a) and a square trajectory is shown in (b) that are both cases of the m-rectangle language. This example shows the scale invariance of the model due to different sized m-rectangles.

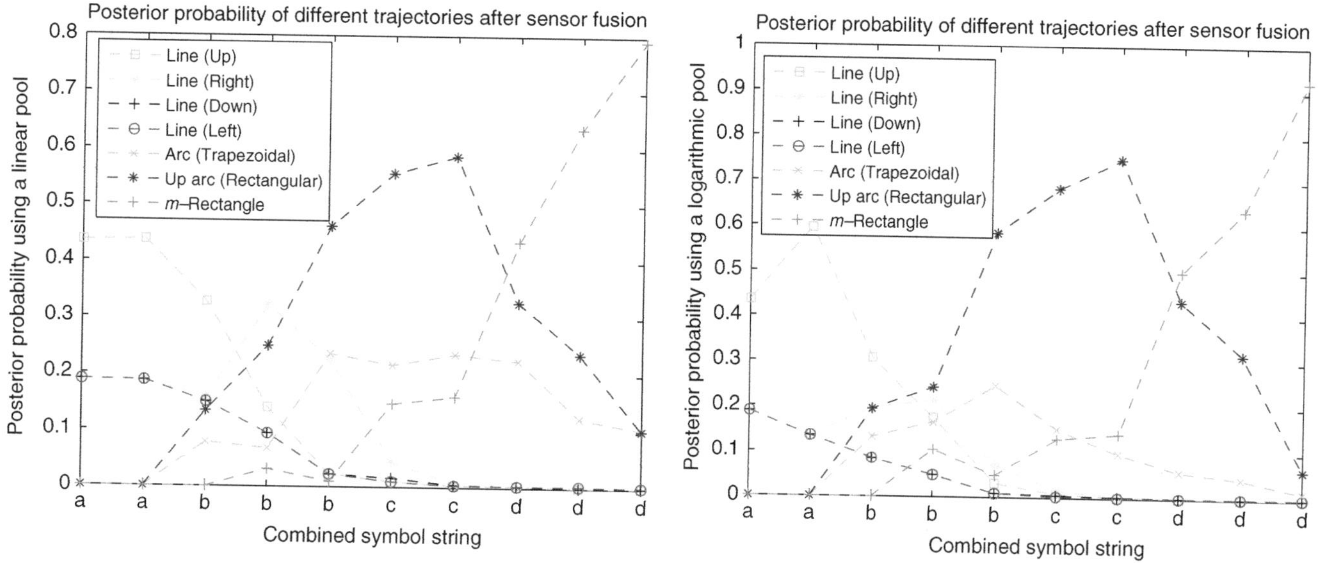

(a)

(b)

Figure 10.12 The noiseless rectangular trajectory in scenario 2. The linearly pooled probabilities are shown in (a) and the logarithmically pooled probabilities are shown in (b).

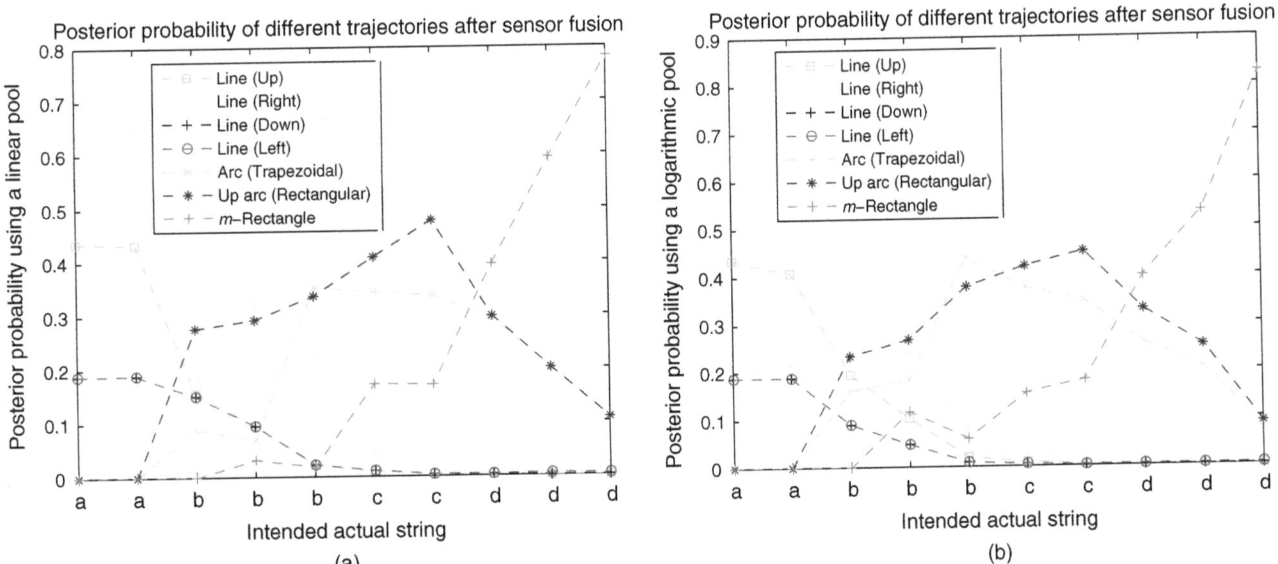

Figure 10.13 The noisy rectangular trajectory in scenario 2. The linearly pooled probabilities are shown in (a) and the logarithmically pooled probabilities are shown in (b).

10.7 CONCLUSION

In this chapter, we considered syntactic (higher level) tracking to assist human radar operators in making inferences about the target behavior given track estimates. Our premise for syntactic signal processing is that the geometric pattern of a target's trajectory can be modeled as "words" (modes) spoken by a SCFG language. The syntactic tracker constructs a parse tree of the geometric patterns that form the target trajectory and provides valuable information about the targets' intent. The parsing of the motion trajectories is implemented with Earley–Stolcke parsing algorithm. In [14], we have fed back this syntactic information to improve the performance of the tracker and demonstrate the performance on GMTI data collected with DRDC Ottawa's XWEAR radar. In related work, [24] presents data fusion algorithms for SCFGs in a camera-network for surveillance.

ACKNOWLEDGMENT

The author thanks Mr. Mustafa Fanaswalla for preparing several of the figures in this chapter and for discussions about stochastic context-free grammars. Also thanks to Prof. Lang White of University of Adelaide for several discussions on reciprocal processes.

REFERENCES

1. S. Blackman and R. Popoli, *Design and Analysis of Modern Tracking Systems*. Artech House, Norwood, MA, USA, 1999.
2. R. Durbin, S. Eddy, A. Krogh, and G. Mitchison, *Biological Sequence Analysis: Probabilistic Models of Proteins and Nucleic Acids*. Cambridge University Press, Cambridge, UK, 1998.
3. N. Visnevski, V. Krishnamurthy, A. Wang, and S. Haykin, Syntactic modeling and signal processing of multifunction radars: A stochastic context free grammar approach. *Proc. IEEE*, 95(5), 1000–1025, 2007.
4. K. Lari and S.J. Young, The estimation of stochastic context free grammars using the inside–outside algorithm. *Comput. Speech Language*, 4, 35–56, 1990.
5. M.I. Miller and A. O'Sullivan, Entropies and combinatorics of random branching processes and context-free languages. *IEEE Trans. Inform. Theory*, 38, 1292–1310, 1992.
6. L. White and F. Carravetta, Optimal smoothing for finite state hidden reciprocal processes. *IEEE Trans. Autom. Control*, 56(9), 2156–2161, 2011.
7. H. Sheng, C. Li, Q. Wei, and Z. Xiong, Real-time detection of abnormal vehicle events with multi-feature over highway surveillance video. In *11th International IEEE Conference on Intelligent Transportation Systems, ITSC 2008.*, October 2008, pp. 550–556.
8. C. Piciarelli, C. Micheloni, and G.L. Foresti, Trajectory-based anomalous event detection. *IEEE Trans. Circuits Syst. Video Technol.*, 18(11), 1544–1554, 2008.

9. C.J. Cohen, F. Morelli, and K.A. Scott, A surveillance system for the recognition of intent within individuals and crowds. *IEEE Conference on Technologies for Homeland Security, 2008*, May 2008, pp. 559–565.
10. S. Hongeng, F. Bremond, and R. Nevatia, Bayesian framework for video surveillance application. *Proceedings of the 15th International Conference on Pattern Recognition, 2000*, Vol. 1, 2000, pp. 164–170.
11. J. Muncaster and Y. Ma. Activity recognition using dynamic Bayesian networks with automatic state selection. *IEEE Workshop on Motion and Video Computing, WMVC '07*, February 2007, pp. 30–34.
12. M.S. Ryoo and J.K. Aggarwal, Recognition of composite human activities through context-free grammar based representation. *CVPR '06: Proceedings of the 2006 IEEE Computer Society Conference on Computer Vision and Pattern Recognition*, IEEE Computer Society, Washington, DC, USA, 2006, pp. 1709–1718.
13. Y.A. Ivanov and A.F. Bobick, Recognition of visual activities and interactions by stochastic parsing. *IEEE Trans. on Pattern Anal. Mach. Intell.*, 22, 852–872, 2000.
14. A. Wang, V. Krishnamurthy, and B. Balaji, Intent inference and syntactic tracking with gmti measurements. *IEEE Trans. Aero. Electron. Syst.*, 47(4), 2824–2843, 2011.
15. A. Stolcke, An efficient probabilistic context-free parsing algorithm that computes prefix probabilities. *Comput. Ling.*, 21(2), 165–201, 1995.
16. B.C. Levy, R. Frezza, and A.J. Krener, Modeling and estimation of discrete-time Gaussian reciprocal processes. *IEEE Trans. Autom. Control*, 35(9), 1013–1023, 1990.
17. M. Fanaswala, V. Krishnamurthy, and L. White, Destination-aware target tracking via syntactic signal processing. *2011 IEEE International Conference on Acoustics, Speech and Signal Processing (ICASSP)*, 2011, pp. 3692–3695.
18. K.S. Fu. *Syntactic Pattern Recognition and Applications*, Prentice-Hall, Englewood Cliffs, NJ, 1982.
19. J.E. Hopcroft, R. Motwani, and J.D. Ullman, *Introduction to Automata Theory, Languages, and Computation*, 3rd edition, Pearson Education, 2007.
20. L.R. Rabiner. A tutorial on hidden Markov models and selected applications in speech recognition. *Proc. IEEE*, 77(2), 257–285, 1989.
21. A.V. Aho and J.D. Ullman, *The Theory of Parsing, Translation and Compiling*, Volume I: Parsing, Prentice-Hall, Englewood Cliffs, NJ, 1972.
22. D. Jurafsky and J.H. Martin. *Speech and Language Processing: An Introduction to Natural Language Processing*, Prentice-Hall, Inc., 2009.
23. B. Jamison, Reciprocal processes. *Prob. Theory Rel. Fields*, 30(1), 65–86, 1974.
24. V. Krishnamurthy and M.H. Fanaswala, Intent inference via syntactic tracking. *Digital Signal Process.*, 21(5), 648–659, 2011.
25. J. Benediktsson and P. Swain, Consensus theoretic classification methods. *IEEE Trans. Syst. Man Cybernet.*, 22(4), 688–704, 1992.

PART III

SENSOR MANAGEMENT AND CONTROL

PART III

SENSOR MANAGEMENT AND CONTROL

CHAPTER 11

Radar Resource Management for Target Tracking—A Stochastic Control Approach

VIKRAM KRISHNAMURTHY

11.1 INTRODUCTION

A typical multitarget tracking system [1, 2] comprises of three subsystems: (i) a track initiation procedure that detects real targets and initializes track state estimators, (ii) a track state estimator that generates estimates of current and future values of the target kinematic parameters (based on the available history of noisy measurements), and (iii) a data association process that allocates measurements to existing tracks and aims to discard false measurements that do not correspond to real targets. Modern, target tracking systems combine these three processes into a single integrated functionality. Target tracking algorithms such as Integrated Probabilistic Data Association (IPDA) [3], MHT [4], HMM-based trackers [5] and particle filter-based trackers [6] are all based on approximate solutions to this complete target and measurement model estimation problem.

This chapter considers a further enhancement called *radar resource management*. The goal is to develop feedback control algorithms that enable a radar to adapt dynamically to its environment. Flexible multifunction radars (e.g., modern electronically scanned array (ESA) radars) are capable of switching between various measurement modes, for example, radar transmit waveforms, beam pointing directions, and so on, so that the tracking system is able to tell the radar which mode to use at the next measurement epoch. The use of such sophisticated radars fundamentally changes the paradigm of how the operator interacts with the radar. Instead of the operator continually changing the radar from mode to mode depending on the environment,

Integrated Tracking, Classification, and Sensor Management: Theory and Applications, Edited by Mahendra Mallick, Vikram Krishnamurthy, and Ba-Ngu Vo.

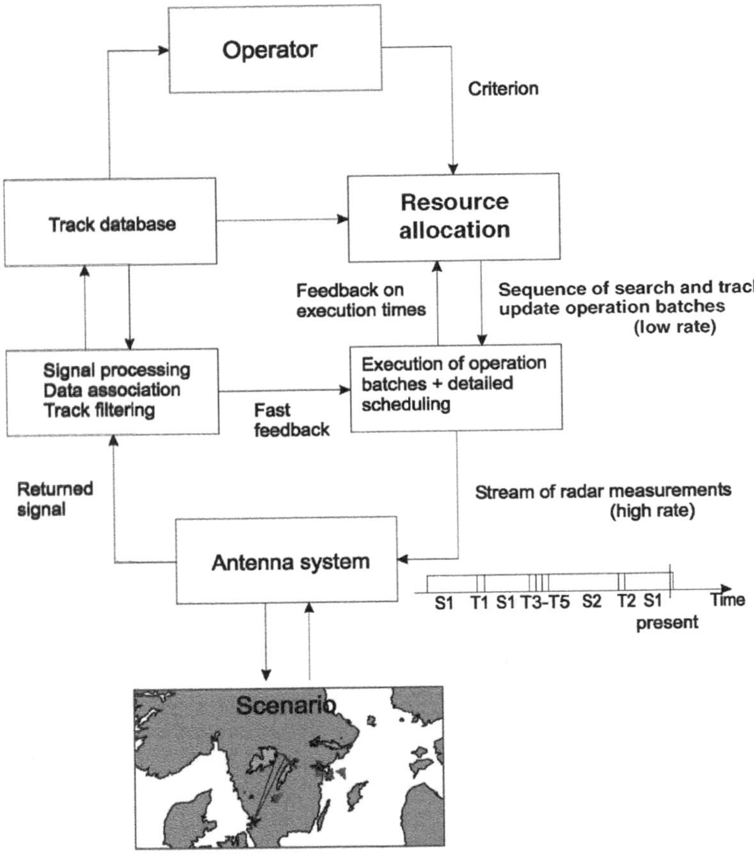

Figure 11.1 Schematic of radar resource management.

the aim is to construct automated resource management algorithms that dynamically adapt where the radar radiates its pulses to achieve the command operator objectives. This results in radars that autonomously switch beams, transmitted waveforms, target dwell and revisit times. An attractive feature of this concept of "inference-based autonomous control" is that a single-phased array radar has the potential to function as multiple conventional radars. Indeed, significant improvements in tracking performance have been demonstrated for radars systems even with relatively simple tracker driven sensor measurement mode selection (control) strategies [7, 8]. Figure 11.1 illustrates the schematic of a radar resource manager in an airborne radar system [9]. As can be seen, the radar manager operates in closed loop with the tracker and radar. (In Figure 11.1, the task S1 means search sector 1, T1 means track target 1, etc.) As is apparent from the figure, radar resource management is a complex task involving prioritizing and servicing various requests for scanning and tracking of targets.

11.1.1 Approaches to Radar Resource Management

There are two broad methodologies in the literature for formulating radar resource management problems:

(i) *Heuristic scheduling based on rule-based design* Under this methodology, a scheduling rule is defined based on descriptive rule-based design, see [2, 10] and also [11, 12], for single target resource allocation algorithms. Detailed scheduling of measurement order, given a set of measurements with specified types and processing time intervals is considered in [13]. Heuristic schedulers are often designed to operate in real time with limited computational requirements. However, since heuristic schedulers are not based on optimizing a cost function, their performance is difficult to predict.

(ii) *Stochastic control-based scheduling* In this chapter, we develop a stochastic control formalism for radar resource management. Due to the stochastic nature of radar detection and target dynamics, scheduling of radar measurements is a *partially observed* stochastic control problem. In this approach, a multistage cost function is minimized over a finite or infinite horizon. Globally optimal stochastic optimization methods such as stochastic dynamic programming (DP) can in principle be used to compute the optimal radar resource management policy. Unfortunately, the curse of dimensionality inherent in DP makes their direct application intractable. One can resort to the myopic case, that is, optimize an instantaneous cost [14], but this is typically inappropriate in surveillance applications. Indeed, optimizing the radar performance over a *long*-term horizon (e.g., one to several minutes) is desirable due to the following reasons:

- Due to the large surveillance volumes to be covered with scarce resources, the number of tracked targets and track load depend dynamically on the search scan allocation. In particular, the long-term dynamical behavior of the number of tracked targets and the system performance as a function of the allocated search resources need to be taken into account. These dynamics also involve future track load and hence the future resource demand.
- It is preferable to avoid repeated target initiations and drops due to optimistically trying to track resource consuming, low-SNR targets in situations where the system is saturated.
- Course changes of the ESA radar platform and the spatially inhomogeneous antenna gain of an ESA lead to dynamically changing resource demand when tracking a set of targets.
- Reacquisitions of targets that reappear after a blindness period (i.e., Doppler blindness, elevation and vegetation mask for ground targets), and tracking a set of interacting targets with possibly mixing tracks, result in time intervals of increased resource demand. Before such time intervals, it is important for the resource management algorithm to decide which measurements are affordable and to what extent other, parallel tasks of the system will be affected.

- Synchronization of search scans and adaptive track updates may reduce the resource demand. However, synchronization requires the radar resource manager to consider a time interval stretching at least over the next search scan pass of the target.

To solve the long-term stochastic optimization problem via DP, simplifying assumptions are needed. In the existing literature, several such assumptions are used as we now briefly describe: In [15–17], it is assumed that all targets are tracked initially and the scheduling problem concerns the ordering of track update measurements. In [16, 17], to facilitate efficient prediction of the tracking dynamics, a regular, discrete timescale is assumed where at each time instant a single measurement occurs. In [16, 18], the multidimensional kinematic state was quantized to a Markov chain, and the beam scheduling problem was formulated as a special type of partially observed Markov decision processes (POMDPs) called a multiarmed bandit. For general POMDPs, the number of states becomes prohibitively large, leading to computational intractability. (POMDPs are appropriate for the classification problem where the state is discrete [19].) In [9], a Lagrangian relaxation approach is developed for radar resource management of airborne radars. More recently, [20, 21] develop structural results for radar resource management formulated as a POMDP. In particular, these papers give conditions under which the optimal radar scheduling policy is characterized by a simple switching curve. Because in general solving stochastic control problems is computationally intractable, such structural results are useful as they characterize the optimal policy without brute force computation. Finally, we also refer the reader to the January 2009 special issue of *IEEE Signal Processing* magazine on waveform-agile radars.

11.1.2 Architecture of Radar Resource Manager

As illustrated in Figure 11.2, radar resource management naturally decomposes into two different scales. At the slow timescale, with clock rate in the order of several seconds, the radar *macromanager* needs to decide on what requests to service, how to prioritize targets, what batch of measurements to make, and how to make them. At this slow timescale, measurements are abstracted into *measurement operations*. A measurement operation is considered to be an algorithm in the radar that generates a sequence of measurements needed for achieving a low-level measurement task such as "update track with repeated update attempts," see for instance the track update algorithm designed in [11]. The physical level aspects of sensor performance are abstracted into quality of service (QoS) measures used in an optimization criterion. The QoS measures are defined targetwise based on concepts such as track accuracy and track continuity, and expressed in terms of tracking utility.

Given the decisions of the radar macromanager, on a fast timescale (clock rate in the order of milliseconds), the radar *micromanager* deals with controlling the tracker to obtain measurements of the various targets. At the fast timescale, multiple targets are tracked/estimated by a Bayesian tracker. The highest priority targets (decided by

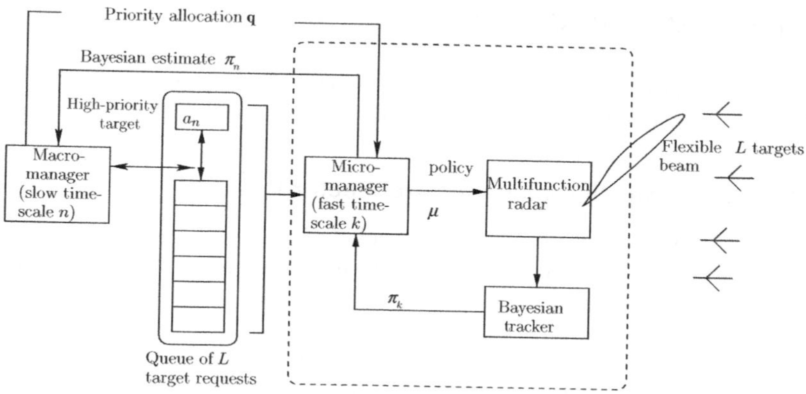

Figure 11.2 Two-timescale radar resource management.

the macromanager) are allocated more resources. Based on the tracker estimates, the micromanager deals with how long to maintain the current priority allocation of targets before returning control to the macromanager to pick a new priority allocation. Apart from such *dwell time management*, the micromanager also interfaces with the tracker to decide when to update a radar waveform (*radar waveform management*), aperture (*aperture management*), and so on. In simple terms, the micromanagement problem can be viewed as a sequential detection problem—based on the track estimates determine the optimal stopping time at which control is handed back to the macromanager to determine a new priority/choice.

Note that the priority allocation from the macromanager, the track variances from the tracker, and the decisions of the micromanager interact with each other. So in its most general form, radar resource management is a complex task involving interacting stochastic control problems at multiple timescales.

11.1.3 Organization of Chapter

The remainder of this chapter is organized as follows. Section 11.2 formulates the sensor management problem as a two-timescale stochastic control problem. Two alternative formulations are given for the micromanagement problem, one formulation deal with maximizing the mutual information of multiple targets, the other formulation deals with a POMDP setup involving a finite state Markov chain. In either case, the micromanagement problem is formulated as a sequential stopping time problem. Section 11.3 gives sufficient conditions under which the optimal micromanagement policy is monotone with respect to the state. As mentioned above, such a qualitative property is of significant importance as it permits design of efficient algorithms to compute the optimal policy. Also the macromanagement problem is discussed briefly. Finally, Section 11.4 presents an alternative formulation of sensor management for a maneuvering target modeled as a jump Markov linear system (JMLS) and several myopic sensor management algorithms are presented. Section11.4 is based on methods developed in [22].

11.2 PROBLEM FORMULATION

To fix ideas, the sensor management formulation in this section focuses on an airborne GMTI radar. The system is a sanitized version of an actual GMTI radar manager.

11.2.1 Macro and Micromanager Architecture

Consider a GMTI radar with an agile beam tracking L ground moving targets indexed by $l \in \{1, \ldots, L\}$. We define three discrete-time indices to reflect the three timescales in a radar management problem:

- Individual observations of target l are obtained on the fast timescale $t = 1, 2, \ldots$. The clock rate at which t ticks is typically 1 ms. At this timescale, ground targets can be considered to be static.
- Decision epoch $k = 1, 2, \ldots, \tau$ is the timescale at which the micromanager and target tracker operate. The clock at which k ticks is typically 0.1 s. Here τ denotes a stopping time at which the micromanager decides to stop the current scheduling interval and return control to the macromanager. At this epoch timescale k, the targets move according to the kinematic model (11.2), (11.3) below. Each epoch k comprises of intervals $t = 1, 2, \ldots, \Delta$ of the fast timescale, where Δ is typically of the order of 100. In other words, 100 observations are integrated at the t-timescale to yield a single observation at the k-timescale.
- The scheduling interval $n = 1, 2 \ldots,$ is the timescale at which the macromanager operates. Each scheduling interval n comprises of τ_n decision epochs. This stopping time τ_n is determined by the micromanager. τ_n is typically in the range 10–50—in absolute time it corresponds to the range 1–5 s. In such a time period, a ground target moving at 50 km h^{-1} moves approximately in the range 14–70 m.

Macro-manager. At the beginning of each scheduling interval n, the radar macromanager allocates the target priority vector $\mathbf{q}_n = (\mathbf{q}_n^1, \ldots, \mathbf{q}_n^L)$. Here the priority of target l is $\mathbf{q}^l \in [0, 1]$ and $\sum_{l=1}^{L} \mathbf{q}^l = 1$. The priority weight $\mathbf{q}^l$ determines how much resources the radar devotes to target l in a given scheduling interval. This affects the track variances as will be described below. The choice $\mathbf{q}_n$ of the radar macromanager is typically rule based, depending on several extrinsic factors. For example, in many GMTI radar systems, the macromanager picks the target priority vector $\mathbf{q}_{n+1}$ based on the track variances (uncertainty) and threat levels

$$\mathbf{q}_{n+1} = \text{priority}\left(\text{track\_variance}_n(l), \text{threat}_n(l), \mathbf{q}_n(l); l = 1, \ldots, L\right). \quad (11.1)$$

Here the track variances of the L targets are determined via the micromanager as will be discussed below.

Micromanager. Once the target priority vector $\mathbf{q}_n$ is chosen, the micromanager is initiated for the nth scheduling interval. The clock on the decision epoch timescale k is reset to $k = 0$ and commences ticking. At this decision epoch timescale, $k = 0, 1, \ldots$, the L targets are tracked/estimated by a Bayesian tracker. Target l with priority $\mathbf{q}_n^l$ is allocated $\mathbf{q}_n^l \Delta$ observations on the fast timescale t. The question we seek to answer is: *How long should the micromanager track the L targets with priority vector* $\mathbf{q}_n$ *before returning control to the macromanager to pick a new priority vector?* We formulate this as a sequential decision problem.

Note that the priority allocation vector $\mathbf{q}$ and track variances of the L targets result in the micro and macromanagement problems interacting with each other.

11.2.2 Target and Measurement Model

On the micromanager timescale, the target dynamics can be viewed as a linear time invariant Gaussian state space model

$$s_{k+1}^l = F s_k^l + G w_k^l, \tag{11.2}$$

$$z_k^l = \begin{cases} H s_k^l + \frac{1}{\mathbf{q}^l} v_k^l & \text{with probability } p_d^l \\ \emptyset & \text{with probability } 1 - p_d^l \end{cases}. \tag{11.3}$$

Actually this model is a linearized version of the actual measurement model that takes into account the dynamics of the radar platform. Since the sensor manager operates on the quality of the track estimates, the actual measurement model is of less importance. The micromanager needs to operate seamlessly with a legacy tracking system. In actual GMTI systems, a Kalman filter tracker is used for state estimation since the observation process is almost linear as discussed above. (Approximate nonlinear filters such as particle filters are unnecessary. In fact, despite the explosion of papers on particle filters, the authors were surprised to find how little such algorithms are used in real GMTI radar systems.)

Define the one-step ahead predicted covariance matrix of target l at time k as

$$P_k^l = \mathbb{E}\Big\{ \big(s_k^l - \mathbb{E}\{s_k^l | z_{1:k-1}^l\}\big)\big(s_k^l - \mathbb{E}\{s_k^l | z_{1:k-1}^l\}\big)' \Big\}.$$

Based on the priority vector $\mathbf{q}$, the covariance of the state estimate of target $l \in \{1, \ldots, L\}$ is computed via the following measurement (z_k^l)-dependent Riccati equation

$$P_{k+1}^l = \mathcal{R}(P_k^l, z_k) \stackrel{\text{def}}{=} F P_k^l F^T + Q - I(z_k^l \neq \emptyset) F P_k^l H^T (H P_k^l H^T + \frac{1}{\mathbf{q}^l} R)^{-1} H P_k^l F^T. \tag{11.4}$$

So in the special case when a target l is allocated zero priority ($\mathbf{q}^{(l)} = 0$), or when $z_k^l = \emptyset$, then (11.4) specializes to the Kalman predictor updated via the Liapunov equation

$$P_{k|k-1}^l = \mathcal{L}(P_k^l) \stackrel{\text{def}}{=} F^l P_{k-1}^l F^{l^T} + Q^l. \tag{11.5}$$

At this stage there are two choices we can choose for formulating the state space of the radar management problem

- *Covariance-based utility function.* Here the micromanagement problem is formulated in terms of the dynamics (11.4) of the covariance matrices P_k^l. Note that due to missing observations, the covariance update is no longer deterministic. So the micromanagement problem is a stochastic control problem. This is discussed in Section 11.2.3.
- *Finite state utility function.* Here the quality of estimates of the targets is mapped to a finite set. For example, a scalar metric based on the covariance matrix and target threat can be mapped into qualitative variables such as

  $$\{\texttt{low\_threat}, \texttt{high\_threat}\} \\ \times\{\texttt{low\_certainty}, \texttt{medium\_certainty}, \texttt{high\_certainty}\}.$$

  Such a finite state process is a random process due to the random evolution of the target (11.2) and can be approximated by a Markov chain. The state of this Markov chain is observed via noisy measurements (tracker estimates) depending on the resource allocation of the macro and micromanager. The resulting stochastic control problem involves a partially observed Markov decision process. This formulation is discussed in Section 11.2.4.

11.2.3 Micromanagement to Maximize Mutual Information of Targets

Here we formulate the sensor micromanagement problem in terms of the dynamics of the Kalman covariance update. As mentioned in Section 11.2.1, once the radar macromanager determines the priority vector $\mathbf{q}$, the micromanager switches on and its clock $k = 1, 2, \ldots$ begins to tick. At each slot k, the micromanager chooses action $u_k \in \{1\ (\text{stop}), 2\ (\text{continue})\}$.

In order to formulate the sequential detection problem, let us specify the costs incurred with choosing actions $u_k \in \{1\ (\text{stop}), 2\ (\text{continue})\}$.

Radar operating cost If the micromanager chooses action $u_k = 2$ (continue), it incurs the radar operating cost denoted as $c_{\mathbf{q}}$. Here $c_{\mathbf{q}} > 0$ depends on the radar operating parameters,

Stopping cost—stochastic observability If the micromanager chooses action $u_k = 1$ (stop), a stopping cost is incurred. In this chapter, we formulate a stopping cost in terms of the stochastic observability of the targets, see also [23, 24]. Define the stochastic observability of each target $l \in \{1, , \ldots, L\}$ as the mutual information

$$I(s_k^l; z_{1:k}^l) = \alpha^l h(s_k^l) - \beta^l h(s_k^l | z_{1:k}^l). \tag{11.6}$$

Recall from elementary information theory [25], that $h(s_k^l)$ denotes the differential entropy of target l at time k and $h(s_k^l | z_{1:k}^l)$ denotes the conditional differential entropy of target l at time k given the observation history $z^l_{1:k}$. The mutual information $I(s^l_k; z^l_{1:k})$ is the average reduction in uncertainty of the targets coordinates s^l_k given measurements $z^l_{1:k}$. In (11.6), α^l and β^l are non-negative constants chosen by the designer. In the standard definition of mutual information $\alpha^l = \beta^l = 1$. However, we are also interested in the special case when $\alpha^l = 0$, in which case, we are considering the conditional entropy for each target.

Recall a denotes the highest priority target. With the above definition, we consider the following stopping cost if the micromanager chooses action $u_k = 1$ at time k

$$\bar{C}(s_k, z_{1:k}) = -I(s^a_k; z^a_{1:k}) + \max_{l \neq a}(I(s^l_k, z^l_{1:k}). \tag{11.7}$$

The stopping cost is the difference of the mutual information between the target with highest mutual information and the target with highest priority. This can be viewed as a stopping cost that discourages stopping too soon. Naturally, several other choices can be considered, including replacing the $\max(\cdot)$ with a $\min(\cdot)$ or the arithmetic average.

The following lemma follows from straightforward arguments in [25].

Lemma 1 *Under the assumption of linear Gaussian dynamics for each target l, the mutual information of target l is*

$$I(s^l_k, z^l_{1:k}) = \alpha^l \log |\bar{P}^l_k| - \beta^l \log |P^l_k|. \tag{11.8}$$

Here $\bar{P}^l_k = \mathbb{E}\{(x_k - \mathbb{E}\{x_k\})(x_k - \mathbb{E}\{x_k\})'\}$ *denotes the predicted (a priori) covariance of target l given no observations, and is computed as the Kalman predictor covariance using (11.5) for k iterations. Recall* $P_k = \mathbb{E}\{(x_k - \mathbb{E}\{x_k|z_{1:k}\})(x_k - \mathbb{E}\{x_k|z_{1:k}\})'\}$ *is the posterior covariance and is computed as the Kalman filter covariance (11.4).*

In light of the above lemma, the stopping cost is completely defined in terms of the Kalman filter and predictor covariances. For notational convenience define the four-tuple

$$P_k = (P^a_k, \bar{P}^a_k, P^{-a}_k, \bar{P}^{-a}_k). \tag{11.9}$$

Therefore, the stopping cost (11.7) can be expressed as

$$\bar{C}(P_k) = -\alpha^a \log |\bar{P}^a_k| + \beta^a \log |P^a_k| + \max_{l \neq a} \left(\alpha^l \log |\bar{P}^l_k| - \beta^l \log |P^l_k| \right). \tag{11.10}$$

Another special case we are also interested in, is when $\alpha^l = 0$. Then the stopping cost is difference in the conditional differential entropy instead of the mutual information. This is particularly useful as a cost when $\mathbf{q}^l = 0$, since then the mutual information of target l is zero (because $\bar{P}_k = P_k$).

With the above stopping and continuing costs, we are now ready to formulate the sequential detection problem that the micromanager needs to solve. Let μ denotes a stationary decision policy of the form (recall P_k is defined in (11.9))

$$\mu : P_k \rightarrow u_{k+1} \in \{1 \text{ (stop)} , 2 \text{ (continue)} \}. \tag{11.11}$$

Let $\boldsymbol{\mu}$ denotes the family of such stationary policies. For any symmetric positive definite matrices P_0, and policy $\mu \in \boldsymbol{\mu}$, define the stopping time $\tau = \inf\{k : u_k = 1\}$. For each initial prior P_0, and policy μ by the micromanager, the following cost is associated with the sequential decision procedure

$$J_\mu(P) = \mathbb{E}^\mu\{(\tau - 1)c_\mathbf{q} + \bar{C}(P_\tau)|P_0 = P\}. \tag{11.12}$$

The goal is to determine the optimal stopping time τ with minimal cost, that is, compute the optimal policy $\mu^* \in \boldsymbol{\mu}$ to minimize (11.12), that is, $J_{\mu^*}(\pi) = \inf_{\mu\in\boldsymbol{\mu}} J_\mu(\pi)$. The existence of an optimal stationary policy μ^* follows from [26, Prop. 1.3, Chapter 3]. Since $c_\mathbf{q}$ is nonnegative, for the conditional entropy cost function, stopping is guaranteed in finite time, that is, τ is finite with probability 1. In general τ is not necessarily finite—however, this does not cause problems from a practical point of view since the micromanager has typically a prespecified upper-bound time at which it always chooses $u_k = 1$ and reverts back to the macromanager.

Considering the above cost (11.12), the optimal stationary policy $\mu^* \in \boldsymbol{\mu}$ and associated value function $\bar{V}(P)$ are the solution of the following "Bellman's dynamic programming equation" (recall our notation $P = (P^a, \bar{P}^a, P^{-a}, \bar{P}^{-a})$)

$$\begin{aligned}
\bar{V}(P) &= \min\left\{\bar{C}(P), c_\mathbf{q} + \mathbb{E}_z\left[V\left(\mathcal{R}(P^a, z^a), \mathcal{L}(\bar{P}^a), \mathcal{R}(P^{-a}, z^{-a}), \mathcal{L}(\bar{P}^{-a})\right)\right]\right\},\\
\mu(P) &= \operatorname{argmin}\left\{\bar{C}(P), c_\mathbf{q} + \mathbb{E}_z\left[V\left(\mathcal{R}(P^a, z^a), \mathcal{L}(\bar{P}^a), \mathcal{R}(P^{-a}, z^{-a}), \mathcal{L}(\bar{P}^{-a})\right)\right]\right\}.
\end{aligned} \tag{11.13}$$

Here $\mathcal{R}(P^{-a}, z^{-a})$ denotes the covariance update for the $L - 1$ lower priority targets according to (11.4). Our goal is to characterize the optimal policy μ^* and optimal stopping set defined as

$$\mathcal{R}_{\text{stop}} = \{(P^a, \bar{P}^a, P^{-a}, \bar{P}^{-a}) : \mu^*(P^a, \bar{P}^a, P^{-a}, \bar{P}^{-a}) = 1\}. \tag{11.14}$$

Of course, in the special case when $\alpha^l = 0$, then $\mathcal{R}_{\text{stop}} = \{(P^a, P^{-a}) : \mu^*(P^a, P^{-a}) = 1\}$.

11.2.4 Formulation of Micromanagement as a Multivariate POMDP

Here we consider the alternative approach of mapping the quality of estimates of the targets to a finite set. For example, a scalar metric based on the covariance matrix and target threat can be mapped into qualitative variables such as

$$\mathbf{S} = \{\texttt{low\_threat}, \texttt{high\_threat}\}$$
$$\times\{\texttt{low\_certainly}, \texttt{medium\_certainty}, \texttt{high\_certainty}\}.$$

Such a finite state process is a random process due to the random evolution of the target (11.2). It can be approximated by a finite Markov chain with state space **S**. The state of this Markov chain is observed via noisy measurements (tracker estimates) depending on the resource allocation of the macro and micromanager. The resulting stochastic control problem involves a partially observed Markov decision process. We refer the reader to [9], a much more detailed justification is given of how to represent the quality of the tracker performance as a finite state Markov chain.

The aim below is to formulate the sensor micromanagement problem for the L targets as a multivariate POMDP. This POMDP comprises of the following six ingredients:

(i) *Markov chain.* X_k models the qualitative threat/accuracy of the tracks of the L evolving targets. For each target l, each process $X_n^{(l)} \in \mathbf{S}$. Define the composite state space of all L targets as

$$\mathbf{S}^{\text{comp}} = \mathbf{S} \times \mathbf{S} \times \cdots \times \mathbf{S}. \tag{11.15}$$

(ii) *Action space.* At the $(k+1)$th time instant within the tth scheduling interval, the micromanager picks action u_{k+1} as a function μ of the Bayesian estimates π_k (defined in (11.21)) of all L targets

$$u_{k+1} = \mu(\pi_k) \in \{\text{continue} = 2, \text{stop} = 1\},$$
$$\text{where } \mu \in \mathcal{A} = \text{stationary scheduling policies}. \tag{11.16}$$

- If $u_k = \text{continue} = 2$, the micromanager continues with the current target priority allocation a_n. So k increments to $k+1$ and the L targets are tracked with target a_t given the highest priority.
- If $u_k = \text{stop} = 1$, the micromanager stops the current scheduling interval n, and returns control to the macromanager to determine a new target a_{n+1} for the next scheduling interval $n+1$.

(iii) *Multivariate target measurements.* Given the composite state X_k of the L targets, the measurement vector Y_k is recorded at time k. Here

$$Y_k = (Y_k^{(1)}, \ldots, Y_k^{(L)}) \text{ is obtained from the multivariate distribution}$$
$$\mathbb{P}_{a_n}(Y_k | X_k, u_k). \tag{11.17}$$

Assume each target's observation $Y_k^{(l)}, l = 1, \ldots, L$, is finite valued, that is,

$$Y_k^{(l)} \in \mathbf{y} = \{O_1, O_2, \ldots, O_M\}, \text{ and so } Y_k \in \mathbf{Y} \stackrel{\triangle}{=} \mathbf{y} \times \cdots \times \mathbf{y}$$
$$= \{O_1, O_2, \ldots, O_{M^L}\}. \tag{11.18}$$

For example in [5, 20, 27], $Y^{(l)}$ models the noisy observed distance of the target to the base-station. In other types of radar resource management [2], $Y^{(l)}$ models the estimated track covariance of target l. In radar resource management, since a_t denotes the high-priority target, if $u_k = 2$, then $Y_k^{(l)}$ is a more accurate measurement of $X_k^{(l)}$ for $l = a_t$ compared to the other $L-1$ targets. Note that since the targets are correlated, observing one target yields information about other targets. Also since the elements of Y_k can be correlated, measurements of one target give information about another target.

(iv) *Multitarget Bayesian tracking and information state.* In scheduling interval t, with priority allocation a_t, at any time k, denote the history of past observations and actions as

$$Z_k^{(l)} = \{a_n, \pi_{t-1}, u_1, Y_1^{(l)}, \ldots, u_k, Y_k^{(l)}\}, \quad Z_k = (Z_k^{(1)}, \ldots, Z_k^{(L)}). \quad (11.19)$$

Here π_{t-1} is the *a posteriori* distribution of the L targets from the macro-manager at the end of the $(t-1)$th scheduling interval. Based on the history Z_{k+1}, the Bayesian tracker computes the *a posteriori* state distribution π_{k+1} of the L targets defined as

$$\pi_{k+1} = \{(\pi_{k+1}(\mathbf{i}), \mathbf{i} \in \mathbf{S}^{\text{comp}}\}, \quad \pi_{k+1}(\mathbf{i}) = \mathbb{P}_{a_n}(X_{k+1} = \mathbf{i} | Z_{k+1}). \quad (11.20)$$

The S^L-dimensional column vector π_k is computed recursively via the Hidden Markov Bayesian filter

$$\begin{aligned}
\pi_{k+1} &= T_{a_n}(\pi_k, u_{k+1}, Y_{k+1}), \text{ where } T_a(\pi, u, Y) = \frac{B_a(u, Y)P'\pi}{\sigma_a(\pi, u, Y)}, \\
&\sigma_a(\pi, u, Y) = \mathbf{1}'_{S^L} B_a(u, Y)P'\pi \\
&\text{and } B_a(u, Y) = \text{diag}(\mathbb{P}_a(Y|1, u), \ldots, \mathbb{P}_a(Y|S^L, u)). \\
&\pi_k \text{ is initialized at } k = 1 \text{ by } \pi_1 = \pi_{t-1}.
\end{aligned} \quad (11.21)$$

Here $\mathbf{1}_{S^L}$ is the S^L dimension vector of ones. π_k is referred to as the *information state*, since (see [28]) it is a sufficient statistic to describe the history Z_k in (11.19). The composite Bayesian estimate π_k in (11.21) lives in an $S^L - 1$ dimensional unit simplex

$$\Pi^{\text{comp}} \triangleq \left\{\pi \in \mathbb{R}^{S^L} : \mathbf{1}'_{S^L}\pi = 1, \quad 0 \le \pi(\mathbf{i}) \le 1 \text{ for all } \mathbf{i} \in \mathbf{S}^{\text{comp}}\right\}. \quad (11.22)$$

The Bayesian *a posteriori* distribution of each individual target l, defined as $\pi_k^{(l)} = \mathbb{P}_{a_t}(X_k^{(l)} | Z_k^{(l)})$, can be computed by marginalizing the joint distribution

π_k. For each target l, the estimate $\pi_k^{(l)}$, lives in an $S-1$ dimensional unit simplex

$$\Pi \triangleq \left\{\pi^{(l)} \in \mathbb{R}^S : \mathbf{1}'_S \pi^{(l)} = 1, \quad 0 \leq \pi^{(l)}(i) \leq 1, \quad i \in \{1 \ldots, S\}\right\}. \tag{11.23}$$

Obviously, $\Pi^{\text{comp}} = \underbrace{\Pi \times \Pi \cdots \times \Pi}$ (L times).

Let e_i denote the unit S^L dimensional vector with 1 in the ith position. Note $e_1, e_2, \ldots, e_{S^L}$ form the S^L corners of the simplex Π^{comp}. These corner points represent the elements of state space $\mathbf{S}^{\text{comp}}$. So,

$$\text{States } (1, 2, \ldots, S^L) \text{ of } X \text{ correspond to the information states } e_1, e_2, \ldots, e_{S^L}, \text{ respectively.} \tag{11.24}$$

(v) *Tracking cost.* At time k, with given current composite state X_k of the L targets, if action $u_{k+1} = \mu(\pi_k) \in \{\text{continue} = 2, \text{stop} = 1\}$ is taken, then the micromanager accrues an instantaneous cost $c_{\mathbf{q}}(X_k, u_{k+1})$. Here $c_{\mathbf{q}}(X, u) \geq 0$ and

$$c_{\mathbf{q}}(X, u) = \begin{cases} u = 2 & \text{cost of continuing with current priority allocation } a \\ & \text{given state } X \\ u = 1 & \text{cost of terminating current priority allocation } a \\ & \text{given state } X. \end{cases} \tag{11.25}$$

In (11.25), the nonnegative L-dimensional vector $\mathbf{q} = (\mathbf{q}(1), \ldots, \mathbf{q}(L))'$ denotes target priority allocations and is set by the macromanager. $\mathbf{q}$ links the micro and macromanagement. The costs $c_{\mathbf{q}}(X, u)$ are chosen as decreasing functions (elementwise) of $\mathbf{q}$ since higher priority targets should incur lower tracking costs. The cost $c_{\mathbf{q}}(X, 1)$ can also be viewed as a *switching cost* incurred by the micromanager when it decides to terminate the current target priority allocation and revert back to the macromanager. In such a case, the higher the target priority, the more it should cost to switch back to the macromanager to request a new high-priority target.

If $u = 1$ is chosen, control reverts back to the macromanager to determine a new target priority allocation. Let τ_n denote the time k (in the nth scheduling interval) at which action $u_k = \text{stop} = 1$ is chosen. The random variable τ_n is a stopping time, that is, the event $\{\tau_n \leq k\}$ for any positive integer k is a function of the history Z_k or equivalently information state π_k (more formally $\{\tau_n \leq k\}$ is measurable with respect to the sigma-algebra generated by Z_k or π_k). Let $0 \leq \rho < 1$ denote a user defined economic discount factor. Then the sample path cost incurred during this interval is

$$\hat{J}_\mu = \sum_{k=1}^{\tau_n} \rho^{k-1} C(\pi_k, u_{k+1}),$$
$$\text{where } u_{k+1} = \mu(\pi_k) \in \{\text{continue} = 2, \text{stop} = 1\} \tag{11.26}$$
$$\text{and } C(\pi_k, u_{k+1}) = c_{\mathbf{q}}'(u_{k+1})\pi_k, \quad c_{\mathbf{q}}(u) \triangleq \left[c_{\mathbf{q}}(1, u) \cdots c_{\mathbf{q}}(S^L, u)\right]'.$$

(vi) *Discounted cost stopping time problem formulation*. The final ingredient in the multivariate POMDP formulation is the optimization objective. Recall that at the fast timescale in a scheduling interval n, the aim is to determine whether to continue or stop with the current target priority a_n. So the objective is to compute the optimal sensor schedule $\mu^*(\pi_k) \in \{\text{continue} = 2, \text{stop} = 1\}$ to minimize the expected discounted cost $J^{\mathbf{q}}(\mu, \pi_{n-1})$ over the set of admissible control laws $\mathcal{A}$ defined in (11.16). That is, with $c_{\mathbf{q}}(X_k, u_{k+1})$ defined in (11.25), and $\mathbb{E}\{\cdot\}$ denoting mathematical expectation with respect to $X_1, \ldots, X_{\tau_n}, Y_1, \ldots, Y_{\tau_n}$

$$J_{\mu^*}(\pi) = \inf_{\mu \in \mathcal{A}} J_\mu(\pi) \text{ where } J_\mu(\pi) \triangleq \mathbb{E}\left\{\sum_{k=1}^{\tau_n} \rho^{k-1} C_{\mathbf{q}}(\pi_k, \mu(\pi_k)) | \pi_{n-1} = \pi\right\}. \tag{11.27}$$

Summary: The above six ingredients constitute a multivariate POMDP for the sensor micromanagement problem for L targets. The dependence of the costs (11.25) on the priority allocation vector $\mathbf{q}$ results in the discounted optimal cost of the POMDP (11.27) depending on $\mathbf{q}$.

Consider the micromanagement problem represented by the multivariate POMDP with objective function (11.27). For fixed priority allocation vector $\mathbf{q}$, the optimal stationary policy $\mu^* : \Pi^{\text{comp}} \to \{1, 2\}$ and associated optimal cost $J_{\mu^*}(\pi)$ are the solution to "Bellman's equation" for $V_a^{\mathbf{q}}(\pi)$

$$J_\mu(\pi) = V^{\mathbf{q}}(\pi) = \min_{u \in \{1,2\}} Q^{\mathbf{q}}(\pi, u), \quad \mu^*(\pi) = \arg\min_{u \in \{1,2\}} Q^{\mathbf{q}}(\pi, u),$$
$$\text{where } Q^{\mathbf{q}}(\pi, 2) = C(\pi, 2) + \rho \sum_{Y \in \mathbf{Y}} V^{\mathbf{q}}(T(\pi, 2, Y))\, \sigma(\pi, 2, Y), \quad Q^{\mathbf{q}}(\pi, 1) = C(\pi, 1). \tag{11.28}$$

Our goal is to characterize the optimal policy μ^* and the optimal stopping set defined as

$$\mathcal{R}_{\text{stop}} = \{\pi : \mu^*(\pi) = 1\}. \tag{11.29}$$

Since the information state space Π^{comp} of a POMDP is an uncountable set, the dynamic programming equations (11.28) do not translate into practical solution methodologies as $V(\pi)$ needs to be evaluated at each $\pi \in \Pi^{\text{comp}}$, an uncountable set, see [28] for details. In our multivariate POMDP, the state space dimension is S^L (exponential in the number of targets) and so applying value iteration is numerically intractable. The optimal policy can in principle be determined by a finite dimensional algorithm. This finite dimensional characterization was first derived in [29], where it was shown that the value function $V_t(\pi)$ is piecewise linear and convex. Several linear programming-based algorithms have been presented in the operations research literature for computing these piecewise linear segments. These include Sondik's algorithm [29], Monahan's algorithm, Cheng's algorithm [28], and the Witness algorithm [27]. We refer the reader to the website `http://www.cs.brown.edu/research/ai/pomdp/index.html` for an excellent tutorial exposition with graphics of these various algorithms. However, from a practical point of view, due to the large state space, such algorithms are not tractable.

11.3 STRUCTURAL RESULTS AND LATTICE PROGRAMMING FOR MICROMANAGEMENT

As mentioned above, solving the dynamic programming equation to compute the optimal policy μ^* is computationally intractable. Can conditions be imposed on the micromanager/target model so that the optimal policy has a "simple structure"? In that case, one only needs to search over the space of policies that satisfy this simple structure. The main idea of this section is to give conditions under which that the optimal micromanagement policy μ^* is monotonically increasing in the state. Of course, the state comprises either of covariance matrices (Section 11.2.3) or posterior distributions (Section 11.2.4)—such states are only partially orderable. Why are such monotone policies defined or partially ordered state spaces useful? Instead of attempting to solve an intractable dynamic programming problem, one simply needs to search for policies that are monotone. Numerically efficient algorithms can be designed to estimate parameterized policies that satisfy this monotone structure. Another appealing property is that the monotone results is that the assumptions under which they hold are set valued. Hence, there is an inherent robustness since even if the underlying model parameters are not exactly specified, the structural results still hold.

The methods we will describe below to prove the existence of a monotone policy fall under the area of lattice programming. We refer to [30–32] for excellent descriptions of lattice programming to establish the existence of monotone policies. In particular, the later two references deal with partially observed stochastic control problems where the states are partially ordered.

In this section, we present the main results regarding the monotonicity of the micromanagement policies for the two cases described above. We also illustrate how to parameterize such a monotone policy, so that the optimal parameterized policy can be computed via simulation-based stochastic optimization.

11.3.1 Monotone Policies for Micromanagement with Mutual Information Stopping Cost

Here we consider the formulation of Section 11.2.3 where the goal was to maximize the mutual information of the targets. Recall in Section 11.2.3 this was formulated in terms of the Kalman filter covariances. We prove below that the optimal policy has a monotone structure with respect to the positive definite matrix ordering.

To introduce the positive definite ordering, we introduce the following notation: Let $\mathcal{M}$ denote the set of all $m \times m$ real-valued, symmetric positive semidefinite matrices. For $P, Q \in \mathcal{M}$ define the positive definite partial ordering $\succeq$ as $P \succeq Q$ if $x'Px \geq x'Qx$ for all $x \neq 0$, and $P \succ Q$ if $x'Px > x'Qx$ for $x \neq 0$. Define $\preceq$ with the inequalities reversed. Note that $[\mathcal{M}, \succeq]$ is a partially ordered set (poset).

Note that ordering positive definite matrices is equivalent to ordering their eigenvalues. To formalize this notation, denote the componentwise partial order on $\mathbb{R}^m$ by $\preceq_l$, that is, for $x = (x_1, \ldots, x_m)$ and $y = (y_1, \ldots, y_m)$, then $x \preceq_l y$ (equivalently, $y \succeq_l x$) if $x_i < y_i$ for all $i = 1, \ldots, m$.

For any matrix $P \in \mathcal{M}$, let $\lambda_P \in \mathbb{R}^m_+$ denotes the eigenvalues of matrix P arranged in decreasing order as a vector. Note $P \succeq Q$ is equivalent to $\lambda_P \succeq_l \lambda_Q$. Clearly, $[\mathbb{R}^m_+, \succeq_l]$ is a poset.

Define f to be increasing if $\lambda_P \preceq_l \lambda_Q$ implies $f(\lambda_P) \leq f(\lambda_Q)$, or equivalently, if $P \preceq Q$ implies $f(P) < f(Q)$. Finally we say that $f(P^{-a})$ is increasing in P^{-a} if $f(\cdot)$ is increasing in each component P^l of $P^{-a}, l \neq a$.

The following result characterizes the monotone structure of the policy $\mu^*(P^a, \bar{P}^a, P^{-a}, \bar{P}^{-a})$. The proof appears in [33].

Theorem 1

(i) *The optimal decision policy $\mu^*(P^a, \bar{P}^a, P^{-a}, \bar{P}^{-a})$ at the radar micromanager is monotone increasing in P^a, decreasing in $\bar{P}^a$, decreasing in P^{-a}, and increasing in $\bar{P}^{-a}$ on the poset $[\mathcal{M}, \succeq]$. Equivalently, $\mu^*(P^a, \bar{P}^a, P^{-a}, \bar{P}^{-a})$ is monotone increasing in λ^a_P, decreasing in $\lambda^a_{\bar{P}}$, decreasing in $\lambda^{-a}_{P^{-a}}$, and increasing in $\lambda^{-a}_{\bar{P}^{-a}}$ on the poset $[\mathbb{R}^m_+, \succeq_l]$.*

(ii) *In the special case when $\alpha^l = 0$ for all $l \in \{1, \ldots, L\}$, that is, the stopping cost is the conditional entropy, the optimal policy $\mu^*(P^a, P^{-a})$ at the radar micromanager is increasing in P^a and decreasing in P^{-a} on the poset $[\mathcal{M}, \succeq]$. Equivalently, $\mu^*(P^a, P^{-a})$ is monotone increasing in λ^a_P, and decreasing in $\lambda^{-a}_{P^{-a}}$ on the poset $[\mathbb{R}^m_+, \succeq_l]$.*

The proof is presented in [33]. The monotone property of the optimal policy μ^* is useful since monotone parameterized policies are readily implementable at the micromanager level and can be adapted in real time. The micromanager simply needs to implement/learn this monotone parameterized policy. Let us give an example of a parameterized policy in terms of covariance matrices that satisfies the above monotone structure. Below θ^l and $\underline{\theta}^l \in \mathbb{R}^m$ are unit-norm vectors, that is, $\theta^{l\prime}\theta^l = 1$ and $\underline{\theta}^{l\prime}\underline{\theta}^l = 1$

for $l = 1, \ldots, L$. Let Θ define the space of all such vectors. Define the parameterized policy μ_θ as

$$\mu_\theta(P^a, P^{-a}) = \begin{cases} 1 \text{ (stop)} & \text{if } -\theta^{a\prime} P^a \theta^a + \underline{\theta}^{a\prime} \bar{P}^a \underline{\theta}^a + \sum_{l \neq a} \theta^{l\prime} P^l \theta^l - \underline{\theta}^{l\prime} \bar{P}^l \underline{\theta}^l \geq 0, \\ 2 \text{ (continue)} & \text{otherwise.} \end{cases} \tag{11.30}$$

It is easily verified that the above policy satisfies the monotone structure of the theorem. Moreover, the parameters $\theta, \underline{\theta}$ of the policy can be estimated efficiently using simulation-based stochastic optimization.

11.3.2 Monotone POMDP Policies for Micromanagement

Here was consider the formulation of Section 11.2.4 where the micromanagement problem was formulated as a multivariate POMDP. The aim is to give conditions on the model to ensure that the optimal policy $\mu^*(\pi)$ is monotone increasing with respect to the multivariate information state π.

To define a monotone policy, we first need to order the information states. How can multivariate information states (posterior distributions) be ordered? Below we use the totally positive (TP2) stochastic order. The TP2 order is preserved under conditional expectations—therefore it is the most convenient way of ordering information states.

To introduce the definition of the TP2 order let $\mathbf{i} = (i_1, \ldots, i_L)$ and $\mathbf{j} = (j_1, \ldots, j_L)$ denote the indices of two L-variate probability mass functions Denote the elementwise minimum and maximum vectors

$$\mathbf{i} \wedge \mathbf{j} = [\min(\mathbf{i}_1, \mathbf{j}_1), \ldots, \min(\mathbf{i}_L, \mathbf{j}_L)]', \qquad \mathbf{i} \vee \mathbf{j} = [\max(\mathbf{i}_1, \mathbf{j}_1), \ldots, \max(\mathbf{i}_L, \mathbf{j}_L)]'. \tag{11.31}$$

Definition 1 (TP2 Ordering and MLR Ordering) *Let P and Q denote any two L-variate probability mass functions. Then $P \underset{TP2}{\geq} Q$ if $P(\mathbf{i})Q(\mathbf{j}) \leq P(\mathbf{i} \vee \mathbf{j})Q(\mathbf{i} \wedge \mathbf{j})$. If P and Q are univariate, then this definition is equivalent to the MLR ordering denoted as $P \succeq Q$.*

MLR *partial order*, and it is not always possible to order any two information states $\pi \in \Pi^{\text{comp}}$. Also, unlike the MLR order, in general the TP2 order is not reflexive, that is, $P \underset{\text{TP2}}{\geq} P$ does not hold. This introduces additional complications when dealing with multivariate POMDPs.

Definition 2 (MTP2 Reflexive Distributions) *A multivariate distribution P is said to be multivariate TP2 (MTP2) if $P \underset{TP2}{\geq} P$ holds, that is, $P(\mathbf{i})P(\mathbf{j}) \leq P(\mathbf{i} \vee \mathbf{j})P(\mathbf{i} \wedge \mathbf{j})$. If $\mathbf{i}, \mathbf{j} \in \{1, \ldots, S\}$ are scalar indices, this is equivalent to saying that a $S \times S$ matrix P is MTP2 if all second-order minors are non-negative.*

First, we list the assumptions for correlated targets with correlated measurements. Assume for any fixed $a \in \{1, \ldots, L\}$, the following conditions hold for the multivariate POMDP (11.27).

(A1) The cost $c_{\mathbf{q}}(X, u) \geq c_{\mathbf{q}}(S^L, u)$ for all $X \in \mathbf{S}^{\text{comp}}$ (defined in (11.15)).

(A2) The composite transition probability matrix P of the L targets is MTP2.

(A3) For each a and u, the L-variate observation probabilities $\mathbb{P}_a(Y|X, u)$ is MTP2 in (Y, X).

(S) The costs are submodular. That is, for $X \in \mathbf{S}^{\text{comp}}$, $c_{\mathbf{q}}(X, 2) - c_{\mathbf{q}}(X, 1) \geq c_{\mathbf{q}}(S^L, 2) - c_{\mathbf{q}}(S^L, 1)$.

As described in [21], these assumptions hold for several sensor management problems.

Theorem 2 (Existence of Threshold Policy for Sensor Micromanagement) *Consider the multivariate POMDP (11.16), (11.17), (11.21), (11.25), (11.27) and any fixed target priority allocation $a \in \{1, \ldots, L\}$. Then under (A1), (A2), (A3), (S), the optimal policy $\mu^*(\pi)$ is TP2 increasing on Π. That is, $\pi \underset{TP2}{\geq} \tilde{\pi}$, implies $\mu^*(\pi) \geq \mu^*(\tilde{\pi})$.*
As a consequence, there exists a curve Γ (which we call a "threshold curve") that partitions information state space Π^{comp} into two individually connected regions $\mathcal{R}_{stop}$, $\Pi^{comp} - \mathcal{R}_{stop}$, such that

$$\textit{Optimal scheduling policy } \mu_a^*(\pi) = \begin{cases} \textit{continue} = 2 & \textit{if } \pi \in \Pi^{comp} - \mathcal{R}_{stop}, \\ \textit{stop} = 1 & \textit{if } \pi \in \mathcal{R}_{stop}. \end{cases} \tag{11.32}$$

Moreover, the region $\mathcal{R}_{stop}$ is convex and therefore the threshold curve Γ is continuous and differentiable almost everywhere (the set of points where it is not differentiable has measure zero).

The proof is in [21]. Under the conditions of Theorem 2, the optimal scheduling policy for the multivariate POMDP is a threshold policy with a threshold curve Γ that partitions the information state space Π^{comp}. Note that without these conditions, the optimal policy of the multivariate POMDP can be an arbitrarily complex partition of the simplex Π^{comp}—and solving such a multivariate POMDP is computationally intractable. The convexity of region $\mathcal{R}_{\text{stop}}$ follows from the clever but elementary observation in [31, Lemma 1]. It is well known [34] that a convex function is continuous and differentiable almost everywhere.

Due to the existence of a threshold curve Γ, computing the optimal policy μ_a^* reduces to estimating this threshold curve. In general, any user-defined basis function approximation can be used to parameterize this curve. However, any such approximation needs to capture the essential feature of Theorem 2: the parameterized optimal policy needs to be TP2 increasing.

In this section, we derive *linear* approximations to Γ; refer to [21] for an extensive treatment where multilinear thresholds are also derived for independent targets. The linear thresholds have two attractive properties: (i) Estimating them is computationally efficient. (ii) We give novel conditions on the threshold coefficients that are necessary and sufficient for the resulting linear/multilinear threshold policy to be TP2 increasing on lines. Due to the necessity and sufficiency of the condition, optimizing over the space of linear/multilinear thresholds yields the "best" linear/multilinear approximation to the threshold curve Γ.

We start with the following definition of a linear threshold policy: For any fixed target priority allocation $a \in \{1, \dots, L\}$, we define the linear threshold policy $\mu_{\theta_a}(\pi)$ as

$$\mu_{\theta_a}(\pi) = \begin{cases} \text{continue} = 1 & \text{if } {\theta_a}'\pi < 1 \\ \text{stop} = 2 & \text{if } {\theta_a}'\pi \geq 1, \end{cases} \qquad \pi \in \Pi^{\text{comp}}. \tag{11.33}$$

Here $\theta_a \in \mathbb{R}_+^{S^L}$ (S^L dimension vector with nonnegative elements) denotes the vector of coefficients of the linear threshold policy.

Theorem 3 below characterizes the best *linear hyperplane approximation* to Γ; the proof is in [21].

Theorem 3 (Best Linear Threshold Policy) *Assume conditions (A1), (A2), (A3), (S) hold for the multivariate POMDP (11.27). Then for any fixed target priority allocation $a \in \{1, \dots, L\}$*

(i) The linear threshold policy $\mu_{\theta_a}(\pi)$ defined in (11.33) is TP2 increasing on lines if and only if

$$\theta_a(S^L) \geq \theta_a(i), \quad i = 1, \dots, S^L - 1. \tag{11.34}$$

(ii) Therefore, the optimal linear threshold approximation to threshold curve Γ of Theorem 2 is the solution of the following constrained optimization problem

$$\theta_a^* = \arg \min_{\theta_a \in \mathbb{R}_+^{S^L}} J^{\mathbf{q}}(\mu_{\theta_a}, \pi), \qquad \textit{subject to } \theta_a(S^L) \geq \theta_a(i), \quad i = 1, \dots, S^L - 1. \tag{11.35}$$

where the cost $J^{\mathbf{q}}(\mu_{\theta_a}, \pi)$ is obtained as in (11.27) by applying threshold policy μ_{θ_a} in (11.33).

The parameters θ_a of the optimal parameterized policy can be estimated via simulation-based stochastic optimization. For example in [21], a policy gradient algorithm that uses the simultaneous perturbation stochastic approximation (SPSA) algorithm is given. We refer the reader to [35] for details of the SPSA algorithm.

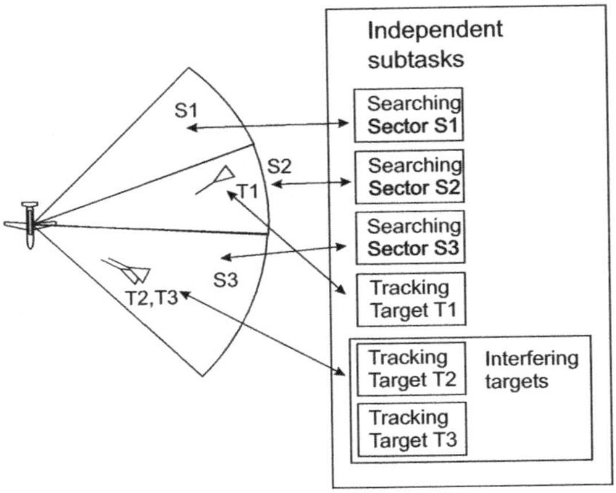

Figure 11.3 Formation of independent subtasks. S1, S2, and S3 represent sectors that should be searched, while T1, T2, and T3 are tracked targets.

11.3.3 Radar Macromanagement

Above, we have formulated the micromanagement problem as a sequential detection problem. We now return to the radar macromanagement problem. At the macromanager level, various subtasks requests need to be prioritized and serviced. These include searching various sectors and tracking various targets. Figure 11.3 illustrates some of these subtasks.

Macromanagement algorithms often involve metalevel rules. However, what is important for our purposes is that macromanagement rules invariably involve the performance (cost) $J^*_\mu(\pi_{n-1})$ of the micromanager from the inner level optimization.

The high-priority target a is chosen according to a metalevel rule based on several extrinsic factors. For example, the macromanager can pick the a based on the track variances (uncertainty), threat levels, and priority allocation vector $\mathbf{q} = (\mathbf{q}(1), \ldots, \mathbf{q}(L))$ of the L targets

$$\text{Macromanager picks target } a_{t+1} = \arg \max_{l \in \{1,2,\ldots,L\}} \text{priority}\Big(\text{track\_variance}(l), \text{threat}(l), \mathbf{q}(l)\Big). \quad (11.36)$$

It is the priority allocation vector $\mathbf{q}$ that couples the micro- and macromanagement problems. For example, the macromanager can optimize the above rule (11.36) with respect to $\mathbf{q}$ as follows

$$\text{Compute } \mathbf{q}^* = \arg\min_{\mathbf{q}} \left[\sum_{l=1}^{L} \text{priority}(\mathbf{q}, l)\, J^{\mathbf{q}}(\mu_l^*, \pi) \right], \quad (11.37)$$

$$\text{subject to constraints } \mathbf{q}(l) \geq 0, \quad \mathbf{A}\mathbf{q} \leq \mathbf{b}. \quad (11.38)$$

Here $\mathbf{A} \in \mathbb{R}^{N \times L}$ and $\mathbf{b} \in \mathbb{R}^N$ denote user-defined macromanager defined parameters that constraint the priority allocation $\mathbf{q}$. In (11.37), $J^{\mathbf{q}}(\mu_l^*, \pi)$ is the optimal cost from the micromanager's multivariate POMDP, see (11.27). The function "priority($\mathbf{q}$, l)" in (11.37) determines how the different targets are allocated priority. Naturally it should be an increasing function of $\mathbf{q}$.

11.4 RADAR SCHEDULING FOR MANEUVERING TARGETS MODELED AS JUMP MARKOV LINEAR SYSTEM

Here we consider an alternative formulation to the two-timescale sensor management of previous sections. We consider a maneuvering target modeled by a jump Markov linear system. Then myopic suboptimal sensor scheduling strategies are presented. The notation in this section is independent of the previous sections.

11.4.1 Formulation of Jump Markov Linear System Model

11.4.1.1 Target Mode. In order to model target maneuvers the underlying target trajectory model jumps between a finite set of known modes representing various trajectory models such as constant speed and heading, constant heading with linear acceleration, coordinated turn, and so on. This mode jumping can be modeled by a finite state Markov chain r_k. Let r_k denotes a discrete-time, time-homogeneous, s-state, first-order Markov chain with state space $\{e_1, \ldots, e_s\}$ where e_i denotes the unit s-dimensional vector with 1 in the ith entry. Denote the transition probabilities as $p_{m,n} = \mathbb{P}\{r_{k+1} = n | r_k = m\}$ for any $m, n \in S$ where $S = \{1, 2, ..., s\}$. The transition probability matrix $\mathcal{P} = [p_{m,n}]$, is thus an $s \times s$ stochastic matrix with elements satisfying $p_{m,n} \geq 0$ and $\sum_{n=1}^{s} p_{m,n} = 1$, for each $m \in S$. Denote the initial probability distribution by $p_m = \mathbb{P}\{r_1 = m\}$, for $m \in S$, such that $p_m \geq 0, \forall m \in S$ and $\sum_{m=1}^{s} p_m = 1$.

11.4.1.2 Target Kinematic State. Target trajectories of interest are modeled by a process $\{x_k\}$ where typically x_k is the target kinematic state (position, velocity) in Cartesian coordinates. Because the target model jumps according to the modes r_k, the target state x_k can be modeled as the following JMLS

$$x_{k+1} = A(r_{k+1})x_k + B(r_{k+1})w_k, \tag{11.39}$$

where $x_t \in \mathbb{R}^{n_x}$ is the system state and $w_k \overset{i.i.d.}{\sim} \mathcal{N}(0, I_{n_w}) \in \mathbb{R}^{n_w}$. $A(\cdot)$ and $B(\cdot)$ are functions of the Markov chain state r_k, that is $(A(\cdot), B(\cdot)) \subset \{(A(m), B(m)); m \in S\}$, and they evolve according to the realization of the finite state Markov chain r_k. We assume that $x_0 \sim \mathcal{N}(\hat{x}_0, P_0)$ where $P_0 > 0$ and let x_0 and w_k be mutually independent for all k.

11.4.1.3 Sensor Choice and Observation Process. Assume that at time k there are L_k possible sensors available that can be used to give measurements of the target's state x_k. However, at each time instant only one of these L_k possible sensors can be chosen. The variable $u_k \in \{1, \dots L_k\}$ denotes the sensor selected at time k from among the L_k possible sensors. u_k will be chosen to minimize a cost function as described later. For notational simplicity we will ignore the time dependence k of L_k.

Let the vector $y_k(u_k) \in \mathbb{R}^{n_y}$ denotes the observation vector at time k if sensor $u_k \in \{1, \dots, L\}$ was chosen. In tracking problems, as well as the "true" observation $y_k(u_k)$ several false measurements are also obtained by sensor u_k. These false measurements are typically modeled as detections that are uniformly distributed throughout the observation span of a particular sensor.

In general, the real target observation probability density $p(y_k|x_k, r_k, u_k)$ of a given sensor at time k could be non-Gaussian. However, to simplify our presentation, in this chapter we will use the following linear Gaussian sensor model

$$y_k(u_k) = C(r_k, u_k)\, x_k + D(r_k, u_k)\, v_k, \tag{11.40}$$

where $v_k \overset{i.i.d.}{\sim} \mathcal{N}(0, I_{n_v}) \in \mathbb{R}^{n_v}$ are i.i.d. Gaussian sequences and

$$D(i, j)D'(i, j) > 0, \forall i \in S,\ j \in \{1, 2, \dots, L\}, \tag{11.41}$$

that is, positive definite. We assume that x_0, w_k, and v_k are mutually independent for all k.

Equations (11.39) and (11.40) denote a controlled partially observed jump Markov linear system. The model parameters

$$\lambda = \{p_m, p_{mn}, A(m), B(m), C(m, u_k), D(m, u_k), \hat{x}_0, P_0; m, n \in S\}$$

are assumed known. This is a reasonable assumption in most tracking problems.

Note that if the realization of the Markov chain sequence r_k is known (or if $s = 1$) and there is only one sensor, then (11.39) and (11.40) reduce to a linear Gaussian state space model and the Kalman filter yields optimal (conditional mean) estimates of the target state. On the other hand, if $L = 1$, $A(r_k) = I$, $B(r_k) = 0$, then y_k in (11.40) specializes to a standard Hidden Markov model, that is, the finite state Markov chain r_k observed in i.i.d. noise.

Define

$$Y_k = \{u_1, u_2, \dots, u_k, y_1(u_1), y_2(u_2), \dots, y_k(u_k)\}, \tag{11.42}$$

so that Y_k represents the information available at time k upon which to base state estimates and sensor selection decisions. The sensor adaptive tracking problem proceeds in three stages for each $k = 0, 1, \dots, N-1$, where N is a fixed positive integer.

(1) *Selection* Based on Y_k we generate $u_{t+1} = \mu_{t+1}(Y_k)$ that determines which sensor is to be used at the next measurement time.

(2) *Observation* We then observe $y_{t+1}(u_{t+1})$, where u_{t+1} is the sensor selected in the previous stage.

(3) *Estimation* After observing $y_{t+1}(u_{t+1})$ we generate our best estimate $\hat{x}_{t+1}$ of the state of the Markov chain x_{t+1} as $\hat{x}_{t+1} = E\{x_{t+1}|Y_{t+1}\}$. Note that the state estimate $\hat{x}_{t+1}$ is dependent on the sequence of sensors selected from time 1 to $k+1$, that is $u_1, \ldots, u_{t+1}$ (since it depends on Y_{t+1}).

We define the sensor sequence

$$\mu = \{\mu_1, \mu_2, \ldots, \mu_N\}$$

and say that sequences are admissible if $u_{t+1} = \mu_{t+1}(Y_k)$, where

$$\mu_{t+1} : Y_k \rightarrow \{1, \ldots, M\}$$

is a sequence of functions.

We assume the following cost is associated with estimation errors and with the particular sensor sequence chosen. If based on the observation at time t, the decision is made to choose $u_{t+1} = l$ (i.e. to choose the lth sensor at time $k+1$ where $l \in \{1, \ldots, L\}$), then the instantaneous cost incurred at time k is

$$\alpha_k(l)\,\|x_k - \hat{x}_k\|^2 + \beta_k(l)\|r_k - \hat{r}_k\|^2 + \gamma_k(x_k, r_k, l). \tag{11.43}$$

Here $\alpha_k(l)$ and $\beta_k(l)$, $l = 1, 2 \ldots, L$ are known nonnegative scalar weights. $\hat{x}_k = E\{x_k|Y_k\}$ and $\hat{r}_k = E\{r_k|Y_k\}$ denote the optimal (conditional mean) estimates of x_k and r_k given the information sequence Y_k—note that these depend on the sequence μ. $\|x_k - \hat{x}_k\|^2$ denotes the mean square state estimation error in the target trajectory at time k due to the choice of sensor sequence $u_1, \ldots, u_k$. Thus, the first two terms denote the square of the tracking error estimate generated by sensor l. Finally, $\gamma_k(x_k, r_k, l) \geq 0$ denotes the instantaneous cost of using the sensor $u_{t+1} = l$ when the target's mode is r_k and the target's state is x_k.

Our aim is to find the sensor sequence that minimizes the total accumulated cost J_μ over the set of admissible sensor selection rules where

$$J_\mu = \mathbb{E}\left\{\sum_{t=1}^{N} \alpha_k(u_{t+1})\|x_k - \hat{x}_k\|^2 + \beta_k(u_{t+1})\|r_k - \hat{r}_k\|^2 + \sum_{t=0}^{N-1} \gamma_k(x_k, r_k, u_{t+1})\right\}, \tag{11.44}$$

where $u_{t+1} = \mu_{t+1}(Y_k)$. Note that the above expectation is with respect to (x_0, r_0) (which has prior distribution $\pi_0^{x_0, r_0}$), $v_1, \ldots, v_N, w_0, \ldots, w_{N-1}, r_0, \ldots, r_{N-1}$, see, for example, [26, pp. 211].

The terms $\alpha_k(l)$ and $\beta_k(l)$ allow different sensors $l \in \{1, 2, \ldots, L\}$ to be weighted differently and the time index in α_k allows us to weight the state estimate errors over time.

Define the conditional density of x_k, r_k given the history Y_k, that is,

$$\pi_k^{x_k,r_k}(x,i) \triangleq p(x_k = x, r_k = e_i | Y_k), \quad \text{where } x \in \mathbb{R}^{n_x}, i \in S = \{1, \ldots, s\}. \tag{11.45}$$

To simplify notation in the sequel define the probability density functions

$$\pi_k^{x_k}(x) \triangleq p(x_k = x | Y_k) = \sum_{i=1}^{s} \pi_k^{x_k,r_k}(x,i),$$

$$\pi_k^{r_k}(i) \triangleq \mathbb{P}(r_k = e_i | Y_k) = \int_{\mathbb{R}^{n_x}} \pi_k^{x_k,r_k}(x,i)dx, \quad \pi_k^{r_k} \triangleq \begin{bmatrix} \mathbb{P}(r_k = e_1|Y_k) \\ \vdots \\ \mathbb{P}(r_k = e_s|Y_k) \end{bmatrix}.$$

Using Bayes' rule it can be shown that the information state (filtered density) evolves as

$$\begin{aligned} \pi_{t+1}^{x_{t+1},r_{t+1}} &= \frac{1}{K_k} p(y_{t+1}|r_{t+1}, x_{t+1}, u_{t+1}) \\ &\int_{\mathbb{R}^{n_x}} \sum_{m=1}^{s} p(x_{t+1}, r_{t+1}|x_k = x, r_k = m)\pi^{x_k,r_k}(x,m)dx, \end{aligned} \tag{11.46}$$

where K_k denotes a normalization term. For ease of notation we express the above filtered density update (11.46) as

$$\pi_{t+1}^{x_{t+1},r_{t+1}} = T_k(\pi_k^{x_k,r_k}, y_{t+1}, u_{t+1}). \tag{11.47}$$

Defining $Q(r_k) = C(r_k)C'(r_k)$ and $\Sigma(r_k, u_k) = D(r_k, u_k)D'(r_k, u_k)$ for the JMLS (11.39) and (11.40), the various terms in (11.46) are

$$\begin{aligned} p(y_{t+1}|r_{t+1}, x_{t+1}, u_{t+1}) &\sim \mathcal{N}(C(r_{t+1}, u_{t+1}), \Sigma(r_{t+1}, u_{t+1})), \\ p(x_{t+1}, r_{t+1}|x_k = x, r_k = m) &= p_{m,r_{t+1}}\mathcal{N}(A(r_{t+1})x, Q(r_{t+1})). \end{aligned} \tag{11.48}$$

For the JMLS (11.39) and (11.40), the information state $\pi_k^{x_k,r_k}$ is a Gaussian mixture with s^k components. Computing $\pi_k^{x_k,r_k}$ and hence also $E\{x_k|Y_k\}$ or $E\{r_k|Y_k\}$ involve prohibitive computational cost that is exponential in k. Thus, in general, the solution to the sensor adaptive tracking problem for a JMLS target model is not practical without some sort of approximation.

11.4.2 Suboptimal Radar Scheduling Algorithms

In this section, we describe a one-step ahead suboptimal solution to the sensor selection problem with multiple jumping models. The idea behind the one-step ahead algorithm is to compute the expected posterior density for the target state for each sensor based on the current posterior density and the known measurement models, then make a measurement using the sensor that gave the best predicted cost. The idea is illustrated below for the no false measurement case. The false measurement case is briefly described in Section 3.3.

11.4.2.1 One-Step Ahead Algorithm. To simplify our notation, we will assume that the weights α and β are constants. The one-step ahead algorithm proceeds recursively as follows: Assume that $\pi_{k-1}^{x_{k-1},r_{k-1}}$ (or a suitable approximate value) has been computed at time $k-1$.

Step 1 Minimize at time k the instantaneous cost defined in (11.43) that can be re-expressed in terms of the information state as

$$J_{\mu_k} = \alpha\left[\langle x^2, \pi_k^{x_k}\rangle - \langle x, \pi_k^{x_k}\rangle^2\right] + \beta\left(1 - \pi_k^{r_k\prime}\pi_k^{r_k}\right) + \langle \gamma_k'(\cdot,\cdot,u_k), \pi_k^{x_k,r_k}\rangle \tag{11.49}$$

subject to $\pi_k^{x_k,r_k} = T(\pi_{k-1}^{x_{k-1},r_{k-1}}, y_k, u_k)$, where $T(\cdot)$ is defined in (11.46).

Using the dynamic programming recursion (11.28), the above cost function is straightforwardly minimized by

$$V_k(\pi^{x_k,r_k}) = \alpha\left[\langle x^2, \pi_k^{x_k}\rangle - \langle x, \pi_k^{x_k}\rangle^2\right] + \beta\left(1 - \pi_k^{r_k\prime}\pi_k^{r_k}\right), \tag{11.50}$$

$$\begin{aligned} u_k^* = \arg\min_u & \left[\langle \gamma_{k-1}'(\cdot,\cdot,u_k), \pi_{k-1}^{x_{k-1},r_{k-1}}\rangle\right. \\ & \left. + E_{y_k}\left\{V_k(T(\pi_{k-1}^{x_{k-1},r_{k-1}}, y_k, u))\right\}\right]. \end{aligned} \tag{11.51}$$

Step 2 Set $\pi_k^{x_k,r_k} = T(\pi_{k-1}^{x_{k-1},r_{k-1}}, y_k, u_k^*)$

Step 3 $k \to t+1$

For a JMLS, the information state $\pi_k^{x_k,r_k}$ is a Gaussian mixture with s^t components. So in principle (although the computational complexity grows exponentially with k), the above one-step ahead algorithm provides an algorithm for the sensor adaptive tracking problem. Below we discuss how an approximation of $\pi_k^{x_k,r_k}$ can be efficiently computed using the IMM algorithm—the resulting algorithm will be termed the sensor adaptive IMM (SA-IMM) algorithm. But first, we briefly consider the linear Gaussian and HMM cases where π_k is easily computable.

Linear quadratic sensor selection Given $P_{k-1|k-1}$, it follows from (11.51) that the one-step ahead algorithm at time k for selecting the optimal sensor is

$$u_k^* = \arg\min_{u\in\{1,\dots,L\}}\left[P_{t|t}(u) + \gamma_{k-1}(u)\right],$$

where $P_{t|t}(u_k)$ evolves according to the Kalman filter covariance update. The one-step ahead algorithm only requires L Kalman covariances to be evaluated at each time.

HMM sensor scheduling The information state evolves according to the well-known HMM filter [36]

$$\pi^{r_k} = T(\pi_{k-1}^{r_{k-1},y_k,u_k}) = \frac{\phi(y_k(u_k))\mathcal{P}'\pi_{k-1}^{r_{k-1}}}{\mathbf{1}'\phi(y_k(u_k))\mathcal{P}'\pi_{k-1}^{r_{k-1}}},$$

where $\mathcal{P}$ denotes the transition probability matrix of r_k (see Section 11.2.2),

$$\phi(y_k(u_k)) = \text{diag}\left[p(y_k|r_k = e_1, u_k), \ldots, p(y_k|r_k = e_s, u_k)\right]$$

and $\mathbf{1}$ denotes the column vector of ones. Given $\pi_{k-1}^{r_{k-1}}$, the one-step ahead adaptive sensor solution for HMMs becomes

$$u_k^* = \arg \min_{u \in \{1,\ldots,L\}} \gamma_{k-1}'(\cdot, u)\pi_{k-1}^{r_{k-1}} + \beta \sum_{m=1}^{M} \left[1 - \frac{\pi_{k-1}^{r_{k-1}\prime}\mathcal{P}\phi^2(y_k(u_k))\mathcal{P}'\pi_{k-1}^{r_{k-1}}}{(\prime\phi(y_k(u_k))\mathcal{P}'\pi_{k-1}^{r_{k-1}})^2}\right].$$

11.4.2.2 Sensor Adaptive IMM (SA-IMM) Algorithm. The IMM algorithm computes the approximate information state for a JMLS in a computational efficient manner by approximating the information state π^{x_k, r_k} (s^t component Gaussian mixture) by an s-component Gaussian mixture at each time instant k. It is worthwhile mentioning that there are several other suboptimal algorithms for approximating the information state of a JMLS—in particular, particle filter-based sequential sampling methods appear to be extremely promising, see [6] for details.

Given an approximate computation of the state $\pi^{x_{k-1}, r_{k-1}}$, the SA-IMM algorithm operates recursively via the steps below.

Step 1—IMM algorithm Compute an approximation of the updated information state $T(\pi^{x_{k-1}, r_{k-1}}, y_k, u_k)$ as follows. We have $\pi^{x_k, r_k} = \pi^{r_k} p(x_k|r_k, Y_k)$. Furthermore, π^{r_k} and $p(x_k|r_k, Y_k)$ can be expressed recursively according to the following two equations:

(i) By elementary application of Bayes' rule we have

$$p(x_k|r_k, Y_k) = p(y_k|x_k, r_k, u_k)p(x_k|r_k, Y_{k-1}) \tag{11.52}$$

$$= p(y_k|x_k, r_k, u_k)\int_{\mathbb{R}^{n_x}} p(x_k|x_{k-1}, r_k)p(x_{k-1}|r_k, Y_{k-1})dx_{k-1} \tag{11.53}$$

$$= p(y_k|x_k, r_k, u_k)\int_{\mathbb{R}^{n_x}} p(x_k|x_{k-1}, r_k) \tag{11.54}$$

$$\times \left[\frac{\sum_{i=1}^{s} p_{i,r_k} p(x_{k-1}|r_{k-1} = e_i, Y_{k-1})p(r_{k-1} = e_i|Y_{k-1})}{\sum_{i=1}^{s} a_{i,r_k} p(r_{k-1} = i|Y_{k-1})}\right] dx_{k-1}. \tag{11.55}$$

(ii) Modal update

$$\pi^{r_k}(j) = \sum_{i=1}^{s} p_{i,j}\pi^{r_{k-1}}(i) \int_{\mathbb{R}^{n_x}} p(y_k|x_k, r_k = e_j, u_k)p(x_k|r_k = e_j, Y_{k-1})dx_k.$$

Consider the term $p(x_k|r_k = e_j, Y_{k-1})$ enclosed in square brackets on the last line of (11.55). The IMM algorithm approximates this by a single Gaussian density with the same mean and variance for each $j = 1, \ldots, s$. Note that the densities $p(y_k|x_k, r_k, u_k)$, $p(x_k|x_{k-1}, r_k)$ are Gaussian as described in (11.48). Thus with the IMM approximation, π^{x_k, r_k} is a Gaussian mixture with s components.

Step 2—sensor adaptive step Compute u_k^* according to (11.51).

Step 3—IMM state update Update the information state π^{x_k,r_k} using the new observation $y_k(u_k^*)$ according to the IMM algorithm. Each of the components $p(x_k|r_k = e_j, Y_k)$, $j = 1, \ldots, s$ can be computed via a Kalman filter—hence the SA-IMM algorithm involves a bank of s parallel Kalman filters, see [1] for details.

Note that the IMM state estimator above, can be replaced with a sequential Monte Carlo (particle filter) state estimator. The algorithm is similar to the sensor adaptive IMM algorithm presented above, except that the information state in Step 1 and Step 3 is updated via a sequential Monte Carlo (particle filtering) algorithm. There are several classes of particle filtering algorithms that can be used, [6] presents several algorithms for jump Markov linear systems together with variance reduction methods.

11.4.2.3 Sensor Adaptive IPDA (SA-IPDA). When false measurements (clutter) are present the usual procedure is to identify which measurements are "close" enough to the predicted measurement for a particular target track then use this set of "validated" measurements in the track update procedure. The nearest-neighbor rule simply selects the validated measurement closest to the predicted measurement position whereas the Probabilistic Data Association (PDA) and IPDA algorithms use all of the validated measurements [1–3]. The track is updated with each measurement in the validated set and the updated estimates are combined in a weighted averaging process to form the new estimate. The weighting is proportional to the likelihood of each measurement scenario. The IPDA algorithm differs from the standard PDA algorithm in that IPDA recursively computes the probability of target existence along with the track estimate. Thus, IPDA recursively computes the probability that the trail of detections currently being followed by the tracker originate from a real target rather than being a random collection of detections arising from clutter that happen to appear target like.

In the SA-IPDA algorithm we must decide on which sensor to use next, based only on the current state estimate and known statistical properties of each sensor and the clutter. Let ρ_k be the false measurements at time k and Ω_k the sequence of all real and false measurements and sensor choices up to and including time k. The SA-IPDA algorithm computes the expected posterior density if sensor l was used by first computing the posterior density if real measurement y_k and clutter ρ_k were obtained and then averaging it over y_k and ρ_k, with x_k, Ω_{k-1}, and u_k fixed, to yield

$$\begin{aligned}
\bar{p}(x_k|\,\Omega_{k-1}, u_k = l) &\triangleq E\{p(x_k|y_k, \rho_k, u_k = l, \Omega_{k-1})|\, x_k, \Omega_{k-1}, u_k = l\} \\
&= \int\int \frac{p(y_k, \rho_k|x_k, u_k = l)p(x_k|\Omega_{k-1})}{p(y_k, \rho_k|\Omega_{k-1}, u_k = l)} p(y_k, \rho_k|x_k, u_k = l)dy_k d\rho_k \\
&= p(x_k|\Omega_{k-1}) \int\int \frac{p(y_k, \rho_k|x_k, u_k = l)^2}{p(y_k, \rho_k|\Omega_{k-1}, u_k = l)} dy_k d\rho_k. \qquad (11.56)
\end{aligned}$$

The sensor l^* that minimizes a specific cost related to this expected density after normalization is chosen and a standard PDA or IPDA update follows.

Instead of minimizing a cost of the expected posterior density, another approach is to minimize the expected cost of the actual posterior density. For instance, we could compute the variance of the posterior density given y_k, ρ_k, $u_k = l$, and Ω_{k-1}, average it over y_k and ρ_k to yield

$$E\{\text{var}\{x_k | y_k, \rho_k, \Omega_{k-1}, u_k = l\} | \Omega_{k-1}, u_k = l\}$$

and select the minimizing sensor. In the linear Gaussian case with uniformly distributed false measurements, this is equivalent to computing a modified matrix Riccati equation update for each sensor, similar to that used in [8].

11.5 SUMMARY

This chapter has presented several methodologies for a stochastic control approach to radar resource management. The radar resource management task was decomposed into micromanagement at the fast timescale and macromanagement at the slow timescale. The micromanagement problem deals with scheduling the optimal Bayesian filter while the macromanagement problem deals with allocating target priority. It was shown that the optimal micromanagement policy has a simple threshold structure thereby permitting the use of simulation-based stochastic optimization algorithms to compute it. Finally, one-step ahead algorithms were given for sensor scheduling involving more general maneuvering target models.

There are other applications of radar resource management that we have not covered in this chapter. For example, [37] deals with electronic counter measures for missile deflection where ships needs to adapt their jamming signals to deflect multiple incoming missiles. In these papers a dynamic game-theoretic setting is developed. Also in a more abstract setting, similar methodologies used in this chapter can be used for managing wireless sensor networks. Extensions involving game theoretic settings are given in [38, 39]. Finally, in recent work [40, 41] we have used similar formulations to examine how local and global decision makers interact in a social learning setup. Such models can capture the interaction between the micro- and macromanagers.

REFERENCES

1. Y. Bar-Shalom and X.R. Li, *Estimation and Tracking: Principles, Techniques and Software*, Artech House, Boston, 1993.
2. S. Blackman and R. Popoli, *Design and Analysis of Modern Tracking Systems*, Artech House, Norwood, MA, USA, 1999.
3. D. Musicki, R. Evans, and S. Stankovic, Integrated probabilistic data association. *IEEE Trans. Auto. Control*, 39(6), 1237–1241, 1994.
4. D. Reid, An algorithm for tracking multiple targets. *IEEE Trans. Auto. Control*, 24(6), 843–854, 1979.

5. V. Krishnamurthy, Algorithms for optimal scheduling and management of hidden Markov model sensors. *IEEE Trans. Signal Proc.*, 50(6), 1382–1397, 2002.
6. A. Doucet, N. Gordon, and V. Krishnamurthy, Particle filters for state estimation of jump Markov linear systems. *IEEE Trans. Signal Proc.*, 49(3), 613–624, 2001.
7. D.E. Kershaw and R.J. Evans, Optimal waveform design for tracking systems. *IEEE Trans. Inform. Theory*, 40(5), 1536–1550, 1994.
8. D. Kershaw and R. Evans. Waveform selective probabilistic data association. *IEEE Trans. Aerosp. Electron. Syst.*, 33(4), 1180–1188, 1997.
9. J. Wintenby and V. Krishnamurthy, Hierarchical resource management in adaptive airborne surveillance radars—a stochastic discrete event system formulation. *IEEE Trans. Aerosp. Electron. Syst.*, 20(2), 401–420, 2006.
10. V.C. Vannicola and J.A. Mineo, Expert system for sensor resource allocation. *Proceedings of the 33rd Midwest Symposium on Circuits and Systems*, 1990.
11. G. Van Keuk and S. Blackman, On phased array radar tracking and parameter control. *IEEE Trans. Aerosp. Electron. Syst.*, 29(1), 186–194, 1993.
12. W. Koch, On adaptive parameter control for phased-array tracking. *Proceedings of the SPIE Conference on Signal and Data Processing of Small Targets*, Vol. 3809, 1999, pp. 444–455.
13. D. Strömberg and P. Grahn, Scheduling of tasks in phased array radar. *IEEE International Symposium on Phased Array Systems and Technology*, 1996.
14. C. Kreucher, A. Hero, and K. Kastella, A comparison of task driven and information driven sensor management for target tracking. *Proceedings of IEEE Control and Decision Conference*, 2005.
15. R. Washburn, M. Schneider, and J. Fox, Stochastic dynamic programming based approaches to sensor resource management. *5th International Conference on Information Fusion*, Annapolis, Maryland, USA, 2002, pp. 608–615.
16. V. Krishnamurthy and R.J. Evans, Hidden Markov model multi-arm bandits: a methodology for beam scheduling in multi-target tracking. *IEEE Trans. Signal Proc.*, 49(12), 2893–2908, 2001.
17. S. Howard, S. Suvorova, and B. Moran, Optimal policy for scheduling of gauss-markov systems. *Proceedings of the 7th International Conference on Information Fusion*, 2004.
18. V. Krishnamurthy, Emission management for low probability intercept sensors in network centric warfare. *IEEE Trans. Aerosp. Electron. Syst.*, 41(1), 133–152, 2005.
19. D. Castañon, Approximate dynamic programming for sensor management. *Proceedings of the 36th Conference on Decision and Control*, 1997.
20. V. Krishnamurthy and D. Djonin, Structured threshold policies for dynamic sensor scheduling—a partially observed Markov decision process approach. *IEEE Trans. Signal Proc.*, 55(10), 4938–4957, 2007.
21. V. Krishnamurthy and D.V. Djonin, Optimal threshold policies for multivariate POMDPs in radar resource management. *IEEE Trans. Signal Process.*, 57(10), 3954–3969, 2009.
22. R. Evans, V. Krishnamurthy, and G. Nair, Networked sensor management and data rate control for tracking maneuvering targets. *IEEE Trans. Signal Proc.*, 53(6), 1979–1991, 2005.
23. R.R. Mohler and C.S. Hwang, Nonlinear data observability and information. *J. Franklin Inst.*, 325(4), 443–464, 1988.

24. A. Logothetis and A. Isaksson, On sensor scheduling via information theoretic criteria. *Proceedings of the American Control Conference*, San Diego, 1999, pp. 2402–2406.

25. T.M. Cover and J.A. Thomas, *Elements of Information Theory*, Second Edition, Wiley-Interscience, New York, 2006.

26. D.P. Bertsekas, *Dynamic Programming and Optimal Control*, Vols 1 and 2, Athena Scientific, Belmont, Massachusetts, 2000.

27. A.R. Cassandra, *Exact and Approximate Algorithms for Partially Observed Markov Decision Process*, Ph.D. thesis, Brown University, 1998.

28. W.S. Lovejoy, A survey of algorithmic methods for partially observed Markov decision processes. *Ann. Operations Res.*, 28, 47–66, 1991.

29. R.D. Smallwood and E.J. Sondik, Optimal control of partially observable Markov processes over a finite horizon. *Operations Res.*, 21, 1071–1088, 1973.

30. D.P. Heyman and M.J. Sobel, *Stochastic Models in Operations Research*, Vol. 2, McGraw-Hill, 1984.

31. W.S. Lovejoy, On the convexity of policy regions in partially observed systems. *Operations Res.*, 35(4), 619–621, 1987.

32. U. Rieder, Structural results for partially observed control models. *Methods Models Operations Res.*, 35, 473–490, 1991.

33. V. Krishnamurthy, R. Bitmead, M. Gevers, and E. Miehling, Maximizing stochastic observability in gmti radar management: A sequential detection approach. *IEEE Trans. Signal Processing*, 60(2), 700–714, 2011.

34. D.G. Luenberger, *Linear and Nonlinear Programming*, 2nd edition, Addison Wesley, Reading, MA, USA, 1984.

35. J. Spall, *Introduction to Stochastic Search and Optimization*, Wiley, New York, NY, USA, 2003.

36. R.J. Elliott, L. Aggoun, and J.B. Moore, *Hidden Markov Models—Estimation and Control*, Springer-Verlag, New York, 1995.

37. M. Maskery and V. Krishnamurthy, Network-enabled missile deflection: games and correlation equilibrium. *IEEE Trans. Aerosp. Electron. Syst.*, 43(3), 843–863, 2007.

38. V. Krishnamurthy, Decentralized spectrum access amongst cognitive radios—an interacting multivariate global game-theoretic approach. *IEEE Trans. Signal Proc.*, 57(10), 3999–4013, 2009.

39. V. Krishnamurthy, M. Maskery, and G. Yin, Decentralized activation in a ZigBee-enabled unattended ground sensor network: A correlated equilibrium game theoretic analysis. *IEEE Trans. Signal Proc.*, 56(12), 6086–6101, 2008.

40. V. Krishnamurthy, Bayesian sequential detection with phase-distributed change time and nonlinear penalty—a lattice programming pomdp approach. *IEEE Trans. Inform. Theory*, 57(3), 2011. http://arxiv.org/abs/1011.5298.

41. V. Krishnamurthy, Quickest detection with social learning: how local and global decision makers interact. *IEEE Trans. Inform. Theory*, 58, 2012. http://arxiv.org/abs/1007.0571.

CHAPTER 12

Sensor Management for Large-Scale Multisensor-Multitarget Tracking

RATNASINGHAM THARMARASA AND THIA KIRUBARAJAN

12.1 INTRODUCTION

12.1.1 Sensor Management

Sensor management has become a crucial part of tracking because of technological advances in recent years [1]. The use of a large number of sensors, which can be deployed all over the surveillance region, in tracking applications has become feasible because of the availability of cheap sensors (e.g., a large number of sonobuoys used in antisubmarine warfare (ASW)). In the context of target tracking, major issues in sensor management are optimal path planning [2–4], optimal sensor placement [5–7], and optimal sensor selection [8–10].

In optimal path planning, the objective is to find the optimal and, in the presence of any threat, safe path of a moving sensor, such as unmanned aerial vehicle (UAV), under the given constraints, such as maximum velocity and turn rate, to monitor the surveillance region [2]. In the optimal sensor placement, the objective is to decide where, when, and how many sensors must be placed such that the target state estimation errors remain below a certain threshold level [5]. In the optimal sensor selection, one has to decide which of the already deployed sensors must be used at each measurement time step in order to maximize the tracking performance [11]. These sensor management problems are a class of partially observed Markov decision processes [12]. In this chapter, the problem of optimal sensor selection is considered in centralized, distributed, and decentralized tracking systems for multitarget tracking purposes [10, 13–15]. Multiple, possibly time varying, number of targets, which may be closely spaced, is considered in large-scale problems. Mathematical formulations of sensor selection problems, and techniques to find suboptimal solutions in real time are also discussed in this chapter.

Integrated Tracking, Classification, and Sensor Management: Theory and Applications, Edited by Mahendra Mallick, Vikram Krishnamurthy, and Ba-Ngu Vo.

In general, sensor management decisions, such as deploying new sensors or activating existing ones must be made with a lead time that is needed to carry out the necessary operations. Hence, sensors must be managed in order to control the estimated/predicted future performance of the trackers. The posterior Cramér–Rao lower bound (PCRLB) gives a measure of the achievable optimum performance. Furthermore, this bound can be calculated predictively [5, 16–18]. The PCRLB is independent of the filtering algorithm employed, and is therefore not constrained by the idiosyncrasies of any particular filtering methodology. In this chapter, the PCRLB is used as the measure of tracking performance. Section 12.3 describes the PCRLB for multitarget tracking. However, the sensor management algorithms proposed in this chapter are not restricted to the PCRLB. The PRCLB can be replaced with any other performance measure such as Bhattacharyya bound, Weiss–Weinstein bound, Kullback–Leibler (KL) divergence, Renyi divergence, and so on [19].

12.1.2 Centralized Tracking

In Section 12.4, the sensor selection problem for centralized tracking system, in which all the deployed sensors are connected to a central fusion center (CFC), is considered [10, 13, 20, 21]. In order to get maximum information, one might want to use as many sensors as possible at each time step. Note that the information from any sensor is always positive as long as the accuracy of its measurements is known. However, physical limitations (e.g., limited frequency channels, limited capacity of processor) make possible for only a subset of sensors to be used at any one time [22]. Then, the problem is to select a subset of sensors that yields optimal tracking performance.

In general once a sensor subset is activated, it remains active for some time until the next sensor subset becomes active. That is, measurements are received from a sensor subset for a number of consecutive sensor revisit times. Although a particular sensor subset may be optimal at the time of initial activation, due to target motion, it may not remain optimal throughout the entire activation period. In order to get a sensor subset that remains optimal during the entire activation interval and not just at the time of activation, the overall tracking performance over the entire activation period is optimized.

In this chapter, two different scenarios are considered under the centralized architecture. In the first scenario, it is assumed that the total number of targets in the surveillance region is known and fixed. The problem is then to select the sensors that allow one to best alleviate the measurement association ambiguities and thereby provide accurate target state estimates. Next, a scenario in which the number of targets in the surveillance region is unknown and time varying is considered. The problem is then to select sensors to both accurately track existing targets and quickly detect new ones. This problem can be formulated as a bicriterion optimization; one objective is to provide accurate state estimates of existing targets and the other is to maximize the probability of detecting new targets.

Strategies for selecting the optimal sensor subset (n out of N) based on the multitarget PCRLB is presented. If the total number of possible combinations $\binom{N}{n}$ is small, then each and every combination can be evaluated and the best one can be selected.

However, in large-scale problems, finding the optimal sensor subset by complete enumeration is not viable. A search technique is presented to obtain a suboptimal solution in real time. Local searches are typically sensitive to the initial guess, and an intelligent choice for the initial guess is also presented.

12.1.3 Distributed Tracking

Sensor selection problem for distributed architecture, which has few local fusion centers (LFCs) and one CFC [14, 21, 23], is discussed in Section 12.5. In distributed architecture, each LFC uses a certain number of sensors at each time steps to obtain information about the surveillance region. LFCs update the estimates of their local tracks using the measurements that are received from their local sensors via wireless channel and then report their estimates to CFC, which will combine all of these information to form the global tracks. Then, the problem is to decide which of the available sensors must be used by each LFC, and decide what frequency channel and transmission power (if controllable) should be used by each active sensor.

In some cases, it is assumed that the LFCs are not fixed, that is, the locations or the number of LFCs can be changed [24, 25]. This might be true if all the deployed sensors can act as a LFC and/or the LFCs are moving platforms. However, this is not always possible. Hence, in this chapter the scenario where the LFCs are different from the sensors is considered. After the deployment of all the sensors and the LFCs, the locations and the number of them are fixed.

In order to reduce the computational load, sensors may be clustered based on target or LFC locations and each cluster is handled separately [26, 27]. However, target-based clustering is not possible for two reasons: first, the number of targets may be different from the number of LFCs; second, targets may be closely spaced. In some cases, it is assumed that a particular number of sensors is required to track each target and one sensor can detect only one target [28–30]. This is not a valid assumption in practice. If a good estimate is available for one target at the current time, that target could be tracked even with one sensor at the next measurement time. In addition, if there are many targets in the coverage region of a sensor, then a sensor can detect all of them with corresponding detection probabilities. LFC-based clustering is also not optimal if many targets are in one LFC's field of view while no or very few targets are in the fields of view of other LFCs. Hence, in this chapter no clustering is used for distributed architecture.

In general, sensors and fusion centers communicate through wireless frequency channels. Hence, the bandwidth limitation, which limits the communication, is an important issue [31–33]. In centralized tracking, it is not necessary to take into consideration the cochannel interference since all the sensors are connected to one fusion center and each sensor uses a different frequency channel. The advantage of distributed tracking is the reusability of the frequency channels. However, there is a trade-off between reusability and reachability. The transmission power of the sensors can be software controllable with a minimum and maximum limits [34]. This will enable us to save the power as well as to increase the reachability of each sensor.

In this chapter, the following scenario is considered for distributed architecture: the sensors and the LFCs are already deployed and their positions are known; each LFC can use a certain maximum number of sensors because of the physical limitation [35]; only a limited number of frequency channels are available; transmission power of each sensor is software controllable within certain lower and upper limits; in order to extract the signal, signal-to-noise ratio (SNR) at each LFC for each frequency must be greater than or equal to a known threshold level; LFCs communicate with CFC at each measurement time steps; active sensors are changed at every measurement step. In order to eliminate the redundancy, an additional constraint that one sensor can be used by at most one LFC is enforced.

12.1.4 Decentralized Tracking

Finally, in Section 12.6, the above two works are extended to decentralized architecture [15, 20, 36]. In the distributed architecture, CFC can perform sensor selection for all the local fusion centers by considering whole surveillance region information and then by instructing the LFC to use the selected sensors. Even though this is the optimal way to select the sensors, this may not always be feasible due to computational and communication constraints. Hence, the decentralized architecture, which does not have any CFC and where each fusion center (FC) can communicate only with its neighbors, is considered [20, 37]. Since each FC communicates only with its neighbors, no FC has the global picture of all the targets and the sensors in the surveillance region. Hence, each FC must select its own sensors by considering which sensors are/may be used by its neighbors.

The main challenge in decentralized tracking is to decide what type of track information to communicate between the FCs so that communication requirement is low, fusion can be made easily and the result will be close to optimal. The possible options are measurements, tracks and tracklets [20, 38]. Even though communicating the measurements is the optimal way, it requires high communication bandwidth. Communicating tracks requires less communication compared to measurements. However, in this architecture a track information from a FC can reach another fusion center via multiple FCs directly or indirectly, and avoiding multiple counting of that information is difficult. Tracklet, which is a track for a target based on only the most recent measurements from its local sensors since the data from the tracker was last communicated for that target, is a good choice for solving the above problem [38]. In addition, tracklets require less communication and easy to fuse. Hence, in this chapter it is assumed that tracklets are communicated among the fusion centers.

Furthermore, two different scenarios are considered under decentralized tracking. In the first scenario, it is assumed that communication and sensor selection are synchronized over the FCs. That is, all the fusion centers change their sensors at the same time, send their tracklets just before sensor selection, and their sensor change intervals are equal and known. In this case, each fusion center has to decide its sensors by considering which sensors may be used by neighbors. Note that no FC precisely knows which sensors will be used by its neighbors.

Even though avoiding clustering, as explained for distributed tracking, will help improve the tracking performance, there is a possibility for a sensor to be selected by more than one fusion center. Using one sensor by more than one fusion center will introduce redundancy as well as correlation between tracklets. Hence, in order to avoid this problem sensors are clustered based on FCs, that is, a sensor can only be used by the nearest FC.

In the second scenario, it is assumed that FCs may change their sensors at different time steps, and exact sensor change times of neighbors are unknown, instead only an interval is known. This in turn introduces additional uncertainties, such as when new sensor subset will be used by neighbors and when tracklets will be received. Therefore, these uncertainties must be incorporated in the objective functions.

12.1.5 Organization of the Chapter

This chapter is structured as follows: Section 12.2 describes the target tracking architectures. Section 12.3 explains the derivation of the PCRLB for multitarget tracking. Sections 12.4–12.6 provide algorithms for sensor subset selection for centralized, distributed, and decentralized architectures, respectively. Conclusions are given in Section 12.7.

12.2 TARGET TRACKING ARCHITECTURES

Three major types of architecture are commonly used in multisensor-multitarget tracking applications: centralized, distributed, and decentralized architectures [20, 37, 39].

12.2.1 Centralized Tracking

In the centralized architecture, shown in Figure 12.1a, more than one sensor are monitoring the region of interest. All the sensors generate the measurements at each revisit time and report those measurements to a central fusion center that fuses all

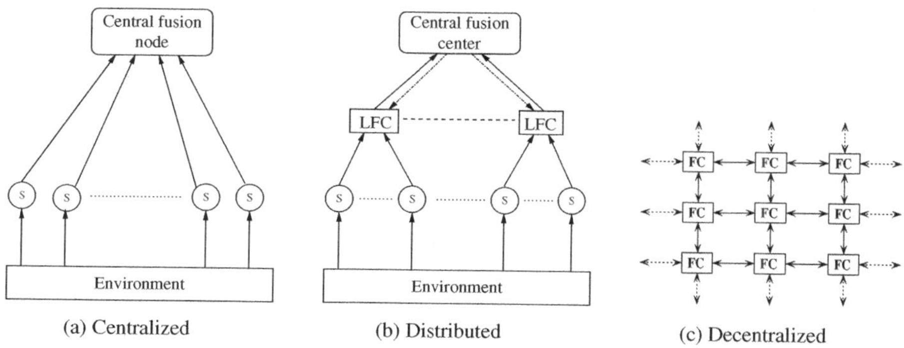

Figure 12.1 Architectures.

the measurements and updates the tracks. This is the optimal architecture in terms of tracking performance. However in a large surveillance region with many sensors, this architecture may not be feasible because of limited resources, that is, communication bandwidth, computation power.

12.2.2 Distributed Tracking

In order to avoid the heavy communication and computation requirement of centralized architecture, distributed architecture, also known as "hierarchical architecture," is used as an alternative, shown in Figure 12.1b [39]. In this architecture, sensors are connected to LFCs and LFCs are in turn connected to a CFC. Each LFC updates its local tracks based on the measurements from the local sensors and sends its tracks to CFC. Then, the CFC performs the track-to-track fusion and may send back the updated tracks to the LFCs if feedback path is available.

12.2.3 Decentralized Tracking

In some cases, neither centralized nor distributed tracking is possible, if there is no central fusion center that can communicate with all the sensors or LFC in a large surveillance region. In such cases, decentralized architecture, shown in Figure 12.1c, which has multiple FCs and no CFC, is used [39]. In this architecture, each FC gets the measurements from one or more sensors that are connected to it, and uses those measurements to update its tracks. In addition, tracks are also updated whenever an FC gets additional information from its neighbors. Even though many FCs are available, each FC can communicate only with its neighbors, the FCs within the communication distance, every few measurement time steps.

12.3 POSTERIOR CRAMÉR–RAO LOWER BOUND

Let X_k be an unknown and random state vector, and let $\hat{X}_k(Z_k)$ be an unbiased estimate of X_k based on the measurement data, Z_k. The PCRLB, which is defined to be the inverse of the Fisher information matrix (FIM), $J(k)$ [40], then gives a lower bound of the error covariance matrix, that is,

$$C(k) \triangleq \mathbb{E}\{[\hat{X}_k(Z_k) - X_k][\hat{X}_k(Z_k) - X_k]'\} \geq J(k)^{-1}, \tag{12.1}$$

where $\mathbb{E}$ denotes expectation over (X_k, Z_k). The inequality in (12.1) means that $C(k) - J(k)^{-1}$ is a positive semidefinite matrix.

A recursive formula for the evaluation of the posterior FIM, $J(k)$, is given by [17]

$$J(k) = \underbrace{D_k^{22} - D_k^{21}\left(J(k-1) + D_k^{11}\right)^{-1} D_k^{12}}_{J_X(k)} + J_z(k), \tag{12.2}$$

where

$$D_k^{11} = \mathbb{E}\left\{-\Delta_{X_{k-1}}^{X_{k-1}} \ln p(X_k|X_{k-1})\right\}, \tag{12.3}$$

$$D_k^{12} = \mathbb{E}\left\{-\Delta_{X_{k-1}}^{X_k} \ln p(X_k|X_{k-1})\right\}, \tag{12.4}$$

$$D_k^{21} = (D_k^{12})', \tag{12.5}$$

$$D_k^{22} = \mathbb{E}\left\{-\Delta_{X_k}^{X_k} \ln p(X_k|X_{k-1})\right\}, \tag{12.6}$$

$$J_z(k) = \mathbb{E}\left\{-\Delta_{X_k}^{X_k} \ln p(z_k|X_k)\right\}, \tag{12.7}$$

and Δ_α^β is a second-order partial derivative operator whose (i, j)th term is given by

$$\Delta_\alpha^\beta(i, j) = \frac{\partial^2}{\partial\alpha(i), \partial\beta(j)}, \tag{12.8}$$

$\alpha(i)$ and $\beta(i)$ are the ith components of vectors α and β, respectively. In the above, $z_k = \left[z_k(1),\ z_k(2),\ \ldots,\ z_k(n)\right]$, where $z_k(i)$ is the measurement vector at sensor i at sampling time k and n is the number of sensors utilized at sampling time k.

12.3.1 Multitarget PCRLB for Centralized Tracking

Let the state vector at time k, obtained by stacking the state vectors of all targets, be denoted by $X_k = \left[x_k^1, x_k^2, \ldots, x_k^T\right]'$, where x_k^t is the state vector of target t and T is the total number of targets in the surveillance region. If we assume that targets are moving independently and the state equation of each target is linear, then the overall state equation is given by

$$X_k = F_{k,k-1}X_{k-1} + \nu_{k,k-1}, \tag{12.9}$$

where

$$F_{k,k-1} = \text{diag}\left(F_{k,k-1}^1,\ F_{k,k-1}^2,\ \ldots,\ F_{k,k-1}^T\right), \tag{12.10}$$

$$\nu_{k,k-1} = \left[\nu_{k,k-1}^1,\ \nu_{k,k-1}^2,\ \ldots,\ \nu_{k,k-1}^T\right]'. \tag{12.11}$$

In the above, $F_{k,k-1}^t$ is the state transition matrix and $\nu_{k,k-1}^t$ is the process noise of target t to sampling time k from sampling time $k-1$. If $\nu_{k,k-1}^t$ is Gaussian with zero mean and covariance $Q_{k,k-1}^t$, then the covariance matrix of $\nu_{k,k-1}$, $Q_{k,k-1}$, is given by

$$Q_{k,k-1} = \text{diag}\left(Q_{k,k-1}^1,\ Q_{k,k-1}^2,\ \ldots,\ Q_{k,k-1}^T\right). \tag{12.12}$$

It can easily be shown that in the case of linear, Gaussian dynamics (e.g., [41]) we have

$$D_k^{11} = F'_{k,k-1} Q_{k,k-1}^{-1} F_{k,k-1}, \tag{12.13}$$

$$D_k^{12} = -F'_{k,k-1} Q_{k,k-1}^{-1}, \tag{12.14}$$

$$D_k^{22} = Q_{k,k-1}^{-1}. \tag{12.15}$$

Using the Matrix inversion lemma and (12.13)–(12.15), it can then be shown that

$$J_X(k) = \left[Q_{k,k-1} + F_{k,k-1} J(k-1)^{-1} F'_{k,k-1}\right]^{-1}. \tag{12.16}$$

The matrix $J_X(k)$ gives the prior information regarding the target states at time k. The measurement contribution is then given by $J_z(k)$. We note that $J_z(k)$ is dependent on the states of the operational sensors. The derivation of $J_z(k)$ is the focus of the next section.

12.3.1.1 Measurement Contribution to the PCRLB. Consider the general case in which there is measurement origin uncertainty, with measurements originating from one of the targets or from clutter. The jth measurement at the ith sensor is given by

$$z_k(j, i) = \begin{cases} h_k^i(x_k^t) + \omega_k^i(j) & \text{if originated from target } t, \\ \upsilon_k^i(j) & \text{if false alarm,} \end{cases} \tag{12.17}$$

where h_k^i is (in general) a nonlinear function, $\omega_k^i(j)$ is a zero mean Gaussian random variable with covariance R_k, and $\upsilon_k^i(j)$ is uniformly distributed across the surveillance region A (with hyper-volume, V). The probability mass function of the number of false alarms, $\mu_{FA}(m)$, which is Poisson distributed with mean λV, is given by

$$\mu_{FA}(m) = \frac{e^{-\lambda V}(\lambda V)^m}{m!}, \tag{12.18}$$

where m is the number of false alarms and λ is the spatial density of the false alarms.

When multiple targets are present, the association between the measurements and the targets is not known and must be considered in the PCRLB calculation. The following assumptions are made regarding the measurements:

- Each measurement can be generated by one of the targets or the clutter.
- Each target can produce zero or one measurement at any one time.

If sensors have independent measurement processes, J_z can be written as

$$J_z(k) = \sum_{i=1}^{n} J_{z_i}(k), \tag{12.19}$$

where

$$J_{z_i}(k) = \sum_{m_k(i)=0}^{\infty} p(m_k(i)) J_{z_i}(m_k(i), k), \tag{12.20}$$

$$J_{z_i}(m_k(i), k) = \mathbb{E}\left\{ -\Delta_{X_k}^{X_k} \ln p(z_k(i)|X_k, m_k(i))|m_k(i) \right\}. \tag{12.21}$$

In the above, n is the number of sensors used at time k and $m_k(i)$ is the number of measurements at sensor i at time k.

The probability of receiving $m_k(i)$ measurements, $p(m_k(i))$, from sensor i is given by

$$p(m_k(i)) = \sum_{d=0}^{\min(T, m_k(i))} \left(\mu_{FA}(m_k(i) - d) \sum_{D_k^i} \left(\prod_{t=1}^{T} \left(P_D^t(i)\right)^{D_k^i(t)} \left(1 - P_D^t(i)\right)^{(1 - D_k^i(t))} \right) \right), \tag{12.22}$$

where D_k^i is the detection vector that indicates which targets are detected at sensor i (at time k).[1] The total number of targets that are detected is d, that is, $\sum_{t=1}^{T} D_k^i(t) = d$. $P_D^t(i)$ is the probability of detection of target t by sensor i.

The probability density function of the measurement $z_k(i)$ conditioned on X_k and $m_k(i)$ is given by

$$p(z_k(i)|X_k, m_k(i)) = \sum_{a_k(i)} p(z_k(i)|X_k, m_k(i), a_k(i)) p(a_k(i)|m_k(i)), \tag{12.23}$$

where $a_k(i)$ is the association vector that indicates which measurement originated from which target. Each element $a_k(j, i)$ of $a_k(i)$ is a random variable that takes a value in $[0,\ 1,\ \ldots,\ T]$, with 0 indicating a false alarm. $a_k(j, i) = t$ indicates that the measurement j originates from target t. If the targets are well separated in the measurement space, there is no measurement origin uncertainty in terms of targets and any one measurement can be originated from a known target or clutter. However, if the targets are closely spaced or cross one another, it is hard to find the association vector and all possible associations must be considered in the calculation of measurement information $J_z(k)$.

[1] $D_k^i(t)$ takes the value 1 if target t is detected and 0 otherwise.

It can be shown that the (t_1, t_2)th block of $J_{z_i}(k)$ is given by [13]

$$\left[J_{z_i}(k)\right]_{t_1 t_2} = \mathbb{E}\left\{\left[H_k^i\right]_{t_1}' \left[\Sigma_k^i\right]_{t_1 t_2} R_k^{-1} \left[H_k^i\right]_{t_2} \middle| X_k\right\}, \tag{12.24}$$

where

$$\left[H_k^i\right]_t(\alpha, \beta) = \frac{\partial h_k^i(\alpha, x_k^t)}{\partial x_k^t(\beta)} \tag{12.25}$$

and $\left[H_k^i\right]_t(\alpha, \beta)$ denotes the (α, β)th element of matrix $\left[H_k^i\right]_t$. Σ_k^i is the information reduction matrix (IRM) for sensor i and $\left[\Sigma_k^i\right]_{t_1 t_2}$ is the (t_1, t_2)th block of the IRM, which is given by

$$\left[\Sigma_k^i\right]_{t_1 t_2} = \sum_{m_k(i)=0}^{\infty} p(m_k(i)) \left[\Sigma_k^i(m_k(i))\right]_{t_1 t_2}, \tag{12.26}$$

where

$$\left[\Sigma_k^i(m_k(i))\right]_{t_1 t_2} = R_k^{-1} \mathbb{E}\left[\frac{G_k^i(t_1) G_k^i(t_2)'}{p(z_k(i)|X_k, m_k(i))^2}\right] \tag{12.27}$$

and

$$\begin{aligned} G_k^i(t) = \frac{1}{p(m_k(i))} \sum_{\substack{a_k(i) \\ D(a_k(i),t)=1}} & \left(\frac{\left(m_k(i) - d(a_k(i))\right)!}{m_k(i)!} \mu_{FA}\left(m_k(i) - d(a_k(i))\right) \right. \\ & \times \left(\prod_{\tau=1}^{T} \left(P_D^\tau(i)\right)^{D(a_k(i),\tau)} \left(1 - P_D^\tau(i)\right)^{(1-D(a_k(i),\tau))} \right) \left(\prod_{j=1}^{m_k(i)} p_{a_k(j,i)}(z_k(j,i)) \right) \\ & \left. \times \left(z_k(j_t, i) - h_k^i(x_k^t)\right) \right). \end{aligned} \tag{12.28}$$

The expectations in (12.24) and (12.27) are over X_k and z_k, respectively. For detailed derivation of multitarget IRM, reader is referred to [13].

No closed form analytical solution exists for the IRM Σ_k^i, which must then be calculated using a numerical integration technique. However, in general, the probability of detection depends on the range of the target from the sensor. As a result, the IRM is also a function of the target range. In the single target case, the relationship between the IRM and range can be determined off-line during a preprocessing stage [5]. Importantly then, real-time sensor management can be performed because the one-time computational "hit" has already be taken, prior to online implementation.

However, in the multitarget case, the IRM depends not only on the range of each target from the sensor but also on the distances between the targets in the measurement

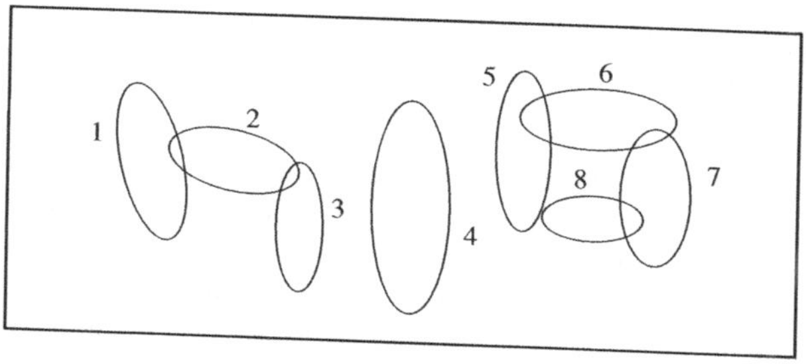

Figure 12.2 Target grouping.

space. Hence, the number of variables on which the IRM depends increases exponentially with the number of targets. Even though off-line calculation is viable in small-scale problems, it is difficult for large-scale problems because of memory limitation. However, even if we have many targets, frequently only a few will be in close proximity to one another (i.e., be in one group), and off-line calculation is again viable. This idea is explored (along with other potential simplifications) in the next section.

12.3.1.2 Reducing the Computational Complexity. We propose the following approximations in order to reduce the computational load of the ensuing calculation.

- *Approximation 1.* Measurements can be restricted to the validation gate of each target. We can then group the targets whose validation gates overlap with one another. After that, each group can be treated separately. For example, in Figure 12.2, there are eight targets and the ellipses indicate the corresponding validation gates. Only targets 2 and 3 affect target 1, hence they are considered as a group. Note that even though target 1 does not overlap with target 3, they are connected by target 2 and they affect each other indirectly. Similarly, targets 5, 6, 7, and 8 are in one group and target 4 is alone in another group. Hence, in order to calculate $\left[\Sigma_k^i\right]_{t_1t_2}$, we check whether the corresponding targets (t_1 and t_2) are in the same group, If they are, then all the members in that group are considered in calculating $\left[\Sigma_k^i\right]_{t_1t_2}$, otherwise $\left[\Sigma_k^i\right]_{t_1t_2} = 0$.
- *Approximation 2.* The IRM is a symmetric matrix, that is, $\left[\Sigma_k^i\right]_{t_1t_2} = \left[\Sigma_k^i\right]_{t_2t_1}$. Hence, we can reduce the computational burden by calculating only the upper or lower triangular entries of Σ_k^i.
- *Approximation 3.* A validation matrix is used to reduce the joint association events, by eliminating the ones with negligible probabilities [20]. The validation matrix is defined by

$$\Omega = [\varpi_{jt}] \quad j = 1,\ 2,\ \ldots,\ m; \quad t = 0,\ 1,\ \ldots,\ T, \tag{12.29}$$

where $\varpi_{jt}(\in \{0, 1\})$ indicates whether measurement j lies inside the validation region of target t.

In addition to the above simplifications, the computation time can be further reduced by calculating the $J_{z_i}(k)$ values for each sensor in parallel.

12.3.1.3 *Scalar Performance Measure.* Since the PCRLB, which is the primary performance criterion in the sensor selection optimization, is a matrix, it needs to be converted into a corresponding scalar quantity for ease of operation. Possible scalar performance metrics based on the PCRLB are its trace, determinant, or the maximum eigenvalue [42]. The trace of PCRLB (i.e., trace($\mathbf{J}^{-1}$)) and the determinant of the PCRLB (i.e., det($\mathbf{J}^{-1}$)) are proportional to the circumference and the volume of the rectangular region enclosing the minimum achievable covariance ellipsoid, respectively. The maximum eigenvalue of PCRLB (i.e., $\lambda_{\max}(\mathbf{J}^{-1})$) is the maximum distance of any point in the minimum achievable covariance ellipsoid from its center.

In situations, where the smallest principal axis shrinks to zero while the uncertainties along the remaining principal axes remain large, the volume of the uncertainty ellipsoid is zero. Hence, the sensor manager with determinant of PCRLB as the scaler performance measure may select sensors, which give poor tracking accuracy. In addition, calculation of the determinant or the maximum eigenvalue of a matrix is computationally more expensive than calculating the trace of a matrix. Hence, in this chapter, the trace of the PCRLB is used as the scalar performance metric. However, the algorithms presented in the following sections are not tied to the trace of the PCRLB or any other scalar metric.

12.4 SENSOR ARRAY MANAGEMENT FOR CENTRALIZED TRACKING

12.4.1 Problem Description

In this section, the sensor subset selection problem for multitarget tracking for centralized architecture is presented [10, 13].

A large number of sensors are already deployed throughout the surveillance region in order to monitor the surveillance region, and it is assumed that the sensor locations are fixed and known. Multiple, possibly time varying, number of targets are present in the surveillance region. A sample scenario, in which three targets are present and each enter the surveillance region at different time steps, is shown in Figure 12.3. Even though large number of sensors available to use at any time, due to physical limitations only a certain number of sensors can be used at any one time step. Then the problem is to select a subset of sensors such that tracking performance is optimized.

Formulation of the above problem based on PCRLB is given in the next section.

12.4.2 Problem Formulation

Suppose, a total of N sensors are available in the surveillance region. Out of these N sensors, only n can be utilized at any one time. Measurements are received at time

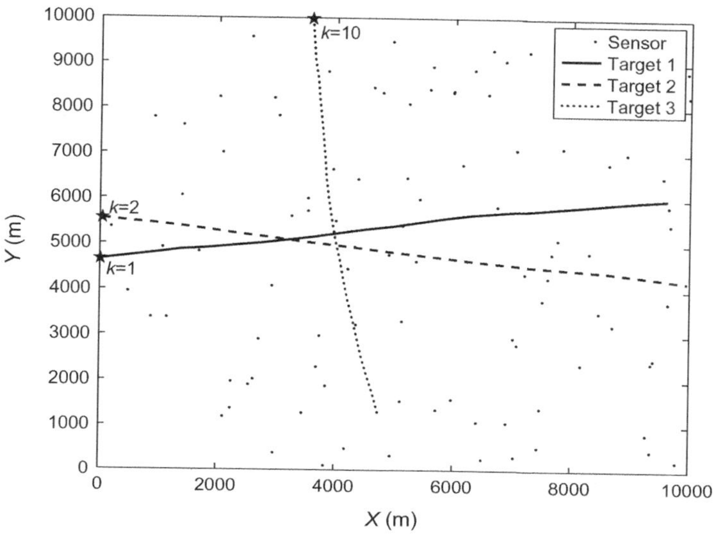

Figure 12.3 A sample scenario.

intervals of T_s. A sensor lead time, T_L $(= \rho \times T_s)$, is needed to activate a sensor subset, that is, the sensor subset that is to be activated at sampling time k has to be selected at or before sampling time $k - \rho$. Furthermore, the active sensor set can only be changed every $T_c = l \times T_s\,(l \geq 1)$ time interval. Hence, once activated, a particular sensor subset remains active for one or more revisits.

12.4.2.1 Sensor Subset Selection for Fixed Number of Targets.

In this section, the sensor selection problem for a fixed and known (say T) number of targets is considered. The problem is then to select a subset consisting of n sensors from the N available in order to optimize the tracking performance of the T targets over the activation period of a particular sensor set.

Sensors are selected based on the predictive PCRLB, which is dependent on the predicted target states. In order to minimize the difference between the predictive PCRLB and the true value given the measurements were available, one needs to wait until the latest possible sampling time, $k - \rho$ before selecting the sensors that are to be activated at sampling time k. Note that the number of targets and their states are determined based on the tracking results at $k - \rho$. The target states in the time interval $k - \rho$ to $k + l - 1$ are then predicted and used to calculate $\mathbf{J}_X(k)$ and $\mathbf{J}_{z_i}(\tau)$ for $i = 1,\ 2,\ \ldots,\ N;\ \tau = k,\ (k+1),\ \ldots,\ (k+l-1)$. The values of $\mathbf{J}_X(k)$ and $\mathbf{J}_{z_i}(\tau)$ do not depend on the sensors that are to be selected at sampling time k. Hence, they need to be calculated only once.

Formulation 1 Let s_{ik} be the indicator of the state of sensor i at sampling time k that takes value 1 if that sensor is active at that time and 0 otherwise. Because the active sensor set is only changed every l time steps, $s_{ik} = s_{i(k+1)} = \cdots = s_{i(k+l-1)}$.

The optimal sensor subset $s^*(k)$ is then given by

$$\mathbf{s}^*(k) = \arg\min_{\{s_{ik}\}} \sum_{\tau=k}^{k+l-1} \text{trace}\left\{ \left[\mathbf{J}_X(\tau) + \sum_{i=1}^{N} s_{ik} \mathbf{D}_{\tau i} \mathbf{J}_{z_i}(\tau) \right]^{-1} \right\}, \quad (12.30)$$

subject to

$$\sum_{i=1}^{N} s_{ik} = n, \quad (12.31)$$

$$s_{ik} \in \{0, 1\}, \quad i = 1, 2, \ldots, N, \quad (12.32)$$

where

$$\mathbf{D}_{\tau i} = \text{diag}[d_{\tau 1i} I_{r\times r}, d_{\tau 2i} I_{r\times r}, \ldots, d_{\tau Ti} I_{r\times r}], \quad (12.33)$$

$$d_{\tau ti} = \begin{cases} 1 \text{ if target } t \text{ is within the coverage region of sensor } i \text{ at time step } \tau \\ 0 \text{ otherwise} \end{cases} \quad (12.34)$$

r is the dimensionality of the state vector of each target. Note that the $d_{\tau ti}$ is determined using the predicted target location and the known sensor location.

$\mathbf{J}_X(k)$ is calculated using the PCRLB at sampling time $k - \rho$, together with the expected measurement information in the time interval $k - \rho$ to k based on the current active sensor set. $\mathbf{J}_X(k+1)$, $\mathbf{J}_X(k+2), \ldots$ are then calculated using the following recursion

$$\mathbf{J}_X(k+1) = \left(Q_{k+1,k} + \mathbf{F}_{k+1,k} \left[\mathbf{J}_X(k) + \sum_{i=1}^{N} s_{ik} \mathbf{D}_{ki} \mathbf{J}_{z_i}(k) \right]^{-1} \mathbf{F}'_{k+1,k} \right)^{-1}. \quad (12.35)$$

In the objective function minimized in (12.30), equal weights are given to all time steps. However, the accuracy with which one can predict the target states decreases with time. Hence, the accuracy with which one can predict the PCRLBs also decreases with time. Therefore, different weights may be given to each time step. These weights may, for example, be inversely proportional to the prediction interval and also dependent on the process noise covariance.

Formulation 2 If the targets are well separated in the state space, even if they are close in the measurement space, the above formulation gives impressive results. However, if the targets are close in the state space, there is a possibility for track swap. In order to reduce the possibility of track swap, we should try to avoid the track confidence region overlap in the state space. Hence, in addition to minimizing the trace of PCRLB, we also attempt to reduce the error covariance in the direction of the vector that connects the state estimates of each two targets.

The vector that connects the predicted states of target t_1 and t_2 at time τ is given by

$$\mathbf{y}_{t_1 t_2}(\tau) = \hat{\mathbf{x}}^{t_1}(\tau|k-\rho) - \hat{\mathbf{x}}^{t_2}(\tau|k-\rho), \tag{12.36}$$

where $\hat{\mathbf{x}}^{t}(\tau|k-\rho)$ is the predicted state of target t at time τ, given estimate at time $k-\rho$.

If the PCRLB at time τ, $\mathbf{B}(\tau)$, is given by

$$\mathbf{B}(\tau) = \left[\mathbf{J}_X(\tau) + \sum_{i=1}^{N} s_{i\tau}\mathbf{D}_{\tau i}\mathbf{J}_{z_i}(\tau)\right]^{-1}, \tag{12.37}$$

then $\mathbf{b}_{t_1 t_2}(\tau)'[\mathbf{B}(\tau)]_{t_1 t_1}\mathbf{b}_{t_1 t_2}(\tau)$ gives the (mean square) error bound in estimating target t_1's state in the direction of target t_2 (at time τ). In the above, $\mathbf{b}_{t_1 t_2}(\tau)$ is a unit vector along $\mathbf{y}_{t_1 t_2}(\tau)$ and $[\mathbf{B}(\tau)]_{tt}$ is the tth block diagonal matrix of $\mathbf{B}(\tau)$ that gives the covariance of target t.

Now, the optimization problem can be modified as follows

$$\mathbf{s}^*(k) = \arg\min_{\{s_{ik}\}} \sum_{\tau=k}^{k+l-1} \left\{ \sum_{t_1=1}^{T} \left[w_{t_1} \operatorname{trace}\left([\mathbf{B}(\tau)]_{t_1 t_1}\right) + \sum_{\substack{t_2=1\\ t_2 \neq t_1}}^{T} w_{t_1 t_2}\, \mathbf{b}'_{t_1 t_2}[\mathbf{B}(\tau)]_{t_1 t_1}\mathbf{b}_{t_1 t_2} \right] \right\}, \tag{12.38}$$

where

$$w_{t_1 t_2} = \min\left\{1, \frac{g \times \left(\sqrt{\mathbf{b}'_{t_1 t_2}[\mathbf{J}_X(\tau)^{-1}]_{t_1 t_1}\mathbf{b}_{t_1 t_2}} + \sqrt{\mathbf{b}'_{t_1 t_2}[\mathbf{J}_X(\tau)^{-1}]_{t_2 t_2}\mathbf{b}_{t_1 t_2}}\right)}{\| \mathbf{y}_{t_1 t_2}(\tau) \|}\right\}, \tag{12.39}$$

$$w_{t_1} = 1 - \max\left\{w_{t_1 t_2} | t_2 \in \{1, 2, \ldots, T\}, t_2 \neq t_1\right\}, \tag{12.40}$$

g is the number of standard deviations of the measurement gate. The constraints are again given by (12.31) and (12.32).

The weights in the objective function, w_{t_1} and $w_{t_1 t_2}$, are selected based on the distances between the targets and the predicted covariances. If there are no targets in close proximity to target t_1, then $w_{t_1 t_2}$ approaches zero and w_{t_1} approaches one. Conversely, if there is a target, t_2, close to target t_1, then w_{t_1} approaches zero and $w_{t_1 t_2}$ approaches one.

If we are only interested in estimating the locations of the targets, then only the PCRLB terms relating to the target locations need to be considered in the above equations.

12.4.2.2 Sensor Subset Selection for a Varying Number of Targets. In the previous section it is assumed that the total number of targets in the surveillance region is fixed and known. However, in most realistic scenarios the above assumption is not valid. In this section a sensor subset selection algorithm is presented for a scenario in which the total number of targets in the surveillance region is not known and may vary with time, that is, either some new targets may enter or some existing targets may leave the surveillance region. The problem is then to select a sensor subset in order to (i) optimize the tracking performance of existing targets and (ii) quickly identify new targets.

The following assumptions are made:

- New targets appear by passing through the perimeter of the surveillance region (i.e., targets do not "pop-up" in the interior of the surveillance region).
- The probability mass function of the number of new targets, $\mu_{NT}(\tau)$, at any one time has a Poisson distribution with mean $\lambda_b V_p$, and given by

$$\mu_{NT}(\tau) = \frac{e^{-\lambda_b V_p}(\lambda_b V_p)^\tau}{\tau!}, \tag{12.41}$$

  where τ is the number of new targets, λ_b is the spatial distribution of the new targets, and V_p is the volume of the perimeter.
- The targets die (disappear) only when they leave the surveillance region.

The sensor selection algorithm presented herein can, with slight modification, cater for a different set of assumptions.

When a target leaves the surveillance region its track is deleted and the PCRLB is adjusted by removing the entries that relate to this target. A bicriterion optimization approach with two objectives, one to accurately track existing targets and the other to quickly identify new ones, is then used to select the sensors.

The first objective is to minimize the PCRLB (see (12.30)) for established (i.e., detected) targets that are still within the surveillance region. If there are T targets in the surveillance region at sampling time $k-1$, then the prior information matrix of those targets at sampling time k, $\mathbf{J}_{X_{\text{old}}}(k)$, is a $Tr \times Tr$ matrix, and can be calculated using the PCRLB at sampling time $k-1$ and the state transition matrices. If a new target is initialized at sampling time k, then the updated prior information matrix is augmented as follows

$$\mathbf{J}_X(k) = \begin{bmatrix} \mathbf{J}_{X_{\text{old}}}(k) & 0 \\ 0 & \mathbf{J}_{X_{\text{new}}} \end{bmatrix}, \tag{12.42}$$

where $\mathbf{J}_{X_{\text{new}}}$ is the prior information matrix of the new target, initialized as $\mathbf{J}_{X_{\text{new}}} = \text{diag}\left\{1/V_{x1}^2,\ 1/V_{x2}^2, \ldots, 1/V_{xr}^2\right\}$. V_{xi} is the surveillance volume, the size of the range of possible values, of the ith component of the state vector. If a target t leaves the surveillance region, then $\mathbf{J}_X(k)$ is adjusted by eliminating the rows and columns corresponding to target t. The adjustments above are modified in an obvious manner if more than one target enters or leaves the surveillance region at any one time.

The second objective is to maximize the probability of detecting new targets. First, let us assume that only one new target may enter the surveillance region at any one time. To find the probability of detecting this new target, P_{new} particles, each of which represents a possible state of the new target, are uniformly distributed along the perimeter of the surveillance region.[2] Note that the particles could be distributed nonuniformly if certain points on the perimeter are more likely entry points for new targets (e.g., based on some prior threat assessment). After selecting a point on the perimeter, the distance that the target could potentially have traveled inside the surveillance region has to be determined. If the velocity of the new target in the direction perpendicular to the perimeter is uniformly distributed between 0 and $V_{\max}$, then the probability density function of the distance of the new target's position from the perimeter, q, is given by

$$\begin{aligned} p(q) &= \int_0^{V_{\max}} p(q|v)\, p(v) dv, \\ &= \int_{q/T_s}^{V_{\max}} \frac{1}{vT_s} \frac{1}{V_{\max}} dv, \\ &= \frac{1}{V_{\max} T_s} \ln\left(\frac{V_{\max} T_s}{q}\right), \end{aligned} \tag{12.43}$$

where T_s is the sampling time. The velocity components of the particles are uniformly distributed between the known lower and upper velocity limits. The probability of detection of the new target at sampling time k, p_D^k, is given by

$$p_D^k = 1 - \frac{1}{P_{\text{new}}} \sum_{j=1}^{P_{\text{new}}} \prod_{i=1}^{N} (1 - s_{ik} P_D(i, j)), \tag{12.44}$$

where $P_D(i, j)$ is the probability of detecting, at sensor i, a target whose state is given by particle j. In general, this probability of detection depends on both the sensor state and the target state.

The performance can be improved even further by considering the possibility that a new target had entered the surveillance region at sampling time $k - 1$ but had not been detected at that time. The probability of not detecting a new target at sampling time $k - 1$ is given by

$$p_M^{k-1} = \frac{1}{P_{\text{new}}} \sum_{j=1}^{P_{\text{new}}} \prod_{i=1}^{N} (1 - s_{i(k-1)} P_D(i, j)). \tag{12.45}$$

[2]If new targets can appear anywhere in the surveillance region, particles should be placed uniformly all over the surveillance region.

In order to incorporate the missed target, while P_{new} particles are used for the new target that enter the region at sampling time k, $p_M^{k-1} \times P_{\text{new}}$ particles are used for the new target that was not detected at sampling time $k-1$. The particles representing the new target that enters the region at sampling time k are uniformly distributed along the perimeter of the surveillance region. However, the particles representing the missed target are distributed according to the probability of missing a new target at each point on the perimeter of the surveillance region. After selecting a point on the perimeter, the probability density function of the depth at which the target could now lie is given by

$$\begin{aligned} p(q) &= \int_{q/2T_s}^{\min(V_{\max},\, q/T_s)} \frac{1}{vT_s}\frac{1}{V_{\max}} dv, \\ &= \begin{cases} \frac{1}{V_{\max}T_s}\ln(2) & \text{if } q < V_{\max}T_s \\ \frac{1}{V_{\max}T_s}\ln\left(\frac{2V_{\max}T_s}{q}\right) & \text{else if } q < 2V_{\max}T_s \\ 0 & \text{otherwise.} \end{cases} \end{aligned} \tag{12.46}$$

Now, the number of particles representing both a new detected target and a missed target (that entered on the previous time step) is $\bar{P}_{\text{new}} = (1 + p_M^{k-1}) \times P_{\text{new}}$. Hence, the second objective function now becomes

$$OBJ(2) = \frac{1}{\bar{P}_{\text{new}}} \sum_{j=1}^{\bar{P}_{\text{new}}} \prod_{i=1}^{N} (1 - s_{ik} P_D(i, j)). \tag{12.47}$$

In the above equations only one new target is considered. However, if many new targets are considered, then under the assumption that the new targets are moving independently, the above objective function can be modified as follows

$$OBJ(2) = \left(\sum_{t_n=1}^{T_N} \mu_{NT}(t_n) \times t_n \right) \times \frac{1}{\bar{P}_{\text{new}}} \sum_{j=1}^{\bar{P}_{\text{new}}} \prod_{i=1}^{N} (1 - s_{ik} P_D(i, j)), \tag{12.48}$$

where T_N is the maximum number of new targets that can enter the surveillance region at any one time.

Finally, the bicriterion optimization problem is given as follows

$$\begin{aligned} \mathbf{s}^*(k) = \arg\min_{\{s_{ik}\}} &\left\{ \left(\sum_{\tau=k}^{k+l-1} \text{trace}\left\{ \left[\mathbf{J}_X(\tau) + \sum_{i=1}^{N} s_{ik} \mathbf{D}_{\tau i} \mathbf{J}_{z_i}(\tau) \right]^{-1} \right\} \right) \right. \\ &\left. + w_b \left(\sum_{t_n=1}^{T_N} \mu_{NT}(t_n) \times t_n \right) \times \frac{1}{\bar{P}_{\text{new}}} \sum_{j=1}^{\bar{P}_{\text{new}}} \prod_{i=1}^{N} (1 - s_{ik} P_D(i, j)) \right\}, \end{aligned} \tag{12.49}$$

In the above, w_b is the weight applied to target birth function. This weight both balances the units and specifies the relative importance of the two objectives.

In the above equations, the effect of missed detections is considered only at sampling time $k-1$. The performance can be improved by considering the effect of missed detections at earlier time steps ($k-2$, $k-3$, etc.) as well. However, the discounted effects do indeed become negligible very quickly, and the above approach is a reasonable approximation.

12.4.3 Solution Technique

The optimization problems formulated in the previous two sections are all NP-hard [43, 44]. Enumerating every possibility in order to find the optimal solution is feasible only in small-scale problems. Complete enumeration is not viable when the number of possibilities is very large. However, for large-scale problems, an iterative local search technique is used to find a near optimal solution [45]. This is given as follows:

- *Step 1* Select an initial solution as follows:
  - Select the sensor that gives optimal performance when only one sensor is used.
  - Select a second sensor that results in optimal performance when only two sensors are used: including the first selected sensor.
  - Continue adding sensors until total number of sensors is n.

  At each stage, a complete enumeration is performed to determine which sensor should be added to the selection.
- *Step 2* Search for better solutions in the neighborhood (until some prespecified time limit is reached). Initially, form the neighborhood by swapping only one sensor from the current solution. If the current solution is optimal in the above neighborhood, then form a new neighborhood by swapping two sensors. Continuously increase the number of sensors that are swapped for forming the neighborhood until the time limit is reached.

The local optimum of a certain neighborhood can be obtained quickly by swapping the sensors in the order of their rank, where the rank of each sensor is based on its individual performance. A detailed explanation of local search technique is given in Appendix 12A.

12.4.4 Simulation

In the following scenarios, it is assumed that the targets move according to independent constant velocity models [46]. The state of target t at time k is given by $\mathbf{x}_k^t = \left[\xi_k^t,\ \dot{\xi}_k^t,\ \eta_k^t,\ \dot{\eta}_k^t\right]'$, where ξ_k^t and η_k^t are the x and y coordinates of target t, respectively, and $\dot{\xi}_k^t$ and $\dot{\eta}_k^t$ are the velocities of target t in the corresponding directions, respectively. The sensors have independent measurement processes. The instantaneous probability of detection of each target at each sensor is given by a Swerling I model [20], that is,

$$P_D^t = P_F^{\frac{R_t^2}{R_t^2 + S_0 R_0^2}}, \tag{12.50}$$

where R_t is the target range from the sensor and S_0 is the signal-to-noise ratio at a nominal range R_0.

In simulations, either an array of bearings-only sensors or an array of range-only sensors is used. The jth measurement at sensor i is given by

$$z_k(j,i) = \begin{cases} \left.\begin{array}{ll} \tan^{-1}\left(\frac{\eta_k^t - \eta_s^i}{\xi_k^t - \xi_s^i}\right) + \omega_k^i(j) & \text{if bearing-only} \\ \sqrt{(\xi_k^t - \xi_s^i)^2 + (\eta_k^t - \eta_s^i)^2} + \omega_k^i(j) & \text{if range-only} \end{array}\right\} & \text{if target } t \text{ generated} \\ \upsilon_k^i(j) & \text{if false alarm,} \end{cases} \tag{12.51}$$

where (ξ_s^i, η_s^i) is the i-th sensor location, $\omega_k^i(j) \sim \mathcal{N}(0, \sigma^2)$ and

$$\upsilon_k^i(j) \sim \begin{cases} U[-\pi, \pi] & \text{if bearing-only} \\ U[0, R_c] & \text{if range-only,} \end{cases} \tag{12.52}$$

R_c is the coverage range of the sensor.

Also a nominal minimum detection range of 1 m, inside which the target cannot be detected, is included. This is necessary because, otherwise, as shown in [8], the bearing-only PCRLB is noninformative and approaches zero as the number of Monte Carlo samples tends to infinity. For brevity, the minimum detection range is not included in the notation.

Because the measurement equation is nonlinear, a stacked single particle filter is used to track the targets. The reader is referred to [47, 48] for details of the approach.

12.4.4.1 Measurement Contribution to FIM. For both types of sensor, the (t_1, t_2)th block matrix of $\mathbf{J}_z(k)$, $\left[\mathbf{J}_z(k)\right]_{t_1 t_2}$, is given by

$$\left[\mathbf{J}_z(k)\right]_{t_1 t_2} = \begin{pmatrix} j_{11}^{t_1 t_2} & 0 & j_{13}^{t_1 t_2} & 0 \\ 0 & 0 & 0 & 0 \\ j_{31}^{t_1 t_2} & 0 & j_{33}^{t_1 t_2} & 0 \\ 0 & 0 & 0 & 0 \end{pmatrix}, \tag{12.53}$$

For bearings-only sensors, $j_{11}^{t_1t_2}$, $j_{13}^{t_1t_2}$, $j_{31}^{t_1t_2}$, and $j_{33}^{t_1t_2}$ are then given by

$$\begin{aligned}
j_{11}^{t_1t_2} &= \frac{1}{\sigma^2}\sum_{i=1}^{n}\mathbb{E}\left[\left[\mathbf{\Sigma}_k^i\right]_{t_1t_2}\frac{(\eta_k^{t_1}-\eta_s^i)(\eta_k^{t_2}-\eta_s^i)}{\left((\xi_k^{t_1}-\xi_s^i)^2+(\eta_k^{t_1}-\eta_s^i)^2\right)\left((\xi_k^{t_2}-\xi_s^i)^2+(\eta_k^{t_2}-\eta_s^i)^2\right)}\right],\\
j_{13}^{t_1t_2} &= -\frac{1}{\sigma^2}\sum_{i=1}^{n}\mathbb{E}\left[\left[\mathbf{\Sigma}_k^i\right]_{t_1t_2}\frac{(\eta_k^{t_1}-\eta_s^i)(\xi_k^{t_2}-\xi_s^i)}{\left((\xi_k^{t_1}-\xi_s^i)^2+(\eta_k^{t_1}-\eta_s^i)^2\right)\left((\xi_k^{t_2}-\xi_s^i)^2+(\eta_k^{t_2}-\eta_s^i)^2\right)}\right],\\
j_{31}^{t_1t_2} &= -\frac{1}{\sigma^2}\sum_{i=1}^{n}\mathbb{E}\left[\left[\mathbf{\Sigma}_k^i\right]_{t_1t_2}\frac{(\xi_k^{t_1}-\xi_s^i)(\eta_k^{t_2}-\eta_s^i)}{\left((\xi_k^{t_1}-\xi_s^i)^2+(\eta_k^{t_1}-\eta_s^i)^2\right)\left((\xi_k^{t_2}-\xi_s^i)^2+(\eta_k^{t_2}-\eta_s^i)^2\right)}\right],\\
j_{33}^{t_1t_2} &= \frac{1}{\sigma^2}\sum_{i=1}^{n}\mathbb{E}\left[\left[\mathbf{\Sigma}_k^i\right]_{t_1t_2}\frac{(\xi_k^{t_1}-\xi_s^i)(\xi_k^{t_2}-\xi_s^i)}{\left((\xi_k^{t_1}-\xi_s^i)^2+(\eta_k^{t_1}-\eta_s^i)^2\right)\left((\xi_k^{t_2}-\xi_s^i)^2+(\eta_k^{t_2}-\eta_s^i)^2\right)}\right],
\end{aligned} \tag{12.54}$$

and for range-only sensors

$$\begin{aligned}
j_{11}^{t_1t_2} &= \frac{1}{\sigma^2}\sum_{i=1}^{n}\mathbb{E}\left[\left[\mathbf{\Sigma}_k^i\right]_{t_1t_2}\frac{(\xi_k^{t_1}-\xi_s^i)(\xi_k^{t_2}-\xi_s^i)}{\sqrt{\left((\xi_k^{t_1}-\xi_s^i)^2+(\eta_k^{t_1}-\eta_s^i)^2\right)\left((\xi_k^{t_2}-\xi_s^i)^2+(\eta_k^{t_2}-\eta_s^i)^2\right)}}\right],\\
j_{13}^{t_1t_2} &= \frac{1}{\sigma^2}\sum_{i=1}^{n}\mathbb{E}\left[\left[\mathbf{\Sigma}_k^i\right]_{t_1t_2}\frac{(\xi_k^{t_1}-\xi_s^i)(\eta_k^{t_2}-\eta_s^i)}{\sqrt{\left((\xi_k^{t_1}-\xi_s^i)^2+(\eta_k^{t_1}-\eta_s^i)^2\right)\left((\xi_k^{t_2}-\xi_s^i)^2+(\eta_k^{t_2}-\eta_s^i)^2\right)}}\right],\\
j_{31}^{t_1t_2} &= \frac{1}{\sigma^2}\sum_{i=1}^{n}\mathbb{E}\left[\left[\mathbf{\Sigma}_k^i\right]_{t_1t_2}\frac{(\eta_k^{t_1}-\eta_s^i)(\xi_k^{t_2}-\xi_s^i)}{\sqrt{\left((\xi_k^{t_1}-\xi_s^i)^2+(\eta_k^{t_1}-\eta_s^i)^2\right)\left((\xi_k^{t_2}-\xi_s^i)^2+(\eta_k^{t_2}-\eta_s^i)^2\right)}}\right],\\
j_{33}^{t_1t_2} &= \frac{1}{\sigma^2}\sum_{i=1}^{n}\mathbb{E}\left[\left[\mathbf{\Sigma}_k^i\right]_{t_1t_2}\frac{(\eta_k^{t_1}-\eta_s^i)(\eta_k^{t_2}-\eta_s^i)}{\sqrt{\left((\xi_k^{t_1}-\xi_s^i)^2+(\eta_k^{t_1}-\eta_s^i)^2\right)\left((\xi_k^{t_2}-\xi_s^i)^2+(\eta_k^{t_2}-\eta_s^i)^2\right)}}\right].
\end{aligned} \tag{12.55}$$

12.4.5 Simulation Results

12.4.5.1 Fixed Number of Targets. In order to test the performance of the formulations presented in Section 12.4.2, the following scenario is considered. A total of 15 sensors are in the surveillance region, but only two of them can be used at any one time. Measurements are available at 1 min intervals. The active sensor subset can be modified at 1 min intervals (i.e., at every revisit). The sensors are selected with a lead time of $T_L = 1$ min, $S_0R_0^2 = 3.2\times 10^9\ \mathrm{m}^2$, $P_F = 10^{-3}$. The particle filter uses 1000 sample points.

First, a scenario with a known number (= 2) of targets is considered. Bearing-only sensors are used with the measurement error of target generated measurements set at

TABLE 12.1 RMSE Values of State Estimates When Ignoring and Accounting for Measurement Origin Uncertainty

	Position RMSE (m)	
Sensor Selection Strategy	Target 1	Target 2
Neglecting measurement origin uncertainty	89.68	86.54
Accounting for measurement origin uncertainty (Formulation 1)	75.11	75.59

0.1 rad (= 5.7^o). The measurement contribution to the PCRLB decreases with range due to both increasing bearings cross section (making targets unobservable at long ranges) and the range dependent P_D. Therefore, Formulation 1 (defined in Section 12.4.2) is likely to select sensors in close proximity to the targets, irrespective of the effect this has on our ability to resolve measurement association ambiguities. Hence, results can be expected to be similar to those obtained using the formulation in [10] where the measurement origin uncertainty is neglected, and this was indeed the case. A better comparison is obtained when using range-only sensors (with a 40 m range error standard deviation). The root mean square error (RMSE) values of the state estimates based on 100 Monte Carlo runs are shown in Table 12.1.

Comparison of the performances of Formulations 1 and 2, in a scenario with two closely spaced, but resolved, targets is considered next. Bearing-only sensors are again used. The sensors selected at sampling time $k = 1$ under these two formulations are shown in Figure 12.4a and b, respectively. The particle distributions obtained subsequent to resampling are then shown in Figure 12.5a and b, respectively. Formulation 2 reduces the association ambiguity, with the particle distributions relating to these two targets well separated after resampling. As a result, association ambiguities and estimation errors will be reduced at subsequent sampling times. Indeed, comparisons (based on 100 Monte Carlo runs each of five time steps) show that Formulation 2 significantly improves performance in tracking both targets (see Table 12.2).

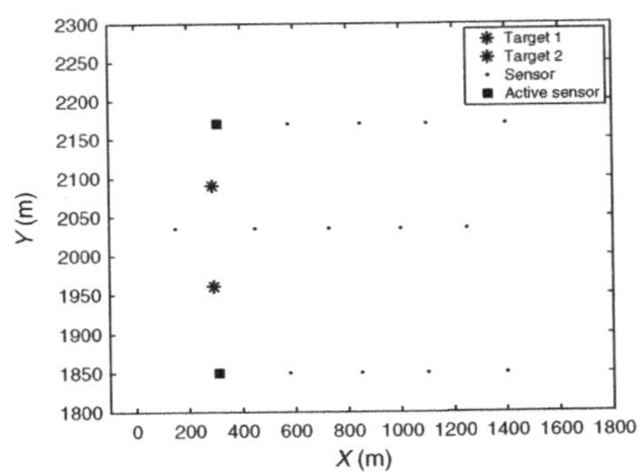

(a) Formulation 1

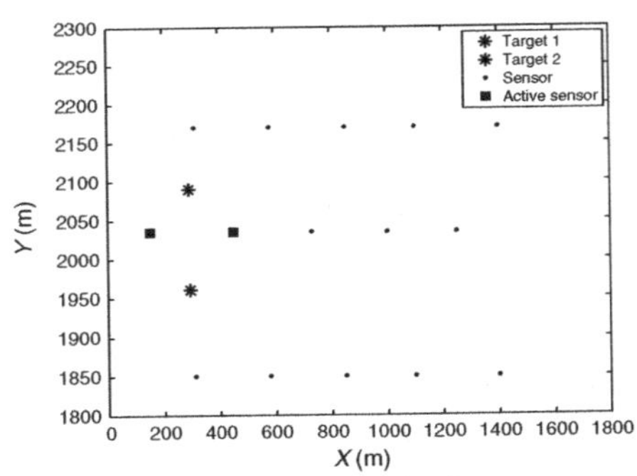

(b) Formulation 2

Figure 12.4 Sensors selected at sampling time $k = 1$.

TABLE 12.2 RMSE Values of State Estimates Using Formulations 1 and 2

	Position RMSE (m)	
Sensor Selection Strategy	Target 1	Target 2
Formulation 1	33.393	33.053
Formulation 2	25.192	23.029

12.4.5.2 Variable Number of Targets. The performance of the formulation that allows for a time-varying number of targets (see Section 12.4.2) is analyzed here. In the focal scenario a total of 100 sensors are in the surveillance region of which only 12 can be used at any one time step. Sensors give the target bearing (with a measurement error standard deviation of 0.05 rad ($\approx 3^o$) for target generated measurements). Measurements are available at 30 s intervals. The active sensor subset is changed at 30 s intervals (i.e., again at every revisit). The sensors are selected with a lead time of $T_L = 30$ s. The sensor field of view $V = 2\pi$ rad, $S_0 R_0^2 = 3.2 \times 10^7$ m^2, $P_F = 10^{-3}$, $\lambda = 0.004$/rad and $\lambda_b = 0.00001$/m. The surveillance region is 5000 m $\times$ 5000 m. The particle filter uses 5000 sample points. The target trajectories and sensor positions are shown in Figure 12.6. There are three targets, and they enter and leave the surveillance region at different times.

Figure 12.7 shows the sensors selected at time steps 13–18. At time step 13 there are two established targets and one new target that has just entered the surveillance region. Because there was no prior information available regarding the new target when the sensor selection was performed, nine sensors are assigned to cover the established targets and three sensors are assigned in an attempt to detect any new ones. Note that the sensors assigned to the established targets can also cover some parts of the perimeter through which new targets can enter. Upon detecting the new target at time step 14, a greater number of sensors are assigned to this target than to each of the other targets since the prior information regarding the state of the new target is minimal.

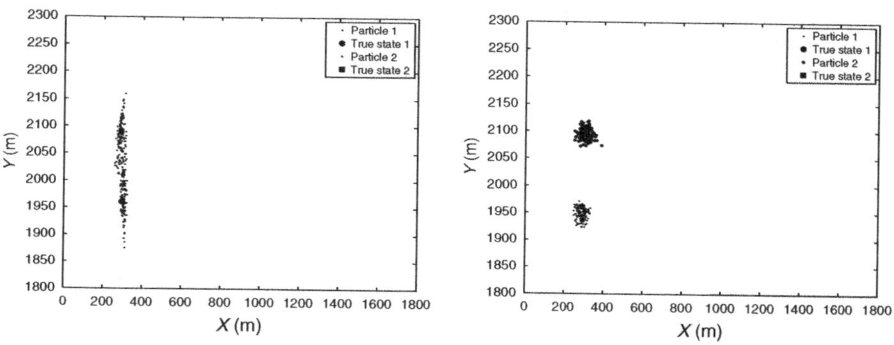

(a) Formulation 1 (b) Formulation 2

Figure 12.5 Resampled particles at sampling time $k = 1$.

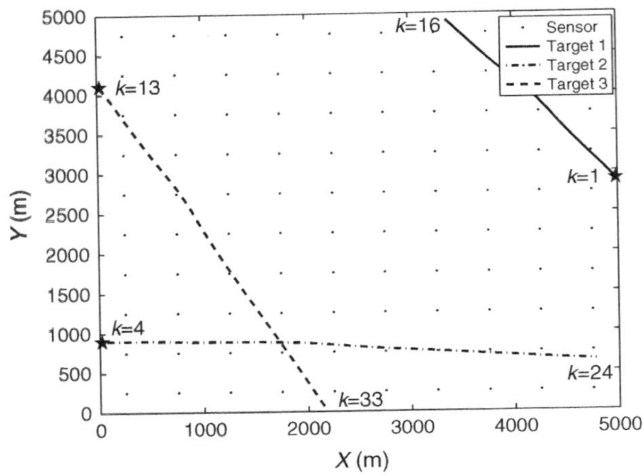

Figure 12.6 Target trajectories (with entering and leaving times) and sensor positions.

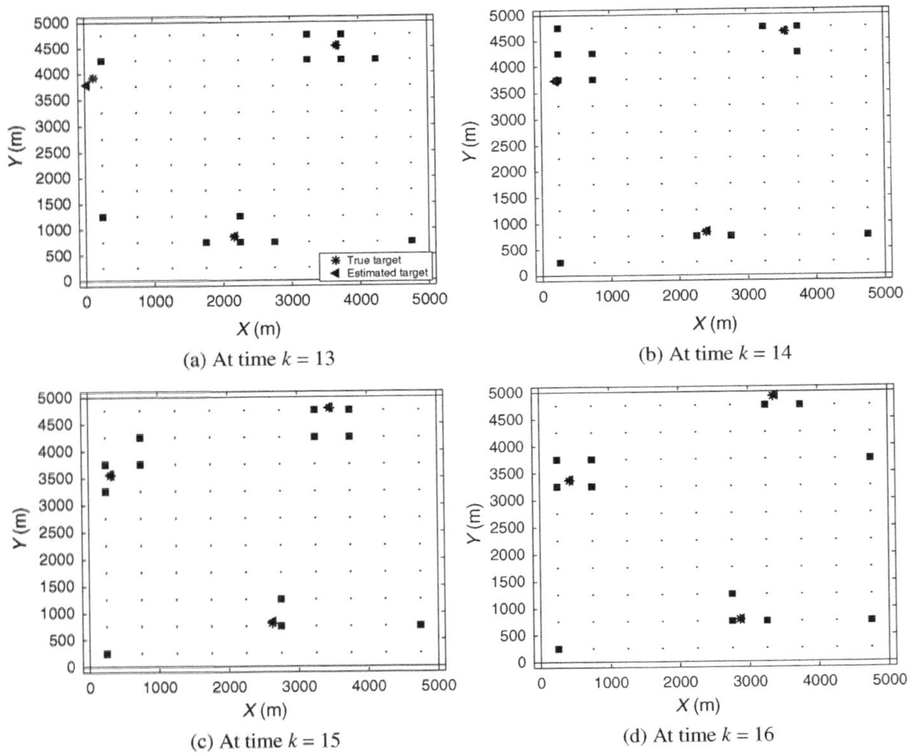

Figure 12.7 Sensors selected in a scenario with a 5000 m × 5000 m surveillance region and 100 sensors.

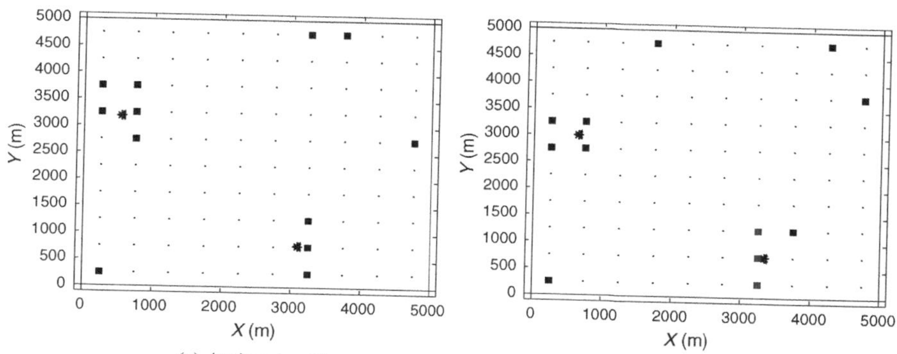

(e) At time $k = 17$

(f) At time $k = 18$

Figure 12.7 *(Continued)*

However, at time steps 15 and 16, the number of sensors that are assigned to the new target is reduced in order to balance the estimation uncertainties of all targets. Target 1 leaves the surveillance region at time step 17. However, two sensors are assigned to the target 1 at that time step, as it was not certain that the target was about to leave the surveillance region. At time step 17, the tracker deletes the track corresponding to target 1, and at time step 18 the sensors that would have been assigned to target 1 return to the surveillance task.

In the above scenario, an average of 0.08 time steps are required to detect each new target. However when using Formulation 2 that does not task assets specifically for surveillance: again see Section 12.4.2, an average of 1.4 time steps are required before a new target track is established. The improvement in performance of the formulation in Section 12.4.2 would have been even more significant had the surveillance region been larger, in which case scheduling some sensors to perform surveillance tasks would have been of even more critical importance. Figures 12.8 and 12.9 are based on a large surveillance region (10,000 m $\times$ 10,000 m) with 400 sensors. The sensors selected using Formulation 2 and the formulation that allows for a time-varying num-

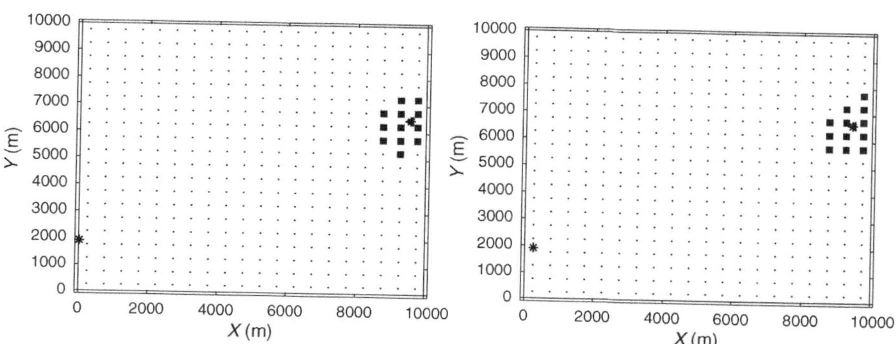

(a) At time $k = 4$

(b) At time $k = 5$

Figure 12.8 Sensors selected using Formulation 2 of Section 12.4.2, in a scenario with a 10,000 m $\times$ 10,000 m surveillance region and 400 sensors.

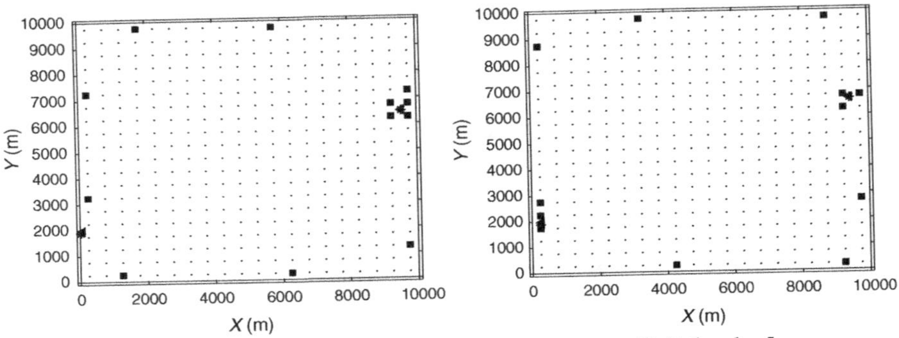

(a) At time $k = 4$ (b) At time $k = 5$

Figure 12.9 Sensors selected using the formulation that allows for a time varying number of targets (see Section 12.4.2), in a scenario with 10,000 m × 10,000 m surveillance region and 400 sensors.

ber of targets are shown in Figures 12.8 and 12.9, respectively. In the first case (Figure 12.8), the second target is never detected, because all active sensors are being used to track the established target, and the second target has entered the surveillance region through a point on the perimeter that is not within the coverage region of any active sensor.

12.4.5.3 *Considering the Effect of Missed Targets.* The effect of taking into consideration the possibility that a target could have entered the surveillance region on the previous time step, but not being detected at that time of entry is analyzed here. A scenario in which there is no established target in the surveillance region and only three sensors can be used at any one time is considered. Figure 12.10 shows the sensors selected at time steps $k - 1$ and k at a point no consideration is given to the possibility that targets could have been entered the surveillance region at time step $k - 1$ and not been detected at that time. However, in Figure 12.11 the effect

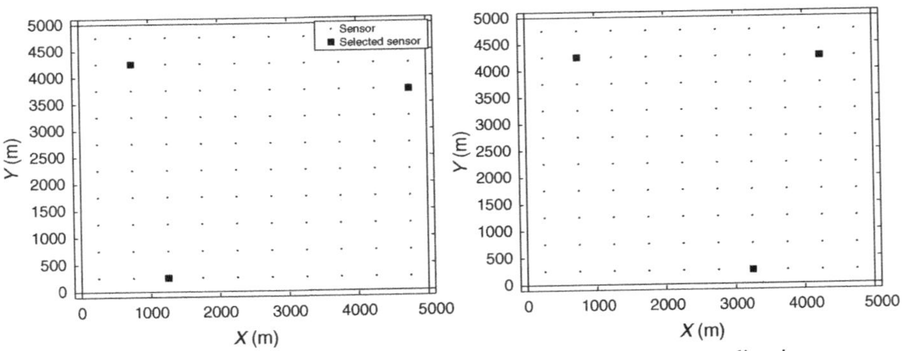

(a) At first sampling time (b) At next sampling time

Figure 12.10 Sensors selected without considering missed targets.

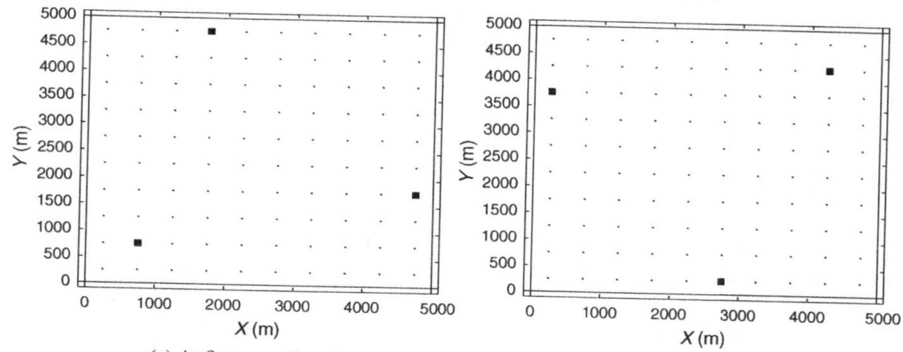

(a) At first sampling time (b) At next sampling time

Figure 12.11 Sensors selected by considering the possibility of missed targets.

of possible missed detections is considered. In Figure 12.10, the sensors selected at time step k are independent of the sensors selected at time step $k-1$. However, in Figure 12.11 the sensors selected at time step k attempt to minimize the probability of again missing any target that could have entered the surveillance region on the previous time step, by prioritizing regions that were not covered well at time step $k-1$.

12.5 SENSOR ARRAY MANAGEMENT FOR DISTRIBUTED TRACKING

In this section, sensor selection algorithm for a distributed tracking system that has a hierarchical level architecture with full feedback, as shown in Figure 12.11b, is presented [14]. A sample scenario with three targets, each enter the surveillance region at different time, 100 sensors and four FCs is shown in Figure 12.12. The

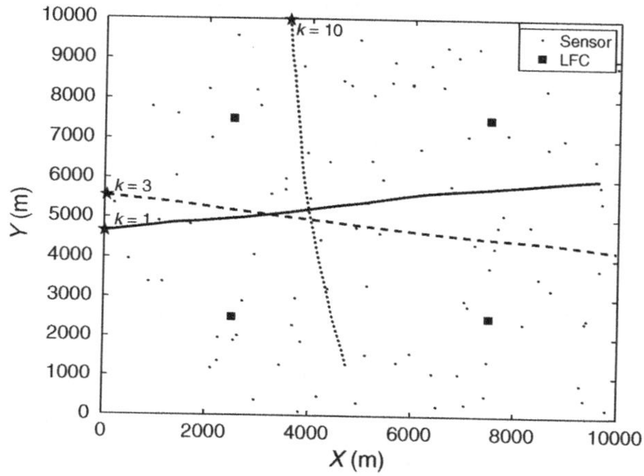

Figure 12.12 A sample scenario.

time steps at which targets enter the surveillance region are shown next to the initial positions (⋆).

12.5.1 Track Fusion

The main challenge in distributed and decentralized tracking is determining what to communicate between the FCs so that communication requirement is low, fusion can be made easily so that the result will be close to optimal. Even though communicating the measurements is the optimal way, in general, it requires a lot of communication. Hence, other approaches that require less communication must be considered.

One option is to communicate the tracks instead of raw measurements. If the tracks are uncorrelated, then they can be fused easily at FCs [20]. However, due to common prior information and common process noise, they are correlated and fusion becomes more complicated. Hence, in order to avoid double counting of common information, the error cross-covariance must be calculated and accounted for during fusion. However, exact error cross-covariance calculation needs the information of when a measurement update is made along with the Kalman gain used at each measurement step by each tracking platform. This imposes additional computational and communication load.

The second option is to communicate tracklets. A tracklet is a track computed so that its errors are not cross-correlated with the errors of any other data in the system for the same target [38]. It is equivalent to a track for a target based on only the most recent measurements since the data from the tracker was last communicated for that target. The main advantages of tracklet are reduced communication and computation requirements. Hence, in this chapter, it is assumed that tracklets are communicated between the fusion centers. The calculation of tracklet is explained below.

Suppose last communication was performed at time $k - l$ and the next communication is at time k. The state estimates and its covariances of a target just after the last communication and just before the next communication are $\hat{\mathbf{x}}_{k-l|k-l}$, $\mathbf{P}_{k-l|k-l}$, and $\hat{\mathbf{x}}_{k|k}$, $\mathbf{P}_{k|k}$. Then the information filter-based tracklet is given by [49]

$$\mathbf{y}_k = \mathbf{P}_{k|k}^{-1}\,\hat{\mathbf{x}}_{k|k} - \mathbf{P}_{k|k-l}^{-1}\,\hat{\mathbf{x}}_{k|k-l}, \tag{12.56}$$

$$\mathbf{Y}_k = \mathbf{P}_{k|k}^{-1} - \mathbf{P}_{k|k-l}^{-1}, \tag{12.57}$$

where

$$\hat{\mathbf{x}}_{k|k-l} = \mathbf{F}_{k,k-l}\,\hat{\mathbf{x}}_{k-l|k-l}, \tag{12.58}$$

$$\mathbf{P}_{k|k-l} = \mathbf{F}_{k,k-l}\,\mathbf{P}_{k-l|k-l}\,\mathbf{F}'_{k,k-l} + Q_{k,k-l}. \tag{12.59}$$

If the particular target has been initialized only after last communication, then

$$\mathbf{y}_k = \mathbf{P}_{k|k}^{-1}\,\hat{\mathbf{x}}_{k|k}, \tag{12.60}$$

$$\mathbf{Y}_k = \mathbf{P}_{k|k}^{-1}. \tag{12.61}$$

Fusion centers must update their tracks upon receiving tracklets from other fusion centers. As the first step, FCs have to associate the tracklets with existing tracks. In order to verify whether tracklet j is corresponding to track i, the errors in the difference between the estimates and the corresponding covariances are calculated as follows [20]

$$\mathbf{\Delta}_{ij} = \hat{\mathbf{x}}^i_{k|k} - \mathbf{Y}^{j^{-1}}_k \mathbf{y}^j_k, \tag{12.62}$$

$$\mathbf{T}_{ij} = \mathbf{P}^i_{k|k} + \mathbf{Y}^{j^{-1}}_k. \tag{12.63}$$

If $\mathbf{\Delta}'_{ij}\mathbf{T}^{-1}_{ij}\mathbf{\Delta}_{ij} \leq D$, where D is the threshold, then both estimates are accepted as belonging to same target.

After a tracklet is associated with a track, the associated track is updated as follows

$$\left(\mathbf{P}^{\text{fused}}_{k|k}\right)^{-1} = \mathbf{P}^{-1}_{k|k} + \mathbf{Y}_k, \tag{12.64}$$

$$\left(\mathbf{P}^{\text{fused}}_{k|k}\right)^{-1} \hat{\mathbf{x}}^{\text{fused}}_{k|k} = \mathbf{P}^{-1}_{k|k}\, \hat{\mathbf{x}}_{k|k} + \mathbf{y}_k. \tag{12.65}$$

If no target is associated to a tracklet, then a new track is formed and is given by

$$\mathbf{P}_{k|k} = \mathbf{Y}^{-1}_k, \tag{12.66}$$

$$\hat{\mathbf{x}}_{k|k} = \mathbf{P}_{k|k}\, \mathbf{y}_k. \tag{12.67}$$

12.5.2 Performance of Distributed Tracking with Full Feedback at Every Measurement Step

In this section, the performance of distributed tracker is studied in a scenario, in which LFCs send their tracks at every measurement steps and CFC send the updated tracks back to LFCs.

If the estimate and corresponding covariance of the jth LFC at time k is given by $\hat{\mathbf{x}}^j_{k|k}$ and $\mathbf{P}^j_{k|k}$, $j = 1, 2, \ldots, M$, respectively, the central estimate in terms of the local estimates is given by [20, 50]

$$\mathbf{P}^{-1}_{k|k}\, \hat{\mathbf{x}}_{k|k} = \mathbf{P}^{-1}_{k|k-1}\, \hat{\mathbf{x}}_{k|k-1} + \sum_{j=1}^{M} \left[\left(\mathbf{P}^j_{k|k}\right)^{-1} \hat{\mathbf{x}}^j_{k|k} - \left(\mathbf{P}^j_{k|k-1}\right)^{-1} \hat{x}^j_{k|k-1}\right], \tag{12.68}$$

and the corresponding covariance is given by

$$\mathbf{P}^{-1}_{k|k} = \mathbf{P}^{-1}_{k|k-1} + \sum_{j=1}^{M} \left(\mathbf{P}^j_{k|k}\right)^{-1} - \left(\mathbf{P}^j_{k|k-1}\right)^{-1}, \tag{12.69}$$

where M is the number of LFCs.

Under the assumption that the LFCs have the feedback from the CFC at every measurement step, $\mathbf{P}^j_{k|k-1} = \mathbf{P}_{k|k-1}$ for all j. Then $(\mathbf{P}^j_{k|k})^{-1} - (\mathbf{P}^j_{k|k-1})^{-1}$ provides the information gained from the jth LFC's sensors.

Hence, the performance of distributed tracking with feedback at every measurement time step is equivalent to the performance of centralized tracking.

12.5.3 PCRLB for Distributed Tracking

For distributed tracking it is assumed that LFCs communicate with CFC at each measurement time steps. In Section 12.5.2, it has been shown that the performance of distributed tracking with communication at every time step is equivalent to the centralized tracking. Hence, the equation derived for centralized architecture in Section 12.3 can still be used for the distributed architecture under the assumption considered in this chapter. An approximate PCRLB for a generalized, decentralized, architecture is given in Section 12.6.1.

12.5.4 Problem Description

There are M LFCs and N sensors. It is assumed that the number of LFCs and sensors, and their locations are fixed and known. Each LFC can handle a certain maximum number of sensors, n_j, because of physical limitations [35]. Sensors transmit their measurements to their LFCs through the allocated frequency channels. The available number of channels, F, is also limited. Thus, even though large number of sensors are available, only a few sensors can be used at any one time. Hence, sensors will be activated or deactivated by a control message whenever the active sensors is changed [51]. Any active sensor will be connected to one of the LFCs through a frequency channel. Note that one sensor can be connected to at most one LFC to avoid the redundancy and to reduce the correlation between the tracks from each LFC. The transmission powers of the sensors are software controllable within certain lower and upper limits. In order to recover the signals, which are sent by the sensors, at the LFCs, the signal-to-noise ratio at each LFC for each channel must be greater than or equal to a known threshold level.

The objective is to maximize the tracking performance of the system. The tracking performance is measured by the accuracies of the existing targets' estimates and the detection probabilities of new incoming targets. Then, the problem is to assign the sensor subset for each LFC, and assign the transmitting frequency and the power to each sensor, such that SNR is above the threshold level, in order to maximize the tracking performance.

Since a hierarchical architecture with feedback is considered, each LFC has the knowledge of the entire system. Therefore, even if sensor management is performed at every LFC, the result will be the same. In order to avoid redundancy and to reduce the workload of the LFCs, sensor management is performed at the CFC and the results, that is, which sensors should be used by each LFC and what frequency channel and how much transmission power should be assigned to each sensor, are sent to all the LFCs. However, if an LFC does not receive any message from the CFC due to communication problems, it will perform the sensor management based on its own knowledge.

12.5.5 Problem Formulation

In Section 12.5.3, it is shown that the PCRLB for the architecture considered in this section is same as the PCRLB of centralized tracking. Hence, centralized PCRLB equations are used in formulation.

12.5.5.1 Objectives. Let s_{ijf} be the indicator function that takes the value 1 if sensor i is assigned to LFC j through the frequency channel f and 0 otherwise.

The first objective is to minimize the PCRLB of existing targets and given by

$$\min \text{trace} \left\{ \left[\mathbf{J}_X(k) + \sum_{i=1}^{N} \sum_{j=1}^{M} \sum_{f=1}^{F} s_{ijf} \mathbf{J}_{z_i}(k) \right]^{-1} \right\}. \tag{12.70}$$

The second objective is to quickly detect the new incoming targets. It is assumed that the new targets appear only through the perimeter of the surveillance region. As explained in Section 12.4.2, P_{new} particles are distributed along the perimeter of the surveillance region to represent the possible states of the new target. Then the second objective, which is similar to (12.47), is given by

$$\min \frac{1}{P_{\text{new}}} \sum_{p=1}^{P_{\text{new}}} \prod_{i=1}^{N} \left(1 - \sum_{j=1}^{M} \sum_{f=1}^{F} s_{ijf} P_d(i, p) \right). \tag{12.71}$$

The objectives are slightly modified in order to avoid wasting a lot of power for marginal improvement in the tracking performance. The modified objectives are

$$\min \sum_{t=1}^{T} \max \left(V_L, \text{ trace} \left\{ \left[\left(\mathbf{J}_X(k) + \sum_{i=1}^{N} \alpha_i \mathbf{J}_{z_i}(k) \right)^{-1} \right]_t \right\} \right), \tag{12.72}$$

$$\min \max \left(M_L, \frac{1}{P_{\text{new}}} \sum_{p=1}^{P_{\text{new}}} \prod_{i=1}^{N} (1 - \alpha_i P_d(i, p)) \right), \tag{12.73}$$

where $\alpha_i = \sum_{j=1}^{M} \sum_{f=1}^{F} s_{ijf}$, V_L is the tolerable variance of the estimation error of a target, M_L is the tolerable missing probability of a new target, and $[\cdot]_t$ is the tth block diagonal matrix that corresponds to target t.

12.5.5.2 Constraints. In this section, the constraints of the above optimization problem are given. First constraint is that a sensor either connected to an LFC through a frequency or not, that is, s_{ijf} can take only one or zero,

$$s_{ijf} \in \{0,\ 1\} \qquad \forall\, i,\, j \text{ and } f. \tag{12.74}$$

Since α_i is introduced in the objective for simplification, the relationship of α_i with the s_{ijf}'s must be added as a constraint and it is

$$\sum_{j=1}^{M}\sum_{f=1}^{F} s_{ijf} = \alpha_i \qquad \forall\, i. \tag{12.75}$$

From (12.74) and (12.75), α_i is an integer, takes zero if it is not used by any fusion center. If a sensor is used by more than one LFC, then duplicate information will be sent to CFC making the information ineffective. Also, the tracks will be correlated and must be considered in fusion. Hence, a sensor can be assigned to at most one LFC

$$\alpha_i \leq 1 \qquad \forall\, i. \tag{12.76}$$

It has been already established that α_i is an integer. However, from the above constraint, α_i is now a binary integer.

Due to physical limitations, a LFC can handle a maximum of certain number, n_j, of sensors

$$\sum_{i=1}^{N}\sum_{f=1}^{F} s_{ijf} \leq n_j \qquad \forall\, j. \tag{12.77}$$

Note that there is no need to use two channels to connect a sensor to an LFC. In the above equations, a sensor is connected to a LFC through at most one frequency channel.

Two sensors that are connected to the same LFC cannot use the same frequency since both information will be interfering with each other. Hence at most one sensor can be connected to one LFC through one frequency

$$\sum_{i=1}^{N} s_{ijf} \leq 1 \qquad \forall\, j \text{ and } f. \tag{12.78}$$

It is assumed that transmitting power of the sensors are software controllable with a minimum and maximum limit. Hence, if a sensor is not active then transmitting power should be zero, otherwise its transmission power should be greater than the lower limit, p_l, and less than the upper limit, p_u

$$P_t^i \leq p_u\, \alpha_i \qquad \forall\, i, \tag{12.79}$$

$$P_t^i \geq p_l\, \alpha_i \qquad \forall\, i, \tag{12.80}$$

where P_t^i is the transmitting power of sensor i.

The last constraint is that, in order to extract the signal, SNR at each LFC for each frequency must be greater than or equal to the threshold level, $\sigma_{\min}$ [52]

$$\frac{P_t^i s_{ijf} r_{ij}^{-\lambda}}{\sum_{\substack{x=1\\ x\neq i}}^{N} \sum_{\substack{y=1\\ y\neq j}}^{M} P_t^x s_{xyf} r_{xj}^{-\lambda} + N_0} \geq s_{ijf}\, \sigma_{\min} \qquad \forall\, i,\, j \text{ and } f, \tag{12.81}$$

where r_{ij} is the distance between sensor i and LFC j, λ is the decaying factor and N_0 is the environmental noise.

As a summary, the optimization problem is

$$\min \sum_{t=1}^{T} \max \left(V_L,\ \text{trace} \left\{ \left[\left(\mathbf{J}_X(k) + \sum_{i=1}^{N} \alpha_i \mathbf{J}_{z_i}(k) \right)^{-1} \right]_t \right\} \right), \tag{12.82}$$

$$\min\ \max \left(M_L,\ \frac{1}{P_{\text{new}}} \sum_{p=1}^{P_{\text{new}}} \prod_{i=1}^{N} (1 - \alpha_i P_d(i, p)) \right). \tag{12.83}$$

Subject to

$$s_{ijf} \in \{0,\ 1\} \qquad \forall\, i,\, j \text{ and } f, \tag{12.84}$$

$$\sum_{j=1}^{M} \sum_{f=1}^{F} s_{ijf} = \alpha_i \qquad \forall\, i, \tag{12.85}$$

$$\alpha_i \leq 1 \qquad \forall\, i, \tag{12.86}$$

$$\sum_{i=1}^{N} \sum_{f=1}^{F} s_{ijf} \leq n_j \qquad \forall\, j, \tag{12.87}$$

$$\sum_{i=1}^{N} s_{ijf} \leq 1 \qquad \forall\, j \text{ and } f, \tag{12.88}$$

$$P_t^i \leq p_u\, \alpha_i \qquad \forall\, i, \tag{12.89}$$

$$P_t^i \geq p_l\, \alpha_i \qquad \forall\, i, \tag{12.90}$$

$$\frac{P_t^i s_{ijf} r_{ij}^{-\lambda}}{\sum_{\substack{x=1\\ x\neq i}}^{N} \sum_{\substack{y=1\\ y\neq j}}^{M} P_t^x s_{xyf} r_{xj}^{-\lambda} + N_0} \geq s_{ijf}\, \sigma_{\min} \qquad \forall\, i,\, j \text{ and } f. \tag{12.91}$$

The solution methodology to this problem is given in the next section.

12.5.6 Solution Technique

The original is a multiobjective NP-hard combinatorial optimization problem [44]. Finding the optimal solution for large-scale problems in real time can be difficult.

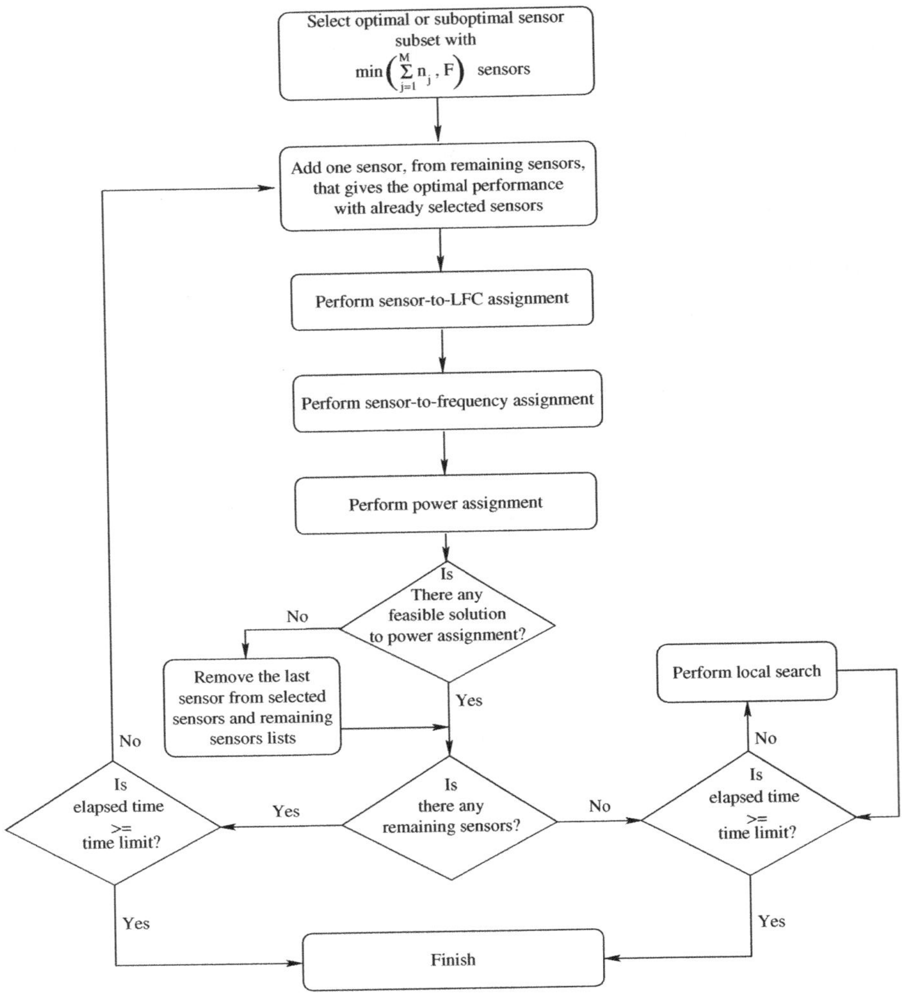

Figure 12.13 Algorithm used to find a suboptimal solution.

Hence, an algorithm is presented to find a suboptimal solution in real time. The flowchart of the algorithm is shown in Figure 12.13.

In the algorithm, first min $\left(\sum_{j=1}^{M} n_j,\ F\right)$ optimal or suboptimal sensors are selected from the available sensors by considering only the objective function. Note that even though the constraints are not considered to select the above sensor subset, this solution will always satisfy all the constraints. Then the sensors are added one by one so that the objective value is minimized and all the constraints are satisfied. After adding a sensor to already selected sensors, feasibility check is performed as follows:

1. Sensors are assigned to LFCs. Note that this assignment has to be performed from the beginning whenever a new sensor is added to the selected sensor subset.
2. After sensors are assigned to LFCs, the transmitting frequencies will be assigned to each sensor.
3. Finally, transmitting power is assigned to each sensor. However, if no feasible solution is found for power assignment, the selected sensor subset is not feasible and the sensor that was added last is removed from the selected sensors. Note that this removed sensor is not needed to check for feasibility in the future after adding one or more sensors to current sensor subset, since this is obviously infeasible.

After adding possible maximum number of sensors, local search is performed to find a better solution by swapping the sensors in and out until the stopping criteria is satisfied. While performing the local search, all the remaining sensors including sensors that produced infeasible solutions should be considered.

Detailed explanations of the above steps are given next. Let $s_{ijf} = \alpha_i * \beta_{ij} * \gamma_{if}$, where α_i takes the value 1 if sensor i is selected and 0 otherwise, β_{ij} takes the value 1 if sensor i is assigned to LFC j and 0 otherwise, and γ_{if} takes the value 1 if sensor i is assigned to frequency f and 0 otherwise.

The original problem is decomposed into four subproblems:

1. Finding α_i (explained in Section 12.5.6)
2. Finding β_{ij} (explained in Section 12.5.6)
3. Finding γ_{if} (explained in Section 12.5.6)
4. Finding P_t^i (explained in Section 12.5.6)

12.5.6.1 Active Sensor Selection. This section briefly describes how to select the sensors that should be active. Handling multiple objectives is difficult. Hence, both objectives are combined by giving suitable weights to each objective to form a single objective, which is easy to handle. The weights can be selected based on the requirements. The combined objective is

$$\min \left\{ \sum_{t=1}^{T} \max \left(V_L,\ \operatorname{trace} \left\{ \left[\left(\mathbf{J}_X(k) + \sum_{i=1}^{N} \alpha_i \mathbf{J}_{z_i}(k) \right)^{-1} \right]_t \right\} \right) + w_d * \max \left(M_L,\ \frac{1}{P_{\text{new}}} \sum_{p=1}^{P_{\text{new}}} \prod_{i=1}^{N} (1 - \alpha_i P_d(i, p)) \right) \right\}, \tag{12.92}$$

where w_d is the weight given to the detection function.

To find α_i, first one has to decide how many sensors to select. The maximum number of sensors that can be active at any one time is $\sum_{j=1}^{M} \min(n_j, F)$. However, if $\sum_{j=1}^{M} \min(n_j, F)$ sensors are selected by considering only the objective, a feasible β_{ij}, γ_{if}, and P_t^i might not be found with selected sensors, α_i. Hence, first $\min\left(\sum_{j=1}^{M} n_j,\ F\right)$ sensors, which definitely give a feasible solution to the original problem, is selected by finding a better initial solution followed by a local search. The solution consists of the following steps:

1. Select the best sensor that gives the minimum objective value when only one sensor is used.
2. Continue adding sensors one by one until $\min\left(\sum_{j=1}^{M} n_j,\ F\right)$ sensors selected. If n sensors are already selected, then the $(n+1)$th sensor is the one that gives the minimum objective value when added with the already selected sensors.
3. Perform a local search by swapping sensors in and out such that the objective value is minimized.

After finding a better solution with $\min\left(\sum_{j=1}^{M} n_j,\ F\right)$ sensors, add sensors one by one until no more sensors can be (or need to be) added. The steps to select the $(n+1)$th sensor are as follows:

1. Rank the remaining sensors based on their performances combined with already selected sensors.
2. Select the best one as the $(n+1)$th sensor and solve for β_{ij}, γ_{if} and P_t^i as explained in Section 12.5.6. If there is no feasible solution for P_t^i, select the next best sensor as the $(n+1)$th sensor and perform the feasibility check again. Keep on changing the sensor in the rank order until get a feasible solution.

After selecting the maximum possible number of sensors, a local search is performed by swapping the sensors in and out such that the objective value is minimized and a feasible solution is obtained for β_{ij}, γ_{if}, and P_t^i.

Note that the algorithm is interrupted at any stage if the time limit is reached.

12.5.6.2 Sensor-to-LFC Assignment.

After selecting the active sensors, corresponding LFCs to which the selected sensors should be connected must be determined. In order to improve the SNR at each LFC for each frequency, the distances of the sensors from the assigned LFCs must be minimized while the distances from other LFCs are maximized.

Then, the sensor-to-LFC assignment, β_{ij}, can be formulated as

$$\max \sum_{i=1}^{N} \sum_{j=1}^{M} \left(\frac{r_{ij}^{-\lambda}}{\sum_{\substack{y=1 \\ y \neq j}}^{M} r_{iy}^{-\lambda}} \right) \beta_{ij}, \tag{12.93}$$

subject to

$$\sum_{i=1}^{N} \beta_{ij} \leq n_j \qquad \forall\, j, \tag{12.94}$$

$$\sum_{j=1}^{M} \beta_{ij} = \alpha_i \qquad \forall\, i, \tag{12.95}$$

$$\beta_{ij} \in \{0,\ 1\} \qquad \forall\, i \text{ and } j. \tag{12.96}$$

The above problem is a linear binary integer programming, and can be solved using CPLEX solver [53].

12.5.6.3 Sensor-to-Frequency Assignment. After finding the β_{ij} values for the selected sensors, α_i, the remaining problem is to assign the transmitting frequencies, γ_{ij}, and transmitting powers, p_t^i, to all the selected sensors. Since the sensor battery energies are limited, the power consumption must be minimized to increase the sensor's lifetime. The remaining problem is

$$\min \sum_{i=1}^{N} P_t^i, \tag{12.97}$$

subject to

$$\gamma_{if} \in \{0,\ 1\} \qquad \forall\, i \text{ and } f, \tag{12.98}$$

$$\sum_{i=1}^{N} \alpha_i\, \beta_{ij}\gamma_{if} \leq 1 \qquad \forall\, j \text{ and } f, \tag{12.99}$$

$$P_t^i \leq p_u\, \alpha_i \qquad \forall\, i, \tag{12.100}$$

$$P_t^i \geq p_l\, \alpha_i \qquad \forall\, i, \tag{12.101}$$

$$\sum_{i=1}^{N} \alpha_i\, \beta_{ij}\, \gamma_{if}\, P_t^i\, r_{ij}^{-\lambda} \geq \sigma_{\min} \left(\sum_{\substack{x=1 \\ x \neq i}}^{N} P_t^x \sum_{y=1}^{M} \gamma_{xf} r_{xj}^{-\lambda} + N_0 \right) \qquad \forall\, j \text{ and } f. \tag{12.102}$$

This is also a NP-hard problem. If the transmission power is fixed, the above problem can be solved easily. However, in order to incorporate the advantage of the variable transmission power, γ_{if} and P_t^i have to be considered together. The following approximation is used to solve the above problem.

1. All the LFCs are arranged in an order. Let the first LFC be the one in any one corner. The next LFC is the closest to the previous LFC form the remaining LFCs.

2. Assign the frequencies to the first ($j = 1$) LFC's sensors. Since the number of sensors assigned to this LFC is less than or equal to F, simply assign any one frequency, but different, to one sensor.
3. Assign the frequencies to the jth LFC's sensors by solving the following optimization problem

$$\min \max_{\substack{i=1,\ldots,N_j \\ f=1,\ldots,F}} C_{if}\,\overline{\gamma}_{if}, \tag{12.103}$$

subject to

$$\overline{\gamma}_{if} \in \{0,\ 1\} \qquad \forall\, i \text{ and } f, \tag{12.104}$$

$$\sum_{f=1}^{F} \overline{\gamma}_{if} = 1 \qquad \forall\, i, \tag{12.105}$$

$$\sum_{i=1}^{N_j} \overline{\gamma}_{if} \leq 1 \qquad \forall\, f, \tag{12.106}$$

where N_j is the number of sensors assigned to the jth LFC and C_{if} is the cost of assigning frequency f to sensor i.

Note that $\overline{\gamma}$ is used instead of γ, since only the subset of sensors that are assigned to the j-th LFC is considered. However after finding the solution to $\overline{\gamma}$, γ will be updated.

To find the value of C_{if}, the cost of assigning frequency f to sensor i, first the transmission powers correspond to the sensors that use frequency f, including sensor i, is found by using the formulation given in the next section (12.5.6), and then calculate C_{if} by

$$C_{if} = \max_{y \in \{i_f\}} \left(\frac{\sum_{\substack{x=1 \\ x \neq y}}^{N_f} P_t^x r_{xj_y}^{-\lambda} + N_0}{P_t^y r_{yj_y}^{-\lambda}} \right), \tag{12.107}$$

where i_f indicates the sensors that are using frequency f including sensor i, N_f is the number of sensors using frequency f, and j_y is the LFC to which sensor y is assigned.

The C_{if} gives the maximum of the noise-to-signal ratio for frequency f at any LFC, if frequency f is assigned to sensor i. In order to improve the possibility of getting a feasible solution for P_t^i, the maximum value of the noise-to-signal ratio over all the frequencies is minimized in (12.103).

The above problem can be reformulated as a linear integer problem

$$\min \quad T, \tag{12.108}$$

subject to (12.104), (12.105), (12.106) and

$$C_{if}\,\overline{\gamma}_{if} \leq T \qquad \forall\, i,\ f. \tag{12.109}$$

This problem can also be solved using CPLEX solver [53].

4. Continue this process until no more LFC remains.

12.5.6.4 Transmission Power Assignment.

After finding the α_i, β_{ij}, and γ_{if}, the remaining problem is to find the transmission power for each active sensor. In order to increase the lifetime of the sensors, transmitting powers must be minimized while satisfying the constraints.

Transmitting power assignment, P_t^i, can be formulated as

$$\min \sum_{i=1}^{N} P_t^i, \tag{12.110}$$

subject to

$$P_t^i \leq \alpha_i\, p_u \qquad \forall\, i, \tag{12.111}$$

$$P_t^i \geq \alpha_i\, p_l \qquad \forall\, i, \tag{12.112}$$

$$\sum_{i=1}^{N} \alpha_i\, \beta_{ij}\, \gamma_{if}\, P_t^i\, r_{ij}^{-\lambda} \geq \sigma_{\min} \left(\sum_{\substack{x=1 \\ x \neq i}}^{N} P_t^x \sum_{y=1}^{M} \gamma_{xf} r_{xj}^{-\lambda} + N_0 \right) \qquad \forall\, j,\, f. \tag{12.113}$$

The above problem is a linear convex problem [54], and can be solved using the CPLEX solver [53].

Note that for some α_i, β_{ij}, and γ_{if}, there will not be any feasible solution for power. In that case, α_i, β_{ij}, or γ_{if} has to be changed so that a feasible solution is found.

12.5.7 Simulation Results

In the simulation scenario, the number of targets $T = 2$; the number of LFCs $M = 4$; the number of sensors $N = 100$; the maximum usable sensors by each LFC $n_j = 4$; the available number of channels $F = 5$; the measurement interval is 30 s; measurements are the bearings from the sensors to targets; measurement standard deviation $\sqrt{R} = 0.05$ radians; the field of view of the sensor $V = 2\pi$ radians; the transmission power bounds $p_l = 0$ and $p_u = 1$ W; the threshold of SNR $\sigma_{\min} = 10$ dB; the probability of false alarm is 0.001.

Since the measurement model is nonlinear, extended Kalman filter (EKF)-based information filter is used to track the targets at each LFC [20, 46]. Because of the existence of multiple targets and false alarms, measurements must be associated to already established tracks, new tracks or a dummy track that corresponds to tracks

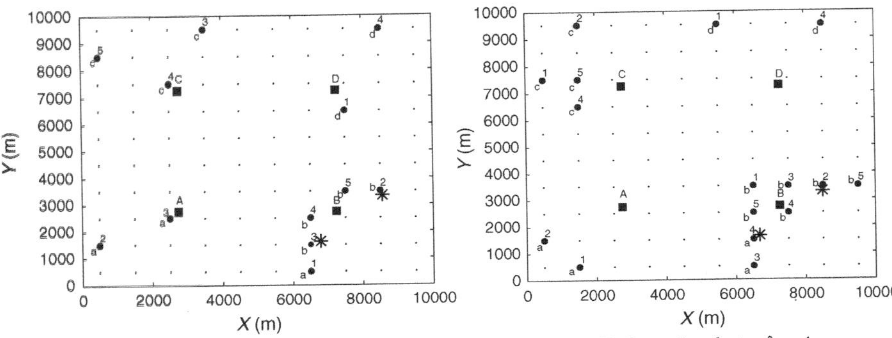

(a) Decaying factor $\lambda = 2$ (b) Decaying factor $\lambda = 4$

Figure 12.14 Selected sensors and their LFC and frequency assignments.

due to false alarms. First, the $(S + 1)D$ (or multiframe) assignment algorithm is used to associate the measurements with the already established tracks [55]. The remaining measurements, which are not associated with any of the already established tracks, are associated with new tracks or dummy track using the S–D algorithm [56]. New tracks are formed with the associated measurements using iterated least squares (ILS) estimator [46].

Estimation at the CFC is performed using (12.68) and (12.69) that consider a single target. However, multiple targets can be tracked by each fusion center, and the estimate of each target has to be updated. Hence, first CFC will group the estimates, that belong to same target, from all the LFCs. Since LFCs get the feedback from the CFC, track ID can be used to identify the target that corresponds to each estimate [57]. However, track ID will be the same only for the already established tracks. If new tracks are formed by LFCs then the technique explained in Section 12.5.1 is used to decide if they are from the same target.

The solution for the sensor management using the algorithm presented in this chapter is shown in Figure 12.14. The selected sensors, their LFC and frequency assignments are shown. In Figure 12.14, "■" indicates the LFC and the capital letter (e.g., A, B, ...) near to it indicates its name. The "·" indicates the sensor, "•" indicates the selected sensor, and the small letter (e.g., a, b, ...) near to it indicates that the sensor is assigned to the corresponding capital letter LFC (e.g., sensor "a" is assigned to LFC "A"). The number near the selected sensor indicates the frequency channel that is assigned to it. The "∗" indicates the target. When the decaying factor is high, the cochannel interference is less. Hence, more sensors could be used with decaying factor 4 than decaying factor 2. However, even with decaying factor 2, the presented algorithm assigns enough sensors to the already exiting targets and other sensors to cover the boundaries through which new targets can enter the surveillance region.

In order to compare the presented method with another method, a slightly different scenario is used. The differences are the sensors are randomly distributed in the surveillance region; the number of targets $T = 3$; the available number of channels $F = 20$. Since the available number of channels (20) is greater than the maximum

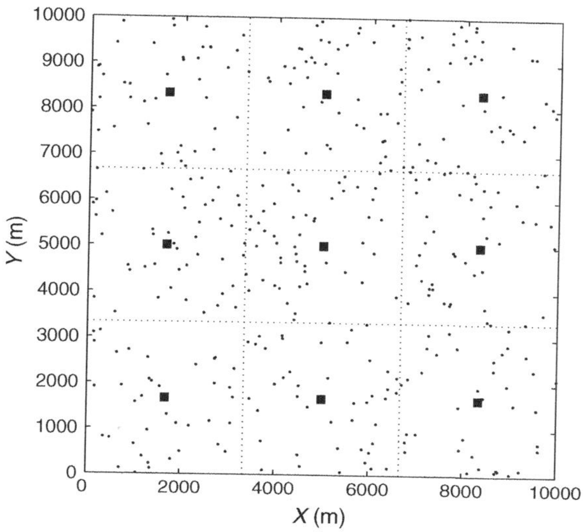

Figure 12.15 A sample scenario with clustered sensors.

usable sensors by all the LFCs (4 × 4), the frequency assignment is not an issue. However, no clustering is used in the presented algorithm.

In the alternate method that is used for comparison, sensors are clustered based on target or fusion center [27]. Target-based clustering cannot be applied to the scenarios considered in this chapter, since number of target is time varying and a sensor can get measurements from more than one target. Fusion center-based clustering is applicable to the above scenario, and an algorithm using clustering is explained next.

First sensors are clustered based on the LFCs, that is, a sensor can be used only by the closest LFC. The clustered sensors are shown in Figure 12.15. Clustering helps to perform the sensor-to-LFC assignment, and it is performed prior to selecting the active sensors. The remaining problem is to select the active sensors.

The comparisons of presented algorithm with the algorithm that used LFC based clustering are shown in Figures 12.16–12.18. Figure 12.16 shows the selected sensors and their LFC and frequency assignments. The symbol notations in this figure are same as in Figure 12.14. There are three targets around the LFC B. In the presented algorithm, enough sensors are assigned to those targets and others sensors are assigned to catch the new targets. In the algorithm with clustering, only four sensors, which are not enough to track all three targets, are assigned to the existing targets. Figures 12.17 and 12.18 show the comparison of the RMSE and the PCRLB, respectively. Presented algorithm gives more than 30% improvement than the algorithm that used clustering.

The average computation time of the presented algorithm for the above scenario is 4.2 s. The corresponding time for LFC-based clustering algorithm is 3.1 s. Both algorithms are coded in *Matlab* and run on a 2.4 GHz Pentium 4 processor. The presented algorithm gives around 30% improvement in the performance than the

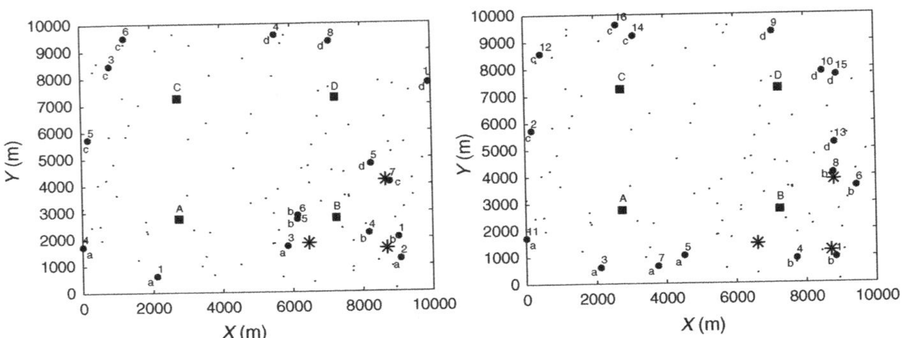

(a) Presented algorithm (b) Algorithm using clustering

Figure 12.16 Sensors selected with presented algorithm and an algorithm that used LFC-based clustering.

algorithm that used clustering with around 35% more computation time. Computation time, however, is not a big concern as long as it is smaller than the allocated time, and can be further reduced by coding the algorithms in C.

To see the effect of clustering on computation and the performance the following scenario is considered: the number of targets $T = 3$; the available number of channels $F = 6$, the maximum usable sensors by each LFC $n_j = 4$; decaying factor $\lambda = 4$. In the algorithm with clustering, LFC-based clustering is used instead of sensor-to-LFC assignment, as in Section 12.5.6. Algorithms are stopped if either allocated time is reached or a local optimum is found with one sensor swapping.

The best objective function values found by both algorithms in the allocated times are shown in Table 12.3. When the allocated time is less than or equal to 15 s, the algorithm that used clustering gives the better solution. However, with more allocated time, presented algorithm finds a better solution that is approximately 30–50% better

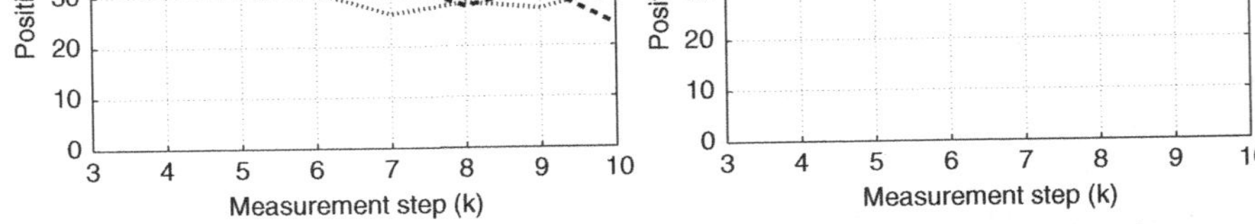

(a) Presented algorithm (b) Algorithm using clustering

Figure 12.17 RMSE comparison of presented algorithm with an algorithm that used clustering.

TABLE 12.3 The Best Objective Function Values Found Using Clustering and Presented Algorithms

	Scenario 1		Scenario 2	
Allocated time (s)	Algorithm with Clustering	Presented Algorithm	Algorithm with Clustering	Presented Algorithm
5	1222	5851	7214	10,966
15	929	5831	6926	10,964
25	929	715	6926	3285
35	929	668	6926	3285

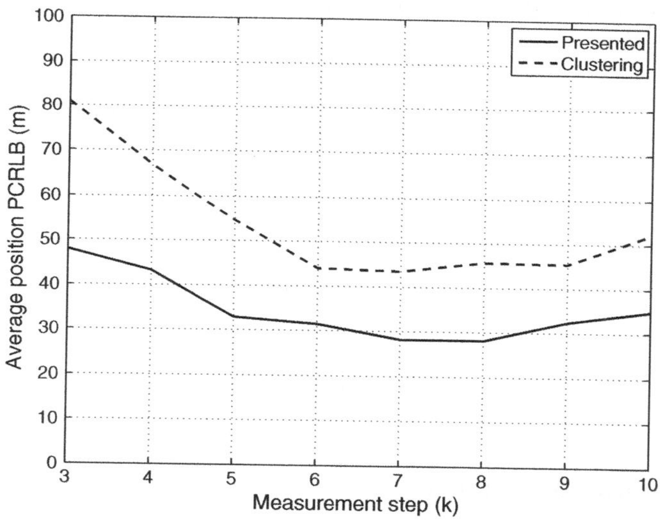

Figure 12.18 PCRLB comparison of presented algorithm with an algorithm that used LFC-based clustering.

than the solution obtained by the algorithm with clustering. Note that the computation times can be reduced significantly by coding the algorithm in *C*, and in most cases there will be sufficient time to reach a better solution using the presented algorithm.

12.6 SENSOR ARRAY MANAGEMENT FOR DECENTRALIZED TRACKING

In this section, the algorithms presented in previous two sections are extended to decentralized architecture, shown in Figure 12.1c, in which there is no central fusion center and each FC gets information only from its neighbors [15, 36].

12.6.1 PCRLB for Decentralized Tracking

In the decentralized tracking, equations derived for centralized tracking in Section 12.3 will be used to update the PCRLB whenever a FC obtains measurements from its local sensors. However, if a FC receives a tracklet from its neighbor, then the equations derived so far cannot be used, and an approximate PCRLB updating equation for track fusion is given below.

Suppose, the neighbors of FC f are FCs $f_{n_1}, f_{n_2}, \ldots, f_{n_j}$, and their last communication time steps are $l_{n_1}, l_{n_2}, \ldots, l_{n_j}$, respectively. The FIMs at last communication time steps are $\mathbf{J}^{f_{n_1}}(l_{n_1}), \mathbf{J}^{f_{n_2}}(l_{n_2}), \ldots, \mathbf{J}^{f_{n_j}}(l_{n_j})$. The FIMs, which are updated using only their local sensor measurements, at current time step k are $\mathbf{J}^{f_{n_1}}(k), \mathbf{J}^{f_{n_2}}(k), \ldots, \mathbf{J}^{f_{n_j}}(k)$. An approximate PCRLB equation, based on information filter (see Section 12.5.1), for track fusion at FC f is

$$\mathbf{J}^{f}_{\text{fused}}(k) = \mathbf{J}^{f}(k) + \sum_{i=1}^{j} c_i \left(\mathbf{J}^{f_{n_i}}(k) - \mathbf{J}^{f_{n_i}}_{X}(k|l_{n_i}) \right), \tag{12.114}$$

where

$$\mathbf{J}^{f_{n_i}}_{X}(k|l_{n_i}) = \left[Q_{k,l_{n_i}} + \mathbf{F}_{k,l_{n_i}} \mathbf{J}^{f_{n_i}}(l_{n_i})^{-1} \mathbf{F}'_{k,l_{n_i}} \right]^{-1}, \tag{12.115}$$

and c_i takes one if FC f received tracklets from neighbor n_i at time k, and zero otherwise.

12.6.2 Problem Description

There are few fusion centers in the surveillance region. Each FC can communicate only with its neighbors. The FCs that are in the communicable distance from an FC are considered as its neighbors. There are a large number of sensors, which are already deployed, and their locations are fixed and known. However, due to physical limitations, only a maximum of certain number of sensors, n_f, can be used by each FC, f, at any one time. In addition, the active sensor subset can only be changed after a certain minimum number of time steps. The number of targets in the surveillance region is time varying, that is, a new target can appear or an already existing target can disappear at any time. A sample scenario is shown in Figure 12.19, where "." indicates sensor, "■" indicates FC, and "★" indicates target.

The problem is to assign a sensor subset to each FC such that the tracking performance is maximized, that is, the already established tracks are maintained accurately and the newly incoming targets are detected immediately and accurately. Because of the additional difficulties due to lack of global knowledge, it is assumed that the frequencies are already allocated to each FC, and only problem is to find the active sensors.

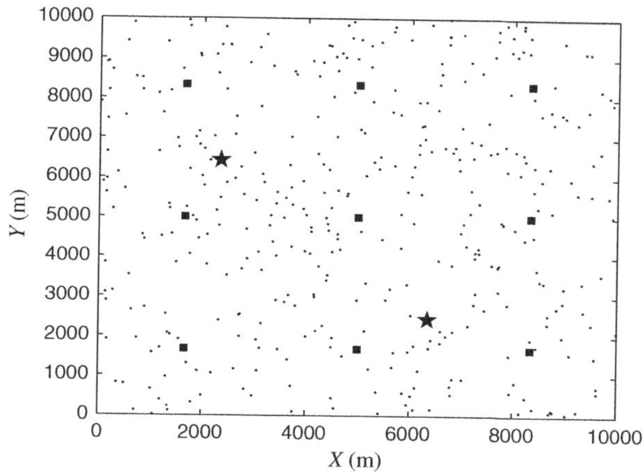

Figure 12.19 A sample scenario.

12.6.3 Problem Formulation

The optimal solution can be obtained, if there is a central fusion center and it performs the sensor selection considering all the possibilities. However, in the decentralized architecture this is not feasible, since FCs may not have the global knowledge, communication might not be reliable, and it is computationally very demanding. Hence, each FC has to decide its own sensors. Even though each FC selects its own sensors, it cannot make the decision based on its region alone. In order to ensure better sensor selection and thus better tracking results, FCs have to consider the sensors that will be used by other FCs, at least the neighboring ones, as well. Considering all other FCs is not feasible due to lack of global knowledge and heavy computation requirements in large-scale problems. Hence, each FC considers only its neighbors while performing the sensor selection. It is assumed that tracklets are communicated among the fusion centers, and each FC has the information of all the sensors in the neighborhood.

First, in order to avoid one sensor being used by more than one fusion center, which will introduce redundancy as well as correlation between tracklets, sensors are clustered based on geographic location. That is, a sensor can only be used by the nearest FC. Then, each FC will select the sensors from its region. However, in the later sections some techniques are presented to avoid clustering, which will help to improve the performance as shown in the previous section.

12.6.3.1 Synchronous Sensor Selection. In this section, it is assumed that FCs exchange their information immediately prior to the selection of sensors, and all the FCs change sensors at the same time. Active sensor subsets of all the FCs are changed after every l measurement time steps. A sample communication and sensor change time steps with $l = 5$ are shown in Figure 12.20, where "□" indicates the tracklet transmission and fusion, and "◇" indicates the sensor change.

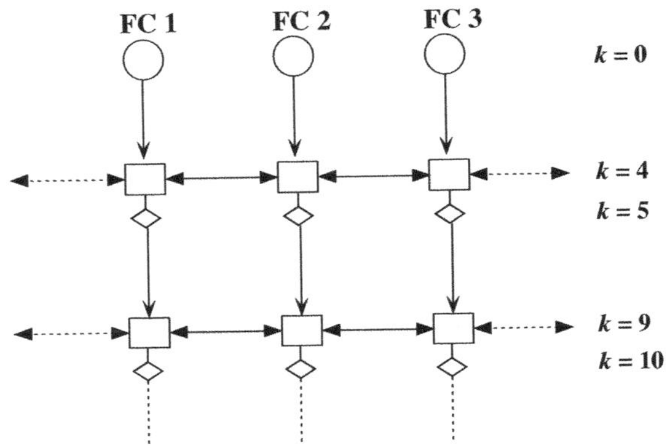

Figure 12.20 A sample sensor change and communication time steps.

Consider the sensor management at FC a. Suppose the next sensor change occurs at time $k+1$, and the FIM, after fusing neighbors' information, at time k is $\mathbf{J}^a(k)$. Suppose FC has M neighbors (including itself). FC has to decide which of the sensors should be used at time steps $k+1, k+2, \ldots, k+l$. Note that the same sensors will be used in all these time steps. Then the objective is to

$$\arg \min_{\{s_i^f; f=1,2,\ldots,M\}} \left(\sum_{j=1}^{l} \left(w_j \sum_{t=1}^{T} \left(w_t \min_{f=1,2,\ldots,M} \operatorname{trace}\left(\left[\mathbf{J}^f(k+j)^{-1} \right]_t \right) \right) \right) \right. \\ \left. + w_{\text{new}} \left(\mathbb{E} \min_{f=1,2,\ldots,M} \operatorname{trace}\left(\mathbf{J}^f_{\text{new}}(\mathbf{x}_{\text{new}})^{-1} \right) \right) \right), \tag{12.116}$$

where

$$\mathbf{J}^f(k+j) = [F_{k+j,k+j-1} \mathbf{J}^f(k+j-1)^{-1} F'_{k+j,k+j-1} + Q_{k+j,k+j-1}]^{-1} \\ + \sum_{i=1}^{N_f} s_i^f \mathbf{J}^f_{z_i}(k+j). \tag{12.117}$$

Note that instead of $\mathbf{J}^f(k+l)$, $\mathbf{J}^f_{\text{fused}}(k+l)$ is used, and it is given by

$$\mathbf{J}^f_{\text{fused}}(k+l) = \mathbf{J}^f(k+l) + \sum_{\substack{b=1 \\ b \neq f}}^{M} \left(\mathbf{J}^b(k+l) - \mathbf{J}^b_X(k+l|k) \right). \tag{12.118}$$

In (12.116), w_{new} is the weight given to the new targets, w_j is the weight given to $(k+j)$th time step and N_f is the number of sensors that belong to FC f. s_i^f takes one if ith sensor of FC f is selected and zero otherwise. The $\mathbb{E}$ denotes the expectation over all possible new target positions. In general, the expectation has to be found numerically, as explained in the previous two sections.

The information about the new target, $\mathbf{J}_{\text{new}}$, is given by

$$\mathbf{J}^f_{\text{new}}(\mathbf{x}_{\text{new}}) = \sum_{i=1}^{N_f} s_i^f \mathbf{J}_{z_i}(\mathbf{x}_{\text{new}}). \tag{12.119}$$

Since no prior information is available about the new target, only the information from the sensors is considered in the above equation. However, if there is any prior information, that also can be added. If the velocity information of the new target is not available in one time step, then only the position PCRLB can be considered.

Note that in Sections 12.4 and 12.5, the probability of detection is used as the cost function for newly incoming targets. In those two sections, it was assumed that the new targets appear only through the boundaries. In that case, even one sensor is enough to initialize the new target. However, if the new targets can appear anywhere in the surveillance region, then, in general, more than one sensor may be necessary to initialize a track. For example, if the measurements are the bearings from sensor to target, then at least two sensors are necessary to initialize the track. In such case, PCRLB of the new targets is the suitable and better choice for cost function, and used in this section. Even in the formulations of Sections 12.4 and 12.5, the probability of detection can be replaced with PCRLB, when new targets can appear anywhere in the surveillance region.

The constrains are

$$\sum_{i=1}^{N_f} s_i^f = n_f \qquad \text{for } f = 1, 2 \ldots, M, \tag{12.120}$$

$$s_i^f \in \{0, 1\} \qquad \forall\, f, i. \tag{12.121}$$

Note that only the targets in the neighboring regions are considered in the above equations. Even for the new target, the $\mathbb{E}$ is taken only over the neighboring regions.

For sensor management purpose, it is assumed that neighbors also have the same amount of information about all the targets at time k. Then, $\mathbf{J}_X^f(k+1) = \mathbf{J}_X^a(k+1)$ for all f.

Sequential Sensor Management In the above algorithm, each FC selects its own sensors by considering which sensors will be used by its neighboring FCs. However, in the end, the neighbors might select different sensors than what a FC thought, since neighbors consider their neighbors that are not considered by the FC, and also they might have different estimates. That is, although a FC considers its neighbors' action in its sensor selection, it does not have all the information nor does it have the correct information of neighbors at all times. The effect of this problem can be mitigated by performing the sensor management in a predefined sequence.

1	2	1	2	1
2	3	2	3	2
1	2	1	2	1
2	3	2	3	2
1	2	1	2	1

Figure 12.21 Predefined sequence of fusion centers.

An example sequence corresponding to the architecture shown in Figure 12.1c is shown in Figure 12.21. First, the FCs with tag 1 will perform the sensor selection by considering which sensors will be used by its neighbors. After selecting their sensors, they will inform their selection to their neighbors. Then the FCs with tag 2 select their sensors. However, now they do not need to find the sensors for neighbors with tag 1. Instead they will use the known values. Hence, computational load of these FCs is less than FCs with tag 1. After they select their sensors, they will inform their selection to their neighbors with tag 3. Finally FCs with tag 3 perform the sensor selection using the information about the neighboring FCs. Computation load of these FCs is very low compared to others, since they have to find only their sensors.

Since the FCs with tag 2 consider the actual sensors that will be used by their neighbors with tag 1, they will select better sensors that will cover the uncovered areas by neighbors. Similarly, FCs with tag 3 will select the sensor subsets that overcome the neighbors' mistakes.

An additional advantage of sequential sensor selection is that clustering is not necessary, since FCs with tag 2 can avoid the sensors selected by FCs with tag 1, similarly FC with tag 3 can avoid the sensors already selected by its neighbors.

However, the disadvantage of sequential sensor selection is that it requires more communication and later FCs must wait until others select their sensors.

12.6.3.2 Asynchronous Sensor Selection. In the previous section, it was assumed that communication and sensor selection are synchronized over the FCs, and sensor change intervals are fixed and known. However, in general scenario FCs may not change their sensors at the same time, and sensor change interval may not be fixed. Hence, in this section a more general scenario is considered with an additional objective to maximize the sensor lifetimes, when sensors have limited energy resources.

The active sensor subset of each FC is changed if the estimation uncertainty of any target goes above a certain limit. However, there is a lower and upper limit for sensor change interval. Assume last sensor change occurred at time step k and current time step is $k + j$, then a binary variable, c, that takes value 1 if need to change the current

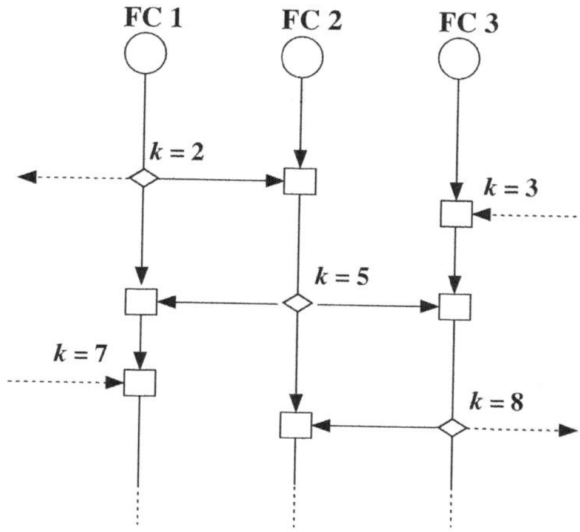

Figure 12.22 A sample sensor change and communication time steps.

active sensor subset, and zero otherwise is given by

$$c = \begin{cases} 0 & \text{if } j \le I_l \text{ or } (j \le I_u \text{ and } U_t < L_u \text{ for all } t) \\ 1 & \text{if } j > I_u \text{ or } (j > I_l \text{ and } U_t > L_u \text{ for any } t), \end{cases} \tag{12.122}$$

where I_l and I_u are lower and upper limits of sensor change interval respectively, U_t is the uncertainty of state estimate of target t and L_u is tolerable estimation uncertainty. Thus, FCs may change their sensor subsets at different time steps and any FC cannot predict the exact next sensor change times of its neighbors. Rather, it will know only an interval, which can be calculated using the last sensor change time, lower and upper limit of sensor change interval of its neighbor.

Fusion centers transmit the tracklets to their neighbors just before they change their active sensor subset. In addition to tracklets, the detail of new sensor subset is also transmitted at that time. Since the sensors are changed asynchronously and FCs send their track information just before they change their sensors, communication times are also asynchronous. A sample sensor change time steps are shown in Figure 12.22, where "□" indicates the track fusion, and "◇" indicates the sensor change and the transmission of tracklets.

Let us consider the sensor selection at FC a at time k, at which sensor selection criteria is satisfied.
The objectives are as follows:

1. *Optimize the tracking performance of existing targets*

$$\min \sum_{j=1}^{I_u} \sum_{t=1}^{T} \left(w_{jt} \min_{f=1,2,\ldots,M} \max_{v=1,2,\ldots,n_v^f(k+j)} \operatorname{trace}\left(\left[\mathbf{J}_v^f(k+j)^{-1} \right]_t \right) \right), \tag{12.123}$$

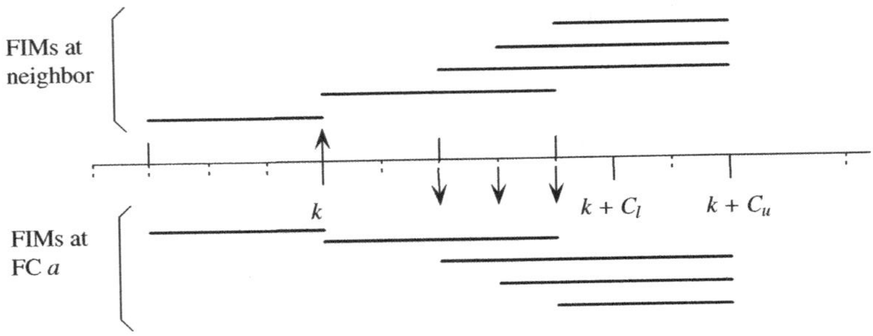

Figure 12.23 Possible versions of FIMs.

where $n_v^f(k+j)$ is the number of possible versions of FIMs at FC f at time step $k+j$ and w_{jt} is the weight given to target t at time step $k+j$. It is possible to have more than one version of FIM at any time because of the uncertainties in sensor change times and tracklet transmission times. For example, assume FC a has only one neighbor, last sensor change of neighbor occurred at $k-3$, $I_l = 5$, and $I_u = 7$.

Figure 12.23 shows the possible versions of FIMs at neighbor and FC a. At time k, FIM of neighbor is updated using the tracklet from FC a, and FC a starts to use the new sensor subset. At k and $k+1$, only one version of FIM is available in both FCs due to no uncertainty in those time steps. However, at $k+2$, $k+3$, and $k+4$ there are possibilities for sensor change and tracklet transmission at neighbor. Hence, there are two possible versions of FIMs at time $k+2$ and 3 versions at $k+2$ to $k+I_u$, as shown in Figure 12.23. Multiple version of FIMs are possible at FC a due to uncertainty in tracklet receive time from neighbor, and at neighbor due to uncertainty in sensor change time. Even though, there is a possibility for sensor change at any time step between $k+I_l$ and $k+I_u$ at FC a, in (12.123) the summation is taken over I_u time steps under the assumption that the next sensor subset may be used until $k+I_u$. However, the possibility of changing another new sensor subset before $k+I_u$ is incorporated in w_{jt}.

Note that only the targets in the neighboring regions are considered in the above equation, under the assumption that other targets will be tracked by some other FCs.

2. *Detect the newly incoming targets quickly and accurately*

$$\min \sum_{j=1}^{I_u} \sum_{p=1}^{P_{\text{new}}} w_{jp} \min_{f=1,2,\ldots,M} \max_{\mathbf{s}^f=\left\{\mathbf{s}_{\text{current}}^f, \mathbf{s}_{\text{new}}^f\right\}} \operatorname{trace}\left(\mathbf{J}_{\text{new}}^f\left(\mathbf{x}_{\text{new}}^p, \mathbf{s}^f, j\right)^{-1}\right), \tag{12.124}$$

where $\mathbf{x}_{\text{new}}^p$ is the possible state of new target, P_{new} is the number of particles used for representing possible new target states, $\mathbf{s}_{\text{current}}^f$ and $\mathbf{s}_{\text{new}}^f$ are current and next sensor subsets of FC f, and w_{jp} is the weight given to particle p at time step j. Note that new target particles are distributed only over the neighboring regions, and prior information about new target states are used in distributing particles. In the $\mathbf{J}^f_{\text{new}}$ calculation, only the $\mathbf{s}^f_{\text{new}}$ is used for FC a, and for others $\mathbf{s}^f_{\text{current}}$ is used if $k + j - l_f < I_l$, both possibilities, that is, $\mathbf{s}^f_{\text{current}}$ and $\mathbf{s}^f_{\text{new}}$, are considered if $I_l \leq k + j - l_f < I_u$, and $\mathbf{s}^f_{\text{new}}$ is used otherwise.

3. *Optimize the sensor lifetimes*

$$\min \sum_{f=1}^{M} \sum_{i=1}^{N_f} C_i^f s_i^f, \tag{12.125}$$

where C_i^f is the cost of ith sensor of FC f, which is the function of its remaining energy and the sensors close to it. If there is another sensor, which can give almost equal performance when replaced for the particular sensor, and with sufficient energy, then the cost of particular sensor must be close to zero. On the other hand, if there is no sensor close by, then there is no replacement for it, and one might want to use it for long time, and the cost must be high. The cost can be calculated as

$$C_i = \max\left(0, \frac{\min(D^i_{\text{neig}}, D_{\max})}{D_{\max}} - \frac{B^i_{\text{rem}}}{B_{\max}}\right), \tag{12.126}$$

where D^i_{neig} is the distance of the nearest sensor, which has more energy than sensor i, B^i_{rem} and $B_{\max}$ are the remaining and initial battery energy of sensor i, respectively. $D_{\max}$ is design parameter, and it should be decided based on sensor type. Note that the cost of each sensor can be calculated prior to sensor selection.

Handling multiple objectives can be challenging. Therefore, a single objective is formed by combining the objectives with suitable weights, and given by

$$\begin{aligned}\arg \min_{\left\{s_i^f; f=1,2,\ldots,M\right\}} & \sum_{j=1}^{I_u} \sum_{t=1}^{T} \left(w_{jt} \min_{f=1,2,\ldots,M} \max_{v=1,2,\ldots,n_v^f(k+j)} \operatorname{trace}\left(\left[\mathbf{J}_v^f(k+j)^{-1}\right]_t\right)\right) \\ & + \sum_{j=1}^{I_u} \sum_{p=1}^{P_{\text{new}}} w_{jp} \min_{f=1,2,\ldots,M} \max_{\mathbf{s}^f=\left\{\mathbf{s}^f_{\text{current}}, \mathbf{s}^f_{\text{new}}\right\}} \operatorname{trace}\left(\mathbf{J}^f_{\text{new}}(\mathbf{x}^p_{\text{new}}, \mathbf{s}^f, j)^{-1}\right) \\ & + w_b \sum_{f=1}^{M} \sum_{i=1}^{N_f} C_i^f s_i^f,\end{aligned} \tag{12.127}$$

where w_b is the weight given to the objective of sensor lifetimes.

The constrains are

$$\sum_{i=1}^{N_f} s_i^f \leq n_f \qquad \text{for } f = 1, 2 \ldots, M, \tag{12.128}$$

$$s_i^f \in \{0, 1\} \qquad \forall\, f, i. \tag{12.129}$$

If w_b is zero, then the solution of the above problem will be same as that of the above problem with modified constraint

$$\sum_{i=1}^{N_f} s_i^f = n_f \qquad \text{for } f = 1, 2 \ldots, M, \tag{12.130}$$

instead of (12.128).

Sensors are clustered based on FCs in order to ensure that no sensor is used by more than one FC. However, an advantage of asynchronous sensor changes is that clustering can be easily avoided if any two neighbors do not change the sensors at the same time. However, if two neighbors decided to change their sensors at the same time, then collaboration is necessary to avoid the clustering.

Weight Selection. This section describes how to select the weights, which are used in the objective functions, scenario dependently.

The weights are as follows:

- *Weight of target t at time $k + j$, w_{jt}.*
  The weight is given to each target according to the target type and the state. One might want to give more weight to enemy targets compare to friendly targets. Similarly, the state of the target is also an important factor in weight selection. Low-speed targets are less dangerous than high-speed targets. In addition to the speed, position of the target should also be considered in weight selection. More weights are given to the targets that are close to important locations.

  PCRLB of each target is calculated based on its predicted states. If accurate prediction are available over all the time steps, then equal weight can be given to all the time steps. However, prediction accuracy of target state diminishes with time, hence the weight of time steps also should be decreased with time.

  In (12.123), it is assumed that the new sensor subset will be used up to $k + I_u$. However, another new sensor subset can be used at anytime after $k + I_l$, and this uncertainty must also be incorporated into the weights.

  Then the weight is given by

$$w_{jt} = w_{jt}^p \times \frac{E_{\max}}{\max\left(E_{\max}, \sqrt{\operatorname{trace}\left(\left[\mathbf{J}_X(k+j|k)^{-1}\right]_t\right)}\right)} \times \frac{1}{\max(1, j - I_l + 1)}, \tag{12.131}$$

where w^p_{jt} is the weight given to target t at time $k + j$ based only on target type and state, as explained above, and $E_{\max}$ is a design parameter. $E_{\max}$ is the maximum error in predicted state that may not effect the sensor selection significantly, and this is a function of measurement models and sensor locations. For an example, if the measurements are angle only and sensors are placed 500 m apart, then a suitable value for $E_{\max}$ can be 125.
Finally, all the weights are normalized so that it is easy to define the weights of other objectives.

- *Weight of new target particle p at time k + j,* w_{jp}.
  Similar to existing targets, more weights can be given to the particles that are close to important locations. In addition, it may be necessary to give less or more priority to new targets compare to already established tracks. Then the weight is given by

$$w_{jp} = w_{\text{new}} \times \frac{w_p \times (1/(\max(1, j - I_l + 1))}{\sum_{p=1}^{P_{\text{new}}} \sum_{j=1}^{I_u} w_p \times (1/(\max(1, j - I_l + 1))}, \tag{12.132}$$

  where w_{new} is weight given to new targets compared to already established tracks, w_p is the weight given to particle p based on its state.

- *Weight of sensor lifetimes,* w_b.
  Since the units of first two objective functions are different from third objective function, a suitable weight is needed to balance them. In addition, this weight can be used to specify how important is the sensors' life times compared to tracking performance. For example, if we expect to have trace of each target PCRLB around 100, and equal importance given to tracking performance and sensor lifetimes, then $w_b = 100$.

Collaboration. In the above formulation, it was assumed that PCRLB at neighbors are the same as at FC a at last communication time steps, since the exact values are not known. Even if the PCRLB is predicted based on the tracklets that are received from the neighbors, the prediction might not be accurate, since neighbors use their neighbors' tracklets, which are not available at FC a, to update their tracks. In addition to PCRLB, the estimates may also be different at FCs. Note that current estimates will affect the future PCRLB calculation, which is a function of predicted states. In order to mitigate the effect of this problem, additional information, such as current estimates, corresponding covariances and any additional information about next sensor change time, may be requested from neighbors just before deciding the new sensor subset.

As mentioned in the previous section, avoiding clustering will help to improve the tracking performance. However, if the clustering is avoided, then there is a possibility that a sensor being used by more than one FC, which change their sensors at the same time. This problem can also be solved by collaboration. FCs can communicate their tentative selections with neighbors, and make the final selection based on neighbor's selection. However, obviously it will require more communication and computation.

12.6.4 Solution Technique

The above are NP-hard combinatorial optimization problems [44]. Finding the optimal solution in real time is difficult, since the complete enumeration is not viable when the number of possibilities is very large. However, for large-scale problems, an iterative local search technique is used to find a near optimal solution [45]:

1. Select an initial solution as follows:
   - Select a sensor for first FC that gives optimal performance when only one sensor is used.
   - Select a second sensor for second FC that gives optimal performance when only two sensors are used including the first selected sensor.
   - Continue to select one sensor for each remaining FCs.
   - Continue to add one sensor at a time in the order described above until all FCs reach their limits.

   At each stage, a complete enumeration is performed to determine which sensor should be added.
2. Search for better solutions in the neighborhood until some pre-specified time limit is reached:
   - *Step 1* Set $f = 1, i_f = 1$ for $\forall f = 1, 2, \ldots, M$
   - *Step 2* Remove the i_fth sensor selected for FC f, and replace with another sensor.
     - If the new sensor set yields better performance, make this as the selected set and
       * if $i_f = N_f$, then set $i_f = 1$, otherwise set $i_f = i_f + 1$,
       * if $f = M$, then set $f = 1$, otherwise set $f = f + 1$
       * go to Step 2.
     - otherwise perform Step 2 with another replacing sensor until no more sensor is remaining.
   - *Step 3* For (12.127) with $w_b > 0$, remove a sensor at a time from the selected sensors and then perform the local search with reduced number of sensors as explained in Step 2. Note that we do not need to search for better solution by removing any sensor with $C_i = 0$.

Initially, form a neighborhood by swapping only one sensor from the current solution. If the current solution is optimal in the above neighborhood, then form a new neighborhood by swapping two sensors. Keep on increasing the number of sensors that are swapped for forming the neighborhood until the time limit is reached.

The local optimum of a certain neighborhood can be obtained quickly by swapping the sensors in the order of their rank, where the rank of each sensor is based on its individual performance.

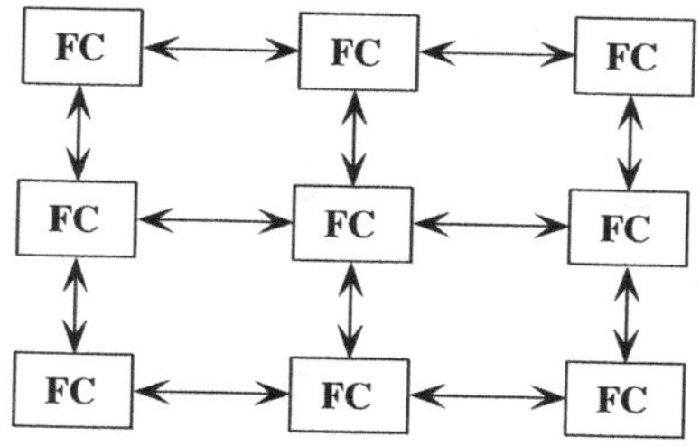

Figure 12.24 Neighbors of FCs.

12.6.5 Simulation Results

12.6.5.1 Synchronous Sensor Selection. In the simulation scenario, the number of FCs $F = 9$; the total number of sensors $N = 400$; each FC's capacity $n_f = 5$; the measurement interval is 30 s; sensor subsets are changed every five measurement steps; the measurements are the bearing from the sensor to target; measurement standard deviation $\sqrt{R} = 0.01$ rad; the field of view of the sensor $V = 2\pi$ rad; new targets can appear anywhere in the surveillance region. Communication range of FCs = 4000 m, and their neighbors are shown in Figure 12.24. The weights are $w_{k+1} = w_{k+2} = w_{k+3} = w_{k+4} = w_{k+5} = 1$; $w_{\text{new}} = 5$.

The sensors selected at the beginning, when no tracks are initialized, are shown in Figure 12.25. Since it is assumed that new targets can appear anywhere in the region, sensors try to cover the whole region. Note that at least two sensors are

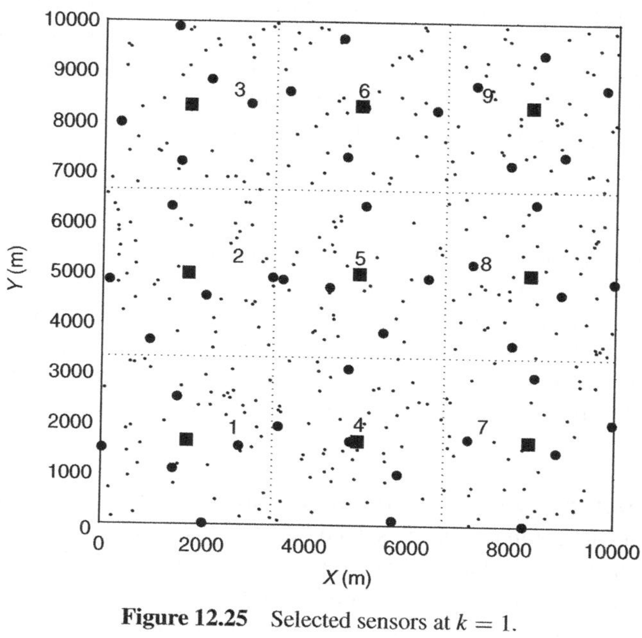

Figure 12.25 Selected sensors at $k = 1$.

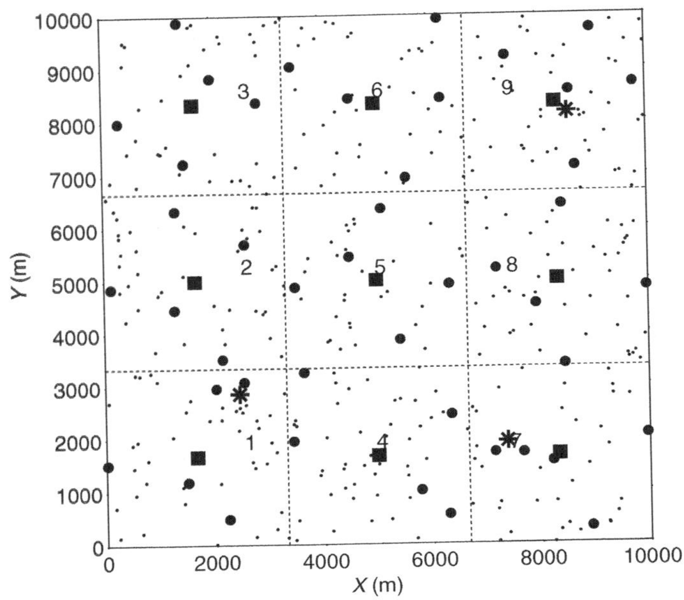

Figure 12.26 Selected sensors at $k = 13$.

needed to initialize a track since the measurement is bearing only. Hence, each point in the surveillance region must be covered by at least two sensors. In addition, for better initialization of X and Y coordinates, sensors have to look from perpendicular directions. Since the PCRLB that explicitly considers new target births is used in sensor selection, the selected sensors, shown in Figure 12.25, provide better coverage of newly emerging targets thereby resulting in superior estimation results later on.

The sensors selected at $k = 13$, with three targets, are shown in Figure 12.26. Few sensors are assigned to track the existing targets while the others are assigned to initialize the new targets immediately and accurately. One of the targets has just appeared and, therefore, no sensor is assigned to that target particularly. However, that target is tracked by the sensors that are assigned for new targets.

The position RMSE of all the targets at each FC is shown in Figure 12.27. The minimum RMSE of each target over all the FCs is shown in Figure 12.28. The RMSEs of some targets are high at some FCs since those targets are far away from those FCs. However, each target is tracked accurately at every measurement step by at least one FC.

12.6.5.2 Comparison of Presented Solution Technique with GA and ACO.

In order to compare the presented solution technique with other heuristic algorithms, genetic algorithm [58] and ant colony optimization algorithm [59] are considered; see Appendices 12B and 12C for detail explanation of GA and ACO algorithm, respectively. The parameters of GA are population size = 50, crossover

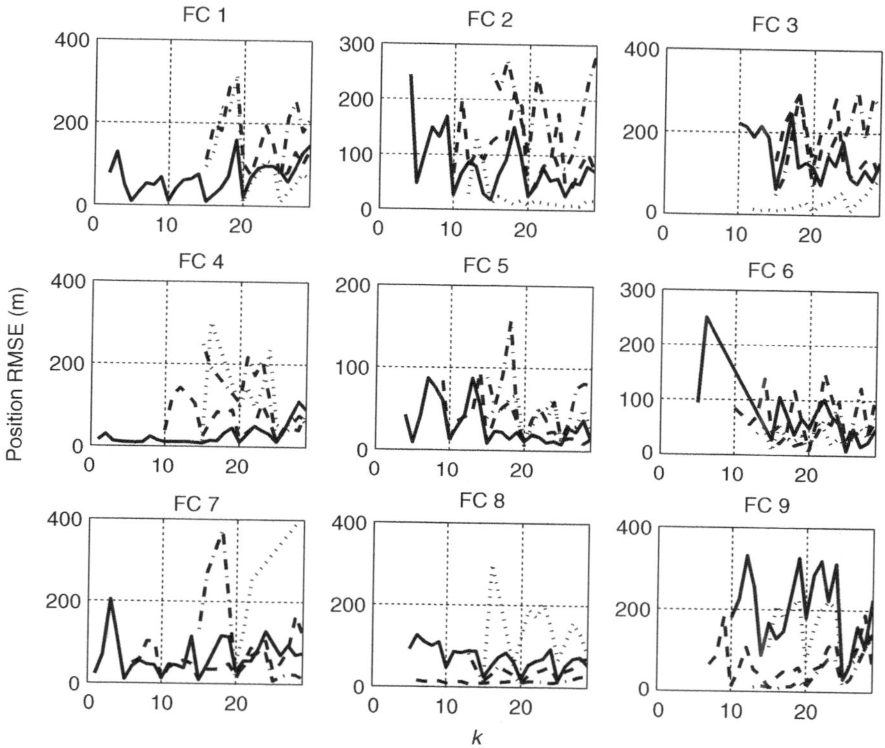

Figure 12.27 RMSEs at all FCs.

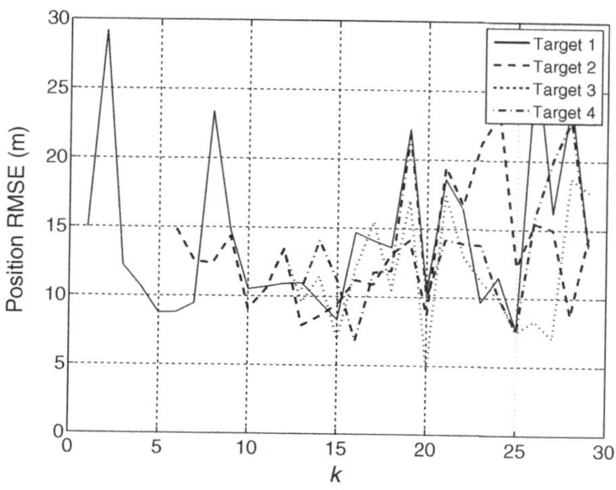

Figure 12.28 The minimum RMSEs over all the FCs.

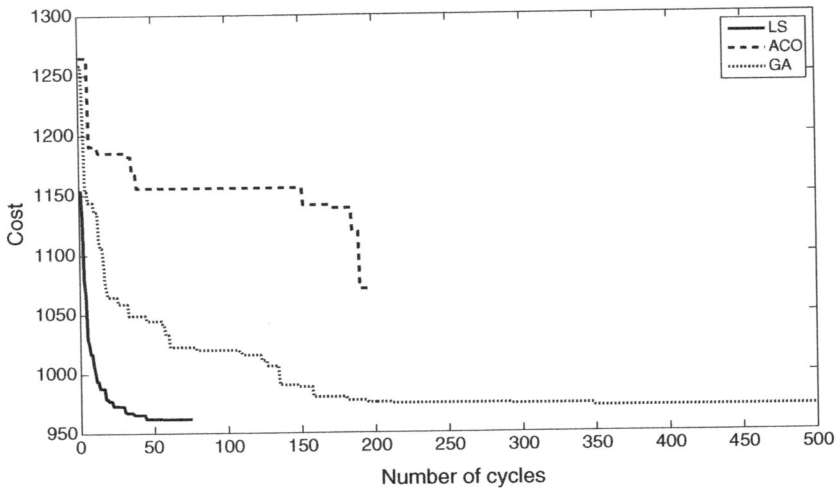

Figure 12.29 Comparison of local search (presented technique), genetic algorithm, and ant colony optimization.

ratio = 0.9, mutation ratio = 0.001, and elitism ratio = 0.04. The parameters used for the ACO algorithm are number of ants, $nAnts = 50$; pheromone trails lower and upper bounds, $\tau_{\min} = 0.01$, $\tau_{\max} = 1$, respectively; pheromone factor weight, $\alpha = 2$; pheromone evaporation rate, $\rho = 0.98$. In order to match the presented algorithm with other two, 50 sensor swaps are represented by one cycle. All the algorithms are stopped after 500 cycles or 30 min, whichever occurs first. In addition, the presented solution technique is stopped if the local optimum is reached with swapping only one sensor at a time. The performances of all the algorithms in a typical run are shown in Figure 12.29. The presented algorithm finds the local optimum in 75 cycles and it takes 2.9 s. GA takes 21.8 s for 500 cycles and ACO takes 30 s for 198 cycles. The suboptimal values found by the presented technique, GA and ACO are 961.9, 1023 and 1156, respectively. Hence, presented algorithm finds a better solution quickly compare to other two algorithms. ACO algorithm proves to be the least optimal, since it consumes more time for each cycle and converges very slowly.

12.6.5.3 Asynchronous Sensor Selection. In the simulation the parameter settings are the number of FCs $F = 9$; the total number of sensors $N = 400$; each FC's capacity $n_f = 5$; the measurement interval is 30 s; lower and upper limits of sensor change interval are $I_l = 6$ and $I_u = 8$ time steps; the measurements are the bearing from the sensor to target; measurement standard deviation $\sqrt{R} = 0.01$ radians; the field of view of the sensor $V = 2\pi$ radians; surveillance region is 10,000 m × 10,000 m; new targets can appear anywhere in the surveillance region; new target weight $w_{\text{new}} = 1$; high-priority area is a circle with center [6000 m 4000 m] and radius 1000 m; communication range of FCs = 4000 m, and their neighbors are shown in Figure 12.30; the weights of the targets in the high-priority area is 4; $D_{\max} = 1200$ m,

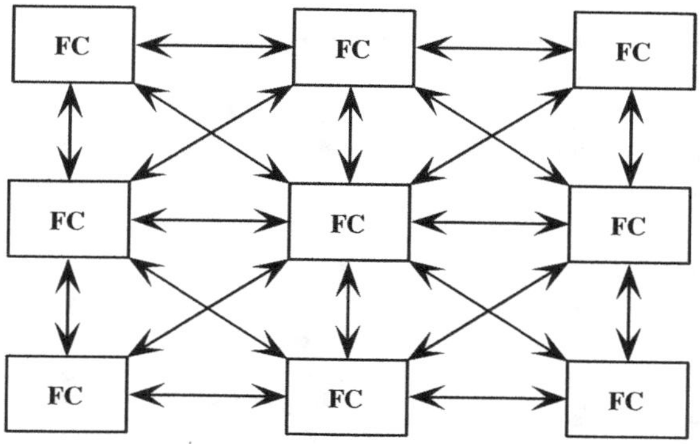

Figure 12.30 Neighbors of FCs.

$B_{max} = 100$ measurement steps; weight of sensor lifetimes $w_b = 0$ (later a different scenario is considered with $w_b > 0$ to see the effect of remaining sensor energies on sensor selection).

Selected sensors at three time steps are shown in Figures 12.31–12.33, where "•" indicates the selected sensors and "◇" indicates the estimated target positions.

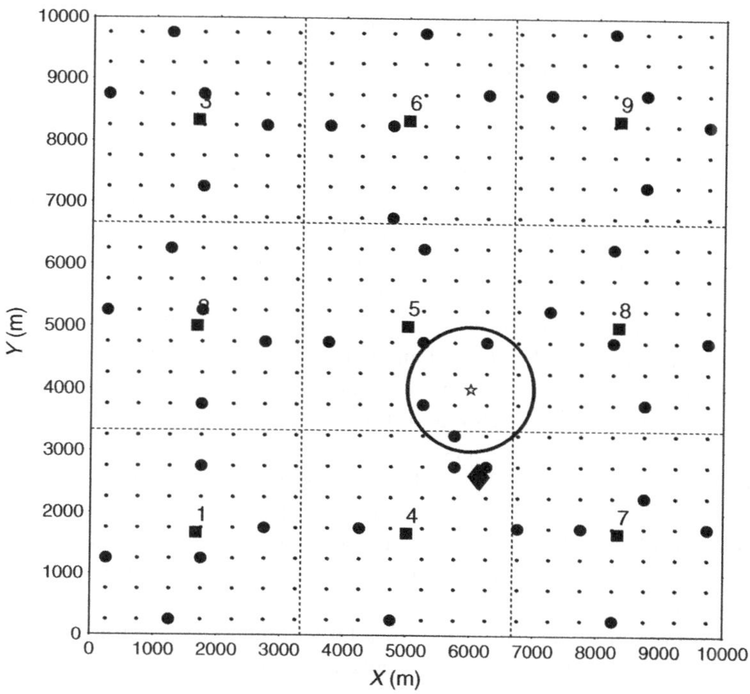

Figure 12.31 Selected sensors at $k = 4$.

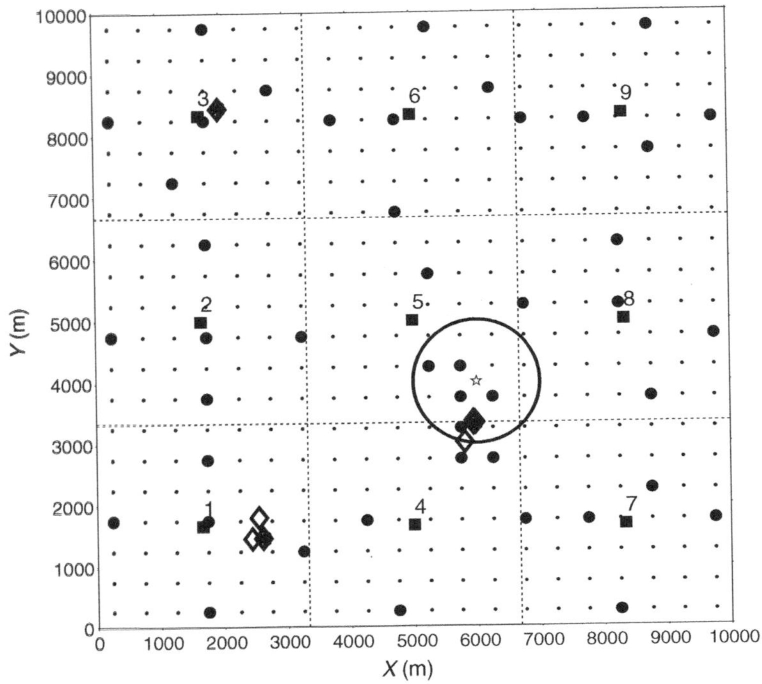

Figure 12.32 Selected sensors at $k = 8$.

In Figure 12.31, at $k = 4$, there is only one target in the surveillance region, and sensors of FC 4 are selected such that tracking performance of that target is optimized in the current and next few time steps. Three sensors of the FC 4 cover the target while other two are monitoring for new targets. There is no other target close to other FCs and sensors of those FCs are monitoring the surveillance region for new targets. Note that most of the sensors of FC 5 are covering high-priority area.

In Figure 12.32, at $k = 8$, FC 5 changes its sensors, and four of its sensors are covering the target that is moving into its high-priority area. Two more new targets appeared in the surveillance region and those are tracked by FCs using the sensors that were monitoring for new targets.

In Figure 12.33, at $k = 11$, FC 4 change its sensors for second time. Earlier, it allocated most of its sensors for first target (see Figure 12.31), which has now moved to the region of FC 5 and tracked by that FC. Currently, there is no target in its area, however there is a target in the area of FC 1, and moving toward its area. Hence, FC 4 allocates three sensors for that target and other two for monitoring new target.

Despite some sensors are allocated for a target and others are for new target, all sensors provide some information about new and already existing target. For example, in Figure 12.33 even though three sensors are mostly monitoring existing target, the same sensors can also monitor some part of area for new targets. Similarly other two

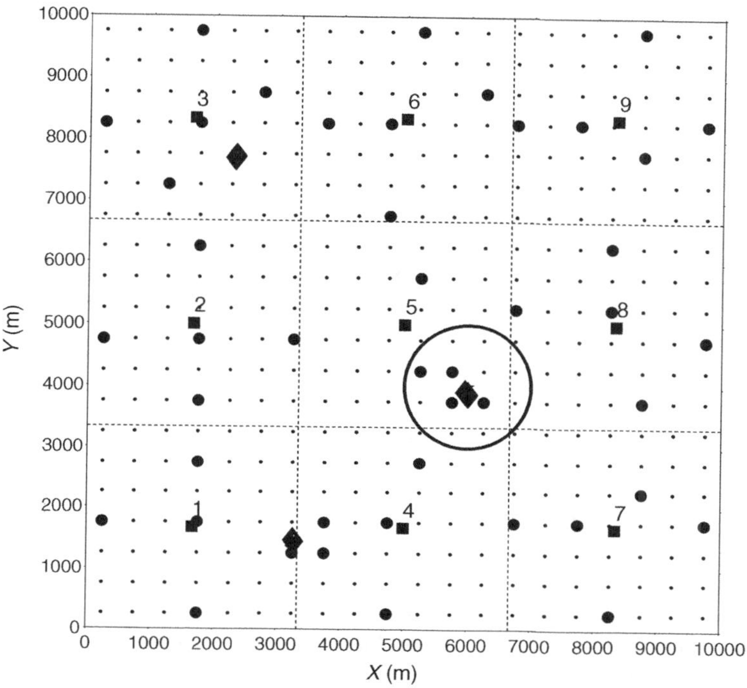

Figure 12.33 Selected sensors at $k = 11$.

sensors, which are mainly monitoring for new targets, can generate measurements for existing target.

In order to see the effect of remaining sensor energies on sensor selection, a different scenario with 40 sensors and 2 FCs is considered. Sensors selected with $w_b = 0$, $w_b = 50$, and $w_b = 100$ are shown in Figures 12.34–12.36, respectively, where "$*$" indicates the target, and the numbers next to sensors indicate the remaining energies in term of number of measurement steps. When $w_b = 0$, sensors are selected to optimize the tracking performance without worrying about the remaining sensor energies (see Figure 12.34). With $w_b = 50$, only one sensor with low energy is selected (see Figure 12.35). However, the selected sensors still provides near optimal coverage. With $w_b = 100$, all the selected sensors have high energy (see Figure 12.36) and the tracking performance degrades slightly.

12.7 CONCLUSIONS

In this chapter, the problem of sensor selection is considered for centralized, distributed, and decentralized architectures for multitarget tracking with unknown associations of measurements to targets as well as with unknown and potentially time-varying number of targets. The posterior Cramér–Rao lower bound, which provides

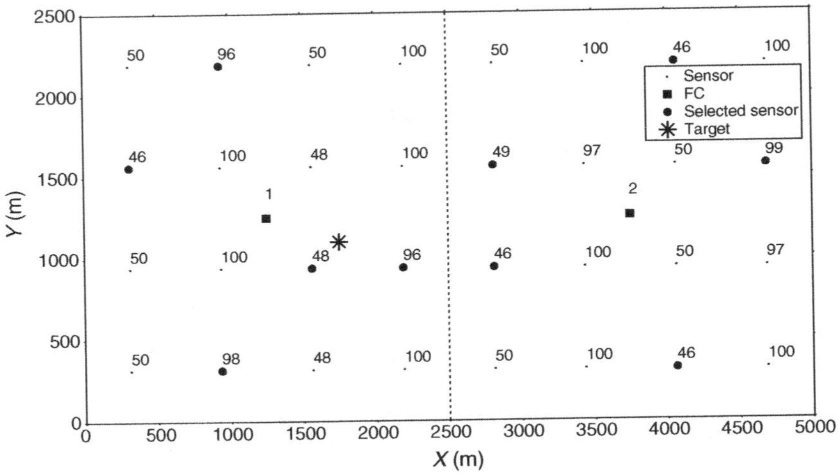

Figure 12.34 Selected sensors with $w_b = 0$.

a measure of the optimal achievable accuracy of target state estimation, was used as the performance measure of the tracking systems.

In the centralized architecture, the problem is to select a sensor subset that gives the optimal tracking performance throughout the measurement steps. Although a particular sensor subset may be optimal at the time of initial activation, due to target motion, it may not remain optimal throughout the entire activation period. In order to get a sensor subset that remains optimal during the entire activation interval and not just at the time of activation, the predicted tracking performance over all the active time

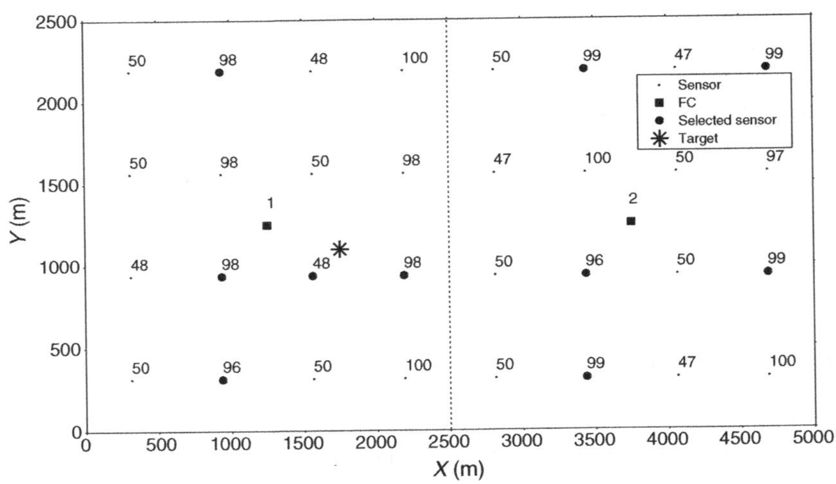

Figure 12.35 Selected sensors with $w_b = 50$.

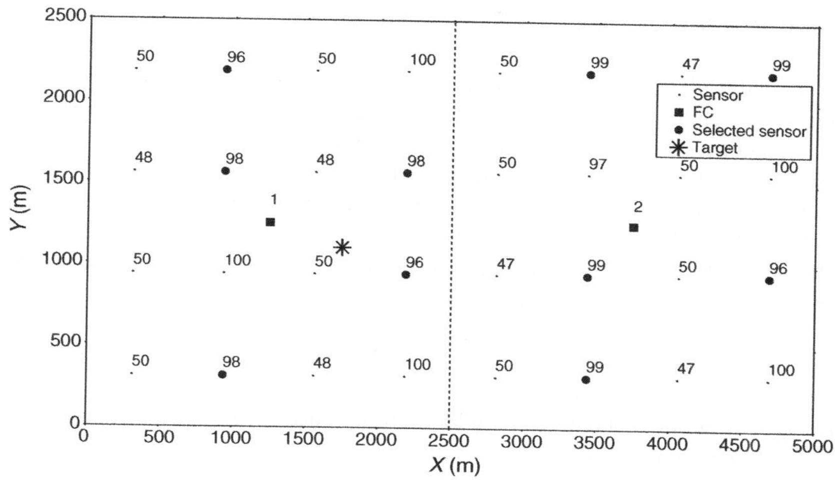

Figure 12.36 Selected sensors with $w_b = 100$.

steps was considered. The varying number of targets was handled using a bicriterion optimization technique, where one objective is to maximize the estimation accuracies of established targets and the other is to maximize the probability of detecting new targets. A search technique was presented to find a suboptimal solution in real time for large-scale scenarios.

In the distributed architecture, the problem is to select subsets of sensors, assign them to local fusion centers, and assign the frequency and transmission power to each active sensor in order to maximize the tracking performance. The optimal formulation was presented for the sensor management for the above problem based on the PCRLB. Finding the optimal solution to the above problem in real time can be challenging in large-scale scenarios. An algorithm was presented to find a suboptimal solution in real time by decomposing the original problem into four subproblems, which are easy to solve, without using the simple clustering algorithms.

In the decentralized architecture, the problem is to select subsets of sensors for each FC in order to maximize the overall tracking performance. In the decentralized architecture, there is no central fusion center and each FC communicates only with its neighbors. Hence, the fusion centers do not have global knowledge, therefore selecting the optimal sensor subsets, by considering all the FCs at once, is not feasible. Algorithms were presented to select suboptimal sensor subsets by considering only the neighboring fusion centers for synchronous and asynchronous sensor selection scenarios. In addition to tracking performance, sensor lifetimes, which are limited, were also considered in sensor selection. Several weights were used in the objectives to specify the importance of each element, and we presented how to select those weights scenario dependently. Finding the optimal solution in real time proved to be a extremely difficult task. As a result, a solution technique based on iterative local search was presented to find a suboptimal solution in real time for large-scale problems.

APPENDIX 12A LOCAL SEARCH

A local search algorithm improves the current estimate of the optimum point by searching for better solutions in a local neighborhood of the current solution and stops if the neighborhood does not contain an improving solution [44, 60]. A local search is sensitive to (the quality of) the initial solution and the choice of the neighborhood. The selection of the initial solution to the sensor subset problem is discussed in *Step 1* of Section 12.4.3. The choice of the neighborhood has a very significant influence on the efficiency of local search heuristics and it also determines the computational time of one iteration [45]. A neighborhood structure to sensor selection problem is given next.

Let $C_s = [C_s(1), C_s(2), \ldots, C_s(n)]$ be the current solution and $N_S(i) = [n_{si}(1), n_{si}(2), \ldots, n_{si}(n)]$ for $i = 1, 2, \ldots, (Nn)$ be the ith neighbor. In addition, q and r are the quotient and the remainder of the division of $(i - 1)$ by n, respectively, and $b_s(q)$ is the qth best sensor. Then

- If $b_s(q + 1) \neq C_s(k)$ for $k = 1, 2, \ldots, n$

$$n_{si}(j) = \begin{cases} b_s(q+1) & \text{if} \quad j = r + 1 \\ C_s(j) & \text{else}, \end{cases} \tag{12A.1}$$

  for $j = 1, 2, \ldots, n$.
- Else

$$N_s(i) = NULL. \tag{12A.2}$$

Here, the $NULL$ set is not considered as an element of the neighborhood. In the above structure, only one sensor is changed from the current solution [61]. The size of the above neighborhood structure is $(N - n)n = O(Nn)$. Another neighborhood structure can be created by changing at most two sensors from the current solution. The size of that structure will be $O(N^2n^2)$. The second neighborhood structure might provide a better local optimum but would take much longer to search. Hence, the first structure is used for real-time capability. However, if time permits the second structure can be used to search for better solution, after finding the possible best solution with one sensor swapping.

There are two possible implementations for a local search procedure [44], namely, (1) the first improving move ("first-fit") implementation moves the solution to an improving neighbor first encountered during the search and (2) the best improving move ("best-fit") implementation, on the other hand, moves the solution to the neighbor that improves the solution the most. In either case, the running time of each iteration is bounded by $O(Nn)$ if the function evaluation time remains constant. The "first-fit" strategy may result in a smaller improvement of the current solution than the "best-fit" strategy. While every neighborhood needs to be searched completely with the "best-fit" strategy, only the final neighborhood needs to be searched with the "first-fit" strategy. Thus, the former is used to find the local optimum, resulting in some computational saving.

The order in which the neighborhood is searched also affects the computational time. Intuitively, the probability of having a high rank (based on the individual performance) sensor in the optimal sensor subset is higher than having a randomly selected sensor. Hence, neighborhood is searched by swapping the sensors in the ranking order.

The complete algorithm is as follows:

- *Step 1* Set the initial solution as a new subset, find the objective function value (LB) and assign $i = 1$, $j = 1$, and $iter = 0$.
- *Step 2* Check whether the ith best sensor is already in the new subset. If it is in the new subset, set $i = i + 1$ and repeat step 2.
- *Step 3* Remove the jth sensor from the new subset and replace it by the ith best sensor and form a temporary subset. Then find the function value (TLB) using the temporary subset.
- *Step 4*
  - If ($iter <$ maximum allowable iterations (a predetermined value))
    - * if ($TLB < LB$)
      - change the temporary subset as the new subset and set $LB = TLB$.
      - $i = 1$, $j = 1$, and $iter = iter + 1$.
      - go to step 2.
    - * else
      - if ($j < n$), set $j = j + 1$ and go to step 3.
      - else if ($i < N$), set $j = 1, i = i + 1$ and go to step 2.
      - else, go to step 5.
- *Step 5* Finish.

The running time of the above algorithm can also be controlled by changing the stopping criterion. For example, the maximum running time can be bounded by a constant, instead of bounding the number of iterations.

The computational time of the above local search algorithm is mainly determined by the number of iterations and the time used to calculate the first improving neighbor of the current solution. The computation time of cost of existing targets is far higher than that of new targets. Hence, a function evaluation can be done in $O(Tln)$ time. But, once the function value of the given initial solution is calculated, the function value of a neighbor, formed by swapping only one sensor, can be found in $O(Tl)$ time. For example, if sensor a is replaced by sensor b and the summation of the values of $J_{z_i}(\tau)$ for the current sensors for all τ is obtained, then the new sum can be calculated by just subtracting $J_{z_a}(\tau)$ and adding $J_{z_b}(\tau)$. Since there are $O(Nn)$ neighbors to test, the running time of each iteration is bounded by $O(TNln)$.

APPENDIX 12B GENETIC ALGORITHM

Genetic algorithm is a class of learning algorithms based on parallel search for an optimal solution, and is inspired by the biological processes of evolution [62, 63]. The parallel searches, which are performed synchronously in time steps, are called generations. In each generation, a certain number of search paths, also known as individual are maintained. The whole set of individuals in a generation is referred to as the population. The main idea in genetic algorithms is to preserve and create variations of the individuals that seem most promising and remove the others.

The general framework of GA is shown in Figure 12B.1 [62], and its steps are explained below with an example:

- *Initialize the population* Individuals of initial population are generated at random. To the problem considered in this chapter, initial individuals, which are strings of bits [62], are found by randomly selecting n_f sensors for each FC. For example, individual v_k, for a scenario with three FCs, five sensors are available for each FC and each FC can handle two sensors, is $00101 \cdot 10010 \cdot 01010$.
- *Find the fitness* A fitness function must be defined to give a score to each individual in the current population. Fitness of each individual, v_k, is given by

$$\frac{1}{\mathrm{func}(\mathrm{v_k}) - 0.5 * \mathrm{bestSoFar}}, \tag{12B.1}$$

where "func" refers the function defined in (12.116) and "bestSoFar" refers to the best function value found so far.

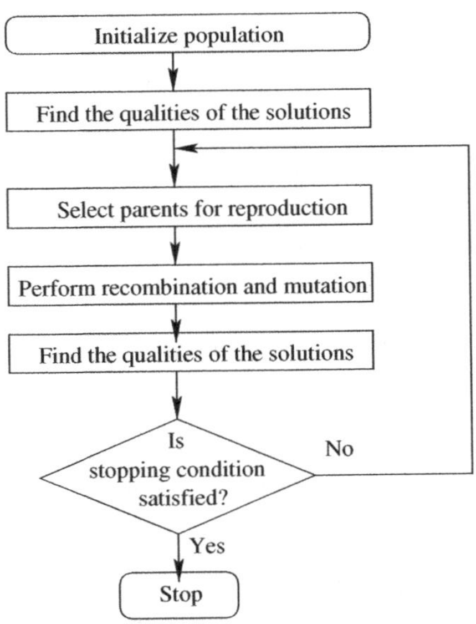

Figure 12B.1 Framework of genetic algorithm.

- *Select parent* Selection is based on the fitness. Individuals that have high score will most likely be selected to reproduce, whereas those with low scores will be discarded. Selection probability p_k for each individual v_k is given by

$$p_k = \frac{f(v_k)}{\sum_{i=1}^{\text{popSize}} f(v_i)}, \tag{12B.2}$$

  where popSize is the size of the population.
- *Crossover* It is an operator that combines two parents to generate an individual. Randomly chooses a locus and exchanges the subsequences before and after that locus between two chromosomes to create two new individuals. For an example, consider the following parents and crossover at position 7

$$\begin{array}{l} 00101 \cdot 10010 \cdot 01010 \\ \mathbf{10010} \cdot \mathbf{11000} \cdot \mathbf{01010} \end{array} \Rightarrow \begin{array}{l} 00101 \cdot 10\mathbf{000} \cdot \mathbf{01010} \\ \mathbf{10010} \cdot \mathbf{11}010 \cdot 01010 \end{array}.$$

  Crossover operation is normally performed with a certain probability, called crossover ratio. For an example, if the crossover ratio is 0.9, 90% of the new individuals are formed by crossover and remaining 10% are just copied.
- *Mutate* Change the new individual by flipping each bit with some probability called mutation ratio. For an example, only one bit is flipped and that is 13th bit

$$00101 \cdot 10000 \cdot 01010 \rightarrow 00101 \cdot 10000 \cdot 01\mathbf{1}10.$$

- *Repair* It is a process that rectifies infeasible individuals that may be produced during crossover and mutation [64]. Flip the random bits from the selected bits such that the individual becomes feasible. For example, in the above individual that was obtained after mutation, only one sensor is selected for second FC and three sensors are selected for third FC. In order to make the individual feasible, randomly select a 0 bit from 10th to 15th bits and change to 1, similarly select a 1 bit from 16th to 20th bits and change to 0

$$00101 \cdot 10000 \cdot 01110 \rightarrow 00101 \cdot 1000\mathbf{1} \cdot 011\mathbf{0}0.$$

Elitism, in which we first copies a few of the top scored individuals to the new population and then continues generating the rest of the population by crossover, can also be used to improve the performance, and prevents loosing the few best found solutions [58].

Choosing the size of the population can be tricky since a small population size provides an insufficient sample size over the space of solutions and large population requires a lot of evaluation and will be slow.

APPENDIX 12C ANT COLONY OPTIMIZATION

Ant colony optimization algorithms are stochastic search procedures [59, 65]. These algorithms are inspired by ant behavior. Ants are capable of finding the shortest path between their nest and a food source. While moving, ants leave a chemical (pheromone) trail on the ground. When choosing their path, they tend to choose the path with strong pheromone. This indirect communication among ants via pheromone trails enables them to find the shortest path effectively. That is, information about an ant's experience is communicated via pheromone, which is utilized by other ants.

The pseudocode of ACO algorithm is given in Figure 12C.1.

The detailed algorithm, that is used for problem considered in this chapter, is explained below:

1. *Initialize the pheromone trails*
   $\tau_{\text{vertex}}(s_i) = \tau_{\max};\ \tau_{\text{clique}}(s_i, s_j) = \tau_{\max}$
2. *Construct the solutions*
   for each ant $a \in [1...n\text{Ants}]$
   Set *Candidates* = all the sensors that belong to itself and the neighbors.
   Choose the first sensor s_i from *Candidates* with probability

$$p(s_i) = \frac{\tau_{\text{vertex}}(s_i)^{\alpha}}{\sum_{j \in Candidates} \tau_{\text{vertex}}(s_j)^{\alpha}}. \tag{12C.1}$$

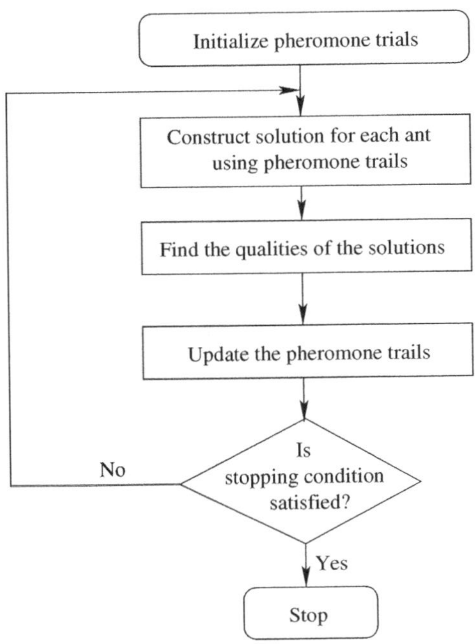

Figure 12C.1 Ant colony optimization algorithm.

Set $S_a = \{s_i\}$ and remove s_i from the *Candidates*.
while *candidates* $\neq \emptyset$
Choose a sensor, s_i, from *Candidates* with probability

$$p(s_i, S_a) = \frac{\zeta(s_i, S_a)^\beta \sum_{j \in S_a} \tau_{\text{clique}}(s_i, s_j)^\alpha}{\sum_{l \in Candidiates} \zeta(s_l, S_a)^\beta \sum_{j \in S_a} \tau_{\text{clique}}(s_l, s_j)^\alpha}. \tag{12C.2}$$

Add s_i to S_a and remove s_i from the *Candidates*.
if s_i belongs to FC f, and the number of sensors belonging to this FC from S_a is equal to n_f
Remove all the sensors that belong to FC f from *Candidates*.
end if
end while
Apply local search (optional).
end for

3. *Trial update*
Order the solutions according to their qualities.
Update the pheromone trails using mBest solutions as follows [66]

$$\tau_{\text{vertex}}(s_i) = (1-\rho)\cdot\tau_{\text{vertex}}(s_i) + \rho \cdot \sum_{\substack{b=1 \\ s_i \in S_b}}^{m\text{Best}} \frac{1/(1+\text{func}(S_b) - \text{func}(S_{\text{BestSoFar}}))}{\sum_{c=1}^{m\text{Best}}(1/(1+\text{func}(S_c) - \text{func}(S_{\text{BestSoFar}})))}, \tag{12C.3}$$

$$\tau_{\text{clique}}(s_i, s_j) = (1-\rho)\cdot\tau_{\text{clique}}(s_i, s_j) + \rho \cdot \sum_{\substack{b=1 \\ s_i \in S_b \\ s_j \in S_b}}^{m\text{Best}} \frac{1/(1+\text{func}(S_b) - \text{func}(S_{\text{BestSoFar}}))}{\sum_{c=1}^{m\text{Best}}(1/(1+\text{func}(S_c) - \text{func}(S_{\text{BestSoFar}})))}. \tag{12C.4}$$

4. *Termination*
If maximum number of cycles or allocated time is reached, return $S_{\text{BestSoFar}}$. Otherwise go to step 2.

In (12C.2), ζ is the heuristic factor that evaluates the promise of sensor s_i based on the sensors selected so far, S_a. This is an optional factor. It is not used always, that is, $\zeta(s_i, S_a) = 1$, in this implementation since it was difficult to find a computationally efficient way to define this factor.

The parameters used in ACO are number of ants, nAnts; pheromone trails upper and lower bounds, $\tau_{\max}$, and $\tau_{\min}$, respectively; pheromone evaporation rate, ρ; pheromone factor weight, α; heuristic factor weight, β.

The quality solution from ACO can be improved by increasing the value of nAnts. However, after a certain limit, quality will not improve significantly, but computation time will increase significantly. In addition, this value must be calculated experimentally. In ACO, $\tau_{\max}$ and $\tau_{\min}$ are used to avoid premature stagnation of search, that is, a situation where all ants construct the same solution repeatedly. Diversification can be emphasized either by decreasing the value of α or by increasing the value of ρ. These values should be set according to the time available to find the solution.

REFERENCES

1. A.O. Hero, D. Castanon, D. Cochran, and K. Kastella, *Foundations and Applications of Sensor Management*, Springer, New York, NY, November 2007.
2. D. Gu, I. Postlethwaite, and Y. Kim, A comprehensive study on flight path selection algorithms, *Proceeding of the IEE Seminar on Target Tracking: Algorithms and Applications*, Austin Court, Birmingham, UK, March 2006, pp. 77–90.
3. Y. Qu, Q. Pan, and J. Yan, Flight path planning of UAV based on heuristically search and genetic algorithms, *Proceedings of the 31st Annual Conference of IEEE Industrial Electronics Society*, Raleigh, NC, November 2005, pp. 45–49.
4. A. Sinha, T. Kirubarajan, and Y. Bar-Shalom, Autonomous search, tracking and classification by multiple cooperative UAVs, *Proceedings of SPIE Conference on Signal Processing, Sensor Fusion, and Target Recognition XV*, Orlando, FL, Vol. 6235, April 2006, 623508-1–623508-12.
5. M.L. Hernandez, T. Kirubarajan, and Y. Bar-Shalom, Multisensor resource deployment using posterior Cramér–Rao bounds, *IEEE Trans. Aerosp. Electron. Syst.*, 40(2), 399–416, 2004.
6. K. Johansson, K. Jored, and P. Svensson, Submarine tracking using multi-sensor fusion and reactive planning for the positioning of passive sonobuoys, *Proceedings of Hydroakustik*, Stockholm, Sweden, September 1997.
7. L.Y. Pao, M. Kalandros, and J. Thomas, Controlling target estimate covariance in centralized multisensor systems, *Colorado Advanced Software Institute (CASI)*, Final Report, 1997.
8. P.R. Horridge and M.L. Hernandez, Performance bounds for angle-only filtering with application to sensor network management, *Proceedings of the 6th International Conference on Information Fusion*, Cairns, Australia, July 2003, pp. 695–703.
9. V. Krishnamurthy, Algorithms for optimal scheduling and management of hidden markov model sensors, *IEEE Trans. Signal Process.*, 50(6), 1382–1397, 2002.
10. R. Tharmarasa, T. Kirubarajan, and M.L. Hernandez, Large-scale optimal sensor array management for target tracking, *IEEE Trans. Syst., Man Cybernet., Part C: Appl. Rev.*, 37(5), 803–814, 2007.

11. A.L. Buczak, Y. Jin, H. Darabi, and M. Jafari, Genetic algorithm based sensor network optimization for target tracking, *Proceedings of the Artificial Neural Networks in Engineering Conference*, St. Louis, MO, Vol. 9, November 1999, pp. 349–354.

12. D. Akselrod, T. Lang, M. Mcdonald, and T. Kirubarajan, Markov decision process based resource and information management for sensor networks, in: *Sensor Networks: Where Theory Meets Practice*, Springer, Berlin, Heidelberg, March 2010.

13. R. Tharmarasa, T. Kirubarajan, M.L. Hernandez, and A. Sinha, PCRLB based multisensor array management for multitarget tracking, *IEEE Trans. Aerosp. Electron. Syst.*, 43(2), 539–555, 2007.

14. R. Tharmarasa, T. Kirubarajan, J. Peng, and T. Lang, Optimization-based dynamic sensor management for distributed multitarget tracking, *IEEE Trans. Syst., Man Cybernet., Part C: Appl. Rev.*, 39(5), 534–546, 2009.

15. R. Tharmarasa, T. Kirubarajan, A. Sinha, and T. Lang, Decentralized sensor selection for large-scale multisensor-multitarget tracking, *IEEE Trans. Aerosp. Electron. Syst.*, 47(2), 1307–1324, 2011.

16. P. Tichavsky, Posterior Cramér–Rao bounds for adaptive harmonic retrieval, *IEEE Trans. Signal Process.*, 43(5), 1299–1302, 1995.

17. P. Tichavsky, C.H. Muravchik, and A. Nehorai, Posterior Cramér-Rao bounds for discrete-time nonlinear filtering, *IEEE Trans. Signal Process.*, 46(5), 1386–1396, 1998.

18. X. Zhang and P.K. Willett, Cramér–Rao bounds for discrete time linear filtering with measurement origin uncertainty, *Proceedings of the Workshop on Estimation, Tracking and Fusion: A Tribute to Yaakov Bar-Shalom*, Monterey, CA, May 2001, pp. 546–561.

19. H. Van Trees and K. Bell, *Bayesian Bounds for Parameter Estimation and Nonlinear Filtering/Tracking*, Wiley-IEEE Press, August 2007.

20. Y. Bar-Shalom and X.R. Li, *Multitarget-Multisensor Tracking: Principles and Techniques*, YBS Publishing, 1995.

21. S. Blackman and R. Popoli, *Design and Analysis of Modern Tracking Systems*, Artech House, 1999.

22. M.L. Hernandez, A.D. Marrs, N.J. Gordon, S. Maskell, and C.M. Reed, Cramér–Rao bounds for non-linear filtering with measurement origin uncertainty, *Proceedings of the 5th International Conference on Information Fusion*, Annapolis, Maryland, Vol. 1, July 2002, pp. 18–25.

23. B. Grocholsky, *Information-theoretic control of multiple sensor platforms*, Ph.D. thesis, Department of Aerospace, Mechatronic and Mechanical Engineering, The University of Sydney, Australia, 2002.

24. M. Howard, D. Payton, and R. Estkowski, Coalitions for distributed sensor fusion, *Proceedings of the 5th International Conference on Information Fusion*, Annapolis, MD, Vol. 1, July 2002, pp. 636–642.

25. J. Liu, J. Reich, P. Cheung, and F. Zhao, Distributed group management in sensor networks: algorithms and applications to localization and tracking, *Proceedings of Second International Workshop on Information Processing in Sensor Networks*, Palo Alto, CA, April 2003, pp. 113–128.

26. A.L. Buczak and V. Jamalabad, Self organization of a heterogeneous sensor network by genetic algorithm, in: C.H. Dagli, C.L. Chen, M. Akay (Eds.), *Intelligent Engineering Systems Through Artificial Neural Networks*, vol. 8, ASME Press, New York, NY, pp. 259–264, 1998.

27. H. Yang and B. Sikdar, A protocol for tracking mobile targets sensor networks using sensor network, *Proceedings of IEEE Workshop on Sensor Network Protocols and Applications*, Anchogare, AK, May 2003, pp. 71–81.
28. R. Bejar, B. Krishnamachari, C. Gomes, and B. Selman, Distributed constraint satisfaction in a wireless sensor tracking system, *Proceedings of the Workshop on Distributed Constraint Reasoning*, Seattle, WA, 2001, pp. 81–90.
29. P.J. Modi, H. Jung, M. Tambe, W.M. Shen, and S. Kulkarni, A dynamic distributed constraint satisfaction approach to resource allocation, *Lecture Notes in Computer Science*, Vol. 2239, pp. 685–700, 2001.
30. A. Petcu and B. Faltings, A value ordering heuristic for local search in distributed resource allocation, *Lecture Notes in Computer Science*, Vol. 3419, Springer Verlag, Heidelberg, Germany, pp. 86–97, 2005.
31. R. Evans, V. Krishnamurthy, G. Nair, and L. Sciacca, Networked sensor management and data rate control for tracking maneuvering targets, *IEEE Trans. Signal Process.*, 53(6), 1979–1991, 2005.
32. J.E. Wieselthier, G.D. Nguyen, and A. Ephremides, Resource management in energy-limited, bandwidth-limited, transceiver-limited wireless networks for session-based multicasting, *Comput. Netw.: Int. J. Comput. Telecommuni. Netw.*, 39(5), 113–131, 2002.
33. J.L. Williams, J.W. Fisher III, and A.S. Willsky, Approximate dynamic programming for communication-constrained sensor network management, *IEEE Trans. Signal Process.*, 55(8), 4300–4311, 2007.
34. N. Bulusu, D. Estrin, L. Girod, and J. Heidemann, Scalable coordination for wireless sensor networks: self-configuring localization systems, *Proceedings of the Sixth International Symposium on Communication Theory and Applications*, Ambleside, UK, July 2001.
35. P. Brown and T. Kirby-Smit, Operational field trials of GPS equipped sonobuoys, *Proceedings of the International Technical Meeting of the Satellite Division of the Institute of Navigation, ION GPS-96*, Kansas City, MO, Vol. 2, September 1996, pp. 1553–1561.
36. R. Tharmarasa and T. Kirubarajan, Collaborative sensor management for decentralized asynchronous sensor networks, *Proceedings of the SPIE Conference on Signal and Data Processing of Small Targets*, San Diego, CA, Vol. 6699, August 2007, 66990Z-1–66990Z-12.
37. M.E. Liggins, C.Y. Chong, I. Kadar, M.G. Alford, V. Vannicola, and S. Thomopoulos, Distributed fusion architectures and algorithms for target tracking, *Proceed. IEEE*, 85(1), 95–107, 1997.
38. O.E. Drummond, A hybrid sensor fusion algorithm architecture and tracklets, *Proceedings of the SPIE Conference on Signal and Data Processing of Small Targets*, San Diego, CA, Vol. 3163, July 1997, pp. 485–502.
39. N. Xiong and P. Svensson, Multi-sensor management for information fusion: issues and approaches, *Inform. Fusion*, 3(1), 163–186, 2002.
40. H. Van Trees, *Detection, Estimation and Modulation Theory*, Vol. I, Wiley, New York, 1968.
41. B. Ristic, S. Zollo, and S. Arulampalam, Performance bounds for maneuvering target tracking using asynchronous multi-platform angle-only measurements, *Proceedings of the 4th International Conference on Information Fusion*, Montreal, Quebec, August 2001.
42. R.A. Horn and C.R. Johnson, *Matrix Analysis*, Cambridge University Press, 1985.

43. K.L. Hoffman, Combinatorial optimization: current successes and directions for the future, *J. Comput. Appl. Math.*, 124(1–2), 341–360, 2000.
44. C.H. Papadimitriou and K. Steiglitz, *Combinatorial Optimization Algorithms and Complexity*, Prentice Hall, 1988.
45. J. Hurink, Solving complex optimization problems by local search, Habilitationsschrift, http://wwwhome.math.utwente.nl/~hurinkjl/papers/habil.pdf, Universität Osnabrück, 1999.
46. Y. Bar-Shalom, X. Li, and T. Kirubarajan, *Estimation with Applications to Tracking and Navigation*, Wiley, New York, NY, 2001.
47. C. Hue and J.-P. Le Cadre, Sequential Monte Carlo methods for multiple target tracking and data fusion, *IEEE Trans. Signal Process.*, 50(2), 309–325, 2002.
48. C. Kreucher, K. Kastella, and A.O. Hero III, Multitarget tracking using the joint multitarget probability density, *IEEE Trans. Aerosp. Electron. Syst.*, 41(4), 1396–1414, 2005.
49. O.E. Drummond, Track and tracklet fusion filtering using data from distributed sensors, *Proceeding of the Estimation, Tracking and Fusion: A Tribute to Yaakov Bar-Shalom*, Monterey, CA, May 2001, pp. 167–186.
50. C.Y. Chong, S. Mori, W.H. Barker, and K.C. Chang, Architectures and algorithms for track association and fusion, *IEEE Aerosp. Electron. Syst. Mag.*, 15(1), 5–13, 2000.
51. X. Yu, K. Niyogi, S. Mehrotra, and N. Venkatasubramanian, Adaptive target tracking in sensor networks, *Proceedings of the Communication Networks and Distributed Systems Modeling and Simulation Conference*, San Diego, CA, January 2004.
52. I. Katzela and M. Naghshineh, Channel assignment schemes for cellular mobile telecommunication systems: a comprehensive survey, *IEEE Pers. Commun.*, 3(3), 10–31, 1996.
53. CPLEX: *ILOG CPLEX 8.0 User's Manual*, ILOG S.A., July 2002.
54. S. Boyd and L. Vandenberghe, *Convex Optimization*, Cambridge University Press, 2006.
55. T. Sathyan, A. Sinha, and T. Kirubarajan, Passive geolocation and tracking of an unknown number of emitters, *IEEE Trans. Aerosp. Electron. Syst.*, 42(2), 740–750, 2006.
56. K.R. Pattipati, R.L. Popp, and T. Kirubarajan, Survey of assignment techniques for multitarget tracking, in: Y. Bar-Shalom and W.D. Blair (Eds.), *Multisensor-Multitarget Tracking: Applications and Advances*, Vol. 3, Artech House, Boston, MA, 2000.
57. J. Shin, L.J. Guibas, and F.A. Zhao, A Distributed algorithm for managing multi-target identities in wireless ad-hoc sensor networks, *Second International Workshop on Information Processing in Sensor Networks*, Palo Alto, CA, April 2003, pp. 223–238.
58. M. Hristakeva and D. Shrestha, Solving the 0–1 Knapsack problem with genetic algorithms, *Proceedings of the 37th Midwest Instruction and Computing Symposium*, Morris, MN, April 2004.
59. C. Solnon and D. Bridge, An ant colony optimization meta-heuristic for subset selection problems, in: N. Nedjah and L. Mourelle (Eds.), *System Engineering using Particle Swarm Optimization*, Nova Science publisher, March 2006.
60. E. Aarts and J.K. Lenstra, *Local Search in Combinatorial Optimization*, John Wiley & Sons, Ltd., Chichester, England, 1997.
61. M.G.C. Resende and R. Werneck, On the implementation of a swap-based local search procedure for the p-median problem, *Technical Report TD-5E4QKA*, AT&T Labs Research, September 2002.

62. Z. Michalewicz, *Genetic Algorithms + Data Structures = Evolution Programs*, 3rd edition, Springer-Verlag, Berlin, NY, 1996.
63. M. Mitchell, *An Introduction to Genetic Algorithms*, MIT Press, Cambridge, MA, 1996.
64. G.R. Raidl, An improved genetic algorithm for the multiconstrained 0–1 knapsack problem, *Proceedings of the IEEE International Conference on Evolutionary Computation*, Anchorage, AK, May 1998, pp. 207–211.
65. M. Dorigo and T. Stutzle, *Ant Colony Optimization*, MIT Press, Cambridge, MA, 2004.
66. C. Blum and M. Dorigo, The hyper-cube framework for ant colony optimization, *IEEE Trans. Syst., Man Cybernet.–Part B: Cybernet.*, 34(2), 1161–1172, 2004.

PART IV

ESTIMATION AND CLASSIFICATION

CHAPTER 13

Efficient Inference in General Hybrid Bayesian Networks for Classification

WEI SUN AND KUO-CHU CHANG

13.1 INTRODUCTION

Bayesian network (BN), also known as probability belief network, causal network [1–3], is a graphical model for knowledge representation under uncertainty and a popular tool for probabilistic inference. It models dependence relationships between random variables involved in the problem domain by conditional probability distributions (CPDs). In the network, CPD is encoded in the directed arc linking the associated random variables. The random variables that have arcs pointing to other random variables are called parent nodes and the random variables that have incoming arcs are called children nodes. The most important property of Bayesian networks is that it fully specifies the joint distribution over all random variables by a product of all CPDs. This is because each random variable is conditionally independent of its nondescendant given its parents. Factoring reduces the numbers of parameters to represent the joint distribution and so saves the computations for reasoning. One of the important tasks after constructing the BN model is to conduct probabilistic inference. However, this task is NP-hard in general [4]. This is true even for the seemly easier task of finding approximate solutions [5]. Nevertheless, for some special classes such as discrete polytree or linear Gaussian polytree networks, there exists an exact inference algorithm using message passing [1] that could be done in linear time. In the past decades, researchers have proposed a great number of inference algorithms for various Bayesian networks in the literature [6]. They can be divided into two basic groups: exact and approximate algorithms. Exact inference only works for very limited types of networks with special structure and CPDs in the model. For example, the most popular exact inference algorithm—the Clique Tree [7–10], also known as the Junction Tree or clustering algorithm [11], only works for discrete network or the simplest hybrid model called conditional linear Gaussian (CLG) [12]. In general, the

Integrated Tracking, Classification, and Sensor Management: Theory and Applications, Edited by Mahendra Mallick, Vikram Krishnamurthy, and Ba-Ngu Vo.

complexity of the exact inference is exponential to the size of the largest clique[1] of the triangulated moral graph in the network. For networks with many loops or general hybrid models that have mixed continuous and discrete variables, the intractability rules out the use of the exact inference algorithms.

For probabilistic inference with hybrid models, relatively little has been developed so far. Unfortunately, we have to deal with heterogeneity in variable types in order to construct the realistic model in most of applications such as classification, sensor fusion, and so on. The simplest hybrid model CLG is the only hybrid model for which exact inference could be done. The state-of-the-art algorithm for exact inference in CLG is Lauritzen's algorithm [9]. It computes the exact answers in the sense that the first two moments of the posterior distributions are correct, while the true distribution might be a mixture of Gaussians. An extended computational scheme of Lauritzen's algorithm could provide full density estimations with extra complexity [10]. In general, the hybrid model may involve arbitrary distributions and arbitrary functional relationships between continuous variables. It is well known that no exact inference is possible in this case. However, approximate methods have been proposed [13, 14] to handle different hybrid models. In recent years, researchers also proposed inference algorithms using mixture of truncated exponentials (MTE) [15, 16] or mixture of Gaussians [17] to approximate arbitrary distributions in order to derive the close form solution for inference in hybrid models.

Generally, there are three main categories of approximate inference methods for Bayesian networks: model simplification, stochastic sampling, and loopy belief propagation. Model simplification methods simplify the model to make the inference algorithm applicable. Some commonly applied simplification methods include the removal of weak dependency, discretization, and linearization. Stochastic sampling is a popular framework including a number of algorithms, such as likelihood weighting (LW) [18, 19] and the state-of-the-art importance sampling algorithm called adaptive importance sampling (AIS-BN) for discrete BNs [20]. The major issue for sampling methods is to find a good sampling distribution. The sampling algorithm could be very slow to converge or in some cases with unlikely evidence, it may not converge even with a huge sample size. In recent years, applying Pearl's message passing algorithm to the network with loops, so-called "loopy belief propagation" (LBP) [21, 22], has become very popular in the literature. Although the propagating messages are not exact, researchers found that loopy belief propagation usually converges and when it converges, it provides good approximate results.

We are particularly interested in the message passing framework because of its simplicity of implementation and good empirical performance. Without the computational burden such as numerical integration in [23], we propose a partitioned message passing algorithm for hybrid models even with nonlinear relationships between continuous variables. First, we need to find a general way to represent messages. Essentially, messages are likelihoods or probabilities. In discrete case, messages are represented and manipulated by probability vectors and conditional probability tables

[1] A fully connected subnetwork.

(CPTs) that is relatively straightforward. For continuous variables, however, it is more complicated for message representation and manipulation as they may have arbitrary distributions. We propose to use the first two moments, mean and variance of a probability distribution, to represent the continuous message regardless of its distribution. This simplification makes message calculation and propagation efficient between continuous variables while keeping the key information of the original distributions. Furthermore, to deal with the possible arbitrary functional relationship between continuous variables, a state estimation method is needed to approximate the distribution of random variable under nonlinear transformation. Several weighted sampling algorithms such as particle filtering [24] and Bayesian bootstrapping [25] for nonlinear state estimation were proposed in the literature. However, we prefer to use unscented transformation [26, 27] due to its computational efficiency and accuracy. Unscented transformation uses a deterministic sampling scheme and can provide good estimates of the first two moments for the continuous variable undergone nonlinear transformation. For arbitrary continuous network, this approach called unscented message passing (UMP-BN) works very well [28]. But in hybrid model, message propagation between discrete and continuous variables is not straightforward due to their different formats. To deal with this issue, we propose to apply the concept called conditioning. First we partition the original hybrid Bayesian networks into separate discrete and continuous network segments by conditioning on discrete parents of continuous variables [29]. We can then process message passing separately for each network segment before final integration.

One of the benefits of partitioning networks is to ensure that there is at least one efficient inference method applicable to each network segment. In hybrid networks, we assume that continuous node is not allowed to have any discrete child node. Therefore, the original networks can be partitioned into separate parts by the discrete parents of continuous variables. We call these nodes the interface nodes. Each network segment separated by the interface nodes consists of purely discrete or continuous variables. By conditioning on interface nodes, the variables in different network segments are independent of each other. We then conduct loopy propagation separately in each subnetwork. Finally, messages computed in different segments are integrated through the interface nodes. We then estimate the posterior distribution of every hidden variable given evidence in all network segments.

The algorithm proposed here aims to tackle nonlinear hybrid models. In particular, we address this problem from the perspective of the general classification model, in which there usually are some categorical variables of our interest, along with both discrete and continuous feature variables. We believe that the proposed combination of known efficient methods and the introduction of interface nodes for hybrid network partition makes the new algorithm a good alternative for inference in nonlinear hybrid models. The remainder of this chapter is organized as follows. Section 13.2 first reviews the Pearl's message passing formulae. We then discuss the message representation and manipulation for continuous variable and how to propagate messages between continuous variables with nonlinear functional relationship. Section 13.3 describes the methods of network partition and message integration by introducing the concept of interface nodes. We show how message passing could be done separately and finally integrated together via the channel of interface nodes. Section 13.4 presents the algorithm of hybrid message passing by conditioning. Several numerical experiments are presented in Section 13.5 using example classification models. Finally, Section 13.6 concludes the research we have done and suggests some potential future work.

13.2 MESSAGE PASSING: REPRESENTATION AND PROPAGATION

Pearl's message passing algorithm [1] is the first exact inference algorithm developed originally for polytree discrete Bayesian networks. Applying Pearl's algorithm in the network with loops usually provides accurate approximate answers and this method is called loopy belief propagation. Recall that in Pearl's message passing algorithm, $\mathbf{e}_X^+$ and $\mathbf{e}_X^-$ are defined as the evidence from the subnetwork "above" a node X, and the subnetwork "below" X, respectively. In a polytree, any node X d-separates (d-separation means given a set of variables, one set of variables is conditionally independent on another set of variables) the set of evidence $\mathbf{e}$ into $\{\mathbf{e}_X^+, \mathbf{e}_X^-\}$. In the algorithm, each node in the network maintains two values called λ value and π value. λ value of a node X, defined as

$$\lambda(X) = P(\mathbf{e}_X^- | X) \tag{13.1}$$

is the likelihood function of X given observations $\mathbf{e}_X^-$. π value of a node X, defined as

$$\pi(X) = P(X|\ \mathbf{e}_X^+) \tag{13.2}$$

is the conditional probability of X given $\mathbf{e}_X^+$.

It is easy to see that the belief of a node X given all evidence is just the normalized product of π value and λ value

$$\begin{aligned} BEL(X) &= P(X|\mathbf{e}) = P(X|\mathbf{e}_X^+, \mathbf{e}_X^-) \\ &= \frac{P(\mathbf{e}_X^-|X, \mathbf{e}_X^+)P(X|\mathbf{e}_X^+)P(\mathbf{e}_X^+)}{P(\mathbf{e}_X^+, \mathbf{e}_X^-)} \\ &= \alpha P(\mathbf{e}_X^-|X)P(X|\mathbf{e}_X^+) \\ &= \alpha\lambda(X)\pi(X), \end{aligned} \tag{13.3}$$

where α is a normalizing constant. In the process of message passing, every node sends λ messages to each of its parents and π messages to each of its children. And based on its received messages, every node updates its λ and π values correspondingly. For a typical node X with m parents $\mathbf{T} = (T_1, T_2, ..., T_m)$ and n children $\mathbf{Y} = (Y_1, Y_2, ..., Y_n)$ as illustrated in Figure 13.1, the conventional propagation equations of Pearl's message passing algorithm can be expressed as the following [1, p. 183]

$$BEL(X) = \alpha\pi(X)\lambda(X), \tag{13.4}$$

$$\lambda(X) = \prod_{j=1}^{n} \lambda_{Y_j}(X), \tag{13.5}$$

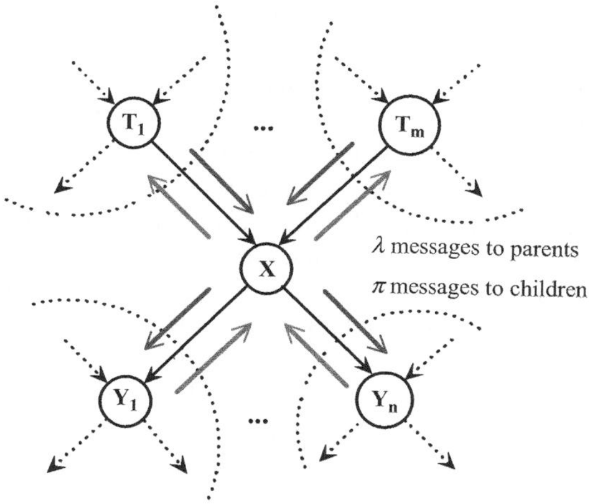

Figure 13.1 A typical node X with m parents and n children.

$$\pi(X) = \sum_{\mathbf{T}} P(X|\,\mathbf{T}) \prod_{i=1}^{m} \pi_X(T_i), \tag{13.6}$$

$$\lambda_X(T_i) = \sum_X \lambda(X) \sum_{T_k:\, k \neq i} P(X|\mathbf{T}) \prod_{k \neq i} \pi_X(T_k), \tag{13.7}$$

$$\pi_{Y_j}(X) = \alpha \left[\prod_{k \neq j} \lambda_{Y_k}(X) \right] \pi(X), \tag{13.8}$$

where $\lambda_{Y_j}(X)$ is the λ message node X receives from its child Y_j, $\lambda_X(T_i)$ is the λ message X sends to its parent T_i, $\pi_X(T_i)$ is the π message node X receives from its parent T_i, $\pi_{Y_j}(X)$ is the π message X sends to its child Y_j, and α is a normalizing constant.

When this algorithm applies to polytree network, the messages propagated are exact and so are the beliefs of all nodes after receiving all messages. For the network with loops, we can still apply this algorithm as the "loopy propagation" mentioned above. In general, loopy propagation will not provide the exact solutions. But empirical investigations on its performance have reported surprisingly good results.

For discrete variables, messages could be represented by probability vectors. And the conditional probability table of node X given its parent $\mathbf{T}$, $P(X|\,\mathbf{T})$, could be represented by a matrix. Therefore, the calculations in the above formulae are product of vectors and multiplication of vector and matrices, which could be carried out easily. However, for continuous variables, message representation and the corresponding calculations are much more complicated. First, integral replaces summation in the above equations. Furthermore, since continuous variable could have arbitrary distribution over the continuous space, in general it is very difficult to obtain exact close-form analytical results when combining multiple continuous distributions. In order to make the computations feasible while keeping the key information, we use the first two moments, mean and variance, to represent continuous message regardless of the original distribution. Then, the product of different continuous distributions could be approximated with a Gaussian distribution. Note that for continuous case, $P(X|\,\mathbf{T})$ is a continuous conditional distribution and it may involve arbitrary function between continuous variables. To integrate the product of continuous distributions as shown in (13.6) and (13.7), it has to take into account the functional transformation of continuous variables. Fortunately, unscented transformation [26, 27] provides good estimates of mean and variance for the continuous variables through nonlinear transformation. In our algorithm, unscented transformation plays a key role for computing continuous messages. Specifically, we use it to formulate and compute the π and λ messages since both computations involve the conditional probability distribution in which nonlinear transformation may be required.

13.2.1 Unscented Transformation

Proposed in 1996 by Julier and Uhlmann [26], unscented transformation is a deterministic sampling method to estimate mean and variance of continuous random variable undergone nonlinear transformation. Consider the following problem: a continuous random variable $\mathbf{x}$ with mean $\bar{\mathbf{x}}$ and covariance matrix $\mathbf{P_x}$ undergoes an arbitrary nonlinear transformation, written as $\mathbf{y} = g(\mathbf{x})$; the question is how to compute the mean and covariance of $\mathbf{y}$?

From probability theory, we have

$$p(\mathbf{y}) = \int_{\mathbf{x}} p(\mathbf{y}|\mathbf{x})p(\mathbf{x})\,dx.$$

However, in general the above integral may be difficult to compute analytically and may not always have a close form solution. Therefore, instead of finding the distribution, we retreat to seek for its mean and covariance. Based on the principle that it is easier to approximate a probability distribution than an arbitrary nonlinear function, unscented transformation uses a minimal set of deterministically chosen sample points called sigma points to capture the true mean and covariance of the prior distribution. Those sigma points are propagated through the original functional transformation individually. According to its formulae, posterior mean and covariance calculated from these propagated sigma points are accurate to the second order for any nonlinearity. In the special case when the transformation function is linear, the posterior mean and variance are exact.

The original unscented transformation encounters difficulties with high-dimensional variables, so the scaled unscented transformation was developed soon afterward [27]. The scaled unscented transformation is a generalization of the original unscented transformation. We will use the two terms interchangeably, but both mean scaled unscented transformation in the remainder of this chapter.

Now let us describe the formulae of unscented transformation. Assume $\mathbf{x}$ is L dimensional multivariate random variable. First, a set of $2L + 1$ sigma points are specified by the following formulae

$$\lambda = \alpha^2(L+\kappa) - L$$

$$\mathcal{X} = \begin{cases} \mathcal{X}_0 &= \bar{\mathbf{x}} & i = 0 \\ \mathcal{X}_i &= \bar{\mathbf{x}} + \left(\sqrt{(L+\lambda)\mathbf{P_x}}\right)_i & i = 1, \ldots, L \\ \mathcal{X}_i &= \bar{\mathbf{x}} - \left(\sqrt{(L+\lambda)\mathbf{P_x}}\right)_i & i = L+1, \ldots, 2L \end{cases} \tag{13.9}$$

and the associated weights for these $2L+1$ sigma points are

$$\begin{aligned} w_0^{(m)} &= \tfrac{\lambda}{L+\lambda} & i = 0 \\ w_0^{(c)} &= \tfrac{\lambda}{L+\lambda} + (1-\alpha^2+\beta) & i = 0 \\ w_0^{(m)} &= w_0^{(c)} = \tfrac{1}{2(L+\lambda)} & i = 1, \ldots, 2L, \end{aligned} \tag{13.10}$$

where α, β, κ are scaling parameters and the superscripts "(m)", "(c)" indicate the weights for computing posterior mean and covariance respectively. The values of scaling parameters could be chosen by $0 \le \alpha \le 1,\ \beta \ge 0$, and $\kappa \ge 0$. It has been shown empirically that the specific values chosen for the parameters are not critical because unscented transformation is not sensitive to those parameters. We choose $\alpha = 0.8,\ \beta = 2$ (optimal for Gaussian prior [27]), and $\kappa = 0$ in all of our experiments.

After selecting sigma points, they are propagated through the functional transformation

$$\mathcal{Y}_i = \mathbf{g}(\mathcal{X}_i), \quad i = 0, \ldots, 2L. \tag{13.11}$$

Finally, the posterior mean and covariance are estimated by combining the propagated sigma points as follows

$$\bar{\mathbf{y}} \approx \sum_{i=0}^{2L} w_i^{(m)} \mathcal{Y}_i, \tag{13.12}$$

$$\mathbf{P_y} \approx \sum_{i=0}^{2L} w_i^{(c)} (\mathcal{Y}_i - \bar{\mathbf{y}})(\mathcal{Y}_i - \bar{\mathbf{y}})^{\mathbf{T}}, \tag{13.13}$$

$$\mathbf{P_{xy}} \approx \sum_{i=0}^{2L} w_i^{(c)} (\mathcal{X}_i - \bar{\mathbf{x}})(\mathcal{Y}_i - \bar{\mathbf{y}})^{\mathbf{T}}. \tag{13.14}$$

In short, we denote the unscented transformation for X undergoing a functional transformation $Y = f(X)$ as the following

$$(Y.mu, Y.cov) = UT(X \xrightarrow{f(X)} Y), \tag{13.15}$$

where $Y.mu,\ Y.cov$ stand for the mean and covariance of Y, respectively.

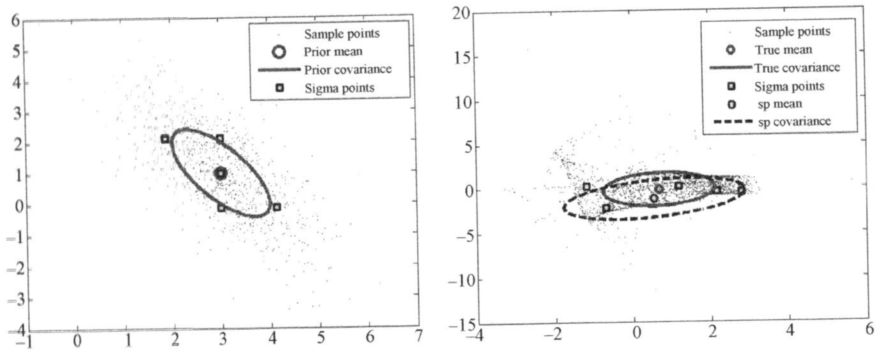

(a) Prior distribution (b) After nonlinear transformation

Figure 13.2 Demonstration of unscented transformation.

We shall demonstrate the unscented transformation by a simple two-dimensional Gaussian example. Let $\mathbf{x} = [x_1\ x_2]$ with mean and covariance matrix given as,

$$\bar{\mathbf{x}} = \begin{bmatrix} 3 \\ 1 \end{bmatrix} \quad \mathbf{P_x} = \begin{bmatrix} 1 & -1 \\ -1 & 2 \end{bmatrix}.$$

In order to show the robustness of unscented transformation, we choose a set of functions with severe nonlinearity shown as below

$$y_1 = \log(x_1^2)\cos(x_2), \qquad y_2 = \sqrt{\exp(x_2)}\sin(x_1 x_2).$$

The true posterior statistics are approximated very closely by brute force Monte Carlo simulation using 100,000 sample points drawn from the prior distribution and then propagated through the nonlinear mapping. We compare them with the estimates calculated by unscented transformation using only five sigma points. In Figure 13.2, it shows that the mean calculated by transformed sigma points is very close to the true mean and the posterior covariance seems consistent and efficient because the sigma-point covariance ellipse is larger but still tightly around the true posterior covariance ellipse.

13.2.2 Unscented Message Passing

Now let us take a closer look at Pearl's general message propagation formulae shown in (13.4)–(13.8). In recursive Bayesian inference, π message represents prior information and λ message represents evidential support in the form of a likelihood function. Combination of different messages by multiplication are represented by (13.4), (13.5), and (13.8). They are similar to the data fusion concept where estimates received from multiple sources are combined.

Under the assumption of Gaussian distribution, the fusion formula is relatively straightforward [30]. Specifically, (13.4), (13.5), and (13.8) can be rewritten in terms of the first two moments of the probability distributions as the following

$$BEL(X)\begin{cases} cov = \left(\frac{1}{\pi(X).cov} + \frac{1}{\lambda(X).cov}\right)^{-1} \\ mu = cov\left[\frac{\pi(X).mu}{\pi(X).cov} + \frac{\lambda(X).mu}{\lambda(X).cov}\right] \end{cases}, \tag{13.16}$$

$$\lambda(X)\begin{cases} cov = \left(\sum_{j=1}^{n} \frac{1}{\lambda_{Y_j}(X).cov}\right)^{-1} \\ mu = cov\left[\sum_{j=1}^{n} \frac{\lambda_{Y_j}(X).mu}{\lambda_{Y_j}(X).cov}\right] \end{cases}, \tag{13.17}$$

$$\pi_{Y_j}(X)\begin{cases} cov = \left(\frac{1}{\pi(X).cov} + \sum_{k \neq j} \frac{1}{\lambda_{Y_k}(X).cov}\right)^{-1} \\ mu = cov\left[\frac{\pi(X).mu}{\pi(X).cov} + \sum_{k \neq j} \frac{\lambda_{Y_k}(X).mu}{\lambda_{Y_k}(X).cov}\right] \end{cases}, \tag{13.18}$$

where mu, cov stand for corresponding mean and covariance, respectively.

Computation of the π message for node X is performed by (13.6). Analytically, this is equivalent to treating X as a functional transformation of $\mathbf{T}$ and the function is the one defined in CPD of X denoted as $h(X)$. Technically, we take $\mathbf{T}$ as a multivariate random variable with a mean vector and a covariance matrix; then by using unscented transformation, we obtain an estimate of mean and variance of X to serve as the π message for node X. In (13.6), $\pi_X(T_i)$ is the π messages sending to X from its parent T_i, which is also represented by "mean" and "covariance." By combining all the incoming $\pi_X(T_i)$ messages, we can estimate the mean vector and covariance matrix of $\mathbf{T}$. Obviously, the simplest way is to view all parents as independent variables; then combine their means into a mean vector, and place their variances at the diagonal positions to form a diagonal covariance matrix.[2] With that, we can compute the π message of node X by

$$(\pi(X).mu, \pi(X).cov) = UT(\mathbf{T} \xrightarrow{h(X)} X). \tag{13.19}$$

Similarly but a bit more complicated, (13.7) computes the λ message sending to its parent (T_i) from node X. Note here that we integrate out X and all of its parents

[2]This is actually how the original loopy algorithm works and why it is not exact. To improve the algorithm, we can estimate the correlations between all parents and include them in the covariance matrix of $\mathbf{T}$.

except the one (T_i) we are sending λ message to. Theoretically, this is equivalent to regarding T_i as the functional transformation of X and $\mathbf{T}\backslash T_i$. It is necessary to mention that the function used for transformation is the inverse function of the original one specified in $P(X|\ \mathbf{T})$ with T_i as the independent variable. We denote this inverse function as $\upsilon(X, \mathbf{T}\backslash T_i)$. Note that in practical problems, the original function may not be invertible or its inverse function may not be unique. In such case, we need additional steps to apply the method. However, without loss of generality, here we assume the inverse function is unique and always available. To compute the message, we first augment X with $\mathbf{T}\backslash T_i$ to obtain a new multivariate random variable called $\mathbf{TX}$; then the mean vector and covariance matrix of $\mathbf{TX}$ are estimated by combining $\lambda(X)$ and $\pi_X(T_k)(k \neq i)$. After applying unscented transformation to $\mathbf{TX}$ with the new inverse function $\upsilon(X, \mathbf{T}\backslash T_i)$, we obtain an estimate of the "mean" and "variance" for T_i serving as the $\lambda_X(T_i)$ message as below,

$$(\lambda_X(T_i).mu, \lambda_X(T_i).cov) = UT(\mathbf{TX} \overset{\upsilon(X,\mathbf{T}\backslash T_i)}{\longrightarrow} T_i). \tag{13.20}$$

With (13.16)–(13.20), we can now compute all messages for continuous variables. As one may notice, unscented transformation plays a key role here. This is why we call it unscented message passing for continuous Bayesian networks [28].

So far, we have summarized message representation and propagation for discrete and continuous variables, respectively. However, for hybrid model, we have to deal with messages passing between both types of variables. Since they are in different formats, it is not easy to integrate them directly. As mentioned in Section 13.1, our approach is to partition the original network before propagating messages between them.

13.3 NETWORK PARTITION AND MESSAGE INTEGRATION FOR HYBRID MODEL

First of all, as mentioned earlier, we assume that discrete node can only have discrete parents in the hybrid models, which implies continuous variable can not have any discrete child node.

Definition 1 (Discrete Parent) *In a hybrid Bayesian network, a discrete variable is called a discrete parent if and only if it has at least one continuous child node.*

It is well known that Bayesian network has an important property that every node is independent of its nondescendant nodes given its parents. Therefore, the following theorem follows:

Theorem 1 *All discrete parents in the hybrid Bayesian network model can partition the network into independent network segments each has either purely discrete or purely continuous variables. We call the set of all discrete parents in the hybrid*

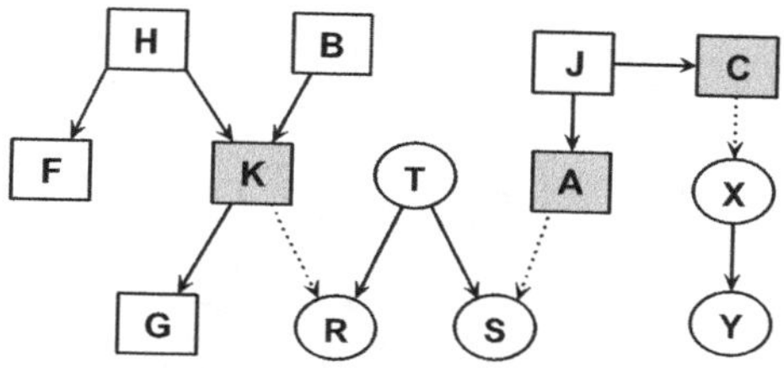

Figure 13.3 Demonstration of interface nodes and network partition.

network the interface nodes. In other words, the interface nodes "d-separate" the network into different network segments.

It is obvious that the variables in different segments of the network are independent of each other given the interface nodes. An example is shown in Figure 13.3 where a 13-node hybrid model is presented. Following the convention, we use square or rectangle to depict the discrete variable and circle or ellipse to depict the continuous variable. As can be seen, K, A, and C are the interface nodes in this example. By representing the arcs between discrete parents and their continuous children as dot lines, four independent network segments are formulated—two discrete parts (H, B, F, K, G and J, A, C) and two continuous parts (T, R, S and X, Y).

After partitioning the network with the interface nodes, we shall choose the most appropriate inference algorithm for each network segment. In fact, we can also combine some segments together if the same algorithm works for all of them. The purpose of introducing the interface nodes is to facilitate the network partition so that at least one algorithm could be applicable to each segment. In general, separate message passing in either discrete or continuous network segment is always doable. Typically, the continuous network segment with nonlinear and/or non-Gaussian CPDs is the most difficult one to deal with. In such case, we shall apply unscented message passing presented in Section 13.2.2 for approximate solutions.

Finally, we need to summarize the prior and evidence information for each network segment and encode it as messages to be passed between network segments through the interface nodes. This is similar to general message passing but requires message integrations between different network segments.

13.3.1 Message Integration for Hybrid Model

For a hybrid model, without loss of generality, let us assume that the network is partitioned into two parts denoted as $\mathcal{D}$ and $\mathcal{C}$. Part $\mathcal{D}$ is a discrete network and it is solvable by appropriate algorithms such as the Junction tree or discrete loopy propagation. Part $\mathcal{C}$ is an arbitrary continuous network. Let us denote the observable evidence in part $\mathcal{D}$ as E_d, and the evidence from $\mathcal{C}$ as E_c. Therefore, the entire evidence set $\mathbf{E}$ consists of E_d and E_c. As mentioned before, given interface nodes, variables from the two network segments are conditional independent of each other.

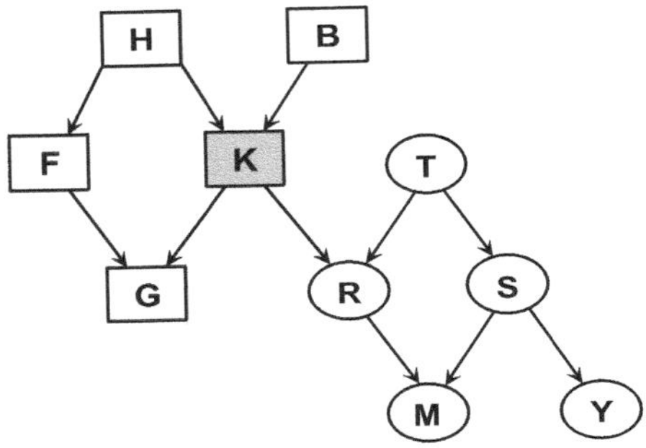

Figure 13.4 A synthetic hybrid Bayesian network model.

The evidence from part $\mathcal{D}$ affects the posterior probability of hidden nodes in part $\mathcal{C}$ and vice versa only through the channel of the interface nodes.

We can therefore summarize the prior and evidence information of each network segment and encode them as either π or λ messages at the interface nodes. Assuming that the set of interface nodes between two network segments is $\mathbf{I}$, then the two messages are $\lambda(\mathbf{I}) = P(E_c|\mathbf{I})$ and $\pi(\mathbf{I}) = P(\mathbf{I}|E_d)$. These messages are to be passed between network segments to facilitate information integration. As in Pearl's algorithm, this approach can be easily integrated with the UMP-BN loopy algorithm mentioned above in a unified manner.

We use the following concrete example to illustrate how to integrate messages from different network segments. As can be seen in Figure 13.4, a synthetic hybrid model has K as the interface node dividing the network into a discrete part consisting of H, B, F, K, G and a continuous part consisting of T, R, S, M, Y. For the purpose of illustration, let us assume all discrete nodes are binary and all continuous nodes are scalar Gaussian variables.

Suppose the leaf nodes G, M, Y are observable evidence. We will first focus on the continuous segment. In this step, we will compute the λ message sending to the interface node K from continuous evidence. And conditioning on each possible state of K, we estimate the posterior distributions for all hidden continuous variables given continuous evidence. Under Gaussian assumption, these posterior distributions are represented by means and variances and they are intermediate results that will be combined after we obtain the *a posterior* probability distribution of the interface node K given all evidence. Probabilities of all possible states of K are served as the mixing weights, similar to computing the mean and variance of a Gaussian mixture.

Given K, it is straightforward to compute the likelihood of continuous evidence $M = m, Y = y$ because we can easily estimate the conditional probability distribution of evidence node given interface nodes and other observations. For example, let

$$P(M = m, Y = y|K = 1) = a,$$

$$P(M = m, Y = y|K = 2) = b,$$

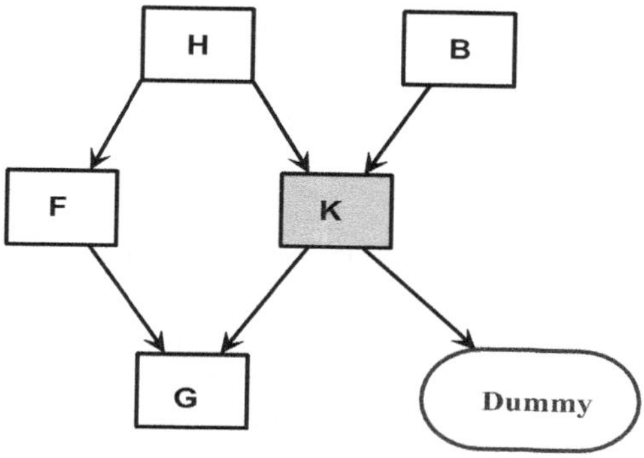

Figure 13.5 Transformed model with dummy node.

then to incorporate the evidence likelihood is equivalent to adding a binary discrete dummy node as the child of the interface node K with the conditional probability table shown as the following

K	Dummy	
	1	2
1	αa	1-αa
2	αb	1-αb

where α is a normalizing constant.

By setting "Dummy" be observed as state 1, the entire continuous segment could be replaced by the node "Dummy." Then the original hybrid Bayesian network can be transformed into a purely discrete model shown in Figure 13.5 in which "Dummy" integrates all of continuous evidence information.

The second step is to compute the posterior distributions for all hidden discrete nodes given $G = g$, $Dummy = 1$. We have several algorithms to choose for inference depending on the complexity of the transformed model. In general, we can always apply discrete loopy propagation algorithm to obtain approximate results regardless of network topology. Note that the posterior distributions of the discrete nodes have taken into account all evidence including the ones from continuous segment via the "Dummy" node. However, we need to send the updated information back to the continuous subnetwork via the set of interface nodes. This is done by computing the joint posterior probability distribution of the interface nodes denoted as $P(\mathbf{I}|\mathbf{E})$. Essentially, it is the π messages to be sent to the continuous network segment.

With the messages encoded in the interface nodes, the last step is to go back to the continuous segment to compute the *a posterior* probability distributions for all hidden continuous variables. Recall that in the first step, for any hidden continuous variable X, we already have $P(X|\mathbf{I}, E_c)$ computed and saved. The following derivation shows how to compute $P(X|\mathbf{E})$

$$
\begin{aligned}
P(X|\mathbf{E}) &= P(X|E_c, E_d) \\
&= \sum_{\mathbf{I}} P(X, \mathbf{I}|E_c, E_d)
\end{aligned}
$$

$$= \sum_{\mathbf{I}} P(X|\mathbf{I}, E_c, E_d) P(\mathbf{I}|E_c, E_d)$$

$$= \sum_{\mathbf{I}} P(X|\mathbf{I}, E_c) P(\mathbf{I}|\mathbf{E}). \tag{13.21}$$

The fourth equality is due to the fact that the set of interface node *d-separate* the node X with E_d.

Assuming given an instantiation of the set of interface nodes $\mathbf{I}=i$, $P(X|\mathbf{I} = i, E_c)$ is a Gaussian distribution with mean $\bar{x}_i$ and variance σ_i^2. Then (13.21) is equivalent to computing the probability density function of a Gaussian mixture with $P(\mathbf{I} = i|\mathbf{E})$ as the weighting factors. Denoting $P(\mathbf{I} = i|\mathbf{E})$ as p_i, the mean $\bar{x}$ and the variance σ_x^2 of $P(X|\mathbf{E})$ can be computed as the following [30, p. 56]

$$\bar{x} = \sum_i p_i \bar{x}_i, \tag{13.22}$$

$$\sigma_x^2 = \sum_i p_i \sigma_i^2 + \sum_i p_i \bar{x}_i^2 - \bar{x}^2. \tag{13.23}$$

Through the above three steps, we successfully integrate messages from different subnetworks to obtain the approximate posterior marginal distribution for both continuous and discrete hidden variables given all evidence. There are two approximations in the algorithm. One is from loopy propagation method itself. Another one is that we approximate continuous variable as Gaussian distributed as we only use the first two moments to represent continuous messages. However, it provides promising performance as will be seen in the numerical experiment results.

13.4 HYBRID MESSAGE PASSING ALGORITHM FOR CLASSIFICATION

We have presented separate message passing in either discrete or continuous network segment and message integration in hybrid model via interface nodes. In this section, we summarize this hybrid message passing algorithm for general mixed Bayesian networks (HMP-BN), as shown in Table 13.1. HMP-BN identifies discrete parent nodes as the interface nodes to partition the original network into separate subnetworks, each subnetwork contains either pure discrete or pure continuous variables. Then separate message passing algorithms are applied to different sub-networks. For continuous subnetworks with nonlinear functions involved, HMP-BN is capable to use unscented transformation to compute messages and propagate. Finally, separate messages are integrated via interface nodes, and the posterior probability distributions are computed based on final messages. HMP-BN can handle hybrid Bayesian networks with nonlinear functions, with the only restriction on network structure that a discrete node is not allowed to have any continuous parent.

TABLE 13.1 Hybrid Message Passing Algorithm for General Mixed Bayesian Network.

Algorithm: Hybrid Message Passing for General Mixed Bayesian Network (HMP-BN).
Input: General hybrid Bayesian network given a set of evidence.
Output: Posterior marginal distributions of all hidden nodes.

1. Determine the interface nodes and partition the network into independent segments with interface nodes. Choose the appropriate inference algorithm for each network segment.
2. Continuous network segment: Compute the λ message sending to the interface nodes and the intermediate posterior distribution of the hidden continuous variables given the interface nodes and the local evidence.
3. Transform the original network into an equivalent discrete model with a dummy node added as a child of the interface nodes. This dummy discrete node carries the λ message from continuous evidence to the interface nodes.
4. Compute the posterior distribution for every hidden discrete variable using the transformed discrete model. The joint posterior probability table of the interface nodes is saved as the π message to be sent back to the continuous network segment.
5. Compute the posterior distribution for every hidden continuous variable given all evidence by integrating the π message using (13.21).

Note that for each connected continuous subnetwork, one "Dummy" node will be added as the only child of the discrete parents to replace the whole continuous component. In order to incorporate evidence information, we allow a node to send a λ message to itself. For discrete network, we initialize the messages by letting all evidence nodes send to themselves a vector of a "1" for observed state and 0's for other states. All other messages are initialized as vectors of 1s. For continuous network, message is represented by "mean" and "variance." We initialize the messages for all continuous evidence nodes sending to themselves as the one with mean equal to the observed value and variance equal to zero. All other messages in continuous network are initialized as uniform, specifically, zero mean and infinity variance (the so-called "diffusion prior"). Then in each iteration, every node computes its own belief and outgoing messages based on the incoming messages from its neighbors. We assess the convergence by checking if any belief change is less than a prespecified threshold (e.g., 10^{-4}). We use parallel updating for each node till the messages are converged.

13.5 NUMERICAL EXPERIMENTS

13.5.1 Experiment Method

We use two notional classification examples as the experimental models for testing the new algorithm. One is called GHM-1, shown in Figure 13.6. GHM-1 has the same network structure as the synthetic model shown in Figure 13.4 but with a

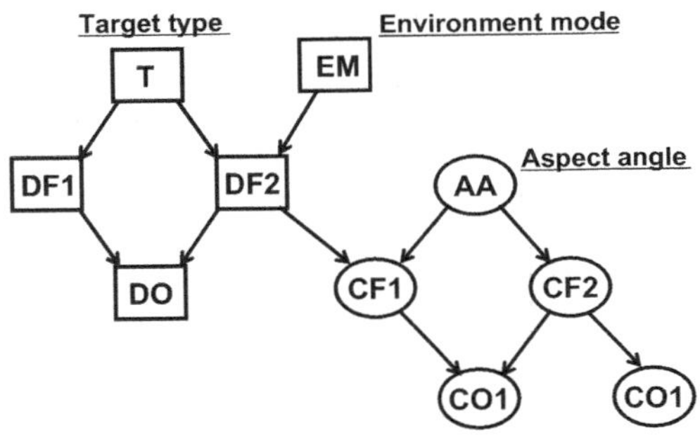

Figure 13.6 GHM-1.

meaningful model. Another classification example model is shown in Figure 13.7 called GHM-2. While GHM-1 has one loop in each subnetwork respectively (partitioned by the interface node $DF2$), GHM-2 has multiple loops in the continuous segment. In both models, besides the obviously suggestive names at the nodes, we like to explain the following names in short—EM (environment mode) includes discrete environment parameters such as weather type, terrain type, sensor mode, and so on; DF stands for discrete features such as class, report, detected/not; likewise, CF means

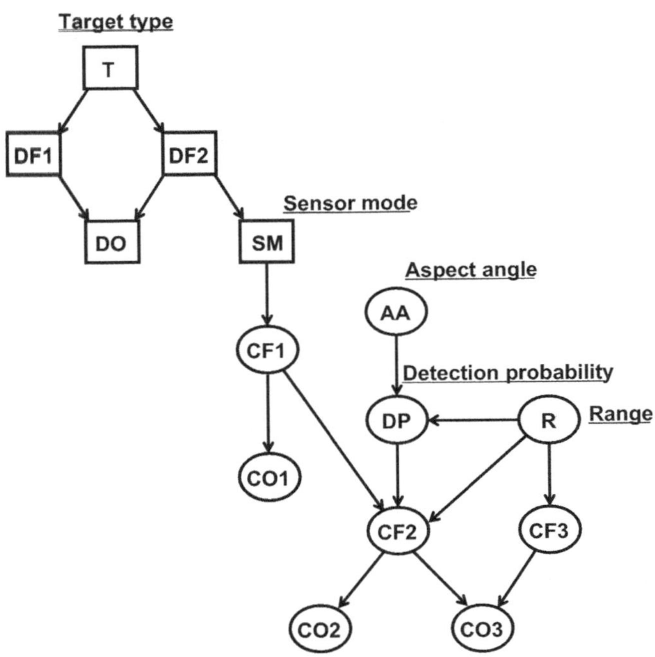

Figure 13.7 GHM-2.

continuous features such as frequency, range, detection probability; and DO/CO stand for discrete/continuous observations, which are directly observable evidence.

For GHM-1, we model its continuous segment as a linear Gaussian network given the interface node $DF2$. Therefore, the original network is a CLG so that the exact inference algorithm (Junction tree) can be used to provide the true answer as a golden standard for performance comparison. The CPTs and CPDs for nodes in GHM-1 are randomly specified.

Note that our algorithm can handle general arbitrary hybrid model, not just CLG. GHM-2 is designed specifically to test the algorithm under the situation where non-linear CPDs are involved in the model. The discrete nodes in GHM-2 are binary and we randomly specify the CPTs for them similar to the one in GHM-1. But the CPDs for the continuous nodes are deliberately specified using severe nonlinear functions shown as below to test the robustness of the algorithm

$$
\begin{aligned}
\mathbf{AA} &\sim \mathcal{N}(-10,\ 3)\\
\mathbf{R} &\sim \mathcal{N}(100,\ 10)\\
\mathbf{CF1|SM = 1} &\sim \mathcal{N}(50,\ 5)\\
\mathbf{CF1|SM = 2} &\sim \mathcal{N}(60,\ 5)\\
\mathbf{DP} &\sim \mathcal{N}(R + 2AA,\ 1)\\
\mathbf{CO1} &\sim \mathcal{N}(e^{\sqrt[3]{CF1}},\ 3)\\
\mathbf{CF2} &\sim \mathcal{N}(\sqrt{R} \times \log(DP) - CF1, 5)\\
\mathbf{CF3} &\sim \mathcal{N}(\sqrt{R} + 6,\ 3)\\
\mathbf{CO2} &\sim \mathcal{N}(-5 \times CF2,\ 5)\\
\mathbf{CO3} &\sim \mathcal{N}(0.5 \times CF2 \times CF3,\ 5)
\end{aligned}
$$

Since no exact algorithm is available for nonlinear model in general, for comparison purpose, we use brute force sampling method, likelihood weighting, to obtain an approximate true solution with a large number of samples (20 million samples).

In our experiments, we first randomly sample the network and clamp the evidence nodes by their sampled value. Then we run the HMP-BN to compute the posterior distributions for the hidden nodes. It is important to mention that in both discrete and continuous network segments, we implement HMP-BN using loopy algorithms to make it general, although Junction tree could be used in network segment whenever it is applicable. In addition, we run LW using as many samples as it can generate within the roughly same amount of time HMP-BN consumes. There are 10 random runs for GHM-1 and 5 random runs for GHM-2. We compare the average Kullback–Leibler (KL) divergences of the posterior distributions obtained by different algorithms.

Given unlikely evidence, it is well-known that the sampling methods converge very slowly even with large sample size. We use GHM-1 to test the robustness of our algorithm in this case because Junction tree can provide the ground true for GHM-1 regardless of the evidence likelihood. We generate 10 random cases with evidence

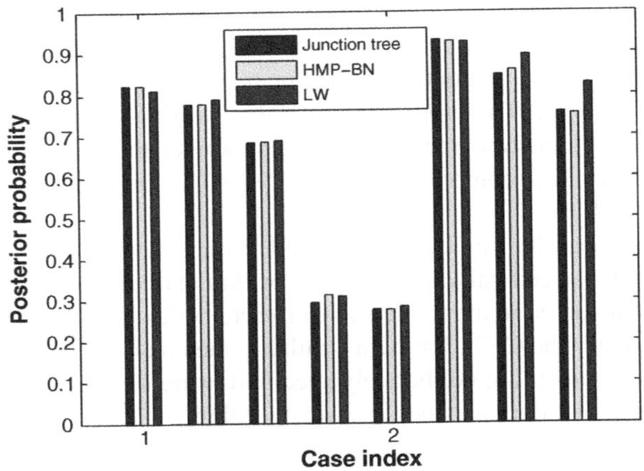

Figure 13.8 Posterior probability of hidden discrete variables in two typical runs.

likelihood between 10^{-5} and 10^{-15} and run both HMP-BN and LW to compare the performances.

13.5.2 Experiment Results

For model GHM-1, there are four hidden discrete nodes and three hidden continuous nodes. Figure 13.8 illustrates the posterior probabilities of hidden discrete nodes computed by Junction tree, HMP-BN, and LW in two typical runs. Since GHM-1 is a simple model and we did not use unlikely evidence, both HMP-BN and LW perform well.

For continuous variables in GHM-1, Figure 13.9 shows the performance comparisons in means and variances of the posterior distributions for the hidden continuous

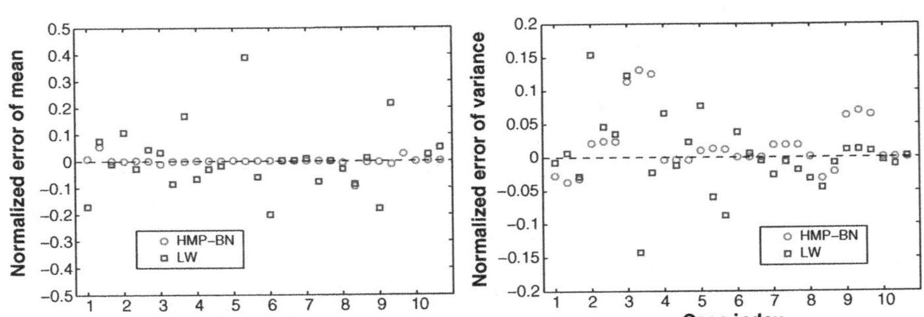

(a)Meancomparison (b)Variancecomparison

Figure 13.9 GHM-1 performance comparison for 10 random runs (the ground true is provided by Junction tree).

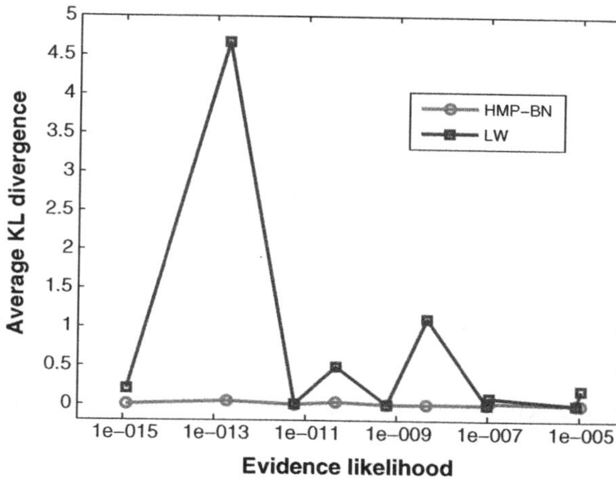

Figure 13.10 GHM-1: performance comparison given unlikely evidence.

nodes in all of the 10 runs. The normalized error is defined as the ratio of the absolute error over the corresponding true value. From the figure, it is evident that HMP-BN provides accurate estimates of means, while the estimated variances deviate from the true somewhat but HMP-BN is still better than LW in most cases.

We then demonstrate the robustness of HMP-BN by testing its performance given unlikely evidence shown in Figure 13.10. In this experiment, 10 random sets of evidence are chosen with likelihood between 10^{-5} and 10^{-15}. As can be seen, HMP-BN performs significantly better than LW in this case. The average KL divergences are consistently small with the maximum value less than 0.05. This is not surprising because LW uses the prior to generate samples so that it hardly hits the area close to the observations.

We summarize the performance results with GHM-1 in Table 13.2. Note that given unlikely evidence, the average KL divergence by HMP-BN is more than one order of magnitude better than LW.

In GHM-2, due to the nonlinear nature of the model, no exact method exists to provide the benchmark. We use LW with 20 millions samples to obtain an approximation to the true value. We implemented five simulation runs with randomly sampled evidence. In this experiment, we adopt our newly developed algorithm UMP-BN for

TABLE 13.2 Average KL Divergence Comparison in Testing GHM-1

Average KL Divergence	Normal Evidence $>10^{-5}$	Unlikely Evidence 10^{-5}–10^{-15}
HMP-BN	0.0011	0.0108
LW	0.0052	0.67

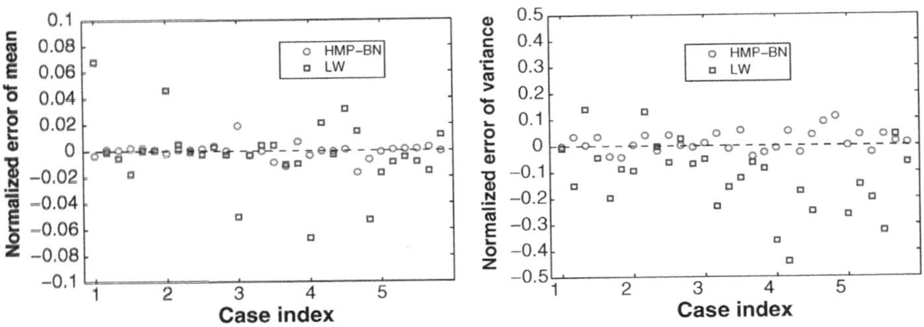

(a) Mean comparison (b) Variance comparison

Figure 13.11 GHM-2 performance comparison for five random runs (the reference base is provided by LW with 20 millions samples).

inference in continuous network segment [28]. Figure 13.11 shows the performance comparison in means and variances of the posterior distribution for the hidden continuous variables. Also, Table 13.3 summarizes the average KL divergences in testing GHM-2. From the data, we see that HMP-BN combining with UMP-BN applied in the continuous subnetwork produces very good results. In this nonlinear model with the normal evidence, the new algorithm performs much better than LW despite its advantages of being a model-free algorithm. However, since there is only one interface nodes in these models, implementing HMP-BN is relatively simple.

13.5.3 Complexity of HMP-BN

In general, when there are multiple interface nodes, HMP-BN computes the posterior distributions of hidden continuous variables given continuous evidence, conditioned on every combination of instantiations of all interface nodes. So the complexity of the algorithm is highly depends on the size of interface nodes. To assess the complexity of HMP-BN, we conducted a random experiment using network structure borrowed from the ALARM model [31] as shown in Figure 13.12 in which there are 37 nodes. We randomly selected each node to be discrete or continuous with only a requirement that continuous variable cannot have any discrete child node. In this experiment, the average number of interface nodes was about 12. HMP-BN still provided

TABLE 13.3 Average KL Divergence Comparison in Testing GHM-2

	Average KL Divergence
HMP-BN	0.0056
LW	0.0639

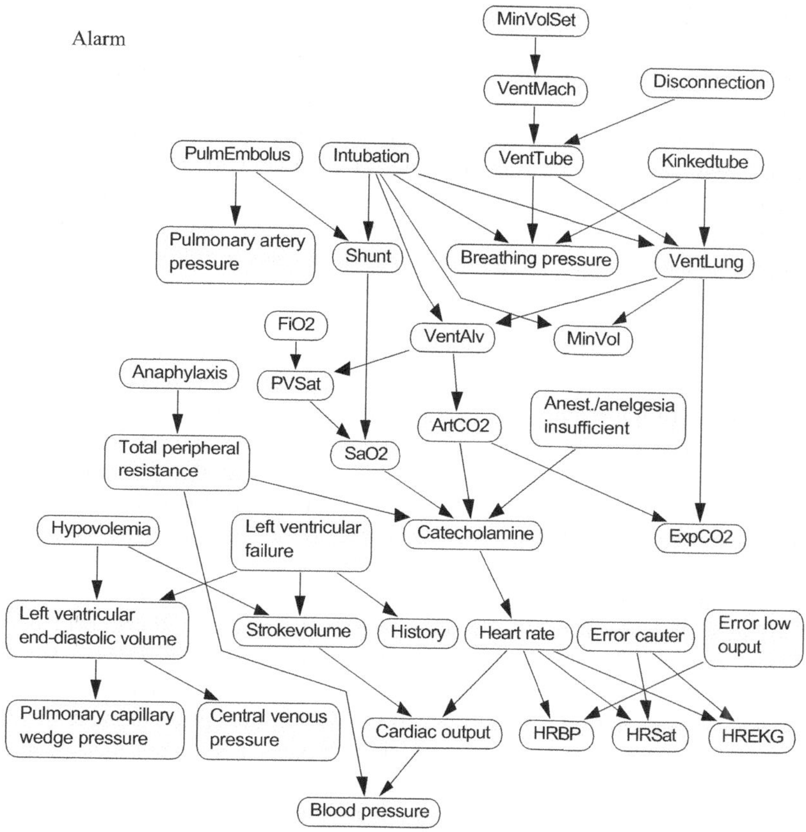

Figure 13.12 ALARM: a network constructed by medical expert for monitoring patients in intensive care.

good estimates of the posterior distributions but it took much longer time than the one with only one interface node. If we have n interface nodes $K_1, K_2, \ldots, K_n$ with number of states $n_1, n_2, \ldots, n_k$, respectively, the computational complexity of HMP-BN is proportional to $\mathcal{O}(n_1 \times n_2 \times n_3 \ldots \times n_k)$. This implies that our algorithm is not scalable for large number of interface nodes. However, our goal is not to propose an algorithm for all models (NP-hard in general) and we suspect that it is rare to have a large number of interface nodes in most practical models. Even with considerable size of interface nodes, HMP-BN provides good results within reasonable time while the stochastic sampling methods could perform very poorly using the same amount of time. In addition, there are several ways to reduce the computational burden such as assuming that some interface nodes with small correlations are independent of each other. Nevertheless, this is beyond the scope of this chapter and could be an interesting topic for future research.

13.6 CONCLUDING REMARKS

In this chapter, we develop a hybrid propagation algorithm for general Bayesian networks with mixed discrete and continuous variables. In the algorithm, we first partition the network into discrete and continuous segments by introducing the interface nodes. We then apply message passing for each network segment and encode the updated information as messages to be exchanged between segments through the set of interface nodes. Finally, we integrate the separate messages from different network segments and compute the *a posterior* distributions for all hidden nodes. The preliminary simulation results show that the algorithm works well for general hybrid Bayesian networks.

The main contribution of this chapter is to provide a general framework for inference in hybrid model. Based on the principle of decomposition and conditioning, we introduce the set of interface nodes to partition the network. Therefore, it is possible to apply exact inference algorithms such as Junction tree to some applicable network segments and it enables the integration of different efficient algorithms from multiple subnetworks. For complicated network segment such as the one with nonlinear and/or non-Gaussian variables, we provide options to use loopy-type message passing algorithm.

Although the bottleneck of our algorithm is the size of interface nodes, we believe that HMP-BN is a good alternative for nonlinear and/or non-Gaussian hybrid models. Another recently developed method termed direct message passing (DMP) can propagate messages between different types of variables directly. DMP has the same complexity when exact solution is requested. But DMP has the option to make the algorithm scalable in trading off the accuracy of the estimates. For details, please see [32].

Note that the focus of this chapter is on developing a unified message passing algorithm for general hybrid networks popularly used in classification problems. While the algorithm works well to estimate the means and variances for the hidden continuous variables, the true posterior distributions may have multiple modes. In practice, it might be more important to know where the probability mass is than just knowing mean and variance. One idea for future research is to utilize the messages computed in HMP-BN to obtain a good importance function and apply importance sampling to estimate the probability distributions.

REFERENCES

1. J. Pearl, *Probabilistic Reasoning in Intelligent Systems: Networks of Plausible Inference*, Morgan Kauffman, San Mateo, 1988.
2. R.E. Neapolitan, *Probabilistic Reasoning in Expert Systems*, John Wiley & Sons, New York, 1990.
3. E. Charniak, Bayesian networks without tears: making bayesian networks more accessible to the probabilistically unsophisticated, *AI Mag.*, 12(4), 50–63, 1991.
4. G.F. Cooper, The computational complexity of probabilistic inference using bayesian belief networks, *Arti. Intell.*, 42, 393–405, 1990.

5. P. Dagum and M. Luby, Approximating probabilistic inference in bayesian belief networks is np–hard, *Artif. Intell.*, 60, 141–153, 1993.

6. H. Guo and W. Hsu, A survey of algorithms for real-time bayesian network inference, in *AAAI/KDD/UAI–2002 Joint Workshop on Real-Time Decision Support and Diagnosis Systems*, Edmonton, Alberta, Canada, 2002.

7. S.L. Lauritzen and D.J. Spiegelhalter, Local computations with probabilities on graphical structures and their applications to expert systems, *Proc. R. Stat. Soc., Ser. B*, 50, 157–224, 1988.

8. P.P. Shenoy and G. R. Shafer, Axioms for probability and belief-function propagation, *Proceedings of the Conference on Uncertainty in Artificial Intelligence*, 1990, Vol. 4, pp. 169–198.

9. S. Lauritzen, Propagation of probabilities, means and variances in mixed graphical association models, *JASA*, 87(420), 1098–1108, 1992.

10. S. L. Lauritzen and F. Jensen, Stable local computations with conditional gaussian distributions, *Stat. Comput.*, 11(2), 191–203, 2001.

11. F.V. Jensen, *An Introduction to Bayesian Networks*, Springer-Verlag, New York, 1996.

12. Uri N. Lerner, *Hybrid Bayesian Networks for Reasoning about Complex Systems*, Stanford University, October 2002.

13. K.C. Chang, Almost instant time inference for hybrid partially dynamic bayesian networks, *IEEE Trans. Aerosp. Electron. Syst.*, 43, 13–22, 2007.

14. U. Lerner, D. Koller, and D. Angelov, A general algorithm for approximate inference and its application to hybrid bayes nets, *Proceedings of the Fifteenth Annual Conference on Uncertainty in Artificial Intelligence*, 1999, pp. 324–333.

15. B.R. Cobb and P.P. Shenoy, Inference in hybrid bayesian networks with mixtures of truncated exponentials, *Int. J. Approx. Reason.*, 41, 257–286, 2006.

16. A. Salmeron, S. Moral, and R. Rumi, Mixtures of truncated exponentials in hybrid bayesian networks, *Symbolic and Quantitative Approaches to Reasoning with Uncertainty. Proceedings of the 6th European Conference, ECSQARU 2001 (Lecture Notes in Artificial Intelligence, Vol. 2143)*, 2001, pp. 156–167.

17. P. Shenoy, Inference in hybrid bayesian networks using mixtures of gaussians, *Proceedings of the 22nd Annual Conference on Uncertainty in Artificial Intelligence (UAI)*, AUAI Press, 2006, pp. 428–436.

18. R. Fung and K.C. Chang, Weighting and integrating evidence for stochastic simulation in bayesian networks, in *Uncertainty in Artificial Intelligence 5*, New York, 1989, Elsevier Science Publishing Company, Inc, pp. 209–219.

19. R.D. Shachter and M.A. Peot, Simulation approaches to general probabilistic inference on belief networks, *Proceedings of the Conference on Uncertainty in Artificial Intelligence*, Elsevier Science, 1989, pp. 311–318.

20. J. Cheng and M.J. Druzdzel, Ais-bn: An adaptive importance sampling algorithm for evidential reasoning in large bayesian networks, *J. Art. Intell. Res. (JAIR)*, 13, 155–188, 2000.

21. K. Murphy, Y. Weiss, and M. Jordan, Loopy belief propagation for approximate inference: an empirical study, *Proceedings of the Fifteenth Conference on Uncertainty in Artificial Intelligence*, Morgan Kaufmann, 1999, pp. 467–476.

22. Y. Weiss and W.T. Freeman, Correctness of belief propagation in gaussian graphical models of arbitrary topology, *Tech. Report CSD-99-1046*, Dept. of Computer Science, University of California at Berkeley, 1999.

23. C. Yuan and M.J. Druzdzel, Hybrid loopy belief propagation, *Proceedings of the Third European Workshop on Probabilistic Graphical Models*, 2006, pp. 317–324.

24. N. Gordon, S. Arulampalam, S. Maskell, and T. Clapp, A tutorial on particle filters for on-line nonlinear/non-gaussian bayesian tracking, *IEEE Trans. Signal Process.*, 50, 174–188, 2002.

25. E.R. Beadle and P.M Djuric, A fast-weighted bayesian bootstrap filter for nonlinear model state estimation, *IEEE Trans. Aerosp. Electron. Syst.*, 33, 338–343, 1997.

26. S.J. Julier and J.K. Uhlmann, A general method for approximating non-linear transformations of probability distribution, *Tech. Report*, RRG, Dept. of Engineering Science, University of Oxford, November 1996.

27. S.J. Julier, The scaled unscented transformation, *Proceedings of the American Control Conference*, Anchorage, Alaska, USA, Vol. 6, May 2002, pp. 4555–4559.

28. W. Sun and K.C. Chang, Unscented message passing for arbitrary continuous bayesian networks, *Proceedings of the 22nd AAAI Conference on Artificial Intelligence*, Vancouver, Canada, July 2007.

29. W. Sun and K.C. Chang, Hybrid message passing for general mixed bayesian networks, *Proceedings of the 10th International Conference on Information Fusion*, Québec City, Canada, 2007, pp. 1300–1307.

30. X.R. Li, Y. Bar-Shalom, and T. Kirubarajan, *Estimation with Applications to Tracking and Navigation*, John Wiley and Sons, Inc., New York, 2001.

31. I. Beinlich, G. Suermondt, R. Chavez, and G. Cooper, The alarm monitoring system: a case study with two probabilistic inference techniques for belief networks, *Proceeding of 2nd European Conference on AI and Medicine*, 1989.

32. W. Sun and K.C. Chang, Scalable inference for hybrid bayesian networks with full density estimations, *Proceedings of the 13th International Conference on Information Fusion*, Edinburgh, UK, 2010.

CHAPTER 14

Evaluating Multisensor Classification Performance with Bayesian Networks

ESWAR SIVARAMAN AND KUO-CHU CHANG

14.1 INTRODUCTION

A common problem in classification is to use one/more sensors to observe repeated measurements of the features or attributes of a prespecified target, and to determine its identity/target type [1]. This is usually obtained by Bayesian updating of the posterior probabilities of the target types, given the sensor observations, and is referred to as multisensor data fusion for ID classification. Data fusion systems are also used extensively for target tracking, situation assessments, and automated reasoning/diagnostics [2]. The challenge in multisensor fusion is to select the best sensor(s) in a sensor suite, specific to an operating scenario, as also the algorithms for fusion; consequently, it is critical that the performance of the fusion system be understood prior to deployment.

The problem of kinematic performance evaluation for continuous state-estimation problems is well studied. However, unlike the covariance matrix for kinematic performance evaluation, there does not exist an equivalent approach for quantifying the long-run performance of multisensor, discrete-state classification systems, or combined classification and kinematic estimation systems, except through Monte Carlo simulations [2–6].

This chapter presents a new analytical approach for quantifying the long-run performance of a multisensor, discrete-state classification system, under the assumption of independent, asynchronous measurements, and consolidates earlier research [5, 7–9]. The questions addressed and answered in this research are as follows:

- Suppose a total of k measurements have been reported by a sensor; what is the accuracy of the posterior assessment of the target's identity? How many measurements must be taken to ensure a certain accuracy in the prediction?

Integrated Tracking, Classification, and Sensor Management: Theory and Applications, Edited by Mahendra Mallick, Vikram Krishnamurthy, and Ba-Ngu Vo.

- Suppose a total of $k_1, k_2, \ldots, k_r$ measurements have been reported by r sensors; what is the accuracy of the prediction obtained after fusing all these measurements?
- How do we quantify the classification performance of a sensor? What metric can we use to compare one sensor over another? How do we *design* into the sensor desirable performance traits?
- How do we make use of multiple sensors to classify a broader set of targets than what they are individually capable of?

The methodology described in this research has been applied to fusion performance evaluation of a practical air-target tracking system involving two types of sensors—electronically scanned radar and IR search and track [6]. It illustrates the use of off-line evaluation to estimate marginal performance gains and sensor mode selection, using measures and metrics derived herein. It also showcases a tool kit for fusion performance evaluation based on a Bayesian-network representation for model-based identification [10].

Our specific contributions include (i) a new methodology for evaluating discrete-state classification performance (Section 14.2.1), (ii) a new metric based on the Kullback–Leibler (KL) divergence for quantifying sensor quality (Section 14.2.4), (iii) an approximate formula based on this metric to compute rate of convergence toward perfect classification (Section 14.2.4), and (iv) a simple methodology for comparing the performance of a discrete-state classification system with multiple sensors of varying quality (Section 14.3).

14.2 SINGLE-SENSOR MODEL

This section outlines the basic notation, and the approach that we develop for quantifying a single sensor's long-run classification performance, and which will be integral to the discussion in the subsequent sections. The extension of this methodology to the case of multiple sensors is presented in Section 14.3.

14.2.1 A New Approach for Quantifying Classification Performance

We define $\mathbf{T} = \{t_1, t_2, \ldots, t_n\}$ to be the set of n objects ("target types") that need to be identified, and specify the prior probability distribution of each target type as $P_0(T = t_i) = P_0(t_i)$. Consider a sensor s that can report any of a total of m_s discrete output measurements from the set $O^s = \{o_1^s, o_2^s, \ldots, o_{m_s}^s\}$.[1] We further specify a sensor's *local confusion matrix* (LCM) as $A^s = [a_{ij}^s]_{n \times m_s}$ where a_{ij}^s specifies the conditional probability that sensor s observes object t_i and reports the output measurement as $o_j \in O^s$. Although the sensor may not observe the target directly, but through several layers of causality linked via features/attributes, the sensor's LCM can be readily derived either by (i) forward sampling of the sensor-target dependency model or

[1] We assume that the sensor's outputs are discrete, or can be suitably discretized.

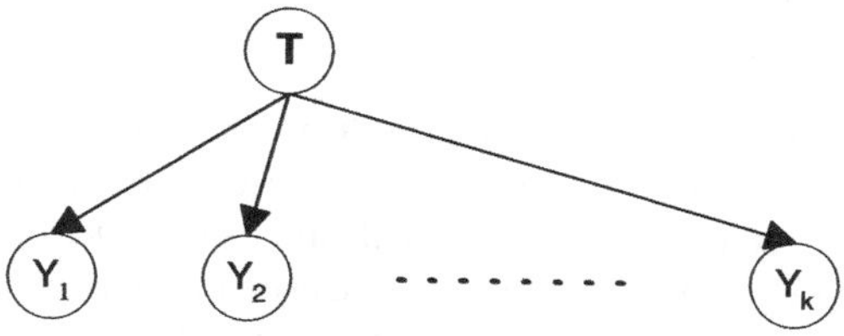

Figure 14.1 Single-sensor classification model.

(ii) from the conditional distributions $P(o_j^s|\mathbf{pa}(S_s))$ that are specified for all $o_j^s \in O^s$ and all combinations of the variables that are **parents** of sensor s [5, 11]. Appendix 14A illustrates the steps involved in calculating a sensor's LCM for different configurations of classification systems.

Consider a single-sensor model as is illustrated in Figure 14.1, where $y_{1:k} = \{y_1, y_2, \ldots, y_k\}$ indicate k successive measurements taken by a sensor with LCM $A = [a_{ij}]_{n \times m}$. Figure 14.1 illustrates the case of sensor measurements that are conditionally independent, given the target. The case of sensor measurements that are not conditionally independent, given the target, is not addressed in this chapter—the issues and challenges associated with dependent measurements are briefly discussed in Appendix 14A.

We now define a discrete metric analogous to covariance calculations for kinematic performance. Consider a single sensor that has observed a sequence $y_{1:k}$ of k measurements, where the target is one of n possible types, with prior probabilities $P_0(t_i); i = 1, 2, \ldots, n$. Suppose the target was actually of type t_i. The posterior assessment of the target's identity being predicted as type t_j is

$$P(\text{pred} = t_j|\text{meas} = y_{1:k}, \text{actual} = t_i) \propto P(y_{1:k}|t_j) \tag{14.1}$$

$$= \prod_{i=1}^{k} P(y_i|t_j), \tag{14.2}$$

where (14.1) uses the fact that the predicted probability is independent of the actual target type, given the sensor measurements, and the multiplicative form of (14.2) arises from the independence of the sensor measurements.

Now, given that the actual target was of type t_i, the expected value of the posterior probability that the target will be (mis)classified as type t_j, after k measurements, is

$$\mathbf{E}[P(\text{pred} = t_j|\text{actual} = t_i)] = \sum_{y_{1:k}} P(t_j, y_{1:k}|t_i)$$

$$= \sum_{y_{1:k}} P(t_j|y_{1:k})P(y_{1:k}|t_i), \tag{14.3}$$

where the summation in (14.3) is over all possible sequences of k sensor measurements. Clearly, a sensor that is capable of discriminating the target's identity would reveal a posterior probability distribution that converges to unity for $t_j = t_i$, and zero otherwise. It is this intuition that forms the basis of this chapter, and the metrics derived herein.

We define the *global classification matrix* (GCM) of a sensor to be an $n \times n$ matrix $\mathbf{G}_k$, whose ith row corresponds to the expected posterior probability distribution for all the target types, averaged over all possible realizations of k measurements, assuming that t_i was the true target type. For a single sensor that has observed a total of k conditionally independent measurements, the expected value of the probability that the sensor will (mis)classify target-type t_i as t_j is

$$g_{ij}^k = \sum_{y_{1:k}} P(t_j|y_{1:k}, t_i)P(y_{1:k}|t_i) \tag{14.4}$$

$$= \sum_{y_{1:k}} \frac{\prod_{c=1}^k P(y_c|j)P(y_c|i)}{\sum_t P_0(t)\prod_{c=1}^k P(y_c|t)} P_0(t_j), \tag{14.5}$$

where the outer summation in (14.5) is over all possible sequences of k measurements, and $P_0(\cdot)$ is the prior probability distribution of the target's types.

The GCM $\mathbf{G}_k$ reflects the expected value of the posterior assessments for the different target types, obtained from a total of k sensor measurements, and is an indicator of the long-run classification accuracy of the sensor. We show in Section 14.2.3 that as long as the rows of the sensor's LCM are different, the GCM will converge to the identity matrix with increasing measurements, that is, the targets will all be perfectly classified. This convergence phenomenon is natural and not surprising as long as the sensors are discriminatory—we show in Sections 14.2.3 and 14.2.4 that a minimal condition for a sensor to be discriminatory is for the rows of the sensor's LCM to be different. However, it is nontrivial to develop an analytical method to predict the convergence rate. The rate at which a sensor's GCM converges to the identity matrix, that is, perfect classification, is an indicator of its classification effectiveness and is a very important parameter when designing a practical multisensor classification system [5]. Since the calculation of the GCM is not dependent on the actual observations of the sensor, its calculation remains *off-line*, and thus, allows us to compare the speed of convergence of one sensor over another before they are even used. This is the thrust of this research, namely, to develop an approach for off-line performance evaluation. However, while the calculation of (14.5) is relatively straightforward, it incurs an explosive growth in the size of the set being enumerated upon, namely, all possible combinations of $y_{1:k}$. The next section presents results that eliminate this problem.

14.2.2 Efficient Estimation of the Global Classification Matrix

This section presents a systematic approach for the efficient calculation of the GCM. We begin with a simple result concerning the off-diagonal elements of the GCM.

Theorem 1 *The ratio of the off-diagonal elements g_{ij}^k and g_{ji}^k of the global classification matrix $\mathbf{G}_k$ is equal to the inverse of the ratio of the a priori probabilities $P_0(t_i)$ and $P_0(t_j)$, that is, $P_0(t_i)g_{ij}^k = P_0(t_j)g_{ji}^k$.*

Proof

$$\begin{aligned}
g_{ij}^k &= \sum_{y_{1:k}} P(t_j|y_{1:k})P(y_{1:k}|t_i) \\
&= \sum_{y_{1:k}} \left[\frac{P(y_{1:k}|t_j)P_0(t_j)}{P(y_{1:k})}\right]\left[\frac{P(t_i|y_{1:k})P(y_{1:k})}{P_0(t_i)}\right], \\
&= \frac{P_0(t_j)}{P_0(t_i)} \sum_{y_{1:k}} P(t_i|y_{1:k})P(y_{1:k}|t_j) = \frac{P_0(t_j)}{P_0(t_i)} g_{ji}^k
\end{aligned}$$

whereupon the result follows. ■

The advantage of Theorem 1 is that once the diagonal elements of $\mathbf{G}_k$ are known, it only remains to compute the upper triangular elements $\{g_{ij}^k : j > i\}$, whose solution reduces to that of simply solving the following linear equations, for $i = 1, 2, \ldots, n$

$$1 - g_{ii}^k = \sum_{j=i+1}^{n} g_{ij}^k + \sum_{j=1}^{i-1} g_{ij}^k = \sum_{j=i+1}^{n} g_{ij}^k + \sum_{j=1}^{i-1} \left[\frac{P_0(t_j)}{P_0(t_i)}\right] g_{ji}^k, \tag{14.6}$$

where we have used the fact that each row of $\mathbf{G}_k$ adds up to 1. A systematic approach to solving these linear equations is presented in Appendix 14B. Clearly, in contrast with the enumerative approach for computing g_{ij}^k with (14.5), we have now reduced the calculations to that of merely solving simple linear equations once the diagonal elements (g_{ii}^k) are computed, which we simplify next.

Note that for $i = 1, 2, \ldots, n$, the diagonal element of the sensor's GCM is

$$\begin{aligned}
g_{ii}^k &= \sum_{y_{1:k}} P(t_i|y_{1:k})P(y_{1:k}|t_i) = \sum_{y_{1:k}} \frac{P(y_{1:k}|i)^2}{P(y_{1:k})} P_0(t_i) \\
&= \sum_{y_{1:k}} \frac{\prod_{c=1}^{k} P(y_c|i)^2}{\sum_t P_0(t) \prod_{c=1}^{k} P(y_c|t)} P_0(t_i),
\end{aligned} \tag{14.7}$$

where $y_{1:k}$ indicates the k successive measurements $\{y_1, y_2, \ldots, y_k\}$ from the sensor. Consider a sequence $y_{1:k}$ of k measurements; for $i = 1, 2, \ldots, m$, let α_i denote the number of times the sensor measurements report output o_i. Clearly, $\alpha_1 + \alpha_2 + \cdots + \alpha_m = k$. Let us now ask another question—how many of these $y_{1:k}$ sequences have the same values of $(\alpha_1, \alpha_2, \ldots, \alpha_m)$? The answer is just the multinomial coefficient

$$\binom{k}{\alpha_1, \alpha_2, \ldots, \alpha_m} = \frac{k!}{\alpha_1! \alpha_2! \ldots \alpha_m!},$$

where $\alpha_1 + \alpha_2 + \cdots + \alpha_m = k$. Consequently, the expression for g_{ii}^k can be readily simplified as

$$g_{ii}^k = \sum_{\alpha_1=0}^{k} \sum_{\alpha_2=0}^{k-\alpha_1} \cdots \sum_{\alpha_{m-1}=0}^{k-(\alpha_1+\cdots+\alpha_{m-2})} \binom{k}{\alpha_{1:m}} \frac{\prod_{j=1}^m (a_{ij})^{2\alpha_j}}{\sum_t P_0(t) \prod_{j=1}^m (a_{tj})^{\alpha_j}} P_0(t_i), \quad (14.8)$$

which follows from the fact that there are exactly $\binom{k}{\alpha_1,\alpha_2,\ldots,\alpha_m}$ different $y_{1:k}$ sequences with the same $(\alpha_1, \alpha_2, \ldots, \alpha_m)$ values, and we are using the equality $P(y_{1:k}|i) = \prod_{j=1}^m (a_{ij})^{\alpha_j}$, where a_{ij} is the probability that the sensor reports output o_j for target-type t_i. Additionally, note that the indices of the summations cover only $\alpha_1, \alpha_2, \ldots, \alpha_{m-1}$, since the choice of values for these parameters automatically fixes $\alpha_m = k - \sum_{i=1}^{m-1} \alpha_i$. Note also that the total number of summations in this reformulation is just $\binom{k+m-1}{k}$, the number of different solutions to the equation $\sum_{i=1}^m \alpha_i = k$, $\alpha_i \geq 0$ [12]. This is a significant reduction compared to the m^k summations that would be required in the enumerative approach of (14.5). The computational savings attained from the reformulation in (14.8) have led to much insight about the classification performance of both single- and multisensor models, the results of which are presented next. Once g_{ii}^k is computed, the calculation of the other elements of the GCM follows readily from the solution to (14.6), using the technique described in Appendix 14B.

In addition to the result in (14.8), we now present an efficient, recursive formulation for computing g_{ii}^{k+1}, using just the results for g_{ii}^k. We begin by specifying a few required definitions.

- Given a set of positive integers $\{\alpha_i\}_1^m$, such that $\sum_{j=1}^m \alpha_j = k$, the $(m-1) \times 1$ vector $[\alpha_1, \alpha_2, \ldots, \alpha_{m-1}]$ will be notated as $\vec{\alpha}$, and the sum $\sum_{r<m} \alpha_r = k - \alpha_m$ will be notated as $\vec{\alpha}_\oplus$.
- The multinomial coefficient $\binom{k}{\alpha_1,\alpha_2,\ldots,\alpha_{m-1},k-\vec{\alpha}_\oplus}$ will be notated as $\binom{k}{\vec{\alpha},k-\vec{\alpha}_\oplus}$.
- The vector δ_j will denote the vector that is 1 in the jth place and is 0 elsewhere, that is, $\delta_j = [0, 0, \ldots, 1, \ldots, 0]$. The dimension of the vector will be obvious from the context in which it appears.

We now reformulate (14.8) as $g_{ii}^k = \sum_{\alpha_m=0}^{k} \sum_{\forall \vec{\alpha}:\vec{\alpha}_\oplus = k-\alpha_m} g_{ii,\vec{\alpha}}^k$, where

$$g_{ii,\vec{\alpha}}^k = \binom{k}{\vec{\alpha}, k-\vec{\alpha}_\oplus} \frac{(a_{im})^{2(k-\vec{\alpha}_\oplus)} \prod_{j=1}^{m-1} (a_{ij})^{2\alpha_i}}{\sum_{t=1}^n P_0(t)(a_{tm})^{(k-\vec{\alpha}_\oplus)} \prod_{j=1}^{m-1} (a_{tj})^{\alpha_j}} P_0(i).$$

Based on this definition, it readily follows that

$$\frac{g_{ii,\vec{\alpha}}^{k+1}}{g_{ii,\vec{\alpha}+\delta_p}^{k+1}} = \frac{(\alpha_p + 1)}{(k+1-\vec{\alpha}_\oplus)} \left(\frac{a_{im}}{a_{ip}}\right)^2 \frac{\sum_{t=1}^n P_0(t) a_{tp} (a_{tm})^{(k-\vec{\alpha}_\oplus)} \prod_{j=1}^{m-1} (a_{tj})^{\alpha_j}}{\sum_{t=1}^n P_0(t)(a_{tm})^{(k+1-\vec{\alpha}_\oplus)} \prod_{j=1}^{m-1} (a_{tj})^{\alpha_j}}. \quad (14.9)$$

Appendix 14C illustrates the intuition underlying the above result with an example based on the simple binomial case, that is, $n = |\mathbf{T}| = 2$, $m = |O| = 2$. The main result underlying the recursion is

Theorem 2 *Given a set of n targets, and a sensor with a range of m outputs and a LCM A, the diagonal elements of the GCM may be recursively computed as*

$$\frac{1}{g_{ii,\vec{\alpha}}^{k+1}} = \frac{(k+1-\vec{\alpha}_\oplus)}{(a_{im})^2}\left[\frac{1}{(k+1)}\frac{1}{g_{ii,\vec{\alpha}}^{k}} - \sum_{p=1}^{m-1}\frac{(a_{ip})^2}{(\alpha_p+1)}\frac{1}{g_{ii,\vec{\alpha}+\delta_p}^{k+1}}\right]. \quad (14.10)$$

The values of $g_{ii,\vec{\alpha}}^{k+1}$ for all $\vec{\alpha}$ such that $\vec{\alpha}_\oplus = k+1$ (i.e., $\alpha_m = 0$) will need to be specified as initial conditions. The total number of initial conditions required for this recursion is equal to the number of integer solutions to the equation $\alpha_1 + \alpha_2 + \cdots + \alpha_{m-1} = k+1$, namely, $\binom{k+m-1}{k+1}$.

Proof Given a set of $\{\alpha_i\}_1^{m-1}$ such that $0 \le \vec{\alpha}_\oplus \le (k+1)$, where $\vec{\alpha} = [\alpha_1, \alpha_2, \ldots, \alpha_{m-1}]$, and $\vec{\alpha}_\oplus = \sum_{r<m}\alpha_r$, the expression for $g_{ii,\vec{\alpha}}^{k+1}$ may be rewritten as

$$g_{ii,\vec{\alpha}}^{k+1} = \binom{k+1}{\vec{\alpha},(k+1-\vec{\alpha}_\oplus)}\frac{P_0(t_i)(a_{im})^{2(k+1-\vec{\alpha}_\oplus)}\prod_{j=1}^{m-1}a_{ij}^{2\alpha_i}}{\sum_{t=1}^{n}P_0(t)(a_{tm})^{(k+1-\vec{\alpha}_\oplus)}\prod_{j=1}^{m-1}a_{tj}^{\alpha_j}}$$

$$= (a_{im})^2\binom{k+1}{\vec{\alpha},(k+1-\vec{\alpha}_\oplus)}\left[\frac{P_0(t_i)(a_{im})^{2(k-\vec{\alpha}_\oplus)}\prod_{j=1}^{m-1}a_{ij}^{2\alpha_i}}{\sum_{t=1}^{n}P_0(t)(a_{tm})^{(k-\vec{\alpha}_\oplus)}\prod_{j=1}^{m-1}a_{tj}^{\alpha_j}}\right]\times$$

$$\left[\frac{\sum_{t=1}^{n}P_0(t)(a_{tm})^{(k-\vec{\alpha}_\oplus)}\prod_{j=1}^{m-1}a_{tj}^{\alpha_j}}{\sum_{t=1}^{n}P_0(t)(a_{tm})^{(k+1-\vec{\alpha}_\oplus)}\prod_{j=1}^{m-1}a_{tj}^{\alpha_j}}\right]$$

$$= (a_{im})^2\frac{(k+1)}{(k+1-\vec{\alpha}_\oplus)}g_{ii,\vec{\alpha}}^{k}\left[\frac{\sum_{t=1}^{n}P_0(t)(a_{tm})^{(k-\vec{\alpha}_\oplus)}\left(\sum_{p=1}^{m}a_{tp}\right)\prod_{j=1}^{m-1}a_{tj}^{\alpha_j}}{\sum_{t=1}^{n}P_0(t)(a_{tm})^{(k+1-\vec{\alpha}_\oplus)}\prod_{j=1}^{m-1}a_{tj}^{\alpha_j}}\right]$$

$$\left(\because \sum_{p=1}^{m}a_{tp} = 1\right)$$

$$= (a_{im})^2\frac{(k+1)}{(k+1-\vec{\alpha}_\oplus)}g_{ii,\vec{\alpha}}^{k}\left[1+\sum_{p=1}^{m-1}\frac{\sum_{t=1}^{n}P_0(t)(a_{tp})(a_{tm})^{(k-\vec{\alpha}_\oplus)}\prod_{j=1}^{m-1}a_{tj}^{\alpha_j}}{\sum_{t=1}^{n}P_0(t)(a_{tm})^{(k+1-\vec{\alpha}_\oplus)}\prod_{j=1}^{m-1}a_{tj}^{\alpha_j}}\right]$$

$$= (a_{im})^2\frac{(k+1)}{(k+1-\vec{\alpha}_\oplus)}g_{ii,\vec{\alpha}}^{k}\left[1+\sum_{p=1}^{m-1}\frac{(k+1-\vec{\alpha}_\oplus)}{(\alpha_p+1)}\left(\frac{a_{ip}}{a_{im}}\right)^2\frac{g_{ii,\vec{\alpha}}^{k+1}}{g_{ii,\vec{\alpha}+\delta_p}^{k+1}}\right],$$

where the last equality follows from (14.9). By rearranging the terms in the last equality above, we get

$$g_{ii,\vec{\alpha}}^{k+1}\left[1-\sum_{p=1}^{m-1}\frac{(k+1)}{(\alpha_p+1)}(a_{ip})^2\frac{g_{ii,\vec{\alpha}}^{k+1}}{g_{ii,\vec{\alpha}+\delta_p}^{k+1}}\right]=\frac{(k+1)}{(k+1-\vec{\alpha}_{\oplus})}(a_{im}^2)g_{ii,\vec{\alpha}}^{k}$$

simplifying which, the result follows. ■

A full illustration of the intuition underlying the recursion of (14.10) is presented in Appendix 14C. Appendix 14C also presents a graph-theoretic formulation for the efficient representation of the recursion presented in (14.10), and which offers further computational efficiency in computing the diagonal elements of the GCM.

14.2.3 The Global Classification Matrix: Some Experiments

We now present an example of the use of the GCM in quantifying the classification performance of a sensor, assuming conditionally independent sensor measurements. This example is used to highlight general principles related to a sensor's GCM that are subsequently used (Section 14.2.4) in the development of metrics to evaluate the long-run performance of a sensor, and their use in comparing one sensor over another. Consider a simple binomial case, that is, $n = |\mathbf{T}| = 2$, $m = |O| = 2$, in which case the calculation of g_{ii}^k, for $i = 1, 2$, simplifies to

$$g_{ii}^k=\sum_{\alpha_1=0}^{k}\frac{k!}{\alpha_1!(k-\alpha_1)!}\left[\frac{(a_{i1})^{2\alpha_1}(a_{i2})^{2(k-\alpha_1)}}{\sum_t P_0(t)(a_{t1})^{\alpha_1}(a_{t2})^{k-\alpha_1}}P_0(t_i)\right],$$

where α_1 is the number of times the sensor reports output 1 (i.e., o_1) in the k measurement readings, and the a_{ij}s are the elements of the sensor's 2×2 LCM. Observe that we require only $(k+1)$ summations. Let us now consider a specific example.

Example 1 Consider the problem of classifying two targets with priors $P_0(t_1) = 1/3$ and $P_0(t_2) = 2/3$, using a sensor with the following LCM.

$$A=\begin{pmatrix}0.4 & 0.6\\ 0.2 & 0.8\end{pmatrix}.$$

We now present the results of a few experiments for the sensor of Example 1. The GCM obtained for this sensor, for increasing k, is presented in Table 14.1. Figure 14.2a illustrates the change in the determinant of the GCM ($\det(\mathbf{G}_k)$) with increase in the number of measurements (k). As k increases, the GCM converges toward the identity matrix, that is, perfect classification of all targets. We will use $\det(\mathbf{G}_k)$ as a summary measure to quantify the rate of convergence toward perfect classification—while the trace of the GCM is an equivalent scalar measure of classification performance

TABLE 14.1 Global Classification Matrix for Example 1

k	$\mathbf{G}_k$	$\det(\mathbf{G}_k)$
2	$\begin{pmatrix} 0.3914 & 0.6086 \\ 0.3043 & 0.6957 \end{pmatrix}$	0.0871
10	$\begin{pmatrix} 0.5587 & 0.4413 \\ 0.2206 & 0.7794 \end{pmatrix}$	0.3381
25	$\begin{pmatrix} 0.7393 & 0.2607 \\ 0.1303 & 0.8697 \end{pmatrix}$	0.6090
50	$\begin{pmatrix} 0.8818 & 0.1182 \\ 0.0591 & 0.9409 \end{pmatrix}$	0.8227
100	$\begin{pmatrix} 0.9724 & 0.0276 \\ 0.0138 & 0.9862 \end{pmatrix}$	0.9587
150	$\begin{pmatrix} 0.9931 & 0.0069 \\ 0.0035 & 0.9965 \end{pmatrix}$	0.9896

[5, 10], the determinant is particularly well suited for deriving off-line metrics to quantify sensor quality (Section 14.2.4).

The general pattern revealed in Table 14.1 and Figure 14.2a, namely, that $\det(\mathbf{G})$ increases at a nondecreasing rate, holds true as long as the rows of the sensor's LCM are different—in the extreme case, the determinant will converge to 1, indicating perfect classification.

Figure 14.2b illustrates the rate of change of $\det(\mathbf{G}_k)$ for different values of a_{11}, while keeping the second row of A fixed with $a_{22} = 0.8$. The main points to be inferred from Figure 14.2b are as follows:

- Observe that when $a_{11} = 0.2$, that is, the two rows of the sensor's LCM are identical, $\det(\mathbf{G}_k)$ remains unchanged at zero, since the GCM is singular with identical rows, each row being the prior probability distribution of the various targets. It can be easily shown that if two or more rows of the sensor's LCM are identical, then the rows of the GCM corresponding to those targets will converge to the ratio of their priors. This is illustrated for the following sensor with $n = 3$, $m = 3$, and prior $P_0 = [\frac{2}{9}\ \frac{3}{9}\ \frac{4}{9}]$. The GCM obtained after 50 measurements is shown alongside.

$$A = \begin{pmatrix} 0.10 & 0.7 & 0.20 \\ 0.10 & 0.7 & 0.20 \\ 0.65 & 0.2 & 0.15 \end{pmatrix} \qquad \mathbf{G}_{50} = \begin{pmatrix} 0.4 & 0.6 & 0 \\ 0.4 & 0.6 & 0 \\ 0 & 0 & 1 \end{pmatrix}$$

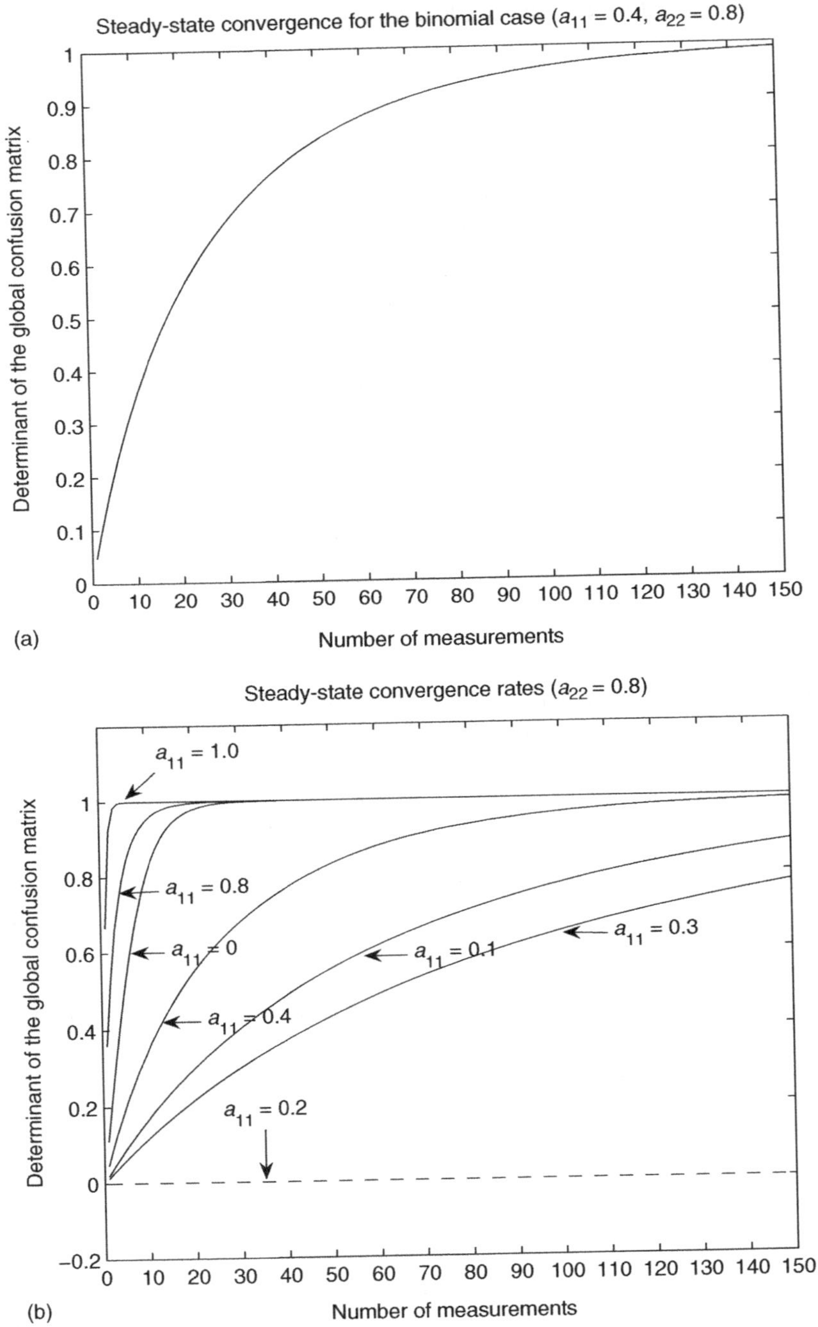

Figure 14.2 Steady-state convergence for the binomial case. (a) Long-run convergence for sensor of Example 1. (b) Convergence rates for binary sensor with varying values of a_{11} ($a_{22} = 0.8$).

TABLE 14.2 Kullback–Leibler Distance for Binomial Sensor Models of Figure 14.2b ($n = 2$, $m = 2$, $a_{22} = 0.8$)

$a_{11} = 1 - a_{12}$	$KL_{2\|\|1}$	$KL_{1\|\|2}$	KL_{12}
0	∞	0.321	0.322
0.10	0.064	0.053	0.029
0.20	0	0	0
0.30	0.037	0.041	0.019
0.40	0.132	0.151	0.070
0.60	0.483	0.551	0.257
0.80	1.200	1.200	0.6
1	∞	2.322	2.322

Since the sensor is unable to distinguish between targets 1 and 2, the GCM converges to identical rows for these targets with values corresponding to the ratio of their priors (e.g., $\frac{2/9}{2/9+3/9} = 0.4$). Consequently, in order for a sensor to be able to classify the targets perfectly, *the only requirement is that the rows of its LCM be different.*

- Figure 14.2b reveals a very interesting observation, namely, that the sensor which is totally incapable of correctly classifying target 1 (i.e., $a_{11} = 0$) converges at a faster rate toward perfect classification than a sensor which apparently is mildly better (e.g., $a_{11} = 0.1, 0.3, 0.4$). We now generalize this observation to quantify the quality of a sensor and which can be used to compare one sensor with another.

Table 14.2 reports both the symmetric and the asymmetric KL distance between the two rows of the sensor's LCM for the different values of a_{11} reported in Figure 14.2b. The KL distance from the first row to the second row is computed as $KL_{2\|1} = \sum_{i=1}^{2} a_{2i}(\log_2 a_{2i} - \log_2 a_{1i})$. Since this metric is asymmetric, we compute a symmetric equivalent KL_{12} with the harmonic mean of $KL_{2\|1}$ and $KL_{1\|2}$ (also known as the resistor average), and which is calculated as [13, 14]

$$KL_{12} = 1/(1/KL_{1\|2} + 1/KL_{2\|1}).$$

Note that the order of increasing convergence ($a_{11} = 0.2, 0.3, 0.1, 0.4, 0.6, 0.8, 1.0$) in Figure 14.2b corresponds to the order of increasing symmetric KL distances in Table 14.2. Thus, the greater the symmetric KL distance between the two rows, the faster the sensor's convergence toward perfect classification. The results of Table 14.2 and Figure 14.2b suggest that a sensor's ability to classify individual targets is not so critical, as is the requirement that it be able to distinguish the targets, that is, the KL distance between the rows of the LCM be nonzero, with higher values leading to faster convergence—we formalize this next.

14.2.4 Sensor Design Quality Metrics

We now present a metric D_s to quantify the discriminatory power of a sensor s with LCM $A^s = [a_{ij}^s]_{n \times m_s}$ (and different rows), computed as the average of the symmetric

KL distance between all pairs of the rows of A^s, that is,

$$D_s = \frac{1}{\binom{n}{2}} \sum_{i=1}^{n} \sum_{j=i+1}^{n} KL_{ij}, \tag{14.11}$$

where $n = |\mathbf{T}|$ is the number of target types to be classified, and KL_{ij} is the harmonic mean of the KL distances $KL_{i\|j}$ and $KL_{j\|i}$. If the rows of the sensor's LCM are all different, then the GCM will converge to the identity matrix—in the discussion below, we will focus on the use of the D_s metric to quantify this rate of convergence toward perfect classification. The case when some of the rows of the sensor's LCM are identical is discussed in Section 14.3.2. The D_s metric offers a simple measure to quantify a sensor's discriminatory power and its speed of convergence—a sensor with a larger D_s value converges faster than a sensor with a lower value, an illustration of which is provided in Figure 14.3a for two sensors with parameters $n = 4$, $m = 4$, and equal priors for all four targets. The LCM for the two sensors is shown alongside—the speed of convergence of Sensor 1 ($D_1 = 0.868$) is clearly much superior to Sensor 2 ($D_2 = 0.22$).

The D_s metric also results in an approximation of a sensor's classification accuracy and its rate of convergence toward perfect classification. Suppose $\mathbf{G}_k^s$ is the GCM obtained from sensor s after k measurements; then, for $n \geq 2$, $|\mathbf{G}_k^s|$ may be approximated as

$$|\mathbf{G}_k^s| \approx 1 - \exp(-C_s k), \quad C_s = \frac{D_s}{n + \frac{(n-2)}{(n-1)} \frac{1}{D_s}}. \tag{14.12}$$

The formula presented in (14.12) is an empirical approximation to the exponential nature of the growth curve that minimizes the mean-squared error between the exact rate of convergence and the approximation. The quality of the approximation presented in (14.12) is illustrated in Figure 14.3b, where the actual and approximated values of $|\mathbf{G}_k^s|$ for $k = 1, 2, \ldots, 100$ are shown for four different sensors—the LCMs for the four sensors and the priors used in the calculations are all shown alongside. The mean-squared error for the four approximations in Figure 14.3b is 0.0005, 0.0007, 0.0006, and 0.0003, respectively. Based on (14.12), it also follows that the minimum number of measurements required for achieving a classification accuracy of $100\beta\%$ is $k_\beta \approx (1/C_s)\ln(1/(1-\beta))$.

The quality of the approximation derived from the formula $|\mathbf{G}_k^s| \approx 1 - \exp(-C_s k)$ was also compared to Monte Carlo simulations of the sensor's GCM. The results for the binomial case of Example 1 are shown in Figure 14.4—each run constituted a simulation of 150 sensor measurements, with results for replications of 10, 25, 50, or 100 runs. The particulars of the simulation are detailed in Appendix 14D.1. As can be seen from Figure 14.4, the alignment of the simulated approximation to the exact convergence curve increases with increasing number of simulation runs. However, the analytical approximation based on (14.12) clearly outperforms the simulation results.

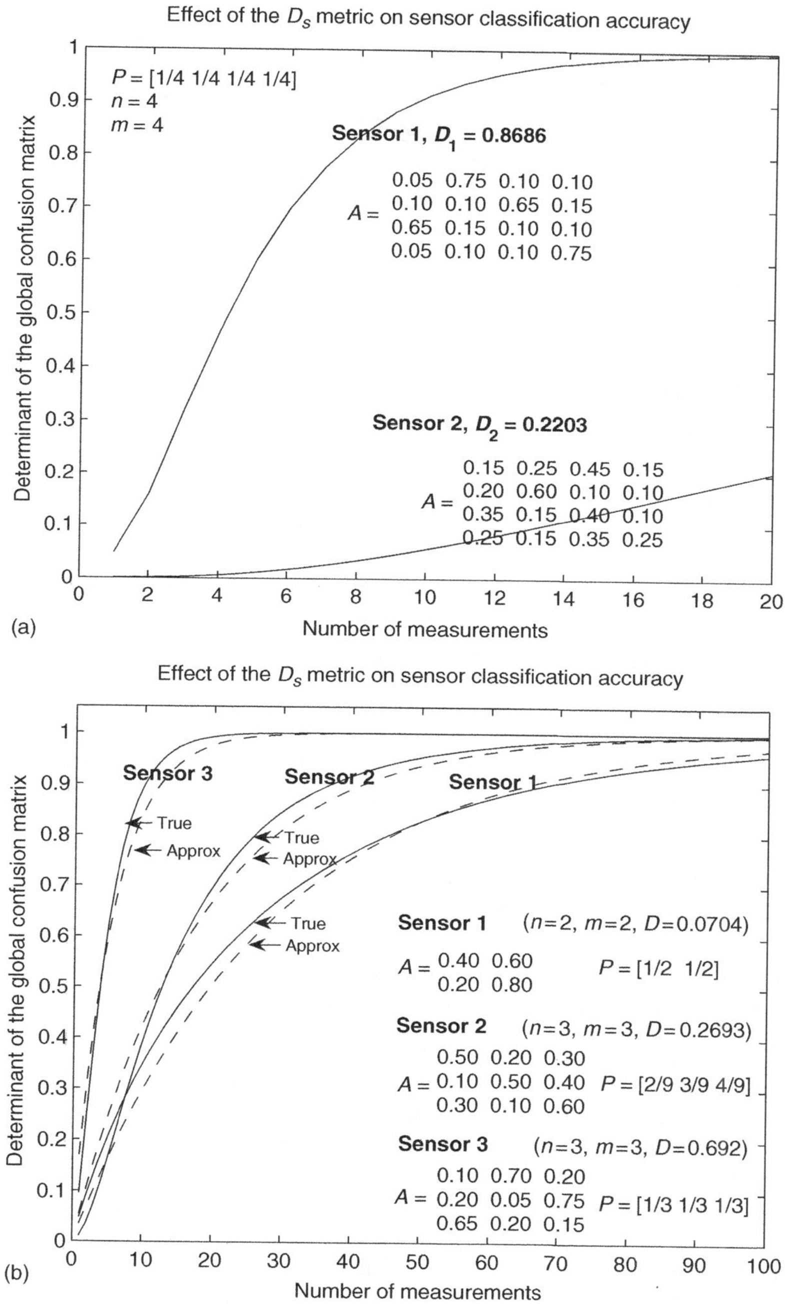

Figure 14.3 Experiments with the D_s metric and sensor performance. (a) Effect of increasing D_s on sensor performance. (b) Approximating classification performance.

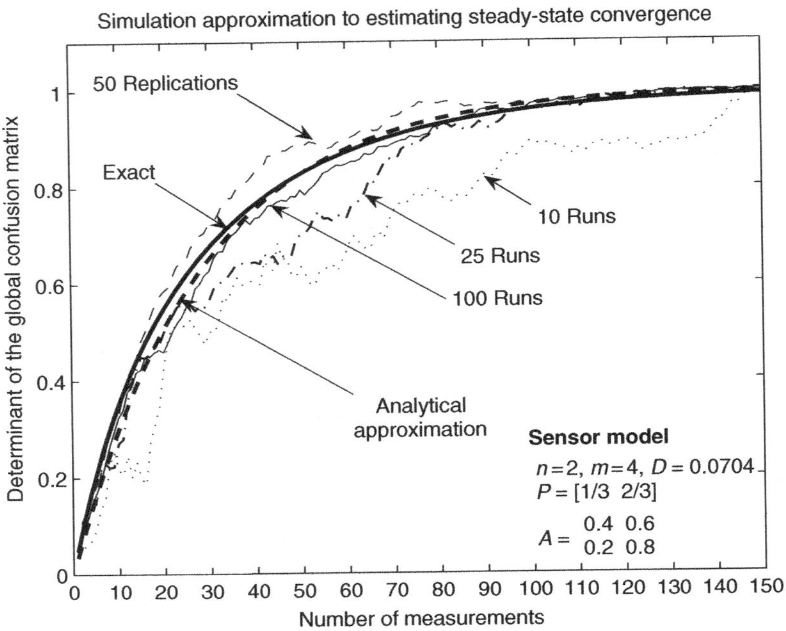

Figure 14.4 Monte Carlo simulation results for Example 1.

In summary, a sensor's classification performance is determined largely by its LCM, and so long as the rows of its LCM are different, the higher the value of D_s, the faster the rate of convergence toward perfect classification.

14.3 MULTISENSOR FUSION SYSTEMS—DESIGN AND PERFORMANCE EVALUATION

We introduced the D_s metric in Section 14.2.4 to quantify a sensor's discriminatory power, and presented an approximation $|\mathbf{G}_k^s| \approx 1 - \exp(-C_s k)$, where $C_s = D_s / \left[n + \frac{(n-2)}{(n-1)} \frac{1}{D_s}\right]$, for estimating the convergence rate of a single sensor. We now extend the use of these results for estimating the performance of multiple sensors, each contributing independent, asynchronous measurements. The discussion in Section 14.3.1 assumes that the sensors are all individually "good," that is, the rows of a sensor's LCM are all different, to enable each sensor to discriminate among all target types—"not-so-good" multisensor configurations are discussed in Section 14.3.2.

14.3.1 Performance Evaluation of Multisensor Models—Good Sensors

In multisensor fusion, we could either have r sensors monitoring the same features (Figure 14A.1c), or different features (Figure 14A.1d)—rules for reducing these configurations to the single-sensor model of Figure 14.1 are discussed in Appendix 14A. In the discussion that follows, we assume that the multisensor model is specified as a collection of r sensors, each contributing conditionally independent, asynchronous measurements. Each sensor s is specified by an output-set O^s, and LCM A^s, for $s = 1, 2, \ldots, r$. In such a case, all of the r sensors can be replaced by a single equivalent sensor with output-set $O^1 \times O^2 \times \cdots \times O^r$, and whose LCM is merely the Kronecker product of each row of the individual sensor LCMs—consider the following example.

Example 2 Given two sensors S_1 and S_2 with LCMs

$$A^1 = \begin{pmatrix} 0.25 & 0.75 \\ 0.55 & 0.45 \end{pmatrix} \qquad A^2 = \begin{pmatrix} 0.3 & 0.6 & 0.1 \\ 0.4 & 0.4 & 0.2 \end{pmatrix},$$

with $D_1 = 0.1403$, $D_2 = 0.0646$. The LCM of the equivalent sensor obtained by taking the Kronecker product of each row of A_1 and A_2 is

$$A^{1,2} = \begin{pmatrix} 0.075 & 0.15 & 0.025 & 0.225 & 0.45 & 0.075 \\ 0.22 & 0.22 & 0.11 & 0.18 & 0.18 & 0.09 \end{pmatrix},$$

whose D value is $D_{1,2} = 0.2049$. The interesting result is that $D_{1,2} \approx D_1 + D_2$. Indeed, this result holds most generally, as formalized next.

The following approximation is required to generalize the result presented in Example 2. An incomplete proof of this approximation is presented in Appendix 14E—the proof description ends with a result, which while trivial to verify, seems intractable to establish formally.

Approximation 1 *Consider two sets of probability distributions* $\{P_i^1, Q_i^1\}_{i=1}^{m_1}$ *and* $\{P_j^2, Q_j^2\}_{j=1}^{m_2}$ *with supports of sizes* m_1 *and* m_2, *and with symmetric KL distances* $KL^1_{P^1,Q^1} = (KL^1_{P^1\|Q^1} * KL^1_{Q^1\|P^1})/(KL^1_{P^1\|Q^1} + KL^1_{Q^1\|P^1})$ *and* $KL^2_{P^2,Q^2}$. *Let* $P^{12} = P^1 \otimes P^2$ *and* $Q^{12} = Q^1 \otimes Q^2$ *be the probability distributions obtained though the corresponding Kronecker products. Then, the symmetric KL distance between* P^{12} *and* Q^{12} *is*

$$KL^{12}_{P^{12},Q^{12}} \approx KL^1_{P^1,Q^1} + KL^2_{P^2,Q^2}.$$

This approximation is very accurate, with near negligible margin of error.

Approximation 1 implies that the D_s metric computed for the Kronecker product of two sensors' LCMs is approximately equal to the sum of the two sensors' D_s values, as illustrated in Example 2. The following result follows directly from Approximation 1—since the result depends on an approximation that is as yet unproven, it is presented as a proposition.

Proposition 1 *If $A_{k_1,k_2,\ldots,k_r}$ is the equivalent sensor obtained from k_s copies of sensor s (by taking the Kronecker product of each row), then*

$$D_{k_1,k_2,\ldots,k_r} \approx \sum_{s=1}^{r} k_s * D_s. \tag{14.13}$$

Proof The result follows directly from a recursive application of Approximation 1. ■

The implication of (14.13) is that the D value of the equivalent sensor obtained by combining k_s copies of sensor s ($s = 1, 2, \ldots, r$) can be readily approximated as the sum of the D values of each sensor type, weighted by its number of copies. We can now use (14.13) to evaluate the performance of a multisensor classification system.

Suppose we have a total of r sensors $S_1, S_2, \ldots, S_r$, with sensor s having reported k_s measurements—can we estimate the value of $|\mathbf{G}_k|$ obtained after fusing all these measurements? Observe that fusing k_s measurements from each sensor s is equivalent to taking *just one measurement from an equivalent sensor constructed from k_s copies of sensor s*. Thus, the estimate of $|\mathbf{G}_k|$ obtained from fusing k_s measurements from sensor s, for $s = 1, 2, \ldots, r$ is

$$\begin{aligned} |\mathbf{G}_{k_1,k_2\ldots,k_r}| &\approx 1 - \exp\left(-C_{k_1,k_2,\ldots,k_r}\right) \\ C_{k_1,k_2,\ldots,k_r} &= \frac{D_{k_1,k_2,\ldots,k_r}}{n + \frac{(n-2)}{(n-1)} \frac{1}{D_{k_1,k_2,\ldots,k_r}}}. \end{aligned} \tag{14.14}$$

Table 14.3 reports the results of the approximation in (14.14) compared with the estimate for $|\mathbf{G}_{k_1,k_2}|$ obtained from Monte Carlo simulations averaged over 150 runs, for different values of k_1 and k_2, for the two sensors of Example 2, using a prior of $\left[\frac{1}{3}, \frac{2}{3}\right]$. The values in Table 14.3 include those computed using the approximation $1 - \exp(-C_{k_1,k_2})$, followed by the simulation results shown within $(\cdot)$. As the values indicate, the approximation is quite good.

The impact of (14.14) is that it allows us to quantify the performance of a multisensor fusion system off-line, and the accuracy of the estimate of the target's identity based on the number of measurements that have been reported thus far. This is very

TABLE 14.3 Example 2: Approximation vs. Simulation

	k_2			
k_1	5	25	50	100
5	0.4008	0.6859	0.8559	0.9721
	(0.4158)	(0.6776)	(0.8690)	(0.9558)
25	0.8527	0.9228	0.9656	0.9931
	(0.8556)	(0.9223)	(0.9646)	(0.9833)
50	0.9745	0.9866	0.9940	0.9988
	(0.9687)	(0.9785)	(0.9967)	(0.9986)
100	0.9992	0.9996	0.9998	1.000
	(0.9997)	(0.9967)	(1.000)	(0.9997)

important for real-time decision-making and inference—for example, incoming hostile missiles are usually monitored by many sensors, each of which report outputs asynchronously, and which must all be fused to estimate posterior probabilities of the missile's identity, which is then acted upon for deciding the response. Comparing with the Kalman filter (KF) equations, (14.14) is analogous to the calculation of the updated covariance matrix in a KF, for computing off-line estimates of the classification accuracy.

14.3.2 Performance Evaluation of Multisensor Fusion Systems—Not-so-Good Sensors

We have thus far focused on quantifying the performance of good sensors, the rows of whose LCMs are all different. What if we have a set of sensors, each of which is selective in the target types that they identify well, but all of which are required to identify all possible targets? Consider the following two sensors whose LCMs, and long-run GCMs, are shown below.

$$A_1 = \begin{pmatrix} .1 & .7 & .1 & .1 \\ .5 & .1 & .3 & .1 \\ .3 & .4 & .2 & .1 \\ .3 & .4 & .2 & .1 \end{pmatrix} \qquad \mathbf{G}_\infty = \begin{pmatrix} 1 & 0 & 0 & 0 \\ 0 & 1 & 0 & 0 \\ 0 & 0 & \dfrac{P_0(t_3)}{P_0(t_3)+(P_0 t_4)} & \dfrac{P_0(t_4)}{P_0(t_3)+P_0(t_4)} \\ 0 & 0 & \dfrac{P_0(t_3)}{P_0(t_3)+P_0(t_4)} & \dfrac{P_0(t_4)}{P_0(t_3)+P_0(t_4)} \end{pmatrix}$$

$$A_2 = \begin{pmatrix} .4 & .2 & .3 & .1 \\ .4 & .2 & .3 & .1 \\ .5 & .1 & .3 & .1 \\ .1 & .7 & .1 & .1 \end{pmatrix} \qquad \mathbf{G}_\infty = \begin{pmatrix} \dfrac{P_0(t_1)}{P_0(t_1)+P_0(t_2)} & \dfrac{P_0(t_2)}{P_0(t_1)+P_0(t_2)} & 0 & 0 \\ \dfrac{P_0(t_1)}{P_0(t_1)+P_0(t_2)} & \dfrac{P_0(t_2)}{P_0(t_1)+P_0(t_2)} & 0 & 0 \\ 0 & 0 & 1 & 0 \\ 0 & 0 & 0 & 1 \end{pmatrix}.$$

Observe that sensors S_1 and S_2 are incapable of distinguishing the targets $\{t_3, t_4\}$, and $\{t_1, t_2\}$, respectively, and so the rows corresponding to these targets in their GCMs converge to the ratio of their priors. In such a case, (14.12) can be used to compute the convergence rate of either sensor for just the targets they are capable of classifying, but with a LCM restricted to just these targets for computing the D metric. In the above example, the D metric for sensor 1 will be computed for a 2×4 matrix restricted to the first two rows of A_1, whose D value is 0.7285 ($= D_1$), and likewise, for sensor 2, using a 2×4 matrix restricted to the last two rows of A_2, whose D value is 0.1640 ($= D_2$). Using (14.12), we obtain $C_1 = 0.3642$ and $C_2 = 0.0820$. Consequently, the two sensors contribute independently to the first two and the last two rows of the GCM, resulting in a block-diagonal structure that will eventually converge to the identity matrix. In this case, the estimate of $|\mathbf{G}|$ obtained after fusing, say, k_1 and k_2 measurements from both sensors, would be approximated as

$$|\mathbf{G}_{k_1,k_2}| \approx [1 - \exp(-C_1 k_1)] * [1 - \exp(-C_2 k_2)]. \tag{14.15}$$

Note, however, that in the example presented above, both sensors were disjoint as regards the targets they were individually capable of classifying, and so, the approximation of (14.15) effectively captures the contributions to the overall determinant from each block of the GCM. The case when we have one/more "poor" sensors which are not disjoint in the sets of targets they are capable of classifying, is more interesting, and requires further research. Likewise, it still requires further research to translate these ideas into a principled approach for evaluating multisensor fusion systems involving "not-so-good" sensors.

14.4 SUMMARY AND CONTINUING QUESTIONS

This chapter addressed the problem of evaluating the classification performance of both single- and multisensor classification systems, where one or more sensors observe repeated measurements of a target's features/attributes and compute posterior probability estimates to aid in target identification. Unlike state-estimation problems, where estimates of kinematic tracking performance are monitored by the covariance matrix, an equivalent discrete analog for classification performance has not been done previously. We have developed a new analytical approach for off-line evaluation of the long-run classification performance of a single- or a multisensor system for the case of independent sensor measurements.

We define a new approach to quantify the classification performance based on the GCM of a classification system. The GCM is analogous to the covariance matrix used in kinematic performance evaluation and is an off-line measure that does not depend on the actual sensor measurements. The rate of convergence of the GCM to the identity matrix is indicative of the ability of a sensor to perfectly classify all target types. It is interesting to note that the only requirement for a sensor to be able to perfectly classify all target types is that the rows of its LCM be different.

Our contributions include (i) an efficient approach for computing the GCM of a sensor, (ii) the D_s metric to quantify a sensor's ability to discriminate between the targets being classified, based on the KL distances between the rows of the sensor's LCM, and its use in comparing multiple sensors, (iii) an approximate formula based on the D_s metric to compute off-line estimates of the rate of convergence toward perfect classification, and the number of measurements required to achieve a desired level of classification accuracy, and (iv) applications of the D_s metric to evaluate the classification performance of multisensor systems. We have also presented a methodology for evaluating the classification performance of multisensor systems that involve sensors of varying quality, each reporting independent measurements—this would aid in the design of a system that is capable of classifying more target types than what each individual sensor is capable of; our results are restricted to the case when the sensors are mutually disjoint in the classes/types of targets they monitor; further research is required to address the case when we have multiple sensors each with capabilities to monitor a subset of the targets, not necessarily disjoint. We contend that a possible solution in this case is to selectively allocate the set of all targets in disjoint clusters to each sensor, such that each sensor is responsible for a subset of the targets, in such a way that no other sensor need track any of its allotted targets. However, the question remains—how do we decide upon such an allocation?

There are still many questions that need to be addressed to fully realize the potential for quantifying the classification performance of a system, and in turn, influencing the choice among sensors, to better adapt to the needs of the problem. The success of the results derived in this chapter relies on the reduction of sensor-target dependency models (or other Bayesian-network representations), to an equivalent single-sensor model, as in Figure 14A.1a, each of whose measurements are temporally independent, and a general approach to handle multiple sensors. It still remains to find a tractable approach for solving sensor models wherein the LCM for the sensor *varies* with each measurement, as would be the case for dependences among measurements arising through time-varying features, as in Figure 14A.1e. We are working to generalize the results of this study by introducing a correlation coefficient between sensor measurements (equal to zero for independent measurements) and incorporating its use in approximations analogous to (14.12), and validating the same with the use of particle-filter-based sequential Monte Carlo simulation studies [15–17].

APPENDIX 14A DEVELOPING A SENSOR'S LOCAL CONFUSION MATRIX

Classification systems are typically designed with sensors observing one/more features of the target than the target type itself. A Bayesian-network representation of a simple sensor-target dependency model is illustrated in Figure 14A.1a, where $\mathbf{F}$ could be a single feature or a set of features specified as a tuple variable, and Y is a sensor output that could either be some $f \in \mathbf{F}$ or $t \in \mathbf{T}$, or some other measurement, all of which is compactly represented by specifying that the output of sensor s be one of $o_1^s, o_2^s, \ldots, o_{m_s}^s$. Figures 14A.1b–e[2] illustrate other possible sensor-target dependency models involving multiple sensors—we now present rules for reducing all these models to the single-sensor model of Figure 14A.1a, by creating a single equivalent sensor, whose LCM can be specified with the set of targets as its domain.

Let us assume that the sensors are indexed S_1, S_2, etc., whose measurements are indexed Y_1, Y_2, etc., and that the conditional distributions for sensor s are specified as $P(o_j^s|\mathbf{pa}(S_s))$ for all $o_j^s \in O^s$ and all combinations of the features that are **par**ents of sensor s.

Single-sensor model of Figure 14A.1a where we have a single sensor that observes a (set of) feature(s) $f \in \mathbf{F}$ and reports an output measurement $o_j \in O$ for some target $t_i \in \mathbf{T}$. In this case, the LCM is $A = [a_{ij}]_{n\times m}$, where $a_{ij} = P(Y = o_j|t_i) = \sum_{f\in\mathbf{F}} P(o_j|f)P(f|t_i)$.

[2]The case in Figure 14A.1e is that of a first-order Markovian dependency among features (static/dynamic) F_1 and F_2–the extension to higher order models is similar.

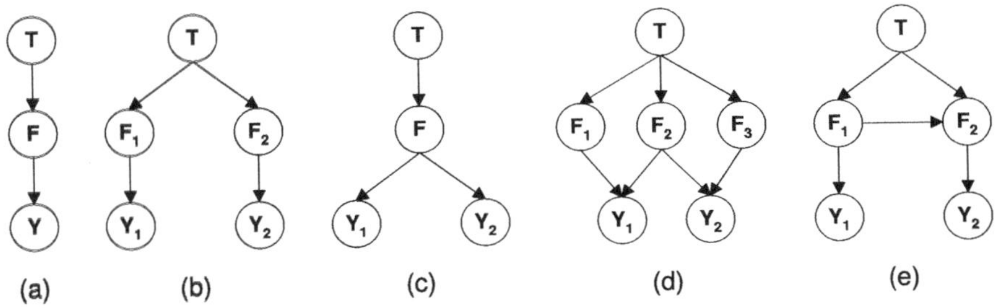

Figure 14A.1 Dependency models between sensors and features in target classification.

Two-sensor model of Figure 14A.1b where two disjoint sets of features are monitored by separate sensors. In this case, the conditional distributions of the two sensors S_1 and S_2 need to be combined to create a single equivalent sensor S with output range $O^1 \times O^2$, and whose LCM is constructed using the relation $P(y_1, y_2|t_i) = \sum_{f_1} P(y_1|f_1)P(f_1|t_i) \sum_{f_2} P(y_2|f_2)P(f_2|t_i)$.

Two-sensor model of Figure 14A.1c where the same set of features is monitored by different sensors (as would be the case in multisensor classification systems—this is discussed further in Section 14.3). As in case (b) above, an equivalent sensor S can be constructed with output range $O \times O$, and whose LCM is constructed using the relation $P(y_1, y_2|t_i) = \sum_f P(y_1|f)P(y_2|f)P(f|t)$, following which the model reduces to that of case (a).

Two-sensor model of Figure 14A.1d where the set of features monitored by the sensors is not disjoint. However, it can be easily reduced to the model of case (c) above, by aggregating the three features F_1, F_2, F_3, into a 3-tuple F [11].

Two-sensor model of Figure 14A.1e where we have two sensors observing disjoint features F_1 and F_2, but where one of them is dependent on the other. This could be a model either of (i) two static features F_1 and F_2 observed by different sensors, or (ii) a single dynamic feature that is observed at different time steps by the same sensor.

In case (i), an equivalent sensor S with output range $O^1 \times O^2$ is constructed, and whose LCM is constructed using the relation

$$P(y_1, y_2|t_i) = \sum_{f_1 \in \mathbf{F_1}} P(y_1|f_1)P(f_1|t_i) \sum_{f_2 \in \mathbf{F_2}} P(y_2|f_2)P(f_2|f_1, t_i). \quad (14A.1)$$

For case (ii), the general form of (14A.1) is exact, and will have to be extended to include as many time steps as are required—a recursive formulation as presented in [18] will need to be derived. The challenge is not in modeling multiple sensors (refer Section 14.3), but in modeling a sensor whose LCM differs for each measurement, and is influenced by previous measurement(s). This is a question meriting further research and is addressed in Section 14.4. It is our belief that all sensor models can be shown to be comprised of the five basic structures illustrated in Figure 14A.1, and that by repeated application of the rules described above, they can be reduced to an equivalent single-sensor model as in Figure 14A.1a, which can be used as the basis for studying the long-run classification accuracy of the system.

APPENDIX 14B SOLVING FOR THE OFF-DIAGONAL ELEMENTS OF THE GLOBAL CLASSIFICATION MATRIX

We now present a systematic approach to solve for the off-diagonal elements of the GCM, as derived in Section 14.2.1, and expressed in (14.6)—the equation is repeated below

$$1 - g_{ii}^k = \sum_{j=i+1}^{n} g_{ij}^k + \sum_{j=1}^{i-1} g_{ij}^k = \sum_{j=i+1}^{n} g_{ij}^k + \sum_{j=1}^{i-1} \left[\frac{P_0(t_j)}{P_0(t_i)} \right] g_{ji}^k. \qquad (14B.1)$$

The solution of (14B.1) will require a few additional definitions, namely,

- The total number of unknown off-diagonal upper triangular elements of **G** is equal to $n(n-1)/2$, namely, $\{g_{12}, g_{13}, \ldots, g_{1n}, g_{23}, \ldots, g_{n-1,n}\}$. This may be represented more compactly as a $\frac{n(n-1)}{2} \times 1$ vector $\mathcal{U} = [u_1|u_2|\cdots|u_{n-1}]^T$, where $u_i = [g_{i,i+1}, g_{i,i+2}, \ldots, g_{i,n}]$ for $i = 1, 2, \ldots, (n-1)$.
- We define an $n \times 1$ vector b containing the L.H.S. of (14.6), that is, $[b]_i = 1 - g_{ii}$. An efficient approach to calculating the diagonal elements (g_{ii}) was derived in Section 14.2.1.
- An $n \times n$ matrix Id will be constructed as follows

```
# Initialize Id to be all zeros
Id ← zeros(n, n)
count ← 1
for row = 1 to n − 1
    for col = row + 1 to n
        Id(row, col) = Id(col, row) = count
        count ← count + 1
```

The matrix Id is essentially a symmetric matrix with zero diagonals, and symmetric upper and lower triangular halves that store the indices for another matrix **S** defined below.

- We now construct an $n \times \frac{n(n-1)}{2}$ matrix $\mathbf{S}$ as follows:

```
# Initialize S to be all zeros
S ← zeros(n, n(n−1)/2)
for row = 1 to n
    # Navigate upper triangular elements of Id
    for col = row + 1 to n
        S(row, Id(row, col)) = 1
    # Navigate lower triangular elements of Id
    for col = 1 to row − 1
        S(row, Id(row, col)) = P(t_col)/P(t_row)
```

Based on the above definitions, (14.6) can now be expressed compactly as $\mathbf{S}\mathcal{U} = b$. The solution of (14.6) follows readily as $\mathcal{U} = \mathbf{S}^{+}b$, where $\mathbf{S}^{+}$ is the generalized pseudo-inverse of $\mathbf{S}$ [19]. Once the upper-diagonal elements of $\mathbf{G}$ have been computed, computing the lower-diagonal elements follows directly from Theorem 1.

Example 3 The technique derived above is best illustrated with an example. Assume that there are $n = 5$ objects to be classified. It follows that $\mathcal{U} = [u_1|u_2|u_3|u_4]^T$, where

$$u_1 = \begin{pmatrix} g_{12} \\ g_{13} \\ g_{14} \\ g_{15} \end{pmatrix}, \qquad u_2 = \begin{pmatrix} g_{23} \\ g_{24} \\ g_{25} \end{pmatrix}, \qquad u_3 = \begin{pmatrix} g_{34} \\ g_{35} \end{pmatrix}, \qquad u_4 = \begin{pmatrix} g_{45} \end{pmatrix}.$$

The other elements of the problem are

$$Id = \begin{pmatrix} 0 & 1 & 2 & 3 & 4 \\ 1 & 0 & 5 & 6 & 7 \\ 2 & 5 & 0 & 8 & 9 \\ 3 & 6 & 8 & 0 & 10 \\ 4 & 7 & 9 & 10 & 0 \end{pmatrix}, \qquad b = \begin{pmatrix} 1 - g_{11} \\ 1 - g_{22} \\ 1 - g_{33} \\ 1 - g_{44} \\ 1 - g_{55} \end{pmatrix}, \text{ and}$$

$$\mathbf{S} = \begin{pmatrix} 1 & 1 & 1 & 1 & 0 & 0 & 0 & 0 & 0 & 0 \\ \frac{P(t_1)}{P(t_2)} & 0 & 0 & 0 & 1 & 1 & 1 & 0 & 0 & 0 \\ 0 & \frac{P(t_1)}{P(t_3)} & 0 & 0 & \frac{P(t_2)}{P(t_3)} & 0 & 0 & 1 & 1 & 0 \\ 0 & 0 & \frac{P(t_1)}{P(t_4)} & 0 & 0 & \frac{P(t_2)}{P(t_4)} & 0 & \frac{P(t_3)}{P(t_4)} & 0 & 1 \\ 0 & 0 & 0 & \frac{P(t_1)}{P(t_5)} & 0 & 0 & \frac{P(t_2)}{P(t_5)} & 0 & \frac{P(t_3)}{P(t_5)} & \frac{P(t_4)}{P(t_5)} \end{pmatrix}.$$

The solution is now very clear, namely, $\mathcal{U} = \mathbf{S}^{+}b$.

APPENDIX 14C A GRAPH-THEORETIC REPRESENTATION OF THE RECURSIVE APPROACH FOR ESTIMATING THE DIAGONAL ELEMENTS OF THE GCM

This section presents a graph-theoretic approach for the efficient representation of the elements involved in the recursion of (14.10). We introduce the general structure of our results by illustrating the recursion of (14.10) for the simple binomial case ($n = |\mathbf{T}| = 2, m = |O| = 2$), and use it to draw parallels with the general case ($n, m > 2$).

Appendix 14C.1 The Binomial Case ($n = 2, m = 2$)

Suppose there are only two targets to be classified (indexed as t_1 and t_2), and the sensor reports one of the two possible outputs o_1, o_2 (indexed as 1 and 2) for each measurement. In this case, for $i = 1, 2$, (14.8) simplifies as

$$g_{ii}^k = \sum_{\alpha_1=0}^{k} \binom{k}{\alpha_1} \left[\frac{P_0(t_i)(a_{i1})^{2\alpha_1}(a_{i2})^{2(k-\alpha_1)}}{P_0(t_1)(a_{11})^{\alpha_1}(a_{12})^{k-\alpha_1} + P_0(t_2)(a_{21})^{\alpha_1}(a_{22})^{k-\alpha_1}} \right], \quad (14C.1)$$

where $P_0(i)$ stands for the prior probability that the target is of type i (either t_1 or t_2), α_1 is the number of times the sensor reports output 1 (that is, o_1) in the k measurement readings, and the a_{ij}s are the elements of the sensor's 2×2 LCM. Observe that in contrast with the 2^k summations that would be required for (14.7), (14C.1) requires only $(k + 1)$ summations. We now reformulate (14C.1) as $g_{ii}^k = \sum_{\alpha_1=0}^{k} g_{ii,\alpha_1}^k$, for $i = 1, 2$, where

$$g_{ii,\alpha_1}^k = \binom{k}{\alpha_1} \frac{P_0(t_i)(a_{i1})^{2\alpha_1}(a_{i1})^{2(k-\alpha_1)}}{P_0(t_1)(a_{11})^{\alpha_1}(a_{12})^{k-\alpha_1} + P_0(t_2)(a_{21})^{\alpha_1}(a_{22})^{k-\alpha_1}}.$$

Based on this definition, it readily follows that

$$\frac{g_{ii,\alpha_1}^{k+1}}{g_{ii,\alpha_1+1}^{k+1}} = \left(\frac{\alpha_1 + 1}{k + 1 - \alpha_1}\right) \left(\frac{a_{i2}}{a_{i1}}\right)^2 \times \left[\frac{P_0(t_1)(a_{11})^{\alpha_1+1}(a_{12})^{k-\alpha_1} + P_0(t_2)(a_{21})^{\alpha_1+1}(a_{22})^{k-\alpha_1}}{P_0(t_1)(a_{11})^{\alpha_1}(a_{12})^{k+1-\alpha_1} + P_0(t_2)(a_{21})^{\alpha_1}(a_{22})^{k+1-\alpha_1}} \right]. \quad (14C.2)$$

The result of (14C.2) is structurally similar to (14.9). The binomial equivalent of the recursion of (14.10) is that the diagonal elements of a sensor's GCM may be recursively computed as

$$\frac{1}{g_{ii,\alpha_1}^{k+1}} = \frac{(k + 1 - \alpha_1)}{(a_{i2})^2} \left[\frac{1}{(k + 1)} \frac{1}{g_{ii,\alpha_1}^k} - \frac{(a_{i1})^2}{(\alpha_1 + 1)} \frac{1}{g_{ii,\alpha_1+1}^{k+1}} \right] \quad (14C.3)$$

for $\alpha_1 = 0, 1, \ldots, k$, where $g_{ii}^{k+1} = \sum_{\alpha_1=0}^{k+1} g_{ii,\alpha_1}^{k+1}$. The initial condition for the recursion is

$$g_{ii,k+1}^{k+1} = \frac{P_0(t_i)a_{i1}^{2(k+1)}}{P_0(t_1)a_{11}^{k+1} + P_0(t_2)a_{21}^{k+1}}.$$

The other values for g_{ii,α_1}^{k+1} are then estimated by setting $\alpha_1 = k, k-1, \ldots, 0$ in (14C.3).

The proof of Theorem 2 is demonstrated for the binomial case below, by simplifying g_{ii,α_1}^{k+1}

$$\begin{aligned}
g_{ii,\alpha_1}^{k+1} &= \binom{k+1}{\alpha_1} \frac{P_0(t_i)(a_{i1})^{2\alpha_1}(a_{i2})^{2(k+1-\alpha_1)}}{P_0(t_1)(a_{11})^{\alpha_1}(a_{12})^{k+1-\alpha_1} + P_0(t_2)(a_{21})^{\alpha_1}(a_{22})^{k+1-\alpha_1}} \\
&= (a_{i2})^2 \binom{k+1}{\alpha_1} \left[\frac{P_0(t_i)a_{i1}^{2\alpha_1}a_{i2}^{2(k-\alpha_1)}}{P_0(t_1)a_{11}^{\alpha_1}a_{12}^{k-\alpha_1} + P_0(t_2)a_{21}^{\alpha_1}a_{22}^{k-\alpha_1}} \right] \times \\
&\qquad \left[\frac{P_0(t_1)a_{11}^{\alpha_1}a_{12}^{k-\alpha_1} + P_0(t_2)a_{21}^{\alpha_1}a_{22}^{k-\alpha_1}}{P_0(t_1)a_{11}^{\alpha_1}a_{12}^{k+1-\alpha_1} + P_0(t_2)a_{21}^{\alpha_1}a_{22}^{k+1-\alpha_1}} \right] \\
&= (a_{i2})^2 \frac{(k+1)}{(k+1-\alpha_1)} g_{ii,\alpha_1}^{k} \left[\frac{P_0(t_1)a_{11}^{\alpha_1}a_{12}^{k+1-\alpha_1} + P_0(t_2)a_{21}^{\alpha_1}a_{22}^{k+1-\alpha_1}}{P_0(t_1)a_{11}^{\alpha_1}a_{12}^{k+1-\alpha_1} + P_0(t_2)a_{21}^{\alpha_1}a_{22}^{k+1-\alpha_1}} + \right. \\
&\qquad \left. \frac{P_0(t_1)a_{11}^{\alpha_1+1}a_{12}^{k-\alpha_1} + P_0(t_2)a_{21}^{\alpha_1+1}a_{22}^{k-\alpha_1}}{P_0(t_1)a_{11}^{\alpha_1}a_{12}^{k+1-\alpha_1} + P_0(t_2)a_{21}^{\alpha_1}a_{22}^{k+1-\alpha_1}} \right] \\
&= (a_{i2})^2 \frac{(k+1)}{(k+1-\alpha_1)} g_{ii,\alpha_1}^{k} \left[1 + \frac{(k+1-\alpha_1)}{(\alpha_1+1)} \left(\frac{a_{i1}}{a_{i2}} \right)^2 \frac{g_{ii,\alpha_1}^{k+1}}{g_{ii,\alpha_1+1}^{k+1}} \right].
\end{aligned}$$

The last term in the second equality above has been modified by expanding the first and the second term in its numerator with the coefficients $a_{11} + a_{12}(= 1)$ and $a_{21} + a_{22}(= 1)$, respectively, to get the two fractions in the last term of the third equality, and we have used (14C.2) in the last equality. The final result follows from simplifying the expression below, which is merely a rearrangement of the last equality above

$$g_{ii,\alpha_1}^{k+1} \left[1 - \frac{(k+1)}{(\alpha_1+1)} (a_{i1})^2 \frac{g_{ii,\alpha_1}^{k}}{g_{ii,\alpha_1+1}^{k+1}} \right] = a_{i2}^2 \frac{(k+1)}{(k+1-\alpha_1)} g_{ii,\alpha_1}^{k}.$$

We now discuss the implementation of the recursion presented above for the binomial case ($n = m = 2$). The key to the recursive relation lies in the fact that there are a total of $(k+1)$ terms in the summation $g_{ii}^k = \sum_{\alpha_1=0}^{k} g_{ii,\alpha_1}^k$, corresponding to $\alpha_1 = 0, 1, 2, \ldots, k$, as opposed to a total of $(k+2)$ terms in the summation for $g_{ii}^{k+1} = \sum_{\alpha_1=0}^{k+1} g_{ii,\alpha_1}^{k+1}$, the extra term accruing from the term $g_{ii,k+1}^{k+1}$ contributed by

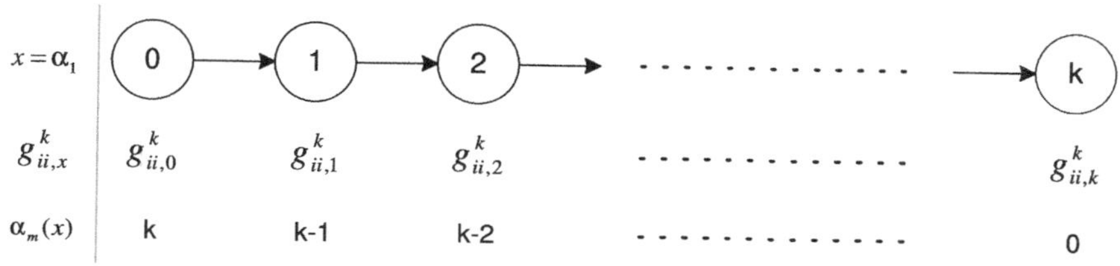

Figure 14C.1 Connected tree representation of the summation $g_{ii}^k = \sum_{\alpha_1=0}^{k} g_{ii,\alpha_1}^k$.

$\alpha_1 = k + 1$. It is this extra term that is required as a boundary condition for the binomial version of Theorem 2, whereupon, we can now compute g_{ii,α_1}^{k+1} as a function of g_{ii,α_1+1}^{k+1} and g_{ii,α_1}^k, by working backward from $\alpha_1 = k, k-1, k-2, \ldots, 0$. We now describe a procedure for programming this recursion using a graph-theoretic representation of the terms in the summation, the use of which will become apparent in the general case $(n, m > 0)$.

Suppose we construct a connected tree $T_m(k)$ with vertices numbered $0, 1, 2, \ldots, k$, corresponding to the values taken by α_1 in the summation for g_{ii,α_1}^k, with each vertex being associated with two values, namely, g_{ii,α_1}^k, and $\alpha_m(x) = (k - \alpha_1)$, and the values stored in the latter term corresponding to the value of $\alpha_m = \alpha_2$ ($m = 2$ in the binomial case). An illustration of the connected tree representation for the binomial case is shown in Figure 14C.1.

More formally, let us define $T_m(k) = (V, E)$, where the vertex set, $V = \{0, 1, 2, \ldots, k\}$, and the edge set, $E = \{(x, y) : x, y \in V, y = x + 1\}$, that is, we connect vertices in the ascending order of their magnitude. We will abuse our notation slightly in using x to indicate both the vertex $x \in V$, and the value of α_1 that vertex x represents, and use $x + 1$ to indicate the child of vertex x. We attach to each vertex $x \in V$, the values $g_{ii,x}^k$ and $\alpha_m = (k - x)$. A graphical depiction of the steps involved in implementing the recursion of (14C.3) is shown in Figure 14C.2.

The binomial version of the recursion of Theorem 2, as presented in (14C.3), essentially updates the tree $T_m(k)$ (with $k + 1$ vertices) to create a new tree $T_m(k + 1)$, each of whose $k + 2$ vertices correspond to the terms in the summation $g_{ii}^{k+1} = \sum_{\alpha_1=0}^{k+1} g_{ii,\alpha_1}^{k+1}$. The key point to note is that the initial condition (i.e., the new node added to $T_m(k + 1)$) corresponds to a value of $\alpha_m = \alpha_2 = 0$, and that the $\alpha_m(\cdot)$ for all the other nodes are all just incremented by one (Step 3 in Figure 14C.2)—this basic idea is generalized for the general case discussed next.

Appendix 14C.2 The Multinomial Case $(n, m > 2)$

We now generalize the graph-theoretic approach presented above for the general case of $n, m > 2$ to efficiently represent the elements involved in the recursion of (14.9). As before, we define $T_m(k)$ to be the connected tree representation of the terms in the summation for $g_{ii}^k = \sum_{\alpha_m=0}^{k} \sum_{\forall \vec{\alpha} : \vec{\alpha}_{\oplus}=k-\alpha_m} g_{ii,\vec{\alpha}}^k$. However, the vertices will not be mere integers like in the binomial case, but, vectors $\vec{\alpha} = (\alpha_1, \alpha_2, \ldots, \alpha_{m-1})$. More

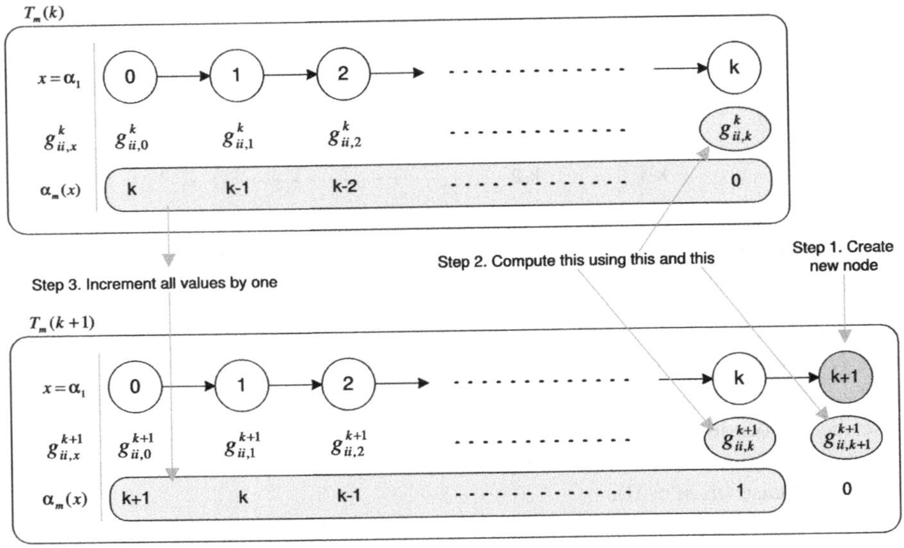

Figure 14C.2 Graphical description of the recursion of (14C.3).

formally, $T_m(k) = (V, E)$, where the vertex set $V = \{\vec{\alpha} : \vec{\alpha}_\oplus = j; j = 0, 1, 2, \ldots, k\}$, and the edge set $E = \{(x, y) : x, y \in V, y = x + \delta_p; p = 1, 2, \ldots, m-1\}$. Additionally, we attach to each vertex $x \in V$ the values $g^k_{ii,x}$ and $\alpha_m(x) = k - \vec{x}_\oplus$. An illustration of the connected tree representation is presented in Figure 14C.3, for $m = 4$ and $k = 2$; observe that the vertices have been ordered ancestrally in the order of decreasing values of $\alpha_m(x)$ – in fact, this is the order in which the tree will grow from k to $k+1$ with the addition of leaf nodes at the end. Much like the binomial case, Figure 14C.4 illustrates the implementation of the recursion in Theorem 2 for the example in Figure 14C.3.

The reader may question the utility of such representations, and the need for the same. Consider for example, the total number of summations that would be required for computing g^k_{ii} for $k = 1, 2, \ldots, K$, for some integer K. An implementation of

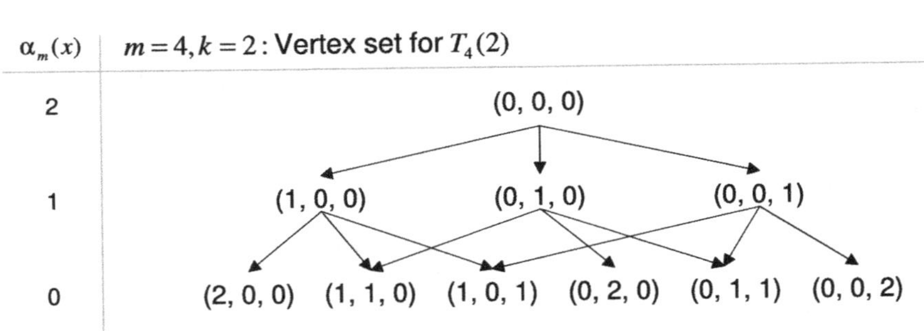

Figure 14C.3 Connected tree representation for the multinomial case.

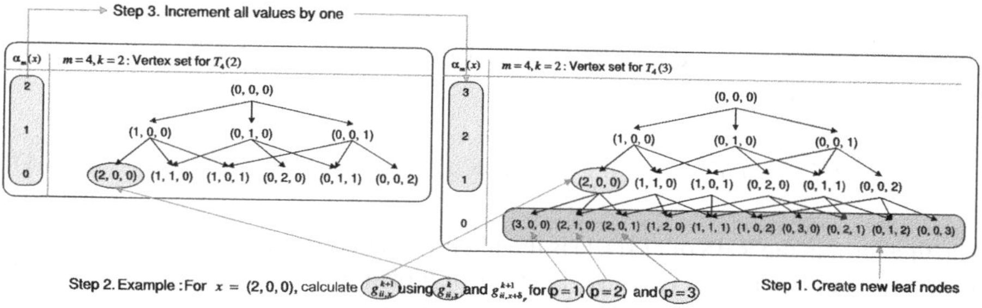

Figure 14C.4 Graphical description of the recursion in Theorem 2.

(14.8) would require a total $\sum_{k=1}^{K}\binom{k+m-1}{k}$ summations for computing all the values of $\{g_{ii}^k\}_1^K$. In contrast with this, the graph-theoretic implementations of (14C.3) and Theorem 2 would require only $\binom{K+m-1}{K}$ summations in total, owing to the recursive implementation that adds only as many leaf nodes to the tree as are required to compensate for the extra terms in the summation for g_{ii}^{k+1} versus g_{ii}^k, thus offering extraordinary computational efficiency.

APPENDIX 14D DESIGNING MONTE CARLO SIMULATIONS OF THE GCM

We now describe an approach to simulate the GCM, for both single- and multisensor systems, to support the discussion in Sections 14.2.3 and 14.3.

Appendix 14D.1 Single-Sensor GCM

Recall that we begin with a sensor's LCM $A = [a_{ij}]_{n\times m}$, and the prior probability distribution $P_0(T = t_i)$ for each target $t_i \in \mathbf{T}$. In order that we may begin to simulate the GCM, it is useful to recall that any row, say, the ith row, of the GCM, signifies the estimates of the posterior probabilities for each target, averaged over all possible sequences of sensor measurements, given that t_i was the true target—this was the basis of the formulations in (14.7) and (14.8). How can we avoid this enumeration and instead use simulation to give us an approximation of $|\mathbf{G}_k|$, perhaps averaged over many runs to improve the approximation?

We now describe a simple approach to generate a single realization of values for $|\mathbf{G}_k|$, for $k = 1, 2, \ldots, K$—an estimate of $|\mathbf{G}_k|$, averaged over several such realizations (i.e., simulation runs), can then be computed to compare with the exact results of (14.8) or the approximations of (14.12), an example of which is presented in Figure 14.4a. In the discussion that follows, we will refer to the ith rows of $\mathbf{G}_k$ and A as $\mathbf{G}_{k\{i.\}}$ and $A_{\{i.\}}$, and the ijth elements as $\mathbf{G}_{k\{ij\}}$ and $A_{\{ij\}}$ respectively.

Suppose the actual target is t_i—what is the probability distribution governing the sensor's outputs? Clearly, it is the values in the ith row of the sensor's LCM, that is,

$A_{\{i,\}}$. Suppose we draw a sensor measurement $y = o_j$ with $y \sim A_{\{i,\}}$ and $o_j \in O$. We can now compute the posterior probabilities for each target type $t \in \mathbf{T}$, given that the true target is t_i, that is,

$$P\left(T^{\text{pred}} = t | T^{\text{actual}} = t_i\right) \propto P(y = o_j | T = t) \times P(T = t). \qquad (14D.1)$$

The first term in the RHS of (14D.1) is the likelihood, derived from the sensor's LCM, and is equal to a_{tj}. The second term is the prior probability that the target is t—recall that we began with the assumption that the true target is t_i, whereupon the current value of the prior probability for target t is obtained from the ith row of the GCM. This is the essence of the idea underlying the simulation, namely, to update the estimates of the posterior probabilities in the ith row of the GCM, and to repeat this for all the rows, that is, assuming that the true target is t_i, for all $i = 1, 2, \ldots, n$.

More generally, given $\mathbf{G}_k$, how do we estimate $\mathbf{G}_{k+1}$? The description above can be formalized thus: for $i = 1, 2, \ldots, n$, (i) simulate a sensor measurement $y = o_j$, with $y \sim A_{\{i,\}}$, (ii) compute $\mathbf{G}_{k+1\{it\}} \propto a_{tj}\mathbf{G}_{k\{it\}}$. The initial value $\mathbf{G}_0$ of the GCM is obtained by setting each of its rows equal to the target's prior probability distribution, that is, $g_{ij} = P\{T = t_j\}, \forall i = 1, 2, \ldots, n$. A single simulation run of length K will thus generate a sequence $\{|\mathbf{G}_k|\}_1^K$, and multiple runs can be averaged to provide Monte carlo estimates of $\{|\mathbf{G}_k|\}_1^K$.

Appendix 14D.2 Multisensor GCM

The approach to computing a Monte Carlo estimate of the GCM obtained after fusing measurements from multiple sensors is identical to that used for a single sensor. Suppose we have two sensors, and we wish to simulate $\mathbf{G}_{k_1 \cdot k_2}$, the GCM obtained after fusing k_1 and k_2 measurements from sensors 1 and 2—since the order in which the measurements are fused is immaterial to the final result, we compute $\mathbf{G}_{k_1}$ for the first sensor (as described above in Appendix 14D.1); we then use this as the prior estimate of the GCM for the second sensor, and simulate another k_2 measurements from the second sensor to obtain an estimate of $\mathbf{G}_{k_1 \cdot k_2}$. The technique readily generalizes for the calculation of $\mathbf{G}_{k_1 \cdot k_2 \cdots k_r}$. The idea is essentially identical to that used for a single-sensor, with the only difference being that while each run of the simulation for a single sensor generates a sequence $\{|\mathbf{G}_k|\}_1^K$, a single run of the simulation for the multisensor model will only generate a single estimate of $\mathbf{G}_{k_1 \cdot k_2 \cdots k_r}$.

APPENDIX 14E PROOF OF APPROXIMATION 1

We now discuss the derivation of inequality 1, discussed in Section 14.3.1, and repeated below:

Inequality 1: Consider two sets of probability distributions $\{P_i^1, Q_i^1\}_{i=1}^{m_1}$ and $\{P_j^2, Q_j^2\}_{j=1}^{m_2}$ with supports of sizes m_1 and m_2, and with symmetric KL distances $KL^1_{P^1,Q^1} = (KL^1_{P^1||Q^1} * KL^1_{Q^1||P^1})/(KL^1_{P^1||Q^1} + KL^1_{Q^1||P^1})$ and $KL^2_{P^2,Q^2}$.

Let $P^{1,2} = P^1 \otimes P^2$ and $Q^{1,2} = Q^1 \otimes Q^2$ be the probability distributions obtained though the corresponding Kronecker products. Then, the symmetric KL distance between $P^{1,2}$ and $Q^{1,2}$ is

$$KL^{1,2}_{P^{1,2},Q^{1,2}} \approx KL^{1}_{P^{1},Q^{1}} + KL^{2}_{P^{2},Q^{2}}.$$

This approximation is very accurate, with near negligible margin of error.

We first show that the asymmetric KL distance between the Kronecker products, $KL^{1,2}_{P^{1,2}||Q^{1,2}}$ and $KL^{1,2}_{Q^{1,2}||P^{1,2}}$, is the sum of the individual asymmetric KL distances. Given two probability distributions defined as above, it follows that

$$\begin{aligned}
KL^{1,2}_{P^{1,2}||Q^{1,2}} &= \sum_{i=1}^{m_1}\sum_{j=1}^{m_2} P_i^1 P_j^2 \log_2\left(\frac{P_i^1 P_j^2}{Q_i^1 Q_j^2}\right)\\
&= \sum_{i=1}^{m_1} P_i^1 \log_2\left(\frac{P_i^1}{Q_i^1}\right)\sum_{j=1}^{m_2} P_j^2 + \sum_{j=1}^{m_2} P_j^2 \log_2\left(\frac{P_j^2}{Q_j^2}\right)\sum_{i=1}^{m_1} P_i^1\\
&= KL^{1}_{P^{1}||Q^{1}} + KL^{2}_{P^{2}||Q^{2}} \qquad \left(\because \sum_{i=1}^{m_1} P_i^1 = \sum_{j=1}^{m_2} P_j^2 = 1\right). \qquad (14E.1)
\end{aligned}$$

Similarly, it can be shown that $KL^{1,2}_{Q^{1,2}||P^{1,2}} = KL^{1}_{Q^{1}||P^{1}} + KL^{2}_{Q^{2}||P^{2}}$.

In order that we may establish the result that $KL^{1,2}_{P^{1,2},Q^{1,2}} \approx KL^{1}_{P^{1},Q^{1}} + KL^{2}_{P^{2},Q^{2}}$, we focus on the error between the exact calculation $KL^{1,2}_{P^{1,2},Q^{1,2}} = 1/(1/KL^{1,2}_{Q^{1,2}||P^{1,2}} + 1/KL^{1,2}_{P^{1,2}||Q^{1,2}})$, and the approximation, $(KL^{1}_{P^{1},Q^{1}} + KL^{2}_{P^{2},Q^{2}})$, as below

$$\begin{aligned}
\text{Error} &= \frac{KL^{1,2}_{P^{1,2}||Q^{1,2}} KL^{1,2}_{Q^{1,2}||P^{1,2}}}{KL^{1,2}_{P^{1,2}||Q^{1,2}} + KL^{1,2}_{Q^{1,2}||P^{1,2}}} - \left(KL^{1}_{P^{1},Q^{1}} + KL^{2}_{P_{2},Q^{2}}\right)\\
&= \frac{\left(KL^{1}_{P^{1}||Q^{1}} + KL^{2}_{P^{2}||Q^{2}}\right)\left(KL^{1}_{Q^{1}||P^{1}} + KL^{2}_{Q^{2}||P^{2}}\right)}{\left(KL^{1}_{P^{1}||Q^{1}} + KL^{2}_{P^{2}||Q^{2}} + KL^{1}_{Q^{1}||P^{1}} + KL^{2}_{Q^{2}||P^{2}}\right)}\\
&\qquad - \frac{KL^{1}_{P^{1}||Q^{1}} KL^{1}_{Q^{1}||P^{1}}}{KL^{1}_{P^{1}||Q^{1}} + KL^{1}_{Q^{1}||P^{1}}} - \frac{KL^{2}_{P^{2}||Q^{2}} KL^{2}_{Q^{2}||P^{2}}}{KL^{2}_{P^{2}||Q^{2}} + KL^{2}_{Q^{2}||P^{2}}}\\
&= \frac{(a+b)(c+d)}{(a+b+c+d)} - \frac{ac}{(a+c)} - \frac{bd}{(b+d)}\\
&= \frac{(ad-bc)^2}{(a+c)(b+d)(a+b+c+d)}\\
&= \frac{a^2}{(a+c)} + \frac{b^2}{(b+d)} - \frac{(a+b)^2}{(a+b+c+d)}, \qquad (14E.2)
\end{aligned}$$

where $a = KL^1_{P^1||Q^1}$, $b = KL^2_{P^2||Q^2}$, $c = KL^2_{Q^1||P^1}$, $d = KL^2_{Q^2||P^2}$. If the two probability distributions are equal, that is, $P^1 = P^2$ and $Q^1 = Q^2$, then, the error in (14E.2) reduces to zero. From Lehmann's inequality [20], which states that if a, b, c, d are positive numbers, then $(a+b)(c+d)/(a+b+c+d) \geq ac/(a+c) + bd/(b+d)$, it follows that Error ≥ 0 (since KL divergences are always positive). Likewise, the error is bounded above by the sum of the first two terms in (14E.2), since the last term is positive. Our experiments indicate that (14E.2) is almost equal to zero, with errors in the order of 10^{-4}, probably since a, b, c, d are not unrestricted positive numbers, but related in that the pairs a, c (and b, d) are asymmetric KL distances of probability distributions. We are unable to derive this result rigorously, except to offer this result as an open conjecture.

REFERENCES

1. S.J. Russell and P. Norvig, *Artificial Intelligence: A Modern Approach*, 2nd edition, Prentice-Hall, Englewood Cliffs, NJ, 2002.
2. J. Llinas, Assessing the performance of multisensor fusion processes, *Handbook of Multisensor Data Fusion*, Second Edition, CRC Press, Boca Raton, FL, USA, 2006.
3. D.L. Hall and J. Llinas, An introduction to multisensor data fusion, *Proceedings of IEEE*, 85(1), 6–23, 1997.
4. K.C. Chang and Z. Tian, Efficient inference for mixed bayesian networks, *Proceedings of the Fifth International Conference on Information Fusion*, Annapolis, Maryland, USA, July 7–11, 2002, pp. 527–534.
5. K.C. Chang, M. Liggins, and Y. Song, Fusion performance modeling for distributed tracking and classification, *Proceedings of the Sixth International Conference on Information Fusion*, Cairns, Queensland, Australia, July 8–11, 2003, pp. 567–574.
6. K.C. Chang and R. Fung, Target identification with bayesian networks in a multiple hypothesis tracking system, *SPIE Opt. Eng. J.*, 36(3), 684–691, 1997.
7. K.C. Chang, E. Sivaraman, and M. Liggins, Performance modeling for multisensor tracking and classification, *Proceedings of SPIE Defense and Security Symposium*, Orlando, FL, 2004, Vol. 5429, pp. 335–342.
8. E. Sivaraman and K.C. Chang, The global confusion matrix: A new approach to evaluating the performance of multisensor classification systems, *Proceedings of the Eighth International Conference on Information Fusion*, Philadelphia, Pennsylvania, USA, June 27- July 1, 2005.
9. E. Sivaraman and K. Chang, Performance evaluation of multisensor classification systems, *IEEE Trans. Aerosp. Electron. Syst.*, 43(4), 1265–1281, 2007.
10. R.N. Carvalho and K.C. Chang, A performance evaluation tool and analysis for multisensor classification systems, *Proceedings of the 12th International Conference on Information Fusion*, Seattle, Washington, USA, July 6–9, 2009, pp. 1123–1130.
11. K.C. Chang and R. Fung, Node aggregation for distributed inference in bayesian networks, *Proceedings of* 11th *IJCAI*, Detroit, MI, 1989, pp. 265–270.
12. R.A. Brualdi, *Introductory Combinatorics*, 3rd edition, Prentice-Hall, Englewood Cliffs, NJ, 1999.

13. D.H. Johnson and S. Sinanoviċ, Symmetrizing the Kullback-Leibler distance, Dept. of Electrical & Computer Engg. Tech. Rep., Rice University, Houston, TX, 2001.
14. D.H. Johnson, C.M. Gruner, K. Baggerly, and C.K. Seshagiri, Information-theoretic analysis of neural coding, *J. Comput. Neurosci.*, 10(1), 47–69, 2001.
15. K.C. Chang and W. Sun, Performance modeling for dynamic bayesian networks, *Proceedings of SPIE'2004*, Orlando, Florida, Vol. 5429, 384–391. 2004.
16. X. Boyen and D. Koller, Tractable inference for complex stochastic processes, *Proceedings of* 14*th UAI*, Madison, Wisconsin, 1998, pp. 33–42.
17. K. Kanazawa, D. Koller, and S. Russell, Stochastic simulation algorithms for dynamic probabilistic networks, *Proceedings of* 11*th UAI*, Montreal, Canada, 1995, pp. 346–351.
18. M. Takikawa, B. D'Ambrosio, and E. Wright, Real-time inference with large-scale temporal bayes nets, *Proceedings of* 18*th UAI*, Edmonton, Canada, 2002, pp. 477–48.
19. L.N. Trefethen and D. Bau, *Numerical Linear Algebra*. Philadelphia, PA: SIAM, 1997.
20. J. Sándor, On Lehmann's inequality and electrical networks, *RGMIA Research Report Collection*, 8(1), 2005.

CHAPTER 15

Detection and Estimation of Radiological Sources

MARK MORELANDE AND BRANKO RISTIC

15.1 INTRODUCTION

Since the terrorist attack on September 11, 2001 there has been widespread concern about the possibility of future attacks involving nuclear weapons. While the threat of an outright terrorist nuclear attack is nonexistent at present, an attack involving an improvised radiological dispersion device, known as a *dirty bomb*, is real. The dirty bomb would consist of waste by-products from nuclear reactors wrapped in conventional explosives, which upon detonation would expel deadly radioactive particles into the environment. Radioactive waste material is relatively easy to obtain with numerous accidents involving its loss or theft reported.

The ability to rapidly detect and localize radioactive sources is important in assisting the emergency responders to disable, isolate, or safely remove the potential threat from dirty bombs. The topic of this chapter is the detection and localization of radiological materials that emit gamma rays, the highly penetrating electromagnetic radiations that can travel large distances through air. The chapter is divided in three main parts. The first (Section 15.2) is devoted to batch estimation of point radiological sources using measurements from a fixed network of sensors. Assuming the availability of measurements from the same sensor network, the second part (Section 15.3) considers the more general problem of estimating the radiation field. This is of interest when sources are distributed and/or the background radiation is unknown. The third part (Section 15.4) is concerned with the search for radiological point sources. This involves sequential detection and estimation of radiological point sources and the control of mobile sensor platforms that collect the measurements.

Throughout the chapter, the focus is on the Bayesian approach. This approach is adopted for two reasons. First, Bayes' rule, which is the basis of the Bayesian

Integrated Tracking, Classification, and Sensor Management: Theory and Applications, Edited by Mahendra Mallick, Vikram Krishnamurthy, and Ba-Ngu Vo.

approach, provides a natural means of updating knowledge of the parameters as more measurements become available. This will be important when search is discussed in Section 15.4. Second, Bayesian methods have been shown to provide accurate inference for up to four sources [1] while the numerical optimization required to compute maximum likelihood estimates (MLE) has produced reliable results only for scenarios with up to two sources [2, 3].

15.2 ESTIMATION OF POINT SOURCES

In this section, radiation is assumed to emanate from an unknown number of sources whose spatial extent is negligible, that is, point sources. The level of background radiation is assumed to be known at each point in the surveillance area. Under these conditions it is desired to estimate the positions and intensities of the sources.

Numerous approaches to radiological source estimation have been proposed in the literature. Recursive least squares estimators were proposed in [3] for position estimation of multiple sources moving with known velocity. In [2, 4], source parameter estimation was performed using maximum likelihood techniques. Bayesian methodologies have been proposed for source parameter estimation in [5], for source detection in [6] and for both source detection and estimation in [1]. In this section, the Bayesian approach to point source parameter estimation is described.

The Bayesian approach to inference assumes that the unknown parameter of interest is a random variable. In this framework, inferential procedures are based on the posterior distribution, that is, the distribution of the parameter conditional on the observations. For the purposes of parameter estimation the posterior distribution is required to compute the minimum mean square error estimator, the posterior mean. For most problems, including point source parameter estimation, the posterior mean cannot be found in closed form but must be approximated. Many methods have been developed for approximating the posterior mean including Laplace approximations [7], grid-based methods [8], and Gaussian quadrature [9]. Monte Carlo methods [10, 11] have recently emerged as a popular approach. Two important advantages of Monte Carlo methods over the alternatives are their ease of implementation and their applicability to high-dimensional problems.

The class of Monte Carlo methods includes a variety of approximations with the common characteristic that they involve the generation of (pseudo-)random variables [11]. The particular method used here is importance sampling that involves drawing random samples from an importance density and weighting them by the ratio of the posterior PDF to the importance density. Importance sampling is not applied here directly to the posterior but rather to a series of target distributions obtained by a factorization of the likelihood [12–14]. The common philosophy of these approaches is to introduce the measurement correction gradually. Their efficiency, in the sense of providing accurate approximation with reasonable computational expense, is demonstrated using simulated and experimental data.

15.2.1 Model

Let $r \geq 0$ denote the known number of sources, in the two-dimensional region of interest. Let $\boldsymbol{\theta}_i = [\mathbf{x}_i', a_i]' \in \mathbb{R}^2 \times \mathbb{R}^+$, $i = 1, \ldots, r$, denote the parameter vector of the ith source where $\mathbf{x}_i$ is the source position in the two-dimensional plane and a_i is the source intensity. The source parameter vectors are collected into a stacked vector $\boldsymbol{\theta} = [\boldsymbol{\theta}_1', \ldots, \boldsymbol{\theta}_r']'$. The availability of a prior distribution for $\boldsymbol{\theta}$, which represents the knowledge of the parameters before any measurements are processed, is assumed. Thus, initially $\boldsymbol{\theta} \sim \pi_0(\cdot)$.

Measurements of radiation dose are collected from m sensors. The statistical model for the measurements assumes that the radiation sources are isotropic point sources, the sensors are omnidirectional, radiation measurements are independent random variables and, for convenience, the exposure time, denoted τ, is the same for all sensors. Let $z_j \in \mathbb{N}$ denote the count obtained at position $\boldsymbol{\rho}_j$ by the jth sensor. Using these definitions and the above assumptions, the likelihood of the parameter value $\boldsymbol{\theta}$ given the measurement vector $\mathbf{z} = [z_1, \ldots, z_m]'$ is [2, 15]

$$\ell(\boldsymbol{\theta}; \mathbf{z}) = \prod_{j=1}^{m} P(z_j; \lambda_j(\boldsymbol{\theta})), \tag{15.1}$$

where $P(z; \alpha) = e^{-\alpha}\alpha^z / z!$ is the Poisson probability density function (PDF) evaluated at $z \in \mathbb{N}$ with parameter α. The mean radiation count $\lambda_j(\boldsymbol{\theta})$ for the jth sensor is the sum of the radiation counts from each source plus background radiation,

$$\lambda_j(\boldsymbol{\theta}) = \left[\mu(\boldsymbol{\rho}_j) + \sum_{i=1}^{r} \frac{a_i}{\|\mathbf{x}_i - \boldsymbol{\rho}_j\|^2}\right] \tau, \tag{15.2}$$

where $\mu(\boldsymbol{\rho})$ is the known count rate due to the background radiation at position $\boldsymbol{\rho}$. Equation (15.2) is valid for short distances where the effect of atmospheric attenuation can be ignored. For larger distances the mean count for the ith source should be scaled by $\exp(-\beta\|\mathbf{x}_i - \boldsymbol{\rho}_j\|)$ where β is the air absorption coefficient.

The problem is to estimate the parameter vector $\boldsymbol{\theta}$ using the vector of measurements $\mathbf{z}$.

15.2.2 Source Parameter Estimation

The minimum mean square error estimator of the random parameter $\boldsymbol{\theta}$ is the posterior mean,

$$\mathrm{E}(\boldsymbol{\theta}|\mathbf{z}) = \int \boldsymbol{\theta}\, \pi(\boldsymbol{\theta})\, d\boldsymbol{\theta}, \tag{15.3}$$

where $\pi(\cdot)$ is the posterior PDF, which can be found using Bayes' rule as

$$\pi(\boldsymbol{\theta}) \propto \ell(\boldsymbol{\theta}; \mathbf{z})\pi_0(\boldsymbol{\theta}). \tag{15.4}$$

The integral (15.3) cannot be evaluated exactly and so must be approximated. Approximation of (15.3) via importance sampling involves drawing samples $\boldsymbol{\theta}^1, \ldots, \boldsymbol{\theta}^n$, where n is the sample size, from an importance density $q(\cdot)$ and approximating the integral (15.3) by

$$\widehat{\mathrm{E}(\boldsymbol{\theta}|\mathbf{z})} = \hat{\boldsymbol{\theta}} = \sum_{i=1}^{n} w^i \boldsymbol{\theta}^i. \tag{15.5}$$

The weights $w^1, \ldots, w^n$ are given by $w^i = \tilde{w}^i / \sum_{j=1}^{n} \tilde{w}^j$ with

$$\tilde{w}^i = \ell(\boldsymbol{\theta}^i; \mathbf{z})\pi_0(\boldsymbol{\theta}^i)/q(\boldsymbol{\theta}^i). \tag{15.6}$$

Let $\{w^i, \boldsymbol{\theta}^i\}$ denote the weighted sample comprised of $\boldsymbol{\theta}^1, \ldots, \boldsymbol{\theta}^n$ and $w^1, \ldots, w^n$. Weighted samples of this sort are referred to as importance samples. The weights are omitted from the notation if they are uniform.

For a large class of importance densities, the approximation (15.5) becomes increasingly accurate as the sample size $n \to \infty$ [16]. However, for a finite n, the accuracy of the approximation depends greatly on the particular importance density. A desirable property of the importance density is that it produce weights with a small sample variance, that is, the weights should be even [17]. An importance sample that satisfies this requirement provides accurate approximation of a wide range of posterior quantities. Equation (15.6) then suggests an importance density that resembles the posterior PDF. Use of the posterior PDF itself as an importance density, which would result in zero variance weights, is impossible as it cannot be calculated for the model described in Section 15.2.1. Constructing a useful approximation to the posterior is difficult because of the potentially large dimension of the parameter vector and the much greater diffuseness of the prior compared to the likelihood. Importance sampling with the prior PDF, as in [5, 6], is inaccurate for any reasonable sample size for the same reasons.

The approach taken here to address these difficulties is to construct a series of target distributions from which it is desired to draw samples sequentially. The first distribution in the series is the prior while the final distribution is the posterior. Consecutive distributions in the series should not differ too greatly so that an importance sample with reasonably even weights can be obtained from the target distribution at a particular stage using the target distribution at the previous stage as an importance distribution. To formalize this idea, the likelihood is decomposed into s factors

$$\ell(\boldsymbol{\theta}; \mathbf{z}) = \prod_{b=1}^{s} \ell_b(\boldsymbol{\theta}; \mathbf{z}). \tag{15.7}$$

The corresponding sequence of target PDFs is defined as, for $b = 1, \ldots, s$,

$$\pi_b(\boldsymbol{\theta}) \propto \prod_{j=1}^{b} \ell_j(\boldsymbol{\theta}; \mathbf{z})\, \pi_0(\boldsymbol{\theta}) \propto \ell_b(\boldsymbol{\theta}; \mathbf{z}) \pi_{b-1}(\boldsymbol{\theta}). \tag{15.8}$$

The idea is to obtain successive sample-based approximations to $\pi_0(\cdot)$, $\pi_1(\cdot)$, and so on up to $\pi_s(\cdot) = \pi(\cdot)$. There are a number of ways of doing this, three of which are considered here. These methods differ in both the manner in which the likelihood is factorized and the methods used to construct the approximations to the target distributions.

The first method, proposed in [12], is to partition the measurement vector as $\mathbf{z} = [\mathbf{z}_1', \ldots, \mathbf{z}_s']'$ and set

$$\ell_b(\boldsymbol{\theta}; \mathbf{z}) = \ell(\boldsymbol{\theta}; \mathbf{z}_b), \qquad b = 1, \ldots, s. \tag{15.9}$$

The likelihood factorization specified by (15.7) and (15.9) allows the correction supplied by the measurement $\mathbf{z}$ to be introduced gradually. Approximation of $\pi_b(\cdot)$ given a sample-based approximation to $\pi_{b-1}(\cdot)$ consists of three steps: weight computation, resampling, and a move step. The move step involves drawing from a transition density with stationary distribution $\pi_b(\cdot)$, that is, a π_b-invariant transition density. The procedure, referred to here as sequential measurement importance sampling (SMIS), is summarized by Algorithm 1. The availability of a procedure for drawing samples from the prior is assumed.

The second method, referred to as annealed importance sampling (AIS) [14], adopts the likelihood factorization of (15.7) with factors

$$\ell_b(\boldsymbol{\theta}; \mathbf{z}) = \ell(\boldsymbol{\theta}; \mathbf{z})^{\gamma_b}, \tag{15.10}$$

Algorithm 1: Sequential Measurement Importance Sampling for Posterior PDF Approximation

select the measurement partition $\mathbf{z}_1, \ldots, \mathbf{z}_s$
draw $\boldsymbol{\theta}^{0,i} \sim \pi_0(\cdot)$ for $i = 1, \ldots, n$
for $b = 1, \ldots, s$ **do**
  compute the weights $w_b^i \propto \ell(\boldsymbol{\theta}^{b-1,i}; \mathbf{z}_b)$, $i = 1, \ldots, n$
  select sample indices $j^1, \ldots, j^n$ such that $\mathrm{P}(j^i = k) = w_b^k$
  draw $\boldsymbol{\theta}^{b,i} \sim g_b(\cdot|\boldsymbol{\theta}^{b-1,j^i})$, $i = 1, \ldots, n$, where $g_b(\cdot|\boldsymbol{\theta})$ is a π_b-invariant transition density
end
compute a state estimate $\hat{\boldsymbol{\theta}}$ based on $\{\boldsymbol{\theta}^{s,i}\}$

Algorithm 2: Annealed Importance Sampling for Posterior PDF Approximation

---

select the correction factors $\gamma_1, \ldots, \gamma_s$
set $w_0^i = 1/n$ and draw $\boldsymbol{\theta}^{0,i} \sim \pi_0(\cdot)$ for $i = 1, \ldots, n$
for $b = 1, \ldots, s$ **do**
  compute the weights $w_b^i \propto w_{b-1}^i \ell(\boldsymbol{\theta}^{b-1,i}; \mathbf{z})^{\gamma_b}$, $i = 1, \ldots, n$
  draw $\boldsymbol{\theta}^{b,i} \sim g_b(\cdot|\boldsymbol{\theta}^{b-1,i})$, $i = 1, \ldots, n$, where $g_b(\cdot|\boldsymbol{\theta})$ is a π_b-invariant transition density
end
compute a state estimate $\hat{\boldsymbol{\theta}}$ based on $\{w_s^i, \boldsymbol{\theta}^{s,i}\}$

---

where the correction factors $\gamma_1, \ldots, \gamma_s$ are positive and sum to one. Equation (15.10) uses a different approach to (15.9) to produce the effect of gradually introducing the correction provided by $\mathbf{z}$. Each incremental correction involves a weight calculation and at least one, but possibly multiple Markov chain Monte Carlo (MCMC) steps. The algorithm is summarised in Algorithm 2.

The final method of Monte Carlo approximation, progressive correction [13], uses the same likelihood factorization (15.10) used by AIS. The difference between progressive correction importance sampling (PCIS) and AIS arises in the manner in which the target PDFs $\pi_1(\cdot), \ldots, \pi_s(\cdot)$ are approximated. The key distinction is the introduction of a resampling step to remove lowly weighted particles. Another difference is that PCIS as proposed in [13], and implemented here, draws samples at each stage from a kernel density rather than a π_b-invariant transition density. However, PCIS can also be implemented with a MCMC step rather than a draw from a kernel density, as in [1]. PCIS is summarized by Algorithm 3. The kernel density $h_b(\cdot)$ can be selected, for instance, using standard results from kernel density estimation [18].

The computational expense and performance of Algorithms 1–3 depend somewhat on the precise manner in which the likelihood is factorized, that is, the number

---

Algorithm 3: Progressive Correction for Posterior PDF Approximation

---

select the correction factors $\gamma_1, \ldots, \gamma_s$
set $w_0^i = 1/n$ and draw $\boldsymbol{\theta}^{0,i} \sim \pi_0(\cdot)$ for $i = 1, \ldots, n$
for $b = 1, \ldots, s$ **do**
  compute the weights $w_b^i \propto \ell(\boldsymbol{\theta}^{b-1,i}; \mathbf{z})^{\gamma_b}$, $i = 1, \ldots, n$
  select sample indices $j^1, \ldots, j^n$ such that $\mathrm{P}(j^i = k) = w_b^k$
  set $\boldsymbol{\theta}^{b,i} = \boldsymbol{\theta}^{b-1,j^i} + \boldsymbol{\epsilon}^{b,i}$, $i = 1, \ldots, n$, where $\boldsymbol{\epsilon}^{b,i} \sim h_b(\cdot)$ with $h_b(\cdot)$ a suitable kernel density
end
compute a state estimate $\hat{\boldsymbol{\theta}}$ based on $\{\boldsymbol{\theta}^{s,i}\}$

---

s of steps and the selection of the partition $\mathbf{z}_1, \ldots, \mathbf{z}_s$ for SMIS and the correction factors $\gamma_1, \ldots, \gamma_s$ for AIS and PCIS. It is obvious that computational expense for a given sample size n is minimized by choosing s as small as possible. However, the requirement that successive target distributions not differ too greatly suggests choosing a large s. Although in Algorithms 1–3 the likelihood factorization is selected *a priori*, an adaptive scheme, such as that proposed in [13], can be used to choose a likelihood factorization that balances the trade-off between performance and computational expense.

If the prior PDF is the same for each source parameter vector then the posterior PDF has $r!$ equally weighted modes, one for each permutation of the labels $1, \ldots, r$ assigned to the sources. Each of these label assignments provides an equally valid solution to the estimation problem since the particular label given to a source is not important in this problem. The posterior PDF approximations computed by Algorithms 1–3 can be forced to concentrate on one of the possible label assignments by performing clustering after each correction. The computationally efficient k-means algorithm is used for this purpose. As an aside, we note that the permutation symmetry inherent in the problem could be characterized by considering the parameter of interest to be a set of individual source parameter vectors instead of a vector formed by concatenating the individual source parameter vectors [19].

15.2.3 Simulation Results

The simulation scenario involves up to four sources distributed in a 150×150 surveillance region and observed by a regular 10×10 grid of sensors. This scenario is depicted in Figure 15.1 with sources indicated by crosses and sensors indicated by dots. The individual source parameters are $\boldsymbol{\theta}_1 = [30, 35, a_1]'$, $\boldsymbol{\theta}_2 = [60, 100, a_2]'$, $\boldsymbol{\theta}_3 = [100, 75, a_3]'$, and $\boldsymbol{\theta}_4 = [30, 125, a_4]'$. Scenarios involving $r \leq 4$ sources contain sources with parameters $\boldsymbol{\theta}_1, \ldots, \boldsymbol{\theta}_r$. The source intensity for each source is such that the mean count at the sensor closest to that source is a set value. This value quantifies the difficulty of the scenario for a given arrangement of sensors and sources. The prior source parameter distribution is $\pi_0(\boldsymbol{\theta}) = \prod_{i=1}^{r}[U_{[0,150]^2}(\mathbf{x}_i)G(a_i; 1, 1000)]$ where $U_A(\cdot)$ is the uniform PDF over the set A and $G(\cdot; \omega, \kappa)$ is the gamma PDF with shape parameter ω and scale parameter κ. The exposure time for the sensors is $\tau = 1$s.

A fair comparison between the SMIS, AIS, and PCIS requires that the algorithms are compared for a roughly equivalent computational expense. All three algorithms have computational expense that is $O(ns)$. However, the number of likelihood computations, which determines the computational expense to a large extent in computer implementations, varies significantly. The AIS and PCIS perform ns likelihood computations while the SMIS performs $ns/2$ likelihood computations, assuming that the measurement partitions are equal length. Thus, the SMIS can use a sample size that is twice that of the AIS and PCIS for an equivalent computational expense. Algorithm performance is measured by the RMS error of the source position estimates. If p

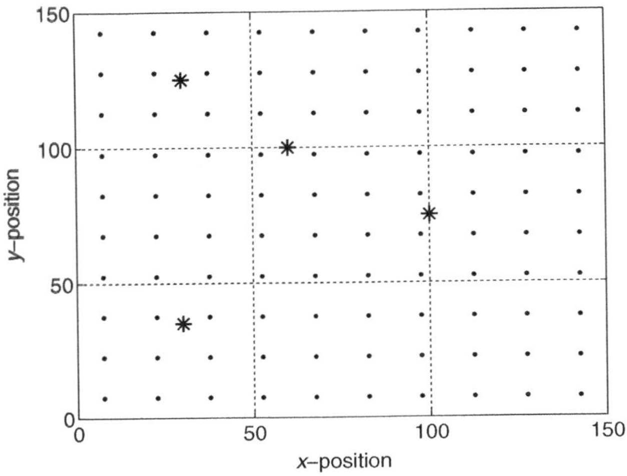

Figure 15.1 Simulation scenario for point source estimation. Sensors are located at the dots and sources are positioned at the crosses.

realizations are generated and $\hat{\mathbf{x}}_i^a$ is the position estimate of the ith source for the ath realization we use

$$\sqrt{1/(rp)\sum_{i=1}^{r}\sum_{a=1}^{p}\|\mathbf{x}_i - \hat{\mathbf{x}}_i^a\|^2} \tag{15.11}$$

as a measure of the source-averaged RMS position error.

The source-averaged RMS position errors of the algorithms with several different sample sizes are computed over $p = 200$ realizations for $r = 1, \ldots, 4$ sources. The source intensities are such that the mean count at the nearest sensor to each source is 10. The likelihood partitionings for each algorithm are selected using an adaptive scheme [13], which limits the number of iterations to $\bar{s} = 30$. The MCMC step in the AIS and SMIS is implemented using a Metropolis–Hastings independence sampler [10]. The results are shown in Figure 15.2.

For sample sizes below a certain threshold, the accuracy of a Monte Carlo procedure is expected to improve as the sample size increases. For sample sizes above the threshold the algorithm operates in the asymptotic region where all sample sizes provide practically exact computation of the required quantities. The threshold sample size for which an algorithm enters the asymptotic region is an important indicator of performance. With this in mind two observations can be made regarding the results of Figure 15.2. First, the threshold sample size for entry into the asymptotic region increases for all of the algorithms as the number of sources, and therefore the number of parameters, increases. This is to be expected. Second, for all source numbers, PCIS has by far the lowest threshold sample size required to enter the asymptotic region. As a result PCIS provides accurate estimation with a significantly smaller computational expense than AIS and SMIS.

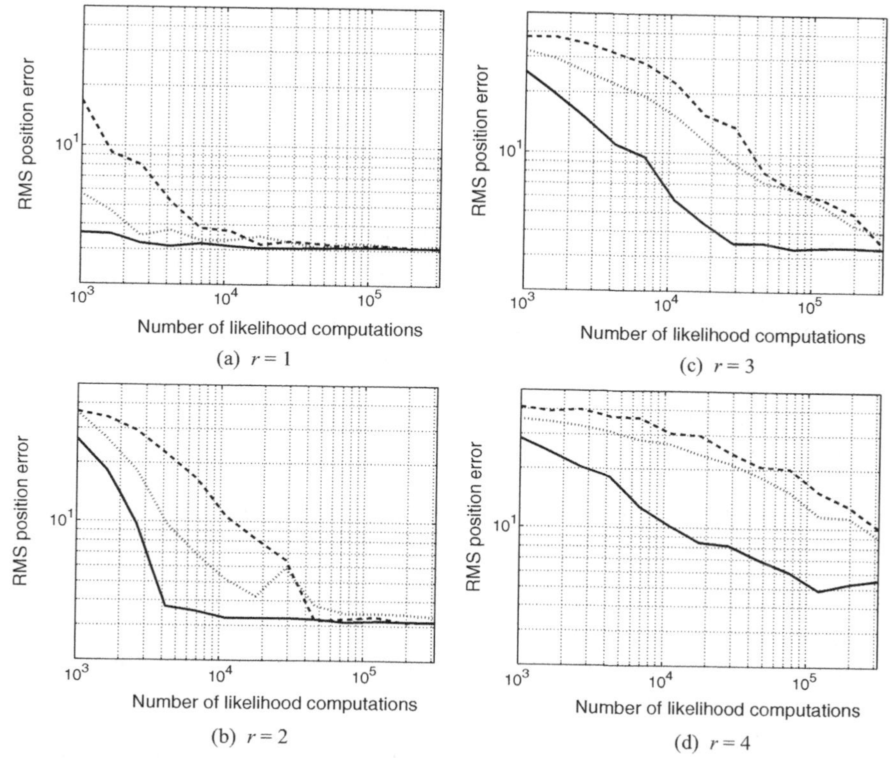

Figure 15.2 Source-averaged RMS position errors for PCIS (black solid), AIS (dotted), SMIS (dashed) plotted against the number of likelihood computations.

15.2.4 Experimental Results

A radiological field trial was conducted on a large, flat, and open area without any obstacles. The data sets were collected using the DSTO-developed low-cost advanced radiological survey (LCAARS) system. The LCAARS survey system consists of an AN/PDR-77 radiation survey meter equipped with an RS232 interface module, a gamma probe and software written in Visual Basic and running on a laptop computer. The gamma probe contains two Geiger–Müller (GM) tubes to cover both low and high ranges of dose rates and, when connected to the AN/PDR-77 radiation survey meter, it is capable of measuring gamma radiation dose rates from background to 9.99 Sv/h without saturating [20]. The gamma probe has a fairly flat response ($\sim \pm 10\%$) from 0.1 to above 1 MeV [20, p. 7]. Measured dose rate data were recorded in μSv/h and converted into raw count measurements $z_j \in \mathbb{Z}^+$ by multiplication with a conversion factor. Approximate omnidirectionality of the gamma probe was verified experimentally.

Three radiation sources were used in the field trial: two cesium sources (^{137}Cs) and one cobalt (^{60}Co) source. The source parameters are given in Table 15.1. While

TABLE 15.1 Radiation Sources Used in the Field Trial

Source	Type	Position	Activity (MBq)	Intensity
1	^{137}Cs	(11,10)	26×10^3	9105
2	^{137}Cs	(3,50)	5×10^3	1868
3	^{60}Co	(41,5)	0.2×10^3	467

the strengths of radiation sources are typically characterized in terms of their activity in GBq, in this work the intensity of a radiation point source is defined as the mean count rate due to that source at a distance of 1 m from the source. This can be seen from (15.1) and (15.2). Both the activities and intensities are given in Table 15.1. The sources were mounted at the same height above the ground as the gamma probe. To ensure that radiation sources appear as isotropic, they were placed in a vertical configuration such that the handling rods were pointing up. Data sets were collected with one, two, and three radiation sources. These data sets are referred to as test set 1, 2, and 3, where test set r contains sources $1, \ldots, r$.

Radiation dose measurements were collected on grid points that were carefully measured and marked beforehand in the local Cartesian coordinate system on the asphalt surface of the airfield. The data were acquired when the trolley-mounted gamma probe was positioned over individual grid points. During data collection at any grid point, the gamma probe was held in place until approximately 60 measurements were acquired. The exposure duration for each radiation dose measurement was kept constant at about 0.8 s. Multiple measurements are collected at each location so that the candidate algorithms can be tested using several different batches of experimental data. The background radiation level was obtained by taking measurements at a number of grid positions in the absence of radiation sources. Based on these measurements the background radiation is taken to be Poisson with mean rate $\mu(\boldsymbol{\rho}) \approx 0.9$ for all $\boldsymbol{\rho}$.

Figure 15.3 is an aerial view of the experimental setup at the Puckapunyal airfield. The three stars show the locations of the three radiation sources and the white discs indicate the points where measurements were collected.

Two approaches to source parameter estimation are considered: Bayesian and maximum likelihood. Bayesian estimation is performed using the PCIS algorithm described in Section 15.2.2 with a sample size $n = 5000$. The correction factors for PCIS are chosen adaptively with the maximum number of factors set to 75. The prior PDF used by PCIS for the source parameter vector $\boldsymbol{\theta}$ is

$$\pi_0(\boldsymbol{\theta}) = \prod_{i=1}^{r} U_S(\mathbf{x}_i) G(a_i; 1.5, 8000), \tag{15.12}$$

where $S \subset \mathbb{R}^2$ is the area over which measurements were taken. The prior distribution for the intensities is such that it is broad enough to cover all likely source strength values. The maximum likelihood estimate is found using the iterative Nelder–Mead method [21]. The Nelder–Mead method is applied with 50 different initial points, each drawn from the prior PDF $\pi_0(\cdot)$. Of the estimates found by the 50 applications of the Nelder–Mead method, the estimate with the highest likelihood is selected as the MLE.

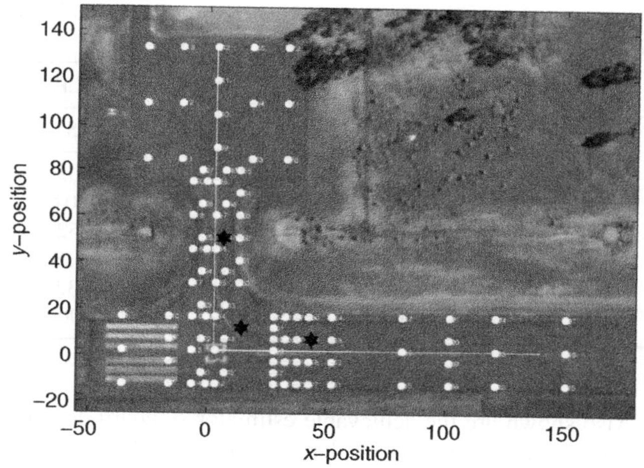

Figure 15.3 Aerial view of the Puckapunyal airfield site where the field trial was conducted. The black stars mark source positions and the white discs indicate measurement positions.

Even multiple optimizations does not guarantee convergence to the global maximum. Convergence to a local maximum can result in estimates that are very far from the true parameters. To obtain a meaningful idea of estimator RMSE it is necessary to exclude these divergent estimates. In this performance analysis an estimate is said to be divergent if its position error is more than 10 m.

Tables 15.2–15.4 list the RMS errors of PCIS and MLE and the number of times the MLE computed by the Nelder–Mead method diverged for test sets 1, 2, and 3,

TABLE 15.2 RMS Errors Averaged Over 100 Realizations for Test 1 Data Set (One Source)

		$\sqrt{\text{CRB}}$	MLE	PCIS
Divergences		–	0	0
Source 1	Position RMSE	0.53	0.70	0.69
	Intensity RMSE	303	392	389

TABLE 15.3 RMS Errors Averaged Over 100 Realizations for Test 2 Data Set (Two Sources)

		$\sqrt{\text{CRB}}$	MLE	PCIS
Divergences		–	0	0
Source 1	Position RMSE	0.55	0.89	0.91
	Intensity RMSE	343	533	575
Source 2	Position RMSE	1.64	2.31	2.26
	Intensity RMSE	393	519	444

TABLE 15.4 RMS Errors Averaged Over 100 Realizations for Test 3 Data Set (Three Sources)

		$\sqrt{\text{CRB}}$	MLE	PCIS
Divergences		–	55	0
Source 1	Position RMSE	0.58	2.63	0.88
	Intensity RMSE	410	1904	805
Source 2	Position RMSE	1.64	5.52	2.29
	Intensity RMSE	395	1518	539
Source 3	Position RMSE	2.67	1.32	0.38
	Intensity RMSE	185	900	410

respectively. Also shown are the achievable estimator standard deviations as determined by the Cramér–Rao bound (CRB), derived in [4]. The RMS errors are computed by averaging over 100 realizations of experimental data. It should be kept in mind that the CRB is computed using an idealized measurement model that is based on several assumptions and approximations, for example, uniform directional response, neglected air attenuation, perfect knowledge of sensor locations, known and constant background radiation. As such the CRBs should be seen only as a guideline for achievable accuracy. Overall the results indicate good performance and validate the measurement model in Section 15.2.1. The results of Tables 15.2 and 15.3 show that both PCIS and MLE provide accurate source parameter estimates for test sets 1 and 2. However, for test set 3, which contains three sources, only PCIS is able to reliably provide accurate parameter estimates. The MLE computed by the Nelder–Mead method fails to converge to the global maximum on most occasions. Even on those occasions where the computed estimate is accepted, it is clear from the RMS errors that convergence to the global maximum has often not been achieved.

15.3 ESTIMATION OF DISTRIBUTED SOURCES

The assumption of point sources on a known background is too restrictive for many practical applications. Instead, sources may be distributed over a finite area and the background radiation may vary spatially in an unknown way. Both of these complications can be accommodated by considering the significantly more difficult problem of estimating the field, that is, the radiation at each point in a surveillance region. In this case, the mean count for a GM counter at a given position is the convolution between a kernel function and the field.

Estimation of the radiation field from measurements collected at a finite number of locations is an ill-posed inverse problem. Such problems arise in a number of areas and have been the subject of wide interest [22–24]. The ill-posed nature of the field estimation problem arises from the smoothing applied by the convolution of the field with the kernel. Any attempt to invert this smoothing will necessarily involve a high-pass inversion filter resulting in field estimates that are highly sensitive to statistical deviations in the measurements. Estimation accuracy essentially depends on what can be assumed about the field. If a parsimonious representation of the field is available then accurate estimation with a reasonable number of measurements can be achieved. An extreme example of this is the point source assumption for which a field composed of r point sources is represented by a vector of $3r$ real-valued parameters.

The approach taken in [25] is to extend the point source model by representing the field by a weighted sum of Gaussians, that is, a Gaussian mixture. This modeling approach is motivated by the fact that a large class of functions, for example, those that are absolutely integrable, can be approximated with arbitrary accuracy by a Gaussian mixture [26]. A similar approach is popular for plume models [27, 28], although in that case only a single plume is present so only a single Gaussian is used.

In this section, the Gaussian mixture model is adopted for the radiation field. Interpreting each mixture component as a distributed source, the Gaussian mixture model requires estimating an additional three parameters for each source. These additional parameters determine the spatial extent of the source. Although the Monte Carlo methods described for point sources in Section 15.2.2 are readily applicable to this situation, the increased dimension of the parameter vector necessarily increases the sample size required for accurate estimation. With this in mind an alternative algorithm, also based on the notion of gradual correction of the prior by the likelihood, is described.

15.3.1 Model

Let $f : \mathbb{R}^2 \to \mathbb{R}$ denotes a field function such that $f(\mathbf{x})$ is the strength of the radiation field at a location $\mathbf{x}$. The field is to be estimated using measurements obtained from a collection of spatially distributed sensors. Approximation of the field by a Gaussian mixture can be justified by theoretical results concerning the approximation of functions by radial basis functions. In particular, the following result is of interest. Let $\mathbf{a} = [a_1, \ldots, a_r]' \in \mathbb{R}^r$, $\boldsymbol{\mu} = [\boldsymbol{\mu}'_1, \ldots, \boldsymbol{\mu}'_r]' \in \mathbb{R}^{2r}$, and $\boldsymbol{\sigma} = [\sigma_1, \ldots, \sigma_r]' \in \mathbb{R}^{+r}$ denote vectors containing the component weights, locations, and scales for the r-element mixture

$$c_r(\mathbf{x}; \mathbf{a}, \boldsymbol{\mu}, \boldsymbol{\sigma}) = \sum_{i=1}^{r} a_i N(\mathbf{x}; \boldsymbol{\mu}_i, \sigma_i^2 \mathbf{I}_2), \qquad \mathbf{x} \in \mathbb{R}^2, \tag{15.13}$$

where $\mathbf{I}_k$ is the $k \times k$ identity matrix. Let $Z = \{c_r(\cdot; \mathbf{a}, \boldsymbol{\mu}, \boldsymbol{\sigma}) : r \in \mathbb{N}, \mathbf{a} \in \mathbb{R}^r, \boldsymbol{\mu} \in \mathbb{R}^{2r}, \boldsymbol{\sigma} \in \mathbb{R}^{+r}\}$ denote the set of mixture functions. Let $L^p(\mathbb{R}^2)$ denotes the set of real-valued functions on $\mathbb{R}^2$ that are pth power integrable. Then the following result holds [26, Theorem 2]:

Theorem 1 *The set Z of mixture functions is dense in the set $L^1(\mathbb{R}^2)$ of absolutely integrable functions.*

Similar results to Theorem 1 exist for pth power integrable functions, $p \geq 1$ and continuous functions [26, 29, 30]. Although we restrict our attention here to Gaussian mixtures the results of [26, 29] hold for a quite general class of mixtures.

Motivated by the results concerning function approximation by Gaussian mixtures, the field is modeled by

$$f(\mathbf{x}) = \sum_{i=1}^{r} a_i N(\mathbf{x}; \boldsymbol{\mu}_i, \Sigma_i), \qquad \mathbf{x} \in \mathbb{R}^2. \tag{15.14}$$

The spreading matrix is represented by three parameters, denoted as $\sigma_i, \varsigma_i > 0$ and $\rho_i \in (-1, 1)$ for the ith component

$$\Sigma_i = \begin{bmatrix} \sigma_i^2 & \rho_i \sigma_i \varsigma_i \\ \rho_i \sigma_i \varsigma_i & \varsigma_i^2 \end{bmatrix}. \tag{15.15}$$

For a given number r of mixture components, the field estimation problem amounts to estimating the $6r$-dimensional parameter vector $\boldsymbol{\theta} = [\boldsymbol{\mu}', \boldsymbol{\sigma}', \mathbf{a}']'$ where $\mathbf{a} = [a_1, \ldots, a_r]$ is the vector of amplitudes, $\boldsymbol{\mu} = [\boldsymbol{\mu}_1', \ldots, \boldsymbol{\mu}_r']'$ is the vector of locations, and $\boldsymbol{\sigma} = [\boldsymbol{\sigma}_1', \ldots, \boldsymbol{\sigma}_r']'$, with $\boldsymbol{\sigma}_i = [\sigma_i, \varsigma_i, \rho_i]'$, is the vector of spreading parameters. The dependence of the field on the mixture parameters is made explicit by writing $f(\mathbf{x}; \boldsymbol{\theta})$ for the value of the field at the point $\mathbf{x}$ for the mixture parameters $\boldsymbol{\theta}$. Since a Bayesian approach is adopted, a prior PDF, denoted $\pi_0(\cdot)$, is assumed to be available for the parameter $\boldsymbol{\theta}$.

The field is observed by a collection of m GM counters distributed throughout the surveillance region. The jth sensor is located at $\boldsymbol{\rho}_j$ and returns the measurement z_j. The sensor measurements are collected into the vector $\mathbf{z} = [z_1, \ldots, z_m]'$. The likelihood of $\boldsymbol{\theta}$ given that $\mathbf{z}$ has been observed is

$$\ell(\boldsymbol{\theta}; \mathbf{z}) = \prod_{j=1}^{m} P(z_j; \lambda_j(\boldsymbol{\theta})), \tag{15.16}$$

where the mean count for the jth sensor is given by

$$\lambda_j(\boldsymbol{\theta}) = \int k(\mathbf{x} - \boldsymbol{\rho}_j) f(\mathbf{x}; \boldsymbol{\theta})\, d\mathbf{x} \tag{15.17}$$

with $k(\cdot)$ the kernel

$$k(\mathbf{x}) = \begin{cases} 1/\upsilon^2, & \|\mathbf{x}\| < \upsilon, \\ 1/\|\mathbf{x}\|^2, & \|\mathbf{x}\| \geq \upsilon, \end{cases} \tag{15.18}$$

where $\upsilon > 0$ is a known constant that limits the response of the sensor to the field at positions close to the sensor. This models the saturation effect present in a real sensor.

15.3.2 Estimation

As with parameter estimation of point sources in Section 15.2.2, the main challenge in Bayesian estimation of a field modelled by a Gaussian mixture is in approximating the posterior mean. The same strategy used in Section 15.2.2 is applied here. The likelihood is factorized as

$$\ell(\boldsymbol{\theta}; \mathbf{z}) = \prod_{b=1}^{s} \ell_b(\boldsymbol{\theta}; \mathbf{z}). \tag{15.19}$$

It is desired to sequentially approximate the series of target distributions $\pi_0(\cdot)$, $\pi_1(\cdot), \ldots, \pi_s(\cdot) = \pi(\cdot)$ where, for $b = 1, \ldots, s$,

$$\pi_b(\boldsymbol{\theta}) \propto \prod_{j=1}^{b} \ell_j(\boldsymbol{\theta}; \mathbf{z})\, \pi_0(\boldsymbol{\theta}) \propto \ell_b(\boldsymbol{\theta}; \mathbf{z})\, \pi_{b-1}(\boldsymbol{\theta}). \tag{15.20}$$

The final estimate of $\boldsymbol{\theta}$ is obtained as

$$\hat{\boldsymbol{\theta}} = \int \boldsymbol{\theta}\, \hat{\pi}_s(\boldsymbol{\theta})\, d\boldsymbol{\theta}, \tag{15.21}$$

where $\hat{\pi}_s(\cdot)$ is an approximation to $\pi_s(\cdot) = \pi(\cdot)$. In Section 15.2.2 three importance sampling algorithms were described for sequentially drawing from the series of target distributions. Although the same algorithms can be applied here an argument against this is the increased dimension of the parameter vector. Parameter estimation of r point sources involves a $3r$-dimensional parameter vector while estimation of a field modelled by a r-component mixture involves a $6r$-dimensional parameter vector.

The idea of sequentially approximating a series of target PDFs, with the ultimate aim of approximating the posterior PDF, is not limited to Monte Carlo methods. In principle, any method of approximating the target PDFs can be applied. If, in the interests of reducing computational expense, Monte Carlo methods are deemed inappropriate, an appealing alternative, provided the posterior is unimodal, is to use Gaussian approximations to the target PDFs. Each target PDF is then represented by a mean and covariance matrix. A simple means of propagating the Gaussian approximation from one target PDF to the next can be achieved by replacing the nonlinear, Poisson likelihood with a linearized Gaussian approximation. In a Gaussian approximation the intermediate posterior PDF at the bth step is written as, for $b = 0, \ldots, s$,

$$\hat{\pi}_b(\boldsymbol{\theta}) = N(\boldsymbol{\theta}; \hat{\boldsymbol{\theta}}^b, \mathbf{P}^b), \tag{15.22}$$

where $\hat{\boldsymbol{\theta}}^b$ and $\mathbf{P}^b$ may be regarded as approximations to the mean and covariance matrix of $\pi_b(\cdot)$. For $b = 0$, the prior PDF approximation is constructed by selecting $\hat{\boldsymbol{\theta}}^0$ and $\mathbf{P}^0$ to match the first two moments of the true prior PDF. For $b > 0$, the bth intermediate posterior PDF is calculated using Bayes' rule with the likelihood replaced by a linearized Gaussian approximation. In the following development the likelihood is factorized as shown in (15.19) with

$$\ell_b(\boldsymbol{\theta}; \mathbf{z}) = \ell(\boldsymbol{\theta}; \mathbf{z})^{\gamma_b}, \tag{15.23}$$

where $\gamma_1, \ldots, \gamma_s$ are positive and sum to one. The linearized Gaussian approximation can also be implemented with the likelihood factorized by partitioning the measurement vector, as in (15.9).

Let $\boldsymbol{\lambda}(\boldsymbol{\theta}) = [\lambda_1(\boldsymbol{\theta}), \ldots, \lambda_m(\boldsymbol{\theta})]'$ denote the vector of mean counts for the parameter vector $\boldsymbol{\theta}$. For the vector $\mathbf{v} = [v_1, \ldots, v_k]'$, define the $k \times k$ matrix $\mathbf{D}(\mathbf{v})$ with (i, j)th element given by, for $i, j = 1, \ldots, k$,

$$D_{i,j}(\mathbf{v}) = \begin{cases} v_i, & i = j, \\ 0, & \text{otherwise.} \end{cases} \tag{15.24}$$

The linearized Gaussian approximation to the likelihood at the bth stage is then given by, for $b = 1, \ldots, s$,

$$\hat{\ell}_b(\boldsymbol{\theta}; \mathbf{z}) = N(\mathbf{z}; \boldsymbol{\lambda}(\hat{\boldsymbol{\theta}}^{b-1}) + \mathbf{L}_b(\boldsymbol{\theta} - \hat{\boldsymbol{\theta}}^{b-1}), \mathbf{D}(\boldsymbol{\lambda}(\hat{\boldsymbol{\theta}}^{b-1})))^{\gamma_b} \tag{15.25}$$

where $\mathbf{L}_b = \nabla_{\boldsymbol{\theta}} \boldsymbol{\lambda}_b(\boldsymbol{\theta})|_{\boldsymbol{\theta}=\hat{\boldsymbol{\theta}}^{b-1}}$ is the $m \times 6r$ gradient matrix. The Gaussian approximation to the bth target PDF is then found as

$$\begin{aligned} \hat{\pi}_b(\boldsymbol{\theta}) &\propto \hat{\ell}_b(\boldsymbol{\theta}; \mathbf{z}) \hat{\pi}_{b-1}(\boldsymbol{\theta}) \\ &\propto N(\mathbf{z}; \boldsymbol{\lambda}(\hat{\boldsymbol{\theta}}^{b-1}) + \mathbf{L}_b(\boldsymbol{\theta} - \hat{\boldsymbol{\theta}}^{b-1}), \mathbf{D}(\boldsymbol{\lambda}(\hat{\boldsymbol{\theta}}^{b-1}))/\gamma_b) N(\boldsymbol{\theta}; \hat{\boldsymbol{\theta}}^{b-1}, \mathbf{P}^{b-1}) \\ &\propto N(\boldsymbol{\theta}; \hat{\boldsymbol{\theta}}^b, \mathbf{P}^b) \end{aligned} \tag{15.26}$$

where the mean and covariance matrix of the Gaussian approximation to the bth target PDF are

$$\hat{\boldsymbol{\theta}}^b = \hat{\boldsymbol{\theta}}^{b-1} + \mathbf{K}_b[\mathbf{z} - \boldsymbol{\lambda}(\hat{\boldsymbol{\theta}}^{b-1})], \tag{15.27}$$

$$\mathbf{P}^b = \mathbf{P}^{b-1} - \mathbf{K}_b \mathbf{L}_b \mathbf{P}^{b-1} \tag{15.28}$$

with $\mathbf{S}_b = \mathbf{L}_b \mathbf{P}^{b-1} \mathbf{L}_b' + \mathbf{D}(\boldsymbol{\lambda}(\hat{\boldsymbol{\theta}}^{b-1}))/\gamma_b$ and $\mathbf{K}_b = \mathbf{P}^{b-1} \mathbf{L}_b' \mathbf{S}_b^{-1}$. Equation (15.26) follows from the previous line by the application of a well-known rule governing products of Gaussian PDFs [31]. The final estimate of $\boldsymbol{\theta}$ is obtained as $\hat{\boldsymbol{\theta}} = \hat{\boldsymbol{\theta}}^s$. The algorithm is referred to as progressive correction with a Gaussian approximation, or PCG.

The linearized Gaussian likelihood approximation is at the heart of perhaps the most well-known nonlinear filtering approximation, the extended Kalman filter (EKF) [32]. The notorious unreliability of the EKF in certain situations may suggest the use of alternative methods of obtaining a Gaussian approximation, for example, Laplace approximations [7] or the unscented transformation [33]. Nevertheless, in a gradual correction framework, it is possible to achieve good results using a linearized Gaussian approximation. This can be explained by considering the effects of likelihood linearization on posterior PDF approximation. When the likelihood is linearized, errors in the posterior PDF approximation can be expected if the likelihood is large in parts of the state space that are not close to the point about which linearization is performed. The approximation errors are exacerbated if the likelihood is large in only a small part of the parameter space, that is, the measurements are informative about the unknown parameters. In the first iteration linearization is performed about a parameter value, derived from the prior, which is likely to be far from parameter values supported by the likelihood. Application of the full measurement correction would then result in gross errors. However, at this stage progressive correction applies only a partial measurement correction so that errors due to linearization are reduced. If the size γ_1 of the correction is sufficiently small then the errors caused by linearization can be made very small. In the latter stages, as the progressive measurement correction approaches the full measurement correction, the parameter value about which linearization is performed should approach the desired part of the parameter space so that errors are minimized even for an informative measurement.

Calculation of the mean count $\lambda_j(\boldsymbol{\theta})$ for a parameter value $\boldsymbol{\theta}$ and the gradient matrix $\mathbf{L}_b$ requires the evaluation of integrals of the form

$$I = \int \mathbf{u}(\boldsymbol{\theta}, \mathbf{x}) k(\mathbf{x} - \boldsymbol{\rho}_j) N(\mathbf{x}; \boldsymbol{\mu}, \Sigma)\, d\mathbf{x}, \tag{15.29}$$

where the function $\mathbf{u}(\bullet)$ depends on the quantity being computed. A Monte Carlo approximation can be obtained by drawing $\mathbf{x}^c \sim N(\boldsymbol{\mu}, \Sigma)$ for $c = 1, \ldots, d$ and computing

$$I \approx 1/d \sum_{c=1}^{d} \mathbf{u}(\boldsymbol{\theta}, \mathbf{x}^c) k(\mathbf{x}^c - \boldsymbol{\rho}_j). \tag{15.30}$$

15.3.3 Simulation Results

Simulations are performed for the scenarios shown in Figure 15.4. The radiation field is a mixture with two components in the first scenario and three components in the second scenario. The plots show the 2-sigma ellipses of the mixture components along with the sensor positions, indicated by dots. The line width of each ellipse is proportional to the peak field contribution of the corresponding component. These scenarios are intended to model a scenario with one reasonably strong source superimposed on an unknown spatially varying background. The second scenario is considerably more challenging than the first because, in addition to there being an extra mixture component and therefore more parameters to estimate, the two background mixture components overlap significantly. The field is observed by a regular grid of 15×15 sensors distributed over the $150\,\mathrm{m} \times 150\,\mathrm{m}$ surveillance region. The distance threshold used in the kernel (15.18) to model saturation is set to $\upsilon = 5$.

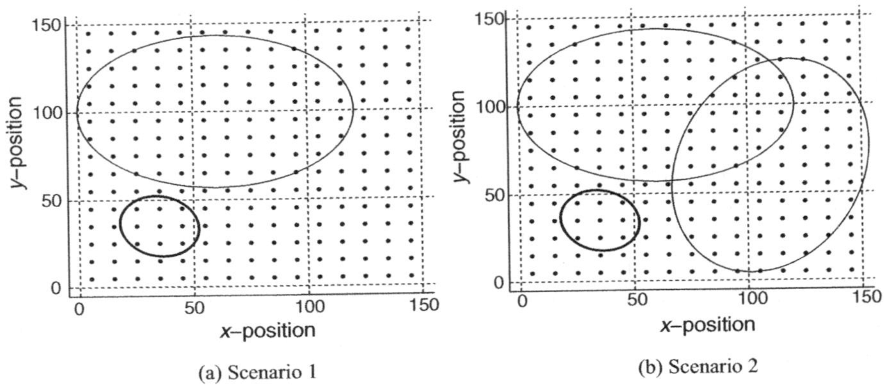

Figure 15.4 Simulation scenarios for distributed source estimation: The ellipses represent the locations and extents of the radiation field mixture components. The line width of each ellipse indicates the strength of the component. The dots are the sensor positions.

The following parameters are used. Progressive correction is implemented with $s = 40$ steps for both the importance sampling approximation described in Section 15.2.2 and the linearized Gaussian approximation of Section 15.3.2. The correction factors are evenly distributed on a log scale with $\gamma_{40}/\gamma_1 = 10^4$. A sample size of 300 is used for the Monte Carlo approximation (15.30) to the integrals required for the mean count and gradient. The prior PDF is

$$\pi_0(\boldsymbol{\theta}) = \prod_{i=1}^{q} G(a_i; 2, 2000) U_S(\boldsymbol{\mu}_i) U_{(-1,1)}(\rho_i) G(\varsigma_i; 5, 20) G(\sigma_i; 5, 20), \qquad (15.31)$$

where $S = [0, 150]^2$ is the surveillance region.

Algorithm performance is measured by the RMSE of the position and spread parameter estimates computed over 100 realizations for each scenario. PCIS is implemented with several sample sizes between 10 and 1000. The results are shown in Figure 15.5 for the two-component mixture field and in Figure 15.6 for the three-component mixture field. The performance achieved by the Gaussian approximation is far better than that achieved by the importance sampling approximation. Moreover, this improvement is achieved with a fraction of the computational expense. For example, for the two-component mixture, PCG has a computational expense, as measured by runtime, which is less than 1% that of PCIS with a sample size of 1000 but achieves superior accuracy. The improvements offered by PCG compared to PCIS are even more dramatic for the three-component mixture. The observed results are remarkable given the relatively poor performance usually achieved by linearized likelihood approximations, particularly in complicated problems. As discussed in Section 15.3.2, an intuitive explanation for this is that progressive correction acts in such a way as to reduce errors caused by linearization.

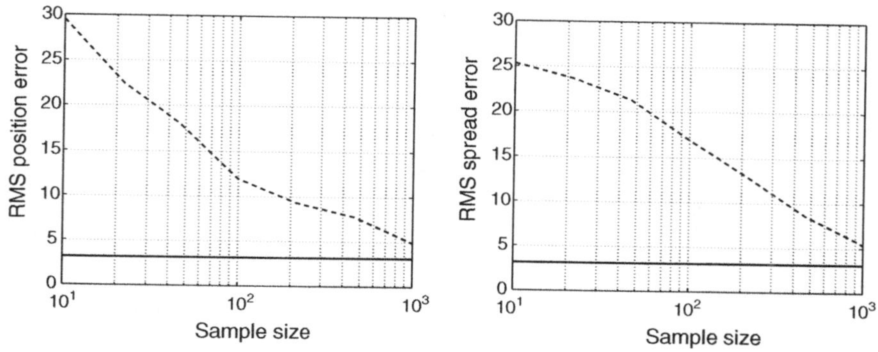

(a) (b)

Figure 15.5 RMS errors of (a) position and (b) spread parameters for PCIS (dashed) and PCG (solid) plotted against sample size for a two-component mixture field.

Since no procedure is given here for determining the number r of mixture components it is of interest to examine the dependence of field estimation accuracy on the assumed value of r. The results of applying PCG with $r = 2$ and 4 to the three-component mixture of Figure 15.4 are shown in Figure 15.7. The plots show how the mixture parameters estimated by PCG adapt to provide a reasonable estimate of the three-component mixture field. A more detailed analysis of the effects of mismatch in the number of components is given in Table 15.5. This shows the RMS error, averaged over 500 realizations, of the PCG field estimator of the three-component mixture field for assumed source numbers of $1, \ldots, 5$. Little difference in the RMS error is observed for $r = 4$ and 5 compared to the matched model. However, for $r = 2$, and particularly for $r = 1$, the RMS error increases significantly compared to matched model. Thus, in this example, underestimating the model complexity is much worse

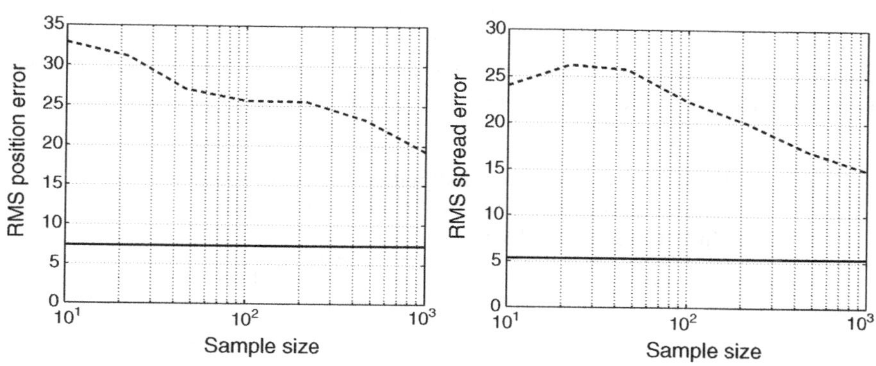

(a) (b)

Figure 15.6 RMS errors of (a) position and (b) spread parameters for PCIS (dashed) and PCG (solid) plotted against sample size for a three-component mixture field.

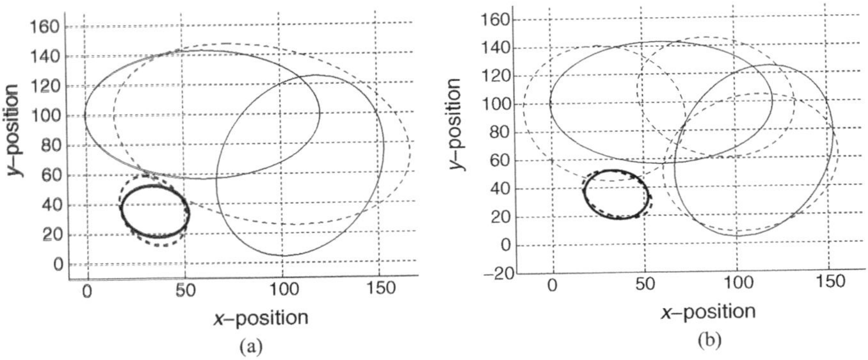

Figure 15.7 PCG estimates (dashed) of the three-component field (solid) using (a) $r = 2$ components and (b) $r = 4$ components. The ellipses represent the locations and extents of the mixture components. The line width of each ellipse indicates the strength of the component.

than overestimating the model complexity. In fact, apart from increased computational expense, the penalty for overestimating model complexity is practically negligible.

15.3.4 Experimental Results

Distributed source estimation is considered for the experimental data described in Section 15.2.4. Although the sources in this experiment are essentially point sources, distributed source estimation is still applicable because it does not require assumptions about the level of background radiation.

Radiation field estimation is performed using the experimental data with minimal assumptions about the nature of the field under consideration. In particular, no assumptions are made about the level of background radiation and the number of components in the Gaussian mixture representation of the field is not matched to the number of point sources. It is assumed only that the field can be accurately represented by a Gaussian mixture with $r = 5$ components. Under these conditions algorithm performance cannot be assessed by comparing parameter estimates to true values. Instead, algorithm performance is measured by how well the estimated field can be used to

TABLE 15.5 RMS Field Estimation Error of PCG with Different Assumed Numbers of Components for the Three-Component Mixture

Assumed r	RMS Error
1	0.321
2	0.155
3	0.102
4	0.103
5	0.104

TABLE 15.6 Mean True and False Source Detections Using Experimental Data

	One Source	Two Sources	Three Sources
True detections	1	2	2.86
False detections	0	0	0.01

find sources. Sources are hypothesized to lie at the locations of peaks of the field estimate with value greater than a specified threshold. Hypothesized sources that are within 10 m of a true source position are classified as true, otherwise they are false. If multiple detected sources are true for the same true source position and no other all but one of these is classified as false. The mean number of true and false sources are compiled over a number of measurement realisations. Only PCG is considered in this performance analysis. It is implemented with $s = 75$ correction steps with the correction factors evenly distributed on a log scale with $\gamma_{75}/\gamma_1 = 10^3$. The Monte Carlo approximation to the integral (15.30) is used with $d = 200$ samples. The prior PDF is

$$\pi_0(\boldsymbol{\theta}) = \prod_{i=1}^{r} G(a_i; 2, 500) U_S(\boldsymbol{\mu}_i) U_{(-1,1)}(\rho_i) G(\varsigma_i; 2, 5) G(\sigma_i; 2, 5), \qquad (15.32)$$

where the surveillance area $S \subset \mathbb{R}^2$ is different for each scenario. The kernel for the measurement model is given by (15.18) with the distance threshold set to $\upsilon = 5$.

The source detection results, averaged over 100 measurement realizations, are shown in Table 15.6. The sources are detected on every occasion in the one and two source scenarios with no false sources. In the three source scenario the true sources are found on most occasions and one false source is found over the 100 realizations. These demonstrate that the radiation field estimation algorithm can be used with real data in a nonparametric fashion, where the term nonparametric is used in the sense that the assumed measurement model need not closely match the way in which measurements are generated.

15.4 SEARCHING FOR POINT SOURCES

Section 15.2 considered parameter estimation of a known number of point sources using a batch of measurements obtained from a fixed network of sensors. This rather restrictive scenario is extended in this section to the case where the number of sources is unknown and the measurements are acquired and processed sequentially. This models the case of an observer equipped with a sensor moving through a surveillance area searching for radiation sources. In addition to the problem of detecting sources and estimating their parameters, it is also necessary to determine the best locations from which to acquire measurements.

The simplistic approach to search is to follow a predefined path that scans the area in a uniform manner [34]. One such method, parallel sweep search, was applied to detect a radioactive source of known strength in [35]. Such methods do not take advantage of the fact that previously acquired measurements can be used to guide future sensing decisions. The value of exploiting the currently available information to manage sensors has been recognized in many fields of engineering, such as bearings-only motion control [36, 37], target tracking [38, 39], classification [40], and robotics [41].

In the context of search for nuclear material, information-driven observer control has been applied in [42] and [43, 44]. Cortez et al. [42] divide the search area into a regular grid of cells. Using a simplified measurement model, where the counts observed in a particular cell depend only on the radiation level in that cell, a search strategy is proposed where (a) the observer stays in one cell until the variance of the measurement falls below a certain threshold; (b) the observer moves to a neighboring cell selected based on its ability to reduce the uncertainty in the estimated radiation field. Ristic and Gunatilaka [43] consider a single-sensor single-source setup where the observer control is based on the Fisher information gain of future measurements. Klimenko et al. [45] also consider a nuclear search strategy along a predefined path, but they adaptively change the exposure time, using ideas from the sequential testing theory.

In [44], the problem of choosing a future search strategy on the basis of previously acquired measurements is formulated as a partially observable Markov decision process (POMDP) [46]. In a POMDP, the information contained in previously acquired measurements is represented by the posterior PDF of the parameters that characterize the process of interest. The aim is to choose a sensing strategy that acts on the posterior PDF in a desirable way. The application of this framework to radiation source search, using information divergence to measure the desirability of candidate sensing actions, is described in this section. Sequential Monte Carlo (SMC) methods for realizing the POMDP are described and applied to simulated and experimental data.

15.4.1 Model

The unknown number of point sources in the surveillance region $S \subset \mathbb{R}^2$ is denoted as $r \in \{0, \ldots, \bar{r}\}$ where $\bar{r}$ is the assumed maximum number of sources. As in Section 15.2.1, the parameter vector for the ith source is $\boldsymbol{\theta}_i = [\mathbf{x}_i', a_i]' \in S \times \mathbb{R}^+$ where $\mathbf{x}_i$ is the source position in Cartesian coordinates and a_i is the source intensity. The source parameter vectors are collected into a stacked vector $\boldsymbol{\theta} = [\boldsymbol{\theta}_1', \ldots, \boldsymbol{\theta}_r']'$. Since the number of sources is unknown the parameter of interest is $\boldsymbol{\vartheta} = [\boldsymbol{\theta}', r]' \in \bigcup_{i=0}^{\bar{r}}(S \times \mathbb{R}^+)^i \times \{i\}$.

Radiation measurements are made by $o \geq 1$ mobile observers using identical low-cost omnidirectional GM counters and transmitted to a fusion center. The sensors use, in general, different exposure durations and transmit measurements as they become available. Thus, at each measurement sampling instant, it is likely that only a subset of the sensors will contribute measurements. Let $\mathbf{z}_k \in \mathbb{N}^{m_k}$ denotes the vector of measurements received at the fusion centre at the kth sampling instant t_k, where $m_k \leq o$ is the number of sensors contributing measurements. The vector of sensing parameters, applied in order to acquire the jth measurement $z_{k,j}$ at time t_k, is denoted as $\boldsymbol{\zeta}_{k,j}$ for $j = 1, \ldots, m_k$. Here $\boldsymbol{\zeta}_{k,j} = [\boldsymbol{\rho}'_{k,j}, \tau_{k,j}]'$ with $\boldsymbol{\rho}_{k,j}$ being the sensor position and $\tau_{k,j}$ the exposure time. The likelihood of $\boldsymbol{\vartheta}$ given that $\mathbf{z}_k$ is observed with sensing parameters $\boldsymbol{\zeta}_{k,1}, \ldots, \boldsymbol{\zeta}_{k,m_k}$ is

$$\ell(\boldsymbol{\vartheta}; \mathbf{z}_k, \boldsymbol{\zeta}_{k,1:m_k}) = \prod_{j=1}^{m_k} P(z_{k,j}; \lambda(\boldsymbol{\theta}; \boldsymbol{\zeta}_{k,j})), \tag{15.33}$$

where, for $\boldsymbol{\zeta} = [\boldsymbol{\rho}', \tau]'$,

$$\lambda(\boldsymbol{\theta}; \boldsymbol{\zeta}) = \left[\mu(\boldsymbol{\rho}) + \sum_{i=1}^{r} \frac{a_i}{\|\boldsymbol{\rho} - \mathbf{x}_i\|^2}\right]\tau \tag{15.34}$$

with $\mu(\boldsymbol{\rho})$ the known count rate due to the background radiation at position $\boldsymbol{\rho}$. In (15.33) $\boldsymbol{\zeta}_{k,1:m_k}$ is the sequence $\boldsymbol{\zeta}_{k,1}, \ldots, \boldsymbol{\zeta}_{k,m_k}$. Similar notation is used for all sequences. Measurements are assumed to be independently distributed. As in Section 15.2 a number of assumptions are inherent in the measurement model (15.33). These include that the point sources are isotropic, the sensors are omnidirectional and that the surveillance region is sufficiently small that atmospheric attenuation can be ignored.

The aim is to determine the number r of sources and accurately estimate their parameters in an efficient manner by controlling the sensing parameters of the observers, that is, the position at which measurements are made and the exposure duration. This search problem is posed as a POMDP in the following section.

15.4.2 Sequential Search Using a POMDP

A summary of the sequential sensor management procedure is given by Algorithm 4. The key is to permit the selection of sensing actions to depend on what has been learnt from previous measurements. Mathematically this can be done using the POMDP framework [46].

Algorithm 4: Sequential Sensor Management

---

establish a prior PDF $\pi_0(\cdot)$
set $k \leftarrow 0$
perform the initial sensing action
while *still searching* **do**
    set $k \leftarrow k + 1$
    collect the measurement z_k.
    compute the posterior PDF $\pi_k(\cdot)$
    select and perform the next sensing action
end

The elements of a POMDP include the posterior PDF, also referred to as the information state, a control vector belonging to a set of admissible controls and a reward function. These elements interact to provide a way of selecting appropriate sensing actions. In particular, the control vector can be viewed as acting on the posterior PDF. The benefits of this action of the control vector on the posterior PDF are measured by the reward function. The aim is to select the control vector from the admissible set that produces the highest reward. In principle, the long-term effects of the current sensing action should be considered by optimizing the rewards accumulated over several measurement samples. This leads to an intractable optimization problem that must be solved approximately. The benefits of long-term scheduling for the problem setup considered here, where sensors are not prevented by the environment from measuring sources or moving to new positions, would not be commensurate with the enormous computational expense required even for a rough approximation solution. Therefore, sensing actions are selected here on the basis of the current reward only. This is referred to as myopic scheduling.

Let J_k denotes the set of sensors contributing measurements at time t_k. After receiving $\mathbf{z}_k$ at time t_k, it is desired to select new sensing actions for sensors J_k. Note that the remaining sensors are in the process of collecting measurements at this time. The actions used by the sensors are determined by control vectors $\zeta_{k,j}$, $j = 1, \ldots, m_k$, composed of the position and exposure duration of the next measurement. Let $Z_{j,k} \subset \mathbb{R}^2 \times \mathbb{R}^+$ denotes the set of admissible control vectors for the jth sensor that produced an observation at time t_k. New control vectors are to be selected jointly from the product space

$$Z_k = \bigtimes_{j=1}^{m_k} Z_{j,k}. \tag{15.35}$$

Let $D(\pi(\cdot), \mathbf{z}, \boldsymbol{\zeta}_{1:m_k}) \geq 0$ denote the reward obtained for observing the measurement vector $\mathbf{z}$ with sensor parameters $\boldsymbol{\zeta}_{1:m_k}$ when the current posterior PDF is $\pi(\cdot)$. Since the actions $\boldsymbol{\zeta}_{1:m_k}$ must be selected without collecting the measurement $\mathbf{z}$ the objective function is the expectation of the reward taken with respect to the prior distribution of $\mathbf{z}$. The sensing actions for the next measurements are then selected as

$$\boldsymbol{\zeta}^*_{1:m_k} = \arg \max_{\boldsymbol{\zeta}_{1:m_k} \in Z_k} \mathrm{E}_{\bar{\ell}_k(\cdot;\boldsymbol{\zeta}_{1:m_k})}[D(\pi_k(\cdot), \boldsymbol{\zeta}_{1:m_k}, \mathbf{z})], \tag{15.36}$$

where $\mathrm{E}_{\bar{\ell}_k(\cdot;\boldsymbol{\zeta}_{1:m_k})}$ denotes expectation with respect to the prior measurement PDF $\bar{\ell}_k(\cdot; \boldsymbol{\zeta}_{1:m_k})$ of the vector of measurements taken with sensing parameters $\boldsymbol{\zeta}_{1:m_k}$. This is given by

$$\bar{\ell}_k(\mathbf{z}; \boldsymbol{\zeta}_{1:m_k}) = \sum_{r=0}^{\infty} \int \ell(\boldsymbol{\theta}; \mathbf{z}, \boldsymbol{\zeta}_{1:m_k}) \pi_k(\boldsymbol{\theta}, r)\, d\boldsymbol{\theta}. \tag{15.37}$$

When sensing actions are selected according to (15.36), the quality of the search depends to a large extent on the reward function. In the detection–estimation problem the reward function should reflect the fact that sensor measurements can provide information about both undetected sources as well as sources that are known to exist with high probability but whose parameters may be poorly known. Because of these dual goals in many cases there is a trade-off between fast detection and accurate estimation. This trade-off is managed here by performing sensing actions that maximize the amount of acquired information. The amount of information provided by a measurement can be measured in a number of ways. It is important to consider not just the quantity of information provided by the measurement but also how it relates to existing knowledge. This can be done using the difference between the posterior PDF, after the measurement is acquired, and the prior PDF. The idea is that a useful measurement will produce a posterior PDF that is different, according to some measure, to the prior PDF.

The difference between two PDFs can be measured in many ways, including by distances, such as the l_p-norm, or divergences, such as the Kullback–Leibler divergence (KLD). Here the Rényi divergence [47] is used to measure the difference between the posterior and prior PDFs. The reward function is then given by, for $\alpha \in \mathbb{R} \setminus \{0, 1\}$,

$$D(\pi_k(\cdot), \mathbf{z}, \boldsymbol{\zeta}_{1:m_k}) = \frac{1}{\alpha - 1} \log \mathrm{E}_{\pi_k(\cdot)} \left[\left(\frac{\pi_{k+1}(\boldsymbol{\vartheta})}{\pi_k(\boldsymbol{\vartheta})} \right)^{\alpha} \right]. \tag{15.38}$$

Although this is not reflected in the notation, the dependence of the reward on the control vectors $\boldsymbol{\zeta}_{1:m_k}$ and the measurement $\mathbf{z}$ enters in the posterior PDF $\pi_{k+1}(\cdot)$ at the next time step. This can be seen more clearly after Bayes' rule is applied to (15.38) to obtain

$$D(\pi_k(\cdot), \mathbf{z}, \boldsymbol{\zeta}_{1:m_k}) = \frac{1}{\alpha - 1} \log \frac{\mathrm{E}_{\pi_k(\cdot)}[\ell(\boldsymbol{\vartheta}; \mathbf{z}, \boldsymbol{\zeta}_{1:m_k})^{\alpha}]}{\{\mathrm{E}_{\pi_k(\cdot)}[\ell(\boldsymbol{\vartheta}; \mathbf{z}, \boldsymbol{\zeta}_{1:m_k})]\}^{\alpha}}. \tag{15.39}$$

The Rényi divergence includes other divergences as special cases. For instance, the KLD is obtained by letting $\alpha \to 1$ and the Bhattacharyya distance is obtained for $\alpha = 1/2$.

There are three computations that must be performed to realize the sequential search: computation of the posterior PDF, computation of the expected reward, and optimal selection of the control vector. None of these computations are tractable. Numerical approximations are described in the following section.

15.4.3 Implementation of the POMDP

Monte Carlo algorithms for approximation of the posterior PDF given a batch of measurements were described in Section 15.2.2. The same basic ideas are applicable here, although some modifications are required to accommodate two important differences between the problems considered in Sections 15.2 and 15.4. First, because the number of sources is unknown, the parameter vector here includes the number of sources in addition to their parameters. Second, measurements are processed sequentially here rather than in a batch, that is, the posterior PDF $\pi_k(\cdot)$ is found, or approximated, using $\mathbf{z}_k$ and $\pi_{k-1}(\cdot)$ rather than the measurement sequence $\mathbf{z}_{1:k}$.

The use of SMC methods to solve sequential Bayesian estimation problems is well established [48]. SMC methods typically involve drawing parameter samples from an importance density and computing sample weights that depend only on the current measurement, not the complete measurement sequence. This works well for dynamic parameters whose evolution in time is subject to randomness. For static parameters, such as the source parameters, the sample values are fixed in time and therefore cannot be adjusted as information is gradually acquired from the measurements. As a result the sample rapidly degenerates as measurements are collected. A number of techniques have been proposed to address this problem, none of which are completely satisfactory. The simplest method is to add process noise to the static parameters. Although, in general, this artificially increases the variance of the prior PDF, it is possible to randomly shift the parameter samples in such a way that the variance is preserved [49]. MCMC methods require reprocessing of the complete measurement sequence and thus cannot be considered to be truly sequential [50]. Methods based on sufficient statistics are not universally applicable [51].

The nature of the measurement model described in Section 15.4.1 prevents the use of sufficient statistics and MCMC methods are cumbersome due to the need to process the entire measurement sequence at each sample time. Therefore, the approach taken here is to simply add a random evolution to the source number and perform progressive correction on the source parameters. The process of perturbing samples of static parameters can be regarded as a form of regularization that reduces approximation variance at the expense of introducing a bias. Note that regularization is an inherent part of progressive correction, as summarized by Algorithm 3, since source parameter samples are perturbed at each stage by drawing from a kernel density.

Regularization is applied to the source number r by permitting new sources to appear and currently existing sources to disappear between measurements. At most one source is permitted to appear with probability ν and sources persist with probability η. Let $b \in \{0, 1\}$ denotes the indicator for the appearance of a new source and $e_i \in \{0, 1\}$ denotes the indicator for the persistence of the ith source. Then, the regularized prior is

$$\varpi_k(\boldsymbol{\vartheta}) = \sum_{s=0}^{\bar{r}} \int f(\boldsymbol{\theta}, r|\boldsymbol{\phi}, s)\pi_{k-1}(\boldsymbol{\phi}, s)\, d\boldsymbol{\phi}. \tag{15.40}$$

The transition density $f(\cdot|\cdot)$ is given by, for $r = 0, \ldots, \bar{r}$,

$$f(\boldsymbol{\theta}, r|\boldsymbol{\phi}, s) = B_s^{-1} \sum_{(b,e_1,\ldots,e_s)\in V_{r,s}} \nu^b(1-\nu)^{1-b} \prod_{i=1}^{s} \eta^{e_i}(1-\eta)^{1-e_i}\, g(\boldsymbol{\theta}|\boldsymbol{\phi}, b, e_1, \ldots, e_s), \tag{15.41}$$

where $V_{r,s} = \{(b, e_1, \ldots, e_s) \in \{0, 1\}^{s+1} : b + e_1 + \cdots + e_s = r\}$ and

$$B_s = \sum_{r=0}^{\bar{r}} \sum_{(b,e_1,\ldots,e_s)\in V_{r,s}} \nu^b(1-\nu)^{1-b} \prod_{i=1}^{s} \eta^{e_i}(1-\eta)^{1-e_i}, \quad (15.42)$$

$$g(\boldsymbol{\theta}|\boldsymbol{\phi}, b, e_1, \ldots, e_s) = \prod_{i=1}^{r-b} \delta(\boldsymbol{\theta}_i - \boldsymbol{\phi}_{c_i})[\pi_0(\boldsymbol{\theta}_r)]^b \quad (15.43)$$

with

$$c_i = \min\left\{p : \sum_{j=1}^{p} e_j = i\right\}. \quad (15.44)$$

The indices $c_1, \ldots, c_{r-b}$ define a relabeling of the sources in (15.43) when sources are removed. For instance, if the jth of s sources is removed then $c_i = i$ for $i = 1, \ldots, j-1$ and $c_i = i+1$ for $i = j, \ldots, s-1$.

Let $\{r^{k-1,i}, \boldsymbol{\theta}^{k-1,i}\}$ denote the collection of samples representing the posterior PDF $\pi_{k-1}(\cdot)$ for the first $k-1$ measurements. Substituting this approximation into (15.40) gives

$$\varpi_k(\boldsymbol{\vartheta}) \approx 1/n \sum_{i=1}^{n} f(\boldsymbol{\theta}, r|\boldsymbol{\theta}^{k-1,i}, r^{k-1,i}). \quad (15.45)$$

A sample from the regularized prior $\varpi_k(\cdot)$ is obtained by drawing $(\tilde{\boldsymbol{\theta}}^{k,i}, \tilde{r}^{k,i}) \sim f(\cdot|\boldsymbol{\theta}^{k-1,i}, r^{k-1,i})$ for $i = 1, \ldots, n$. Let $R_k = \bigcup_{i=1}^{n}\{\tilde{r}^{k,i}\}$ denote the set of source numbers supported by the samples. For $r \in R_k$, let $S_{k,r} = \{i : \tilde{r}^{k,i} = r\}$ denote the collection of samples with r sources. The number of such samples is denoted as $h_{k,r}$. PCIS is used to find a $h_{k,r}$ sample approximation to the posterior PDF of the source parameters conditional on each source number $r \in R_k$. Let $\breve{\boldsymbol{\theta}}^{k,i}$, $i \in S_{k,r}$ denote the source parameter samples drawn by PCIS for approximation of $\pi_k(\cdot|r)$. For $r \in R_k$, the posterior PDF is then approximated by

$$\begin{aligned} \pi_k(\boldsymbol{\theta}, r) &= \mathrm{P}(r|\mathbf{z}_{1:k})\pi_k(\boldsymbol{\theta}|r) \\ &\approx \mathrm{P}(r|\mathbf{z}_{1:k})/h_{k,r} \sum_{i\in S_{k,r}} \delta\left(\boldsymbol{\theta} - \breve{\boldsymbol{\theta}}^{k,i}\right). \end{aligned} \quad (15.46)$$

Note that the weights of the source parameter samples found by PCIS are uniform because resampling is performed during the PCIS iterations. The uniform weights are appropriate for the conditional posterior PDF $\pi_k(\cdot|r)$. The weights associated with these samples for the posterior PDF $\pi_k(\cdot)$ must take into account the posterior source number probability, that is, the weight of the ith sample for $i \in S_{k,r}$, is

$$w_k^i = \mathrm{P}(r|\mathbf{z}_{1:k})/h_{k,r}. \quad (15.47)$$

Using Bayes' rule and expanding the prior PDF of the measurement conditional on the source number gives

$$\begin{aligned} \mathrm{P}(r|\mathbf{z}_{1:k}) &\propto \int \ell(\boldsymbol{\theta}, r; \mathbf{z}_k)\varpi_k(\boldsymbol{\theta}|r)\, d\boldsymbol{\theta}\, \mathrm{P}(r|\mathbf{z}_{1:k-1}) \\ &\approx \kappa_{k,r}\, \mathrm{P}(r|\mathbf{z}_{1:k-1}), \end{aligned} \tag{15.48}$$

where

$$\kappa_{k,r} = 1/h_{k,r} \sum_{i \in S_{k,r}} \ell(\tilde{\boldsymbol{\theta}}^{k,i}, \tilde{r}^{k,i}; \mathbf{z}_k). \tag{15.49}$$

Thus, for $i \in S_{k,r}$, $r \in R_k$, the sample weights are

$$w_k^i \propto \kappa_{k,r}\, \mathrm{P}(r|\mathbf{z}_{1:k-1})/h_{k,r}. \tag{15.50}$$

A recursion of the SMC procedure for approximating the posterior PDF is summarized by Algorithm 5. Note that, since resampling is performed at the end of each recursion, the final samples are evenly weighted so that $\mathrm{P}(r|\mathbf{z}_{1:k-1}) \approx h_{k,r}/n$. Substituting this approximation to the source number posterior into (15.50) gives the weight calculation of step 5.

Algorithm 5: Sequential Monte Carlo Approximation of the Posterior PDF

---

draw $(\tilde{\boldsymbol{\theta}}^{k,i}, \tilde{r}^{k,i}) \sim f(\cdot|\boldsymbol{\theta}^{k-1,i}, r^{k-1,i})$ for $i = 1, \ldots, n$

find $R_k = \bigcup_i \{\tilde{r}^{k,i}\}$, $S_{k,r} = \{i : \tilde{r}^{k,i}\}$ and $h_{k,r} = |S_{k,r}|$ for $r \in R_k$

for $r \in R_k$ **do**

draw $\breve{\boldsymbol{\theta}}^{k,i} \sim \pi_k(\cdot|r)$ for $i \in S_{k,r}$ using PCIS (Algorithm 3)

compute the unnormalized weights, for $i \in S_{k,r}$,

$$\tilde{w}_k^i = 1/h_{k,r} \sum_{j \in S_{k,r}} \ell(\tilde{\boldsymbol{\theta}}^{k,j}, \tilde{r}^{k,j}; \mathbf{z}_k)$$

end

compute normalized weights, for $i = 1, \ldots, n$,

$$w_k^i = \tilde{w}_k^i \Big/ \sum_{j=1}^{n} \tilde{w}_k^j$$

resample by selecting, for $i = 1, \ldots, n$, $j^i = c$ with probability w_k^c and setting $\boldsymbol{\theta}^{k,i} = \breve{\boldsymbol{\theta}}^{k,j^i}$, $r^{k,i} = \tilde{r}^{k,j^i}$.

Given a sample-based approximation to the posterior PDF an approximation to the reward can be found quite easily using (15.39). Replacing $\pi_k(\cdot)$ with the sample approximation represented by $\{r^{k,i}, \boldsymbol{\theta}^{k,i}\}$ gives

$$D(\pi_k(\cdot), \mathbf{z}, \boldsymbol{\zeta}_{1:m_k}) \approx \frac{1}{\alpha - 1} \log(\gamma_\alpha(\mathbf{z}, \boldsymbol{\zeta}_{1:m_k})/\gamma_1(\mathbf{z}, \boldsymbol{\zeta}_{1:m_k})^\alpha), \tag{15.51}$$

where

$$\gamma_\alpha(\mathbf{z}, \boldsymbol{\zeta}_{1:m_k}) = 1/n \sum_{i=1}^{n} \ell(\boldsymbol{\theta}^{k,i}, r^{k,i}; \mathbf{z}, \boldsymbol{\zeta}_{1:m_k})^\alpha. \tag{15.52}$$

For $\alpha = 1$, $\gamma_\alpha(\cdot, \boldsymbol{\zeta})$ is an approximation to the prior measurement PDF. Use of the sample-based approximation (15.51) of the Rényi divergence results in the following approximation to the expected reward

$$\begin{aligned} &\mathrm{E}_{\bar{\ell}_k(\cdot;\boldsymbol{\zeta}_{1:m_k})}[D(\pi_k(\cdot), \mathbf{z}, \boldsymbol{\zeta}_{1:m_k})] \\ &\approx \frac{1}{\alpha - 1} \sum_{\mathbf{z} \in \mathbb{N}^{m_k}} \log(\gamma_\alpha(\mathbf{z}, \boldsymbol{\zeta}_{1:m_k})/\gamma_1(\mathbf{z}, \boldsymbol{\zeta}_{1:m_k})^\alpha)\gamma_1(\mathbf{z}, \boldsymbol{\zeta}_{1:m_k}). \end{aligned} \tag{15.53}$$

Note that the unavailable prior measurement PDF $\bar{\ell}(\cdot; \boldsymbol{\zeta}_{1:m_k})$ has been replaced in (15.53) with the sample approximation (15.52). Equation (15.53) cannot be computed exactly and so must be approximated. One possibility is to limit the infinite summation over the measurements by considering only measurements for which the prior measurement PDF is non-negligible. Since the spread of the prior measurement PDF tends to increase with the measurement value, this strategy is computationally efficient if small values are expected for the measurements but will involve many terms if large measurements are probable. Instead a Monte Carlo approximation is used with measurements drawn from the sample-based approximation $\gamma_1(\cdot, \boldsymbol{\zeta}_{1:m_k})$ of the measurement prior. For $j = 1, \ldots, d$, draw $\mathbf{z}^j \sim \gamma_1(\cdot, \boldsymbol{\zeta}_{1:m_k})$, then approximate the mean reward as

$$\mathrm{E}_{\bar{\ell}_k(\cdot;\boldsymbol{\zeta}_{1:m_k})}[D(\pi_k(\cdot), \mathbf{z}, \boldsymbol{\zeta}_{1:m_k})] \approx \frac{1}{d(\alpha - 1)} \sum_{j=1}^{d} \log(\gamma_\alpha(\mathbf{z}^j, \boldsymbol{\zeta}_{1:m_k})/\gamma_1(\mathbf{z}^j, \boldsymbol{\zeta}_{1:m_k})^\alpha). \tag{15.54}$$

The computational expense of the approximation (15.54) is $O(dn)$. The choice of d is important in determining the computational expense of the sequential search procedure. The goal is not accurate approximation of the mean reward *per se*, but rather to produce an approximation that retains as closely as possible the ordering of the admissible control vectors. For these purposes, a sufficiently accurate approximation seems to be achievable with relatively small d. A reduction in computational expense can be obtained by computing the reward with a decimated sample representation of the posterior PDF when. Thus for the purposes of reward calculation the posterior

PDF $\pi_k(\cdot)$ is represented by the collection $\{r^{k,ei}, \theta^{k,ei}\}$ of $\lfloor n/e \rfloor$ samples, where e is a positive integer. The computational expense is then $O(dn/e)$.

The final implementation issue is selection of the optimal control vector. In principle, there are infinitely many positions and exposure times available to the sensor for its next measurement. In the practical implementation the set of available control vectors is reduced to a finite set by discretizing the set of allowable sensor positions and reducing the exposure times to a singleton. Discretization of the control space seems reasonable when the necessity of approximating the expected reward prevents fine distinctions being drawn between the merits of neighboring control vectors. The finite set of allowable control vectors for the jth sensor at time t_k is then

$$\begin{aligned} Z_{k,j} \\ &= (S \cap \{\rho_{k,j} + b\Delta/u[\cos(2\pi i/v), \sin(2\pi i/v)]' : b=1,\ldots,u, i=1,\ldots,v\}) \times \{\tau\}, \end{aligned} \tag{15.55}$$

where Δ is the maximum distance between consecutive measurements from a given sensor. Recall that S is the search region.

15.4.4 Simulation Results

The simulation scenario contains $r = 2$ sources that are searched for by $o = 2$ observers. The scenario, including the source positions, initial observer positions and surveillance area, is shown in Figure 15.8.

The set of allowable control vectors is given by (15.55) with $\Delta = 20$, $u = 2$, and $v = 6$. The exposure time is fixed to $\tau = 2$. Since the exposure time is the same at all

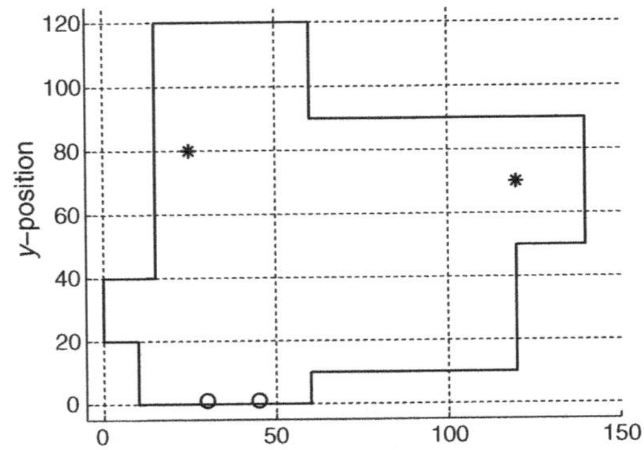

Figure 15.8 Simulation scenario for source search. The asterisks indicate source positions and the circles are the initial observer positions. The solid line indicates the search area.

times for both observers, the measurement vector contains measurements from both observers at each sampling instant.

At time zero no sources have been detected. Thus, the source number samples are $r^{0,i} = 0$ for $i = 1, \ldots, n$. At each sampling instant, after the measurements have been processed, the sample representation $\{r^{k,i}, \boldsymbol{\theta}^{k,i}\}$ of the posterior PDF $\pi_k(\cdot)$ is used to generate source parameter estimates as follows. Let $S_{k,r} = \{i : r^{k,i} = r\}$ denotes the set of samples with r sources and $h_{k,r} = |S_{k,r}|$. Note that the same notation was used in Section 15.4.3 for the prior sample. Since the samples are evenly weighted, the MAP estimate of the source number is $\hat{r}^k = \arg\max_r h_{k,r}$. The source parameters are then estimated as, for $i = 1, \ldots, \hat{r}^k$,

$$\hat{\boldsymbol{\theta}}_i^k = 1/h_{k,r} \sum_{j \in S_{k,r}} \boldsymbol{\theta}^{k,j}. \tag{15.56}$$

In forming the estimate (15.56) it is assumed that clustering is performed so that, for $i = 1, \ldots, \hat{r}^k$, the ith partition of the samples with indices in the set $S_{k,r}$ corresponds to the same source. The computationally efficient k-means algorithm can be used for this purpose. Two measures of performance are used to assess the quality of the source number and source parameter estimates. The first performance measure is the number of sources found, where a source is deemed found if there is a source parameter estimate within 5 m. The second measure of performance is the mean squared error of the position estimates of found sources.

Three simulation experiments are performed using the scenario of Figure 15.8. The first experiment examines the effect of different approximations of the mean reward. The second experiment considers the effect of the Rényi divergence parameter α on search performance while the third experiment compares the performance with multiple observers to performance with a single observer. In all of these experiments the sample size is $n = 5000$ and algorithm performance is determined by averaging over 100 realizations. The maximum number of sources is set to $\bar{r} = 3$. For the regularized prior of (15.40), the source appearance probability is set to $\nu = 0.005$ and the source persistence probability is $\eta = 0.997$. The prior PDF used for the proposal of added sources in the regularized prior is

$$\pi_0(\mathbf{x}, a) = U_S(\mathbf{x}) G(a; 5, 2000). \tag{15.57}$$

For a given sample size n, the computational expense of approximating the mean reward using (15.54) depends on the number d of measurement samples drawn from the prior and the decimation factor e applied to the sample representation of the posterior PDF. The purpose of the first experiment is to examine the effects of these parameters. Simulations are performed for two combinations of e and d: $e = 50, d = 100$, and $e = 2, d = 4$. The computational expense for these combinations is approximately equal. The Rényi divergence parameter is set to $\alpha = 0.5$. The mean number of sources found and RMS position error are shown in Figure 15.9. These results clearly show that decimation of the posterior PDF representation is a better way of reducing computational expense than using a smaller sample size for the Monte Carlo approximation

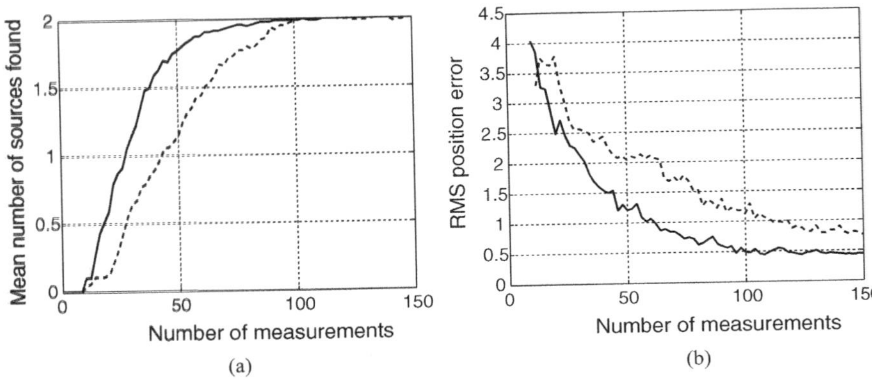

Figure 15.9 Plots of (a) the mean number of sources found and (b) the RMS position error of found sources for source search in the scenario of Figure 15.8. Results are shown for mean reward approximation with decimation factor $e = 50$ and prior measurement sample size $d = 100$ (solid) and $e = 2$, $d = 4$ (dashed).

of the mean reward. This conclusion is based on the supposition that more accurate approximation of the mean reward results in better observer trajectories, that is, trajectories that provide quicker source detection and more accurate source parameter estimation.

The reward function used here is actually a class of reward functions parameterized by the divergence parameter α. It is not clear which value of α is most suitable for source search. Figure 15.10 shows the search performance for $\alpha = 0.1, 0.5,$ and 0.999. For $\alpha = 0.999$, the Rényi divergence is almost the same as the Kullback–Leibler divergence. No significant differences in performance are evident for the various values of α.

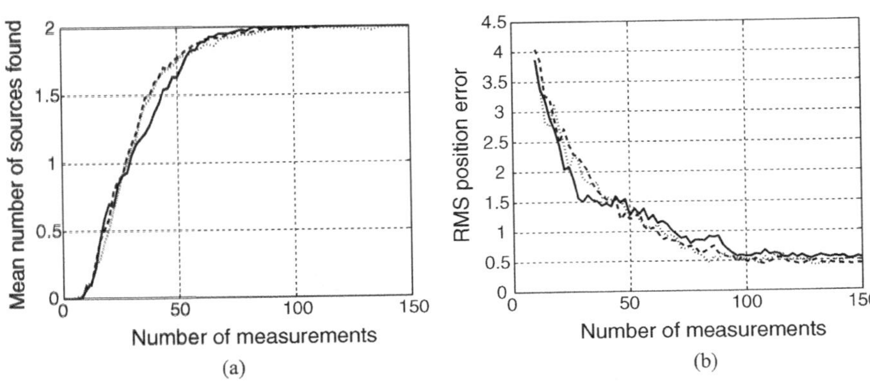

Figure 15.10 Plots of (a) the mean number of sources found and (b) the RMS position error of found sources for source search in the scenario of Figure 15.8. Results are shown for $\alpha = 0.999$ (solid), $\alpha = 0.5$ (dashed), and $\alpha = 0.1$ (dotted).

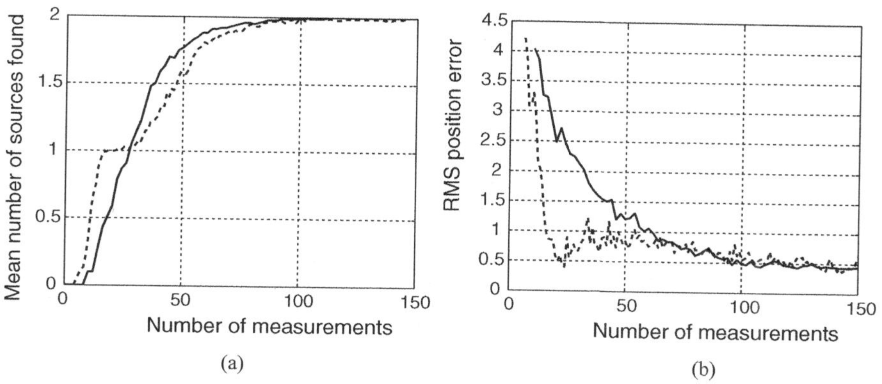

Figure 15.11 Plots of (a) the mean number of sources found and (b) the RMS position error of found sources for source search in the scenario of Figure 15.8 with two observers (solid) and one observer (dashed).

The final experiment compares multiple and single observer search performance. The reward function is the Rényi divergence with $\alpha = 0.5$ and the mean reward is approximated with $d = 100$ and $e = 50$. The results are shown in Figure 15.11. The single observer detects one of the sources with fewer measurements than the two observers. This happens because the single observer uses all of its measurements for detecting one of the sources while the measurements of the two observers are divided between the two sources. The latter strategy results in a much faster detection of the both sources. The reduced rate of source detection for the single observer after the first source is detected indicates a tendency for the observer to linger about a known source instead of searching for new sources. This results in accurate estimation of the parameters of the detected source but delays detection of the second source. Note that the performance measures are plotted against the number of measurements. It should be kept in mind that two observers collect a given number of measurements in half the time of a single observer. Thus, while the number of measurements required to detect one of the sources is greater for two observers than for one observer, the actual time taken is significantly less.

15.4.5 Experimental Results

In this section, the search algorithm is applied to the experimental data described in Section 15.2.4. The data consists of three test sets each containing radiation counts taken at a number of positions throughout a surveillance area. These measurement positions form the candidates for the search algorithm. To mimic the case of an observer moving through the area, after a measurement from a given position has been processed, the candidates for the next measurement are constrained to those positions that lie within 35 m of the current position. The second test set, in which there are two radiation sources, is used here. The scenario is shown in Figure 15.12, with the source positions, initial and candidate observer positions and search area displayed.

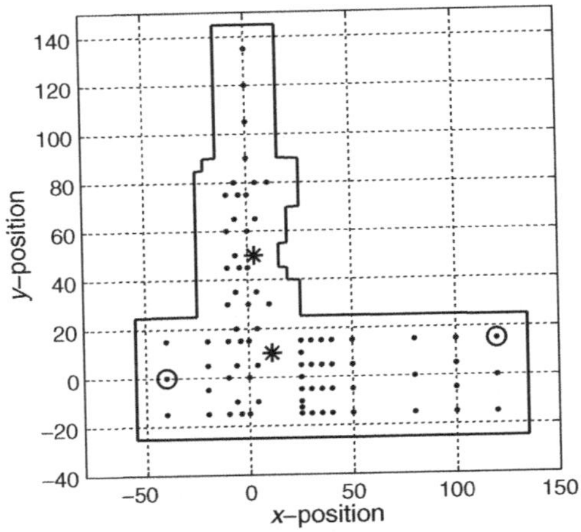

Figure 15.12 Scenario for source search using experimental data. The asterisks indicate source positions, the circles are the initial observer positions and the dots are candidate observer positions. The solid line indicates the boundary of the search area.

Recall that the experimental data includes multiple measurements at each position. This allows performance measures to be computed by averaging over multiple realizations. The search algorithm is applied with a sample size of $n = 5000$. A decimation factor of $e = 50$ and $d = 100$ prior measurement samples are used for approximation of the mean reward. The Rényi divergence parameter is $\alpha = 1/2$. Equation (15.40) is used for regularization of the prior with appearance probability $\nu = 0.005$, persistence probability $\eta = 0.995$ and added source prior PDF

$$\pi_0(\mathbf{x}, a) = U_S(\mathbf{x})G(a; 10, 1000). \tag{15.58}$$

The same measures of performance described in Section 15.4.4 are used here, that is, mean number of sources found and RMS position error. These statistics are found by averaging over 100 realizations. Figure 15.13 shows the results obtained for one and two observers. Strong similarities can be seen to results shown in Figure 15.11 for simulated data, that is, the single observer quickly finds one of the sources but is slower to find both sources. These experimental results demonstrate the practical validity of the assumed measurement model and the proposed search algorithm.

15.5 CONCLUSIONS

In this chapter several inferential problems regarding radiological sources were considered, including estimation of point and distributed sources using a collection of static observers and source search and estimation using mobile observers. The Bayesian approach has been emphasized. Bayesian estimation was favored because of the possibility of developing accurate and computationally efficient approximations

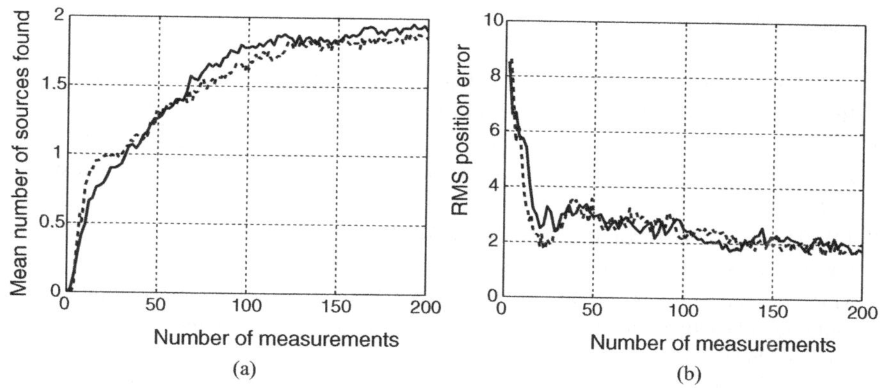

Figure 15.13 Plots of (a) the mean number of sources found and (b) the RMS position error of found sources for source search using experimental data with two observers (solid) and one observer (dashed).

of optimal Bayesian estimators for large dimensional parameters. Another advantage of the Bayesian approach is that it provides a probabilistic summary of what is known about the parameters, that is, the posterior distribution, which can be used as the basis for sequential search strategies. The methods described here were shown to work well with both simulated and experimental data. Several limitations of the methods described here, as well possible solutions to these limitations, are discussed below.

The simplified measurement model assumed here is not applicable to all situations. Perhaps the most critical simplification is that no attenuation-producing obstacles are present. The amount of attenuation caused by an obstacle depends on the type of radiation as well as the composition of the obstacle. In principle, a measurement model can be developed if this knowledge is available. In the more likely event that the required knowledge is unavailable, or only partially available, one possible approach is to seek a parametric model to describe the attenuation due to obstacles and estimate the model parameters along with the source parameters.

The search strategy is based on a reward that measures the difference between the posterior and the prior using the Rényi divergence. The Rényi divergence is only one of a number of candidates for measuring the difference between two distributions. The undesirable behavior of scheduling based on Rényi divergence in certain situations, for example, the tendency of an observer to linger about a known source rather than explore new areas, suggests that alternatives should be considered.

ACKNOWLEDGMENTS

The authors would like to thank A. Gunatilaka, A. Skvortsov, and R. Gailis (HPP division of DSTO, Australia) for motivation, technical discussions and experimental data. The work of Mark Morelande was partially funded by the HPP division of DSTO, Australia.

REFERENCES

1. M.R. Morelande and B. Ristic, Radiological source detection and localization using Bayesian techniques, *IEEE Trans. Signal Process.*, 57(11), 4220–4231, 2009.
2. A. Gunatilaka, B. Ristic, and R. Gailis, On localisation of a radiological point source, *Proceedings of the Conference on Information, Decision and Control*, Adelaide, Australia, 2007.
3. J.W. Howse, L.O. Ticknor, and K.R. Muske, Least squares estimation techniques for position tracking of radioactive sources, *Automatica*, 37, 1727–1737, 2001.
4. M.R. Morelande, B. Ristic, and A. Gunatilaka, Detection and parameter estimation of multiple radioactive sources, *Proceedings of the International Conference on Information Fusion*, Quebec, Canada, 2007.
5. R.J. Nemzek, J.S. Dreicer, D.C. Torney, and T.T. Warnock, Distributed sensor networks for detection of mobile radioactive sources, *IEEE Trans. Nucl. Sci.*, 51(4), 1693–1700, 2004.
6. S.M. Brennan, A.M. Mielke, and D.C. Torney, Radioactive source detection by sensor networks, *IEEE Trans. Nucl. Sci.*, 52(3), 813–819, 2005.
7. L. Tierney, R.E. Kass, and J.B. Kadane, Fully exponential Laplace approximations to expectations and variances of nonpositive functions. *J. Am. Stat. Assoc.*, 84(407), 710–716, 1989.
8. P.M. Reilly, The numerical computation of posterior distributions in Bayesian statistical inference, *Appl. Stat.*, 25(3), 201–209, 1976.
9. J.C. Naylor and A.F.M. Smith, Applications of a method for the efficient computation of posterior distributions, *Appl. Stat.*, 31(3), 214–225, 1982.
10. J.S. Liu, *Monte Carlo Strategies in Scientific Computing*, Springer, 2008.
11. C.P. Robert and G. Casella, *Monte Carlo Statistical Methods*, Springer-Verlag, New York, 1999.
12. N. Chopin, A sequential particle filter method for static models, *Biometrika*, 89(3), 539–551, 2002.
13. C. Musso, N. Oudjane, and F. Le Gland, Improving regularised particle filters, in: A. Doucet, N. de Freitas, and N. Gordon (Eds.), *Sequential Monte Carlo Methods in Practice*, Springer-Verlag, New York, 2001.
14. R.M. Neal, Annealed importance sampling, *Stat. Comput.*, 11, 125–139, 2001.
15. N. Tsoulfanidis, *Measurement and Detection of Radiation*, Taylor & Francis, Washington, DC, 1995.
16. M. Evans and T. Swartz, *Approximating Integrals via Monte Carlo and Deterministic Methods*, Oxford, 2000.
17. M.-S. Oh, Monte Carlo integration via importance sampling: dimensionality effect and an adaptive algorithm, *Contemp. Math.*, 115, 165–187, 1991.
18. B.W. Silverman, *Density Estimation for Statistics and Data Analysis*, Chapman and Hall, 1986.
19. R.P.S. Mahler, *Statistical Multisource Multitarget Information Fusion*, Artech House, 2007.
20. *AN/PDR-77 User's Guide*, CT, USA.
21. J.A. Nelder and R. Mead, A simplex method for function minimization, *Comput. J.*, 7(4), 308–313, 1965.

22. M.R. Banham and A.K. Katsaggelos, Digital image restoration, *IEEE Signal Process. Mag.*, 24–41, 1997.

23. D.L. Donoho, Nonlinear solution of linear inverse problems by wavelet-vaguelette decomposition, *Appl. Comput. Harmonic Anal.*, 2, 101–126, 1995.

24. F. O'Sullivan, A statistical perspective on ill-posed inverse problems, *Stat. Sci.*, 1(4), 502–527, 1986.

25. M.R. Morelande and A. Skvortsov, Radiation field estimation using a Gaussian mixture, *Proceedings of the International Conference on Information Fusion*, Seattle, USA, 2009.

26. J. Park and I.W. Sandberg, Approximation and radial-basis-function networks, *Neural Comput.*, 5, 305–316, 1993.

27. M. Drews, B. Lauritzen, H. Madsen, and J.Q. Smith, Kalman filtration of radiation monitoring data from atmospheric dispersion of radioactive materials, *Radiat. Protect. Dosimetry*, 111(3), 257–269, 2004.

28. H.J. Jeong, M.H. Han, W.T. Hwang, and E.H. Kim, Application of data assimilation to improve the forecasting capability of an atmospheric dispersion model for a radioactive plume, *Ann. Nucl. Energy*, 35(5), 838–844, 2008.

29. J. Park and I.W. Sandberg, Universal approximation using radial-basis-function networks, *Neural Comput.*, 3, 246–257, 1991.

30. I.W. Sandberg, Gaussian radial basis functions and inner product spaces, *Circuits Syst. Signal Process.*, 20(6), 635–642, 2001.

31. Y.C. Ho and R.C.K. Lee, A Bayesian approach to problems in stochastic estimation and control, *IEEE Trans. Autom. Control*, 9, 333–339, 1964.

32. A.H. Jazwinski, *Stochastic Processes and Filtering Theory*, Academic Press, 1970.

33. S. Julier, J. Uhlmann, and H.F. Durrant-Whyte, A new method for the nonlinear transformation of means and covariances in filters and estimators, *IEEE Trans. Autom. Control*, 45(3), 477–482, 2000.

34. B.O. Koopman, *Search and Screening*, Pergamon Press, Elmsford, NY, 1980.

35. K.P. Ziock and W.H. Goldstein, The lost source, varying backgrounds and why bigger may not be better, in: J.I. Trombka, D.P. Spears, and P.H. Solomon (eds.), *Unattended Radiation Sensor Systems for Remote Applications*, American Institute of Physics, 2002.

36. J.-P. Le Cadre and S. Laurent-Michel, Optimizing the receiver maneuvers for bearings-only tracking, *Automatica*, 35(4), 591–606, 1999.

37. J.M. Passerieux and D. Van Cappel, Optimal observer maneuver for bearings-only tracking, *IEEE Trans. Aerosp. Electron. Syst.*, 34(3), 777–788, 1998.

38. A. Doucet, B.-N. Vo, C. Andrieu, and M. Davy, Particle filtering for multi-target tracking and sensor management. *Proceedings of the International Conference on Information Fusion*, 2002, pp. 474–481.

39. C.M. Kreucher, A.O. Hero, K.D. Kastella, and M.R. Morelande, An information-based approach to sensor management in large dynamic networks, *Proc. IEEE*, 95(5), 978–999, 2007.

40. K. Kastella, Discrimination gain to optimize detection and classification, *IEEE Trans. Syst., Man Cybernet., Part A*, 27(1), 112–116, 1997.

41. F. Furukawa, H.F. Durrant-Whyte, and B. Lavis, The element-based method—Theory and its application to Bayesian search and tracking, *Proceedings of the IEEE International Conference on Intelligent Robots and Systems*, San Diego, CA, 2007, pp. 2807–2912.

42. R.A. Cortez, X. Papageorgiu, H.G. Tanner, A.V. Klimenko, K.N. Borozdin, R. Limia, and W.C. Priedhorsky, Smart radiation sensor management, *IEEE Robot. Autom. Mag.*, 85–93, 2008.
43. B. Ristic and A. Gunatilaka, Information driven localisation of a radiological point source, *Inform. Fusion*, 9(2), 317–326, 2007.
44. B. Ristic, M.R. Morelande, and A. Gunatilaka, Information driven search for point sources of gamma radiation, *Signal Process.*, 90(4), 1225–1239, 2010.
45. A.V. Klimenko, W.C. Priedhorsky, N.W. Hengartner, and K.N. Borozin, Efficient strategies for low-level nuclear searches, *IEEE Trans. Nucl. Sci.*, 53(3), 1435–1442, 2006.
46. E.K.P. Chong, C.M. Kreucher, and A.O. Hero, Partially observable Markov decision process approximations for adaptive sensing, *Discrete Event Dyn. Syst.*, 19(3), 377–422, 2009.
47. A. Rényi, On measures of entropy and information, *Proceedings of the Fourth Berkeley Symposium on Mathematics, Statistics and Probability*, 1961, pp. 547–561.
48. A. Doucet, N.J. Gordon, and V. Krishnamurthy, Particle filters for state estimation of jump Markov linear systems, *IEEE Trans. Signal Process.*, 49(3), 613–624, 2001.
49. J. Liu and M. West, Combined parameter and state estimation in simulation-based filtering, in: A. Doucet, N. de Freitas, and N. Gordon (Eds.), *Sequential Monte Carlo Methods in Practice*, Springer, New York, 2001.
50. W.R. Gilks and C. Berzuini, Following a moving target- Monte Carlo inference for dynamic Bayesian models, *J. R. Stat. Soc., Ser. B*, 63(1), 127–146, 2001.
51. G. Storvik, Particle filters for state-space models with the presence of unknown static parameters, *IEEE Trans. Signal Process.*, 50(2), 281–289, 2002.

PART V

DECISION FUSION AND DECISION SUPPORT

CHAPTER 16

Distributed Detection and Decision Fusion with Applications to Wireless Sensor Networks

QI CHENG, RUIXIN NIU, ASHOK SUNDARESAN, AND PRAMOD K. VARSHNEY

16.1 INTRODUCTION

There are many practical situations in which one is faced with a decision-making problem, that is, the problem of choosing a course of action from several possibilities. These examples include radar detection, signal detection in digital communications, image classification, and pattern recognition. The branch of statistics dealing with these types of problems is known as statistical decision theory or hypothesis testing. In the context of radar and communication theory, it is known as detection theory [1–4].

With advances in sensing technology and wireless communications, multiple sensors can be deployed in situ to monitor phenomena of interest with increased precision and coverage. This has given rise to detection problems involving multiple sensors. In classical multisensor detection, local sensors transmit their raw observations to a central processor where optimal detection is carried out based on conventional statistical techniques. However, centralized processing is neither efficient nor necessary in many practical applications. It consumes excessive energy and bandwidth in communications, imposes a heavy computation burden at the central processor, and may cause a single-point-of-failure problem. In distributed processing [1, 5, 6], on the other hand, local sensors carry out preliminary processing of data and only communicate with each other and/or the fusion center with the most informative information relevant to the global objective. The advantages of distributed processing are obvious, that is, reduced communication bandwidth requirement and energy consumption, increased reliability, and robustness. This chapter is focused on distributed detection and decision fusion problems, which involve the design of decision rules at the local

Integrated Tracking, Classification, and Sensor Management: Theory and Applications, Edited by Mahendra Mallick, Vikram Krishnamurthy, and Ba-Ngu Vo.

sensors and at the fusion center to optimize detection performance, under either a Neyman–Pearson or a Bayesian criterion.

In a conventional multisensor distributed detection system, it has been usually assumed that the local decisions' performance indices, including both probabilities of detection and false alarm at local sensors, are known, and there are no limitations on the sensors. As will be discussed later in Section 16.4, in a wireless sensor network (WSN), the assumptions made for the conventional distributed detection problems may not be valid any more. In WSNs, many sensors have limits on their sensing and communication abilities due to severe constraints on sensor size, cost, energy, and communication bandwidth. This changes the basic distributed detection problem to some extent. It is, therefore, important to develop distributed detection approaches specifically for WSNs, by taking into account the unique characteristics and constraints of WSNs. Recently, distributed detection in WSNs has become a very active research area and is discussed in this chapter.

The chapter is organized as follows. Section 16.2 briefly introduces the fundamentals of detection theory. The conventional distributed detection problem with multiple sensors is covered in Section 16.3. There, under the conditional independence assumption, the optimal design of decision rules at the local sensors and at the fusion center is discussed for both parallel and serial network structures, using both Bayesian and Neyman–Pearson formulations. Furthermore, research work in distributed detection with dependent observations is reviewed in Section 16.3. In Section 16.4, distributed detection in WSNs is studied. For nonidentical local observations and without the knowledge of local decisions' performance indices, a counting fusion rule is proposed and its exact performance is analyzed. Some computationally efficient approximations to system-level probability of detection are also presented. The problem of distributed detection in the presence of dependent observations is considered in Section 16.5. Assuming local sensor decision rules to be based on simple binary quantization of sensor observations, a novel method for fusion of correlated decisions using copula theory is discussed. Finally, the summary of the chapter and some challenging issues for distributed detection are presented in Section 16.6.

16.2 ELEMENTS OF DETECTION THEORY

The basic components of a hypothesis testing problem are shown in Figure 16.1. The source is not directly observable, otherwise there would be no decision problem. A possible source of observations can be represented by a hypothesis. If the set of hypotheses is only 2, the problem is a binary hypothesis testing problem. We

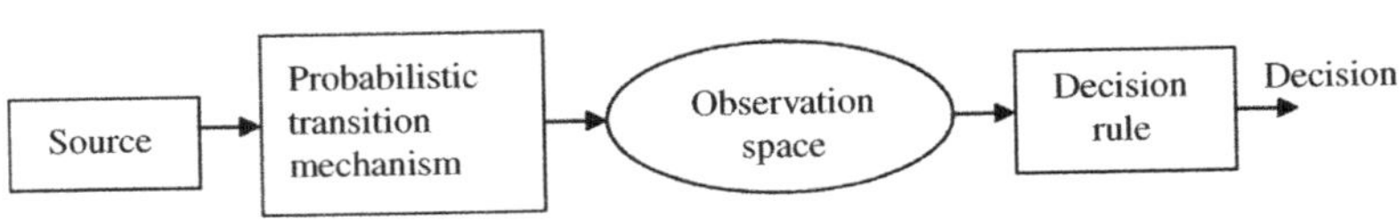

Figure 16.1 Components of a hypothesis testing problem.

label the two possible choices as H_0 and H_1. Hypothesis H_0 usually represents the transmission of symbol zero or the absence of a target. Hypothesis H_1 corresponds to the transmission of symbol 1 or the presence of a target. If there are M hypotheses with $M > 2$, it is a multiple hypothesis testing problem or M-ary detection problem. Corresponding to each hypothesis, an observation, which is in general a vector-valued random variable, is generated in the observation space according to some probabilistic law. By examining the observation, we try to infer which hypothesis is the correct one based on a certain decision rule. Usually, a decision rule partitions the observation space into decision regions corresponding to the different hypotheses. The hypothesis corresponding to the decision region where the observation falls is declared true. Whenever a decision does not match the true hypothesis, an error occurs. To obtain the fewest errors (or least cost), the decision rule plays an essential role and should be designed according to the optimization criterion in use.

There are several different optimization criteria for a detection problem depending on different knowledge of the prior corresponding to the problem. A simple binary hypothesis testing problem is considered here for illustration purposes and the results can be extended to M-ary hypothesis testing problems. In a binary hypothesis testing problem, two types of errors can occur in decision making. When H_0 is the true underlying hypothesis and H_1 is declared true, an error of the first kind is said to occur. In radar terminology, it is known as a false alarm, and the associated conditional probability of error is known as the probability of false alarm, P_f. An error of the second kind, a miss, is said to occur if H_0 is declared true when H_1 is present. The associated conditional probability of error is known as the probability of miss, P_M. Let $P(D_i|H_j)$ denote the conditional probability that we decide H_i (we denote this event as D_i) given the true hypothesis H_j, $i, j = 0, 1$. By integrating the conditional density function of observation $\mathbf{y}$ given hypothesis H_j over Z_i, the decision region corresponding to hypothesis H_i, we have

$$P(D_i|H_j) = \int_{Z_i} p(\mathbf{y}|H_j)d\mathbf{y}. \tag{16.1}$$

By definition, $P_f = P(D_1|H_0)$ and $P_M = P(D_0|H_1)$. The probability of detection or the power P_d is defined as the probability of declaring H_1 true when H_1 is present, that is, $P_d = P(D_1|H_1) = 1 - P_M$.

In the Bayesian formulation, C_{ij} is defined to be the cost of declaring H_i true when H_j is present to reflect different consequences of all decisions. The decision rule is determined by minimizing the average cost or Bayes risk function, $R = \sum_{i=0}^{1} \sum_{j=0}^{1} C_{ij} P(D_i|H_j)P_j$. Here, P_0 and P_1 are the a priori probabilities of the two hypotheses H_0 and H_1, respectively. If $C_{00} = C_{11} = 0$ and $C_{01} = C_{10} = 1$, we have the minimum probability of error criterion, that is, $R = Pe = P(D_1|H_0)P_0 + P(D_0|H_1)P_1$. If the knowledge of a priori probabilities is not readily available, one approach to obtain the decision rule is to use the Bayes solution corresponding to the value of P_1 for which the average cost is maximum, and this criterion is referred to as the minimax criterion. In the Neyman–Pearson formulation, an acceptable value of P_f, α (also called the size of the test) is chosen and a decision rule γ is determined by maximizing P_d while constraining P_f to this value, that is, $\max_\gamma P_d$ subject to $P_f \leq \alpha$. The decision rule based on the above criteria results in a likelihood ratio test (LRT) that involves the comparison of the likelihood ratio $\Lambda(\mathbf{y})$ with a threshold η

$$\Lambda(\mathbf{y}) = \frac{p(\mathbf{y}|\mathrm{H}_1)}{p(\mathbf{y}|\mathrm{H}_0)} \underset{D_0}{\overset{D_1}{\gtrless}} \eta. \tag{16.2}$$

The thresholds, in general, are different for different criteria. The performance of LRTs is best represented in terms of the receiver operating characteristic (ROC) curve since it depends only on the probability density functions of the observation under the two hypotheses. This is a plot of the probability of detection P_d against the probability of false alarm P_f. The ROCs of all continuous LRTs are concave downward and lie above the $P_d = P_f$ line. Details regarding LRTs and ROCs are provided in [2–4].

So far, we have assumed that the parameters characterizing a hypothesis, $\boldsymbol{\theta}$, are fixed and known and the corresponding detection problem is known as simple hypothesis testing. In many situations, however, these parameters can take unknown values or a range of values. Such hypotheses are called composite hypotheses and the corresponding detection problem is known as composite hypothesis testing. If $\boldsymbol{\theta}$ is characterized as a random vector with known probability densities under the two hypotheses, the LRT can be extended to composite hypothesis testing in a straightforward manner

$$\Lambda(\mathbf{y}) = \frac{\int_{\Theta_1} p(\mathbf{y}|\boldsymbol{\theta}, \mathrm{H}_1) p(\boldsymbol{\theta}|\mathrm{H}_1) d\boldsymbol{\theta}}{\int_{\Theta_0} p(\mathbf{y}|\boldsymbol{\theta}, \mathrm{H}_0) p(\boldsymbol{\theta}|\mathrm{H}_0) d\boldsymbol{\theta}}. \tag{16.3}$$

If $\boldsymbol{\theta}$ is nonrandom, one would like to be able to obtain uniformly most powerful (UMP) results for an optimum scheme based on a Neyman–Pearson test. If a UMP test does not exist, we can use the maximum likelihood estimates (MLEs) of its value under the two hypotheses as the true values in an LRT, resulting in a so-called generalized likelihood ratio test (GLRT)

$$\Lambda_g(\mathbf{y}) = \frac{\max_{\boldsymbol{\theta} \in \Theta_1} p(\mathbf{y}|\boldsymbol{\theta}, \mathrm{H}_1)}{\max_{\boldsymbol{\theta} \in \Theta_0} p(\mathbf{y}|\boldsymbol{\theta}, \mathrm{H}_0)} \underset{D_0}{\overset{D_1}{\gtrless}} \eta. \tag{16.4}$$

When the number of observations approaches infinity, we expect that any reasonable test will tend to making perfect inferences of the corresponding hypotheses. That is, the probability of error of the test will approach zero. Consider the Neyman–Pearson test of the form

$$\Lambda_n = \frac{\prod_{i=1}^n p(y_i|\mathrm{H}_1)}{\prod_{i=1}^n p(y_i|\mathrm{H}_0)} \underset{D_0}{\overset{D_1}{\gtrless}} \eta_n, \tag{16.5}$$

where n represents the number of observation samples. Let $\epsilon \in (0, 1)$ be the specified constraint on the probability of false alarm, that is, the test must be designed so that

$\alpha_n \leq \epsilon$. Define the minimum probability of miss to be

$$\beta_n(\epsilon) := \min_{\eta_n, \alpha_n \leq \epsilon} \beta_n. \tag{16.6}$$

The minimum probability of miss $\beta_n(\epsilon)$ converges to zero exponentially as n goes to infinity. The rate of convergence is provided by Stein's lemma given as follows [7]

$$\lim_{n \to \infty} \frac{1}{n} \log \beta_n(\epsilon) = -D, \tag{16.7}$$

where D is the relative entropy or the Kullback–Leibler informational divergence between observation distributions conditioned on the two hypotheses, defined as

$$D := \int p(\mathbf{y}|\mathrm{H}_0) \log \frac{p(\mathbf{y}|\mathrm{H}_0)}{p(\mathbf{y}|\mathrm{H}_1)} \mathrm{d}\mathbf{y}. \tag{16.8}$$

If a Bayesian test is used with the threshold of the test chosen to be $\eta_n = 1$ (this corresponds to the minimum probability of error criterion for equally likely hypotheses), it can be shown that the probability of error approaches zero as n approaches infinity and the rate of convergence is the Chernoff information [7], which is defined as

$$C := - \min_{0 \leq \lambda \leq 1} \log \left(\int p^{\lambda}(\mathbf{y}|\mathrm{H}_0) p^{1-\lambda}(\mathbf{y}|\mathrm{H}_1) d\mathbf{y} \right). \tag{16.9}$$

In addition to the above conventional detection problems, strategies have also been developed for some special problems under other related performance criteria, for modified test structures, and with fewer assumptions on the signals. In radar target detection, for example, it is desirable to maintain a constant probability of false alarm even in a nonstationary noise and clutter background. Adaptive threshold techniques such as those based on cell averaging and order statistics are employed by learning the background noise power level online. The resulting detectors are the so-called constant false alarm rate (CFAR) detectors. More details regarding CFAR detection can be found in [1, 8, 9]. As opposed to fixed number of observations cases, there exist many practical situations where observations become available as time progresses. In such cases, it may be desirable to process the observations sequentially and stop taking new observations whenever the predefined detection performance is reached. The main advantage of such sequential hypothesis testing is that it requires, on an average, fewer observations to achieve the same probability of error as for a fixed-sample-size-test. A modified Neyman–Pearson type of test, also known as Wald's sequential probability ratio test can be found in [1, 4] and references therein. For detecting weak signals in noise, the locally optimum test has been developed where the slope of P_d when the signal approaches zero is maximized under the constraint $P_f = \alpha$ [1, 4]. If the probabilistic transition mechanism is not known, detection techniques that are applicable regardless of the form of the observation distributions are desired.

These techniques are called nonparametric methods and include the sign detector, the Wilcoxon detector, the normal scores detector, and the polarity coincidence correlator [3, 10]. These methods have shown fairly good performance under some assumptions on underlying probability distributions.

16.3 DISTRIBUTED DETECTION WITH MULTIPLE SENSORS

Distributed detection has been an active research field for more than two decades, following the seminal work of Tenney and Sandell [11]. The main task in distributed detection is to design decision rules at local sensors and the fusion center to optimize certain objective functions subject to certain constraints, for example, energy, bandwidth, latency, and false alarm rate. Extension of conventional detection theory to the distributed setting with multiple sensors is nontrivial. Next, we present several important factors that affect the design of distributed detection algorithms. They are the topology of the sensor networks, relation between sensor observations (conditionally independent versus correlated), optimization criteria, and quantization levels.

16.3.1 Topology

There are several topologies used for distributed signal processing. Parallel configuration (Figure 16.2) is the most common topological structure that has been studied quite extensively. In this configuration, different sensors make observations concurrently. The sensors may be spatially separated or co-located depending upon the application. Another popular structure is the serial, or tandem, topology (Figure 16.3). In this case,

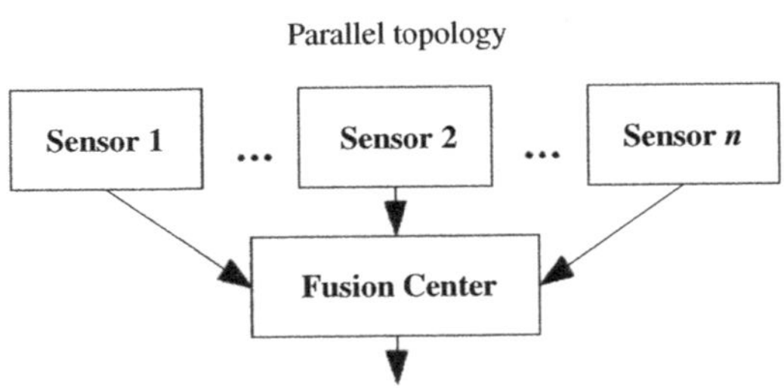

Figure 16.2 Parallel configuration.

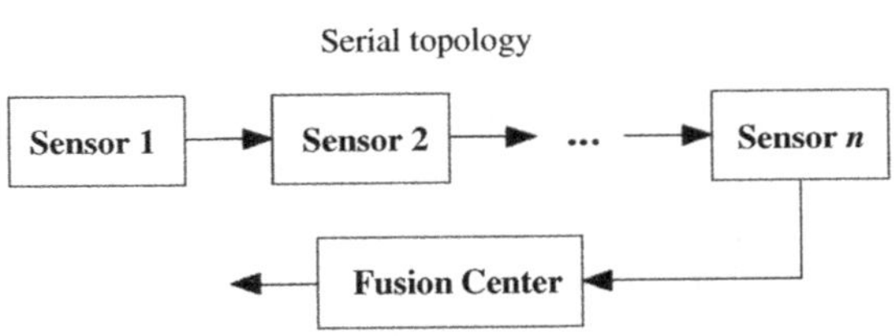

Figure 16.3 Serial configuration.

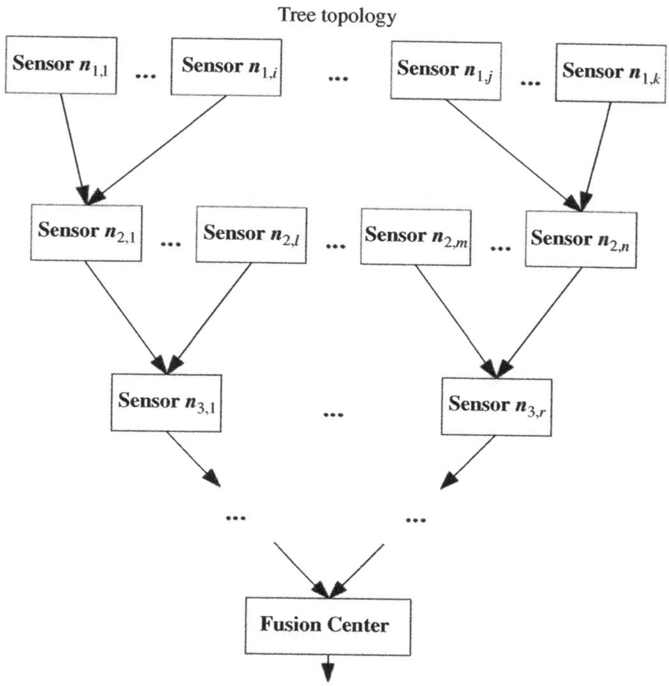

Figure 16.4 Tree configuration.

sensor observations are temporally separated. This separation is due to the nature of the problem under consideration, for example, effect of a phenomenon becomes "visible" to different sensors at different times or there is a hierarchical organizational structure. In addition to these basic configurations, one can envisage configurations that are combinations of these basic topologies, for example, a tree or hierarchical network (Figure 16.4) or a complete network (Figure 16.5). One may also incorporate feedback in the system configuration. In most cases, the structure is dictated by the application and it has a profound impact on the achievable performance.

Let us first focus on a simple binary hypothesis testing problem in which the observations at N sensors either correspond to the presence of a signal (hypothesis H_1) or to the absence of a signal (hypothesis H_0). Let y_i denote the observation that is available at the ith sensor, $i = 1, \ldots, N$, and $p(y_i|\mathrm{H}_j)$ denote the marginal distribution of the observation conditioned on hypothesis H_j, $j = 0, 1$. Based on its own observation y_i and quantized information from a subset of neighboring sensors $\mathbf{u}_{\{i\}} = \{u_j\}_{j \in \mathcal{N}_i}$, the ith sensor employs the mapping rule $u_i = \gamma_i(y_i, \mathbf{u}_{\{i\}})$ and passes the quantized information u_i to another subset of neighboring sensors, depending on the network structure. For general network structures, the optimal solution to the distributed detection problem, that is, the optimal decision rules $(\gamma_1, \ldots, \gamma_N)$, is NP complete [12–14]. Nonetheless, under certain assumptions and specific network topologies, the optimum solution may be tractable.

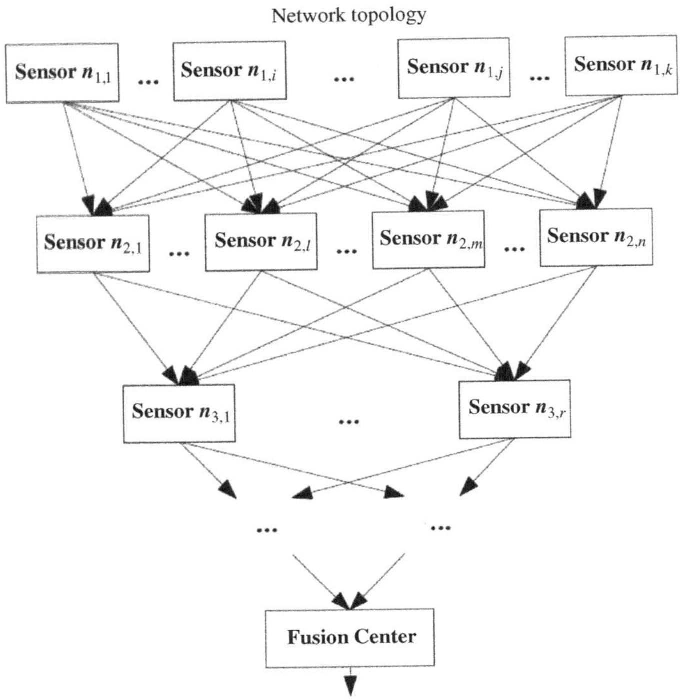

Figure 16.5 Network structure.

16.3.2 Conditional Independence Assumption

The first assumption that we make is the conditional independence of sensor observations. This implies that the joint density of the observations obeys

$$p(y_1, \ldots, y_N|H_j) = \prod_{i=1}^{N} p(y_i|H_j), \qquad \text{for } j = 0, 1. \tag{16.10}$$

Both parallel and serial configurations under the Bayesian and Neyman–Pearson formulations will be discussed, respectively. Solutions for arbitrary topologies have been derived and are discussed in [1, 15].

16.3.2.1 Parallel Configuration. In the parallel configuration, the sensors do not communicate with each other and there is no feedback from the fusion center to any sensor. Thus, the mapping rule at the ith sensor is employed only on the local observation, $u_i = \gamma_i(y_i)$. Based on the received information $\mathbf{u} = [u_1, \ldots, u_N]$, the fusion center arrives at the global decision $u_0 = \gamma_0(\mathbf{u})$ that favors either H_1 (decides $u_0 = 1$) or H_0 (decides $u_0 = 0$). The goal is to obtain the optimal set of decision rules $\Gamma = (\gamma_0, \gamma_1, \ldots, \gamma_N)$ according to the objective function under consideration.

Bayesian Formulation. As discussed in Section 16.2, in a Bayesian formulation, costs are assigned to different courses of action and knowledge of prior probabilities, and the objective is to minimize the Bayesian risk of the overall system operation. Hence, the Bayesian formulation of the distributed detection problem can be expressed as

$$\min_{\Gamma=(\gamma_0,\gamma_1,\ldots,\gamma_N)} R, \tag{16.11}$$

where the Bayes risk R is

$$\begin{aligned} R &= \textstyle\sum_{i=0}^{1}\sum_{j=0}^{1} C_{ij} P(u_0 = i|\mathrm{H}_j) P_j \\ &= C + C_F \textstyle\sum_u P(u_0 = 1|\mathbf{u}) P(\mathbf{u}|H_0) - C_D \sum_u P(u_0 = 1|\mathbf{u}) P(\mathbf{u}|H_1). \end{aligned} \tag{16.12}$$

Here, $C_F = P_0(C_{10} - C_{00})$, $C_D = (1 - P_0)(C_{01} - C_{11})$, and $C = C_{01}(1 - P_0) + C_{00} P_0$. $\sum_{\mathbf{u}}$ indicates summation over all possible values of $\mathbf{u}$. It can be shown that the sensor decision rules and the fusion rule are LRTs given by [1]

$$\frac{p(y_k|\mathrm{H}_1)}{p(y_k|\mathrm{H}_0)} \underset{u_k=0}{\overset{u_k=1}{\gtrless}} \frac{\sum_{\mathbf{u}^k} C_F A(\mathbf{u}^k) \prod_{i=1, i\neq k}^{N} P(u_i|\mathrm{H}_0)}{\sum_{\mathbf{u}^k} C_D A(\mathbf{u}^k) \prod_{i=1, i\neq k}^{N} P(u_i|\mathrm{H}_1)} \quad \text{for } k = 1, \ldots, N, \tag{16.13}$$

and

$$\prod_{k=1}^{N} \frac{P(u_k|\mathrm{H}_1)}{P(u_k|\mathrm{H}_0)} \underset{u_0=0}{\overset{u_0=1}{\gtrless}} \frac{C_F}{C_D}, \tag{16.14}$$

where

$$\begin{aligned} \mathbf{u}^k &= [u_1, \ldots, u_{k-1}, u_{k+1}, \ldots, u_N], \\ A(\mathbf{u}^k) &= P(u_0 = 1|\mathbf{u}^{k1}) - P(u_0 = 1|\mathbf{u}^{k0}), \\ \mathbf{u}^{kj} &= [u_1, \ldots, u_{k-1}, u_k = j, u_{k+1}, \ldots, u_N], \ j = 0, 1. \end{aligned}$$

Given the decision rules, the right-hand side of (16.13) is a constant. Thus, the local decision rules are threshold tests. To find the optimal set of decision rules amounts to simultaneously solving the above $(N + 2^N)$ coupled nonlinear equations.

One approach to solve the problem is by an exhaustive search over all possible decision strategies. For each fusion rule, one may obtain a set of local thresholds and the associated Bayes risk of the system. The final solution is the fusion rule and corresponding set of local thresholds that yield the minimum cost. This approach can be employed for relatively small networks, but it becomes impractical as N becomes large. In fact, the computational complexity is exponential in the number of sensors N [14]. A popular method to design distributed detection systems is to employ a person-by-person optimization (PBPO) technique. This technique consists of optimizing the decision rule of one sensor at a time while keeping the decision rules of the remaining sensors fixed. The overall performance at the fusion center is guaranteed to improve (or, at least, to not worsen) with every iteration of the PBPO algorithm. However, system design equations resulting from this PBPO procedure represent necessary but not, in general, sufficient conditions to determine the globally optimum solution. The Gauss–Seidel cyclic coordinate descent algorithm has been proposed in the literature [16, 17] to obtain the PBPO solution satisfying the necessary conditions of optimality in an iterative manner.

Neyman–Pearson Formulation. The Neyman–Pearson formulation of the distributed detection problem can be stated as follows: for a prescribed bound on the global probability of false alarm, P_f, find (optimum) local and global decision rules that maximize the global probability of detection P_d. Under the conditional independence assumption, the mapping rules at the sensors as well as the decision rule at the fusion center are threshold rules based on the appropriate likelihood ratios [18, 19]

$$\frac{p(y_i|\mathrm{H}_1)}{p(y_i|\mathrm{H}_0)} \begin{cases} > t_i, & \text{then } u_i = 1 \\ = t_i, & \text{then } u_i = 1 \text{ with probability } \epsilon_i \\ < t_i, & \text{then } u_i = 0 \end{cases} \tag{16.15}$$

for $i = 1, \ldots, N$, and

$$\prod_{i=1}^{N} \frac{P(u_i|\mathrm{H}_1)}{P(u_i|\mathrm{H}_0)} \begin{cases} > \lambda_0, & \text{decide } \mathrm{H}_1 \text{ or set } u_0 = 1 \\ = \lambda_0, & \text{randomly decide } \mathrm{H}_1 \text{ with probability } \epsilon\ . \\ < \lambda_0, & \text{decide } \mathrm{H}_0 \text{ or set } u_0 = 0 \end{cases} \tag{16.16}$$

If the likelihood ratio in (16.15) is a continuous random variable with no point mass, then the randomization is unnecessary and ϵ_i can be assumed to be zero without losing optimality. The threshold λ_0 in (16.16) as well as the local thresholds t_i in (16.15) need to be determined so as to maximize P_d for a given $P_f = \alpha$. This can still be quite difficult even though the local decision rules and the global fusion rule are LRTs. Since (16.16) is known to be a monotone fusion rule, one can solve for the set of optimal local thresholds $\{t_i, i = 1, \ldots, N\}$ for a given monotone fusion rule and compute the corresponding P_d. One can then successively consider other possible monotone fusion rules and obtain the corresponding detection probabilities. The final optimal solution is the one monotone fusion rule and the corresponding local decision rules that provide the largest P_d. An iterative gradient method was proposed in [20] to find the thresholds satisfying the preassigned false alarm probability. Finding the optimal solution in this fashion is possible only for very small values of N. The complexity increases with N because (1) the number of monotone rules grows exponentially with N and (2) finding the optimal $\{t_i, i = 1, \ldots, N\}$ for a given fusion rule is an optimization problem involving an $N - 1$ dimensional search (it is one dimension less than N because of the constraint $P_f = \alpha$).

Special Case 1: The Decision Fusion Problem. Given the local detectors, the problem is to determine the fusion rule to combine local decisions optimally. Let us first consider the case where local detectors make only hard decisions, that is, u_i can take only two values 0 or 1 corresponding to the two hypotheses H_0 and H_1. Then, the fusion rule is essentially a logical function with N binary inputs and one binary output. There are 2^{2^N} possible fusion rules in general and an exhaustive search strategy is not feasible for large N. Let P_{fi} and P_{di} denote the probabilities of false alarm and detection of detector i respectively, that is, $P_{fi} = P(u_i = 1|H_0)$ and $P_{di} = P(u_i = 1|H_1)$. As we know, the optimum fusion rule is given by the LRT

$$\prod_{i=1}^{N} \frac{P(u_i|H_1)}{P(u_i|H_0)} \underset{u_0=0}{\overset{u_0=1}{\gtrless}} \lambda. \tag{16.17}$$

Here, λ is determined by the optimization criterion in use. The left-hand side of (16.17) can be written as

$$\begin{aligned} \prod_{i=1}^{N} \frac{P(u_i|H_1)}{P(u_i|H_0)} &= \prod_{i=1}^{N} \left(\frac{P(u_i{=}1|H_1)}{P(u_i{=}1|H_0)}\right)^{u_i} \left(\frac{P(u_i{=}0|H_1)}{P(u_i{=}0|H_0)}\right)^{1-u_i} \\ &= \prod_{i=1}^{N} \left(\frac{P_{di}}{P_{fi}}\right)^{u_i} \left(\frac{1-P_{di}}{1-P_{fi}}\right)^{1-u_i}. \end{aligned} \tag{16.18}$$

Taking the logarithm of both sides of (16.17), we have the Chair–Varshney fusion rule [21]

$$\sum_{i=1}^{N} \left[u_i \log \frac{P_{di}}{P_{fi}} + (1-u_i) \log \frac{1-P_{di}}{1-P_{fi}} \right] \underset{u_0=0}{\overset{u_0=1}{\gtrless}} \log \lambda. \tag{16.19}$$

This rule can also be expressed as

$$\sum_{i=1}^{N} \left[\log \frac{P_{di}(1-P_{fi})}{P_{fi}(1-P_{di})} \right] u_i \underset{u_0=0}{\overset{u_0=1}{\gtrless}} \log \lambda + \sum_{i=1}^{N} \log \frac{1-P_{fi}}{1-P_{di}}. \tag{16.20}$$

Thus, the optimum fusion rule can be implemented by forming a weighted sum of the incoming local decisions and comparing it with a threshold. The weights and the threshold are determined by the local probabilities of detection and false alarm. If the local decisions have the same statistics, that is, $P_{fi} = P_{Fj}$ and $P_{di} = P_{Dj}$ for $i \neq j$, the Chair–Varshney fusion rule reduces to a K-out-of-N form or a counting rule, that is, the global decision $u_0 = 1$ if K or more sensor decisions are 1. This structure of the fusion rule reduces the computational complexity considerably.

If the local detectors are allowed to make multilevel or soft decisions, the observation space at each local detector is partitioned into L mutually exclusive regions with $L > 2$. If the observation at detector i lies in the partition l, we set $u_i = l$,

$l = 0, \ldots, L-1$. Define $\alpha_i^l = P(u_i = l|\mathrm{H}_0)$ and $\beta_i^l = P(u_i = l|\mathrm{H}_1)$. The likelihood ratio in (16.17) can be written as

$$\prod_{i=1}^{N} \frac{P(u_i|\mathrm{H}_1)}{P(u_i|\mathrm{H}_0)} = \prod_{l=0}^{L-1} \prod_{S_l} \frac{\beta_i^l}{\alpha_i^l}, \tag{16.21}$$

where S_l is the set of local decisions u_i that are equal to l. Taking the logarithm of both sides, the optimal fusion rule is as follows

$$\sum_{l=0}^{L-1} \sum_{S_l} \log \frac{\beta_i^l}{\alpha_i^l} \underset{u_0=0}{\overset{u_0=1}{\gtrless}} \log \lambda. \tag{16.22}$$

This fusion rule is a generalization of the fusion rule for the hard decision case.

Special Case 2: Identical Sensors. Consider a scenario in which the observations at the sensors are conditionally independent as well as identically distributed. The symmetry in the problem suggests that the decision rules at the sensors should be identical. But counterexamples have been found in which nonidentical decision rules are optimal [14, 22–24]. Interestingly, identical decision rules are optimal in the asymptotic regime where the number of sensors increases to infinity, that is, the identical decision rule assumption often results in little or no loss of optimality [14, 25]. Therefore, identical local decision rules are frequently assumed in many situations, which reduce the computational complexity considerably.

For any reasonable collection of decision rules Γ, the probability of error at the fusion center goes to zero exponentially as the number of sensors N grows unbounded. It is then adequate to compare collections of decision rules based on their exponential rate of convergence to zero,

$$\lim_{N \to \infty} \frac{\log Pe(\Gamma)}{N}. \tag{16.23}$$

It was shown that, for the binary hypothesis testing problem, using identical local decision rules for all the sensor nodes is asymptotically optimal in terms of the error exponent [25]. In [26], the exact asymptotics of the minimum error probabilities achieved by the optimal parallel fusion network and the system obtained by imposing the identical decision rule constraint were investigated. It was shown analytically that the restriction of identical decision rules leads to little or no loss of performance. Asymptotic regimes applied to distributed detection are convenient because they capture the dominating behaviors of large systems. This leads to valuable insights into the problem structure and its solution.

In [27], by studying in the asymptotic regime, it is shown that binary sensors are optimal if there exists a binary quantization function γ_b whose Chernoff information exceeds half of the information contained in an unquantized observation. The requirement is fulfilled by many practical applications [28] such as the problem of detecting deterministic signals in Gaussian noise and the problem of detecting fluctuating signals in Gaussian noise using a square-law detector. In these scenarios, the gain offered by having more sensor nodes outperforms the benefits of getting detailed information from each sensor.

16.3.2.2 Serial Configuration. In a serial or tandem configuration of N sensors, the $(i-1)$th sensor passes its quantized information to the ith sensor which generates its quantized information based on its own observation and the quantized data received from the previous sensor. The first sensor in the network uses only its observation to derive its quantized data for use by the next sensor. The last sensor in the network makes a decision as to which one of the two possible hypotheses the observations at the sensors correspond to.

Bayesian Formulation. Let us consider the Bayesian formulation of distributed detection for a serial network. For simplicity, we limit our attention to a two-detector serial network. More general results are available in [1]. In this network, the first detector makes the decision $u_1 = i$ based on y_1. The second detector makes the final decision u_2 based on u_1 and y_2. Let C_{jk} denote the cost of deciding $u_2 = j$ when H_k is present. The goal is to derive decision rules at both detectors so that the average cost of making decision u_2 is minimized. The Bayes risk in this case can be expressed as

$$\begin{aligned} R &= \sum_{i,j,k} \int_{y_1,y_2} C_{jk} P(u_1, u_2, y_1, y_2, \mathrm{H}_k) \mathrm{d}y_1 \mathrm{d}y_2 \\ &= \sum_{i,j,k} \int_{y_1,y_2} P_k C_{jk} P(u_2|u_1, y_2) P(u_1|y_1) p(y_1|\mathrm{H}_k) p(y_2|\mathrm{H}_k) \mathrm{d}y_1 \mathrm{d}y_2. \end{aligned} \tag{16.24}$$

As shown in [1], the decision rules at both detectors are LRTs. A single threshold is used at the first detector whereas two thresholds depending upon the decision of the first detector are used at the second detector. t_2^i denotes the threshold at the second detector when $u_1 = i$. The three thresholds are given by

$$t_1 = \frac{C_F(P_{F2}(t_2^1) - P_{F2}(t_2^0))}{C_D(P_{D2}(t_2^1) - P_{D2}(t_2^0))}, \tag{16.25}$$

$$t_2^1 = \frac{C_F P_{F1}}{C_D P_{D1}}, \tag{16.26}$$

$$t_2^0 = \frac{C_F(1 - P_{F1})}{C_D(1 - P_{D1})}, \tag{16.27}$$

where $P_{F2}(t_2^i)$ and $P_{D2}(t_2^i)$ represent the probabilities of false alarm and detection at detector 2 based on threshold value t_2^i, $i = 0, 1$. These coupled equations provide necessary conditions for an optimal solution, and need to be solved simultaneously to determine the three thresholds. System optimization can be carried out based on the PBPO methodology.

For an N-detector serial network, $(2N - 1)$ coupled equations need to be solved to determine the thresholds. For large N, a more convenient approach is to represent these equations recursively [1]. An important question is whether the parallel topology or the serial topology performs better under Bayes criterion. It has been shown that for the two-detector case, the serial network performs at least as well as the parallel network; while the parallel network is better than the serial network in the asymptotic sense [14, 29]. Another interesting issue is the ordering of nonidentical detectors in serial networks. It might be intuitively appealing to put the better detectors toward the end. However, there exist counterexamples in which placing the better detectors toward the end need not always be optimal [30].

Neyman–Pearson Formulation. Let us consider the case when each sensor in the serial configuration makes a binary decision, that is, $u_i \in \{0, 1\}$, for $i = 1, 2, \ldots, N$. Let P_{fi} and P_{di} denote the false alarm and the detection probabilities of the ith sensor, respectively. The Neyman–Pearson problem can be stated as follows: subject to the constraint that $P_{FN} \leq \alpha$, find the decision rules at all the sensors so that these rules together achieve the maximum possible P_{DN}. It is shown in [31] that the solution to the above problem yields likelihood ratio based threshold rule for the ith sensor, with the likelihood ratios being computed using the observations at the ith sensor and the decision from the $(i - 1)$th sensor. The solution is based on the following observation. From the fundamental NP lemma it is clear that the last (Nth) sensor must employ an LRT to achieve the maximum possible P_{DN}. The proof is then completed by showing that the probability of detection at the ith sensor, for any given P_{fi} and $P_{F(i-1)}$, is a monotonic increasing function of $P_{D(i-1)}$ and hence the test at the $(i - 1)$th sensor must be a likelihood ratio threshold test. As in the parallel case, tests at all the sensors are LRTs, but solving for the optimal thresholds that need to be employed in these tests is, in general, quite difficult. However, for the serial case, there exist algorithms that can obtain optimal thresholds with complexity that is linear in N [32].

16.3.3 Dependent Observations

An important result in distributed detection is that for the classical framework, LRTs at the local sensors are optimal if observations are conditionally independent given each hypothesis [14]. This property drastically reduces the search space for an optimal set of local decision rules. Although the resulting problem is not necessarily easy, it is amenable to analysis in many contexts. In general, it is reasonable to assume conditional independence across sensor nodes if the uncertainty comes mainly from device and ambient noise. However, it does not necessarily hold for arbitrary sensor systems. For instance, when sensors lie in close proximity of one another, we expect their observations to become strongly correlated. If the observed signal is random in nature or the sensors are subject to common external noise, conditional independence assumption may also fail. Without the conditional independence assumption, the joint density of the observations, given the hypothesis, cannot be written as the product of the marginal densities, as in (16.10). The optimal tests at the sensors are no longer of the threshold type based solely on the likelihood ratio of the observations at the individual sensors. In general, finding the optimal solution to the distributed detection problem becomes intractable [12].

One may restrict attention to the set of likelihood ratio based tests and employ algorithms to determine the best solution from this restricted set. The resulting system may yield acceptable performance. This approach has been adopted in [33] where detection of known and unknown signals in correlated noise was considered. For the case of two sensors observing a shift-in-mean of Gaussian data, Chen and Papamarcou [34] develop sufficient conditions for the optimality of each sensor implementing a local LRT. Aalo and Viswanathan [35] assume local LRTs at multiple sensors and study the effect of correlated noise on the performance of a distributed detection system. The detection of a known signal in additive Gaussian and Laplacian noise is considered. System performance deteriorates when the correlation increases. In [36], two correlation models are considered. In one, the correlation coefficient between any two sensors decreases geometrically as the sensor separation increases. In the other model, correlation coefficient between any two sensors is a constant. Asymptotic performance with Gaussian noise when the number of sensors goes to infinity is examined. In [37], Blum et al. study distributed detection of known signals in correlated non-Gaussian noise, where the noise is restricted to be circularly symmetric. Lin and Blum examine two-sensor distributed detection of known signals in correlated t-distributed noise in [38]. Simulation results show that in some specific cases the optimum local decision rules are better than LRTs. A distributed M-ary hypothesis testing problem when observations are correlated is examined from a numerical perspective in [39]. Willett et al. study the two detector case with dependent Gaussian observations, the simplest meaningful problem one can consider, in [40]. They discover that the nature of the local decision rules can be quite complicated.

Constraining the local sensor decision rules to be suboptimal binary quantizers for the dependent observations problem, improvement in the global detection performance can still be attained by taking into account the correlation of local decisions while designing the fusion rule. Towards this end, design of fusion rules using correlated decisions has been proposed in [41] and [42]. In [41], Drakopoulos and Lee have developed an optimum fusion rule based on the Neyman–Pearson criterion for correlated decisions assuming that the correlation coefficients between the sensor decisions are known and local sensor thresholds generating the correlated decisions are given. Using a special correlation structure, they studied the performance of the detection system versus the degree of correlation and showed how the performance advantage obtained by using a large number of sensors degrades as the degree of correlation between local decisions increases. In [42], the authors employed the Bahadur–Lazarsfeld series expansion of probability density functions to derive the optimum fusion rule for correlated local decisions. By using the Bahadur–Lazarsfeld expansion of probability density functions, the pdf of local correlated binary decisions can be represented by the pdf of independent random variables multiplied by a correlation factor. In many practical situations, conditional correlation coefficients beyond a certain order can be assumed to be zero. Thus, computation of the optimal fusion rule becomes less burdensome. When all the conditional correlation coefficients are zero, the optimal fusion rule reduces to the Chair–Varshney rule. Here, the implementation of the fusion rule was carried out assuming that the joint density of sensor observations is multivariate Gaussian which takes into consideration the linear dependence of sensor observations by using the Pearson correlation coefficient in the covariance matrix. An implicit assumption is that individual sensor observations are also Gaussian distributed.

In many applications, the dependence can get manifested in many different nonlinear ways. As a result, more general descriptors of correlation than the Pearson correlation coefficient, which only characterizes linear dependence, may be required [43]. Moreover, the marginal distributions of sensor observations characterizing their univariate statistics may also not be identical. Here, emphasis should be laid on the fact that multivariate density (or mass) functions do not necessarily exist for arbitrary marginal density (or mass) functions. In other words, given arbitrary marginal distributions, their joint distribution function cannot be written in a straightforward manner.

In this chapter, another interesting approach for the fusion of correlated decisions, that does not necessarily require prior information about the joint statistics of the sensor observations or decisions, is described. Its novelty lies in the usage of *copula theory* [44]. The application of copula theory is widespread in the fields of econometrics and finance. However, its use for signal processing applications has been quite limited. In this chapter, we employ it for a signal detection problem in a distributed setting. Using copulas, approximate joint distribution functions can be constructed from *arbitrary* marginal distributions. Hence, the fusion strategy described in this chapter is particularly attractive in those cases where the underlying distributions of sensor observations are non-Gaussian that may arise in many practical problems and for which standard expressions for the multivariate distribution do not exist. Moreover, by using different copula functions, nonlinear dependence between sensor observations can also be accounted for. The copula-based fusion method is discussed in detail in Section 16.5.

16.3.4 Discussion

The multiple sensor systems discussed in this section involved the surveillance of an event by multiple sensors. It was implicitly assumed that the knowledge of local sensors' performance indices is available at the fusion center, and no limitations on the sensors were placed. As will be discussed in Section 16.4, many sensors have limits on their sensing and communication abilities and this changes the basic distributed detection problem to some extent. The distributed detection problem for WSNs with some sensor limitations is discussed next.

16.4 DISTRIBUTED DETECTION IN WIRELESS SENSOR NETWORKS

With significant advances in the areas of computer/communication networking, wireless communications, micro-fabrication and integrated circuits (IC) technology,

WSNs have become a fast-growing research area and received considerable attention in recent years [45–48]. Due to their high flexibility, enhanced surveillance coverage, robustness, mobility, and cost effectiveness, WSNs have great potential in environmental monitoring, battlefield surveillance, structural health management, among others. Each sensor node is generally composed of four components, that is, sensing, computation, communication, and power. To monitor phenomena of interest, a very large number of sensor nodes are deployed in situ to achieve coverage and robustness, while without the need to be physically tethered to a wired network. However, to successfully deploy a truly ubiquitous WSN places severe restrictions on power, size, and cost of individual sensor nodes.

Sensors typically work in dangerous or hostile environments with little maintenance. For self-contained sensor nodes powered by onetime batteries or energy scavenged from the environment, it is required that the average power of the node be extremely low to extend the lifetime of sensor nodes. Another important reason for low power operations is the need for a low probability of intercept/detection (LPI/LPD) in a military context. Due to these unique characteristics and requirements of WSNs, distributed detection framework is a suitable choice for the detection task in WSNs. In a distributed detection scheme, sensors make local decisions, which are transmitted through wireless links to a fusion center. Since only discrete decisions, rather than continuous observations, are transmitted, the energy consumption incurred by communications in the network is reduced significantly. Distributed detection problems become quite interesting due to limited channel bandwidth, channel imperfections, limited sensing range, local signal processing capabilities, and energy resources. How to optimize detection performance under these constraints is a challenging problem that is currently being investigated quite extensively.

In a traditional distributed detection paradigm, the signal processing algorithms, including local quantizers at the sensors and the fusion rule at the fusion center, and the communication aspects of the sensor networks are typically designed separately, irrespective of each other. In such a distributed detection system, usually the channels are deemed error-free, since one can always design a communication system with negligible communication error probabilities, by adopting a high transmission power and/or employing powerful error correction codes. In a WSN with severe constraints on energy, bandwidth, and delay, it has been illustrated in [49] that a joint design philosophy integrating communication design and distributed detection design leads to much improved system detection performance. The distributed detection problem in the presence of nonideal communication channels has been studied quite extensively recently. Under the Bayesian criterion, the optimality of the LRT for local sensor decisions has been established for a binary hypothesis testing problem in a sensor network with nonideal channels [50]. For a finite number of sensors, [51] provides the conditions under which the channel outputs no longer reduce the error probability at the fusion center. Channel aware decision fusion algorithms have been investigated with different degrees of channel state information for both single-hop [52–54] and multihop WSNs [55], while channel-optimized local quantizer design methods are provided in [56, 57]. To counter sensor or channel failures, robust binary quantizer design has been proposed in [58]. Channel aware distributed detection has also been studied in the context of cooperative relay network [59, 60].

In the canonical parallel fusion structure, the channels between sensors and the fusion center have been assumed as mutually independent. In sensor networks with a large number of sensors, this assumption is often violated. An alternative is to utilize the multiple access channel (MAC), by taking advantage of the shared nature of the wireless medium. Decentralized detection and estimation have been considered using a MAC channel model in [61–66].

16.4.1 Counting Rule in a Wireless Sensor Network with Signal Decay

Distributed detection has been discussed both in its classical formulation and in the context of the WSN. However, most of the existing results are based on the assumption that the local sensors' detection performances, namely either the local sensors' signal to noise ratio (SNR) or their probability of detection and false alarm rate, are known to the fusion center. For a WSN consisting of passive sensors, it is very difficult to estimate local sensors' performances via experiments because sensors' distances from the target are unknown to the fusion center and the sensors. Even if the local sensors can somehow estimate their detection performances in real time, it will be very expensive to transmit them to the fusion center, especially for a WSN with very limited system resources. Hence, the knowledge of the local sensors' performances cannot be taken for granted in a WSN and a fusion rule that does not require local sensors' performances is highly preferable. To make things worse, in many applications, the sensors are deployed randomly in a surveillance area, and their exact positions are difficult to estimate. Without the knowledge of local sensors' detection performances and their positions, an approach at the fusion center is to treat every sensor equally. An intuitive solution is to use the total number of "1"s as a statistic since the information about which sensor reports a "1" is of little use to the fusion center. In [67, 68], a fusion rule is proposed, which uses the total number of detections ("1"s) transmitted from local sensors as the statistic,

$$\Lambda = \sum_{i=1}^{N} u_i \underset{u_0=0}{\overset{u_0=1}{\gtrless}} T, \tag{16.28}$$

where T is the threshold at the fusion center, which can be decided by a prespecified probability of false alarm P_F. This fusion rule is called the counting rule. It is an attractive solution, since it is quite simple to implement, and achieves very good performance in a WSN with densely deployed low-cost sensor nodes.

16.4.2 Performance Analysis: Sensors with Identical Statistics

First, let us study the special case where the measurements $y_1, \ldots, y_N$ follow independent and identical distributions. It is reasonable to adopt identical local decision rules, since the identical decision rules are asymptotically optimal as the number of sensors in a sensor network approaches infinity, as proved in [25] and also discussed in Section 16.3. A direct result of these assumptions is that $P_{fi} = P_f$ and $P_{di} = P_d$, for $i = 1, \ldots, N$, and that

$$\Lambda | H_0 \sim \text{Binomial}(N, P_f), \tag{16.29}$$

$$\Lambda | H_1 \sim \text{Binomial}(N, P_d). \tag{16.30}$$

Now it is straightforward to obtain the system level probability of detection (P_D) and probability of false alarm (P_F) for the counting rule as follows

$$P_D = \sum_{i=T}^{N} \binom{N}{i} P_d^i (1 - P_d)^{N-i}, \tag{16.31}$$

$$P_F = \sum_{i=T}^{N} \binom{N}{i} P_f^i (1 - P_f)^{N-i}. \tag{16.32}$$

When N is large enough, the distribution function of a Binomial random variable can be approximated by a Gaussian distribution. Therefore, P_D and P_F can be calculated by using the Laplace–DeMoivre approximation [69]

$$P_D \approx Q\left((T - NP_d)/\sqrt{NP_d(1 - P_d)}\right), \tag{16.33}$$

$$P_F \approx Q\left((T - NP_f)/\sqrt{NP_f(1 - P_f)}\right), \tag{16.34}$$

where $Q(\cdot)$ is the complementary distribution function of a standard Gaussian, that is,

$$Q(x) = \int_x^{\infty} \frac{1}{\sqrt{2\pi}} e^{-\frac{t^2}{2}} dt. \tag{16.35}$$

However, for a large N, the Laplace–DeMoivre approximation may not give sufficient relative accuracy for tail probability estimation, especially for a tail region far from the mean of the decision variable [70]. In this case, the saddle point approximation, which is based on large deviation theory, is computationally efficient and can provide very accurate solutions [70].

16.4.3 Performance Analysis: Sensors with Nonidentical Statistics

In a realistic sensor network system, under hypothesis H_1, the spatially dispersed sensors typically have different statistics, since the signal emitted by the target decays as a function of the distance. One thus needs to analyze the counting rule's performance for nonidentical sensor observations.

The distributed detection problem with nonidentical sensor observations is formulated as follows. As shown in Figure 16.6, a total of N sensors are randomly deployed

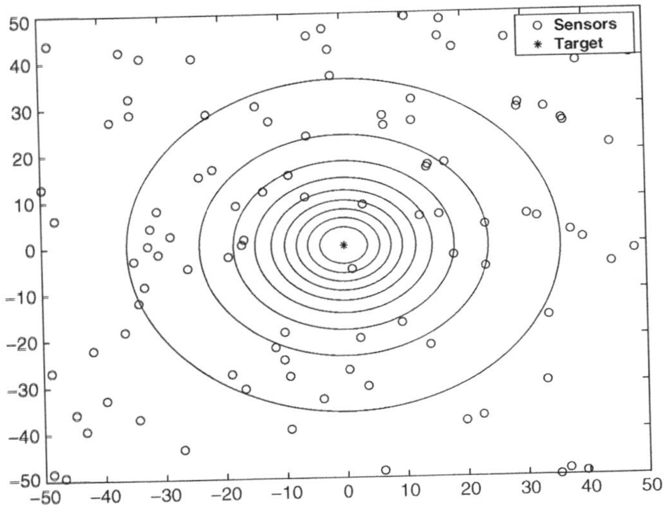

Figure 16.6 A sensor deployment example.

in a region of interest (ROI), which is a square with area b^2. The locations of sensors are unknown to the fusion center, and they are i.i.d. and follow a uniform distribution in the ROI

$$f(x_i, y_i) = \begin{cases} 1/b^2 & -\frac{b}{2} \le x_i,\ y_i \le \frac{b}{2} \\ 0 & \text{otherwise} \end{cases} \tag{16.36}$$

for $i = 1, \ldots, N$, where (x_i, y_i) are the coordinates of sensor i. The location of the target, represented by its coordinates (x_t, y_t), is independent of the positions of sensors, and follows the same uniform distribution within the ROI as that described in (16.36).

It is assumed that the signal power emitted by the target decays isotropically as a function of the distance

$$P_i = P_0 \begin{cases} 1 & 0 < d_i \le d_0 \\ d_0^n / d_i^n & d_0 < d_i \end{cases}, \tag{16.37}$$

where P_i is the signal power received by sensor i, P_0 is the signal power measured at a reference distance d_0, and d_i is the Euclidean distance between the target and local sensor i

$$d_i = \sqrt{(x_i - x_t)^2 + (y_i - y_t)^2} \tag{16.38}$$

and n is the signal decay exponent. Note that the model described in (16.37) is adopted, because the receivers at local sensors have a limited dynamic range, and the amplifier will saturate when the received signal strength is beyond a certain large value. The power decay model of (16.37) is quite general and widely adopted. For example, for both an isotropically radiated electromagnetic wave propagating in free space [71], and a spherical acoustic wave radiated by a simple source [72], the signal power decays at a rate inversely proportional to the square of the distance.

For a local sensor i, the binary hypothesis testing problem is

$$\begin{aligned} H_1: &\quad y_i = \sqrt{P_i} + n_i, \\ H_0: &\quad y_i = n_i, \end{aligned} \tag{16.39}$$

where n_i is a standard Gaussian random variable with zero mean and unit variance. For simplicity, a common threshold τ is adopted by all the sensors, leading to a false alarm rate of

$$P_f = Q(\tau) \tag{16.40}$$

and a probability of detection at local sensor i

$$P_{d_i} = Q\left(\tau - \sqrt{P_i}\right). \tag{16.41}$$

16.4.3.1 Exact Performance Analysis. Under hypothesis H_0, the system level probability of false alarm can be obtained using (16.32), due to the i.i.d. assumption of n_is. Under hypothesis H_1, u_is are not independent of each other, since they are all dependent on the target's coordinates (x_t, y_t), which are RVs. Hence, the distribution of $\Lambda = \sum_{i=1}^{N} u_i$ cannot be obtained by the saddle point approximation technique proposed in [70], since it requires u_is to be i.i.d. Instead, a new method is needed to derive the distribution of Λ. Based on the assumption that the noises at local sensors are i.i.d. and the locations of local sensors are i.i.d., it is shown in [73] that under hypothesis H_1 and conditioned on the target's location (x_t, y_t), Λ follows a Binomial distribution

$$\Lambda|(H_1, x_t, y_t) \sim \text{Binomial}\ (N, \bar{P}_d(x_t,\ y_t)), \tag{16.42}$$

where

$$\bar{P}_d(x_t, y_t) = \frac{1}{b^2} \int_{-\frac{b}{2}}^{\frac{b}{2}} \int_{-\frac{b}{2}}^{\frac{b}{2}} P_d(x, y, x_t, y_t) dx dy. \tag{16.43}$$

(16.42) provides a closed-form solution for the probability mass function (pmf) of $\Lambda|(H_1, x_t, y_t)$

$$Pr(\Lambda = k|H_1, x_t, y_t) = \binom{N}{k} [\bar{P}_d(x_t, y_t)]^k [1 - \bar{P}_d(x_t, y_t)]^{N-k} \quad (k = 0, \ldots, N) \tag{16.44}$$

and it is straightforward to obtain the pmf of $\Lambda|H_1$

$$Pr(\Lambda = k|H_1) = \frac{1}{b^2}\binom{N}{k}\int_{-\frac{b}{2}}^{\frac{b}{2}}\int_{-\frac{b}{2}}^{\frac{b}{2}}[\bar{P}_d(x_t, y_t)]^k[1 - \bar{P}_d(x_t, y_t)]^{N-k}dx_t dy_t. \tag{16.45}$$

The system level probability of detection is therefore

$$P_D = \sum_{k=T}^{N} Pr(\Lambda = k|H_1). \tag{16.46}$$

16.4.3.2 Performance Evaluation Via Approximations. The exact formulas to evaluate the system detection performance, namely (16.45) and (16.46), involve the computationally intensive fourfold integration. Let us introduce some approximation methods that require much less computational load than the exact evaluation of P_D.

Under hypothesis H_1, $\Lambda = \sum_{i=1}^{N} u_i$ is the sum of N independent Bernoulli RVs, and its distribution is often called Poisson Binomial distribution [74], which has a very complicated structure if the P_{d_i}s are nonidentical. Many publications have addressed the problem of approximating Poisson Binomial distribution by either a Poisson or a Binomial distribution [74–77]. However, in the problem formulated here, $\Lambda|H_1$ does not follow a Poisson Binomial distribution, since sensors' locations (x_i, y_i)s, and the target's location (x_t, y_t) are unknown, and u_is are dependent on each other. Nevertheless, it is still interesting to find approximations of the distribution of Λ, based on Binomial or Poisson distribution. The details of various approximations and their performances in terms of the total variation distance (TVD) are provided in the Appendix of this Chapter.

The accuracies of different approximation methods, and their requirements for computation load are summarized in Table 16.1. When N is small, Binomial I approximation is a good choice, considering its accuracy and its small amount of computation involving only a single-fold integration. When N is large, however, the Binomial II and DeMoivre–Laplace Approximation (DLAs) are better choices, since they need only one fourfold integration, compared with $N + 1$ fourfold integrations required by the exact evaluation of the pmf of $\Lambda|H_1$.

TABLE 16.1 Comparison of Different Approximation Methods

Approximation Method	Computation Load	Approximation Accuracy
Binomial I	1 single-fold integration	High for small N
Binomial II	1 single-fold integration and 1 fourfold integration	High for any N
DLA	1 single-fold integration and 1 fourfold integration	High for large N
Exact evaluation	$N + 1$ fourfold integrations	Exact results

16.4.3.3 Random Number of Sensors and Noisy Channels. In a typical WSN, due to the limited energy and/or bandwidth budget for each sensor node, increasing power and/or employing powerful error correction codes may not always be feasible. Furthermore, in a hostile environment, the power of transmitted signal should be kept to a minimum to attain an LPI/LPD. As a result, the data transmitted over the wireless channel between a sensor and the fusion center may endure errors. The local decisions are assumed to be transmitted through binary symmetric channels (BSCs) [7] to the fusion center. The BSCs have been used to model the noisy channels in the context of distributed detection problems by many researchers [50, 78], since this simple model effectively characterizes the loss of information caused by nonideal channels. The channels corresponding to different sensors are assumed i.i.d., with the same crossover probability p_c

$$p_c \triangleq Pr\{u_i' \neq u_i\}, \tag{16.47}$$

where $u_i' = \{0, 1\}$ is the ith sensor's decision received at the fusion center, which may not be the same as u_i, due to the nonzero channel error rate. Note that as long as $p_c < \frac{1}{2}$, it is very easy to show that u_i' is an MLE of u_i. The statistic that the counting rule uses at the fusion center becomes

$$\Lambda_2 = \sum_{i=1}^{N} u_i'. \tag{16.48}$$

As shown later, the nonideal channels only change the probabilities of "1"s and "0"s received at the fusion center, and lead to performance degradation. We keep using the same decision fusion rule, the counting rule.

In many applications, the sensors are deployed randomly in and around the ROI, and oftentimes some of them are out of the communication range of the fusion center, malfunctioning, or out of battery. Therefore, at a particular time, the total number of sensors that work properly in the ROI is a RV. First, let us study the case where this RV follows a Binomial distribution. One may assume that in a network consisting of L sensors, each sensor successfully sends a binary local decision to the fusion center with a probability p_b. As a result, at a particular time, the number of sensors (N) that can communicate with the fusion center successfully is a Binomial (L, p_b). Given this assumption and the BSC communication channels, the distribution of $\Lambda_2|(H_1, x_t, y_t)$ can be derived as

$$\Lambda_2 \sim \text{Binomial}\ (L, p_b\mu(x_t, y_t)), \tag{16.49}$$

where

$$\mu(x_t, y_t) := p_c + (1 - 2p_c)\bar{p}_d(x_t,\ y_t).$$

Now it is easy to show that

$$Pr(\Lambda_2 = k|H_1) = \frac{1}{b^2}\binom{L}{k}\int_{-b/2}^{b/2}\int_{-b/2}^{b/2}[p_b\mu(x_t, y_t)]^k[1 - p_b\mu(x_t, y_t)]^{L-k}dx_t dy_t \tag{16.50}$$

for $k = 0, \ldots, L$. Similarly, it can be shown that

$$\Lambda_2|H_0 \sim \text{Binomial}(L, p_b[p_c + (1 - 2p_c)p_{fa}]). \tag{16.51}$$

In addition to the case where the number of sensors follows a Binomial distribution, the case of Poisson random number of sensors is investigated. The Poisson point process has been adopted widely to model the randomly distributed sensors or wireless nodes [62, 68, 79]. Let us assume that the total number (N) of sensors within a ROI is a Poisson (λ) RV

$$f(N) = \frac{e^{-\lambda}\lambda^N}{N!} \quad (N = 0, \ldots, \infty), \tag{16.52}$$

where λ is the average number of sensors deployed in the ROI. The total number of detections received at the fusion center is denoted as $\Lambda_3 = \sum_{i=1}^{N} u_i'$. Given the above assumptions and notations, the distribution is derived as $\Lambda_3|(H_1, x_t, y_t)$

$$\Lambda_3 \sim \text{Poisson }(\lambda\mu(x_t, y_t)) \tag{16.53}$$

and

$$Pr(\Lambda_3 = k|H_1) = \frac{1}{b^2}\int\int\frac{e^{-\lambda\mu(x_t,y_t)}\left[\lambda\mu(x_t, y_t)\right]^k}{k!}dx_t dy_t \tag{16.54}$$

for $k = 0, \cdots, \infty$. Note that in practice, to get the pmf of $\Lambda_3|H_1$, only a finite number of terms, which have nonnegligible values, are calculated. Similarly, it can be proved that under hypothesis H_0, Λ_3 follows a Poisson ($\lambda[p_c + (1 - 2p_c)p_{fa}]$) distribution.

Approaches for approximating values of P_D for the counting rule have also been derived [73], in a manner analogous to those for the case of a known number of sensors and perfect channels.

16.4.3.4 Performance Analysis for a Large WSN. The performance approximation methods listed in Table 16.1, namely Binomial I and II and DLA, are very useful tools to evaluate P_D. However, when N is large, there will be a large number of terms that need to be estimated. Also, both Binomial II and DLA need a fourfold integral. A very simple alternative is to evaluate P_d by using the central limit theorem (CLT) for a very large sensor network. Conditioned on hypothesis H_1, $\{u_1, \ldots, u_N\}$ are not mutually independent, since they are all dependent on the target's coordinates (x_t, y_t), which are random variables. However, one can assume that the ROI is very large and the signal power of the target decays very fast as the distance increases.

Based on these assumptions, only within a very small fraction of the ROI, which is the area surrounding the target, the received signal power is significantly larger than 0. In a very large ROI, since the target will be located near the border of the ROI with a negligible probability, the border effect can be ignored and the ROI could be deemed as an infinite field. With such an assumption, the target's location is not relevant any more, and u_is can be deemed as independent under hypothesis H_1. Further, $\bar{P}_d(x_t, y_t)$ defined in (16.43) becomes approximately invariant to x_t and y_t

$$\begin{aligned}\bar{P}_d(x_t, y_t) &\approx \bar{P}_d(0, 0) \\ &= \frac{1}{b^2}\int_{-\frac{b}{2}}^{\frac{b}{2}}\int_{-\frac{b}{2}}^{\frac{b}{2}} P_d(x, y, 0, 0)dxdy,\end{aligned} \tag{16.55}$$

where

$$P_d(x, y, 0, 0) = \begin{cases} Q(\tau - \sqrt{P_0}) & 0 < x^2 + y^2 \leq d_0^2 \\ Q\left(\tau - \sqrt{P_0 d_0^n / \left(x^2 + y^2\right)^{\frac{n}{2}}}\right) & d_0^2 < x^2 + y^2 \end{cases}. \tag{16.56}$$

When N is large enough, and there does not exist a dominant u_i, according to CLT [69], the distribution function of Λ approaches a Gaussian distribution with mean η and variance σ^2, where

$$\eta = E\{\Lambda = \sum_{i=1}^{N} u_i | H_1\} = \sum_{i=1}^{N} P_{d_i} \tag{16.57}$$

and

$$\sigma^2 = \text{var}\{\Lambda = \sum_{i=1}^{N} u_i | H_1\} = \sum_{i=1}^{N} P_{d_i}(1 - P_{d_i}). \tag{16.58}$$

Remember that P_{d_i} is a function of $[x_i,\ y_i]^T$. When N is large, the summation in (16.57) can be approximated by

$$\begin{aligned}\eta(x_t, y_t) &= \sum_{i=1}^{N} P_{d_i}(x_i, y_i, x_t, y_t) \\ &\approx N\bar{P}_d(0, 0).\end{aligned} \tag{16.59}$$

The integration in (16.55) can be divided into two parts: one is the integration over a circle with radius $\frac{b}{2}$, and the other is over the rest of the area of the ROI. Part one is much easier if the integration is performed in polar coordinates

$$\begin{aligned}\bar{P}_d &\approx \frac{1}{b^2}\int_0^{2\pi}\left[\int_0^{d_0} Q\left(\tau-\sqrt{P_0}\right) rdr + \int_{d_0}^{\frac{b}{2}} Q\left(\tau-\sqrt{\frac{P_0 d_0^n}{r^n}}\right) rdr\right] d\boldsymbol{\theta}\\ &\quad + \frac{1}{b^2}\left(b^2-\frac{\pi b^2}{4}\right)\gamma\\ &= \frac{2\pi}{b^2}\left[Q\left(\tau-\sqrt{P_0}\right) d_0^2/2 + \int_{d_0}^{\frac{b}{2}} Q\left(\tau-\sqrt{\frac{P_0 d_0^n}{r^n}}\right) rdr\right]\\ &\quad + \left(1-\frac{\pi}{4}\right)\gamma. \end{aligned} \tag{16.60}$$

Note that in Part two integration, namely the integration over the area outside of the circle, d is very large and P_d is quite close to P_f. Therefore, it is reasonable to make the following approximation

$$\begin{aligned} P_d &\approx Q\left(\tau-\sqrt{P_0(\sqrt{2}d_0/b)^n}\right)\\ &:= \gamma. \end{aligned} \tag{16.61}$$

This has been used in the derivation of (16.60). Note that this approximation is conservative, because d is replaced by the largest value ($\frac{\sqrt{2}b}{2}$) it can take. Similarly, one can derive

$$\begin{aligned}\bar{\sigma^2} &\approx \frac{2\pi}{b^2}\left[\left(1-Q\left(\tau-\sqrt{P_0}\right)\right) Q\left(\tau-\sqrt{P_0}\right) d_0^2/2\right. \\ &\quad + \left.\int_{d_0}^{\frac{b}{2}}\left(1-Q\left(\tau-\sqrt{P_0(d_0/r)^n}\right)\right) Q\left(\tau-\sqrt{P_0(d_0/r)^n}\right) rdr\right]\\ &\quad + \left(1-\frac{\pi}{4}\right)\gamma(1-\gamma). \end{aligned} \tag{16.62}$$

Finally, employing CLT, the system level P_D can be written as

$$P_D = Q\left(\frac{T-N\bar{P}_d}{\sqrt{N\bar{\sigma^2}}}\right). \tag{16.63}$$

In [68], the performance of the counting rule for a large sensor network has been analyzed with uncertainty regarding the total number of sensors, which has been assumed to follow a Poisson distribution.

16.5 COPULA-BASED FUSION OF CORRELATED DECISIONS

The focus of this section is design of an optimal fusion rule when sensor observations and hence the sensor decisions are correlated. It is assumed that local sensor decision rules are binary quantizers of the sensor observations. The fusion rule is designed using copulas which characterize the joint statistics of sensor decisions.

16.5.1 Copula Theory

Copulas are basically functions that relate multivariate distribution functions to their corresponding one-dimensional marginal distribution functions [44] and are powerful tools for modeling dependencies between several random variables.

An important theorem that is central to the theory of copulas is the Sklar's theorem [44] given as follows.

Sklar's Theorem: Consider an m-dimensional distribution function F with marginal distribution functions $F_1, \ldots, F_m$. Then there exists a copula C, such that for all $x_1, \ldots, x_m$ in $(-\infty, \infty)$

$$F(x_1, x_2, \ldots, x_m) = C(F_1(x_1), F_2(x_2), \ldots, F_m(x_m)). \tag{16.64}$$

If F_k is continuous for $1 \leq k \leq m$, then C is unique, otherwise it is determined uniquely on $Ran F_1 \times \cdots \times Ran F_m$ where $Ran F_k$ is the range of CDF F_k. Conversely, given a copula C and univariate CDFs $F_1, \ldots, F_m$, then F as defined in equation (16.64) is a valid multivariate cdf with marginals $F_1, \ldots, F_m$.

A detailed proof of Sklar's theorem can be found in [44]. As a direct consequence of Sklar's theorem, for continuous distributions, the joint probability density function is obtained by differentiating both sides of (16.64),

$$f(x_1, \ldots, x_m) = \left(\prod_{i=1}^{m} f(x_i)\right) c(F_1(x_1), \ldots, F_m(x_m)), \tag{16.65}$$

where c is termed as the copula density and is given by

$$c(\mathbf{k}) = \frac{\partial^m (C(k_1, \ldots, k_m))}{\partial k_1, \ldots, \partial k_m}, \tag{16.66}$$

where $k_j = F_j(x_j)$.

Thus, given the knowledge of the marginal density functions of the random variables, $x_1, \ldots, x_m$, and a suitable copula density function, $c(.)$, the joint density function of the random variables, $f(x_1, \ldots, x_m)$, can be constructed as in equation (16.65). Note that, equation (16.65) achieves separation of the dependence information (modeled by the copula density c) from the choice of the marginal densities. This makes the approach well suited for problems with disparate or heterogenous marginal distributions.

Various families of copula functions exist in the literature and the choice of different copula functions may lead to the dependence information getting manifested in different ways. A detailed review on these functions and how they are derived can be found in [44]. In this chapter, for modeling purposes, the Gaussian (or Normal) copula and the t-copula with different degrees of freedom have been utilized.

16.5.2 System Design Using Copulas

The problem of designing the optimal test at the fusion center is now considered. It is assumed that the copula function together with its parameters is known in advance, possibly by performing a training procedure prior to deployment. Signal processing using unknown copula functions is dealt with in [80]. To simplify the presentation of the derivation, the design of the optimal distributed detection system is considered assuming only two sensors (i.e., $i = 1, 2$). Extension to the case when the number of sensors is arbitrary is discussed in [81].

As mentioned previously, local one bit sensor decisions are obtained by using a binary quantizer at each sensor. Let τ_i be the quantizer threshold at the ith sensor. Then the sensor decisions, during any time interval $1 \leq n \leq N$, are quantized versions of sensor observations defined as

$$u_{in} := \mathbb{Q}(z_{in}) = \begin{cases} 0 & \text{if } -\infty < z_{in} < \tau_i, \\ 1 & \text{if } \tau_i \leq z_{in} < \infty. \end{cases} \tag{16.67}$$

Letting $\mathbf{u_1}$ and $\mathbf{u_2}$ be the vectors of sensor decisions obtained over N time instants from sensors 1 and 2 respectively, the optimal test statistic at the fusion center is the likelihood ratio [1] given by

$$\Lambda(\mathbf{u}) = \frac{P(\mathbf{u_1}, \mathbf{u_2}|H_1)}{P(\mathbf{u_1}, \mathbf{u_2}|H_0)}, \tag{16.68}$$

where $P(\mathbf{u_1}, \mathbf{u_2}|H_k)$ is the joint pmf of the sensor decisions under the kth hypothesis, $k = 0, 1$.

Assuming that sensor observations and sensor decisions are i.i.d over time, the optimal fusion statistic becomes

$$\Lambda(\mathbf{u}) = \frac{\prod_{n=1}^{N} P(u_{1n}, u_{2n}|H_1)}{\prod_{n=1}^{N} P(u_{1n}, u_{2n}|H_0)}. \tag{16.69}$$

Let $P_{jk} = Pr(u_{1n} = j, u_{2n} = k|H_1)$, $Q_{jk} = Pr(u_{1n} = j, u_{2n} = k|H_0)$ for all $j = 0, 1;\quad k = 0, 1$. The joint pmf of u_{1n} and u_{2n} at any time instant, $1 \leq n \leq N$, under H_1 and H_0 respectively is given as follows

$$P(u_{1n}, u_{2n}|H_1) = P_{00}^{(1-u_{1n})(1-u_{2n})} P_{01}^{(1-u_{1n})u_{2n}} P_{10}^{u_{1n}(1-u_{2n})} P_{11}^{u_{1n}u_{2n}}, \tag{16.70}$$

$$P(u_{1n}, u_{2n}|H_0) = Q_{00}^{(1-u_{1n})(1-u_{2n})} Q_{01}^{(1-u_{1n})u_{2n}} Q_{10}^{u_{1n}(1-u_{2n})} Q_{11}^{u_{1n}u_{2n}}. \tag{16.71}$$

Substituting equation (16.70) and equation (16.71) in equation (16.69), taking logarithm on both sides and retaining only the data-dependent terms, we get

$$\log \Lambda_1(\mathbf{u}) = C_1 \sum_{n=1}^{N} u_{1n} + C_2 \sum_{n=1}^{N} u_{2n} + C_3 \sum_{n=1}^{N} u_{1n}u_{2n}, \tag{16.72}$$

where

$$C_1 = \log \frac{P_{10}Q_{00}}{P_{00}Q_{10}}, \qquad C_2 = \log \frac{P_{01}Q_{00}}{P_{00}Q_{01}}, \qquad C_3 = \log \frac{P_{00}P_{11}Q_{01}Q_{10}}{P_{01}P_{10}Q_{00}Q_{11}}.$$

When sensor decisions are conditionally independent, the term C_3 equals zero and the optimal fusion rule reduces to the Chair–Varshney fusion rule statistic (i.e., weighted sum of sensor decisions [21]) as shown in equation (16.73).

$$\log \Lambda_2(\mathbf{u}) = C_1 \sum_{n=1}^{N} u_{1n} + C_2 \sum_{i=n}^{N} u_{2n}. \tag{16.73}$$

Thus, for the case when sensor decisions (or observations) are correlated, the optimal fusion rule will not only depend on the weighted sum of sensor decisions but also on the weighted sum of the cross-products of sensor decisions. This is intuitively satisfying because the sum of the cross-products of the sensor decisions is a measure of how many sensors are jointly declaring the presence of a source under H_1. The higher the correlation between sensor observations, the larger this number will be.

For performing the test at the fusion center for the two sensor problem, the set of probabilities P_{00}, P_{01}, P_{10}, P_{11} and Q_{00}, Q_{01}, Q_{10}, Q_{11} that completely specify the joint probability mass function of the sensor decisions u_{1n} and u_{2n} under hypotheses H_1 and H_0, respectively, are required. From equation (16.67), these probabilities can be evaluated by integrating the joint pdf of sensor observations under both hypotheses. For example,

$$P_{00} = \int_{z_{1n}=-\infty}^{\tau_1} \int_{z_{2n}=-\infty}^{\tau_2} f(z_{1n}, z_{2n}|H_1)\mathrm{d}z_{1n}\mathrm{d}z_{2n},$$

$$Q_{00} = \int_{z_{1n}=-\infty}^{\tau_1} \int_{z_{2n}=-\infty}^{\tau_2} f(z_{1n}, z_{2n}|H_0)\mathrm{d}z_{1n}\mathrm{d}z_{2n}.$$

With the knowledge of the copula function and its parameters together with the marginal densities of sensor observations, the joint density of sensor observations can be constructed using equation (16.65). The joint probabilities of sensor decisions that determine the weights in the optimal test statistic can then be obtained by solving double integrals of the joint density of the corresponding sensor observations. The optimal fusion rule is now given by

$$\log \Lambda_1(\mathbf{u}) \underset{H_0}{\overset{H_1}{\gtrless}} \gamma', \tag{16.74}$$

where γ' is the threshold for test at the fusion center.

To derive the fusion rule performance measures, namely, the system probability of detection (P_D) and system probability of false alarm (P_F), we consider the asymptotic distribution of the optimal fusion rule statistic under H_0 and H_1. It can be shown that the optimal fusion test statistic denoted by $\log \Lambda_1(\mathbf{u})$ is *asymptotically* (when N is large) Gaussian [81]. Letting the first- and second-order statistics of $\log \Lambda_1(\mathbf{u})$ be denoted by μ_0 and σ_0^2 under H_0 and μ_1 and σ_1^2 under H_1, the system P_D and system P_F are now given by

$$P_D = Q\left(\frac{\gamma' - \mu_1}{\sigma_1}\right), \tag{16.75}$$

$$P_F = Q\left(\frac{\gamma' - \mu_0}{\sigma_0}\right), \tag{16.76}$$

where $Q(.)$ is the complementary cumulative distribution function of the Gaussian distribution. Under the Neyman–Pearson criterion, γ' can be obtained by constraining $P_F = \alpha$ as below

$$\gamma' = \sigma_0 Q^{-1}(P_F) + \mu_0. \tag{16.77}$$

Recall that the local sensor decision rules were constrained to be binary quantizers and the corresponding sensor thresholds were assumed to be $\tau_i, i = 1, 2$. Let $\boldsymbol{\tau}$ be the vector of sensor thresholds. Clearly, P_D and P_F are functions of the local sensor thresholds. Constraining $P_{FA} = \alpha$, P_D can be written as

$$P_D(\boldsymbol{\tau}) = Q\left(\frac{\sigma_0(\boldsymbol{\tau})Q^{-1}(\alpha) + \mu_0(\boldsymbol{\tau}) - \mu_1(\boldsymbol{\tau})}{\sigma_1(\boldsymbol{\tau})}\right). \tag{16.78}$$

The sensor thresholds are chosen to maximize P_D at a particular value of P_F. The optimal sensor thresholds are obtained by solving the following optimization problem

$$\boldsymbol{\tau}^* = \arg\min_{\boldsymbol{\tau}} \left(\frac{\sigma_0(\boldsymbol{\tau})Q^{-1}(\alpha) + \mu_0(\boldsymbol{\tau}) - \mu_1(\boldsymbol{\tau})}{\sigma_1(\boldsymbol{\tau})}\right). \tag{16.79}$$

16.5.3 Illustrative Example: Application to Radiation Detection

In this section, the application of copula-based fusion for distributed detection of a radiation source is considered. Two sensors are assumed to be monitoring a region for the possible presence of a radioactive source. The radiation sensors receive radiation counts embedded in additive white Gaussian noise and a decision regarding the presence or absence of a radioactive source is made based on a local threshold. The observations received by the local decision makers under both hypotheses are as follows

$$\begin{aligned} H_0 &: z_{in} = s_{in}^0 + w_{in} \quad i = 1, 2; n = 1, \ldots, N, \\ H_1 &: z_{in} = s_{in}^1 + w_{in} \quad i = 1, 2; n = 1, \ldots, N, \end{aligned} \tag{16.80}$$

where s_{in}^0 and s_{in}^1 are the radiation counts under H_0 and H_1, respectively, and w_{in} is the measurement noise. The variable z_{in} represents the observation received at sensor i located at (x_i, y_i) during the nth time interval.

The count variable s_{in}^0, due to background radiation, received during the time interval $(0, t]$ is assumed to be Poisson distributed with known rate λ_b. The count variable under H_1, s_{in}^1, representing source plus background radiation, at sensor i located at (x_i, y_i) is assumed to be Poisson distributed with rate $\lambda_{ci} + \lambda_b$, where λ_{ci} is a function of the source intensity A_0, and distance of the ith sensor from the source. Under the assumption of isotropic behavior of radiation in the presence of the source, the rate λ_{ci} is given by

$$\lambda_{ci} = \frac{A_0}{1 + (x_0 - x_i)^2 + (y_0 - y_i)^2}, \tag{16.81}$$

where (x_0, y_0) represent the source coordinates. The measurement noise w_{in} at the local detectors is Gaussian distributed with a known variance σ_w^2 and is assumed to be spatially and temporally independent. For this example, the radiation counts s_{in}^1 are assumed to be spatially correlated but temporally independent conditioned on the H_1 hypothesis. This implies that under the H_1 hypothesis, sensor observations are dependent and hence are samples drawn from a bivariate distribution $f_{Z_{1n},Z_{2n}}(z_{1n}, z_{2n})$. Under the H_0 hypothesis, the sensor observations s_{in}^0 are assumed to be spatially and temporally independent although in general this may not be true if the sensors are observing a common background. Here, the above problem is solved for a known signal case, that is, values of source intensity A_0 and source coordinates (x_0, y_0) are assumed to be known, although in general this may not be the case.

The marginal pdfs of the sensor observations under both hypotheses at any time instant n are infinite Gaussian mixtures [81]. Let the local sensor probability of detection and local sensor probability of false alarm be represented by p_i and q_i, respectively, for $i = 1, 2$. Once the pdfs of individual sensors' observations under both hypotheses are known, p_i and q_i can be readily evaluated using individual sensor thresholds.

Using copulas, the set of probabilities $P_{jk} = Pr(u_{1n} = j, u_{2n} = k|H_i)$, for all $j = 0, 1; k = 0, 1$ that are required to specify the joint pmf of sensor decisions under the ith hypothesis are given as follows

$$Pr(u_1 = 0, u_2 = 0|H_1) = C(1 - p_1, 1 - p_2), \tag{16.82}$$

$$Pr(u_1 = 0, u_2 = 1|H_1) = 1 - p_1 - C(1 - p_1, 1 - p_2), \tag{16.83}$$

$$Pr(u_1 = 1, u_2 = 0|H_1) = 1 - p_2 - C(1 - p_1, 1 - p_2), \tag{16.84}$$

$$Pr(u_1 = 1, u_2 = 1|H_1) = p_1 + p_2 + C(1 - p_1, 1 - p_2) - 1, \tag{16.85}$$

where C is the copula function used. Joint probabilities under H_0 are obtained similarly with p_i replaced by q_i.

Bivariate Normal (C_ρ^N) and Student-t copula ($C_{\rho,\nu}^t$) functions that are used for determining the joint probabilities of sensor decisions are given as follows

$$C_\rho^N(u, v) = \int_{-\infty}^{\phi^{-1}(u)} \int_{-\infty}^{\phi^{-1}(v)} \frac{1}{2\pi\sqrt{(1-\rho^2)}} \exp\left\{-\frac{s^2 + 2\rho st + t^2}{2(1-\rho^2)}\right\} dsdt, \tag{16.86}$$

$$C_{\rho,\nu}^t(u, v) = \int_{-\infty}^{t_\nu^{-1}(u)} \int_{-\infty}^{t_\nu^{-1}(v)} \frac{1}{2\pi\sqrt{(1-\rho^2)}} \left\{1 + \frac{s^2 + 2\rho st + t^2}{\nu(1-\rho^2)}\right\} dsdt. \tag{16.87}$$

For simulation purposes, the source count rate is determined by using the following source parameters: $A_0 = 10$, $(x_0, y_0) = (10, 10)$. The sensors are assumed to be equidistant from the source located at $(x_1, y_1) = (10, 14)$ and $(x_2, y_2) = (10, 6)$ resulting in equal values of λ_{ci} ($\lambda_{c1} = \lambda_{c2} = \lambda_c = 0.625$). The background radiation count is assumed to be $\lambda_b = 10$ and measurement noise with variance $\sigma_w^2 = 10$ is considered. It is assumed that the sensors are observing the phenomena over $N = 100$ time intervals. The correlation parameter in the copula function (denoted by ρ in equations (16.86) and (16.87)) is assumed to be known under both hypotheses. For convenience of notation, it is denoted as simply ρ_0 under H_0 and ρ_1 under H_1.

In one experiment for performance evaluation, it is assumed that $\rho_0 = 0$ and $\rho_1 = 0.4$. By varying P_F, the maximum attainable P_D and values of the local thresholds, τ_1 and τ_2, that maximize the P_D are determined for the Chair–Varshney rule, the Normal copula-based fusion rule, and the Student-t copula-based fusion rule. The resulting theoretical ROC curves for the three fusion rules are shown in Figure 16.7. Recall that the use of the Student-t copula also requires specification of the number of degrees of freedom ν. A Student-t copula with $\nu = 3$ is considered to obtain the ROC curve in Figure 16.7. From Figure 16.7, it is clear that the performance of the copula-based fusion rules is superior to that of the Chair–Varshney fusion rule. Moreover, for this particular signal model, the Gaussian copula-based fusion rule outperforms the Student-t copula-based fusion rule, especially at a low system probability of false alarm P_F.

In Figure 16.8, the performance comparison of the copula-based fusion rules and the Chair–Varshney fusion rule with varying number of sensor observations (N) is shown with $P_F = 0.001$. The correlation parameters within the copula functions and the signal parameters under both hypotheses are maintained the same. Again, the performance of the proposed copula-based fusion methods is better than the Chair–Varshney fusion rule, and they require fewer sensor observations than the Chair–Varshney rule to attain the same value of P_D.

16.5.4 Remark

The results presented so far assumed perfect knowledge of the copula function parameters, namely, ρ_1, ρ_0, and ν (for Student-t copula function). Any deviation of the

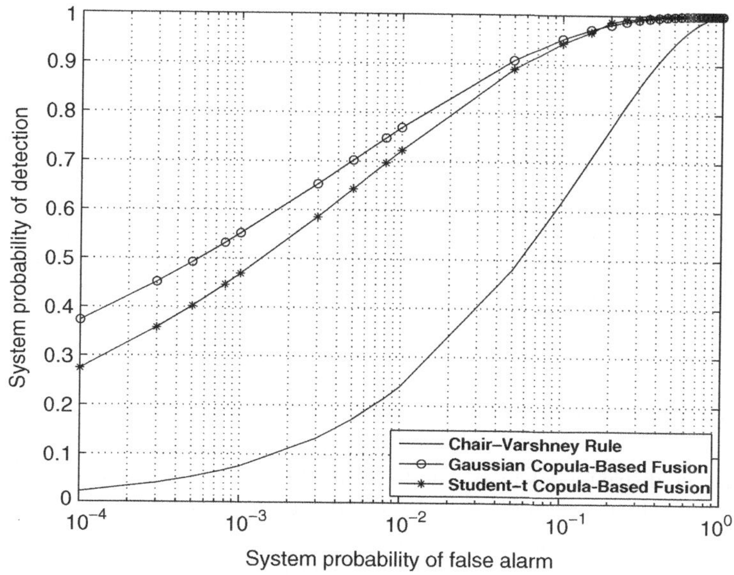

Figure 16.7 Theoretical ROC curves comparing the Chair–Varshney fusion rule and the Copula-based fusion rules.

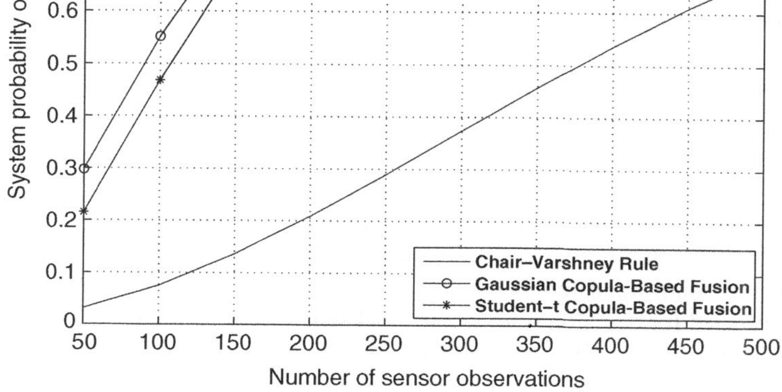

Figure 16.8 Detection performance comparison of the Chair–Varshney test and copula-based tests with increase in the number of sensor observations.

assumed parameter values from the true ones will result in performance degradation and may result in a performance even worse than the Chair–Varshney fusion rule that assumes independence of sensor observations. This is due to the mismatch between the actual multivariate pdf of sensor observations and the pdf specified by using copula functions with the incorrect parameter values. The same argument holds true while selecting a suitable copula function to construct the joint distribution of sensor observations also. The copula-function-based fusion will fail to perform better than the Chair–Varshney rule if the constructed joint distribution using a particular parametric copula function does not adequately model the underlying joint distribution of the sensor observations. The topic of copula function selection for the distributed detection problem is considered in [80].

16.6 CONCLUSION

In this chapter, distributed detection and decision fusion for a multisensor system have been discussed. In a conventional distributed detection framework, it is assumed that local sensors' performance indices are known and communication channels between the sensors and fusion center are perfect. Under these assumptions, the design for optimal decision fusion rule at the fusion center and the optimal local decision rules at sensors was discussed. Various network topologies, including parallel and serial structures, have been considered. Both Bayesian and Neyman–Pearson criteria have been used for designing optimal decision rules.

For the emerging WSNs, some of the assumptions made for the conventional sensor networks may be violated due to severe constraints on sensors' sensing and communication capabilities. In this chapter, the sensors were assumed to have nonidentical views of the phenomenon due to power decay during signal's propagation. Further, the performance indices of local sensors were assumed to be unknown to the detection system. An intuitive fusion rule based on simply counting the total number of detections (i.e., "1"s) was discussed. The exact detection performance of the counting rule in a WSN has been derived. Also, several computationally efficient approaches, using Binomial distribution or DLA approximation, have been proposed for approximating the detection performance. While deriving the detection performance, an isotropic signal decay model, uniformly distributed target location, and sensor location models have been adopted so that the randomness brought by signal decay, sensor, and target locations can be averaged out. However, these models, especially the isotropic signal decay model, may not accurately reflect the reality. Performance evaluation for a sensor network with more realistic models is a challenging problem.

Under the conditional independence assumption, the optimality of the LRT has been established for decision rules at the local sensors and at the fusion center. However, the determination of the optimal LRT thresholds at the local sensors is still quite computationally complex, especially for a large scale sensor network. Further, as discussed in Section 16.3, distributed detection with correlated observations remains a very difficult problem, since in such cases, the LRT test based solely on local observation is no longer an optimal test at sensors, and the optimal solution is intractable in general. Suboptimal solutions that can still exploit the dependence information are required. In this chapter, for the dependent observations problem, a framework for fusion of correlated decisions using copula theory has been described. The local sensor decision rules are assumed to be based on simple binary quantization of the sensor observations. The described method is particularly useful when the marginal densities of sensor observations are non-Gaussian (and potentially nonidentical) and when dependence between sensor observations can get manifested in several different nonlinear ways.

While many advances have been made in the area of distributed detection and decision fusion, many open and challenging problems remain that need further research. For example, distributed detection in a sensor network, where the signal decays following a nonisotropic model, is a very difficult problem. Another challenging problem is the design of the optimal local sensor decision rules when observations from different sensors are dependent conditioned on either hypothesis. Copula theory has provided one approach to treat dependence, further work on copula-based methods and other approaches will be quite valuable.

ACKNOWLEDGMENTS

Qi Cheng's work was supported by the National Science Foundation under grant 0932297 and the US Department of Transportation, Research and Innovative Technology Administration under grant DTRT06-G-0016. Ruixin Niu, Ashok Sundaresan, and Pramod K. Varshney's work was supported in part by Air Force Office of Scientific Research under grant FA9550-10-1-0263 and the National Science Foundation under grant ECCS-0925854.

APPENDIX 16A PERFORMANCE ANALYSIS OF A NETWORK WITH NONIDENTICAL SENSORS VIA APPROXIMATIONS

In Section 16.4, exact performance analysis is computationally intensive. This appendix provides details of some approximation methods that require much less computational load.

Through experiments, the Poisson distribution has been shown to be a poor approximation for the distribution of $\Lambda|H_1$, and is not discussed here.

Appendix 16A.1 Binomial I Approximation

Inspired by the fact that conditioned on (x_t, y_t), $\Lambda|H_1$ is a Binomial distributed RV, a Binomial (N, p_1) distribution is used to approximate the distribution of $\Lambda|H_1$. The parameter of the Binomial distribution is chosen such that its mean matches the mean of $\Lambda|H_1$, namely $p_1 = \bar{P}_d$, where $\bar{P}_d$ is the expected value of P_d

$$\bar{P}_d = \frac{1}{b^4}\int_{-\frac{b}{2}}^{\frac{b}{2}}\int_{-\frac{b}{2}}^{\frac{b}{2}}\int_{-\frac{b}{2}}^{\frac{b}{2}}\int_{-\frac{b}{2}}^{\frac{b}{2}} P_d\left(P_0, \sqrt{(x-x_t)^2+(y-y_t)^2}, \tau\right) dx dy dx_t dy_t. \tag{16A.1}$$

Again, the evaluation of $\bar{P}_d$ requires a fourfold integration. However, given P_0 and τ, P_d is only a function of $d^2 = (x - x_t)^2 + (y - y_t)^2$ (see (16.37) and (16.41)). As a result, once the distribution of $v \triangleq d^2$ is known, the evaluation of $\bar{P}_d$ is reduced to a single-fold integration over v, instead of a fourfold integration. In [82], the distribution of v has been derived

$$f_V(v) = \begin{cases} \frac{\pi}{b^2} + \frac{v}{b^4} - \frac{4\sqrt{v}}{b^3} & 0 < v \le b^2 \\ h(v) & b^2 < v \le 2b^2 \\ 0 & \text{otherwise} \end{cases}, \tag{16A.2}$$

where

$$h(v) \triangleq \frac{2}{b^2}\arcsin\left(\frac{2b^2 - v}{v}\right) - \frac{v}{b^4} + \frac{4\sqrt{v - b^2}}{b^3} - \frac{2}{b^2}.$$

Given $f_V(v)$, the average P_d can be easily calculated as follows

$$\bar{P}_d = \int_0^{2b^2} P_d\left(P_0, \sqrt{v}, \tau\right) f_V(v)dv. \tag{16A.3}$$

Appendix 16A.2 Binomial II Approximation

The Binomial I approximation only matches the mean of $\Lambda|H_1$. It will be of interest to study the approximation that matches the variance of $\Lambda|H_1$ as well, which can be derived as

$$\mathrm{Var}(\Lambda|H_1) = N\bar{P}_d - N^2\bar{P}_d^2 + (N^2 - N)E[\bar{P}_d^2(x_t, y_t)]. \tag{16A.4}$$

From (16A.4), it is clear that the evaluation of $E[\bar{P}_d^2(x_t, y_t)]$, and hence that of $\mathrm{Var}(\Lambda|H_1)$ requires a fourfold integration. Similar to the method proposed in [77], matching a Binomial (M, p_2) RV's mean Mp_2 and variance $Mp_2(1 - p_2)$ to $N\bar{P}_d$ and $\mathrm{Var}(\Lambda|H_1)$ respectively, it follows that

$$\begin{aligned} M &= \mathrm{Round}\left\{(N\bar{P}_d)^2/[N\bar{P}_d - \mathrm{Var}(\Lambda|H_1)]\right\}, \\ p_2 &= N\bar{P}_d/M. \end{aligned} \tag{16A.5}$$

In addition, it can be readily shown that $M \ge N$.

Appendix 16A.3 DeMoivre–Laplace Approximation

According to DeMoivre–Laplace Theorem [69], if $Mp_2(1 - p_2) >> 1$, then

$$\binom{M}{k} p_2^k(1 - p_2)^{M-k} \approx \frac{1}{\sqrt{2\pi Mp_2(1 - p_2)}} e^{-(k-Mp_2)^2/[2Mp_2(1-p_2)]}.$$

As a result, when M is large, the pmf of a Binomial (M, p_2) distribution can be approximated by the samples of a Gaussian pdf, whose mean and variance are $N\bar{P}_d$ and $\text{Var}(\Lambda|H_1)$ respectively

$$Pr(\Lambda = k|H_1) \approx \frac{1}{\sqrt{2\pi \text{Var}(\Lambda|H_1)}} e^{-(k-N\bar{P}_d)^2/[2\text{Var}(\Lambda|H_1)]}. \tag{16A.6}$$

Note that when using DLA, terms in (16A.6) are normalized so that the sum of the pmf values is 1.

Appendix 16A.4 Total Variation Distance

The TVD [75] is a very useful tool to quantify the difference of two distributions. Let r_1 and r_2 be two integer-valued RVs, which follow two distributions with pmfs f_1 and f_2, respectively. The TVD between these two distributions is defined as

$$d_{TV}(f_1, f_2) := \sup_A |Pr(r_1 \in A) - Pr(r_2 \in A)|, \tag{16A.7}$$

where A is any subset of the set of nonnegative integers and it is well known that [76]

$$d_{TV}(f_1, f_2) = \frac{1}{2} \sum_{k=0}^{\infty} |f_1(k) - f_2(k)|. \tag{16A.8}$$

Since by definition, the TVD is the maximum absolute difference of probability mass on any subset A between two distributions, and P_D is equivalent to a complementary cumulative distribution function (see (16.46)), the absolute difference between the true P_D and its approximated value is also bounded by the TVD.

To compare the accuracies of different approximations, their TVDs relative to the true pmf of $\Lambda|H_1$ are shown in Figure 16A.1, as a function of N. It is clear that the accuracy of Binomial I degrades rapidly as N increases. The variance of a Binomial I distribution, $N\bar{P}_d(1 - \bar{P}_d)$ is linear in N, and $\text{Var}(\Lambda|H_1)$ is a quadratic function of N, according to (16A.4). As N increases, their difference, $(N^2 - N)[E[\bar{P}_d^2(x_t, y_t) - \bar{P}_d^2]]$, increases, and the Binomial I approximation becomes less accurate, due to the mismatch in variance.

On the other hand, the Binomial II approximation has the smallest TVD among the three approximations, because both its mean and variance have been matched to those of $\Lambda|H_1$. The jumpy behavior of its TVD curves is due to the fact that M can only take an integer value and hence its variance is close but not identical to the true variance of $\Lambda|H_1$ in most cases. As N increases, the accuracy of DLA improves and converges to that of the Binomial II quickly. This is because as N increases, $M \geq N$ increases too, and the DLA becomes a more accurate approximation of Binomial (M, p_2) distribution.

Another observation is that when the size of the ROI b is large, the TVDs corresponding to all the approximations are small. This can be explained as follows.

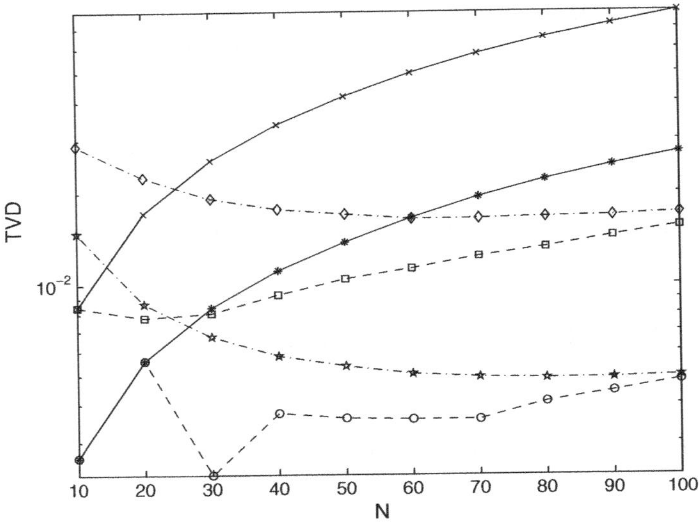

Figure 16A.1 The TVD between the exact pmf of $\Lambda|H_1$, and its approximations based on Binomial I and II, and DLA. $P_0 = 25$. Solid line + x-mark: Binomial I, $b = 20$; solid line + star: Binomial I, $b = 40$; dashdot line + diamond: DLA, $b = 20$; dashdot line + pentagram: DLA, $b = 40$; dashed line + square: Binomial II, $b = 20$; dashed line + circle: Binomial II, $b = 40$.

When the ROI is large, only within a small fraction of the ROI surrounding the target, received signal strength is significantly larger than zero. As a result, for most (x_t, y_t) pairs, except those near the border of the ROI, $\bar{P}_d(x_t, y_t)$ in Equation (16.43) is almost a constant, independent of (x_t, y_t), and $\bar{P}_d(x_t, y_t) \approx \bar{P}_d$. In such cases, (16.45) becomes

$$Pr(\Lambda = k|H_1) \approx \binom{N}{k} \bar{P}_d^k (1 - \bar{P}_d)^{N-k}$$

and Binomial distribution is a very good approximation to the exact distribution of $\Lambda|H_1$.

REFERENCES

1. P.K. Varshney, *Distributed Detection and Data Fusion*, Springer, New York, 1997.
2. H.L. Van Trees, *Detection, Estimation and Modulation Theory*, Vol. 1, Wiley, New York, 1968.
3. H.V. Poor, *An Introduction to Signal Detection and Estimation*, Springer-Verlag, New York, 1988.
4. C.W. Helstrom, *Elements of Signal Detection and Estimation*, Prentice-Hall, Englewood Cliffs, NJ, 1995.

5. R. Viswanathan and P.K. Varshney, Distributed detection with multiple sensors: Part I—Fundamentals, *Proc. IEEE*, 85(1), 54–63, 1997.

6. R.S. Blum, S.A. Kassam, and H.V. Poor, Distributed detection with multiple sensors: Part II—Advanced topics, *Proc. IEEE*, 85(1), 64–67, 1997.

7. T.M. Cover and J.A. Thomas, *Elements of Information Theory*, Wiley, New York, 1991.

8. H.M. Finn and R.S. Johnson, Adaptive detection mode with threshold control as function of spatially sample clutter-level estimates, *RCA Rev.*, 29, 414–464, 1968.

9. H. Rohling, Radar CFAR thresholding in clutter and multiple target situations, *IEEE Trans. Aerosp. Electron. Syst.*, 19(4), 608–621, 1983.

10. H.V. Poor and J.B. Thomas, *Advances in Statistical Signal Processing: Vol. 2. Signal Detection*, JAI Press, Greenwich, CT, 1993.

11. R.R. Tenney and N.R. Sandell Jr., Detection with distributed sensors, *IEEE Trans. Aerosp. Electron Syst.*, 17(4), 501–510, 1981.

12. J. Tsitsiklis and M. Athans, On the complexity of decentralizd desicion making and detection problems, *IEEE Trans. Automat. Contr.*, 30(5), 440–446, 1985.

13. N.S.V. Rao, Computational complexity issues in synthesis of simple distributed detection networks, *IEEE Trans. Syst., Man., Cybern.*, 21(5), 1071–1081, 1991.

14. J.N. Tsitsiklis, Decentralized detection, in: H.V. Poor and J.B. Thomas (Eds.), *Advances in Statistical Signal Processing*, JAI Press, Greenwich, CT, 2, 297–344, 1993.

15. S. Alhakeem and P.K. Varshney, A unified approach to the design of decentralized detection systems, *IEEE Trans. Aerosp. Electron. Syst.*, 31(1), 9–20, 1995.

16. Z.B. Tang, K.R. Pattipati, and D. Kleinman, An algorithm for determining the detection thresholds in a distributed detection problem, *IEEE Trans. Syst., Man., Cybern.*, 21(1), 231–237, 1991.

17. Z.B. Tang, Optimization of detection networks, *PhD thesis*, University of Connecticut, Storrs, CT, Dec. 1990.

18. A.R. Reibman, Performance and fault-tolerance of distributed detection networks, *PhD thesis*, Duke University, Durham, NC, 1987.

19. S.C.A. Thomopoulos, R. Viswanathan, and D.K. Bougoulias, Optimal distributed decision fusion, *IEEE Trans. Aerospace Electron. Syst.*, 25(5), 761–765, 1989.

20. C.W. Helstrom, Gradient algorithms for quantization levels in distributed detection systems, *IEEE Trans. Aerosp. Electron. Syst.*, 31(1), 390–398, Jan. 1995.

21. Z. Chair and P.K. Varshney, Optimal data fusion in multiple sensor detection systems, *IEEE Trans. Aerosp. Electron. Syst.*, 22(1), 98–101, 1986.

22. J.N. Tsitsiklis, On threshold rules in decentralized detection, *Proceedings of the 25th IEEE Conference on Decision and Control.*, Athens, Greece, 1986, pp. 232–236.

23. P. Willet and D. Warren, Decentralized detection: When are identical sensors identical, in *Proceedings of the Conference on Information Science and System*, 1991, pp. 287–292.

24. M. Cherikh and P.B. Kantor, Counterexamples in distributed detection, *IEEE Trans. Inform. Theor.*, 38(1), 162–165, 1992.

25. J.N. Tsitsiklis, Decentralized detection with a large number of sensors, *Math. Contr., Signals Syst.*, 1, 167–182, 1988.

26. P. Chen and A. Papamarcou, New asymptotic results in parallel distributed detection, *IEEE Trans. Inform. Theor.*, 39(6), 1847–1863, 1993.

27. J.Chamberland and V.V. Veeravalli, Decentralized detection in sensor networks, *IEEE Trans. Signal Process.*, 51(2), 407–416, 2003.
28. J.F. Chamberland and V.V. Veeravalli, Asymptotic results for decentralized detection in power constrained wireless sensor networks, *IEEE J. Sel. Areas Commun.*, 22(6), 1007–1015, 2004.
29. J.D. Papastavrou and M. Athans, The team ROC curve in a binary hypothesis testing environment, *IEEE Trans. Aerosp. Electron. Syst.*, 31(1), 96–105, 1995.
30. J.D. Papastavrou, Decentralized decision making in a hypothesis testing environment, *PhD thesis*, MIT, May 1990.
31. R. Viswanathan, S.C.A. Thomopoulos, and R. Tumuluri, Optimal serial distributed decision fusion, *IEEE Trans. Aerosp. Electron. Syst.*, 24(4), 366–376, 1988.
32. Z.B. Tang, K. Pattipati, and D.L. Kleinman, Optimization of detection networks: Part I—Tandem structures, *IEEE Trans. Syst. Man Cybern.*, 21(5), 1044–1059, 1991.
33. G.S. Lauer and N.R. Sandell Jr., Distributed detection with waveform observations: correlated observation processes, *Proceedings of the 1982 American Controls Conference*, Vol. 2, 1982, pp. 812–819.
34. P. Chen and A. Papamarcou, Likelihood ratio partitions for distributed signal detection in correlated Gaussian noise, *Proceedings of the IEEE International Symposium on Information Theory*, Oct. 1996, p. 118.
35. V. Aalo and R. Viswanathan, On distributed detection with correlated sensors: Two examples, *IEEE Trans. Aerosp. Electron. Syst.*, 25(3), 414–421, 1989.
36. V. Aalo and R. Viswanathan, Asymptotic performance of a distributed detection system in correlated Gaussian noise, *IEEE Trans. Signal Process.*, 40(1), 211–213, 1992.
37. R. Blum, P. Willett, and P. Swaszek, Distributed detection of known signals in nonGaussian noise which is dependent from sensor to sensor, *Proceedings of the Conference on Information Science and Systems*, Mar. 1997, pp. 825–830.
38. X. Lin and R. Blum, Numerical solutions for optimal distributed detection of known signals in dependent t-distributed noise—The two-sensor problem, *Proceedings of the Asilomar Conference on Signals, Syst., Comput.*, Nov. 1998, pp. 613–617.
39. Z. Tang, K. Pattipati, and D. Kleinman, A distributed M-ary hypothesis testing problem with correlated observations, *IEEE Trans. Automat. Contr.*, 37(7), 1042–1046, 1992.
40. P.K. Willett, P.F. Swaszek, and R.S. Blum, The good, bad, and ugly: distributed detection of a known signal in dependent Gaussian noise, *IEEE Trans. Signal Process.*, 48(12), 3266–3279, 2000.
41. E. Drakopoulos and C.-C. Lee, Optimum multisensor fusion of correlated local decisions, *IEEE Tran. Aerosp. Electron. Syst.*, 27(4), 593–606, 1991.
42. M. Kam, Q. Zhu, and W.S. Gray, Optimal data fusion of correlated local decisions in multiple sensor detection systems, *IEEE Trans. Aerosp. Electron. Syst.*, 28(3), 916–920, 1992.
43. D.D. Mari and S. Kotz, *Correlation and Dependence*, Imperial College Press, London, 2001.
44. R.B. Nelsen, *An Introduction to Copulas*, Springer-Verlag, New York, 1999.
45. S. Kumar, F. Zhao, and D. Shepherd Eds., Special issue on collaborative signal and information processing in microsensor networks, *IEEE Signal Process. Mag.*, 19(2), 2002.

46. H. Gharavi and S. Kumar (Eds.), Special issue on sensor networks and applications, *Proc. IEEE*, 91(8), 2003.

47. A. Sayeed, D. Estrin, G. Pottie, and K. Ramchandran (Eds.), Special Issue on self-organizing distributed collaborative sensor networks," *IEEE J. Sel. Areas Commun.*, 23(4), 2005.

48. Z.Q. Luo, M. Gastpar, J. Liu, and A. Swami (Eds.), Special Issue on distributed signal processing in sensor networks, *IEEE Signal Process. Mag.*, 2006.

49. B. Chen, L. Tong, and P.K. Varshney, Channel aware distributed detection in wireless sensor networks," *IEEE Signal Process. Mag. (Special Issue on Distributed Signal Processing for Sensor Networks)*, 23(4), 16–26, 2006.

50. B. Chen and P.K. Willett, On the optimality of likelihood ratio test for local sensor decision rules in the presence of nonideal channels, *IEEE Trans. Inform. Theor.*, 51(2), 693–699, 2005.

51. Q. Cheng, B. Chen, and P.K. Varshney, Detection performance limits for distributed sensor networks in the presence of nonideal channels," *IEEE Trans. Wireless Commun.*, 5(11), 3034–3038, 2006.

52. B. Chen, R. Jiang, T. Kasetkasem, and P.K. Varshney, Channel aware decision fusion for wireless sensor networks, *IEEE Trans. Signal Process.*, 52(12), 3454–3458, 2004.

53. R. Niu, B. Chen, and P.K. Varshney, Fusion of decisions transmitted over Rayleigh fading channels in wireless sensor networks, *IEEE Trans. Signal Process.*, 54(3), 1018–1027, 2006.

54. R. Jiang and B. Chen, Fusion of censored decisions in wireless sensor networks, *IEEE Trans. Wireless Commun.*, 4(6), 2668–2673, 2005.

55. Y. Lin, B. Chen, and P.K. Varshney, Decision fusion rules in multihop wireless sensor networks, *IEEE Trans. Aerosp. Electron. Syst.*, 51(2), 475–488, 2005.

56. B. Liu and B. Chen, Channel optimized quantizers for decentralized detection in wireless sensor networks, *IEEE Trans. Inform. Theor.*, 52(7), 3349–3358, 2006.

57. B. Liu and B. Chen, Decentralized detection in wireless sensor networks with channel fading statistics, *EURASIP J. Wireless Commun. Netw.*, 1–8, 2007.

58. Y. Lin, B. Chen, and B. Suter, Robust binary quantizers for detection in sensor networks, *IEEE Trans. Wireless Commun.*, 6, 2172–2181, 2007.

59. B. Liu, B. Chen, and R.S. Blum, Minimum error probability cooperative relay design," *IEEE Trans. Signal Process.*, 55(2), 656–664, 2007.

60. H. Chen, P.K. Varshney, and B. Chen, Cooperative relay for decentralized detection, in *Proceedings of the 2008 IEEE International Conference on Acoustics, Speech and Signal Processing*, Las Vegas, Nevada, March 2008, pp. 2293–2296.

61. T.M. Duman and M. Salehi, Decentralized detection over multiple-access channels, *IEEE Trans. Aerosp. Electron. Syst.*, 34(1), 469–476, 1998.

62. Y. Sung, L. Tong, and A. Swami, Asymptotic locally optimal detector for large-scale sensor networks under the Poisson regime, *IEEE Trans. Signal Process.*, 53(6), 2005–2017, 2005.

63. G. Mergen and L. Tong, Type based estimation over multiple access channels, *IEEE Trans. Signal Process.*, 54(2), 613–626, 2006.

64. G. Mergen, V. Naware, and L. Tong, Asymptotic detection performance of type-based multiple access over multiaccess fading channels, *IEEE Trans. Signal Process.*, 55(3), 1081–1092, 2007.

65. K. Liu and A.M. Sayeed, Type-based decentralized detection in wireless sensor networks, *IEEE Trans. Signal Process.*, 55(5), 1899–1910, 2007.

66. K. Liu, H. El Gamal, and A.M. Sayeed, Decentralized inference over multiple-access channels, *IEEE Trans. Signal Process.*, 55(7), 3445–3455, 2007.

67. R. Niu, P.K. Varshney, and Q. Cheng, Distributed detection in a large wireless sensor network, *Int. J. Inform. Fusion*, 7(4), 380–394, 2006.

68. R. Niu and P.K. Varshney, Distributed detection and fusion in a large wireless sensor network of random size, *EURASIP J. Wireless Commun. Netw.*, 2005(4), 462–472, 2005.

69. A. Papoulis, *Probability, Random Variables, and Stochastic Processes*, McGraw-Hill, New York, NY, 1984.

70. S.A. Aldosari and J.M.F. Moura, Detection in sensor networks: The saddlepoint approximation, *IEEE Trans. Signal Process.*, 55(1), 327–340, 2007.

71. N. Levanon, *Radar Principles*, John Wiley & Sons, Inc., New York, NY, 1988.

72. L.E. Kinsler and A.R. Frey, *Fundamentals of Acoustics*, John Wiley & Sons, Inc., New York, NY, 1962.

73. R. Niu and P.K. Varshney, Performance analysis of distributed detection in a random sensor field, *IEEE Trans. Signal Process.*, 56(1), 339–349, 2008.

74. L. Le Cam, An approximation theorem for the Poisson Binomial Distribution, *Pacific J. Math.*, 10(4), 1181–1197, 1960.

75. A.D. Barbour, L. Holst, and S. Janson, *Poisson Approximation, Oxford Studies in Probability 2*, Oxford: Clarendon Press, Oxford, England, 1992.

76. B. Roos, Binomial approximation to the Poisson Binomial Distribution: The Krawtchouk expansion, *Theory Probab. Appl.*, 45(2), 258–272, 2000.

77. K.P. Choi and A. Xia, Approximating the number of successes in independent trials: Binomial versus Poisson, *Ann. Appl. Probab.*, 12(4), 1139–1148, 2002.

78. S.C.A. Thomopoulos and L. Zhang, Distributed decision fusion with networking delays and channel errors, *Inform. Sci.*, 66(1-2), 91–118, 1992.

79. M. Haenggi, On distances in uniformly random networks, *IEEE Trans. Inform. Theor.*, 51(10), 3584–3586, 2005.

80. A. Sundaresan, Detection and source location estimation of random signal sources using sensor networks, *PhD Thesis*, Syracuse University, Syracuse, NY, December, 2010.

81. A. Sundaresan, P.K. Varshney, and N.S.V. Rao, Copula based fusion of correlated decisions, *IEEE Trans. Aerosp. Electron. Syst.*, 47(1), 454–471, 2011.

82. R. Niu and P.K. Varshney, Target location estimation in sensor networks with quantized data, *IEEE Trans. Signal Process.*, 54(12), 4519–4528, 2006.

CHAPTER 17

Evidential Networks for Decision Support in Surveillance Systems

ALESSIO BENAVOLI AND BRANKO RISTIC

17.1 INTRODUCTION

The amount of data potentially relevant and available to a decision maker in a modern warfare by far exceeds the human ability to review and comprehend it in a timely manner. Moreover, the decisions usually have to be taken under very stressful conditions that adversely affect humans and make them prone to error. All this leads to a need for the development of an automatic knowledge-based information fusion system that will support the decision process in a reliable, timely, and consistent manner [1]. Similar problems also exist in other fields of human endeavor (e.g., management of commercial enterprises, medical diagnosis, etc.), although the military and homeland security domains are particularly challenging due to the inherently incomplete, uncertain and imprecise evidence and domain knowledge.

A review of the early (pre-1990s) attempts at building knowledge-based expert systems is presented in [2, Chapter 9]. The main drawback with these early attempts was the lack of a means and associated difficulties in handling the uncertain domain knowledge and imprecise evidence. The invention of *Bayesian networks* [3] in the mid-1980s for knowledge representation and probabilistic inference represented the next important stepping stone in the development of expert systems. Bayesian networks are based on the assumption that all data (domain knowledge and accumulated evidence) can be conveniently represented by probability functions. In reality, this may not always be the case and so, as alternative to the Bayesian networks, other network-based systems [4, 5] employing alternative uncertainty formalisms, such as possibility theory [6–8] and the theory of evidence [9, 10], have been developed.

Integrated Tracking, Classification, and Sensor Management: Theory and Applications, Edited by Mahendra Mallick, Vikram Krishnamurthy, and Ba-Ngu Vo.

In the Bayesian network context, several architectures [11] have been proposed for exact computation of marginals of multivariate discrete probability distributions. One of the pioneering architectures for computing marginals was proposed by Pearl [3] for multiply connected Bayesian networks. In 1988, Lauritzen and Spiegelhalter [12] proposed an alternative architecture for computing marginals in join trees (also known as junction or clique trees) that applies to any Bayesian network. Subsequently, Jensen et al. [13, 14] proposed a modification of the Lauritzen–Spiegelhalter architecture, which is known as the Hugin architecture, since this architecture is implemented in Hugin, a software tool developed by the same group. This architecture has been generalized by Lauritzen and Jensen [15] so that it applies more generally to other domains including the Dempster–Shafer's belief function theory. Inspired by the work of Shafer et al. [5] first adapted and generalized Pearl's architecture to the case of finding marginals of joint Dempster–Shafer belief functions in join trees. Later, motivated by the work of Lauritzen and Spiegelhalter [12] for the case of probabilistic reasoning, they proposed the valuation algebra (VA) framework for computing marginals in join trees and established the set of axioms that combination and marginalization need to satisfy in order to make the local computation concept applicable [16]. These axioms are satisfied for all three major theories of uncertainty (probability theory, possibility theory and the theory, of evidence). In 1997, Shenoy [17] proposed a refinement of junction trees, called binary join trees, designed to improve the computational efficiency of the Shenoy–Shafer architecture.

This chapter will focus on valuation algebras in the context of the theory of evidence, due to the expressive power of belief functions that can represent both classical probability functions and possibility/necessity functions [18]. The resulting network for reasoning under uncertainty is referred to as the *evidential network*.

The chapter is organized as follows. Section 17.2 introduces the concept of valuation algebra for knowledge representation and reasoning under uncertainty. Section 17.3 describes the algorithms for local computation in a valuation algebra: the fusion algorithm, the binary join tree, and inward propagation. Section 17.4 presents the basic tools of the theory of evidence as a valuation algebra: the belief functions, the combination and marginalization operations, and the approaches to decision making. Finally, Section 17.5 illustrates the theoretical concepts by two examples: decision support systems for target identification and threat assessment.

17.2 VALUATION ALGEBRAS

There are many different formalisms for representing knowledge or information. However, there are some elementary features that can be found in every formalism [19].

Example 1 A cyclist lives in a country with four distinct climatic regions. N is a mountainous region in the north, W is a plain in the west, E is a hilly region in the east, and S is a plain in the south. The cyclist is planning a training ride in region N tomorrow and wants to know if it will rain. He/she found three local weather forecasts. The first predicts "sunshine in regions N and W" with a 60% chance, the second "overcast with 30% chance of rain showers in regions W and S" and the third predicts rain showers with a 60% chance in regions S and E. From his experience, the cyclist knows that when it rains in region E, then, with chance of 50–70%, it also rains in the mountains N.

Example 1 shows the main aspects of knowledge representation and processing:

- Information is a statement about something of interest.
- Information comes piecewise from different sources.
- Information generally includes the elements that are not of interest.
- Information is often uncertain.

A reasoning (decision making) system must be able to combine the information coming from different sources and to extract from it those parts that are relevant to specific questions. This must be accomplished by taking into account the underlying uncertainty. Various methods for combining and extracting the desired information are potentially available, but only those that obey certain rules of "coherence" are of interest. The rules are as follows:

1. The result of combining two pieces of information must be information.
2. The result of extracting a piece of information must be information.
3. The combination process must not add information that is not contained in the pieces of information being combined.
4. The result of the combination of two or more pieces of information must be independent of the order in which the pieces of information are combined.
5. Extracting the information of interest after combining two pieces of information must be equivalent to extracting the information of interest from each piece of information separately followed by the combination of the resulting information.

Example 2 Consider rules 4 and 5 in the context of Example 1. Rule 4 is obvious; it means that the decision of the cyclist is based on the combination of the information coming from the local forecasts and his experience. The order in which this information is combined does not matter. Rule 5 is more subtle. It states that the following procedures are equivalent:

- The cyclist can combine all available information and then extract the forecast for the region of interest N.
- The cyclist can extract the forecast for the region of interest N from available pieces of information (local forecasts and his experience) and then combine them.

Rules 1–5 are well known in mathematics. A system that satisfies the above rules is called an algebraic system [19]. In this chapter, the focus is on a particular algebraic system, a system made up of the pieces of information. The concept of an algebra of the information was introduced by Kohlas and Stark [19] under the name of *information algebra*. The interested here is in a particular *information algebra* referred to as the *valuation algebra* [19].

17.2.1 Mathematical Definitions and Results

This section presents the mathematical definition of a valuation algebra. For an in-depth study of valuation algebras the reader is referred to [19].

The primitive elements of valuation algebras are valuations. Intuitively, a valuation represents some knowledge about the possible values of a set of variables. Each valuation φ refers thus to a finite set of variables $d(\varphi)$ called its domain. Given an arbitrary finite set D of variables, Φ_D denotes the set of all valuations φ with $d(\varphi) = D$. Let V be the set of all available variables. Notation $\Phi = \cup\{\Phi_D : D \subseteq V\}$ is used to denote the set of all valuations φ with $d(\varphi) \subseteq V$. Note that, $d : \Phi \to 2^V$, where 2^V denotes the power set of V, is sometimes called the labeling operation.

Example 3 Consider again Example 1. In this case, the set of variables is $V = \{N, W, E, S\}$ and the set of valuations is $\Phi = \{\varphi_1, \varphi_2, \varphi_3, \varphi_4\}$, where

- φ_1: sunshine with 60% chance in regions N and W, $d(\varphi_1) = \{N, W\}$;
- φ_2: overcast with 30% chance of rain showers in regions W and S, $d(\varphi_2) = \{W, S\}$;
- φ_3: rain showers with 60% chance in regions S and E, $d(\varphi_3) = \{E, S\}$;
- φ_4: rain in region E implies rain in the mountains N with chance between 50% and 70%, $d(\varphi_4) = \{E, N\}$.

The model is shown in the form of a valuation network in Figure 17.1, where the variables are represented by circular nodes and the valuations by diamond shapes.

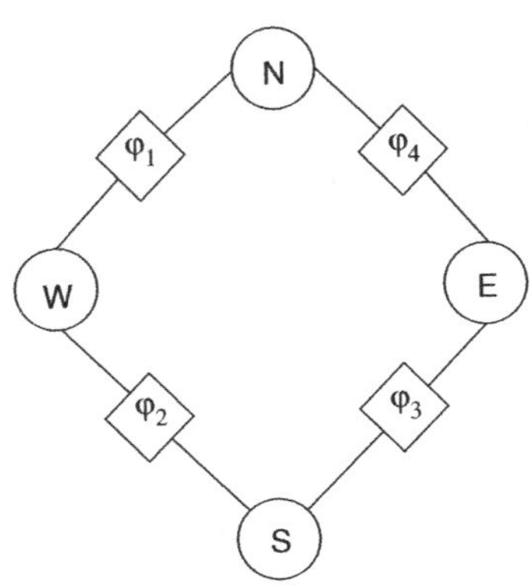

Figure 17.1 Valuation network representing the cyclist example.

Each valuation node is connected by edges to the subset of variables that define its domain. For example, the domain of valuation φ_1 consists of variables N and W.

There are two basic operations for valuations.

Combination is a binary operation $\oplus : \Phi \times \Phi \to \Phi$ that, intuitively, represents aggregation of knowledge. Thus, if $\varphi_1, \varphi_2 \in \Phi$ are two valuations, then the combined valuation $\varphi_1 \oplus \varphi_2$ represents the aggregated knowledge from φ_1 and φ_2.

Marginalization is also a binary operation $\downarrow: \Phi \times 2^V \to \Phi$ that, intuitively, represents focusing the knowledge to a smaller domain. Thus, if $\phi \in \Phi$ is a valuation and $D \subseteq d(\phi)$, then the marginalized valuation $\phi^{\downarrow D}$ represents the knowledge obtained by focusing ϕ from $d(\phi)$ to D.

Instead of marginalization another basic operation called *variable elimination* can be defined as $\varphi^{-x} \triangleq \varphi^{\downarrow d(\varphi)\setminus\{x\}}$ with $x \in V$, where symbol $\setminus$ denotes the set difference. Note that $x \notin d(\varphi)$ implies $\varphi^{-x} = \varphi$.

17.2.2 Axioms

Given a finite collection $\Phi = \{\varphi_1, \ldots, \varphi_r\}$ of valuations, the problem of inference is to marginalize the joint valuation $\oplus\Phi = \varphi_1 \oplus \cdots \oplus \varphi_r$ to a subset of variables $D^o \subseteq V$ with $V = d(\varphi_1) \cup \cdots \cup d(\varphi_r)$ (in the cyclist example $D^o = \{N\}$). The straightforward approach is to compute the joint valuation first and to marginalize to D^o afterward. The problem is that the number of variables increases with each combination. Since the complexity grows exponentially with the domain size, this becomes soon intractable, even if all the given valuations are defined on small domains. For instance, if there are n variables and each variable can assume m different values (or configurations), then there are m^n configurations in the joint domain of all variables.

However, by imposing certain axioms for the operations of *labeling*, *combination*, and *marginalization* [16, 19–21], it is possible to compute the marginal $(\oplus\Psi)^{\downarrow D^o}$ on local domains, that is without explicitly computing the joint valuation. The list of axioms is as follows.

(A1) *Commutative semigroup.* Φ is commutative and associative under combination. Thus, $(\Phi, \oplus)$ forms a commutative semigroup (monoid). Similarly, $(\Phi_D, \oplus)$ forms a commutative semigroup for each domain $D \subseteq V$.

(A2) *Labeling.* if φ_1, φ_2 are valuations, then $d(\varphi_1 \oplus \varphi_2) = d(\varphi_1) \cup d(\varphi_2)$. Furthermore, if $\varphi \in \Phi$ and $D \subseteq d(\varphi)$, then $d(\varphi^{\downarrow D}) = D$.

(A3) *Stability of marginalization.* If $\varphi \in \Phi$ is a valuation, then $\varphi^{\downarrow d(\varphi)} = d(\varphi)$.

(A4) *Transitivity of marginalization.* If $\varphi \in \Phi$ is a valuation and $D \subseteq D_1 \subseteq d(\varphi)$, then $\varphi^{\downarrow D} = (\varphi^{\downarrow D_1})^{\downarrow D}$.

(A5) *Partial distributivity of marginalization over combination.* If $\varphi_1, \varphi_2 \in \Phi$ defined on the domains D_1 and D_2, respectively, then $(\varphi_1 \oplus \varphi_2)^{\downarrow D_1} = \varphi_1 \oplus \varphi_2^{\downarrow D_1 \cap D_2}$.

(A6) *Identity element.* For each domain $D \subset V$, there is an identity valuation $e_D \in \Phi_D$ such that $e_D \oplus \varphi = \varphi$ for all $\varphi \in \Phi_D$. Furthermore, $e_{D_1} \oplus e_{D_2} = e_{D_1 \cup D_2}$ for domains $D_1, D_2 \subset V$.

Axioms (A1) states that the order in which valuations are combined does not matter. Axiom (A2) states that the domain of the combination of two pieces of information is the union of the two domains. Axiom (A3) states that the marginalization of a valuation to its domain does not change the domain. Axiom (A4) states that the marginalization operation is transitive. From (A5), it follows that combination and marginalization satisfy a rule of distributivity. This means that if a piece of information is combined with a second information, and the combination is marginalized to the domain of the first piece of information, then this is the same as the combination of the first piece of information with the marginalization of the second piece of information to the intersection of the two domains. Axiom (A6) says that the combination has a neutral element; in other words, there exists an empty piece of information that, combined with any other information, does not add anything to it.

Definition 1 *A system $\{V, \Phi, d, \oplus, \downarrow\}$ of valuations Φ and variables V is called valuation algebra if the operations of labeling d, combination $\oplus$, and marginalization $\downarrow$ satisfy the above system of axioms [19].*

The transitivity and distributivity axioms can easily be translated for the case where variable elimination is used instead of marginalization.

(A4′) *Transitivity of elimination.* $\varphi \in \Phi$ is a valuation and $x, y \in V$ $(\phi^{-x})^{-y} = (\phi^{-y})^{-x}$.

(A5′) *Partial distributivity of elimination over combination.* If $\varphi_1, \varphi_2 \in \Phi$ are valuations with $x \notin d(\varphi_1)$, then $(\varphi_1 \oplus \varphi_2)^{-x} = \varphi_1 \oplus \varphi_2^{-x}$.

Transitivity of elimination makes it possible to write φ^{-X} for the elimination of several variables $X \subset V$, since the result is independent of the order of elimination. As a consequence, marginalization can be expressed in terms of variable eliminations by $\varphi^{-D} = \varphi^{\downarrow d(\varphi) \setminus D}$. Therefore, the operations of marginalization and variable elimination together with their respective systems of axioms are equivalent.

A definition that will be useful in the sequel is introduced next.

Definition 2 *The set of possible values of a variable x is called a frame (or the state space) of x and will be denoted by Θ_x. Hence, for any subset of variables $D \subseteq V$, frame Θ_D denotes the Cartesian product of the variables $x \in D$, that is*

$$\Theta_D \triangleq \times\{\Theta_x : x \in D\},$$

$\times$ denoting the Cartesian product. The elements of Θ_D are referred to as configurations.

Example 4 Consider again Example 1 and focus on the subset of variables $D = \{N, W\}$. Let their frames be specified as follows: $\Theta_N = \{r_n, \bar{r}_n\}$ and $\Theta_W = \{r_w, \bar{r}_w\}$ (i.e., rain or no rain in the region N and, respectively, W). Then the frame of D consists of $2 \cdot 2 = 4$ configurations and is given by

$$\Theta_D = \{(r_n, r_w), (r_n, \bar{r}_w), (\bar{r}_n, r_w), (\bar{r}_n, \bar{r}_w)\}.$$

The concept of VA is very general and has a wide range of possible instantiations. Some of them, such as discrete probability mass functions or Gaussian potentials, are based on probability theory [19]. But the examples also include nonprobabilistic systems [19] such as relational algebra, propositional logic, constraint systems, systems of linear equations and inequalities, Dempster–Shafer belief functions, Spohns disbelief functions, possibility functions and many others [19]. Two particular formalisms will be discussed in detail here, that is, the discrete probability mass functions in Section 17.2.3 and belief functions in Section 17.4.

17.2.3 Probability Mass Functions as a Valuation Algebra

Probability mass functions are a particular instantiation of VA. Let $D = \{x_1, x_2, \ldots, x_n\}$ be a set of variables with a common frame Θ_D. In the context of probability, valuations are probability mass functions (PMF), $p^D : \Theta_D \to [0, 1]$

$$\begin{cases} p^D(\theta_i) \geq 0 \text{ for each } \theta_i \in \Theta_D, \\ \sum_i p^D(\theta_i) = 1, \\ P^D(A \cup B) = P^D(A) + P^D(B) \text{ for any } A, B \subseteq \Theta_D \text{ with } A \cap B = \emptyset, \end{cases}$$

where

$$P^D(A) = \sum_{\theta \in A} p^D(\theta), \quad \forall\, A \subseteq \Theta_D.$$

In probability theory, the combination operator is *multiplication*.

Definition 3 *If $p_1^{D_1}$ is a PMF over the set of variables D_1 and $p_2^{D_2}$ is a PMF over D_2, then for $x \in \Theta_D = \Theta_{D_1} \times \Theta_{D_2}$ one gets*

$$(p_1 \oplus p_2)^D(x) = \frac{1}{k}\, p_1^{D_1}(x^{\downarrow D_1})\, p_2^{D_2}(x^{\downarrow D_2}), \tag{17.1}$$

where $k > 0$ is the normalization constant.

The neutral element for Θ_D is $p_e(x) = 1/|\Theta_D|$ for all $x \in \Theta_D$.

Marginalization is a projection of a PMF defined on domain D into a PMF defined on a coarser domain $D_0 \subseteq D$.

Definition 4 *If p^D is a PMF defined on a domain D then its marginalization to the domain $D_0 \subseteq D$ is*

$$p^{D\downarrow D_0}(x) = \sum_{y:\ y\downarrow x} p^D(y), \tag{17.2}$$

where the summation in (17.2) is over all $y \in \Theta_D$ such that the configurations y reduce to the configurations $x \in \Theta_{D_0}$ by the elimination of variables $D \setminus D_0$.

Example 5 Let $D = \{N, W\}$ and $D_0 = \{N\}$, with the frames of variables $\Theta_N = \{r_n, \bar{r}_n\}$ and $\Theta_W = \{r_w, \bar{r}_w\}$. Assume that the probability function p^D is

$$p^D(r_n, r_w) = 0.1, \quad p^D(r_n, \bar{r}_w) = 0.15, \quad p^D(\bar{r}_n, r_w) = 0.15, \quad p^D(\bar{r}_n, \bar{r}_w) = 0.6.$$

Then $p^{D\downarrow D_0}(r_n) = 0.25$ and $p^{D\downarrow D_0}(\bar{r}_n) = 0.75$.

Theorem 1 *Probability mass functions together with the above definitions of combination, marginalization, and the neutral element form a VA.*

The previous definitions of combination and marginalization can also be extended to conditional PMFs and, thus, it can be proved that valuation algebra is strictly related to the probability theory and Bayesian networks [19].

17.3 LOCAL COMPUTATION IN A VA

Let $\Psi = \{\varphi_1, \varphi_2, \ldots, \varphi_r\} \subseteq \Phi$ be a given set of valuations and $D^o \subseteq V$, with $V = d(\varphi_1) \cup d(\varphi_2) \cup \cdots \cup d(\varphi_r)$, the domain of interest for decision making. The problem of inference consists in marginalizing the joint valuation $\oplus\Psi = \varphi_1 \oplus \cdots \oplus \varphi_r$ to the subset of variables D^o. The straightforward approach is to compute the joint valuation first and to marginalize to D^o afterward. In Section 17.2.2, it has been observed that, by imposing Axioms (A1)–(A6) for the operations of *labeling*, *combination*, and *marginalization*, it is possible to compute the marginal $(\oplus\Psi)^{\downarrow D^o}$ on local domains, that is without explicitly computing the joint valuation. This is the goal of the so-called *fusion algorithm* [20, 21], which permits to perform inference in a VA using local computation.

17.3.1 Fusion Algorithm

The fundamental operation of the fusion algorithm is to successively delete all variables $x \in \Delta$, where $\Delta \triangleq V \backslash D^o$ is the set of variables of no interest in the VA. The variables can be deleted in any sequence, since according to Axiom (A4) (or, equivalently, (A4′)) all deletion sequences lead to the same result. However, different deletion sequences can have different computational costs. Finding an optimal elimination sequence is an NP-complete problem [20], but there exist several heuristics for finding a good elimination sequence [22–24].

In the fusion algorithm, the marginal of the joint valuation is computed by successively eliminating all the variables in Δ. With respect to the variable $x \in \Delta$ to be eliminated, two subsets of valuations can be defined

$$\Psi_x \triangleq \{\varphi \in \Psi : x \in d(\varphi)\} \text{ and } \Psi_{\bar{x}} \triangleq \{\varphi \in \Psi : x \notin d(\varphi)\}.$$

As a consequence of Axiom (A3) of Section 17.2.2, only the valuations in Ψ_x are affected by the elimination of x. Thus, the remaining set of valuations after eliminating x from Ψ is

$$Fus_x\{\varphi_1, \varphi_2, \ldots, \varphi_r\} \triangleq \{\oplus\Psi_x\}^{\downarrow(S\backslash\{x\})} \cup \Psi_{\bar{x}} \triangleq \{\varphi_{r+1}\} \cup \Psi_{\bar{x}}, \tag{17.3}$$

where $S \triangleq \cup_{\varphi_i \in \Psi_x} d(\varphi_i)$. Note that the Fus_x operation in (17.3) amounts to the union of all valuations not involving x together with the single valuation φ_{r+1}. The latter is obtained by combining all valuations involving x and then marginalizing the resulting valuation to $S\backslash\{x\}$. The valuation on the domain of interest D^o can thus be obtained by recursively applying the fusion algorithm and deleting all variables in $\Delta = \{x_1, x_2, \ldots, x_m\}$, that is,

$$(\varphi_1 \oplus \varphi_2 \oplus \cdots \oplus \varphi_r)^{\downarrow D^o} = \oplus\left\{Fus_{x_m}\left\{Fus_{x_{m-1}}\left\{\ldots Fus_{x_1}\{\varphi_1, \varphi_2, \ldots, \varphi_r\}\right\}\right\}\right\}. \tag{17.4}$$

This technique allows a reduction in the computational load for two reasons: the beliefs are combined on local domains and the variable elimination keeps the domains of the combined beliefs, that is, $d(\varphi_1 \oplus \varphi_2) = d(\varphi_1) \cup d(\varphi_2)$, to a reasonably small size.

Example 6 Consider Example 1 where the set of valuations $\{\varphi_1, \varphi_2, \varphi_3, \varphi_4\}$ is defined respectively on the domains $d(\varphi_1) = \{N, W\}$, $d(\varphi_2) = \{W, S\}$, $d(\varphi_3) = \{S, E\}$, $d(\varphi_4) = \{E, N\}$. Assuming that N is the decision variable, that is, $D^o = \{N\}$, then $\Delta = \{W, S, E\}$ is the set of variables of no interest. The objective is to apply the fusion algorithm to compute the combined valuation $(\varphi_1 \oplus \varphi_2 \oplus \varphi_3 \oplus \varphi_4)^{\downarrow D^o}$. The steps of the fusion algorithm are the following:

1. Elimination of W

$$\varphi_5 = (\varphi_1 \oplus \varphi_2)^{\downarrow(d(\varphi_1)\cup d(\varphi_2))\backslash\{W\}}, \; Fus_W = \{\varphi_5\} \cup \Psi_{\overline{W}} = \{\varphi_3, \varphi_4, \varphi_5\},$$

where $d(\varphi_5) = (d(\varphi_1) \cup d(\varphi_2))\backslash\{W\} = \{N, S\}$ is the domain of φ_5. The original valuation network from Figure 17.1 after this step is modified according to Figure 17.2a.

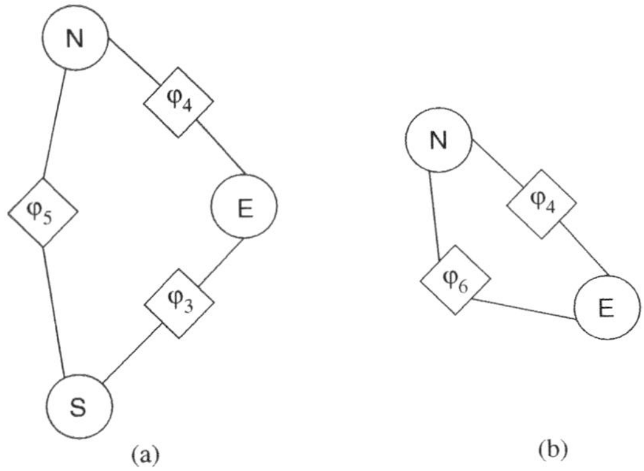

Figure 17.2 Valuation network from Example 1 after the elimination of (a) variable W; (b) variables W and S.

2. Elimination of S

$$\varphi_6 = (\varphi_3 \oplus \varphi_5)^{\downarrow(d(\varphi_3)\cup d(\varphi_5))\setminus\{S\}}, \; Fus_S = \{\varphi_6\} \cup \Psi_{\overline{S}} = \{\varphi_4, \varphi_6\},$$

where $d(\varphi_6) = (d(\varphi_3) \cup d(\varphi_5))\setminus\{S\} = \{N, E\}$. The valuation network after this step is shown in Figure 17.2b.

3. Elimination of E

$$\varphi_7 = (\varphi_4 \oplus \varphi_6)^{\downarrow(d(\varphi_4)\cup d(\varphi_6))\setminus\{E\}}, \; Fus_E = \{\varphi_7\} \cup \Psi_{\overline{E}} = \{\varphi_7\},$$

where $d(\varphi_7) = (d(\varphi_4) \cup d(\varphi_6))\setminus\{E\} = \{N\}$.

At the end of the last step the valuation Fus_E, defined on the domain of interest $\{N\}$, represents the solution to the problem.

17.3.2 Construction of a Binary Join Tree

The fusion algorithm (17.4) is appropriate if the valuations are static (invariant with time). In most practical applications, however, the valuations in the VA change with time. In surveillance, for example, new pieces of evidence (sensor measurements, intelligence reports) constantly arrive and one needs to update the inference every time this happens. Application of the fusion algorithm to dynamic systems would require that every time a valuation in Ψ changes, computations in (17.4) are repeated. This is inefficient because it results in a substantial amount of repeated computations. Instead, it is more efficient to represent the fusion algorithm in the form of a *binary join tree* (BJT) and then to recompute only the relevant changes.

A BJT is a binary tree (N, E) of nodes $N = \{n_1, n_2, \ldots, n_f\}$ and edges $E = \{(n, m) : n, m \in N, n \neq m\}$, where each node has at most three neighbors, one parent and two children. A node without children is called a *leaf*. A node without a parent is called a *root*. As such, a BJT is effectively a graphical representation of the fusion algorithm [20]. For this reason, like the fusion algorithm, the structure of the BJT (i.e., nodes and edges) strongly depends on the elimination sequence Δ.

A BJT has the following characteristics:

- A subset of variables $D_i \subseteq V$ and a valuation $\varphi(n_i)$, such that $d(\varphi(n_i)) = D_i$, are associated to each node n_i.
- The domain of the root of the BJT is such that $D^o \subseteq d(root)$.
- Edges represent the order in which the valuations must be combined (in order to calculate the valuation of the root on D^o).
- Nodes and edges represent the steps in the fusion algorithm.
- A BJT has to satisfy the Markov property, which means that $D_i \cap D_j \subseteq D_k$ for every pair of nodes n_i and n_j and for every node $n_k \in Path(n_i, n_j)$, where $Path(n_i, n_j)$ denotes the set of nodes on the path between n_i and n_j.

The Markov property is a fundamental property of the BJT, as will be discussed in Section 17.5.2. An algorithm for building a BJT is given in Appendix 17A. The BJT constructed for Example 1 is shown in Figure 17.3. This BJT is a result of application of the fusion algorithm described in Example 6, with the elimination sequence: W, S, E. The nodes in the BJT are labeled by integer numbers from 1 to 7. The leaves of the tree (nodes 1–4) represent the original valuations $\varphi_1, \varphi_2, \varphi_3, \varphi_4$. The remaining nodes in the BJT represent the intermediate steps of the fusion algorithm; as such they specify the order in which the valuations must be combined in order to calculate the valuation for the variable N. The vertical labels next to the nodes of the BJT denote the domains (the subsets of variables) of the nodes.

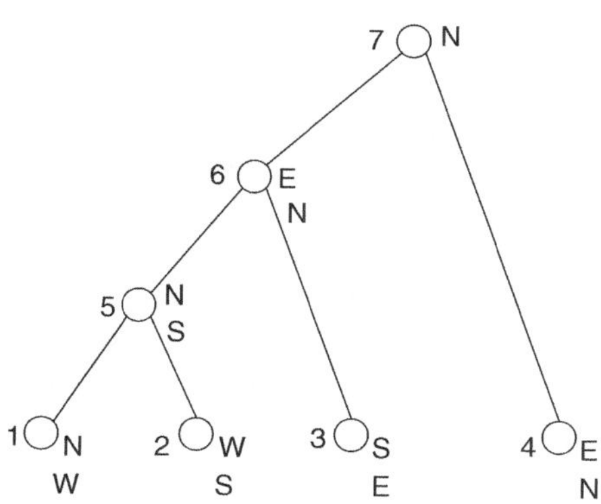

Figure 17.3 BJT representing the cyclist example with $D^o = \{N\}$.

17.3.3 Inward Propagation

In a BJT, marginals are computed by means of a message-passing scheme among the nodes. Initially only the valuations of the leaves of the BJT are specified. The process of propagating the valuations from the leaves toward the root of a BJT is called *inward propagation* [20, 21] and can be implemented using the algorithm reported in Appendix 17B. The key feature of the BJT and inward propagation is that the combination operator is applied only at the nonleaf nodes of the tree, between their left and right children. The advantage of using inward propagation on a BJT instead of the fusion algorithm lies in the ability to reuse the computations of the inward phase if the marginals need to be recomputed. In this way, every time one or more valuations of the leaves of the BJT change, the inward phase recalculates the valuations for all the nodes in the BJT that are affected by the change. More precisely, if n_i is the leaf whose valuation has changed, then the inward phase recomputes the valuations of all the nodes of the BJT along $Path(n_i, root)$.

Suppose that the BJT has been constructed for the domain of interest D^o and the inward propagation has been carried out. Let us also assume that the domain of interest has changed. One way to carry out the inference would be to create a new BJT and to perform inward propagation again. However, there is a more efficient alternative, the so-called *outward propagation* [21]. Outward propagation distributes the knowledge from the root to the leaves of the tree by reversing the direction in which the messages are passed between nodes [21].

In summary, a BJT can be seen as a data structure that allows the intermediate results of the combination process to be saved and the marginals to be computed efficiently.

17.4 THEORY OF EVIDENCE AS A VALUATION ALGEBRA

The term *Theory of evidence* was coined by Glenn Shafer in his book "A Mathematical Theory of Evidence," published in 1976 [9]. In the preface of the book, Shafer writes about the theory "*. . . It offers a reinterpretation of Dempster's work, a reinterpretation that identifies his lower probabilities as epistemic probabilities or degrees of belief, takes the rule for combining such degrees of belief as fundamental, and abandons the idea that they arise as lower bounds over classes of Bayesian probabilities.*" Thus, the theory of evidence is a new formalism that can be used for modeling uncertainty instead of classical probability. Furthermore, as it will be shown in the sequel, evidence theory comprises standard probability as a particular case, being therefore a more general theory. When the theory of evidence is applied for reasoning under uncertainty (instead of, for example, the probability theory), its higher computational complexity must be taken into account. The theory of evidence satisfies Axioms (A1)–(A6) for local computation listed in Section 17.2.2 and thus it is a valuation algebra [19]. This is important because one can exploit local computations to reduce the computational complexity.

Let the frame $\Theta = \{h_1, h_2, \ldots, h_n\}$ define a finite set of possible values of variable h in an evidential network. Elementary values h_i $(i = 1, \ldots, n)$ of the frame Θ are assumed to be mutually exclusive and exhaustive, with $n = |\Theta|$ being the cardinality of the frame. The beliefs about the actual value of the variable h are expressed on the subsets of Θ. The set containing all possible subsets of Θ, that is the power set of Θ, is denoted by $2^{\Theta} = \{H : H \subseteq \Theta\}$; its cardinality is 2^n. In this formalism, the belief is represented by the so-called basic belief assignment (BBA).

Definition 5 *A function $m : 2^{\Theta} \to [0, 1]$ is referred to as a basic belief assignment if*

$$m(\emptyset) = 0, \tag{17.5}$$

$$\sum_{A \subseteq \Theta} m(A) = 1. \tag{17.6}$$

Thus for $A \subset \Theta$, $m(A)$ is the part of belief that supports exactly A, that is, the fact that the true value of x is in A, but due to the lack of further information, does not support any strict subset of A. Condition (17.5) reflects the fact that no belief should be committed to empty set $\emptyset$ and condition (17.6) reflects the convention that the total belief has measure one. A subset A of the frame Θ is called a *focal element* of m if $m(A) > 0$. The union of all the focal elements is called the *core* $\mathcal{K}(m)$ of m. Note that, for any $A, B \subset \Theta$, $m(A)$ and $m(B)$ can be both equal to zero even if $m(A \cup B) \neq 0$. Furthermore, m is not monotone under inclusion, that is, $A \subset B$ does not imply $m(A) < m(B)$.

The BBA m represents the atomic information in the theory of evidence. To obtain the measure of the total belief committed to A, one must add to $m(A)$ the BBAs of all proper subsets B of A. This suggests the following definition.

Definition 6 *A function $Bel : 2^{\Theta} \to [0, 1]$ is called a belief function over Θ if it satisfies the following relationship*

$$Bel(A) = \sum_{B \subseteq A} m(B). \tag{17.7}$$

Bel(A) summarizes all our reasons to believe in A, hence it is the "lower probability" to believe in A. *Bel*(A) can be used to represent all degrees of belief on the elements of the frame, from the complete knowledge to the total ignorance. One can also define the "upper probability" to believe in A, which is known as the *plausibility function*, $Pl : 2^{\Theta} \to [0, 1]$, and it is given by

$$Pl(A) = 1 - Bel(\overline{A}), \tag{17.8}$$

where $\overline{A}$ is the complement of A.

Example 7 Consider again Example 1. The cyclist wants to know if it is going to rain in region N. Assume that, after having processed all the available information, the cyclist ends up with the BBA $m : 2^{\Theta_n} \rightarrow [0, 1]$, where $\Theta_n = \{r_n, \bar{r}_n\}$. From (17.7) and (17.8) it follows

$$\begin{aligned}
Bel(r_n) &= m(r_n),\\
Bel(\bar{r}_n) &= m(\bar{r}_n),\\
Bel(r_n \cup \bar{r}_n) &= m(r_n) + m(\bar{r}_n) + m(r_n \cup \bar{r}_n) = 1,\\
Pl(r_n) &= m(r_n) + m(r_n \cup \bar{r}_n),\\
Pl(\bar{r}_n) &= m(\bar{r}_n) + m(r_n \cup \bar{r}_n),\\
Pl(r_n \cup \bar{r}_n) &= m(r_n) + m(\bar{r}_n) + m(r_n \cup \bar{r}_n) = 1.
\end{aligned} \tag{17.9}$$

Note from this example that $Bel(r_n \cup \bar{r}_n) = Pl(r_n \cup \bar{r}_n) = 1$. This follows from the assumption of exhaustiveness of the frame of discernment, that is, there are only two possibilities, rain or no rain. Observe that if $m(r_n \cup \bar{r}_n) > 0$ then $Bel(r_n) + Bel(\bar{r}_n) < 1$. This property allows to model a wide range of possible situations. Thus, in Example 7 one can distinguish at least five situations.

1. There is a strong belief that it is going to rain. In this case $m(r_n) = 1, m(\bar{r}_n) = 0$ and $m(r_n \cup \bar{r}_n) = 0$. In the formalism of the probability theory, this belief can be represented using the PMF (Section 17.2.3) as $p(r_n) = 1$ and $p(\bar{r}_n) = 0$.
2. It is going to rain with a belief α and it is not going to rain with a belief $1 - \alpha$. In this case $m(r_n) = \alpha, m(\bar{r}_n) = 1 - \alpha$, and $m(r_n \cup \bar{r}_n) = 0$. This statement can be represented using the PMF as $p(r_n) = \alpha$ and $p(\bar{r}_n) = 1 - \alpha$.
3. It is going to rain with a belief of at least α. In this case $m(r_n) = \alpha, m(\bar{r}_n) = 0$ and $m(r_n \cup \bar{r}_n) = 1 - \alpha$. In the formalism of the probability theory, the equivalent representation involves all PMFs that satisfy $\alpha \leq p(r_n) \leq 1$ and $0 \leq p(\bar{r}_n) \leq 1 - \alpha$ with the constraint $p(r_n) + p(\bar{r}_n) = 1$.
4. There is a belief on propositions "it is going to rain" and "it is not going to rain," but there is also some ignorance ("don't know") factor. This is modeled by $m(r_n) = \alpha$, $m(\bar{r}_n) = \beta$ and $m(r_n \cup \bar{r}_n) = 1 - \alpha - \beta$. In the probability formalism, this is equivalent to considering all PMFs such that $\alpha \leq p(r_n) \leq 1 - \beta$ and $\beta \leq p(\bar{r}_n) \leq 1 - \alpha$ with the constraint $p(r_n) + p(\bar{r}_n) = 1$.
5. There is a complete lack of information about weather in N. This is modeled by $m(r_n) = 0, m(\bar{r}_n) = 0$ and $m(r_n \cup \bar{r}_n) = 1$. In the probability formalism, this is equivalent to considering all PMFs such that $p(r_n) + p(\bar{r}_n) = 1$.

The cases 1–5 illustrate the main difference between the standard theory of probability and the theory of evidence. In the former, knowledge is represented by a single PMF. In the latter, when knowledge is not trustworthy enough to elicit a single PMF, it can be represented by a closed-convex set of PMFs, or equivalently, by a BBA. $Bel(\cdot)$ gives the lower envelope of the closed-convex set of probability mass functions, that is $Bel(r_n)$ and $Bel(\bar{r}_n)$ are the lower probabilities of rain and no rain, respectively.

Conversely, $Pl(\cdot)$ is the upper envelope of the closed-convex set of probability mass functions, that is $Pl(r_n)$ and $Pl(\bar{r}_n)$ are the upper probabilities of rain and no rain, respectively. Mathematically this is written as [25]

$$\min_{P \in \mathcal{P}} P(A) = Bel(A), \quad \max_{P \in \mathcal{P}} P(A) = Pl(A), \tag{17.10}$$

where $\mathcal{P}$ denotes the closed-convex set of distributions.

A generic closed-convex set of probability mass functions is not a BBA in general. To be a BBA, its lower envelope (i.e., its lower probability function) must satisfy a particular property specified by the following theorem.

Theorem 2 *If Θ is a frame of discernment, then a function $Bel : 2^{\Theta} \rightarrow [0, 1]$ is a belief function over Θ, if and only if it satisfies the following conditions:*

1. $Bel(\emptyset) = 0$,
2. $Bel(\Theta) = 1$,
3. *For every positive integer n and every collection $A_1, A_2, \ldots, A_n$ of subsets of Θ,*

$$Bel(A_1 \cup A_2 \cup \ldots, A_n) \geq \sum_i Bel(A_i)$$

$$- \sum_{i<j} Bel(A_i \cap A_j) + \cdots + (-1)^{n+1} Bel(A_1 \cap \cdots \cap A_n)$$

or equivalently

$$Bel(A_1 \cup A_2 \cup \cdots \cup A_n) \geq \sum_{\substack{I \subseteq \{1,\ldots,n\} \\ I \neq \emptyset}} (-1)^{|I|-1} Bel\left(\bigcap_{i \in I} A_i\right), \tag{17.11}$$

where $|I|$ denotes the cardinality of the set I.

The third property is commonly referred to as ∞-monotonicity. Only lower probability functions that satisfy this property are belief functions [26]. When symbol $\geq$ in (17.11) is replaced by equality, then ∞-monotonicity reduces to the well-known additivity rule for nonexclusive events in standard probability. Probability functions are therefore a special case of belief functions, as stated by the following theorem.

Definition 7 *If Θ is a frame of discernment, then a function $Bel : 2^{\Theta} \rightarrow [0, 1]$ defined by*

1. $Bel(\emptyset) = 0$,
2. $Bel(\Theta) = 1$,
3. $Bel(A \cup B) = Bel(A) + Bel(B)$ *for each $A, B \subseteq \Theta$ and $A \cap B = \emptyset$*

is a probability function.

Theorem 3 *A belief* $Bel : 2^{\Theta} \to [0, 1]$ *is a probability function if and only if there exists a function* $p : \Theta \to [0, 1]$ *such that*

$$\sum_{\theta \in \Theta} p(\theta) = 1,$$

and

$$P(A) \triangleq Bel(A) = \sum_{\theta \in A} p(\theta), \quad \forall\, A \subseteq \Theta.$$

This function p is unique and is given by $p(\theta) = m(\{\theta\})$ *and it is a PMF.*

In this case, it can easily be proved that $Pl(A) = Bel(A) = P(A)$ and, thus, the convex set of PMFs associated to $Bel(\cdot)$ includes a single PMF. Hence, the standard probability functions can bee seen as the most informative belief functions. The least informative belief function, representing total ignorance, is the so-called *vacuous* belief function (i.e., the convex set includes all the possible PMFs).

Definition 8 *A belief function on frame* Θ *such that*

$$m(\Theta) = 1,\ m(A) = 0\ \forall A \neq \Theta \tag{17.12}$$

or equivalently

$$Bel(\Theta) = 1,\ Bel(A) = 0\ \forall A \neq \Theta \tag{17.13}$$

is referred to as the vacuous belief function.

17.4.1 Combination

The combination operator in the theory of evidence is carried out using the Dempster's rule of combination. Let the BBA $m_1^{D_1}$ be defined on a domain $d(m_1) = D_1 \subseteq V$. Similarly let $m_2^{D_2}$ be another BBA defined on a domain $d(m_2) = D_2 \subseteq V$. Furthermore, suppose that $m_1^{D_1}$ and $m_2^{D_2}$ represent beliefs obtained from two independent sources of information. If $d(m_1) \equiv d(m_2) = D$, the two BBAs can be combined directly using the Dempster's rule [9]

$$(m_1^D \oplus m_2^D)(A) = \frac{\sum_{B \cap C = A} m_1^D(B)\, m_2^D(C)}{1 - \sum_{B \cap C = \emptyset} m_1^D(B)\, m_2^D(C)}, \tag{17.14}$$

where A, B, C are subsets of the frame defined by the Cartesian product of the variables in D; that is, $A, B, C \subseteq \Theta_D$. If the two domains are different, $D_1 \neq D_2$, then before applying the Dempster's rule, one must extend both BBAs to the joint domain $D_1 \cup D_2$ in such a way that they express the same information before and after the extension (hence referred to as the *vacuous extension* and denoted by $\uparrow$). The vacuous extension of $m_1^{D_1}$ to $D_1 \cup D_2$ is defined as [20]

$$m_1^{D_1 \uparrow (D_1 \cup D_2)}(C) = \begin{cases} m_1^{D_1}(A) & \text{if } C = A \times \Theta_{D_2}, A \subseteq \Theta_{D_1}, \\ 0 & \text{otherwise.} \end{cases} \tag{17.15}$$

Example 8 Suppose $V = \{N, W, S\}$ with frames $\Theta_n = \{r_n, \bar{r}_n\}$, $\Theta_w = \{r_w, \bar{r}_w\}$ and $\Theta_s = \{r_s, \bar{r}_s\}$. Let $D_1 = \{N\}$, and $D_2 = \{W, S\}$, that is, $\Theta_{D_1} = \Theta_n$ and $\Theta_{D_2} = \{(r_w, r_s), (r_w, \bar{r}_s), (\bar{r}_w, r_s), (\bar{r}_w, \bar{r}_s)\}$. Let the BBA $m_1^{D_1}$ be defined as $m_1^{D_1}(\{r_n\}) = 0.7$ and $m_1^{D_1}(\{r_n, \bar{r}_n\}) = 0.3$. Then the vacuous extension of $m_1^{D_1}$ to $D_1 \cup D_2$ is given by

$$m_1^{D_1 \uparrow (D_1 \cup D_2)}(\{(r_n, r_w, r_s), (r_n, r_w, \bar{r}_s), (r_n, \bar{r}_w, r_s), (r_n, \bar{r}_w, \bar{r}_s)\}) = 0.7$$

with the remaining belief mass of 0.3 assigned to

$$\begin{aligned} \Theta_{D_1 \cup D_2} = \{&(r_n, r_w, r_s), (r_n, r_w, \bar{r}_s), (r_n, \bar{r}_w, r_s), (r_n, \bar{r}_w, \bar{r}_s), \\ &(\bar{r}_n, r_w, r_s), (\bar{r}_n, r_w, \bar{r}_s), (\bar{r}_n, \bar{r}_w, r_s), (\bar{r}_n, \bar{r}_w, \bar{r}_s)\}. \end{aligned}$$

In the general case of possibly nonidentical domains, the Dempster's rule is defined as

$$m_1^{D_1} \oplus m_2^{D_2} = m_1^{D_1 \uparrow (D_1 \cup D_2)} \oplus m_2^{D_2 \uparrow (D_1 \cup D_2)}. \tag{17.16}$$

The Dempster's rule of combination has been criticized for many years because of a perceived paradox, known as the "Zadeh's paradox" [7]. This "paradox," however, can be easily resolved by careful modeling, as described in [27]. The Dempster's rule needs to be applied with caution, though, because of the assumption of independence among the sources to be combined [25].

Finally, we point out that the neutral element for the Dempster's combination rule is the vacuous belief function.

17.4.2 Marginalization

Marginalization is a projection of a BBA defined on domain D onto a BBA defined on a coarser domain $D' \subseteq D$. Formally this is expressed as

$$m^{D \downarrow D'}(A) = \sum_{B \downarrow A} m^D(B), \tag{17.17}$$

where the summation in (17.17) is over all $B \subseteq \Theta_D$ such that the configurations in B reduce to the configurations in $A \subseteq \Theta_{D'}$ by the elimination of variables $D \setminus D'$.

Example 9 Let $D = \{N, W, S\}$ and $D' = \{N, S\}$, with the frames of variables $\Theta_n = \{r_n, \bar{r}_n\}$, $\Theta_w = \{r_w, \bar{r}_w\}$ and $\Theta_s = \{r_s, \bar{r}_s\}$. Suppose BBA m^D has three focal sets

$$m^D(\{(r_n, r_w, r_s)\}) = 0.6,$$
$$m^D(\{(r_n, r_w, r_s), (r_n, \bar{r}_w, \bar{r}_s)\}) = 0.3,$$
$$m^D(\{(r_n, r_w, r_s), (r_n, \bar{r}_w, r_s), (r_n, \bar{r}_w, \bar{r}_s)\}) = 0.1.$$

Then

$$m^{D\downarrow D'}(\{(r_n, r_s)\}) = 0.6,$$
$$m^{D\downarrow D'}(\{(r_n, r_s), (r_n, \bar{r}_s)\}) = 0.4.$$

Remark 1 Marginalization is the inverse operation of the vacuous extension, but, in general, the vacuous extension is not the inverse of marginalization. For instance, consider a valuation φ and three generic sets D_1, D_2, and D_3 such that $d(\varphi) = D_2$ and $D_1 \subseteq D_2 \subseteq D_3$; then it turns out that $(\varphi^{\uparrow D_3})^{\downarrow D_2} = \varphi$ but, in general, $(\varphi^{\downarrow D_1})^{\uparrow D_2} \neq \varphi$.

Theorem 4 *Belief functions together with the above definitions of combination, marginalization and a neutral element form a valuation algebra.*

The proof is given in [19].

17.4.3 Inferring and Eliciting the Evidential Model

In standard probability, uncertainty is captured by means of a single probability distribution. This means that all uncertainties, whether of a random or an epistemic nature, are represented in the same way. However, it is often very difficult to model epistemic uncertainty by using a single probability distribution. For instance, it is quite common for an expert to represent his/her uncertain knowledge by means of probability intervals (confidence intervals). Since defining a probability interval is equivalent to defining a set of distributions, this knowledge cannot be adequately represented by a single distribution without introducing (consciously or unconsciously) some additional (unjustified) information. In general, this is done on the basis of a certain criterion, such as the *maximum entropy*. For instance, suppose an expert gives his/her opinion on the value of parameter x [28]: "I only know that the value of x lies in an interval A." According to the *maximum entropy* criterion, this statement can be represented by the uniform probability with support A. But this choice introduces information that is not available in the original statement, and may seriously bias the outcome of probabilistic model in a nonconservative manner [28]. This is so because there exists an infinity of probability distributions with support in A, and the uniform distribution is just one of them. A more faithful representation of this statement on parameter x is the set of probability distributions with support A, which is actually a belief function (vacuous belief). Thus, the theory of evidence and, in general, a set of distributions, provide a more powerful framework for representing uncertain knowledge.

In the standard probability theory, inferring the model from data and elicitation from human experts have been widely explored and many techniques are available. On the contrary, this is still an open research area for the theories based on the sets of distributions. From a practical point of view, this is currently the main weakness of these theories. Although this aspect needs further research, there are already several established ways to tackle the inference and elicitation. Hereafter, a brief summary of the available approaches is presented. A significant contribution to this field cames from the Bayesian robustness community, for more details see, for example, the Berger's book [29] (the next section will present the so-called ϵ-contamination model [29]). To build BBAs, it is worth mentioning the multivalued mapping approach proposed by Dempster [9, 25], which gives another possible probabilistic interpretation of a BBA. A discussion about elicitation in evidence and possibility theory is also presented in [28]. For inference in classification problems, an extension to convex set of distributions of the Naive Bayes classifier has been proposed in [30], by using the so-called "Imprecise Dirichlet Model" developed by Walley [26] to model the prior belief. Since the corresponding convex set of distributions is not representable by a BBA (it is actually more general), this model will not be discussed here. Finally, this section describes how to represent an uncertain implication rule by a BBA [31–34].

17.4.3.1 Linear Vacuous Mixture Model. The *linear-vacuous mixture* [26] or ϵ-contamination model [29] is the family of all convex combinations of a known nominal PMF with arbitrary PMFs. This family can be used to address problems in which one takes into account that the model (nominal distribution) may be inexact. Hence, the model is perturbed (contaminated) in order to reflect this modeling uncertainty. It can be easily verified that the lower envelope of the resulting convex set of PMFs is a belief function. In fact, let $p : \Theta \to [0, 1]$ with $\Theta = \{\theta_1, \ldots, \theta_n\}$ denote the nominal PMF, then the BBA associated to the *linear-vacuous mixture* is given by

$$\begin{cases} m(\theta_i) = \epsilon\, p(\theta_i), & \forall \theta_i \in \Theta, \\ m(\Theta) = 1 - \epsilon, \end{cases} \tag{17.18}$$

where $0 \leq \epsilon \leq 1$ is the contamination parameter (which is a design parameter). The BBA of (17.18) is referred to as a simple support function focused on θ_i [9]. An application of this model to dynamic state estimation is presented in [35]. The resulting filter can be seen as a generalization of Kalman filter to the case in which the measurement and process noises are modeled as a linear-vacuous mixture of a nominal Gaussian density and the set of all possible arbitrary probability density functions.

17.4.3.2 Representation of Uncertain Implication Rules. Often knowledge from human experts is expressed in the form of uncertain implication rules, such as "if A then B," with a certain degree of confidence. There are essentially two types of uncertain implication rules. An implication rule "if A then B," which holds with a probability p such that $p \in [\alpha, 1]$ and $0 \leq \alpha \leq 1$, will be referred to as *1dof* uncertain implication rule. $p \in [\alpha, 1]$ means that the implication rule is believed true with probability α and $1 - \alpha$ represents the uncertainty. The term *1dof* means that uncertainty can be represented by means of a single parameter (1 degree of freedom), that is, α. Conversely, an implication rule "if A then B," which holds with a probability p such that $p \in [\alpha, \beta]$ with $0 \leq \alpha \leq \beta \leq 1$, will be referred to as *2dof* uncertain implication rule. $p \in [\alpha, \beta]$ means that the implication rule is believed true with degree α, false with degree $1 - \beta$ and uncertain with degree $\beta - \alpha$. Naturally, if $\beta = 1$, the *2dof* uncertain implication rule reduces to the *1dof* rule.

Since human experts represent their knowledge in the form of uncertain implication rules [31–33], it is important to include such a representation of knowledge into the process of reasoning under uncertainty. Incorporating inference rules based on the material implication of propositional logic into evidence theory is conceptually simple, which is not the case with classical probability theory. Assuming that the probability of the implication "if A then B" is equal to the conditional probability of B given A leads to the trivialization results of Lewis, that is, the probability can then only take three values [36]. Under this assumption, probability theory degenerates into a useless theory. In the evidence theory framework, on the contrary, one can assume the equality between the implication and the conditional belief and still avoid the trivialization [32].

An implication rule can be expressed in evidence theory by means of a BBA function using the principle of minimum commitment [32] and its instantiation is referred to as the ballooning extension [32]. Transformations to convert *1dof* and *2dof* uncertain implication rules into the evidence theory framework have been presented in [32] and, respectively, [22, 34].

In the following, the transformation presented in [34] for *2dof* uncertain implication rules is discussed, since this transformation satisfies three main properties of logical implication, namely the *reflexivity*, *contrapositivity*, and *transitivity* [34]. The fulfillment of these properties guarantees that the transformation is "coherent," in the sense that the logical axioms continue to hold also in the BBA function domain and, thus, that the transformation does not alter the information (see [34] for further details). Maintaining coherence of information is a fundamental requirement in any data fusion process.

Suppose there are two disjoint domains, D_1 and D_2, with associated frames Θ_{D_1} and Θ_{D_2}, respectively. Formally, an implication rule is an expression of the form

$$A \subseteq \Theta_{D_1} \Rightarrow B \subseteq \Theta_{D_2}. \tag{17.19}$$

Furthermore, let us assume that this implication rule is valid only in a certain percentage of cases, that is, with a probability (confidence) p such that $p \in [\alpha, \beta]$, with $0 \leq \alpha \leq \beta \leq 1$.

The implication rule of (17.19) can be expressed by a BBA consisting of three focal sets on the joint domain $D_1 \cup D_2$ [34]

$$m^{D_1 \cup D_2}(C) = \begin{cases} \alpha, & \text{if } C = (A \times B) \cup (\overline{A} \times \Theta_{D_2}) \\ 1 - \beta, & \text{if } C = (\overline{A} \times \overline{B}) \cup (A \times \Theta_{D_2}) \\ \beta - \alpha, & \text{if } C = \Theta_{D_1 \cup D_2}, \end{cases} \tag{17.20}$$

where $\overline{A}$ is the complement of A in Θ_{D_1}, and similarly $\overline{B}$ is the complement of B in Θ_{D_2}. If $\beta = 1$, one obtains a *1dof* implication rule with two focal sets.

Example 10 Let $D_1 = \{E\}$, $D_2 = \{N\}$, $\Theta_n = \{r_n, \bar{r}_n\}$, $\Theta_e = \{r_e, \bar{r}_e\}$, $A = \{r_e\}$, and $B = \{r_n\}$. Then the BBA representation of the statement (see Example 3) "rain in region E implies rain in the mountains N with chance between 50% and 70%," that is, $A \Rightarrow B$ with confidence $p \in [0.5, 0.7]$ is given by

$$\begin{aligned} m^{\{E,N\}}\ (\{(r_e, r_n), (\bar{r}_e, r_n), (\bar{r}_e, \bar{r}_n)\}) &= 0.5, \\ m^{\{E,N\}}\ (\{(\bar{r}_e, \bar{r}_n), (r_e, r_n), (r_e, \bar{r}_n)\}) &= 0.3, \\ m^{\{E,N\}}\ (\{(r_e, r_n), (r_e, \bar{r}_n), (\bar{r}_e, r_n), (\bar{r}_e, \bar{r}_n)\}) &= 0.2. \end{aligned} \tag{17.21}$$

17.4.4 Decision Making

This section presents a brief discussion on decision making when uncertainty is represented by belief functions. The Bayesian methodology to decision making provides the action that maximizes the expected value of some utility function. If $h_{a_i}(x)$ is the considered utility function, which depends on possible actions a_i and on the unknown value of the variable x, then a_j is preferred to a_i [29] if and only if

$$E(h_{a_i}) < E(h_{a_j}). \tag{17.22}$$

where $E(\cdot)$ denotes expectation w.r.t. x.

For example in classification, x is a discrete-valued class variable $x \in \{x_1, \ldots, x_n\}$ and action a_i represents the selection of class x_i. The utility function can be adopted to be the indicator function $h_{a_i}(x) = I_{x_i}(x)$, where $I_{x_i}(x) = 1$ if $x = x_i$ and zero otherwise. In this case $E(h_{a_i}) = \sum_j I_{x_i}(x_j)\, p(x_j) = p(x_i)$ and hence (17.22) becomes $p(x_i) < p(x_j)$. This means that one selects the class with the highest probability.

In the theory of evidence one deals with lower and upper probability and, thus, with interval-valued expectations, leading to the problem of decision making under imprecision [26]. A consequence of imprecision is that, when the lower and upper probabilities are substantially different, then choosing a unique action may not be possible. Instead, a set of possible actions may be put forward to a decision maker. There are various ways for decision making with the convex set of probability functions, such as interval dominance, maximality, minimax, and so on [26, 29]. Before discussing these methods, it is necessary to generalize the notion of expected value of a given function, suitable for the theory of evidence.

17.4.4.1 Choquet Integral. Uncertainty theories generalize the classical probability theory by considering intervals of probability. The classical notion of expected value of a given function must also be generalized. Instead of a single expected value, a range of expected values will be considered. This range is captured by lower and upper expected values, which are based on lower and upper probability functions, respectively. Since lower and upper probability functions are in general non additive, the classical expectation is not applicable. In order to calculate an expected value of a function with respect to an ∞-monotone measure, a generalization of the classical expectation, called the *Choquet integral* [37], has to be introduced.

Definition 9 *Let $h : \Theta \to \mathbb{R}$ be a utility function, $\Theta = \{\theta_1, \theta_2, \ldots, \theta_n\}$ and m a BBA on Θ. The Choquet integral of h with respect to m is defined by*

$$\begin{aligned}\underline{E}[h] &= \sum_{A \subseteq \Theta} m(A) \inf\{h(\theta) : \theta \in A\}, \\ \overline{E}[h] &= \sum_{A \subseteq \Theta} m(A) \sup\{h(\theta) : \theta \in A\}.\end{aligned} \tag{17.23}$$

Example 11 Consider Table 17.1 of lower and upper probabilities. Assuming that $h(\theta) = I_{\{\theta_1\}}(\theta) - I_{\{\theta_3\}}(\theta)$, that is, $h(\theta_1) = 1$, $h(\theta_2) = 0$ and $h(\theta_3) = -1$, then it follows that

$$\begin{aligned}\underline{E}[h] &= m(\theta_1) \cdot 1 + (m(\theta_2) + m(\{\theta_1, \theta_2\}) \cdot 0 + (1 - m(\theta_1) - m(\theta_2) - m(\{\theta_1, \theta_2\})) \cdot (-1) \\ &= 1 \cdot 0.4 + 0 \cdot 0.25 - 1 \cdot 0.35 = 0.05,\end{aligned}$$

$$\begin{aligned}\overline{E}[h] &= m(\theta_3) \cdot (-1) + (m(\theta_2) + m(\{\theta_2, \theta_3\}) \cdot 0 + (1 - m(\theta_2) - m(\theta_3) - m(\{\theta_2, \theta_3\})) \cdot 1 \\ &= -1 \cdot 0.05 + 0 \cdot 0.45 + 1 \cdot 0.5 = 0.45.\end{aligned}$$

The expected value is included in [0.05, 0.45].

TABLE 17.1 Example of Lower and Upper Probabilities

	Bel	*Pl*	*m*
$\{\emptyset\}$	0.0	0.0	0.0
$\{\theta_1\}$	0.4	0.5	0.4
$\{\theta_2\}$	0.25	0.45	0.25
$\{\theta_3\}$	0.05	0.35	0.05
$\{\theta_1, \theta_2\}$	0.65	0.65	0
$\{\theta_1, \theta_3\}$	0.55	0.55	0.1
$\{\theta_2, \theta_3\}$	0.5	0.5	0.2
$\{\theta_1, \theta_2, \theta_3\}$	1.0	1.0	0

17.4.4.2 Interval Dominance. *Interval dominance* is the weakest decision criterion for a convex set of PMFs. Under the interval dominance criterion, a_j dominates (is preferred to) a_i if and only if

$$\overline{E}_x(h_{a_i}) < \underline{E}_x(h_{a_j}). \tag{17.24}$$

In classification, where x represents the class variable and the action a_i represents one of the possible categories of the class, that is, the utility function is $h_{a_i}(x) = I_{x_i}(x)$, the criterion (17.24) simply becomes $\overline{P}(x_i) < \underline{P}(x_j)$, where $\underline{P}(x_j)$ is the lower probability of x_j (the belief of x_j in evidence theory) and $\overline{P}(x_i)$ is the upper probability of x_i (the plausibility of x_i in evidence theory). This means that the category x_j is preferred to x_i if its lower probability is greater than the upper of x_i. Notice that, while for standard expectation the criterion (17.22) determines a total order between actions, (17.24) determines only a partial order. Thus, it can happen that (17.24) returns a set of actions that are undominated, that is, there is no preference between the actions in this set. A way to robustly tackle this case is to consider all consequences implied by these actions.

Example 12 Consider again the example in Table 17.1. In the case $h_{a_i}(\theta) = I_{\theta_i}(\theta)$, and thus

$$\begin{aligned}
\underline{E}[I_{\theta_1}] &= Bel(\theta_1) = 0.4,\\
\underline{E}[I_{\theta_2}] &= Bel(\theta_2) = 0.25,\\
\underline{E}[I_{\theta_3}] &= Bel(\theta_3) = 0.05,\\
\overline{E}[I_{\theta_1}] &= Pl(\theta_1) = 0.5,\\
\overline{E}[I_{\theta_2}] &= Pl(\theta_2) = 0.45,\\
\overline{E}[I_{\theta_3}] &= Pl(\theta_3) = 0.35.
\end{aligned}$$

Since $\overline{E}_x(I_{\theta_i}) < \underline{E}_x(I_{\theta_j})$ is verified only $j = 1$ and $i = 3$, then category θ_1 dominates θ_3 and, thus, the set of undominated categories is $\{\theta_1, \theta_2\}$. If this was a classification problem, two class categories would be put forward as a solution, that is, $\{\theta_1, \theta_2\}$.

17.4.4.3 Maximality. The maximality criterion was proposed by Walley [26]. Recall from the discussion preceding (17.10) that any BBA defines a convex set of distributions $\mathcal{P}$. Then, under maximality, a_j dominates (is preferred to) a_i if for any distribution p in the convex set $\mathcal{P}$, it holds that

$$E_p(h_{a_i}) < E_p(h_{a_j}), \tag{17.25}$$

or, equivalently, if

$$E(h_{a_j} - h_{a_i}) > 0 \;\; \forall p \in \mathcal{P}. \tag{17.26}$$

For any p in $\mathcal{P}$ from the definition of the Choquet integral it follows that

$$E_p(h_{a_j} - h_{a_i}) \geq \inf_{p \in \mathcal{P}} E_p(h_{a_j} - h_{a_i}) = \underline{E}(h_{a_j} - h_{a_i}).$$

Hence, a necessary and sufficient condition for (17.26) to be satisfied is that

$$\underline{E}(h_{a_j} - h_{a_i}) > 0. \tag{17.27}$$

In the interval dominance case, the dominance is verified by considering the best (upper) and worst (lower) distribution in the convex set. Conversely, in the maximality criterion, actions are compared w.r.t. the same distribution, and thus a_j is said to dominate a_i if (17.25) is satisfied for each distribution in the convex set. This is a straightforward generalization of the Bayesian decision criterion (17.22) to the set of distributions. For this reason maximality is our preferred decision criterion. Finally, notice that the interval dominance implies maximality, but the converse is not true, as shown in the following example.

Example 13 Consider again the example in Table 17.1. In the case $h_{a_j}(\theta) - h_{a_i}(\theta) = I_{\theta_j}(\theta) - I_{\theta_i}(\theta)$, by using the Choquet integral (as in Example 11) it follows that

$$\underline{E}[I_{\theta_1} - I_{\theta_2}] = 0.05,$$
$$\underline{E}[I_{\theta_1} - I_{\theta_3}] = 0.05,$$
$$\underline{E}[I_{\theta_2} - I_{\theta_3}] = -0.1.$$

Since $\underline{E}[I_{\theta_1} - I_{\theta_2}], \underline{E}[I_{\theta_1} - I_{\theta_3}] > 0$ this implies that the category θ_1 dominates both θ_2 and θ_3 and, thus, the set of undominated categories is $\{\theta_1\}$. Recall from the previous example that using the interval dominance, the set of undominated categories was $\{\theta_1, \theta_2\}$.

17.4.4.4 Maxmin. In some applications it may be a requirement to select a single action even when dealing with a set of distributions (i.e., returning a set of actions is not useful). In Bayesian robustness, a well-known criterion for this purpose is the so-called *maxmin* (*minmax*) principle [29]. Under maxmin, the best action is determined as follows

$$\max_{a_j} \underline{E}(I_{\theta_j}). \tag{17.28}$$

The idea of maxmin is to select the action with the maximum lower utility. Clearly this is a very pessimistic criterion.

Example 14 Consider again the example in Table 17.1. In this case, $\{\theta_1\}$ is the best maxmin action since $Bel(\theta_1) > Bel(\theta_2) > Bel(\theta_3)$.

17.4.4.5 Pignistic Transformation. There are other criteria that return a single action. One frequently used in the theory of evidence is based on the so-called *pignistic transformation* [38]. The pignistic transformation defines a mapping of a belief measure to a probability measure, which is equivalent to the "center of mass" of the convex set of distributions associated to the BBA. Let m^D be a BBA defined on a subset of variables D with corresponding frame Θ_D. The pignistic transform of m^D is defined for every element of the frame $\theta \in \Theta_D$ as follows [38]

$$BetP(\theta) = \sum_{\theta \in A \subseteq \Theta_D} \frac{m^D(A)}{|A|}. \tag{17.29}$$

Being a probability measure, one can use $BetP$ in (17.22) for decision making.

Example 15 Consider again the example in Table 17.1. The pignistic probability is

$$BetP(\theta_1) = 0.45,$$
$$BetP(\theta_2) = 0.35,$$
$$BetP(\theta_3) = 0.2,$$

and, thus, the best action is again $\{\theta_1\}$ if $h_{a_j} = I_{\theta_j}$.

17.5 EXAMPLES OF DECISION SUPPORT SYSTEMS

17.5.1 Target Identification

Surveillance systems typically involve multiple sensors of different types whose role is to observe physical entities in the region of interest. Decision-level identity fusion [39, Chapter 6] performs fusion of uncertain identity declarations from multiple sensors in order to achieve a joint declaration of target identity (ID). In doing so, the assumption is that all ID declarations to be fused have been associated with the same target (via an association preprocessing step).

Sensors that are useful in creating target ID declarations in the air surveillance context include radar (provides target radar-cross section, its kinematic features, range profile, jet-engine modulation signature, etc.), ESM sensor (target radar frequency, pulse width, platform type, emitter type), IFF sensor (positive ID of friendly platforms), flight/mission plans (indicators of agreement), infra-red sensor (target shape, area), region-based indicators (e.g., originating airfield), and intelligence reports.

Note that there are three basic categories of target ID: allegiance, platform class, and platform type. The frame (state-space) of category allegiance typically includes values such as friend, assumed friend, neutral, suspect (assumed hostile), hostile, and so on. Platform class can take values such as commercial, transporter, fighter plane,

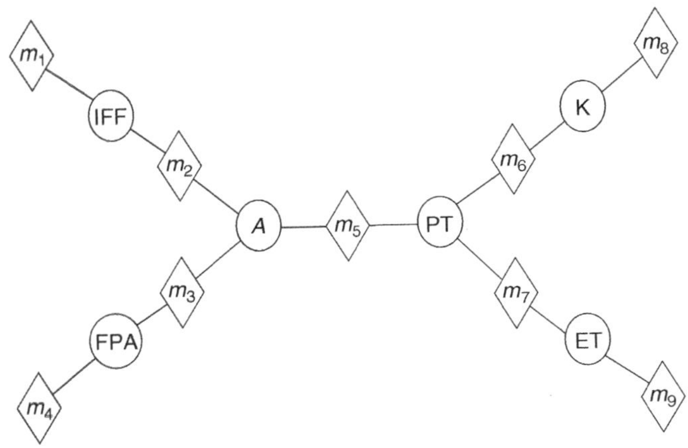

Figure 17.4 Evidential network for target identification.

bomber, helicopter, missile, and so on. Each platform class can be further divided into various types. For example, a fighter plane can be of type F/A-18, Tornado, Mig-29, Eurofighter, and so on.

17.5.1.1 Problem Specification. Figure 17.4 shows a simple evidential network that models the fusion of ID declarations for the purpose of determining the allegiance (denoted by A) and platform type (PT). Valuations are now the basic belief assignments and therefore denoted by m. The list of variables and their frames is given in Table 17.2. For simplicity let us adopt that A is a binary variable (allegiance is friend or hostile), with only six platform types in the frame of PT variable.

Variable IFF is true if the target responds correctly to IFF interrogation. In order to define the valuation m_2 on domain $\{A, \text{IFF}\}$, suppose that (based on prior/expert knowledge) we are confident that in 95–100% of the cases, if IFF is true, than the target is indeed a friend (i.e., $A = 0$). The lack of response to IFF interrogation ($\text{IFF} = 0$) from prior experience is due to a hostile target ($A = 1$) only in 10% to 30% of the cases. One can then summarize the "expert" knowledge about the domain $\{A, \text{IFF}\}$

TABLE 17.2 Variables and Their Frames for the Evidential Network in Figure 17.4

Variable	Description	Frame	Explanation
A	Allegiance	$\{0, 1\}$	0 is friend, 1 is hostile
PT	Platform type	$\{0, 1, \ldots, 5\}$	E.g. 0 is EuroFighter, 1 is FA-22 raptor, and so on
IFF	Correct IFF squawking	$\{0, 1\}$	0 is false, 1 is true
FPA	Flight plan agreement	$\{0, 1\}$	0 is false, 1 is true
K	Kinematic agility	$\{0, 1, 2\}$	0 is low, 1 is medium, 2 is high
ET	Emitter type	$\{0, 1, \ldots, 5\}$	

by the following set of independent rules

$$(\text{IFF} = 1) \Rightarrow (A = 0) \text{ with confidence between } 0.95 \text{ and } 1, \tag{17.30}$$

$$(\text{IFF} = 0) \Rightarrow (A = 1) \text{ with confidence between } 0.10 \text{ and } 0.30, \tag{17.31}$$

Then according to Section 17.4.3, each of the rules above can be represented by a BBA; when the two BBAs are combined by the Dempster's rule, the following valuation on the product space $A \times \text{IFF}$ is obtained

$$\begin{array}{llllllll} m_2 & (\{ & & (0,1), & (1,0) & & \}) = & 0.0950, \\ m_2 & (\{ & (0,0), & (0,1), & (1,0) & & \}) = & 0.89, \\ m_2 & (\{ & & (0,1), & (1,0), & (1,1) & \}) = & 0.0050, \\ m_2 & (\{ & (0,0), & (0,1), & (1,0), & (1,1) & \}) = & 0.0100. \end{array} \tag{17.32}$$

Flight plans are plans filed by pilots with the local aviation authority prior to flying. They generally include basic information such as departure and arrival points, estimated travel time, and so on. If there is evidence that an air target is flying in accordance with a flight plan (FPA $= 1$), then this is a strong indication that it is a friend, that is, $A = 0$. Suppose one can summarize expert knowledge about the domain $\{\text{FPA}, A\}$ by the following two rules

$$(\text{FPA} = 1) \Rightarrow (A = 0) \text{ with confidence between } 0.95 \text{ and } 1, \tag{17.33}$$

$$(\text{FPA} = 0) \Rightarrow (A = 1) \text{ with confidence between } 0.10 \text{ and } 0.30. \tag{17.34}$$

As described above, these two rules can be translated to the corresponding BBA m_3 on its domain $\{\text{FPA}, A\}$.

Variable m_6 represents the relationship between the kinematic agility of the target (observed maximum speed, acceleration, maneuverability) and the platform type. Suppose m_6 is specified by the following set of rules

$$(K = 0) \Rightarrow \text{PT} \in \{0, 1\} \text{ with confidence between } 0.60 \text{ and } 1.0,$$

$$(K = 1) \Rightarrow \text{PT} \in \{2, 3\} \text{ with confidence between } 0.60 \text{ and } 1.0,$$

$$(K = 2) \Rightarrow \text{PT} \in \{4, 5\} \text{ with confidence between } 0.60 \text{ and } 1.0.$$

Variable m_7 represents the relationship between the emitter type and the platform type. Suppose this relationship is given in the form of an ϵ-contamination ($\epsilon = 0.9$) of a normalized confusion matrix whose elements are conditional probabilities $\pi_{ij} = P\{\text{PT} = \{j\}|\text{ET} = \{i\})$, where the diagonal elements of the confusion matrix are all equal to 0.50, while the off-diagonal elements are equal to 0.1. According to the model in Section 17.4.3, one can represent this knowledge via the following BBA on the product space PT × ET

$$\begin{aligned} m_7(\{j, i\}) &= \epsilon\, 0.5 \quad \forall\, i = j, \;\; i, j \in 0, 1, \ldots, 5, \\ m_7(\{j, i\}) &= \epsilon\, 0.1 \quad \forall\, i \neq j, \;\; i, j \in 0, 1, \ldots, 5, \\ m_7(\Theta) &= 1 - \epsilon, \end{aligned} \tag{17.35}$$

where

$$\Theta = \{(0, 0), (0, 1), \ldots, (0, 5), (1, 0), (1, 1), \ldots, (5, 4), (5, 5)\}.$$

Valuation m_5 captures the expert knowledge that relates the PT to variable A. Suppose this knowledge is represented by the following implication rules

$$(A = 0) \Rightarrow (\mathrm{PT} \in \{0, 1, 2\}) \text{ with confidence between } 0.99 \text{ and } 1,$$
$$(A = 1) \Rightarrow (\mathrm{PT} \in \{3, 4, 5\}) \text{ with confidence between } 0.30 \text{ and } 1.$$

This rule represents our prior knowledge that friendly platforms are 0, 1, and 2, while (from unreliable intelligence sources) we anticipate that nonfriendly aircraft in the battle-space are of type 3, 4, or 5.

17.5.1.2 Numerical Results. Suppose that our sources of information provide us with the following valuations: $m_1(\{0\}) = 1$ indicated no response to IFF squawking (defined on domain IFF); m_4 reflects the lack of evidence that the target flies in accordance with a flight plan; $m_4(\{0\}) = 0.9$; the observed kinematic agility is $m_8(\{0, 1\}) = 0.9$; and finally an ESM supplies the information about the emitter type $m_9(\{3\}) = 0.7$, $m_9(\{2\}) = 0.3$.

The set of variables consists of six elements, $V = \{\mathrm{IFF}, \mathrm{FPA}, A, \mathrm{PT}, K, \mathrm{ET}\}$; the set of all valuations (BBAs) consists of nine elements, $\Phi_V = \{m_1, m_2, \ldots, m_9\}$; the domain of interest is $D^o = \{A, \mathrm{PT}\}$, but also its marginals A and PT, separately.

The following steps describe the process for solving the problem in the VA framework: (1) construct the BJT; (2) initialize the leaves of the BJT with the BBAs; (3) apply inward propagation; (4) marginalize the belief of the root of the BJT to D^o; (5) apply a decision criterion (maximality and the pignistic transformation).

Only three pieces of information are necessary to build a BJT: the set of variables of interest for decision making D^o; the set of variables to be eliminated Δ; and the set of the valuations Φ_V with associated domains. The BJT for the described problem was constructed using the algorithm in Appendix 17A with the elimination sequence $\Delta = \{4, 3, 5, 6\}$, and is shown in Figure 17.5.

The output of *inward propagation* is the BBA of node 17, defined on domain $D^o = \{A, \mathrm{PT}\}$. This BBA is then marginalized to domain $\{A\}$ and, respectively, $\{\mathrm{PT}\}$ and finally the corresponding belief and plausibility functions and pignistic probability are computed for the variables $\{A\}$ and $\{\mathrm{PT}\}$. Figure 17.6 shows the belief and plausibility functions and (pignistic) PMF for the singleton values $A = 0$ and

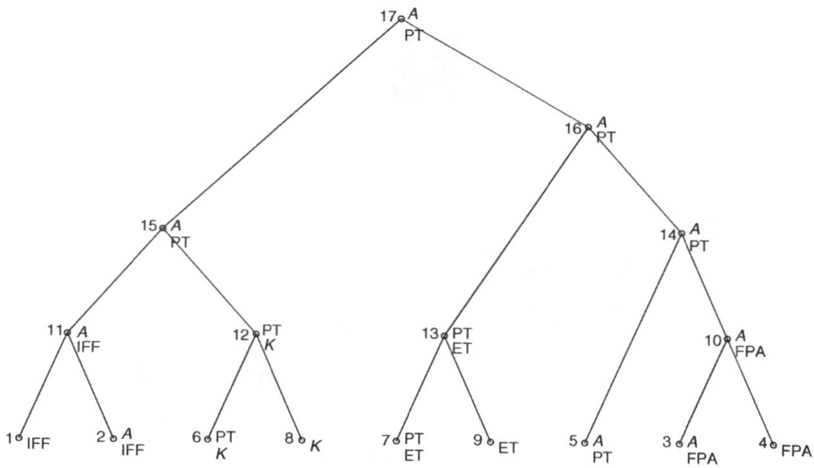

Figure 17.5 Binary joint tree for the fusion of target ID declaration.

$A = 1$ of the "allegiance" variable. Conversely, results for the PT variable are shown in Figure 17.7

From Figure 17.6, it can be noticed that the values of the belief, pignistic, and plausibility functions for $A = 0$ are 0.2, 0.45, and 0.6, respectively, while for $A = 1$ they are 0.3, 0.55, and 0.6, respectively. The pignistic probability supports $A = 1$ more than $A = 0$, but from the values of the belief and plausibility functions, one can see that there is a high level of uncertainty. In fact, in this case, by computing the maximality criterion (computed as described in Example 13), we obtain that both

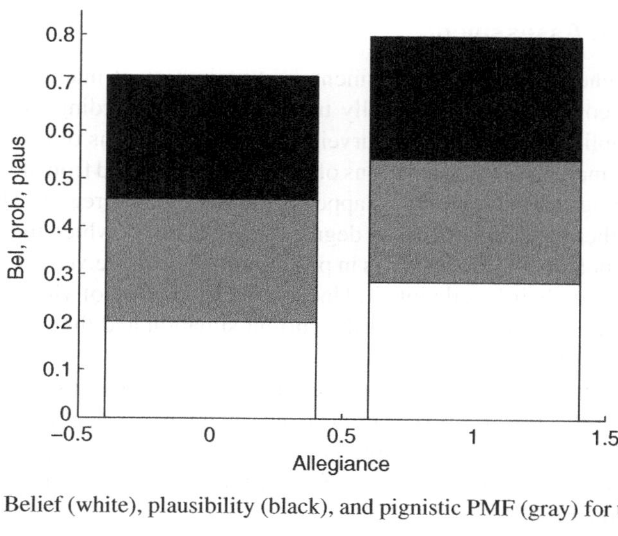

Figure 17.6 Belief (white), plausibility (black), and pignistic PMF (gray) for the variable A.

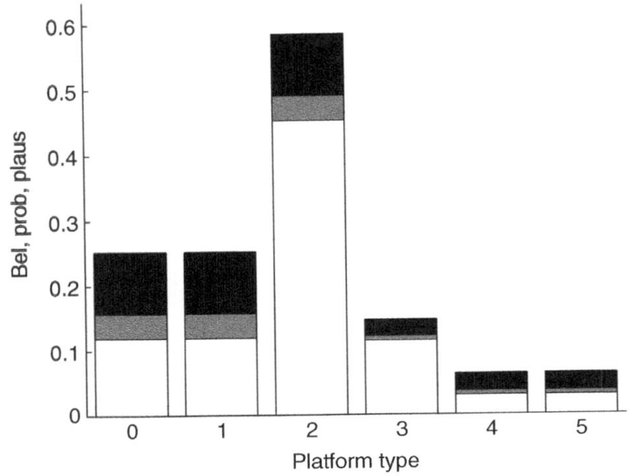

Figure 17.7 Belief (white), plausibility (black), and pignistic PMF (gray) for the variable PT.

values $\{0, 1\}$ for A are undominated. This means that there is not enough information to decide the allegiance type.

Conversely, from Figure 17.7, since the value of the belief function for PT$= 2$ is greater than the values of the plausibility functions for the other types of platforms, based on the interval dominance criterion (and, thus, also on maximality, since it is implied), it can be concluded that the target platform is of type 2. This holds with a probability of at least 0.45 (belief), the baseline being $1/6 \approx 0.17$ (i.e., the uniform distribution over the values of PT, which corresponds to the equiprobability case).

17.5.2 Threat Assessment

Situation awareness and threat assessment are two important interdependent information fusion concepts that are usually treated jointly. According to [2], situation awareness establishes a view of the surveillance region in terms of the observed activities, events, maneuvres, and locations of potential threats, and from this view infers what is happening or what is going to happen in the surveillance region. Threat assessment, on the other hand, estimates the degree of severity with which the engagement events will occur and its significance is in proportion to the perceived capability of the threats to carry out their hostile intent. Under a decision point of view, the objective of threat assessment is to be aware of the current situation and threat in order to act properly and in a timely fashion.

Bayesian networks have been the prevailing technique for constructing situation assessment [40–42], threat assessment [43], and intent assessment [44, 45] systems. This section demonstrates by example how to build a decision support system for threat assessment using the framework of evidential networks mainly following the material presented in [31, 46].

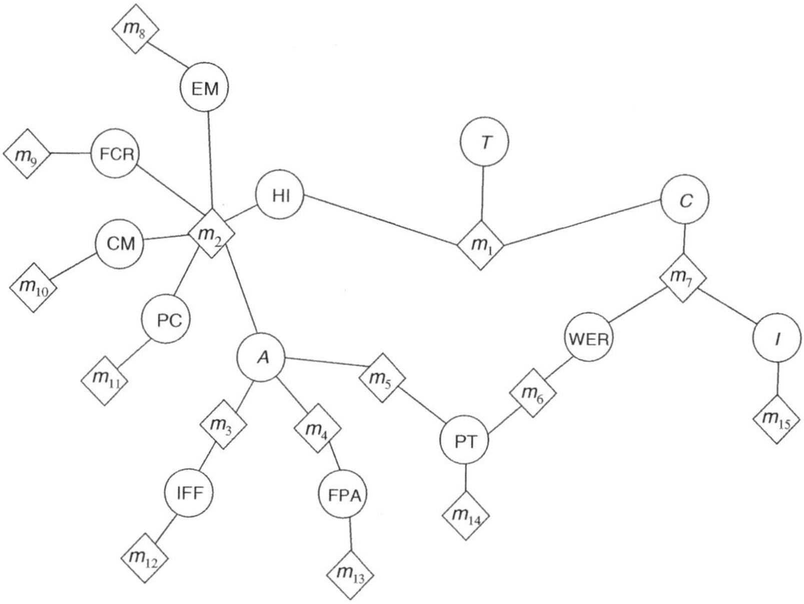

Figure 17.8 Evidential network for threat assessment.

17.5.2.1 Problem Specification. The model of threat, which draws on ideas from [2, 47], is shown in the form of an evidential network in Figure 17.8. The list of variables with explanations and frame definitions is given in Table 17.3. The domain of interest for decision making is the singleton $D^o = \{T\}$.

TABLE 17.3 Variables of the Threat Assessment Model

Variable	Description	Frame	Explanation
T	Threat	$\{0, 1, \ldots, 10\}$	0 none, 10 highest degree of T
HI	(Hostile) intent	$\{0, 1, \ldots, 6\}$	0 none (benign), 6 highest degree of HI
C	Capability	$\{0, 1, 2, 3, 4\}$	0 none, 4 highest degree of C
EM	Evasive maneuvre	$\{0, 1\}$	0 is false, 1 is true
FCR	Fire control radar	$\{0, 1\}$	0 is OFF, 1 is ON
CM	Countermeasures	$\{0, 1\}$	0 is false, 1 is true
PC	Political climate	$\{0, 1\}$	0 is peace, 1 is war
A	Allegiance	$\{0, 1\}$	0 is friend, 1 is not friend (hostile)
IFF	Correct IFF squawking	$\{0, 1\}$	0 is false, 1 is true
FPA	Flight plan agreement	$\{0, 1\}$	0 is false, 1 is true
PT	Platform type	$\{0, 1, \ldots, 5\}$	E.g. 0 is EuroFighter, 1 is FA-22 raptor, etc.
WER	Weapon engagement range	$\{0, 1, 2\}$	0 is small, 1 medium, 2 long range
I	Imminence	$\{0, 1, 2\}$	0 is low, 1 medium, 2 is high

According to the threat model in Figure 17.8, variable T (threat) depends on the degree of hostile intent (HI) of the opponent and on its capability (C). Assuming the threat is linearly related to both HI and C, one may choose to represent the valuation m_1 by the following rule: $T = \text{HI} + C$. Consider in the Cartesian product space $T \times \text{HI} \times C$ the set of triples (t, h, c), such that $t = h + c$, where according to the frames of the variables in Table 17.3, $t \in \{0, \ldots, 10\}$, $h \in \{0, \ldots, 6\}$ and $c \in \{0, \ldots, 4\}$. Then the rule $T = \text{HI} + C$ can be represented by the following BBA

$$\begin{aligned} m_1(\{&(0,0,0),(1,0,1),\ldots,(4,0,4),\\ &(1,1,0),(2,1,1),\ldots,(5,1,4),\\ &\ldots\\ &(6,6,0),(7,6,1),\ldots,(10,6,4)\}) = 1. \end{aligned} \tag{17.36}$$

This BBA has a single focal set consisting of 35 triples (t, h, c).

The degree of hostile intent (HI) is proportional to the evidence that the target (opponent) behaves in a hostile manner. In particular, the target may perform evasive maneuvres (EM), it may employ countermeasures (CM), such as deception jamming or chaff, we may have evidence that it is hostile ($A = 1$) platform, and most importantly, its fire-control radar (FCR) could be turned on (meaning it intends to fire a weapon soon). In addition, the political climate (PC) has an influence on the HI variable in the sense that the climate of political tension means that the target is more likely to have a hostile intent. The relationship between the six variables mentioned (HI, EM, FCR, CM, PC, A) is captured by the valuation m_2. How this relationship may be represented by m_2 depends on many factors (doctrine, engagement rules, etc.), but for the sake of illustration let us adopt the following simple rule: $\text{HI} = \text{EM} + 2 \cdot \text{FCR} + \text{CM} + \text{PC} + A$. This rule reflects the fact that the FCR variable is weighted higher than other variables in contributing to the HI. The adopted rule is represented by the BBA m_2 defined on a six-dimensional product space $\text{HI} \times \text{EM} \times \text{FCR} \times \text{CM} \times \text{PC} \times \text{A}$ as follows

$$\begin{aligned} m_2(\{&(0,0,0,0,0,0),(1,0,0,0,0,1),(1,0,0,0,1,0),\\ &(2,0,0,0,1,1),(1,0,0,1,0,0)\ldots,(2,0,1,0,0,0),\ldots,(6,1,1,1,1,1)\}) = 1. \end{aligned} \tag{17.37}$$

Thus, m_2 has a single focal set consisting of 32 six-tuples.

Valuation m_3 on domain $\{A, \text{IFF}\}$ is defined by the uncertain implication rules given earlier in (17.30) and (17.31), thus m_3 in this evidential network is identical to m_2 of (17.32). Valuation m_4 on domain {FPA, A} is defined by the uncertain implication rules given by (17.33) and (17.34).

As in the target ID fusion example, m_5 represents our knowledge of platform types PT in relation to allegiance A. Suppose this knowledge is represented by the following implication rule

$$(A = 1) \Rightarrow (\text{PT} \in \{3, 4, 5\}) \text{ with confidence between 0.50 and 1.}$$

According to this (uncertain) rule, the hostile aircraft in the battlespace of interest are of type 3, 4, or 5, with confidence higher than 50%.

For each PT, it is usually known *a priori* what types of weapons it carries [48]. Variable m_6 represents the relationship between the weapons engagement range (WER) variable and the PT. Suppose m_6 is defined by the following set of rules

$$(\text{PT} \in \{0, 1\}) \Rightarrow (\text{WER} = 0) \text{ with confidence between 0.40 and 1,}$$
$$(\text{PT} \in \{2, 3\}) \Rightarrow (\text{WER} \in \{1, 2\}) \text{ with confidence between 0.40 and 1,}$$
$$(\text{PT} \in \{4, 5\}) \Rightarrow (\text{WER} = 2) \text{ with confidence between 0.40 and 1.}$$

Variable C (capability) in our threat model is related to the WER and to the imminence (I) of an attack. The degree of imminence is measured by the distance, heading and speed of the target, and according to Table 17.3 can be low, medium, or high. We define valuation m_7 by the following rule on the product space $C \times \text{WER} \times I$: $C = \text{WER} + I$. This rule captures the simple notion that the capability is high if the WER is large and the imminence is high. Thus m_7 is a BBA given by

$$m_7(\{(0,0,0), (1,0,1), (2,0,2), (1,1,0), (2,1,1), (3,1,2),$$
$$(2,2,0), (3,2,1), (4,2,2)\}) = 1.$$

Valuations $m_1, m_2, \ldots, m_7$ represent our prior domain knowledge of the problem. The remaining valuations $m_8, m_9, \ldots, m_{15}$, referred to as *input valuations*, are the drivers of this evidential network for threat assessment. Input valuations are initially represented by vacuous BBAs. As more evidence about the intruder and the situation become available (e.g., from the surveillance sensors and other external sources), input valuations change and become more informative.

17.5.2.2 Numerical Results with Analysis. The evidential network consists of 13 variables and 15 valuations (BBAs) in $\Theta_V = \{m_1, \ldots, m_{15}\}$. The BJT constructed for the threat assessment problem using the variable elimination sequence: IFF, FPA, I, C, EM, FCR, CM, PC, PT, WER, HI, A, is shown in Figure 17.9. The leaves of the tree (the nodes labeled from 1 to 15) represent the original valuations $m_1, \ldots, m_{15}$. The remaining nodes in the BJT represent the intermediate steps of the fusion algorithm.

The BJT satisfies the Markov property defined in Section 17.3.2. Consider for example the subtree of nodes $\{1, 7, 15, 20, 21\}$. Note that variable C is contained in the domains of nodes 1 and 7, but also in the domain of all nodes in the path between 1 and 7, that is, $Path(1, 7) = \{1, 7, 20, 21\}$. A BJT that does not satisfy this property cannot be a representation of the fusion algorithm. For example, let us assume that the domain of node 20 does not include C; this would mean that C has been eliminated

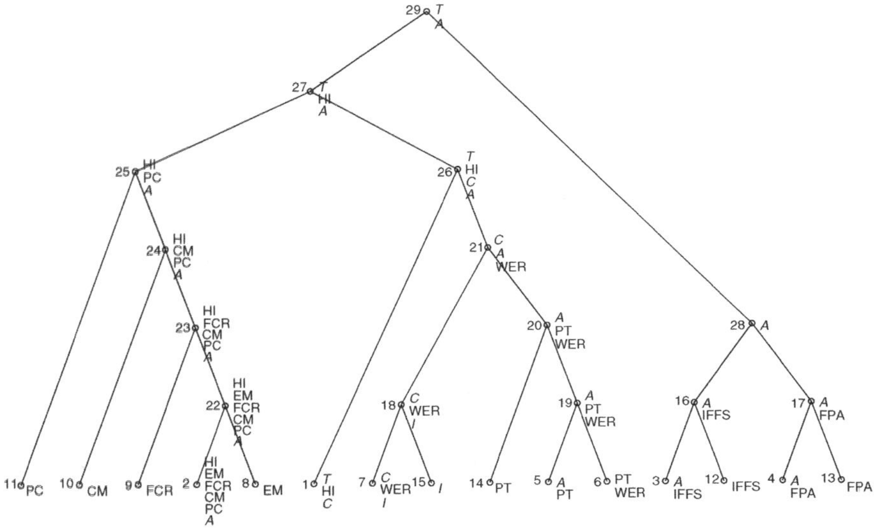

Figure 17.9 Binary join tree for the threat assessment model obtained by applying the OSLA-SCFF heuristic.

during the combination of the BBAs of nodes 7 and 15. If this were true, before combining the valuations at node 20 with node 1 to produce the BBA for node 21 (the domain of 1 contains C), one should again extend the domain of node 20 to a new domain containing C. Since marginalization produces a loss of information (coarsening) that cannot be recuperated with the extension operation (see Remark 1 at the end of Section 17.4.2), the BBA of node 21 would be incorrect, that is, it would be different from $(m_1 \oplus m_7 \oplus m_{15})^{\downarrow\{T,HI,C,WER\}}$.

As mentioned in Section 17.3.1, finding the optimal elimination sequence is an NP-complete problem. The above elimination sequence has been calculated by means of the *One Step Look Ahead—Smallest Clique, Fewest Focal Sets* (OSLA-SCFF) Heuristic [23, p. 61]. This heuristic chooses the variable to be eliminated by minimizing the cardinality of the domain and the number of focal sets associated with the nodes of the BJT. Note that a different elimination sequence would result in a different BJT. For example, the BJT in Figure 17.10 was obtained using the elimination sequence IFF, FPA, *I*, EM, FCR, CM, PC, PT, WER, *A*, *C*, HI, which has been calculated via the *One Step Look Ahead - Fewest Fill-ins* (OSLA-FFI) heuristic [23, p. 60]. Note that the final result of the application of inward propagation is independent of the elimination sequence and, thus, of the structure of the BJT. The difference is only in the computational time required to calculate the result.

Three Extreme Cases.. To apply the inward propagation algorithm, the valuations of the leaves of the tree must be initialized first. The BBAs $m_1, \ldots, m_7$ have been defined earlier. For the input valuations $m_8, \ldots, m_{15}$ (nodes 8–15 in the BJT), consider now the three "extreme" cases: (1) total ignorance, (2) the highest degree of threat,

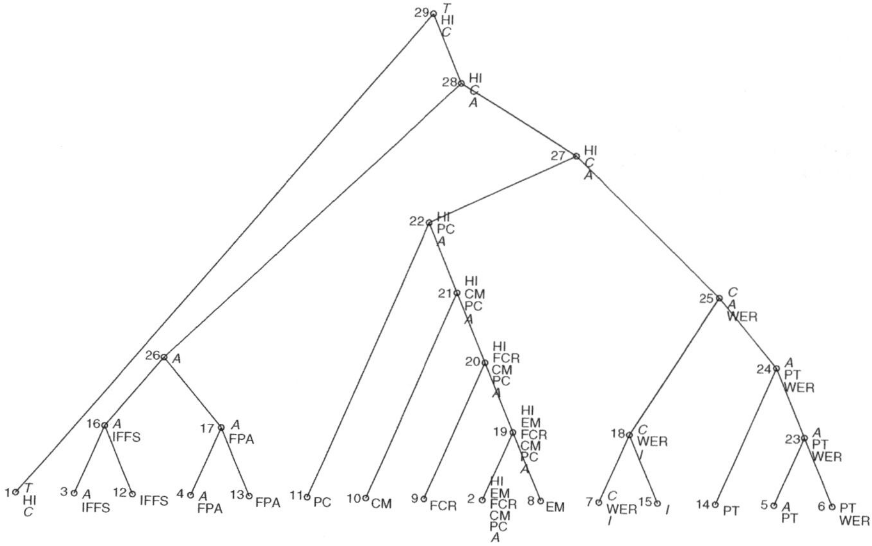

Figure 17.10 Binary join tree for the threat assessment model obtained by applying the OSLA-FFI heuristic.

and (3) the lowest degree of threat. The BBAs for the input valuations in three extreme cases are given in Table 17.4. For the case of the total ignorance, all input valuations are represented by vacuous BBAs. For the case of a highest (lowest) threat, all BBAs are singletons taking high (low) threat values.

The output of *inward propagation* is the BBA of node 29, defined on domain $\{T, A\}$. This BBA is then marginalized to domain $\{T\}$ and finally the corresponding belief, plausibility functions and pignistic probability are computed for the variable $\{T\}$. Figure 17.11 shows the resulting belief, plausibility functions and (pignistic) PMF for the degrees of threat (from 0 to 10) in all three cases. From this figure

TABLE 17.4 Input Valuations for Three Extreme Cases

		No information		High Threat		Low Threat	
BBA	Domain	Focal Set	Mass	Focal Set	Mass	Focal Set	Mass
m_8	EM	{0, 1}	1	{1}	1	{0}	1
m_9	FCR	{0, 1}	1	{1}	1	{0}	1
m_{10}	CM	{0, 1}	1	{1}	1	{0}	1
m_{11}	PC	{0, 1}	1	{1}	1	{0}	1
m_{12}	IFF	{0, 1}	1	{0}	1	{1}	1
m_{13}	FPA	{0, 1}	1	{0}	1	{1}	1
m_{14}	PT	{0, 1, 2, 3, 4, 5}	1	{5}	1	{0}	1
m_{15}	I	{0, 1, 2}	1	{2}	1	{0}	1

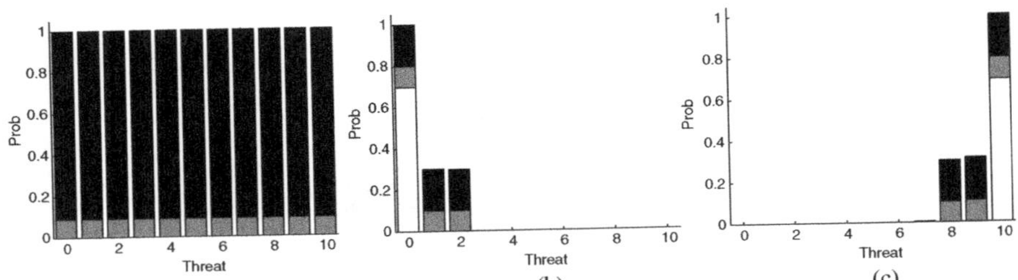

Figure 17.11 Belief (white), Plausibility (black), and Pignistic PMF (gray) for variable T (threat) in three extreme cases: (a) total ignorance; (b) lowest threat; (c) highest threat.

it can be seen that the results are in agreement with the inputs and our intuition. When there is no information (total ignorance), the resulting BBA on domain $\{T\}$ is a vacuous BBA and hence the belief (lower probability) is zero, the plausibility (upper probability) is 1 and, finally, the pignistic probability is a uniform. This means that the prior valuations $m_1, \ldots, m_7$ are balanced, that is they assume that all the degrees of threat are initially equally probable. Finally, based on maximality (computed as described in Example 13), one concludes that all degrees of threat $\{0, 1, \ldots, 10\}$ are undominated in this case (all one can say is that the degree of threat is between 0 and 10). This means that there is not enough information to decide the value of the degree of threat. For the low and high threat cases, the results are also in agreement with intuition. Observe that in the low (high) threat case, the value of the belief function for $T = 0$ ($T = 10$) is greater than the plausibility of all the other values and, thus, based on the interval dominance criterion (and, thus, also on maximality, since it is implied), one can conclude with high confidence that the degree of threat is 0 (10). Note, however, that in the low (high) threat case the plausibility of the degree 0 (10) is less than 1.0. This is due to the intrinsic uncertainty in the prior expert knowledge, expressed by valuations m_1 to m_7.

Dynamic Reasoning Example. Input valuations $m_8, \ldots, m_{15}$ change over time as the new pieces of evidence (from surveillance sensors and other external sources) become available. Whenever this happens, it is necessary to re-evaluate the level of threat. Initially all input valuations $m_8, \ldots, m_{15}$ are vacuous BBAs, representing the initial state of ignorance. When an input valuation is changed, the network recomputes the valuations of the nodes of the BJT along the affected path of the tree. For example, if the belief mass m_{15} changes, only the masses of the nodes 20, 21, 26, 27, and 29 must be recomputed (see Figure 17.9). Consider an example of a sequence of incoming evidence shown in Table 17.5. At time t_1, the information about the current state of the political climate (PC), represented by BBA m_{11}, is supplied. For argument's sake, let this BBA reflect a state of political tension in the region, so that the belief mass given to the state of war is 0.7, while the remaining 0.3 is assigned to ignorance. Then at time t_2 some evidence about the EM variable becomes available; it appears that the target is performing an evasive maneuvre, and therefore a belief mass of 0.8 is assigned to true and 0.2 to the state of ignorance. Each time a new piece of evidence is

TABLE 17.5 The Sequence of Incoming Evidence Driving the Evidential Network

Time	BBA	Domain	Focal Set	Mass
t_1	m_{11}	PC	{1}	0.7
			{0, 1}	0.3
t_2	m_8	EM	{1}	0.8
			{0, 1}	0.2
t_3	m_{15}	I	{0, 1}	0.7
			{0, 1, 2}	0.3
t_4	m_{13}	FPA	{1}	0.9
			{0, 1}	0.1
t_5	m_{15}	I	{1}	0.8
			{0,1,2}	0.2
t_6	m_{14}	PT	{2}	0.8
			{3}	0.1
			{4}	0.1
t_7	m_{12}	IFF	{0}	0.9
			{0, 1}	0.1
t_8	m_{10}	CM	{1}	0.99
			{0, 1}	0.01
t_9	m_9	FCR	{1}	0.99
			{0, 1}	0.01

available, the situation becomes more informative (less uncertain) that is reflected by the pignistic PMF of threat, shown in Figure 17.12. Note how this PMF evolves from being totally uninformative at time t_0 to becoming concentrated ("peaky") at time t_9. At this last time instant the degree of threat with the highest pignistic probability is 3 (on the scale from 0 to 10), but there is still a high uncertainty (plausibilities of degrees 3, 4, 5, 6 are equal). In fact, by computing the maximality decision criterion, the undominated degrees of threat are {3, 4, 5, 6}. Thus, at this point, one can only say that the degree of threat is between 3 and 6. More information is needed to reduce this uncertainty.

Sensitivity Analysis. Sensitivity analysis studies the effect of the changes in the input valuations on the valuation of the output (decision) variable. In this way, sensitivity analysis helps us to identify which inputs are more influential on decision making and how they affect the decision process. Sensitivity analysis was applied to the described threat assessment evidential network [31, 49] and revealed that the most influential variable in decision making was FCR.

Computational Complexity. Threat assessment can be carried out without local computations in valuation algebra, that is, by directly computing the joint belief on the domain Θ_V followed by marginalization of the resulting belief to the domain of T. From Table 17.3 it can be seen that the number of configurations in the joint

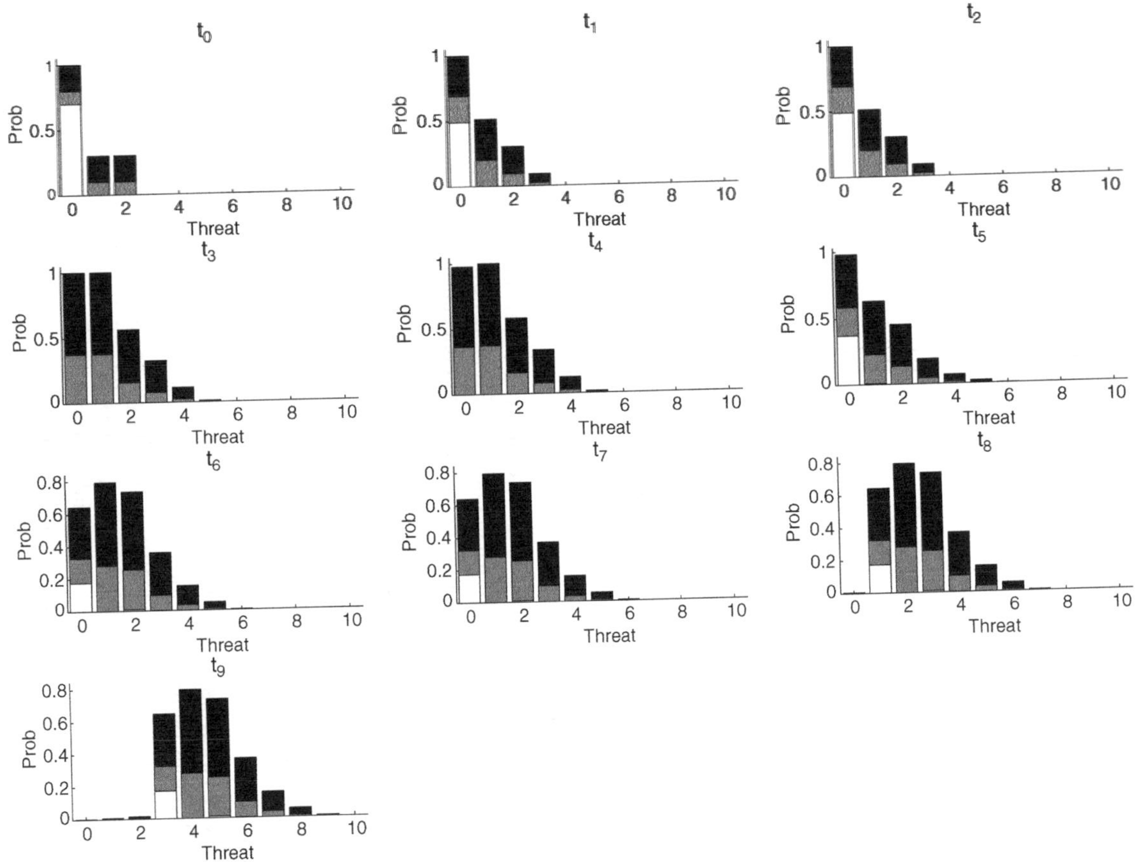

Figure 17.12 Belief (white), plausibility (black), and pignistic PMF (gray) for variable T (threat) in a dynamic situation (from time t_0 to t_9).

frame of Θ_V is 2, 661, 120 (i.e., the Cartesian product of the frames of the single variables). This is a huge number compared with the number of elements of the maximum domains in the two BJTs shown in Figures 17.9 and 17.10. In the BJT obtained by applying the OSLA-SCFF heuristic (Figure 17.9), the number of elements of the maximum domain is only 1155 (for node 21). This number is even lower for the BJT obtained by applying the OSLA-FFI heuristic (Figure 17.10). In this case, the maximum domain has only 385 elements (for nodes 1 and 29). Since the joint belief for Θ_V is defined on the power set of Θ_V, for computing the joint belief it is necessary to calculate, in the worst case, the masses for all the $2^{2661120}$ elements of the power set. This "brute force" approach has been attempted for threat assessment on the joint domain, but the computer could not complete the task even after 48 h of processing. By contrast, using local computations framework, the threat assessment was carried out on the same computer in just 5 s for the BJT obtained by applying the OSLA-SCFF heuristic and 3 s for the BJT obtained by applying the OSLA-FFI heuristic. We point out that although for the adopted threat assessment model the OSLA-FFI heuristic allows to compute the solution faster than OSLA-SCFF, in general this may not be true: the computational complexity and the effectiveness of the heuristic for sequence elimination depend strongly on the structure of the network and the number of focal elements in belief functions. The general rule is the more complex the interdependencies are among the variables, the smaller the advantage in using the VA. Real complex reasoning systems, with hundreds or even thousands of variables, are usually characterized by very localized structures. As the computational complexity grows exponentially with the domain size, the VA framework can solve problems that otherwise would be computationally intractable [31, 49].

APPENDIX 17A CONSTRUCTION OF A BJT

Let $\Psi = \{\varphi_1, \varphi_2, \ldots, \varphi_r\}$ be a given set of valuations and $D^o \subseteq V$, with $V = d(\varphi_1) \cup d(\varphi_2) \cup \cdots \cup d(\varphi_r)$, the domain of interest. Let us introduce the following notation [21]:

- $L(n)$: left child of node n, or *nil* if n is a leaf;
- $R(n)$: right child of node n, or *nil* if n is a leaf;
- $F(n)$: parent of node n, or *nil* if n is the root of the tree;
- $d(n)$: domain of the valuation for the node n;
- *root*: root of the BJT.

The algorithm for constructing a BJT is as follows [21].

Initialization:
Define the initial set of node $N_\psi = \{n_1, n_2, \ldots, n_r\}$ with $d(n_i) = d(\varphi_i)$, $L(n_i) = nil$, $R(n_i) = nil$ and $F(n_i) = nil$.
Fix the set of variables to be eliminated $\Delta = V - D^0$.

4: **function** CONSTRUCT A BJT(N_ψ,Δ)
5:     $N = \emptyset$; $\overline{\Delta} = \emptyset$; $root = nil$;
6:     **repeat**
7:         **if** $\Delta = \emptyset$ **then**
8:             $N_x = N_\psi$;
9:         **else**
10:             select the next variable to be eliminated, $x \in \Delta$, using some heuristic;
11:             $N_x = \{n \in N_\psi : x \in d(n)\}$;
12:         **end if**
13:         **while** $|N_x| > 1$ **do** $\triangleright$ while the cardinality of N_x is greater than 1
14:             generate a new node n with $F(n) = nil$;
15:             select distinct $n_1, n_2 \in N_x$;
16:             $F(n_1) = n$; $F(n_2) = n$;
17:             $L(n) = n_1$; $R(n) = n_2$;
18:             $d(n) = (d(n_1) \cup d(n_2)) - \overline{\Delta}$;
19:             $N_x = (N_x \backslash \{n_1, n_2\}) \cup \{n\}$;
20:             $N = N \cup \{n_1, n_2\}$;
21:         **end while**
22:         **if** $\Delta = \emptyset$ **then**
23:             $root = n$;
24:         **else**
25:             $\Delta = \Delta \backslash \{x\}$; $\overline{\Delta} = \overline{\Delta} \cup \{x\}$;
26:             $N_\psi = \{n \in N_\psi : x \notin d(n)\} \cup \{n\}$;
27:         **end if**
28:     **until** $root \neq nil$
29:     $N = N \cup \{n\}$;
30:     **return** N
31: **end function**

The tree resulting from this procedure is a BJT with $2r - 1$ nodes, $N = \{n_1, n_2, \ldots, n_{2r-1}\}$, such that $D^o \subseteq d(root)$. The only degree of freedom in the BJT construction algorithm is the order in which the variables are eliminated (Step 10).

APPENDIX 17B INWARD PROPAGATION

The objective of the inward propagation algorithm is to compute the valuations for the variables of interest. Consider again the set of valuations $\Psi = \{\varphi_1, \varphi_2, \ldots, \varphi_r\}$, the domain of interest D^o and the set $N = \{n_1, n_2, \ldots, n_{2r-1}\}$ of nodes of the BJT constructed by the algorithm given in Appendix 17A. Inward propagation is performed in two stages. In Stage 1, which is executed only once, the valuations are propagated from the leaves toward the root of the BJT [21]. Stage 2 is performed every time the valuations of one or more leaves of the BJT change. In this case, inward propagation recomputes only the valuations of those nodes of the BJT that belong to the path connecting the leaves with changed valuations to the root of the BJT. The steps of the algorithm for inward propagation are as follows.

```
Initialization:
Initialize Leaves = {n ∈ N : n is a leaf}.
if stage=1 then
    φs = nil;
    UI = Leaves;
else
    UI = {n ∈ Leaves : valuation is changed w.r.t. the previous time};
end if
function INWARD PROPAGATION(Leaves, N, stage, UI, φ, φs)
    if stage=1 then    ▷ It is the first time that inward propagation is performed
        Set next = {n ∈ N : L(n) ∈ Leaves and R(n) ∈ Leaves}.
        for n ∈ Leaves do
            φs(n) = φ(n)^{↓d(F(n))};
        end for
    else    ▷ inward propagation has been already performed at least one time
        next = ∅;
        for n ∈ N do
            if L(n) ∈ UI or R(n) ∈ UI then
                Set next = next ∪ n.
            end if
        end for
    end if

    visitN = ∅    ▷ indicates the nodes visited during the inward propagation
    while |next| > 0 do    ▷ while next is not empty
        extract an element n from next;
        visitN = visitN ∪ n; next = next − n;
        φ(n) = φs(L(n)) ⊕ φs(R(n));
        φs(n) = φ(n)^{↓d(F(n))};
        if n ≠ root then
            DomF = d(F(n));
        else
            DomF = d(d_v)
        end if
        φs(n) = φ(n)^{↓DomF};
        if stage=1 then ▷ It is the first time that inward propagation is performed
            for n ∈ N do
                if (n ∉ Leaves) and (n ∉ next) and (n ∉ visitN) then
                    if (L(n) ∈ Leaves or L(n) ∈ visitN) and (R(n) ∈ Leaves or
                        R(n) ∈ visitN) then
                        next = next ∪ n;
```

```
                    end if
                end if
            end for
        else
            for n ∈ N do
                if (n ∉ UI) and (n ∉ next) and (n ∉ visitN) then
                    if (L(n) ∈ visitN) or (R(n) ∈ visitN) then
                        next = next ∪ n;
                    end if
                end if
            end for
        end if
    end while
    return φ, φ_s                ▷ φ_s(root) is the valuation for the decision variables
end function
```

REFERENCES

1. A. Kott, M. Pollack, and B. Krogh, The situation assessment problem: toward a research agenda. *Proceedings of the DARPA-JFACC Symposium on Advances in Enterprise Control*, San Diego, CA, Nov. 1999.
2. E. Waltz and J. Llinas, *Multisensor Data Fusion*, Artech House, Norwood, 1990.
3. J. Pearl, *Probabilistic Reasoning in Intelligent Systems*, Morgan Kaufmann Publishers, San Francisco, 1988.
4. J.D. Lowrance, T.D. Garvey, and T.M. Strat, A framework for evidential reasoning systems, *Proceedings of the 5th National Conference on AI (AAAI-86)*, Philadelphia, 1986, pp. 896–901.
5. G. Shafer, P.P. Shenoy, and K. Mellouli, Propagating belief functions in qualitative Markov trees, *Int. J. Approx. Reason.*, 1, 349–400, 1987.
6. L.A. Zadeh, Fuzzy sets as a basis for a theory of possibility, *Fuzzy Sets Syst.*, 3–28, 1978.
7. L.A. Zadeh, On the validity of Dempster rule of combination, *Memo M 79/24*, University of California, Berkeley, 1979.
8. D. Dubois and H. Prade, *Possibility Theory: An Approach to Computerised Processing of Uncertainty*. Plenum Publishers, 1988.
9. G. Shafer, *A Mathematical Theory of Evidence*, Princeton University Press, 1976.
10. P. Smets, Data fusion in the transferable belief model, *Proceedings of the 3rd International Conference on Information Fusion*, Paris, France, 2000, PS21–PS33.
11. V. Lepar and P.P. Shenoy, A comparison of lauritzen-spiegelhalter, hugin and shenoy-shafer architectures for computing marginals of probability distributions. *Proceedings of the 14th Conference on Uncertainty in Artificial Intelligence (UAI-98)*, 1998, pp. 328–337.
12. S.L. Lauritzen and D.J. Spiegelhalter, Local computations with probabilities on graphical structures and their application to expert systems, *J. R. Stat. Soc.*, 50, 157–224. 1988.

13. K.G. Olesen, F.V. Jensen, and S.K. Andersen, An algebra of bayesian belief universes for knowledge-based systems, *Network*, 20, 637–659. 1990.
14. S.L. Lauritzen, F.V. Jensen, and K.G. Olesen, Bayesian updating in causal probabilistic networks by local computation, *Comput. Stat. Q.*, 4, 269–282. 1990.
15. S.L. Lauritzen and F.V. Jensen. Local computation with valuations from a commutative semigroup, *Ann. Math. Artif. Intell.*, 21, 51–69. 1997.
16. P.P. Shenoy and G. Shafer, Axioms for probability and belief-function propagation, in: J. Pearl and G. Shafer (Eds.), *Readings in uncertain reasoning*, Morgan Kaufmann, San Mateo, CA, pp. 575–610, 1990.
17. P.P. Shenoy, Binary join trees for computing marginals in the shenoy-shafer architecture, *Int. J. Approx. Reason.*, 17, 239–263, 1997.
18. G.J. Klir, *Uncertainty and Information*, Wiley, New Jersey, 2006.
19. J. Kohlas, *Information Algebras: Generic Structures for Inference*, Springer-Verlag, London, 2003.
20. P.P. Shenoy, Valuation based systems: a framework for managing uncertainty in expert systems, in: L.A. Zadeh and J. Kacprzyk (Eds.), *Fuzzy Logic and the Management of Uncertainty*, Wiley, New York, Chapter 4, pp. 83–104, 1992.
21. R. Haenni, Ordered valuation algebras: a generic framework for approximate inference, In *Int. J. Approx. Reason.*, 37, 1–41, 2004.
22. R.G. Almond, *Graphical Belief Modeling*, Chapman and Hall, 1995.
23. N. Lehmann, *Argumentation Systems and Belief Functions, Tech. Report 01-30*, Department of Informatics, University of Fribourg, Switzerland, 2001.
24. R. Haenni and N. Lehmann, Probabilistic argumentation systems: a new perspective on Dempster-Shafer theory, *Int. J. Intell. Syst.*, 18, 93–106, 2003.
25. A.P. Dempster, Upper and lower probabilities induced by a multivalued mapping, *Ann. Math. Stat.*, 38, 325–339, 1967.
26. P. Walley, *Statistical Reasoning with Imprecise Probabilities*, Chapman and Hall, London, 1991.
27. R. Haenni, Shedding new light on Zadeh's criticism of Dempster's rule of combination, *Proceedings of the 8th International Conference on Information Fusion*, contribution No. C8-1, Philadelphia, USA, 2005.
28. C. Baudrit and D. Dubois, Practical representations of incomplete probabilistic knowledge, *Comput. Stat. Data Anal.*, 51(1), 86–108, 2006.
29. J.O. Berger, *Statistical Decision Theory and Bayesian Analysis*, Springer Series in Statistics, New York, 1985.
30. M. Zaffalon, Statistical inference of the naive credal classifier, *ISIPTA '01: Proceedings of the Second International Symposium on Imprecise Probabilities and Their Applications*, Shaker Publishing, The Netherlands, 2001, pp. 384–393.
31. A. Benavoli, B. Ristic, A. Farina, M. Oxenham, and L. Chisci, An approach to threat assessment based on evidential networks, *Proceedings of the 10th International Conference on Information Fusion*, Ville de Québec, Canada, July 2007.
32. B. Ristic and P. Smets, Target identification using belief functions and implication rules, *IEEE Trans. Aerosp. Electron. Syst.*, 41, 1097–1103, 2005.
33. B. Ristic and P. Smets, Target classification approach based on the belief function theory, *IEEE Trans. Aerosp. Electron. Syst.*, 41, 574–583, 2005.

34. A. Benavoli, L. Chisci, B. Ristic, and A. Farina, Modelling uncertain implication rules in evidence theory, *Proceedings of the 11th International Conference on Information Fusion*, Cologne, Germany, June 30–July 3, 2008, pp. 1–8.
35. A. Benavoli, M. Zaffalon, and E. Miranda, Robust filtering through coherent lower previsions, *IEEE Trans. Autom. Control*, 56(7), 1567–1581, 2011.
36. D. Lewis, Probabilities of conditionals and conditional probabilities, 85, 297–315, 1976.
37. G. Choquet, *Theory of Capacities*, Annales de l'Institut Fourier, 1953–1954.
38. P. Smets, Decision making in the TBM: the necessity of the pignistic transformation, *Int. J. Approx. Reas.*, 38, 133–147, 2005.
39. D.L. Hall and S.A.H. McMullen, *Mathematical Techniques in Multisensor Data Fusion*, Artech House, 2nd edition, 2004.
40. S. Das, R. Grey, and P. Gonsalves, Situation assessment via Bayesian belief networks, *Proceedings of the 5th International Conference on Information Fusion*, Annapolis, MD, USA, July 2002, pp. 664–671.
41. P. Bladon, R.J. Hall, and W.A. Wright, Situation assessment using graphical models, *Proceedings of the 5th International Conference on Information Fusion*, Annapolis, MD, July 2002, pp. 886–893.
42. F. Mirmoeini and V. Krishnamurthy, Reconfigurable Bayesian networks for adaptive situation assessment in battlespace, *Proc. IEEE Conf. Netwo., Sensing Control*, 2005, 810–815.
43. N. Okello and G. Thomas, Threat assessment using Bayesian networks, *Proceedings of the 6th International Conference on Information Fusion*, Cairns, Australia, July 2003, pp. 1102–1109.
44. X.T. Nguyen and P. Heuer, Automated intent assessment simulation environment, *Proc. Information, Decision and Control*, Adelaide, Australia, Feb. 2002, pp. 395–400.
45. K.D. Lee and J. Llinas, Hybrid model for intent estimation, *Proceedings of the 6th Internationa Conference on Information Fusion*, Cairns, Australia, July 2003, pp. 1215–1222.
46. A. Benavoli, B. Ristic, A. Farina, M. Oxenham, and L Chisci, An Application of Evidential Networks to Threat Assessment, *IEEE Trans. Aerosp. Electron. Syst.*, 45(2), 620–639, 2009.
47. M. Oxenham, Enhancing situation awareness for air defence via automated threat analysis, *Proceedings of the 6th International Conference on Information Fusion*, Cairns, Australia, July 2003.
48. Fighter planes. Web: http://www.fighter-planes.com/.
49. A. Benavoli, *Modelling and efficient fusion of uncertain information: beyond the classical probability approach*, Ph.D. thesis, University of Florence, December, 2007.

INDEX

Integrated Tracking, Classification, and Sensor Management: Theory and Applications, Edited by Mahendra Mallick, Vikram Krishnamurthy, and Ba-Ngu Vo.

Printed and bound by CPI Group (UK) Ltd, Croydon, CR0 4YY
16/09/2026
14963417-0006